Geosystems

Combination of satellite *Terra* MODIS sensor image and *GOES* satellite image produces a true-color view of North and South America from 35,000 km (22,000 mi) in space. [Courtesy of MODIS Science Team, Goddard Space Flight Center, NASA and NOAA.]

The top predator in the Arctic region is a marine mammal, the polar bear (*Ursus maritimus*). This female bear is raising and traning her cubs to hunt, feed, and get about on Arctic Ocean pack ice. The cubs are 8-months old and will stay with their mother for more than two years. Changes in climate and pack ice conditions are affecting the survivability of these bears. The International Polar Year 2007–2008 of science and research will focus on Earth's polar regions and the impact of these changes. [Photo by Bobbé Christopherson.]

Sixth Edition

Geosystems

An Introduction to Physical Geography

Robert W. Christopherson

PEARSON

Prentice
Hall

Upper Saddle River, New Jersey 07458

Library of Congress Cataloging-in-Publication Data

Christopherson, Robert W.
  Geosystems : an introduction to physical geography / Robert W. Christopherson.—
    6th ed.
      p. cm.
  ISBN 0-13-153117-4
  1. Physical geography. I. Title.

  GB54.5.C48 2006
  910'.02—dc22

                                                        2004030652

Executive Editor: *Daniel E. Kaveney*
Editor in Chief: *John Challice*
Editor in Chief, Development: *Carol Trueheart*
Development Editor: *Ginger Birkeland*
Executive Managing Editor: *Kathleen Schiaparelli*
Assistant Managing Editor: *Beth Sweeten*
Production Management: *GGS Book Services*
Production Editor: *Cindy Miller*
Production Assistant: *Nancy Bauer*
Production Assistant to the Author: *Bobbé Christopherson*
Manufacturing Buyer: *Alan Fischer*
Associate Editor: *Amanda Griffith*
Marketing Manager: *Robin Farrar*
Media Editor: *Chris Rapp*
Managing Editor, Science Media: *Nicole M. Jackson*
Editorial Assistant: *Margaret Ziegler*
Art Director: *Heather Scott*
Interior Designer: *Tamara Newnam*
Cover Design: *Tamara Newnam*
AV Editor: *Jessica Einsig*
Art Studio: *Precision Graphics*
Photo Editor: *Debbie Latronica*
Director of Creative Services: *Paul Belfanti*
Composition: *Interactive Composition Corporation*
Front Cover Photo: *Grounded iceberg in Scoresby Sund, east-central Greenland—60-m (200-ft) tall "ice tower"
    by Bobbé Christopherson*
Back Cover Photos: *Arctic and Antarctic scenes by Bobbé Christopherson*

Pearson Education LTD., *London*
Pearson Education Singapore, Pte. Ltd
Pearson Education, Canada, Ltd, *Toronto*
Pearson Education–Japan, *Tokyo*
Pearson Education Australia PTY, Limited, *Sydney*
Pearson Education North Asia Ltd, *Hong Kong*
Pearson Educación de Mexico, S.A. de C.V.
Pearson Education Malaysia, Pte. Ltd

To all the students and teachers of Earth,
our home planet, and a sustainable future.

And to Rinie van Meurs,
and the other expedition leaders who teach us the polar regions.

The land still provides our genesis,
however we might like to forget
that our food comes from dank,
muddy Earth, that the oxygen in our lungs was
recently inside a leaf, and that every newspaper or
book we may pick up is made from the hearts of
trees that died for the sake of our imagined lives.
What you hold in your hands right now,
beneath these words,
is consecrated air and time and sunlight.

—Barbara Kingsolver

Brief Contents

Contents

PART II
The Water, Weather, and Climate Systems 174

PART III

The Earth–Atmosphere Interface 318

11 The Dynamic Planet 321

Preface

During the life of this sixth edition of *Geosystems*, important scientific research and global trends in Earth systems science will unfold. As pointed out on the back cover of the text, climate change and its many impacts are hitting the high latitudes at a faster pace than elsewhere on Earth. To reflect global system linkages with these regions, this edition of *Geosystems* weaves high-latitude themes through the chapters.

The scientific community is responding with the comprehensive Arctic Climate Impact Assessment (ACIA) Report, published in late 2004. This historic scientific effort is detailed in Chapter 21. The International Polar Year (IPY) for science and research in the polar regions is set for March 2007 to March 2009. The author and his photographer wife completed four polar expeditions in 2003 and 2004, gathering materials and photos for this edition. As you study physical geography you will find your textbook responsive to the latest science such as the IPY process. Welcome to physical geography!

Geosystems presents introductory physical geography as a contemporary spatial science, integrating the latest science across the broad sweep of this discipline. The sixth edition builds on the widespread success of the first five editions in the United States, Canada, and elsewhere, as well as its companion text, *Elemental Geosystems*, now in its fourth edition. Students and teachers alike continue to express their appreciation of the up-to-date content, systems organization, scientific accuracy, clarity of the summary and review sections, and overall relevancy.

Armed with the spatial analysis tools of geographic science, physical geographers are well equipped to participate in a planetary understanding of environmental conditions. U.N. Secretary General Kofi Annan, recipient of the 2001 Nobel Peace Prize, spoke to the Association of American Geographers annual meeting in 2001, stating,

> As you know only too well the signs of severe environmental distress are all around us. . . . The idea of interdependence is old hat to geographers, but for most people it is a new garment they are only now trying on for size. . . . I look forward to working with you in that all-important journey.

New to the Sixth Edition

Space does not permit a listing of all the changes and additions to *Geosystems*, 6/E, however, here are a few comments. Begin by examining all 21 chapter-opening and 4 part-opening photos and read their captions—all but two are new. These are content-specific and begin the learning process from the start of the chapter. Throughout the text there are hundreds of new photos to illustrate concepts, many integrated in compound figures with arts and maps.

New to this edition is the "High Latitude Connection" feature, linking chapter material to the polar regions. The world is turning its attention to the high latitudes to better understand present global climate change. These High Latitude Connection items include:

- 1.1: "Meltponds as positive feedback,"
- 3.1: "Arctic Haze,"
- 5.1: "Overview of Trends in the Polar Regions,"
- 10.1: "Climate Change in the Polar Regions,"
- 11.1: "Isostatic Rebound in Alaska,"
- 16.1: "A Rebounding Shoreline and Coastal Features,"
- 17.1: "Climate Change Impacts Arctic Ice Shelf,"
- 21.1: "Report from Reykjavik—Arctic Climate Impact Assessment."

New polar region temperature maps are in Figures 5.14 and 5.16. The Chapter 21 opening photo was made at the ACIA Symposium in Iceland in November 2004. The report that emerged from this scientific process is summarized in this book. For beauty as you learn, check out the "Arctic and Antarctic Iceberg Gallery" in Figure 17.36, among many other examples. There are two additional photo galleries on the *Student Animation CD*.

One example of new materials in this edition: Chapter 1 treats the engineering feat of moving the Cape Hatteras Lighthouse 884 m (2900 ft) inland, in terms of its changed latitude and longitude grid coordinates. A compound map, satellite image, and photos cover the same area in Figure 16.16. In addition barrier islands along the Texas coast (16.17), Louisiana (16.18), Florida (16.19), further highlight coastal concepts. Focus Study 16.1, Figure 1 presents the coastal planning process with seven photos keyed to the illustration. You find this same thoroughness of treatment for many different topics in the book.

Other examples include, many with new photos, satellite images, maps, and arts: expanded coverage of jet-contrail impacts on Earth–atmosphere energy budgets; analysis of "earthshine" as a diagnostic tool; four new polar-region temperature maps; new treatment of North Atlantic, Arctic, and Pacific Decadal Oscillations; rewritten section on humidity for clarity of concepts; revised midlatitude weather teaching map; new discussion of *derechos*—straight-line intense winds; mention of the 2004 Atlantic hurricane season and the plight of Florida, with satellite-loop coverage on the text CD; new genetic climate classification map and discussion, with causitive forces portrayed on the climate map; updated section on global climate change including the latest science; discussion of the Loess Hills of Iowa; a revised and expanded glacial and periglacial chapter; new section on evolution and the tremendous biodiversity that results from the process; new treatment of polar climates and biomes; an updated capstone Chapter 21; among many new or updated items.

Geosystems Communicates the Science of Physical Geography

The goal of physical geography is to explain the spatial dimension of Earth's dynamic systems—its energy, air, water, weather, climate, tectonics, landforms, rocks, soils, plants, ecosystems, and biomes. Understanding human-Earth relations is part of the challenge of physical geography—to create a holistic (or complete) view of the planet and its inhabitants.

Geosystems analyzes the worldwide impact of environmental events, synthesizing many physical factors into a complete picture of Earth-system operations. A good example is the eruption of Mount Pinatubo in the Philippines. The global implications of this major event (one of the largest eruptions in the 20th century) are woven through seven chapters of the book (see Figure 1.6 for a summary). Our update on global climate change and its related potential effects is part of the fabric in these seven chapters. These content threads weave together the variety of interesting and diverse topics crucial to a thorough understanding of physical geography.

To assist you in physical geography, this edition of *Geosystems* features more than 500 photographs from across the globe and more than 100 remote-sensing images from a wide variety of orbital platforms. For spatial analysis and location support, 125 maps are utilized, and more than 300 illustrations locate and explain concepts.

Systems Organization Makes *Geosystems* Flow

Each section of this book is organized around the flow of energy, materials, and information. *Geosystems* presents subjects in the same sequence in which they occur in nature. In this way you and your teacher logically progress through topics that unfold according to the flow of individual systems, or in accord with time and the flow of events. See Figure 1.7 in the text for an illustration of this systems organization.

For flexibility, *Geosystems* is divided into four parts, each containing chapters that link content in logical groupings. The diagram from Figure 1.8 illustrates our part structure. A quick check of the Table of Contents and this illustration shows you the order of chapters within these four parts.

The text culminates with Chapter 21, "Earth and the Human Denominator," a unique capstone chapter that summarizes physical geography as an important discipline in understanding Earth's present status and possible future. Think of the world's population and the totality of our impact as the *human denominator*. Just as the denominator in a fraction tells how many parts a whole is divided into, so the growing human population and the increasing demand for resources and rising planetary impact suggest how much the whole Earth system must adjust. High Latitude Connection 21.1 takes you to Iceland for a firsthand account from the important Arctic Climate Impact Assessment

Symposium. This chapter is sure to stimulate further thought and discussion, dealing as it does with the most profound issue of our time, Earth's stewardship.

Geosystems Is a Text That Teaches

Teaching and learning begin with the front and back cover photographs. The grand "ice tower" iceberg stands tall against the sky in east Greenland. On the back cover, photos from the Arctic Ocean region and the waters near Antarctica frame an introduction to this edition. *Geosystems* is written to assist you in the learning process.

Three heading levels are used throughout the text and precise topic sentences begin each paragraph to help you outline and review material. **Boldface** words are defined where they first appear in the text. These terms and concepts are collected alphabetically in the Glossary, with a chapter-number reference. Our complete Glossary appears on the *Student Animation CD* that accompanies this text. *Italics* are used in the text to emphasize other words and phrases of importance. Every figure has a title that summarizes the caption. Also, in the introduction to each chapter, a feature called "In this chapter" gives you an overview.

An important continuing feature is a list of *Key Learning Concepts* that opens each chapter, stating what you should be able to do upon completing the chapter. These objectives are keyed to the main headings in the chapter. At the end of each chapter is a unique *Summary and Review* section that corresponds to the *Key Learning Concepts*. Grouped under each learning concept is a narrative review that defines the boldfaced terms, a key terms list with page numbers, and specific review questions for that concept. You can conveniently review each concept, test your understanding with review questions, and check key terms in the Glossary, then return to the chapter and the next learning concept. In this way, the chapter content is woven together using specific concepts.

A *Critical Thinking* section ends each chapter, challenging you to take the next step with information from the chapter. The key learning concepts help you determine what you want to learn, the text guides you in developing information and more questions, the summary and review assesses what you have learned and what more you might want to know about the subject, and the critical thinking section provokes action and application.

Career Link essays, including several new ones, feature geographers and other scientists in a variety of professional fields practicing their spatial analysis craft. You will read about an astronaut with over 1200 hours in orbit, a weather forecaster at the Forecast Systems Lab, an environmental scientist, a hydrologist with the National Weather Service, a geographer who specializes in risk analysis, a high-latitude researcher who combines physical geography and archeology to understand the impacts of climate change, a snow avalanche specialist, and a researcher on biogeography and paleoenvironments, among others.

FIGURE 1.8 **Earth's four spheres.**
Each sphere is a model of vast Earth systems. This four-part structure is the organizational
framework for Geosystems. [Photos by Bobbé Christopherson.]

  Part 1—Atmosphere: The Energy–Atmosphere System
        (Rainbow and thunderstorm squall line.)
  Part 2—Hydrosphere: The Water, Weather, and Climate Systems
        (Cascading stream in southwestern Iceland.)
  Part 3—Lithosphere: The Earth–Atmosphere Interface
        (Cape Renaud, Graham Land, Antarctica.)
  Part 4—Biosphere: Soils, Ecosystems, and Biomes
        (Humpback whales in the Atlantic Ocean.)

Despite the publication of our new *Geosystems Canadian Edition*, coverage of Canadian physical geography in the text and figures continues throughout this book. Canadian data on a variety of subjects are portrayed on 30 different maps in combination with the United States—physical geography does not stop at the United States–Canadian border!

Focus Study essays, some completely revised and several new to this edition, provide additional explanation of key topics. A few examples from this diverse collection include the stratospheric ozone predicament, solar energy collection and wind power, the newly (2001) calibrated wind-chill chart, forecasting the near-record 1995–2004 hurricane seasons, the 1997–1998 El Niño phenomenon, geothermal energy development, status of the High Plains Aquifer, floodplain strategies, an environmental approach to shoreline planning, the Mount St. Helens eruption, the status of the Colorado River and the Western drought, the continuing global loss of biodiversity, and the dramatic ice cores in Greenland and in Antarctica at Dome C that look back almost a million years into past climates.

News Reports relate topics of special interest; for example, GPS, careers in GIS, a 34-kilometer sky dive to study the atmosphere, new studies of "earthshine," jet streams and airline flight times, how one culture harvests fog, the UV Index, coordination of global climate-change research (accompanied by many URLs), the disappearing Nile Delta, water issues in the Middle East, artificial scouring experiments in the Grand Canyon to restore beaches and habitats, survival issues in the sinking Mississippi River Delta, how sea turtles read Earth's magnetic field, alien and exotic plant and animal invasions, and threats to the Arctic National Wildlife Refuge.

We now live on a planet served by the Internet and its World Wide Web, a resource that weaves threads of information from around the globe into a vast fabric. The fact that we have Internet access into almost all the compartments aboard Spaceship Earth is clearly evident in *Geosystems*. Many entry points link directly from the words in a chapter to an Internet source, allowing you to be up-to-the minute. You will find more than 200 URLs (Internet addresses) in the body of the text (printed in blue and boldface). Given the fluid nature of the Internet, URLs were rechecked at press time for accuracy. If some URLs changed since publication, you can most likely find the new location using elements of the old address. This Internet link begins with a new Table 1.1 presenting the URLs for major geography organizations.

The Geosystems Learning/Teaching Package

The sixth edition provides a complete physical geography program for you and your teacher.

For You, the Student:

- *Student Animation CD*, by Robert Christopherson, is included in each copy of this book. This exciting learning tool contains 48 animations, satellite loops (including the 2004 Florida hurricanes), three exclusive photo galleries, a map reference library, and a searchable version of the text Glossary. Each animation features a self-test with pop-up reinforcement for correct answers. Text references provided throughout help you relate the CD to your use of the book.
- *Student Study Guide*, Sixth Edition (ISBN 0-13-133092-6), by Robert Christopherson and Charlie Thomsen. The study guide includes additional learning objectives, a complete chapter outline, critical thinking exercises, problems and short essay work using actual figures from the text, and a self-test with answer key in the back.
- *Geosystems Online Study Guide* (http://www.prenhall.com/christopherson): This site gives you the opportunity to further explore topics presented in the book using the Internet. The site contains numerous review exercises (from which you get immediate feedback), exercises to expand students' understanding of physical geography, and resources for further exploration. This web site provides an excellent opportunity from which to start using the Internet for the study of geography.
- *Student Lecture Notebook* (ISBN 0-13-186353-3). All of the line art from the transparency set is reproduced in this full-color notebook. Students can now fully focus on the lecture and not be distracted by replicating drawings. Each page is three-hole punched for easy integration with other course materials.

For You, the Teacher:

Geosystems is designed to give you flexibility in presenting your course. The text is comprehensive in that it is true to each scientific discipline from which it draws subject matter. This diversity is a strength of physical geography, yet makes it difficult to cover the entire book in a school term. *Geosystems* is organized to help you customize your presentation. You should feel free to use the text based on your specialty or emphasis, rearranging parts and chapters as desired. The four-part structure of chapters, systems organization within each chapter, Focus Study and News Report features—all will assist you in sampling some chapters while covering others in greater depth. The following materials are available to assist you—have a great class!

- *Instructor Resource Center CD-ROM (IRC)*, (ISBN: 0-13-133096-9), contains everything you need for efficient course preparation. Find all your digital resources in one, well-organized, easy-to-access place. Included on this CD-ROM are:

Figures—JPEGs of all illustrations and photos from the book.

Animations—All the animations ready-to-use in your classroom presentations as PPT slides or Flash files.

PowerPoint™—Pre-authored slides outline the concepts of each chapter with embedded art for use in the classroom or to customize with other items for your own presentation needs.

TestGen—A complete TestGen-EQ software featuring a thorough and revised test bank, with questions and answers.

Electronic files—The complete Instructor's Resource Manual and Test Item File in one easily accessible location.

- *Instructor's Resource Manual,* Sixth Edition (ISBN: 0-13-133099-3), by Robert Christopherson and Charlie Thomsen. The Instructor's Resource Manual, intended as a resource for both new and experienced teachers, includes lecture outlines and key terms, additional source materials, teaching tips, complete annotation of chapter review questions, and a list of overhead transparencies.

- *Annotated Instructor's Edition* (ISBN 0-13-133103-5) of *Geosystems'* sixth edition contains complete annotations of all media and other available resources in the text to save time and make for easier classroom preparation.

- *Geosystems Test Bank* (ISBN: 0-13-133098-5), by Robert Christopherson and Charlie Thomsen. This collaboration has produced the most extensive and fully revised test item file available in physical geography. This test bank employs TestGen-EQ software. TestGen-EQ is a computerized test generator that lets you view and edit test bank questions, transfer questions to tests, and print customized formats. Included is the QuizMaster-EQ program that lets you administer tests on a computer network, record student scores, and print diagnostic reports. Mac and IBM/DOS computer formats are served. Order the TestGenEQ version of the text bank using ISBN 0-13-133097-7.

- *Overhead Transparencies* (ISBN: 0-13-133095-0) includes more than 350 illustrations from the text on 300 transparencies, all enlarged for excellent classroom visibility.

- *Applied Physical Geography—Geosystems in the Laboratory,* Sixth Edition (ISBN: 0-13-133093-4), by Robert Christopherson and Charlie Thomsen of American River College. The new sixth edition is the result of a careful revision. Twenty-one lab exercises, divided into logical sections, allow flexibility in presentation. Each exercise comes with a list of learning concepts. Our manual is the only one that comes with its own complete glossary and stereolenses and stereomaps for viewing photo stereopairs in the manual. A complete *Solutions and Answers Manual* is available to teachers (ISBN: 0-13-133094-2).

Acknowledgments

As in all past editions, I recognize my family, for they have endured our workload, yet never wavered in supporting *Geosystems'* goals—both our moms, my sister Lynne, brothers Randy and Marty, and our children, Keri, Matt, Reneé, and Steve. And now the next generation: Chavon, Bryce, Payton, Brock, Trevor, Blake, Chase, and our newest, Téyenna. When I look into our grandchildren's faces it tells me why we need to work toward a sustainable future—one for the children.

I give special gratitude to all the students, and my colleagues, over 29 years at American River College for defining the importance of Earth's future, for their questions, and their enthusiasm. To all students and teachers, and to our expedition leader Rinie van Meurs on three polar voyages, we dedicate this book. A special mentor, friend, and my life coach since 1960, I thank Dick Vermeil for his inspiration and spirit; the intensity he exemplifies drives this author toward the horizon.

My thanks go to the many authors and scientists who published research articles and books that enrich my work; to all the colleagues who served as reviewers on one or more editions of each book, who participated in our focus groups, or who offered helpful suggestions at our national and regional geography meetings; and, although unnamed here, to all the correspondence received from students and teachers from across the globe who shared with me over the Internet, e-mail, FAX, and phone—a continuing appreciated dialogue. I am grateful to all of them for their generosity of ideas and sacrifice of time. Here is a master list of all our reviewers of the *Geosystems* books, including those that reviewed the *Student Animation CD.*

Ted J. Alsop, *Utah State University*
Ward Barrett, *University of Minnesota*
David Berner, *Normandale Community College*
Franco Biondi, *University of Nevada, Reno*
Peter D. Blanken, *University of Colorado, Boulder*
David R. Butler, *Southwest Texas State University*
Mary-Louise Byrne, *Wilfred Laurier University*
Ian A. Campbell, *University of Alberta–Edmonton*
Randall S. Cerveny, *Arizone State University*
Fred Chambers, *University of Colorado, Boulder*
Muncel Chang, *Butte College and California State University–Chico*
Andrew Comrie, *University of Arizona*
C. Mark Cowell, *Indiana State University*
Richard A. Crooker, *Kutztown University*
Armando M. da Silva, *Towson State University*
Dirk H. de Boer, *University of Saskatchewan*
Dennis Dahms, *University of Northern Iowa*
Mario P. Delisio, *Boise State University*
Joseph R. Desloges, *University of Toronto*
Lee R. Dexter, *Northern Arizona University*
Don W. Duckson, Jr., *Frostburg State University*
Christopher H. Exline, *University of Nevada–Reno*
Michael M. Folsom, *Eastern Washington University*
Mark Francek, *Central Michigan University*

Glen Fredlund, *University of Wisconsin–Milwaukee*
David E. Greenland, *University of North Carolina–Chapel Hill*
Duane Griffin, *Bucknell University*
John W. Hall, *Louisiana State University–Shreveport*
Barry N. Haack, *George Mason University*
Roy Haggerty, *Oregon State University*
Vern Harnapp, *University of Akron*
Jason "Jake" Haugland, *University of Colorado, Boulder*
Gail Hobbs, *Pierce College*
Thomas W. Holder, *University of Georgia*
David A. Howarth, *University of Louisville*
Patricia G. Humbertson, *Youngstown State University*
David W. Icenogle, *Auburn University*
Philip L. Jackson, *Oregon State University*
J. Peter Johnson, Jr., *Carleton University*
Guy King, *California State University–Chico*
Ronald G. Knapp, *SUNY–The College at New Paltz*
Peter W. Knightes, *Central Texas College*
Thomas Krabacher, *California State University–Sacramento*
Richard Kurzhals, *Grand Rapids Junior College*
Steve Ladochy, *California State University, Los Angeles*
Charles W. Lafon, *Texas A & M University*
Paul R. Larson, *Southern Utah University*
Robert D. Larson, *Southwest Texas State University*
Elena Lioubimtseva, *Grand Valley State University*
Joyce Lundberg, *Carleton University*
W. Andrew Marcus, *Montana State University*
Elliot G. McIntire, *California State University, Northridge*
Norman Meek, *California State University, San Bernardino*
Sherry Morea-Oaks, *Boulder, CO*
Patrick Moss, *University of Wisconsin, Madison*
Lawrence C. Nkemdirim, *University of Calgary*
Andrew Oliphant, *San Francisco State University*
John E. Oliver, *Indiana State University*
Bradley M. Opdyke, *Michigan State University*
Richard L. Orndorff, *University of Nevada, Las Vegas*
Patrick Pease, *East Carolina University*
James Penn, *Southeastern Louisiana University*
Greg Pope, *Montclair State University*
Robin J. Rapai, *University of North Dakota*
Philip D. Renner, *American River College*
William C. Rense, *Shippensburg University*
Dar Roberts, *University of California–Santa Barbara*
Wolf Roder, *University of Cincinnati*
Robert Rohli, *Louisiana State University*
Bill Russell, *L.A. Pierce College*
Dorothy Sack, *Ohio University*
Randall Schaetzl, *Michigan State University*
Glenn R. Sebastian, *University of South Alabama*

Daniel A. Selwa, *U.S.C. Coastal Carolina College*
Thomas W. Small, *Frostburg State University*
Daniel J. Smith, *University of Victoria*
Stephen J. Stadler, *Oklahoma State University*
Michael Talbot, *Pima Community College*
Susanna T. Y. Tong, *University of Cincinnati*
Suzanne Traub-Metlay, *Front Range Community College*
David Weide, *University of Nevada–Las Vegas*
Thomas B. Williams, *Western Illinois University*
Brenton M. Yarnal, *Pennsylvania State University*
Stephen R. Yool, *University of Arizona*

I extend my continuing gratitude to the editorial, production, and sales staff of Prentice Hall. Thanks to ESM President Paul Corey for his leadership strength in the publishing industry, friendship, and guidance since before the first edition. Thanks to Dan Kaveney, Geosciences Executive Editor; every author should have such an editor— tireless, energetic, willing to risk new ideas and technologies, and a lifetime friend—for his role in both *Geosystems* and *Elemental Geosystems* and in geographic education in general. Compliments to Amanda Griffith who oversees my ancillaries with such cheer, and Chris Rapp for expertise and creativity on the CD projects, and to Margaret Zeigler for keeping the *Geosciences* office organized and efficient, and to all the staff for allowing us to participate in the entire publishing process.

My thanks to Ginger Birkeland, developmental editor for her work on the sixth edition; her careful reading, assessment, and guidance assisted me in expressing the research and content for student use. Her grounding influence kept me in sight of the goal. And thanks to a tireless production editor, Cindy Miller of GGS Book Services, for such expertise and care in converting all my materials into this textbook. Thanks to the art and design team at Prentice Hall for this powerful cover using Bobbé's photo, beautiful text design, and for letting us in on many decisions. The many sales representatives who spend months in the field communicating the *Geosystems* approach are a tremendous asset to the book—thanks and safe travels to them—and to our son with pride, a Pearson Prentice Hall sales rep who has risen to be one of the best!

My continuing partnership with a special collaborator, photographer, and production assistant, Bobbé Christopherson, is at the heart of the success of *Geosystems* and of my life's work. My wife has worked tirelessly on all the *Geosystems*' projects. You see 350 of her photographs in the sixth edition, all carefully made to illustrate physical geography principles. Please visit the three photo galleries on the *Student Animation CD* and learn more from her artistry with a camera. She prepares our extensive figure and photo logs, processes satellite imagery and all photos, copyedits, obtains permissions, and assists me in proofing art and final pages. Bobbé's love for the smallest of living things is evident in the beautiful and insightful photographs she has in this edition. And, she continues as my best friend and colleague and expedition companion.

Physical geography teaches us a holistic view of the intricate supporting web that is Earth's environment and our place in it. Dramatic changes that demand our understanding are occurring in many human-Earth relations, as we alter physical, chemical, and biological systems. Our attention to the polar regions is in recognition of the linkages between all planetary systems and the high latitudes. All things considered, this is a critical time for you to be enrolled in a physical geography course! The best to you in your studies—and *carpe diem!*

Robert W. Christopherson
Geography Professor Emeritus,
American River College
P. O. Box 128
Lincoln, California 95648
E-mail: bobobbe@aol.com
Web site: http://www.prenhall.com/christopherson

CAPE HATTERAS LIGHT HO[USE]

ERECTED A. D. 1870.

LATITUDE 35° 15′ 14[″]

LONGITUDE 75° 30′ 5[″]

NEW COORDINATES

LATITUDE

35° 15′ 2″ NORTH

LONGITUDE

75° 31′ 44″ WEST

Increasing losses from coastal barrier islands necessitated the dramatic 1999 move of the Cape Hatteras lighthouse more than 884 m (2900 ft) inland. Ironically, this is about the same distance the lighthouse was from the ocean when first built in 1870 (lower-left photo shows path along which it was moved). This move places the lighthouse in a new location relative to Earth's surface coordinates. Can you tell by comparing the new and old coordinates in the photo the direction the lighthouse was moved? More on this, and the answer, in Chapter 16, Figure 16.16. Location is important in geographic science. [Photos by Bobbé Christopherson.]

Essentials of Geography

▶ Key Learning Concepts

After reading the chapter, you should be able to:

▶ **Define** geography and physical geography in particular.

▶ **Describe** systems analysis, open and closed systems, feedback information, and system operations and **relate** those concepts to Earth systems.

▶ **Explain** Earth's reference grid: latitude, longitude, latitudinal geographic zones, and time.

▶ **Define** cartography and mapping basics: map scale and map projections.

▶ **Describe** remote sensing and **explain** geographic information system (GIS) methodology as a tool used in geographic analysis.

During this sixth edition of *Geosystems*, we celebrate the International Polar Year (IPY) in 2007. This interdisciplinary research effort combines many scientific fields that are at the core of physical geography. Bringing together various disciplines is essential to understanding the ongoing dynamics of Earth's systems. The polar regions, which link to all systems across the globe, are experiencing profound change and much preparation is underway toward IPY research, exploration, and discovery (see: **http://dels.nas.edu/us-ipy/index.html**).

Therefore, this is a scientifically exciting time to be enrolled in a physical geography course, and perhaps to be considering it for a major—we'll see. Physical geography deals with our environment and the powerful Earth systems that influence our lives and the many ways humans are altering Earth's systems. Most natural ecosystems now bear the imprint of civilization. This book is an assessment of the operation of Earth's physical systems, in both their natural state and as they are changed by human actions.

Physical geography provides essential information needed to assess Earth systems and their operational status. United Nations Secretary General and recipient of the 2001 Nobel Peace Prize, Kofi Annan, speaking to the Association of American Geographers (AAG) annual meeting in March 2001, offered this assessment:

As you know only too well the signs of severe environmental distress are all around us. Unsustainable practices are woven deeply into the fabric of modern life. Land degradation threatens food security. Forest destruction threatens biodiversity. Water pollution threatens public health, and fierce competition for freshwater may well become a source of conflict and wars in the future.... the overwhelming majority of scientific experts have concluded that climate change is occurring, that humans are contributing, and that we cannot wait any longer to take action ... environmental problems build up over time, and take an equally long time to remedy.

Scientists and governments scramble to understand the worldwide impact of global changes in weather, climate, water resources, and the landscape. We are experiencing unprecedented environmental conditions: increasing air and ocean temperatures, stratospheric ozone losses, increased rates of sea-level rise, heightened thunderstorm intensity, large losses of glacial ice worldwide, increasing extinctions of plant and animal species, weather-related damage with massive floods in one region contrasted with severe droughts elsewhere, and related wildfires. As an example, weather-related damage in 2002 topped $70 billion, many times greater than the previous annual average.

Why do all of these conditions occur? How are these events different from past experience? Why does the environment and rates of global change vary from equator to midlatitudes, between deserts and polar regions? How does solar energy influence the distribution of trees, soils, climates, and lifestyles? How does energy produce the patterns of wind, weather, and ocean currents? How do natural systems affect human populations, and, in turn, what impact are humans having on natural systems? Why are the countries of the world joining in the IPY to study the polar regions? In this book, we explore those questions, and more, through geography's unique perspective. Welcome to an exploration of physical geography!

We live in an extraordinary era of **Earth systems science**. This science contributes to our emerging view of Earth as a complete entity—an interacting set of physical, chemical, and biological systems that produce a whole Earth. Physical geography is at the heart of Earth systems science as we answer the *spatial* questions concerning Earth's physical systems and their interaction with living things.

Physical geographers analyze interactions and changes that are occurring in natural systems to better quantify Earth systems over diverse *temporal* (time) and *spatial* (space or area) scales, and to address the question of how these changes might affect life on Earth. In an editorial in the journal *Science*, John Lawton stated:

One of the great scientific challenges of the twenty-first century is to forecast the future of planet Earth ... we find ourselves, literally, in uncharted territory, performing an uncontrolled experiment with planet Earth.... Wrestling to understand these challenges is the young, and still emerging, discipline of Earth systems science.*

*John Lawton, "Earth systems science," *Science* 292 (June 15, 2001): 1965.

In this chapter: Our study of *Geosystems*—Earth systems—begins in this chapter with a look at the science of physical geography and the geographic tools we use. Physical geography is key to studying entire Earth systems because of its integrative spatial approach.

Physical geographers analyze systems to study the environment. Therefore, we discuss systems and the feedback mechanisms that influence system operations. We then consider location, a key theme of geographic inquiry—the latitude, longitude, and time coordinates that inscribe Earth's surface and the new technologies in use to measure them. The study of longitude and a universal time system provide us with interesting insights into geography. Next, we examine maps as critical tools that geographers use to portray physical and cultural information. This chapter concludes with an overview of the technology that is adding exciting new dimensions to geography: remote sensing from space and computer-based geographic information systems (GIS).

The Science of Geography

Geography (from *geo*, "Earth," and *graphein*, "to write") is the science that studies the relationships among natural systems, geographic areas, society, cultural activities, and the interdependence of all of these *over space*. The term **spatial** refers to the nature and character of physical space, its measurement, and the distribution of things within it. For example, think of your own route to the classroom or library today or the way you get to work and how you used your knowledge of street patterns, traffic trouble spots, one-way streets, parking spaces, or bike rack locations to minimize walking distance. All these are spatial considerations.

Humans are spatial actors. We profoundly influence vast areas because of our mobility and access to energy and technology. In turn, Earth's systems influence our activities in a most obvious way—these systems give us life.

To guide geographic education, *Geography for Life* (1994), prepared by the Association of American Geographers, National Council for Geographic Education, American Geographical Society, and National Geographic Society, subdivides geographic science into essential elements, critical skills, and geographic standards. For a listing of some important geography organizations and their URLs, see Table 1.1.

We simplify geographic science using five spatial themes: **location**, **region**, **human–Earth relationships**, **movement**, and **place**, illustrated and defined in Figure 1.1. *Geosystems* draws on each theme.

Geographic Analysis

Within these five geographic themes, a *method* rather than a specific body of knowledge governs geography. The method is **spatial analysis**. Using this method, geography synthesizes (brings together) topics from many fields, integrating information to form a whole-Earth concept. Geographers view phenomena as occurring in spaces and

Location

Absolute and relative location on Earth. Location answers the question *Where?*—the specific planetary address of a location. This road sign is posted on Interstate 5 freeway in Oregon telling drivers their position on Earth.

Region

Areas having uniform characteristics; how they form and change; their relations to other regions. In the Arctic region a scene from eastcentral Greenland, at Sydkap on Hall Bredning. The Arctic region of the Northern Hemisphere is experiencing dynamic change as temperatures increase.

Human–Earth Relationships

Humans and the environment: resource exploitation, hazard perception, and environmental pollution and modification. The western U.S. drought, into its sixth year in 2004, drastically reduced flows in the Colorado River, leaving Lake Mead behind Hoover Dam at only 50% of capacity.

Movement

Communication, movement, circulation, migration, and diffusion across Earth's surface. Global interdependence links all regions and places. Hurricane Isabel is a product of atmospheric circulation. In 2003, Category 5 Isabel slammed into Cape Hatteras and the Outer Banks, North Carolina, causing $4 billion in damage.

Geographic Science

Place

Tangible and intangible living and nonliving characteristics that make each place unique. No two places on Earth are exactly alike. Patagonia in southern Argentina is such a place. The braided channel of the Rio Gallegos laces across the arid and semiarid landscape.

FIGURE 1.1 Five themes of geographic science.
Definitions of five fundamental themes in geographic science with examples of each—location, region, human–Earth relationships, movement, and place. Drawing from your own experience, can you think of several examples of each theme? [Regions, Place, Human–Earth relationships photos by Bobbé Christopherson; Location photo by author; and Movement *Terra* image from MODIS Land Rapid Response Team, NASA.]

Table 1.1	
A Few Geography Organizations	**URL Addresses**
American Geographical Society	http://www.amergeog.org/
Association of American Geographers*	http://www.aag.org/
National Council for Geographic Education	http://www.ncge.org/
National Geographic Society	http://www.nationalgeographic.com/
Canadian Association of Geographers	http://venus.uwindsor.ca/cag/cagindex.html
Royal Canadian Geographical Society	http://www.rcgs.org/
Institute of Australian Geographers	http://www.iag.org.au/
Australian Geography Teachers Association	http://www.agta.asn.au/
European Geography Association	http://egea.geog.uu.nl/
Royal Geographical Society, Institute of British Geographers	http://www.rgs.org/
Complete global listing of geography organizations	http://www.geoggeol.fau.edu/Links/GeographicalSites.htm

*Includes nine regional divisions: East Lakes, Great Plains/Rocky Mountain, Middle Atlantic, Middle States, New England–St. Lawrence Valley, Pacific Coast Regional, Southeastern, Southwestern, West Lakes.

areas having distinctive characteristics. The language of geography reflects this spatial view: *space, territory, zone, pattern, distribution, place, location, region, sphere, province,* and *distance.* Geographers analyze the differences and similarities among places and locations.

Process, a set of actions or mechanisms that operate in some special order, is central to geographic analysis. As examples in *Geosystems*, numerous processes are involved in Earth's vast water–atmosphere–weather system, or in continental crust movements and earthquake occurrences, or in ecosystem functions. Geographers use spatial analysis to examine how Earth's processes interact over space or area.

Therefore, **physical geography** is the *spatial analysis of all the physical elements and processes that make up the environment: energy, air, water, weather, climate, landforms, soils, animals, plants, microorganisms, and Earth itself.* We add to this the oldest theme in the geographic tradition, that of human activity. As a science, physical geographers employ the **scientific method**. Focus Study 1.1, beginning on page 6, explains this essential process of science.

The Geographic Continuum

Geography is eclectic, integrating a wide range of subject matter from diverse fields; virtually any subject can be examined geographically. Figure 1.2 shows a continuous distribution—a continuum—along which the content of geography is arranged. Disciplines in the physical and life sciences are at one end, and those in the human and cultural sciences are at the other. As the figure shows, various specialties within geography draw from these subject areas.

The continuum in Figure 1.2 reflects a basic duality, or split, within geography—*physical geography* versus *human/cultural geography.* Society parallels this duality. Humans sometimes think of themselves as exempt from physical Earth processes—like actors not paying attention to their stage, props, and lighting. We all depend on Earth's systems to provide oxygen, water, nutrients, energy,

and materials to support life. The growing complexity of the human–Earth relation requires that we shift our study of geographic processes, and perhaps our link to Earth systems, toward the center of the continuum in Figure 1.2 to attain a more *holistic,* or balanced, perspective—such is the thrust of Earth systems science.

Earth Systems Concepts

The word *system* pervades our lives daily: "Check the car's cooling system"; "Is a broadband system available for my Internet browser?"; "How does the grading system work?"; "There is a weather system approaching." Systems of many kinds surround us. Not surprisingly, *systems analysis* has moved to the forefront as a method for understanding operational behavior in many disciplines. The technique began with studies of energy and temperature (thermodynamics) in the nineteenth century and was further developed in engineering during World War II. Today, geographers use systems methodology as an analytical tool. In this book's 4 parts and 21 chapters, content is organized along logical flow paths consistent with systems thinking.

Systems Theory

Simply stated, a **system** is any ordered, interrelated set of things and their attributes, linked by flows of energy and matter, as distinct from the surrounding environment outside the system. The elements within a system may be arranged in a series or interwoven with one another. A system comprises any number of subsystems. Within Earth's systems, both matter and energy are stored and retrieved and energy is transformed from one type to another. (Remember: *Matter* is mass that assumes a physical shape and occupies space; *energy* is a capacity to change the motion of, or to do work on, matter.)

FIGURE 1.2 The content of geography.
Geography derives subject matter from many different sciences. The focus of this book is physical geography, but also we integrate some human and cultural components. Movement toward the middle of the continuum suggests a synthesis of Earth topics and human topics. Examine the subjects listed in the two boxes for any course you completed. Think for a moment and recall any spatial aspects you remember studying in that course.

Open Systems Systems in nature are generally not self-contained: Inputs of energy and matter flow into the system, and outputs of energy and matter flow from the system. Such a system is called an **open system**. Within a system, the parts function in an interrelated manner, acting together in a way that gives each system its character. Earth is an open system *in terms of energy*, for solar energy enters freely and heat energy leaves back into space. Most natural systems are open in terms of energy. Figure 1.3 schematically illustrates an open system and presents the inputs and outputs of an automobile as an example.

FIGURE 1.3 An open system.
In an open system, inputs of energy and matter undergo conversions and are stored as the system operates. Outputs include energy and matter and heat energy (waste) that flow from the system. See how the various inputs and outputs are related to the operation of a car: Matter and energy are converted, stored, and produced in an automobile—an open system. Expand your viewpoint to the entire system of auto production, from raw materials, to assembly, to sales, to car accidents, to junkyards. Can you identify other open systems that you encounter in your daily life?

The Scientific Method

The term *scientific method* may have an aura of complexity that it should not. The scientific method is simply the application of common sense in an organized and objective manner. A scientist observes, makes a general statement to summarize the observations, formulates a hypothesis, conducts experiments to test the hypothesis, and develops a theory and governing scientific laws. Sir Isaac Newton (1642–1727) developed this method of discovering the patterns of nature, although the term *scientific method* was applied later.

Scientists are curious about nature and appreciate the challenge of problem solving. Society depends on the discoveries of science to decipher mysteries, find solutions, and foster progress. *Complexity* dominates nature, making several outcomes possible as a system operates. Science serves an important function to reduce such uncertainty. Yet, the more knowledge we have, the more the uncertainty and awareness of other possible scenarios (outcomes and events) increases. This in turn demands more precise and aggressive science. A problem for society occurs when scientific uncertainty fuels an antiscience viewpoint, whereas defining uncertainty is a by-product of discovery. As we realize scientific principles of complexity and chaos in natural and human-made systems, the need for critical thinking and the scientific method deepens in all aspects of life.

Steps in the Scientific Method

Follow the scientific method illustration in Figure 1.1.1 as you read. The scientific method begins with our perception of the real world and a determination of what we know, what we want to know, and the many unanswered questions that exist. Scientists who study the physical environment turn to nature for clues that they can observe and measure. They discern what data are needed and begin to collect those data. Then, these observations and data are analyzed to identify coherent patterns that may be present. This search for patterns requires *inductive reasoning*, or the process of drawing generalizations from specific facts. This step is important in modern Earth systems science, in which the goal is to understand *a whole functioning Earth*, rather than isolated, small compartments of information. Such understanding allows the scientist to construct models that simulate general operations of Earth systems.

If patterns are discovered, the researcher may formulate a *hypothesis*—a formal generalization of a principle. Examples include the planetesimal hypothesis, nuclear-winter hypothesis, and moisture-benefits-from-hurricanes hypothesis. Further observations are related to the general principles established by the hypothesis. Further data gathered may support or refute the hypothesis, or predictions made according to it may prove accurate or inaccurate. All these findings provide feedback to adjust data collection and model building and to refine the hypothesis statement. Verification of the hypothesis after exhaustive testing may lead to its elevation to the status of a *theory*.

A theory is constructed on the basis of several extensively tested hypotheses. Theories represent truly broad general principles—unifying concepts that tie together the laws that govern nature (for example, the theory of relativity, theory of evolution, atomic theory, Big Bang theory, stratospheric ozone depletion theory, or plate tectonics theory). A theory is a powerful device with which to understand both the order and chaos (disorder) in nature. Using a theory allows predictions to be made about things not yet known, the effects of which can be tested and verified or disproved through tangible evidence. The value of a theory is the continued observation, testing, understanding, and pursuit of knowledge that the theory stimulates. A general theory reinforces our perception of the real world, acting as positive feedback.

Pure science does not make value judgments. Instead, pure science provides people and their institutions with objective information on which to base their own value judgments. Social and political judgments about the applications of science are increasingly critical as Earth's natural systems respond to the impact of modern civilization. From Jane Lubchenco's 1997 AAAS Presidential Address:

> Science alone does not hold the power to achieve the goal of greater sustainability, but scientific knowledge and wisdom are needed to help inform decisions that will enable society to move toward that end.

The Method and Applied Science

The growing awareness that human activity is producing global change places increasing pressure on scientists to participate in decision making. Numerous editorials in scientific journals have called for such *applied science* involvement. F. Sherwood Rowland and Mario Molina provide an example, discussed in Chapter 3, for they first proposed a hypothesis that certain human-made chemicals caused damaging reactions in the stratosphere. They hypothesized in 1974—along with Paul Crutzen, who also contributed to these discoveries—that chlorine-containing products such as chlorofluorocarbons (CFCs), commonly used as aerosol propellants, refrigerants, foaming agents, and cleaning solvents, were depleting our protective stratospheric ozone (O_3) layer.

Subsequently, surface, atmosphere, and satellite measurements confirmed the photochemical reactions and provided data to map the real losses that were occurring. International treaties and agreements to ban the chemical culprits followed. In 1995, the Royal Swedish Academy of Sciences awarded these three scientists the Nobel Prize for chemistry for their pioneering work. Such successful applied science is strengthening soci- ety's resolve and provoking treaties to ban problem chemicals and practices and introduce sustainable substitutes.

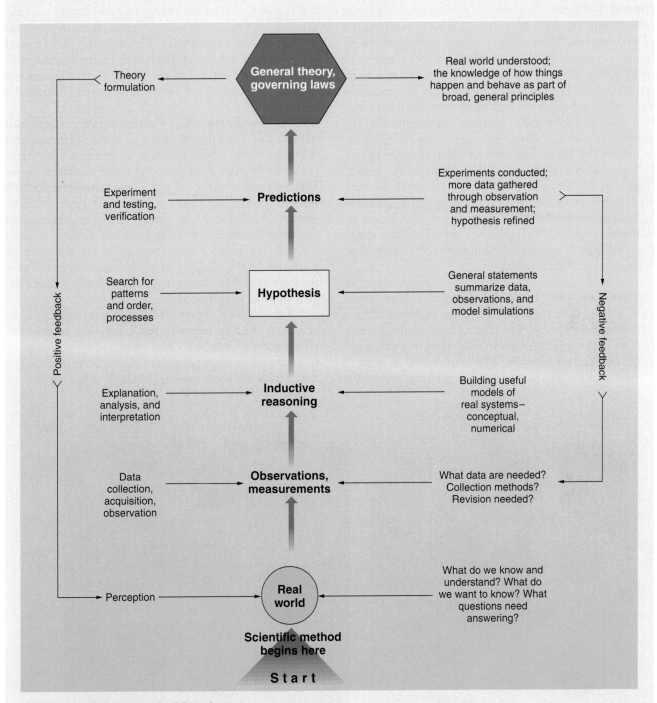

FIGURE 1.1.1 Scientific method flow chart.
The scientific method process: from perceptions, to observations, reasoning, hypothesis, predictions, and possibly to general theory and natural laws.

Most Earth systems are dynamic (energetic, in motion) because of the tremendous infusion of radiant energy from thermonuclear reactions deep within the Sun. This energy comes through the outermost edge of Earth's atmosphere to power terrestrial systems, being transformed along the way into various forms of energy, such as *kinetic energy* (of motion), *potential energy* (of position), or as chemical or mechanical energy—setting a fluid atmosphere and ocean in motion. Eventually, Earth radiates this energy back to the cold vacuum of space as heat energy. Earth's interior is likewise a dynamic system, with a fluid core and physically convecting mantle that influences crustal plates that are in motion. Researchers are examining Earth's energy system to distinguish natural operations from those changes being forced by human activities.

Closed Systems A system that is shut off from the surrounding environment so that it is self-contained is a **closed system**. Although such closed systems are rarely found in nature, Earth is essentially a closed system *in terms of physical matter and resources*—air, water, and material resources. The only exceptions are the slow escape of lightweight gases (such as hydrogen) from the atmosphere into space and the input of frequent but tiny meteors and cosmic and meteoric dust.

Since the initial formation of our planet, no significant quantities of new resources have entered the system. This is it. Earth's physical materials are limited. No matter how numerous and daring the technological reorganizations of matter become, our physical base is, for all practical purposes, fixed. The fact that Earth is a closed material system makes recycling efforts inevitable if we want a sustainable economy.

System Example Figure 1.4 illustrates a simple open-flow system, using plant photosynthesis and respiration as an example. In *photosynthesis* (Figure 1.4a), plants use sunlight as an energy input and material inputs of water, nutrients, and carbon dioxide. The photosynthetic process converts these inputs to stored chemical energy in the form

of plant sugars (carbohydrates). The process also releases an output from the plant system: the oxygen we breathe.

Reversing the process, plants derive energy for their operations from respiration. In *respiration*, the plant consumes inputs of chemical energy (carbohydrates) and oxygen and releases outputs of carbon dioxide, water, and heat energy into the environment (Figure 1.4b). Thus, a plant acts as an open system, in which both energy and materials freely flow into and out of the plant. (Photosynthesis and respiration processes are discussed further in Chapter 19.)

System Feedback As a system operates, it generates outputs that influence its own operations. These outputs function as "information" that is returned to various points in the system via pathways called **feedback loops**. Feedback information can guide, and sometimes control, further system operations. In the plant's photosynthetic system (see Figure 1.4), any increase or decrease in daylength (sunlight availability), carbon dioxide, or water will produce feedback that causes specific responses in the plant. For example, decreasing the water input will slow the growth process; increasing daylength will increase the growth process, within limits.

If the feedback *information* discourages response in the system, it is called **negative feedback**—like bad reviews affecting ticket sales for a film or play. *Further production in the system decreases the growth of the system.* Such negative feedback causes self-regulation in a natural system,

(a) Plant photosynthesis

(b) Plant respiration

FIGURE 1.4 A leaf is a natural open system.
A plant leaf provides an example of a natural open system. (a) In the process of photosynthesis, plants consume light, carbon dioxide (CO_2), nutrients, and water (H_2O), and produce outputs of oxygen (O_2) and carbohydrates (sugars) as stored chemical energy. (b) Plant respiration, illustrated here at night, approximately reverses this process and produces outputs of carbon dioxide (CO_2) and water (H_2O), using oxygen and consuming (oxidizing) carbohydrates to produce energy for cell operations.

stabilizing the system. Think of a weight-loss diet—the human body is an open system. When you stand on the scales, the good or bad news reported about your weight acts as negative feedback to you. This, in turn, provides you with guidance (negative feedback) as to how inputs such as food Calories need to be adjusted, or how exercise and increased metabolism is needed to burn excess Calories.

If feedback *information* encourages increased response in the system, it is called **positive feedback**—like good reviews affecting ticket sales for a film. *Further production in the system stimulates the growth of the system.* In finance, a compound-interest-bearing account provides an example: The larger the account becomes, the more interest it earns, thus the larger the account becomes, and so on. Unchecked positive feedback in a system can create a runaway ("snowballing") condition. In natural systems, such unchecked growth will reach a critical limit, leading to instability, disruption, or death of organisms.

Think of the numerous devastating wildfires that occurred in Indonesia, Mexico, and Florida in 1998, and the western United States between 1999 and 2003, as examples of positive feedback. As the fires burned, they dried wet shrubs and green wood around the fire, thus providing more fuel for combustion. The greater the fire, the greater becomes the availability of fuel, and thus more fire is possible—a positive feedback for the fire. Control of the input of flammable fuel and oxygen is the key to extinguishing such fires. Knowing this process allows us to design strategies for controlling fuel availability, regulating excessive landscaping in vulnerable urban areas and practicing "control burns." Global climate change is creating an important example of positive feedback, as described in our first "High Latitude Connection" feature.

Can you describe possible negative and positive feedback in the automobile illustration in Figure 1.3? For instance, how would the inputs and outputs be affected if tailpipe exhaust emission standards are weakened or strengthened? What if the automobile is operated at high elevation? Or if the car is a sports utility vehicle (SUV) that weighs 2300 kg (5070 lb)? Or if the price of gasoline increases or decreases? Take a moment to assess the affect of such changes on automobile system operations.

System Equilibrium   Most systems maintain structure and character over time. An energy and material system that remains balanced over time, in which conditions are constant or recur, is in a *steady-state condition.* When the rates of inputs and outputs in the system are equal and the amounts of energy and matter in storage within the system are constant (or more realistically, as they fluctuate around a stable average, such as a person's body weight), the system is in **steady-state equilibrium**.

However, a steady-state system may demonstrate a changing trend over time, a condition described as

High Latitude Connection 1.1

Meltponds as Positive Feedback

Another example of positive feedback is the significant increase in meltponds in the polar regions—on icebergs, ice shelves, and the Greenland ice sheet. As explained in News Report 17.3, Chapter 17, meltponds are darker and reflect less sunlight (have a lower albedo) and therefore absorb more solar energy, which in turn melts more ice, which forms more meltponds, and so forth. Thus, a positive feedback system is in operation, further enhancing the effects of higher temperatures and warming trends. Based on satellite images and aerial surveys, scientists determined a spatial increase in meltpond distribution of 400% between 2001 and 2003 across the Arctic (Figure 1.1.1).

Note that throughout *Geosystems* you find these brief High Latitude Connection features. These regions of Earth are experiencing dynamic changes in system operations. Many people think of these polar environs as our "miner's canary," that is, our early warning system. Watch for these points of applied geographic science in the chapters ahead.

FIGURE 1.1.1 High-latitude meltpond increase results from positive feedback.
Meltponds on the western region of the Greenland ice sheet in 2003. The darker water-saturated surface layer from melt and blue-colored meltponds indicate significant change in surface energy budgets on the ice, indicating significant increase in temperatures. [Image by MODIS sensor, *Terra* satellite, courtesy of MODIS Rapid Response Team, NASA/GSFC.]

dynamic equilibrium. These changing trends of either increasing or decreasing system operations may appear gradual. Figure 1.5 illustrates these two equilibrium conditions, steady-state and dynamic.

Examples of dynamic equilibrium include long-term climatic changes and the present pattern of increasing temperatures in the atmosphere and ocean or the ongoing loss of barrier islands along the East and Gulf coastlines. The present rate of species extinction exhibits a downward trend in numbers of living species—for example, Hawai'i has had 114 extinctions since 1980, nearly half the U.S. total.

Note that systems try to maintain their functional operations; they tend to resist abrupt change. However, a system may reach a **threshold** at which it can no longer maintain its character, so it lurches to a new operational level. This abrupt change places the system in a *metastable equilibrium*. An example of such a condition is a landscape, such as a hillside or coastal bluff, which adjusts after a sudden landslide. A new equilibrium is eventually achieved among slope, materials, and energy over time.

This threshold concept raises concern in the scientific community, especially if some natural systems reach their threshold limits. The relatively sudden collapse of ice shelves surrounding a portion of Antarctica or the Ward Hunt Ice Shelf's sudden collapse in 2003 on the north coast of Ellesmere Island, Canada, serve as examples of systems reaching a threshold and changing to a new status, that of disintegration. The bleaching (death) of living coral reefs worldwide accelerated dramatically during 1997 to 2004. Warming conditions and some pollution in the ocean led to such a threshold and coral system collapse. A sudden change to a new equilibrium arrangement may not be as desirable or supportive to us as present conditions (see inset photo in Figure 1.5).

Mount Pinatubo—Global System Impact A particularly dramatic example of interactions between volcanic eruptions and Earth systems illustrates the strength of spatial analysis in physical geography and the systems organization of this textbook. Mount Pinatubo in the Philippines erupted violently in 1991, injecting 15–20 million tons of ash and sulfuric acid mist into the upper atmosphere (Figure 1.6). The smoke and ash soared 40 km (25 mi) into the atmosphere and spread out 500 km wide (310 mi) in

System fluctuates around a stable average and maintains same operation level

(a) Passage of time ⟶

Coastal systems pass a threshold.

System fluctuates around a stable average, but exhibits a trend overall

Increasing system operations

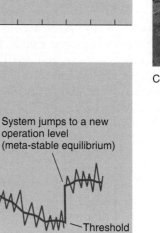

System jumps to a new operation level (meta-stable equilibrium)

Decreasing system operations

Threshold

(b) Passage of time ⟶

FIGURE 1.5 System equilibria: steady-state and dynamic.
(a) Some systems exhibit a steady-state equilibrium over time; system operations fluctuate around a stable average. Other systems are in a condition of dynamic equilibrium, with an increasing or decreasing operational trend as in (b). Rather than changing gradually, some systems may reach a *threshold* at which system operations lurch (change abruptly) to a new set of relations. Landslides along the Pacific Coast south of San Francisco destroyed homes and property as a threshold was reached and the bluffs collapsed. [Inset photo by Sam Sargent/Liaison Agency Inc.]

its mushroom-cloud top. This was the second greatest eruption during the twentieth century; Mount Katmai in Alaska (1912) was the only one greater. The eruption materials from Mount Pinatubo affected Earth systems in several ways, which are noted on the map.

As you progress through this book, you will see the story of Mount Pinatubo and its implications woven through seven chapters: Chapter 1 (systems theory), Chapter 4 (effects on energy budgets in the atmosphere), Chapter 6 (satellite images of the spread of debris by atmospheric winds), Chapter 10 (temporary effect on global atmospheric temperatures), Chapters 11 and 12 (volcanic processes), and Chapter 17 (past climatic effects of volcanoes). This illustrates the value of a systems approach, where instead of description only we can see the linkages and global impacts of such an event.

Systems in *Geosystems*   To further illustrate the conceptual systems organization used in this textbook, notice how chapters and portions of chapters are presented in simple sequences organized around the flow of energy, materials, and information (Figure 1.7). This book presents subjects in the same sequence in which they occur in nature. In this way, a logical progression is followed through topics that unfold according to the flow within individual systems or with time and the flow of events.

The sequence (Figure 1.7a)—*input* (components and driving force), *actions* (movements, processes, and storage

changes), *outputs* (results and consequences), and *human impacts/impact on humans* (measure of relevance)—is seen in Part 1 as an example. The Sun begins Chapter 2, the energy flows across space to the top of the atmosphere, through the atmosphere to the surface and surface energy budgets (Chapters 3 and 4). Then we look at the outputs of temperature (Chapter 5) and winds and ocean currents (Chapter 6). Note the same logical systems flow in the other three parts and in each chapter of this text—Part 2 and Chapter 16 also are shown as examples.

Models of Systems   A **model** is a simplified, idealized representation of part of the real world. Models are designed with varying degrees of generalization. You may have built or drawn a model of something in your life that was a simplification of the real thing—such as a model airplane or model house. Physical geographers use simple system models to demonstrate complex associations in the environment.

The simplicity of a model makes a system easier to understand and to simulate in experiments (for example, the leaf model in Figure 1.4). A good example is a model of the *hydrologic system*, which models Earth's entire water system, its related energy flows, and atmosphere, surface, and subsurface environments through which water moves (see Figure 9.1 in Chapter 9).

Adjusting the variables in a model produces differing conditions and allows predictions of possible system

Eruption cloud spread to 500-kilometer diameter, 35-km high.

GLOBAL IMPACT OF A VOLCANIC ERUPTION

Diffuse sunlight increases; slight enhancement of photosynthesis

Colorful twilight and dawn skies are seen worldwide

12 km³ of material is spread by atmospheric winds

Mount Pinatubo (15° N 120° E) violently erupts June 15, 1991

42% of globe is affected by a thin aerosol cloud (20° S to 30° N)

Atmospheric absorption of solar energy increases (2.5 W/m²)

Atmospheric albedo (reflectivity) increases 1.5% (4.3 W/m²)

15–20 million tons of ash and sulfuric acid mist blasted into the atmosphere

Net radiation at Earth's surface temporarily decreases average temperatures in the Northern Hemisphere by 0.5 C° (0.9 F°)

FIGURE 1.6 The eruption of Mount Pinatubo.
The 1991 Mount Pinatubo eruption, one of the largest volcanic eruptions in the twentieth century, affected the Earth–atmosphere system on a global scale. Geographers and other scientists use the latest technology to study how such eruptions affect the atmosphere's dynamic equilibrium. For a summary of the impacts from this eruption, refer to Chapter 12. [Inset photo by Van Cappellen/REA/SABA.]

Systems in *Geosystems*

(a)

Examples of systems organization in text:

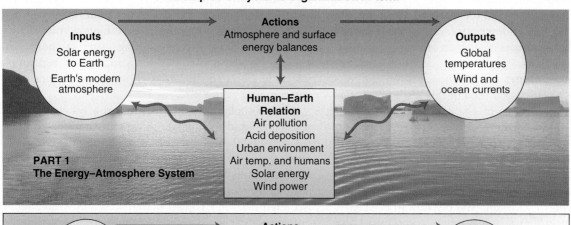

PART 1
The Energy–Atmosphere System

PART 2
The Water, Weather, and Climate Systems

Chapter 16
Oceans, Coastal Processes, and Landforms

(b)

FIGURE 1.7 The systems in *Geosystems*.
(a) Chapters, sections, and topics are organized around simple flow systems, or around time and the flow of events, at various scales. (b) This systems structure as it is applied to Parts 1 and 2. As a chapter example, turn to Chapter 16 and coastal processes. Note the outline headings used: "Coastal System Components," "Coastal System Actions," "Coastal System Outputs," and "Human Impact on Coastal Environments." Thus, Chapter 16 is organized along the systems flow shown in (a).

operations. A general circulation model (GCM) of the atmosphere (see Figure 10.30), operated by the Goddard Institute, accurately predicted the effect of Mount Pinatubo's ash on the atmosphere, the lowering and subsequent recovery of global air temperatures. However, predictions are only as good as the assumptions and accuracy built into the model. It is best to view a model for what it is—a simplification to help us understand complex processes.

We discuss many *system models* in this text, including the hydrologic system, water balance, surface energy budgets, earthquakes and faulting as outputs of Earth systems, river drainage basins, glacier mass budgets, soil profiles, and various ecosystems. Computer-based models are in use to study most natural systems—from the upper atmosphere, to climate change, to Earth's interior. Each of Earth's four "spheres" represents such a model of Earth's systems.

Earth's Four "Spheres"

Earth's surface is a vast area of 500 million square kilometers (193 million square miles) where four immense open systems interact. Figure 1.8 shows a simple model of three

abiotic (nonliving) systems overlapping to form the realm of the **biotic** (living) system. The abiotic spheres are the *atmosphere*, *hydrosphere*, and *lithosphere*. The biotic sphere is called the *biosphere*. Because these four systems are not independent units in nature, their boundaries are really transition zones rather than sharp barriers. As noted in the figure, these four spheres form the part structure in which chapters are grouped in this book. The various arrows in the figure show that content in each part and among chapters interrelate as we build our discussion of Earth's abiotic and biotic systems.

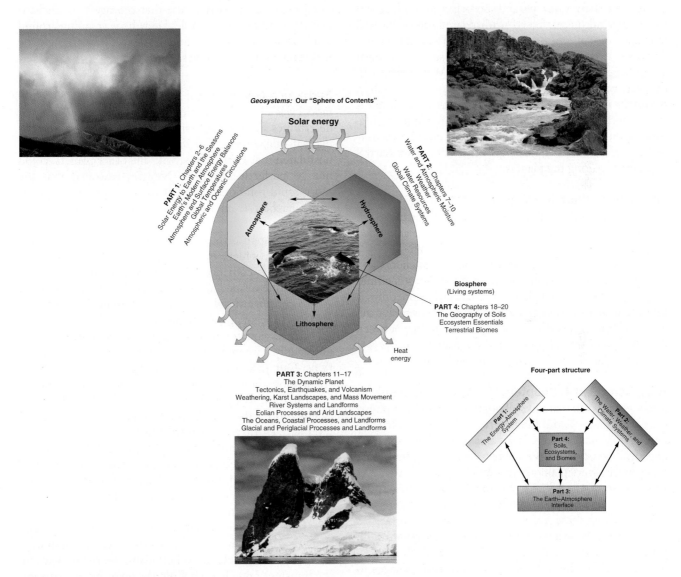

FIGURE 1.8 Earth's four spheres.
Each sphere is a model of vast Earth systems. This four-part structure is the organizational framework for *Geosystems*. [Photos by Bobbé Christopherson.]

> Part 1—Atmosphere: The Energy–Atmosphere System (Rainbow and thunderstorm squall line.)
> Part 2—Hydrosphere: The Water, Weather, and Climate Systems (Cascading stream in southwestern Iceland.)
> Part 3—Lithosphere: The Earth–Atmosphere Interface (Cape Renaud, Graham Land, Antarctica.)
> Part 4—Biosphere: Soils, Ecosystems, and Biomes (Humpback whales in the Atlantic Ocean.)

Atmosphere (Part 1, Chapters 2–6)   The **atmosphere** is a thin, gaseous veil surrounding Earth, held to the planet by the force of gravity. Formed by gases arising from within Earth's crust and interior and the exhalations of all life over time, the lower atmosphere is unique in the Solar System. It is a combination of nitrogen, oxygen, argon, carbon dioxide, water vapor, and trace gases.

Hydrosphere (Part 2, Chapters 7–10)   Earth's waters exist in the atmosphere, on the surface and in the crust near the surface. Collectively, these quantities form the **hydrosphere**. That portion of the hydrosphere that is frozen is the **cryosphere**—ice sheets, ice caps and fields, glaciers, ice shelves, sea ice, and subsurface ground ice and frozen ground (permafrost)—with related topics covered in chapters in all four parts of the book. Water of the hydrosphere exists in all three states: liquid, solid (the frozen cryosphere), and gaseous (water vapor). Water occurs in two general chemical conditions, fresh and saline (salty). It exhibits important heat-storage properties. And water is an extraordinary solvent. Water is the medium of life. Among the planets in the Solar System, only Earth possesses surface water in such quantity, adding to Earth's uniqueness among the planets.

Lithosphere (Part 3, Chapters 11–17)   Earth's crust and a portion of the upper mantle directly below the crust form the **lithosphere**. The crust is quite brittle compared with the layers deep beneath the surface, which move slowly in response to an uneven distribution of heat energy and pressure. In a broad sense, the term *lithosphere* sometimes refers to the entire solid planet. The soil layer sometimes is referred to as the *edaphosphere* and generally covers Earth's land surfaces. In this text, soils represent a bridge between the lithosphere and biosphere and are featured in the first chapter in Part 4, "Soils, Ecosystems, and Biomes."

Biosphere (Part 4, Chapters 18–20)   The intricate, interconnected web that links all organisms with their physical environment is the **biosphere**. Sometimes called the **ecosphere**, the biosphere is the area in which physical and chemical factors form the context of life. The biosphere exists in the overlap among the abiotic spheres, extending from the seafloor, and even the upper layers of the crustal rock, to about 8 km (5 mi) into the atmosphere. Life is sustainable within these natural limits. In turn, life processes have powerfully shaped the other three spheres through various interactive processes. The biosphere evolves, reorganizes itself at times, faces extinctions, and manages to flourish. Earth's biosphere is the only one known in the Solar System; thus, life as we know it is unique to Earth.

A Spherical Planet

We have all heard that some people believed Earth to be flat. Yet Earth's *sphericity* is not as modern a concept as many think. For instance, more than two millennia ago, the Greek mathematician and philosopher Pythagoras (ca. 580–500 B.C.) determined through observation that Earth is spherical. We do not know what observations led Pythagoras to this conclusion. Can you guess at what he saw to deduce Earth's roundness?

He might have noticed ships sailing beyond the horizon and apparently sinking below the water's surface, only to arrive back at port with dry decks. Perhaps he noticed Earth's curved shadow cast on the lunar surface during an eclipse of the Moon. He might have deduced that the Sun and Moon are not just the flat disks they appear to be in the sky, but are spherical, and that Earth must be a sphere as well.

Earth's sphericity was generally accepted by the educated populace as early as the first century A.D. Christopher Columbus, for example, knew he was sailing around a sphere in 1492; that is one reason why he thought he had reached the East Indies.

Earth as a Geoid   Until 1687, the spherical-perfection model was a basic assumption of **geodesy**, the science that determines Earth's shape and size by surveys and mathematical calculations. But in that year, Sir Isaac Newton postulated that Earth, along with the other planets, could not be perfectly spherical. Newton reasoned that the more-rapid rotational speed at the equator—the equator being farthest from the central axis of the planet and therefore moving faster—would produce an equatorial bulge in response to a greater centrifugal (pulling outward) force, which, in effect, pulls Earth's surface outward. He was convinced that Earth is slightly misshapen into what he termed an *oblate spheroid*, or more correctly, an *oblate ellipsoid* (*oblate* means "flattened"), with the oblateness occurring at the poles.

Earth's equatorial bulge and its polar oblateness are universally accepted and confirmed by satellite observations. Our modern era of Earth measurement is one of tremendous precision and is called the "geoidal epoch" because Earth is a **geoid**, meaning literally that "the shape of Earth is Earth-shaped." Imagine Earth's geoid as a sea-level surface that is extended uniformly worldwide, beneath the continents. Both heights on land and depths in the oceans are measured from this hypothetical surface. Think of the geoid surface as a balance between the gravitational attraction of Earth's mass and the centrifugal pull caused by Earth's rotation.

Today's satellite-based measurements are so precise that scientists know how Earth's shape is changing in ways that are unnoticeable to us. Following the last ice age and the transfer of much water from the cryosphere to the hydrosphere, Earth's polar oblateness was decreasing. This reversed in 1997 and the oblateness is now increasing this trend, with the mass redistribution of water as lower-latitude subpolar glaciers melt and the ocean warms. In other words, Earth's equatorial girth is expanding slightly. Also, Earth's shape changes slightly with the seasons as the crust compresses a little during each hemisphere's winter. Such is the excitement in modern geodesy!

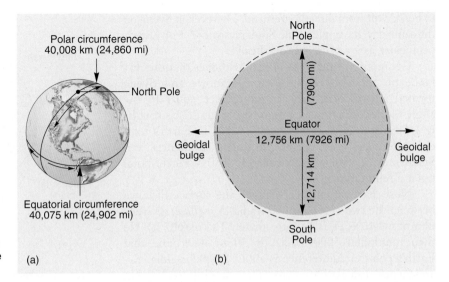

FIGURE 1.9 Earth's dimensions.
Earth's circumference (a) and diameter
(b)—equatorial and polar—are shown. The
dashed line is a perfect circle for reference
to Earth's geoid.

Figure 1.9 gives Earth's average polar and equatorial circumferences and diameters. The Greek geographer, astronomer, and librarian Eratosthenes (ca. 276–195 B.C.) first measured Earth's polar circumference more than 2200 years ago. The ingenious reasoning by which he arrived at his calculation follows.

Measuring Earth in 247 B.C.

Foremost among early geographers, Eratosthenes served as the librarian of Alexandria in Egypt during the third century B.C. He was in a position of scientific leadership, for Alexandria's library was the finest in the ancient world. Among his achievements was calculation of Earth's polar circumference to a high level of accuracy, quite a feat for 247 B.C. Here's how he did it. As you read, follow along on Figure 1.10.

Travelers told Eratosthenes that on June 21 they saw the Sun's rays shine directly to the bottom of a well at Syene, the location of present-day Aswan, Egypt. This meant that the Sun was directly overhead on that day. North of Syene in Alexandria, Eratosthenes knew from his own observations that the Sun's rays never were directly overhead, even at noon on June 21. This day is the longest day of the year and the day on which the noon Sun is at its northernmost position in the sky. Unlike objects in Syene, he knew objects in Alexandria always cast a noontime shadow. Using his knowledge of geometry, Eratosthenes conducted an experiment.

In Alexandria at noon on June 21, he measured the angle of a shadow cast by an obelisk (a perpendicular column). Knowing the height of the obelisk and measuring the length of the Sun's shadow from its base, he solved the triangle for the angle of the Sun's rays, which he determined

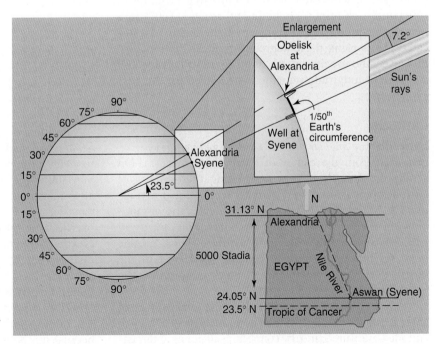

FIGURE 1.10 Eratosthenes' calculation.
Eratosthenes' work teaches the value of
observing carefully and integrating all observations with previous learning. Calculating Earth's circumference required
application of his knowledge of Earth–Sun
relationships, geometry, and geography to
his keen observations. His estimate was remarkably close to modern measurements.

to be 7.2° off from directly overhead. However, at Syene on the same day, the angle of the Sun's rays was 0° from a perpendicular, as it was directly overhead.

Geometric principles told Eratosthenes that the distance on the ground between Alexandria and Syene formed an arc of Earth's circumference equal to the angle of the Sun's rays at Alexandria. Since 7.2° is roughly 1/50 of the 360° (360° ÷ 7.2° = 50) in Earth's total circumference, the distance between Alexandria and Syene must represent approximately 1/50 of Earth's total polar circumference.

Next, Eratosthenes determined the surface distance between the two cities as 5000 stadia. A *stadium*, a Greek unit of measure, equals approximately 185 m (607 ft). He then multiplied 5000 stadia by 50 to determine that Earth's polar circumference is about 250,000 stadia. Eratosthenes' calculations convert to roughly 46,250 km (28,738 mi), which is remarkably close to the correct value of 40,008 km (24,860 mi) for Earth's polar circumference. Not bad for 247 B.C.! (Several values can be used for the distance of a Greek stadium—the 185 m used here represents an average value.)

Location and Time on Earth

An essential for geographic science is an internationally accepted, coordinated grid system to determine location on Earth. You can see the importance in the chapter-opening photo showing the Cape Hatteras Lighthouse location change. The terms *latitude* and *longitude* were in use on maps as early as the first century A.D., with the concepts themselves dating back to Eratosthenes and others.

The geographer, astronomer, and mathematician Ptolemy (ca. A.D. 90–168) contributed greatly to modern maps, and many of his terms and configurations are still used today. Ptolemy divided the circle into 360 degrees (360°), with each degree comprising 60 minutes (60′) and each minute including 60 seconds (60″) in a manner adapted from the ancient Babylonians. He located places using these degrees, minutes, and seconds. However, the precise length of a degree of latitude and a degree of longitude remained unresolved for the next 17 centuries.

Latitude

Latitude *is an angular distance north or south of the equator*, measured from the center of Earth (Figure 1.11a). On a map or globe, the lines designating these angles of latitude run east and west, parallel to the equator (Figure 1.11b). Because Earth's equator divides the distance between the North Pole and the South Pole exactly in half, it is assigned the value of 0° latitude. Thus, latitude increases in value from the equator northward to the North Pole, at 90° north latitude, and southward to the South Pole, at 90° south latitude.

A line connecting all points along the same latitudinal angle is called a **parallel**. In the figure, an angle of 49° north latitude is measured, and, by connecting all points at this latitude, we have the 49th parallel. Thus, *latitude* is the name of

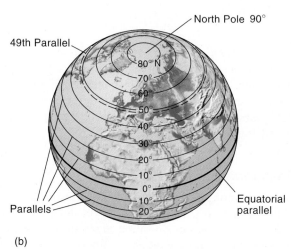

FIGURE 1.11 Parallels of latitude.
(a) Latitude is measured in degrees north or south of the equator, which is 0°. Each pole is at 90°. Note the measurement of 49° latitude. (b) These angles of latitude determine parallels along Earth's surface. Do you know your present latitude?

the angle (49° north latitude), *parallel* names the line (49th parallel), and both indicate distance north of the equator. The 49th parallel is a significant one in the Western Hemisphere, for it forms the boundary between Canada and the United States from Minnesota to the Pacific Ocean.

Latitude is readily determined by observing *fixed celestial objects* such as the Sun or the stars, a method dating to ancient times. During daylight hours, the angle of the Sun above the horizon indicates the observer's latitude, after adjustment is made for the season and for the time of day. Because Polaris (the North Star) is almost directly overhead at the North Pole, persons anywhere in the Northern Hemisphere can determine their latitude at night simply by sighting Polaris and measuring its angle above the local horizon (Figure 1.12). The angle of elevation of Polaris above the horizon equals the latitude of the observation point.

In the Southern Hemisphere, Polaris cannot be seen, because it is below the horizon. Instead, latitude measurement south of the equator is accomplished by sighting on

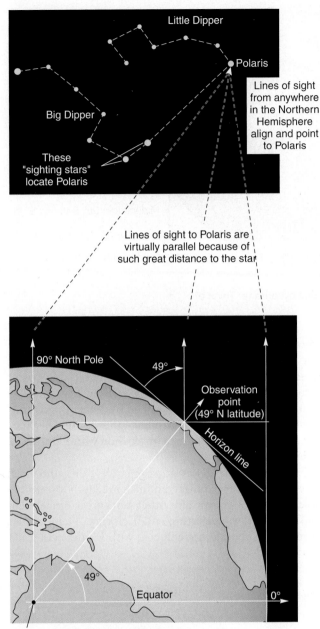

FIGURE 1.12 Determining latitude by using Polaris (the North Star).
To locate Polaris from anywhere in the Northern Hemisphere, you can use the "sighting stars" in the Big Dipper constellation. Next, measure the angular distance between Polaris and the local horizon. This angular distance of Polaris above the horizon is the same as your latitude. In this case, Polaris appears 49° above the horizon, so you are standing at 49° north latitude. (Note that Polaris is at such a great distance from Earth that lines of sight from anywhere in the Northern Hemisphere are considered parallel.) On the next clear night, take a protractor and stick or ruler and sight on Polaris to determine your latitude manually.

a constellation that points to a celestial location above the South Pole. This indicator constellation is the Southern Cross (*Crux Australis*).

Latitudinal Geographic Zones Natural environments differ dramatically from the equator to the poles,

in both their processes and their appearance. These differences result from the amount of solar energy received, which varies by latitude and season of the year. As a convenience, geographers identify *latitudinal geographic zones* as regions with fairly consistent qualities. Figure 1.13 portrays these zones, their locations, and their names: *equatorial* and *tropical*, *subtropical*, *midlatitude*, *subarctic* or *subantarctic*, and *arctic* or *antarctic*. "Lower latitudes" are those nearer the equator, whereas "higher latitudes" are those nearer the poles. These generalized latitudinal zones are useful for reference and comparison, but they do not have rigid boundaries, as one zone transitions into another.

The *Tropic of Cancer* (23.5° north parallel) and the *Tropic of Capricorn* (23.5° south parallel) are the most extreme northern and southern parallels that experience perpendicular (directly overhead) rays of the Sun at local noon. When the Sun arrives overhead at these tropics, it marks the first day of summer in each hemisphere. (The tropics are discussed further in Chapter 2.) The *Arctic Circle* (66.5° north parallel) and the *Antarctic Circle* (66.5° south parallel) are the parallels farthest from the poles that still experience 24 uninterrupted hours of night during local winter or of day during local summer.

Longitude

Longitude *is an angular distance east or west of a point on Earth's surface*, measured from the center of Earth (Figure 1.14a). On a map or globe, the lines designating these angles of longitude run north and south (Figure 1.14b). A line connecting all points along the same longitude is a **meridian**. In the figure a longitudinal angle of 60° E is measured. These meridians run at right angles (90°) to all parallels, including the equator. Thus, *longitude* is the name of the angle, *meridian* names the line, and both indicate distance east or west of an arbitrary **prime meridian**—a meridian designated as zero degrees (Figure 1.14b). Earth's prime meridian passes through the old Royal Observatory at Greenwich, England, as set by treaty—the *Greenwich prime meridian*.

Determination of Latitude and Longitude Table 1.2 compares the length of latitude and longitude degrees. Because meridians of longitude converge toward the poles, the actual distance on the ground spanned by a degree of longitude is greatest at the equator (where meridians separate to their widest distance apart) and diminishes to zero at the poles (where meridians converge). Note the consistent distance represented by a degree of latitude from equator to poles, yet the decreasing value a degree of longitude covers as meridians converge toward each pole.

We have noted that latitude is easily determined by sighting the Sun or the North Star or by using the Southern Cross as a pointer. In contrast, a method of accurately determining longitude, especially at sea, remained a major difficulty in navigation until the late 1700s. The key to measuring the longitude of a place lies

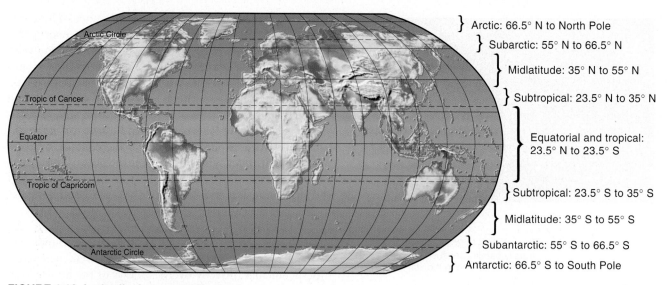

} Arctic: 66.5° N to North Pole

} Subarctic: 55° N to 66.5° N

] Midlatitude: 35° N to 55° N

} Subtropical: 23.5° N to 35° N

] Equatorial and tropical:
23.5° N to 23.5° S

} Subtropical: 23.5° S to 35° S

} Midlatitude: 35° S to 55° S

} Subantarctic: 55° S to 66.5° S

} Antarctic: 66.5° S to South Pole

FIGURE 1.13 Latitudinal geographic zones.
Geographic zones are generalizations that characterize various regions by latitude. Think of these as transitional into one another over broad areas.

in accurately knowing time. The relation between time and longitude and an exciting chapter in human discovery is the topic of Focus Study 1.2 on page 20.

Today, using a handheld instrument that reads radio signals from satellites, you can accurately calibrate latitude, longitude, and elevation. This is the **Global Positioning System (GPS)** technology. News Report 1.1 discusses the dramatic applications of GPS on page 21.

Great Circles and Small Circles

Great circles and small circles are important concepts that help summarize latitude and longitude (Figure 1.15). A **great circle** is any circle of Earth's circumference whose center coincides with the center of Earth. An infinite number of great circles can be drawn on Earth. Every meridian is one-half of a great circle that passes through the poles. On flat maps, airline and shipping routes appear to arch their way across oceans and landmasses. These are *great circle routes*, the shortest distance between two points on Earth (discussion with Figure 1.22 is just ahead).

Only one parallel is a great circle—the equatorial parallel. All other parallels diminish in length toward the poles and, along with any other non-great circles that one might draw, constitute **small circles**. These circles have centers that do not coincide with Earth's center.

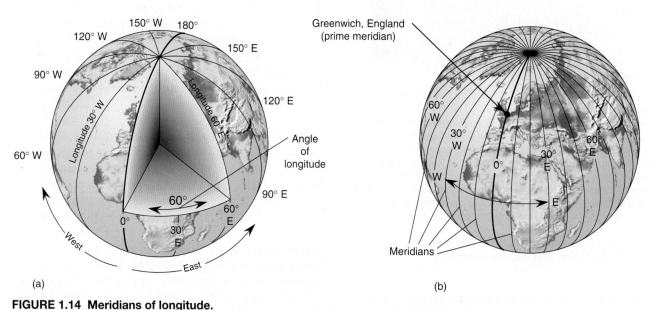

(a)

(b)

FIGURE 1.14 Meridians of longitude.
(a) Longitude is measured in degrees east or west of a 0° starting line, the prime meridian. Note the measurement of 60° E longitude. (b) Angles of longitude measured from the prime meridian determine other meridians. The prime meridian is drawn from the North Pole through the Royal Observatory in Greenwich, England, to the South Pole. North America is west of Greenwich; therefore, it is in the Western Hemisphere. Do you know your present longitude?

Table 1.2 Physical Distances Represented by Degrees of Latitude and Longitude

Latitudinal Location	Latitude Degree Length km	(mi)	Longitude Degree Length km	(mi)
90° (poles)	111.70	(69.41)	0	(0)
60°	111.42	(69.23)	55.80	(34.67)
50°	111.23	(69.12)	71.70	(44.55)
40°	111.04	(69.00)	85.40	(53.07)
30°	110.86	(68.89)	96.49	(59.96)
0° (equator)	110.58	(68.71)	111.32	(69.17)

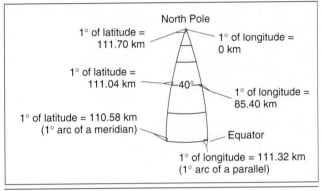

Figure 1.16 combines latitude and parallels with longitude and meridians to illustrate Earth's complete coordinate grid system. Note the dot that marks our measurement of 49° N by 60° E, a location in western Kazakstan. Next time you look at a world globe, follow the parallel and meridian that converge on your location.

Prime Meridian and Standard Time

Coordination of international trade, airline schedules, business and agricultural activities, and daily living depends on a worldwide time system. Today, we take for granted standard time zones and an agreed-upon prime meridian, but such a standard is a relatively recent development.

Setting time was not a great problem in small European countries, most of which are less than 15° wide. But in North America, which spans more than 90° of longitude (the equivalent of six 15° time zones), the problem was serious. In 1870, railroad travelers going from Maine to San Francisco made 22 adjustments to their watches to stay consistent with local time.

In Canada, Sir Sanford Fleming led the fight for standard time and for an international agreement on a prime meridian. His struggle led the United States and Canada to adopt a standard time in 1883. Today, only three adjustments are needed in the continental United States—from Eastern Standard Time to Central, Mountain, and Pacific—and four changes across Canada.

FIGURE 1.15 Great circles and small circles.
(a) Examples of great circles and small circles on Earth. (b) Any plane that divides Earth into equal halves will intersect the globe along a great circle; this great circle is a full circumference of the globe and is the shortest distance between any two surface points. (c) Any plane that splits the globe into unequal portions will intersect the globe along a small circle.

Focus Study 1.2

The Timely Search for Longitude

Unlike latitude, longitude cannot be determined readily from fixed celestial bodies. The problem is Earth's rotation, which constantly changes the apparent position of the Sun and stars. Determining longitude is particularly critical at sea, where no landmarks are visible.

In his historical novel *Shogun*, author James Clavell expressed the frustration of the longitude problem through his pilot, Blackthorn: "Find how to fix longitude and you're the richest man in the world.... The Queen, God bless her, 'll give you ten thousand pound and a dukedom for the answer to the riddle.... Out of sight of land you're always lost, lad."*

In the early 1600s, Galileo explained that longitude could be measured by using two clocks. Any point on Earth takes 24 hours to travel around the full 360° of one rotation (1 day). If you divide 360° by 24 hours, you find that any point on Earth travels through 15° of longitude every hour. Thus, if there were a way to measure time accurately at sea, a comparison of two clocks could give a value for longitude. One clock would indicate the time back at home port (Figure 1.2.1). The other clock would be reset at local noon each day, as determined by the highest Sun position in the sky (solar zenith). The time difference then would indicate the longitudinal difference traveled: 1 hour for each 15° of longitude. The principle was sound; all that was needed were accurate clocks. Unfortunately, the pendulum clock invented by Christian Huygens in 1656 did not work on the rolling deck of a ship at sea!

In 1707, the British lost four ships and 2000 men in a sea tragedy that was blamed specifically on the longitude problem. In response, Parliament passed an act in 1714—"Publik Reward ... to Discover the Longitude at Sea"—and authorized a prize

FIGURE 1.2.1 Clock time determines longitude. Using two clocks on a ship to determine longitude. For example, if the shipboard clock reads local noon and the clock set for home port reads 3:00 P.M., ship time is 3 hours earlier than home time. Therefore, calculating 3 hours at 15° per hour puts the ship at 45° west longitude from home port.

worth more than $2 million in today's dollars to the first successful inventor of an accurate seafaring clock. The Board of Longitude was established to judge any devices submitted.

John Harrison, a self-taught country clockmaker, began work on the problem in 1728 and finally produced his brilliant marine chronometer, known as Number 4, in 1760. The clock was tested on a voyage to Jamaica in 1761. When taken ashore and compared to land-based longitude, Harrison's ingenious Number 4 was only 5 seconds slow, an error that translates to only 1.25′, or 2.3 km (1.4 mi), well within Parliament's standard. After many delays, Harrison finally received most of the prize money in his last years of life.

With his marine clocks, John Harrison tested the waters of space–time. He succeeded, against all odds, in using the fourth—temporal—dimension to link points on the three-dimensional globe. He wrested the world's whereabouts from the stars, and locked the secret in a pocket watch.[†]

From that time it was possible to determine longitude accurately on land and sea, as long as everyone agreed upon a meridian to use as a reference for time comparisons—the Royal Observatory in Greenwich, England. In this modern era of atomic clocks and satellites in mathematically precise orbits, we have far greater accuracy available for the determination of longitude on Earth's surface and a basis for precise navigation.

*From J. Clavell, *Shogun*, Copyright © 1975 by James Clavell (New York: Delacorte Press, a division of Dell Publishing Group, Inc.), p. 10.

[†]From Dava Sobel, *Longitude, The Story of a Lone Genius Who Solved the Greatest Scientific Problem of His Time* (New York: Walker and Co., 1995), p. 175.

News Report 1.1

GPS: A Personal Locator

The Global Positioning System (GPS) comprises 24 orbiting satellites, in six orbital planes, that transmit navigational signals for Earth-bound use (backup GPS satellites are in orbital storage as replacements). Originally devised in the 1970s by the Department of Defense for military purposes, the present system is commercially available worldwide.

A small receiver, some about the size of a pocket radio, receives signals from four or more satellites at the same time, calculates latitude and longitude within 10-m accuracy (33 ft) and elevation within 15 m (49 ft), and displays the results. With the shutdown in 2000 of the Pentagon Selective Availability, commercial resolution is the same as for military applications and its Precise Positioning Service (PPS). *Differential GPS (DGPS)* increases accuracy by comparing readings with another base station (reference receiver) for differential correction (Figure 1.1.1).

You already read how satellite measurements refine our knowledge of Earth's exact shape and how this shape is changing. GPS is useful for diverse applications, such as ocean navigation, land surveying, tracking small changes in Earth's crust, managing the movement of fleets of trucks, mining and resource mapping, tracking wildlife migration and behavior, and environmental planning.

Relative to earthquakes in southern California, Jet Propulsion Laboratory (JPL) operates the GPS Observation Office that monitors a network of 250 seismic stations of the Southern California Integrated GPS Network. This system can record fault movement as small as 1 mm (0.04 in.). Measuring and mapping an ominous bulge on an Oregon volcano is another present-day application. GPS also is useful to the backpacker and sportsperson.

Commercial airlines use GPS to improve accuracy of routes flown and thus increase fuel efficiency. Scientists used GPS to accurately determine the height of Mount Everest in the Himalayan Mountains—now 8850 m

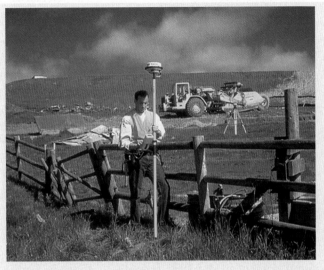

FIGURE 1.1.1 GPS in action.
A GPS unit in operation for surveying and location analysis in an environmental study. [Photo courtesy of Trimble Navigation Ltd., Sunnyvale, California.]

compared to the former 8848 m (29,035 ft; 29,028 ft) (Figure 1.1.2). In contrast, GPS measurements of Mount Kilimanjaro lowered its summit from 5895 m to 5892 m (19,340 ft; 19,330 ft).

In "precision agriculture," farmers use GPS to determine crop yields on specific parts of their farms. A detailed plot map is made to guide the farmer to where more fertilizer, proper seed distribution, irrigation applications, or other work is needed. A computer and GPS unit on board the farm equipment guides the work. This is the science of *variable-rate technology*, made possible by GPS. As an example among others, Hawkeye Community

College in Waterloo, Iowa, offers several degree programs that use GPS technology and GIS methods.

The importance of GPS to geography is obvious because this precise technology reduces the need to maintain ground control points for location, mapping, and spatial analysis. Instead, geographers working in the field can determine their position accurately as they work. Boundaries and data points in a study area are easily determined and entered into a data base, reducing the need for traditional surveys. For this and myriad other applications, GPS sales are exceeding $10 billion a year. Additional frequencies were added in 2003 and 2006, which increased accuracy significantly. Also, the European Union is launching its own GPS system of 20 satellites beginning in 2005. (For a GPS overview, see http://www.colorado.edu/geography/gcraft/notes/gps/gps_f.html.)

FIGURE 1.1.2 GPS used to measure Everest's summit height.
Installation of a Trimble 4800 GPS unit at Earth's highest benchmark, only 18 m (60 ft) below the 8850 m (29,035 ft) summit of Mount Everest. This is a GPS-based revised height measurement announced in 1999. Scientists are using the data collected from a regional GPS network to accurately analyze the height of the world's tallest mountain and the rate that the mountain range is moving due to tectonic forces.
[Photo and GPS installation by Wally Berg, May 20, 1998.]

FIGURE 1.16 Earth's coordinate grid system.
Latitude and parallels, longitude and meridians, allow all places on Earth to be precisely located. The dot is at 49° N latitude by 60° E longitude.

Twenty-seven countries attended the 1884 International Meridian Conference in Washington, DC. Before that year, most nations used their own national capital as a prime meridian for their land maps, whereas more than 70% of the world's merchant ships were using Greenwich as a prime meridian on marine charts. After lengthy

debate at the conference, most participating nations chose the highly respected Royal Observatory at Greenwich, London, England, as the place for the prime meridian of 0° longitude for all maps. Thus, a world standard was set—**Greenwich Mean Time (GMT)**—and a consistent Universal Time was established. (See http://www.gmt2000.co.uk/meridian/place/plco0a1.htm.)

The basis of time is that Earth revolves 360° every 24 hours, or 15° per hour (360° ÷ 24 = 15°). Thus, a time zone of 1 hour is established for each increment of longitude, or 7.5° on either side of a *central meridian*. Assuming it is 9:00 P.M. in Greenwich, then it is 4:00 P.M. in Baltimore (+5 hr), 3:00 P.M. in Oklahoma City (+6 hr), 2:00 P.M. in Salt Lake City (+7 hr), 1:00 P.M. in Seattle and Los Angeles (+8 hr), noon in Anchorage (+9 hr), and 11:00 A.M. in Honolulu (+10 hr). To the east, it is midnight in Ar Riyāḍ, Saudi Arabia (−3 hr). (The designation A.M. is for *ante meridiem*, "before noon," whereas P.M. is for *post meridiem*, "after noon." A 24-hour clock avoids the use of these designations.)

As you can see from the modern international time zones in Figure 1.17, national boundaries and political considerations distort time boundaries. For example, China spans four time zones, but its government decided to keep the entire country operating at the same time. Thus, in some parts of China clocks are several hours off from what the Sun is doing. In the United

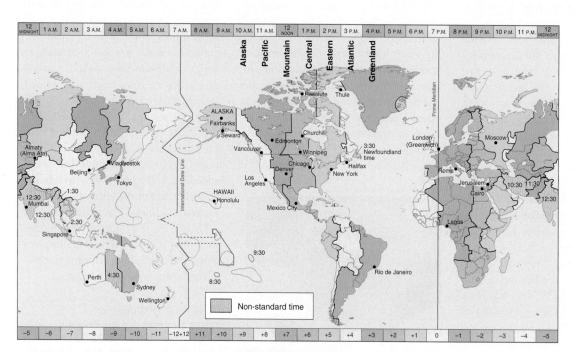

FIGURE 1.17 Modern international standard time zones.
Numbers along the bottom of the map indicate how many hours each zone is earlier (plus sign) or later (minus sign) than Coordinated Universal Time (UTC) at the prime meridian. The United States has five time zones; Canada is divided into six. If it is 7 P.M. in Greenwich, determine the present time in Moscow, London, Halifax, Chicago, Winnipeg, Denver, Los Angeles, Fairbanks, Honolulu, Tokyo, and Singapore. The island country of Kiribati placed itself east of the International Date Line so it is first to experience each new day. These distortions of the IDL only apply to the countries and their territorial waters and not to international waters between them and the 180th meridian. [Adapted from Standard Time Zone Chart of the World, Defense Mapping Agency, Bethesda, Maryland. See http://aa.usno.navy.mil/faq/docs/world_tzones.html.]

States, parts of Florida and west Texas are in the same time zone.

International Date Line

An important corollary of the prime meridian is the 180° meridian on the opposite side of the planet. This meridian is called the **International Date Line** and marks the place where each day officially begins (at 12:01 A.M.). From this "line" the new day sweeps westward. This *westward* movement of time is created by Earth's turning *eastward* on its axis.

At the International Date Line, the west side of the line is always one day ahead of the east side. No matter what time of day it is when the line is crossed, the calendar changes a day (Figure 1.18). Note in the illustration the departures from the IDL and the 180° meridian; this deviation is due to local administrative and political preferences. The political control of island nations extends only 3.5 nautical miles (4 mi) offshore. So in 1995 when Kiribati moved the IDL designation to east of their islands so they would experience "first dawn" on Earth, the political move did not apply to the ocean between their territorial waters and the 180th meridian.

Locating the date line in the sparsely populated Pacific Ocean minimizes most local confusion. However, early explorers before the date-line concept were "lost." For example, Magellan's crew returned from the first circumnavigation of Earth in 1522, confident from their ship's log that the day of their arrival home was Wednesday, September 7. They were shocked when informed by insistent local residents that it was actually Thursday, September 8. Of course, without an International Date Line they had no idea that they must advance a day somewhere when sailing around the world in a westward direction. Imagine the confusion as the crew accounted for each day in their log!

Coordinated Universal Time

For decades, Greenwich Mean Time from the Royal Observatory's astronomical clocks was the world's Universal Time (UT) standard for accuracy. GMT was broadcast using radio time signals as early as 1910. The French government took the initiative in 1912 and called a gathering of nations to better coordinate the various radio time signals from many countries. At this conference, GMT was made standard, and a new organization established to be the custodian of the most "exact" time—the International Bureau of Weights and Measures (BIPM) outside Paris (see http://www.bipm.fr/). Progress in accurately measuring time was made rapidly with the invention of a quartz clock in 1939 and atomic clocks in the early 1950s.

The **Coordinated Universal Time (UTC*)** time signal system replaced GMT universal time in 1972 and became the legal reference for official time in all countries. Although the prime meridian still runs through Greenwich, UTC is based on average time calculations collected by the BIPM near Paris and broadcast worldwide. You might still see official UTC referred to as GMT or Zulu time.

Regular vibrations (natural frequency) of cesium atoms in primary standard clocks measure the length of a second and UTC—accuracies now range down to 15 decimal places, or nanoseconds of time. Time and Frequency Services of the National Institute for Standards and Technology (NIST), U.S. Department of Commerce, operates several of the most advanced clocks. (For more on time, call 303-499-7111 or 808-335-4363, or see http://nist.time.gov/ for UTC.) In Canada, the Institute for Measurement Standards, National Research Council Canada, participates in determining UTC (English, 613-745-1578; French, 613-745-9426; for more see http://www.nrc.ca/).

Daylight Saving Time

In many countries, time is set ahead 1 hour in the spring and set back 1 hour in the fall—a practice known as **daylight saving time**. The idea to extend daylight for early evening activities (at the expense of daylight in the morning) was first proposed by Benjamin Franklin. It was not adopted until World War I and again in World War II, when Great Britain, Australia, Germany, Canada, and the United States used the practice to save energy (one less hour of artificial lighting needed).

In 1986, the United States and Canada increased daylight saving time. Time now "springs forward" 1 hour on the first Sunday in April and "falls back" an hour on the last Sunday in October, except in a few places that do not use daylight saving time (Hawai'i, Arizona, portions of Indiana, and Saskatchewan). In Europe, the last Sundays in March and October are used to begin and end what is called "summer time." (See http://webexhibits.org/daylightsaving/.)

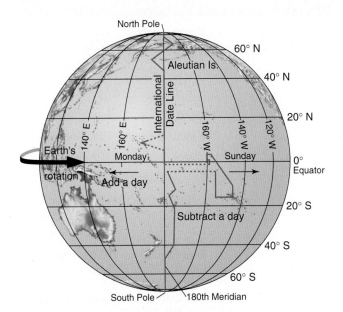

FIGURE 1.18 International Date Line.
International Date Line location, approximately along the 180th meridian (see the IDL location on Figure 1.17).
Note that it is officially one day later west of the IDL.

*UTC is in use because agreement was not reached on whether to use the English word order, CUT, or the French order, TUC. UTC was the compromise and is recommended for all timekeeping applications; use of the term GMT is discouraged.

Maps, Scales, and Projections

The earliest known graphic maps date to 2300 B.C., when the Babylonians used clay tablets to record information about the region of the Tigris and Euphrates Rivers (the area of modern-day Iraq). Today, the making of maps and charts is a specialized science as well as an art, blending aspects of geography, engineering, mathematics, graphics, computer science, and artistic specialties. It is similar in ways to architecture, in which aesthetics and utility combine to produce a useful product.

A **map** is a generalized view of an area, usually some portion of Earth's surface, as seen from above and greatly reduced in size. **Cartography** is that part of geography that involves mapmaking. Maps are critical tools with which geographers depict spatial information and analyze spatial relationships.

We all use maps at some time to visualize our location and our relationship to other places, or maybe to plan a trip, or to coordinate commercial and economic activities. Have you found yourself looking at a map, planning real and imagined adventures to far-distant places? Maps are wonderful tools! Understanding a few basics about maps is essential to our study of physical geography.

The Scale of Maps

Architects, toy designers, and mapmakers have something in common: They all create scale models. They reduce real things and places to the more convenient scale of a drawing, a model car, train, or plane, a diagram, or a map. An architect renders a blueprint of a structure to guide the building contractors, selecting a scale so that a centimeter (or inch) on the drawing represents so many meters (or feet) on the proposed building. Often, the drawing is 1/50 to 1/100 of real size.

The cartographer does the same thing in preparing a map. The ratio of the image on a map to the real world is called **scale**; it relates a unit on the map to a similar unit on the ground. A 1:1 scale means that a centimeter on the map represents a centimeter on the ground (although this is an impractical map scale, for the map is as large as the area mapped!). A more appropriate scale for a local map is 1:24,000, in which 1 unit on the map represents 24,000 identical units on the ground.

Cartographers express map scale as a written scale, a representative fraction, or a graphic scale (Figure 1.19). A *written scale* simply states the ratio—for example, "1 centimeter to 1 kilometer" or "1 inch to 1 mile." A *representative fraction* (RF, or *fractional scale*) is expressed with either a colon or a slash, as in 1:125,000 or 1/125,000. No actual units of measurement are mentioned because any unit is applicable as long as both parts of the fraction are in the same unit: 1 cm to 125,000 cm, 1 in. to 125,000 in., or even 1 arm length to 125,000 arm lengths, and so on.

A *graphic scale*, or bar scale, is a bar graph with units to allow measurement of distances on the map. An important advantage of a graphic scale is that if the map is enlarged or reduced, the graphic scale enlarges or reduces

FIGURE 1.19 Map scale.
Three common expressions of map scale—written scale, representative fraction, and graphic scale.

along with the map. In contrast, written and fractional scales become incorrect with enlargement or reduction. As an example, you can shrink a map from 1:24,000 to 1:63,360, but the scale will still say "1 in. to 2000 ft," instead of the new correct scale of 1 in. to 5280 ft (1 mi).

Scales are *small*, *medium*, and *large*, depending on the ratio described. In relative terms, a scale of 1:24,000 is a large scale, whereas a scale of 1:50,000,000 is a small scale. The greater the denominator in a fractional scale (or the number on the right in a ratio expression), the smaller the scale and the more abstract the map is in relation to what is being mapped. Examples of selected representative fractions and written scales are listed in Table 1.3 for small-, medium-, and large-scale maps.

Table 1.3	Sample Representative Fractions and Written Scales for Small-, Medium-, and Large-Scale Maps		
System	**Scale Size**	**Representative Fraction**	**Written System Scale**
English	Small	1:3,168,000	1 in. = 50 mi
		1:2,500,000	1 in. = 40 mi
		1:1,000,000	1 in. = 16 mi
		1:500,000	1 in. = 8 mi
		1:250,000	1 in. = 4 mi
	Medium	1:125,000	1 in. = 2 mi
		1:63,360	1 in. = 1 mi
		(or 1:62,500)	
		1:31,680	1 in. = 0.5 mi
		1:30,000	1 in. = 2500 ft
	Large	1:24,000	1 in. = 2000 ft
System		**Representative Fraction**	**Written System Scale**
Metric		1:1,000,000	1 cm = 10.0 km
		1:50,000	1 cm = 0.50 km
		1:25,000	1 cm = 0.25 km
		1:20,000	1 cm = 0.20 km

If there is a globe or map available in your library or classroom, check to see the scale at which it was drawn. See if you can find examples of written, representative, and graphic scales on wall maps, highway maps, and in atlases. In general, do you think a world globe is a small- or a large-scale map of Earth's surface?

Map Projections

A globe is not always a helpful map representation of Earth. When you go on a trip, you need more detailed information than a globe can provide. Consequently, to provide local detail, cartographers prepare large-scale flat maps, which are two-dimensional representations (scale models) of our three-dimensional Earth. Unfortunately, such conversion from three dimensions to two causes distortion.

A globe is the only true representation of *distance, direction, area, shape,* and *proximity.* A flat map distorts those properties. Therefore, in preparing a flat map, the cartographer must decide which characteristic to preserve, which to distort, and how much distortion is acceptable.

To understand this problem, consider these important properties of a *globe*:

- Parallels always are parallel to each other, always are evenly spaced along meridians, and always decrease in length toward the poles.
- Meridians converge at both poles and are evenly spaced along any individual parallel.
- The distance between meridians decreases toward poles, with the spacing between meridians at the 60th parallel equal to one-half the equatorial spacing.
- Parallels and meridians always cross each other at right angles.

The problem is that all these qualities cannot be reproduced on a flat surface. Simply taking a globe apart and laying it flat on a table illustrates the challenge faced by cartographers (Figure 1.20). You can see the empty spaces that open up between the sections, or gores, of the globe. This reduction of the spherical Earth to a flat surface is called a **map projection.** Thus, no flat map projection of Earth can ever have all the features of a globe. Flat maps always possess some degree of distortion—much less for large-scale maps representing a few kilometers; much more for small-scale maps covering individual countries, continents, or the entire world.

Properties of Projections There are many projections, four of which are shown in Figure 1.21. *The best projection is always determined by its intended use.* The major decisions in selecting a map projection involve the properties of **equal area** (equivalence) and **true shape** (conformality). If a cartographer selects equal area as the desired trait—for example, for a map showing the distribution of world climates—then true shape must be sacrificed by *stretching* and *shearing*, which allows parallels and meridians to cross at other than right angles. On an equal-area map, a coin covers the same amount of surface area

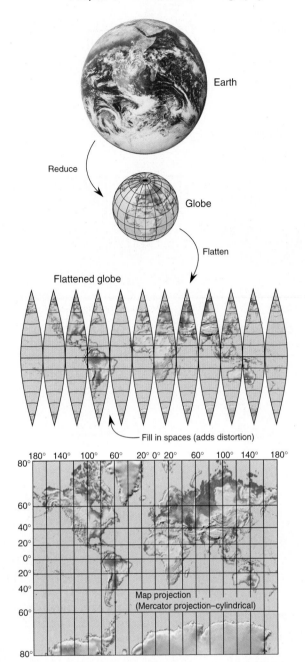

FIGURE 1.20 From globe to flat map.
Conversion of the globe to a flat map projection requires decisions about which properties to preserve and the amount of distortion that is acceptable. [NASA astronaut photo.]

no matter where you place it on the map. In contrast, a cartographer selects the property of true shape, as for a map used for navigational purposes, then equal area must be sacrificed, and the scale will actually change from one region of the map to another.

The Nature and Classes of Projections Despite the fact that modern cartographic technology uses mathematical constructions and computer-assisted graphics, the word *projection* is still used. The term comes from times past, when geographers actually projected the shadow of a wire-skeleton globe onto a geometric surface. The wires

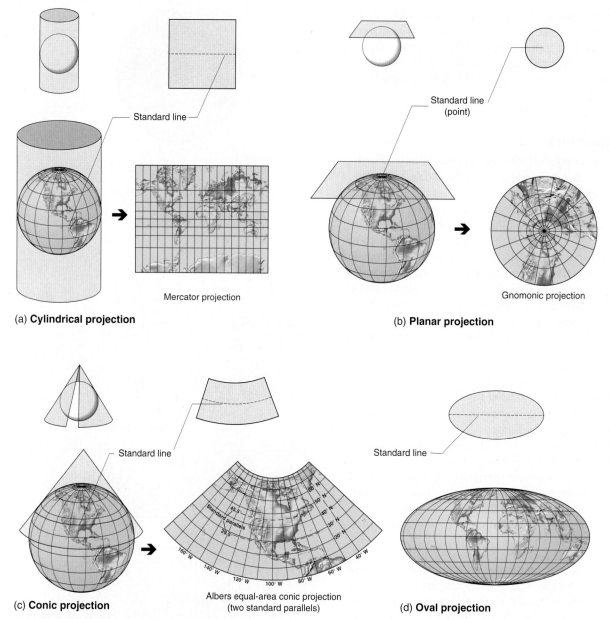

FIGURE 1.21 Classes of map projections.
Four general classes and perspectives of map projections—cylindrical, planar, conic, and oval projections.

represented parallels, meridians, and outlines of the continents. A light source then cast a shadow pattern of latitude and longitude lines from the globe onto various geometric surfaces, such as a *cylinder*, *plane*, or *cone*.

Figure 1.21 illustrates the general classes of map projections and the perspectives from which they are generated. The classes shown include the cylindrical, planar (or azimuthal), and conic. Another class of projections, which cannot be derived from this physical-perspective approach, is the nonperspective oval shape. Still other projections are derived from purely mathematical calculations.

With projections, the contact line or contact point between the wire globe and the projection surface—called a *standard line* or *standard point*—is *the only place where all globe properties are preserved.* Thus, a *standard parallel* or *standard meridian* is a standard line true to scale along its

entire length without any distortion. Areas away from this critical tangent line or point become increasingly distorted. Consequently, this area of optimum spatial properties should be centered on the region of interest so that greatest accuracy is preserved there.

The commonly used **Mercator projection** (from Gerardus Mercator, A.D. 1569) is a cylindrical projection (Figure 1.21a). The Mercator is a true-shape projection, with meridians appearing as equally spaced straight lines and parallels appearing as straight lines that are spaced closer together near the equator. The poles are infinitely stretched, with the 84th north parallel and 84th south parallel fixed at the same length as that of the equator. Note in Figures 1.20 and 1.21a that the Mercator projection is cut around the 80th parallel in each hemisphere because of the severe distortion at higher latitudes.

(a) Gnomonic Projection

(b) Mercator Projection (conformal, true shape)

FIGURE 1.22 Determining great circle routes.
A gnomonic projection (a) is used to determine the shortest distance—great circle route—between San Francisco and London, because on this projection the arc of a great circle is a straight line. This great circle route is then plotted on a Mercator projection (b), which has true compass direction. Note that straight lines of constant direction (or bearing) on a Mercator projection—rhumb lines—are not the shortest route in terms of distance.

Unfortunately, Mercator classroom maps present false notions of the size (area) of midlatitude and poleward land-masses. A dramatic example on the Mercator projection is Greenland, which looks bigger than all of South America. In reality, Greenland is only one-eighth the size of South America and is actually 20% smaller than Argentina alone.

The advantage of the Mercator projection is that lines of constant direction, called **rhumb lines**, are straight and thus facilitate plotting directions between two points (see Figure 1.22). Thus, the Mercator projection is useful in navigation and is the standard for nautical charts prepared by the National Ocean Service since 1910 (formerly U.S. Coast and Geodetic Survey).

The *gnomonic*, or *planar, projection* in Figure 1.21b is generated by projecting a light source at the center of a globe onto a plane that is tangent to (touching) the globe's surface. The resulting severe distortion prevents showing a full hemisphere on one projection. However, a valuable feature is derived: All great circle routes, which are the shortest distance between two points on Earth's surface, are projected as straight lines (Figure 1.22a). The great circle routes plotted on a gnomonic projection then can be transferred to a true-direction projection, such as the Mercator, for determination of precise compass headings (Figure 1.22b).

For more information on maps used in this text and standard map symbols, turn to Appendix A, "Maps in This Text and Topographic Maps." Topographic maps are essential tools of landscape analysis. Geographers, other scientists, travelers, and anyone visiting the outdoors may use topographic maps. U.S. Geological Survey (USGS) topographic maps appear in several chapters of this text. Perhaps you have used a "topo" map in planning a hike. Coming on line is *The National Map* that the USGS says is "The Nation's Topographic Map for the 21st Century," also discussed in this appendix (see **http://nationalmap.usgs.gov/**).

Remote Sensing and GIS

Geographers probe, analyze, and map our home planet through remote sensing and geographic information systems (GIS). These technologies enhance our understanding of Earth. Among many subjects, geographers use remote-sensing data to study humid and arid lands, natural and economic vegetation, snow and ice, Earth energy budgets, seasonal variation of atmospheric and oceanic circulation, sea-level measurements, atmospheric chemistry, geologic features and events, changes in the timing of seasons, and the human activities that produce global change.

Remote Sensing

In this era of observations from orbit outside the atmosphere, from aircraft within it, and from remote submersibles in the oceans, scientists obtain a wide array of remotely sensed data (Figure 1.23). Remote sensing is nothing new to humans; we do it with our eyes all the time. When we scan the environment with our eyes, we are sensing the shape, size, and color of objects from a distance, registering energy from the visible-wavelength portion of the electromagnetic spectrum. Similarly, when a camera views the wavelengths for which its film or sensor is designed (visible light or infrared), it remotely senses energy that is reflected or emitted from a scene.

Our eyes and cameras are familiar means of obtaining **remote-sensing** information about a distant subject without having physical contact. Aerial photographs have been used for years to improve the accuracy of surface maps faster and more cheaply than can be done by on-site surveys. Deriving accurate measurements from photographs is the realm of **photogrammetry**, an important application of remote sensing.

Remote sensors on satellites, the International Space Station, and other craft sense a broader range of

A sample of orbital platforms

Envisat: ESA environment-monitoring satellite; 10 sensors, including a next-generation radar.

ERBS: Earth Radiation Budget Satellite.

GOES: Weather monitoring and forecasting.

Landsat: First in 1972 to *Landsat-7* in 1999, millions of images for Earth systems science and global change.

NOAA*:* First in 1978 through *NOAA-15* and *NOAA-16*, global data, short- and long-term weather forecasts.

RADARSAT-1, -2: Synthetic Aperature Radar in near polar orbit operated by Canada Space Agency.

SciSat-1: Analyzes trace gases, thin clouds, atmospheric aerosols with Arctic focus.

SeaStar: Carries the SeaWiFS (Sea-viewing Wide Field-of-View instrument) to observe Earth's oceans and microscopic marine plants.

Terra **and** ***Aqua:*** Environmental change, error-free surface images, cloud properties, through five instrument packages.

TOMS-EP*:* Total Ozone Mapping Spectrometer, monitoring stratospheric ozone, similar instruments on *NIMBUS-7* and *Meteor-3.*

TRMM*:* Tropical Rainfall Measuring Mission, includes lightning detection, and global energy budget measurements.

UARS*:* Since 1991 measuring atmospheric chemistry and ozone layer changes.

For more info see:
http://www.gsfc.nasa.gov/indepth/earth_esm.html

FIGURE 1.23 Remote-sensing technologies.
Remote-sensing technology measures and monitors Earth's systems from orbiting spacecraft, aircraft in the atmosphere, and ground-based sensors. Various wavelengths (bands) are collected from sensors. Computers process the data to produce digital images for analysis. A sample of remote-sensing platforms is in the margin. The Space Shuttle is shown in its inverted orbital flight mode. (Illustration is not to scale.)

wavelengths than can our eyes. They can be designed to "see" wavelengths shorter than visible light (ultraviolet) and wavelengths longer than visible light (infrared and microwave radar).

Satellites do not take conventional-film photographs. Rather, they record *images* that are transmitted to Earth-based receivers in a way similar to television satellite transmissions or a digital camera. A scene is scanned and broken down into pixels (*pic*ture *el*ements), each identified by coordinates named *lines* (horizontal rows) and *samples* (vertical columns). For example, a grid of 6000 lines and 7000 samples forms 42,000,000 pixels, providing great detail. The large amount of data needed to produce a single image requires computer processing and data storage at ground stations.

Digital data are processed in many ways to enhance their utility: simulated natural color, "false" color to highlight a particular feature, enhanced contrast, signal filtering, and different levels of sampling and resolution. Active and passive are two types of remote-sensing systems.

Active Remote Sensing   Active systems direct a beam of energy at a surface and analyze the energy reflected back. An example is *radar* (*r*adio *d*etection *a*nd *r*anging). A radar transmitter emits short bursts of energy that have relatively long wavelengths (0.3 to 10 m) toward the subject terrain, penetrating clouds and darkness. Energy re-flected back to a radar receiver for analysis is known as *backscatter*. An example is the computer image of wind and sea-surface patterns over the Pacific in Figure 6.6a, developed from 150,000 radar-derived measurements made on a single day by the *Seasat* satellite.

In addition, NASA sent imaging radar systems into orbit on several Space Shuttles. The subjects of study included oceanography, landforms and geology, and biogeography. Shuttle missions in 1994 by *Endeavour* and *Atlantis* marked dramatic contributions to Earth observations using radar and other sensors to study stratospheric ozone, weather, volcanic activity, earthquakes, and water resources, among many subjects.

One Space Shuttle mission in September 1994 was appropriately loaded with radar sensors to study volcanoes. Only 8 hours after launch, the Kliuchevskoi Volcano on the Kamchatka Peninsula of Russia erupted unexpectedly. Previously, this volcano had erupted in 1737 and 1945. The Shuttle radar was able to see through ash and smoke and expose lava flows and the volcanic eruption in dramatic images (Figure 1.24). Astronaut Mission Specialist Dr. Thomas Jones operated the radar and camera to make the image and photo in the figure. He is profiled in a Career Link at the end of this chapter. Chapter 12 discusses volcanic processes.

Several radar-imaging satellites are operating. Canada operates *RADARSAT–1, –2* and the Japanese have *JERS-1*

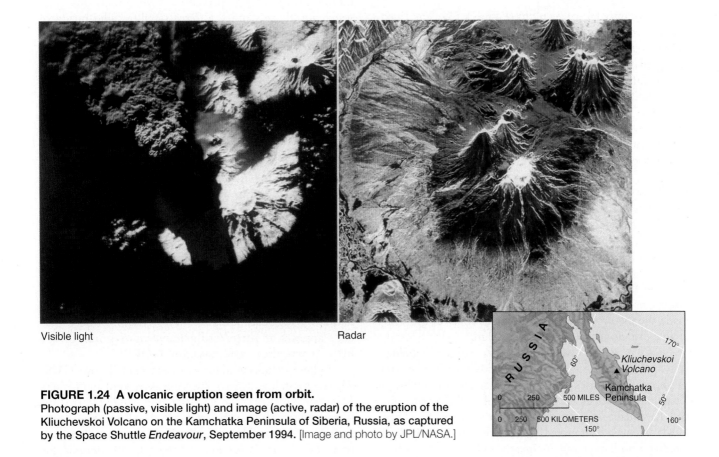

Visible light   Radar

FIGURE 1.24 A volcanic eruption seen from orbit.
Photograph (passive, visible light) and image (active, radar) of the eruption of the Kliuchevskoi Volcano on the Kamchatka Peninsula of Siberia, Russia, as captured by the Space Shuttle *Endeavour*, September 1994. [Image and photo by JPL/NASA.]

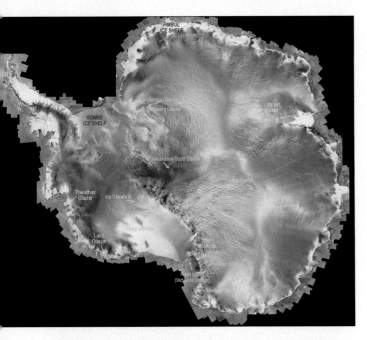

FIGURE 1.25 *RADARSAT–1* **composite of Antarctica.** A mosaic of *RADARSAT–1* images made over 18-days in the fall of 1997. Resolution can be as fine as 8 m (26 ft) with this microwave Synthetic Aperture Radar (SAR) system. The platform is placed in a near-polar orbit, approximately 800 km (497 mi) above the Earth. Satellite coverage is important to the International Polar Year of intense scientific research in the high latitudes. [Image courtesy of Canada Space Agency.]

that image in radar wavelengths (Figure 1.25). The European Space Agency (ESA) now operates two Earth resource satellites (*ERS 1* and *2*). They work in tandem, producing a spectacular 10-cm (3.9-in.) resolution, imaging the same area at different times. Pairs of images produce a digital three-dimensional data set.

ESA's newest environment-monitoring satellite, *Envisat*, went into service in 2002. In addition to more sophisticated radar than *ERS*, *Envisat* carries nine passive sensor packages, making it a multitalented craft. It monitors ocean temperatures, sea level, and wave patterns, polar ice, forests, biological activity in the oceans and on land, cloud heights, atmospheric ozone and pollution, and carbon dioxide concentrations (see http://envisat.esa.int/).

Passive Remote Sensing Passive remote-sensing systems record energy radiated from a surface, particularly visible light and infrared. Our own eyes are passive remote sensors, as was the *Apollo 17* astronaut camera that made the picture of Earth on the back cover of this book from a distance of 37,000 km (23,000 mi).

Passive remote sensors on five *Landsat* satellites, launched by the United States, provide a variety of data, as shown in images of the Appalachian Mountains in Chapter 12, river deltas in Chapter 14, and the Malaspina and Kuskulana glaciers in Alaska in Chapter 17. Three *Landsats* remain operational (4, 5, and the latest, 7), although *Landsat 4* no longer gathers images and is used for orbital tests. See http://geo.arc.nasa.gov/sge/landsat/landsat.html for the *Landsat* Home Page and other links.

The National Oceanic and Atmospheric Administration's (NOAA, see http://www.noaa.gov/) *Polar Operational Environmental Satellites* (*POES*) complete nearly polar orbits 14 times a day. The *advanced very high resolution radiometer* (*AVHRR*) sensors are on *NOAA-14, NOAA-15,* and *NOAA-16.* These sensors are sensitive to visible and infrared wavelengths. The incredible images of Hurricane Andrew (Chapter 8), among others in this text, are from this AVHRR system. These examples of resource analysis were impossible to perform at such a scale just a few years ago. In addition, these satellites measure ozone concentrations and temperatures in the stratosphere.

Key to NASA's Earth Observing System (EOS) is satellite *Terra*, which began beaming back data and images in 2000 (see http://terra.nasa.gov/), followed by another satellite in the series called *Aqua*. Five instrument packages observe Earth systems in detail, exploring the atmosphere, landscapes, oceans, environmental change, and climate, among other abilities. For example, the Clouds and the Earth's Radiant Energy System (CERES) instruments aboard *Terra* monitor the Earth's energy balance, giving new insights into climate change (see Chapter 4). These monitors offer the most accurate global radiation and energy measurements ever available.

Another instrument set, the Moderate-resolution Imaging Spectroradiometer (MODIS), sees Earth's entire surface every 1–2 days in 36 spectral bands, thereby expanding on AVHRR capabilities. Of more than 100 remote-sensing images presented throughout this text, watch for those from the *Terra* satellite. The image of Earth on the half-title page of this book includes a cloud snapshot from *GOES* added to *Terra* MODIS images over a 16-day period in 2000.

The *Geostationary Operational Environmental Satellites*, known as *GOES*, became operational in late 1994, providing frequent infrared and visible images—the ones you see on television weather reports. Geostationary satellites stay in semipermanent positions because they keep pace with Earth's rotational speed at their altitude of 35,400 km (22,000 mi).

GOES-10, online in 1998, operates above 135° W longitude to monitor the West Coast and the eastern Pacific Ocean. *GOES-12* sits above 75° W longitude to monitor central and eastern North America and the western Atlantic and replaces the aging (low maneuvering fuel) *GOES-8* (Figure 1.26). *GOES-11* is in orbital storage for future weather-satellite replacement as needed. See the Geostationary Satellite Server at http://www.goes.noaa.gov/. The *GOES* Project Science appears at http://rsd.gsfc.nasa.gov/goes/ or see http://www.ghcc.msfc.nasa.gov/GOES/.

Other satellites used for weather include Japan's *GMS-5* weather satellite (*GOES-9* serving as backup) and China's *Feng Yun-2* covering the Far East, and *METEOSAT-7* for Europe and Africa, operated by the European Space Agency. (See the Remote Sensing Virtual Library at http://www.vtt.fi/tte/research/tte1/tte14/virtual/ for remote-sensing links; click on "Satellite Data" for specific coverage.)

FIGURE 1.26 *GOES-12* **first image.**
The first image from *GOES-12*, a new environmental satellite in 2001, demonstrates excellent quality from its 37,000 km (23,300 mi) orbital post. This satellite, along with *GOES-11*, is stored in orbit to replace either of the existing *GOES* satellites as needed. [Image courtesy of NOAA.]

FIGURE 1.27   EROS Data Center (EDS).
USGS operates the EROS (Earth Resources Observation Systems) Data Center, outside Sioux Falls, South Dakota, as the main repository of satellite data and images and aerial photography. This center is the location of one of the largest computer complexes dedicated to science (see http://edc.usgs.gov/). [Photo by Bobbé Christopherson.]

One of two commercial systems includes the three French satellites (numbered 1, 2, and 4) called *SPOT* (Systeme Probatoire d'Observation de la Terre; see http://spot4.cnes.fr/waiting.htm) that resolve objects on Earth down to 10 to 20 m (33 to 66 ft), depending on which sensors are used. The other commercial system is Space Imaging, Inc., that offers 1–4 meter resolution from its *Ikonos–2* satellite in a Sun-synchronous orbit at 680 km (420 mi) altitude (http://www.spaceimaging.com/).

In addition, previously unavailable intelligence ("spy") satellite images are now available. These images from a CIA program code-named *Corona* were actually photographs taken by satellite cameras with film returned from orbit. Comparative analysis of Earth's surface over time is a potentially valuable use of these photographs that go back to 1960. These images are released through the National Reconnaissance Office, Department of Defense. This declassified satellite imagery is available through the USGS Eros Data Center, Sioux Falls, South Dakota, where the films are stored (Figure 1.27).

Geographic Information Systems (GIS)

Techniques such as remote sensing, which are important tools for acquiring large volumes of spatial data, are most valuable when that data can be stored, processed, and retrieved in useful ways. A **geographic information system (GIS)** is a computer-based, data-processing tool for gathering, manipulating, and analyzing geographic information. Today's sophisticated computer systems allow the integration of geographic information from direct surveys (on-the-ground mapping) and remote sensing in complex ways never

before possible. Through a GIS, Earth and human phenomena are analyzed over time. This can include the study and forecasting of diseases and epidemics such as West Nile virus, SARS, and the vicious Ebola virus, to name a few recent examples. An example is shown in Figure 1.28b and c.

GIS is a rapidly expanding career field in many sectors of the economy. Regardless of your academic major, the ability to analyze data spatially is important. Be sure to check some of the URLs listed in News Report 1.2 for information on professional career directions in this exciting field or how to acquire this tool for use in your chosen career.

The beginning component for any GIS is a coordinate system such as latitude–longitude, which establishes reference points against which to position data. The coordinate system is digitized, along with all areas, points, and lines. Remotely sensed imagery and data are then added on the coordinate system.

A GIS is capable of analyzing patterns and relationships within a single data plane, such as the floodplain or soil layer in Figure 1.28a. The GIS also can generate an overlay analysis where two or more data planes interact. When the layers are combined, the resulting synthesis—a *composite overlay*—is a valuable product, ready for use in analyzing complex problems. A research study may follow specific points or areas through the complex of overlay planes. The utility of a GIS compared with that of a fixed map is the ability to manipulate the variables for analysis and to constantly change the map—the map is alive!

Before the advent of computers, an environmental-impact analysis required someone to gather data and painstakingly hand-produce overlays of information to determine positive and negative impacts of a project or event. Today, this layered information is handled by a computer-driven GIS, which assesses the complex interconnections

(a)

Topographic base

Parcels

Zoning

Floodplains

Wetlands

Land cover

Soils

Survey control

Composite overlay of all data layers

(b) Land-cover map, Washington state

(c) Probability map for spread of West Nile virus

FIGURE 1.28 A geographic information system (GIS) model.
(a) Layered spatial data in a geographic information system (GIS) format. (b) This comprehensive land-cover map is an important component of a statewide GIS analysis for Washington state. The goal is to evaluate the protection of species and biodiversity. (c) A NASA GIS study uses satellite-derived temperature, humidity, and vegetation variables from 1997 to 2000 in a GIS overlay of environmental conditions for a probability map of the spread of West Nile virus. Various colors indicate temperatures and temperature ranges, moisture or dryness, and vegetation as an indicator of humidity. For more information go to **http://www.gsfc.nasa.gov/topstory/20020828phap.html** and **http://www.cdc.gov/ncidod/dvbid/westnile/.** [(a) After USGS. (b) GIS map courtesy of Dr. Kelly M. Cassidy, *Gap Analysis of Washington State*, v. 5, Map 3, University of Washington, 1997. (c) Composite image from *NOAA-15* AVHRR data, NASA/GSFC.]

among different components. In this way, subtle changes in one element of a landscape may be identified as having a powerful impact elsewhere.

The applications of GIS are wide-ranging. Scientists at NASA's Goddard Space Flight Center completed a comprehensive GIS of Brazil in an effort to better understand land-use patterns—specifically, loss of the rain forest. In Bangladesh, where extensive flooding causes destruction of property and loss of life, a GIS analysis should help identify the most flood-prone areas and mitigate flooding effects. GIS models also are used to analyze regions vulnerable to wildfire. The GIS discloses those high-risk areas that can guide officials in the placement of fire-fighting equipment and crews. (For GIS resources and information see **http://www.geo.ed.ac.uk/home/giswww.html** and **http://erg.usgs.gov/isb/pubs/gis_poster/.**)

One of the most extensive and longest-operating systems is the Canada Geographic Information System (CGIS). Environmental data about natural features, resources, and land use were taken from maps, aerial photographs, and orbital sources, reduced to map segments, and entered into the CGIS. The development of this system has progressed with the ongoing Canada land-inventory project (see the Canadian Land Inventory Web page at **http://geogratis.cgdi.gc.ca/CLI/frames.html**).

USGS Geography (formerly Mapping) is in charge of building and maintaining a bold new national map. The value of the new online, interactive system is that geospatial data is portrayed in a manner that can be readily updated—a dynamic mapping program, involving the principles of GIS. For more on this work in progress see **http://nationalmap.usgs.gov/.**

News Report 1.2

Careers in GIS

Geographic information system (GIS) methodology offers great career opportunities in industry, government, business, marketing, teaching, sales, military, and other fields. Right now, geographers trained in GIS analyze ozone depletion, deforestation, soil erosion, and acid deposition. They map ecosystems and monitor the declining diversity of plant and animal species. Geographers plan, design, and survey urban developments, follow the trends of global warming, study the impact of human population, and analyze air and water pollution.

We are in the midst of a growing GIS revolution. Potent new careers are emerging in almost every academic area as GIS programs are implemented at many universities, colleges, and community colleges. GIS applications are used in environmental analysis, weather forecasting, natural hazard assessment, business and marketing, industry location analysis, criminal justice, and natural resources exploration, among many examples. (For a list of current trends in GIS, simply enter this topic in your search engine; for an alphabetized GIS resources list, check out **http://www. geo.ed.ac.uk/home/giswww.html**.)

GIS degree programs are available at many colleges and universities. GIS curriculum and certificate programs are now available at many community colleges. For example, at American River College, Sacramento, California, a curriculum offering 11 GIS courses fuels two certificate programs (see **http://www.arc.losrios.edu/~earthsci/ GIS/GIS.html**).

A consortium of three universities forms the National Center for Geographic Information and Analysis (NCGIA) for GIS education, research, outreach, and model generation. These GIS centers are: Department of Geography, University of California Santa Barbara, Santa Barbara, CA 93106; Department of Surveying and Engineering, University of Maine, Orono, ME 04669; and State University of New York–Buffalo, Buffalo, NY 14260. (See **http://www.ncgia.ucsb. edu/** or **http://www.ncgia.maine. edu/** or **http://www.geog.buffalo. edu/ncgia**.)

Summary and Review—Essentials of Geography

To assist you, here is a review in handy summary form of the Key Learning Concepts listed on this chapter's title page. Each concept review concludes with a list of the key terms from the chapter, their page numbers, and review questions. Such summary and review sections follow each chapter in the book.

◗ *Define* geography and physical geography in particular.

Geography brings together disciplines from the physical and life sciences with the cultural and human sciences to attain a holistic view of Earth—physical geography is an essential aspect of the emerging **Earth systems sciences**. **Geography** is a science of method, a special way of analyzing phenomena over space; **spatial** refers to the nature and character of physical space. Geography integrates a wide range of subject matter. Geographic education recognizes five major themes: **location**, **region**, **human–Earth relationships**, **movement**, and **place**. Geography's method is **spatial analysis**, used to study the interdependence among geographic areas, natural systems, society, and cultural activities over space. **Process**—that is, analyzing a set of actions or mechanisms that operate in some special order—is central to geographic synthesis.

Physical geography applies spatial analysis to all the physical elements and processes that make up the environment: energy, air, water, weather, climate, landforms, soils, animals, plants, microorganisms, and Earth itself. Understanding the complex relations among these elements is important to human survival because Earth's physical systems and human society are so intertwined. The development of hypotheses and theories about the Universe, Earth, and life involves the **scientific method**.

Earth systems science (p. 2)
geography (p. 2)
spatial (p. 2)
location (p. 2)
region (p. 2)
human–Earth relationships (p. 2)
movement (p. 2)
place (p. 2)
spatial analysis (p. 2)
process (p. 4)
physical geography (p. 4)
scientific method (p. 4)

1. What is unique about the science of geography? On the basis of information in this chapter, define physical geography and review the geographic approach.

2. In general terms, how might a physical geographer analyze water pollution in the Great Lakes?

3. Assess your geographic literacy by examining atlases and maps. What types of maps have you used—political? Physical? Topographic? Do you know what projections they employed? Do you know the names and locations of the four oceans, seven continents, and most individual countries? Can you identify the new countries that have emerged since 1990?

4. Suggest a representative example for each of the five geographic themes and use that theme in a sentence.

5. Have you made decisions today that involve geographic concepts discussed within the five themes presented? Explain briefly.

◗ *Describe* systems analysis, open and closed systems, feedback information, and system operations and *relate* those concepts to Earth systems.

A **system** is any ordered, related set of things and their attributes, as distinct from their surrounding environment. Systems analysis is an important organizational and analytical tool used by geographers. Earth is an **open system** in terms of energy, receiving energy from the Sun, but it is essentially a **closed system** in terms of matter and physical resources.

As a system operates, "information" is returned to various points in the system via pathways called **feedback loops**. If the feedback information discourages response in the system, it is called **negative feedback**. (Further production in the system decreases the growth of the system.) If feedback information encourages response in the system, it is called **positive feedback**. (Further production in the system stimulates the growth of the system.) When the rates of inputs and outputs in the system are equal and the amounts of energy and matter in storage within the system are constant (or as they fluctuate around a stable average), the system is in **steady-state equilibrium**. A system that demonstrates a steady increase or decrease in system operations—a trend over time—is in **dynamic equilibrium**. A **threshold** is the moment in which a system can no longer maintain its character, so it lurches to a new operational level, one which may not be compatible with previous conditions. Geographers often construct a simplified **model** of natural systems to better understand them.

Four immense open systems powerfully interact at Earth's surface: three nonliving **abiotic** systems [**atmosphere**, **hydrosphere** (including the **cryosphere**), and **lithosphere**] and a living **biotic** system (**biosphere**, or **ecosphere**).

system (p. 4)
open system (p. 5)
closed system (p. 8)
feedback loops (p. 8)
negative feedback (p. 8)
positive feedback (p. 9)
steady-state equilibrium (p. 9)
dynamic equilibrium (p. 10)
threshold (p. 10)
model (p. 11)
abiotic (p. 13)
biotic (p. 13)
atmosphere (p. 14)
hydrosphere (p. 14)
cryosphere (p. 14)
lithosphere (p. 14)
biosphere (p. 14)
ecosphere (p. 14)

6. Define systems theory as an organizational strategy. What are open systems, closed systems, and negative feedback? When is a system in a steady-state equilibrium condition? What type of system (open or closed) is the human body? A lake? A wheat plant?

7. Describe Earth as a system in terms of both energy and matter; use simple diagrams to illustrate your description.

8. What are the three abiotic spheres (nonliving) that make up Earth's environment? Relate these to the biotic (living) sphere: the biosphere.

◗ *Explain* Earth's reference grid: latitude, longitude, and latitudinal geographic zones and time.

The science that studies Earth's shape and size is **geodesy**. Earth bulges slightly through the equator and is oblate (flattened) at the poles, producing a misshapen spheroid called a **geoid**. Absolute location on Earth is described with a specific reference grid of **parallels** of **latitude** (measuring distances north and south of the equator) and **meridians** of **longitude** (measuring distances east and west of a prime meridian). A historic breakthrough in navigation and timekeeping occurred with the establishment of an international **prime meridian** (0° through Greenwich, England) and the invention of precise chronometers that enabled accurate measurement of longitude. Latitude, longitude, and elevation are accurately calibrated using a handheld **global positioning system (GPS)** instrument that reads radio signals from satellites. A **great circle** is any circle of Earth's circumference whose center coincides with the center of Earth. Great circle routes are the shortest distance between two points on Earth. **Small circles** are those whose centers do not coincide with Earth's center.

The prime meridian provided the basis for **Greenwich Mean Time (GMT)**, the world's first universal time system. A corollary of the prime meridian is the 180° meridian, the **International Date Line**, which marks the place where each day officially begins. Today, **Coordinated Universal Time (UTC)** is the worldwide standard and the basis for international time zones. **Daylight saving time** is a seasonal change of clocks by 1 hour in summer months.

geodesy (p. 14)
geoid (p. 14)
latitude (p. 16)
parallel (p. 16)
longitude (p. 17)
meridian (p. 17)
prime meridian (p. 17)
Global Positioning System (GPS) (p. 18)
great circle (p. 18)
small circles (p. 19)
Greenwich Mean Time (GMT) (p. 22)
International Date Line (p. 23)
Coordinated Universal Time (UTC) (p. 23)
daylight saving time (p. 23)

9. Draw a simple sketch describing Earth's shape and size.

10. How did Eratosthenes use Sun angles to figure out that the 5000-stadia distance between Alexandria and Syene was 1/50 of Earth's circumference? Once he knew this fraction of Earth's circumference, how did he calculate the distance of Earth's circumference?

11. What are the latitude and longitude coordinates (in degrees, minutes, and seconds) of your present location? Where can you find this information?

12. Define latitude and parallel and define longitude and meridian using a simple sketch with labels.

13. Define a great circle, great circle routes, and a small circle. In terms of these concepts, describe the equator, other parallels, and meridians.

14. Identify the various latitudinal geographic zones that roughly subdivide Earth's surface. In which zone do you live?

15. What does timekeeping have to do with longitude? Explain this relationship. How is Coordinated Universal Time (UTC) determined on Earth?

16. What and where is the prime meridian? How was the location originally selected? Describe the meridian that is opposite the prime meridian on Earth's surface.

17. What is GPS and how does it assist you in finding location and elevation on Earth? Give a couple of examples

where it was utilized to correct heights for some famous mountains.

▶ *Define* cartography and mapping basics: map scale and map projections.

A **map** is a generalized view of an area, usually some portion of Earth's surface, as seen from above, and greatly reduced in size. The science and art of mapmaking is called **cartography**. For the spatial portrayal of Earth's physical systems geographers use maps. **Scale** is the ratio of the image on a map to the real world; it relates a unit on the map to an identical unit on the ground. Cartographers create **map projections** for specific purposes, selecting the best compromise of projection for each application. Compromise is always necessary because Earth's round, three-dimensional surface cannot be exactly duplicated on a flat, two-dimensional map. **Equal area** (equivalence), **true shape** (conformality), true direction, and true distance are all considerations in selecting a projection. The **Mercator projection** is in the cylindrical class; it has true-shape qualities and straight lines that show constant direction. **Rhumb lines** are lines of constant direction and appear as straight lines on the Mercator.

    map (p. 24)
    cartography (p. 24)
    scale (p. 24)
    map projections (p. 25)
    equal area (p. 25)
    true shape (p. 25)
    Mercator projection (p. 26)
    rhumb lines (p. 27)

18. Define cartography. Explain why it is an integrative discipline.

19. What is map scale? In what three ways may it be expressed on a map?

20. State whether each of the following ratios is a large scale, medium scale, or small scale: 1:3,168,000; 1:24,000; 1:250,000.

21. Describe the differences between the characteristics of a globe and those that result when a flat map is prepared.

22. What type of map projection is used in Figure 1.13? In Figure 1.17? (See Appendix A.)

▶ *Describe* remote sensing and *explain* geographic information system (GIS) methodology as a tool used in geographic analysis.

Orbital and aerial **remote sensing** analyze the operation of Earth's systems. Satellites do not take photographs, but record images that are transmitted to Earth-based receivers. Satellite images are recorded in digital form for later processing, enhancement, and generation. Aerial photographs have been used for years to improve the accuracy of surface maps. This is the realm of **photogrammetry**, an important application of remote sensing.

The mountain of data already collected has led to the development of **geographic information system (GIS)** technology. Computers process geographic information from direct surveys and remote sensing in complex ways never before possible. GIS methodology is an important step in better understanding Earth's systems and is a vital career opportunity for geographers.

The science of physical geography is in a unique position to synthesize the spatial, environmental, and human aspects of our increasingly complex relationship with our home planet—Earth.

    remote sensing (p. 27)
    photogrammetry (p. 27)
    geographic information system (GIS) (p. 31)

23. What is remote sensing? What are you viewing when you observe a weather satellite image on TV or in the newspaper? Explain.

24. Describe the *Terra*, *Landsat*, *GOES*, and *NOAA* satellites and explain them using several examples.

25. If you were in charge of planning for development of a large tract of land, how would GIS methodologies assist you? How might planning and zoning be affected if a portion of the tract in the GIS is a floodplain or prime agricultural land?

NetWork

The *Geosystems* Home Page provides online resources for this chapter on the World Wide Web. To begin: Once at the Home Page, click on the cover of this textbook, scroll the Table of Contents menu, select this chapter, and click "Begin." You will find self-tests that are graded, review exercises, specific updates for items in the chapter, and in "Destinations" many links to interesting related pathways on the Internet. *Geosystems* Home Page is found at **http://www.prenhall.com/christopherson**.

Critical Thinking

A. Select a location (for example, your campus, home, workplace, a public place, or a city) and determine the following: latitude, longitude, and elevation. Describe the resources you used to gather this geographic information. Have you ever used a GPS unit to determine these aspects of your location?

B. Let's say there is a world globe in the library or geography department that is 61 cm (24 in.) in diameter. We know that Earth has an equatorial diameter of 12,756 km (7926 mi), so the scale of the globe is the ratio of 61 cm to 12,756 km. We divide Earth's actual diameter by the globe's diameter (12,756 km ÷ 61 cm) and determine that 1 cm of the globe's diameter equals about _____ cm of Earth's actual diameter. Thus, the representative fraction for the globe is expressed in centimeters as 1/_____. (*Hint*: 1 km = 1000 m, 1 m = 100 cm; therefore, Earth's diameter of 12,756 km represents 1,275,600,000 cm.)

C. The various geographic information technologies discussed in your text (GIS, GPS, remote sensing) promise to revolutionize many aspects of modern life. How are they used now, and what sorts of changes might we expect in the near future? Use the Net Search and Destinations sections to find examples and write a brief description of what you find. Be sure to include any URLs for the sites you visit.

Career Link 1.1

Thomas D. Jones, Ph.D., Astronaut, Earth Observer, and Geographer

I first met Dr. Thomas Jones at an annual meeting of the National Council for Geographic Education (NCGE) in Indianapolis. Earth was the feature as the audience orbited the planet through NASA photos and imagery piloted by his enthusiastic geographic analysis. At one point the photo and radar image of the Kliuchevskoi volcanic eruption, on the Kamchatka Peninsula, appeared on the screen. The same photo and image are in Figure 1.24 in *Geosystems*. Afterward, I asked astronaut Tom Jones to sign the figure in a copy of the book—a real thrill.

His enthusiasm for Earth observation and his spatial analysis of natural and human phenomena—from cities, to rock formations, to hydrology, to weather—makes him a real friend of geographic education. I interviewed Tom in June 2001 at the Johnson Space Center (JSC), in Building 9, where the simulators for both the Shuttle and components of the International Space Station (ISS) are housed for training (Figure 1).

Tom was born and raised in Maryland, living there through high school graduation and becoming a National Merit Scholar. I asked him if he had an early interest in geography. Tom said, "Yes, elementary school geography was fascinating. Even as early as first grade, I can remember looking at maps with my older second-grade friend. In Boy Scouts, I loved compass work and map reading, and all the outdoor activities." Tom said, "These early interests feed

right into piloting where you have a map right on your knee and must observe geography."

He received his B.S. degree as a Distinguished Graduate from the U.S. Air Force Academy and served for 6 years on active duty as pilot and commander on B-52D strategic bombers. Tom completed more than 2000 hours of flying time, achieving the rank of Captain. He stated, "Spending thousands of hours in aircraft is a good way to learn how to look carefully at landscapes from above. I have always thought it valuable in my life to have so many hours viewing Earth from above, first in planes, then from orbit."

Tom Jones continued his education at the University of Arizona, earning a Ph.D. in planetary science. He said, "I used remote sensing to study the water and mineral content of asteroids—those chunks of rock between Mars and Jupiter that are the source of many meteors that Earth encounters."

Tom joined NASA in 1990 and became an astronaut in 1991. He flew on Space Shuttle *Endeavour* as a Mission Specialist on STS-59 (April 1994) and STS-68 (October 1994). This later flight is when he captured the volcanic eruption image and photo in Figure 1.24. Tom next went into space aboard *Columbia* in STS-80 (1996), where he operated the robotic arm to launch a satellite and, incidentally, made the photo of Mount Everest at dawn shown in News Report 12.1. On his fourth flight, aboard *Atlantis* in STS-98 (February 2001), he worked installing the Destiny Laboratory Module for the International Space Station (ISS).

I inquired about the *Endeavour* flight in October 1994. Tom said, "We were going to study volcanoes with the SIR-C/X-SAR (Synthetic Aperture Radar), including the volcanic complex on Kamchatka. So it was on our charts and flight plan. We never expected that the biggest volcano in Asia would erupt right after launch and present us with such a wonderful opportunity. We saw it on our first orbit! This huge smudge on the horizon looked like the strangest thunderstorm. Then we realized that Kliuchevskoi had blown."

Tom added, "Earth gave us a gift. We were supposed to fly six weeks earlier and would have missed this chance to see an active eruption."

When working outside the Shuttle above the protective layers of the atmosphere, an astronaut must wear a spacesuit to regulate the temperature differences experienced, from nearly 120°C (156°F) where the Sun strikes the spacesuit, to −250°C (−156°F) when in shadows. Oxygen and water

FIGURE 1 Astronaut Thomas Jones.
Astronaut Mission Specialist Thomas Jones stands in front of a Space Shuttle simulator at the Johnson Space Center, Houston, Texas. Dr. Jones logged 1272 hours in orbit aboard four Shuttle flights. [Photo by Bobbé Christopherson.]

must be provided and carbon dioxide buildup managed. Here on Earth's surface the atmosphere does all this for us. Imagine designing a spacesuit that does everything portrayed in Figure 3.2 in this book.

Tom Jones completed three extravehicular activities (EVAs) on his last flight, totaling 19 hours (Figure 2a). He and his partner Bob Curbeam installed the U.S. Destiny Laboratory Module on the ISS.

I asked him about any feeling of vulnerability during his EVA while he was an Earth satellite drifting in orbit at more than 28,100 kmph (17,500 mph). Tom answered, "Intellectually you are aware of that, and you can stop and think about your independence from things as you look over the brilliant Earth, but most of the time you are focused on the work and the spacesuit is almost invisible to you. It works so well that you get quite comfortable, you forget that you are next to a vacuum traveling ten times faster than a bullet, or in such a harsh environment."

He continued, "Despite the thermal protection of the EMU spacesuit, when the Sun rises, which it does every 90 minutes in orbit, I could feel the warmth when the light hit me. The thermal tubing in the suit that regulates temperature goes throughout the suit. If it does get warm in the suit, you can adjust your thermostat. With sunset, I felt that warmth drop away and when my feet were on a work platform, I felt the coldness of the metal plate and the heat energy conduct out through the feet of the suit, despite the bulky socks I wore."

I asked about Earth observation time. "Everyone gets training for Earth observing, including classroom work in physical and human geography and geology. For my last mission we attended lectures and got experience doing actual camera practice. We have NASA's *Space Shuttle Physiographic Atlas* (1:10,000,000 scale), from the Earth Observation Project, organized along flight paths (west to east). Clear, bold labels used in the atlas help us orient the camera to specific photographic targets. Decal black circles are placed on charts so you know what needs to be photographed."

I asked, "After 1272 hours in Earth orbit, more than 52 days total, what are some of your thoughts about Earth?" Tom answered, "Earth never fails to amaze me and captivate me with the beauty it presents, the ever-changing aspects of light, the changing vision, for you are always seeing Earth in a new way. Every time I looked out the window I saw a different angle or lighting that changes something I might have seen many times before. I always found something new to be amazed at! Looking at Earth is relaxing and enjoyable; it refreshes you instantly."

Tom added, "My days in orbit are a privilege, to see Earth from that distance is a lifelong memory and one that will never leave me. Yet, it comes with a sense of regret

FIGURE 2 Space walk February 2001.
(a) Astronaut Thomas Jones completed three space walks, totaling 19 hours, working on the International Space Station *Destiny* module installation. He is waving to crew members inside *Atlantis*. His partner in the space walks was Robert L. Curbeam. (b) Insignia for the *Atlantis* mission; note the reflection of Earth in the large window in the *Destiny* module. See **http://www.spaceflight. nasa.gov/gallery/images/shuttl e/sts-98/ndxpage1.html**. [Space Shuttle photograph STS98-E-5195 and mission insignia courtesy of NASA.]

because on this last mission I know there was so much I could have seen, but we had an important mission to complete and there wasn't time to linger by the windows. This is the mixed blessing of working in space."

As to the future, Tom wants to focus on further research related to his dissertation topic and nonfiction writing for the general public on space travel and possible missions to asteroids. He is the author of *The Complete Idiot's Guide to NASA* (Indianapolis: Alpha Books, 2002). He said, "The Moon is a close-by testing place for equipment, so we will no doubt be returning. We need to build support for future space exploration."

In closing, I asked Tom about the insignia patch for STS-98 (Figure 2b). He answered, "The crew thought it was important to show the Earth-observing window in the Destiny Laboratory Module, with Earth reflected in it. Uncovering the protective coating to begin operations of this crystal-clear viewing portal was a real thrill for Bob and me on our EVA." This says it all about Tom and the enrichment he gives to geographic education—he has opened a portal for us to better see Earth, our Home Planet.

The Energy–Atmosphere System

PART I

O ur planet and our lives are powered by radiant energy from the star that is closest to Earth—the Sun. For more than 4.6 billion years, solar energy has traveled across interplanetary space to Earth, where a small portion of the solar output is intercepted. Because of Earth's curvature, the energy at the top of the atmosphere is unevenly distributed, creating imbalances from the equator to each pole—the equatorial region experiences energy surpluses; the polar regions experience energy deficits. Also, the annual pulse of seasonal change varies the distribution of energy during the year.

Earth's atmosphere acts as an efficient filter, absorbing most harmful radiation, charged particles, and space debris so that they do not reach Earth's surface. In the lower atmosphere the unevenness of daily energy receipt empowers atmospheric and surface energy balances, giving rise to global patterns of temperature and circulation of wind and ocean currents. Each of us depends on many systems that are set into motion by energy from the Sun. These systems are the subject of Part 1, Chapters 2 through 6.

Sunrise in the Scoresby Sund fjord system, East Greenland, September 2003, at approximately 71° N latitude. The extended length of dawn and twilight in the high latitudes produces wonderful colors and long shadows from transient icebergs. These icebergs probably calved and set sail from glaciers in northern Greenland, were ocean going driven by winds, and floated into this fjord system—all involving physical principles in PART One, The Energy–Atmosphere System. [Photo by Bobbé Christopherson.]

Sunset in Antarctica, February 2004, across the snow and ice of Hovgaard Island and the waters of Penola Strait, along the Antarctic Peninsula. The temperature was −1°C (30°F). [Photo by Bobbé Christopherson.]

CHAPTER 2

Solar Energy to Earth and the Seasons

▶ Key Learning Concepts

After reading the chapter, you should be able to:

▶ **Distinguish** among galaxies, stars, and planets and **locate** Earth.

▶ **Overview** the origin, formation, and development of Earth and **construct** Earth's annual orbit about the Sun.

▶ **Describe** the Sun's operation and **explain** the characteristics of the solar wind and the electromagnetic spectrum of radiant energy.

▶ **Portray** the intercepted solar energy and its uneven distribution at the top of the atmosphere.

▶ **Define** solar altitude, solar declination, and daylength and **describe** the annual variability of each—Earth's seasonality.

The Universe is populated with billions of galaxies. One of these is our own Milky Way Galaxy, consisting of billions of stars. Among these stars is an average yellow star we call the Sun, although the dramatic *SOHO* satellite image in Figure 2.2 seems anything but average! Our Sun radiates energy in all directions and upon its family of orbiting planets. Of special interest to us is the solar energy that falls on the third planet from the Sun.

In this chapter: Incoming solar energy that arrives at the top of Earth's atmosphere establishes the pattern of energy input that drives Earth's physical systems and that daily influences our lives. This solar energy input to the atmosphere, plus Earth's tilt and rotation, produce daily, seasonal, and annual patterns of changing daylength and Sun angle. The Sun is the ultimate energy source for most life processes in our biosphere.

The Solar System, Sun, and Earth

Our Solar System is located on a remote, trailing edge of the **Milky Way Galaxy**, a flattened, disk-shaped mass estimated to contain nearly 400 billion stars (Figure 2.1a, b). Our Solar System is embedded more than halfway out from the galactic center, in one of the Milky Way's spiral arms, called the Orion Arm. A supermassive black hole some 2 million solar masses in size, named Sagittarius A*, sits in the center (pronounced "Sagittarius A Star"). From our Earth-bound perspective in the Milky Way, the Galaxy appears to stretch across the night sky like a narrow band of hazy light. On a clear night, the unaided eye can see only a few thousand of these billions of stars, gathered about us in our "neighborhood."

Solar System Formation and Structure

According to prevailing theory, our Solar System condensed from a large, slowly rotating and collapsing cloud of dust and gas called a *nebula*. **Gravity**, the mutual attracting force exerted by the mass of an object upon all other objects, was the key force in this condensing solar nebula. As the nebular cloud organized and flattened into a disk shape, the early *protosun* grew in mass at the center, drawing more matter to it. Small accretion (accumulation) eddies swirled at varying distances from the center of the solar nebula; these were the *protoplanets*.

The beginnings of the Sun and its Solar System are estimated to have occurred more than 4.6 billion years ago. The early protoplanets, called *planetesimals*, orbited at approximately the same distances from the Sun as the planets are today. The **planetesimal hypothesis**, or *dust-cloud hypothesis*, explains how suns condense from nebular clouds with planetesimals forming in orbits about the central mass. Astronomers study this formation process in other parts of the Galaxy, where planets are observed orbiting distant stars. By the end of 2004, astronomers discovered more than 230 planets orbiting around other stars. Closer to home in our Solar System, 65 new moons (planetary satellites) were discovered since 1999. As of 2004, the new satellite count for the four outer planets is: Jupiter, 61 moons; Saturn, 31 moons; Uranus, 22 moons; and Neptune, 11 moons.

Dimensions and Distances   The **speed of light** is 300,000 kmps (kilometers per second), or 186,000 mps (miles per second)—in other words, about 9.5 trillion kilometers per year, or nearly 6 trillion miles. (In more precise numbers, light speed is 299,792 kmps, or 186,282 mps.) This tremendous distance that light travels in a year is known as a *light-year*, and it is used as a unit of measurement for the vast Universe.

The known Universe that is observable from Earth stretches approximately 12 billion light-years in all directions. The Milky Way Galaxy is about 100,000 light-years from edge to edge (Figure 2.1b). For spatial comparison, our entire Solar System of nine planets is approximately 11 hours in diameter, as measured by light speed (Figure 2.1c). The Moon is an average distance of 384,400 km (238,866 mi) from Earth, or about 1.28 seconds from Earth in terms of light speed—for the *Apollo* astronauts this was a three-day journey. (See a Solar System simulator at **http://space.jpl.nasa.gov/**.)

Earth's Orbit   Earth's orbit around the Sun is presently *elliptical*—a closed, oval path (Figure 2.1d). Earth's average distance from the Sun is approximately 150 million km (93 million mi), which means that light reaches Earth from the Sun in an average of 8 minutes and 20 seconds. Earth is at **perihelion** (its closest position to the Sun) during the Northern Hemisphere winter (January 3 at 147,255,000 km, or 91,500,000 mi). It is at **aphelion** (its farthest position from the Sun) during the Northern Hemisphere summer (July 4 at 152,083,000 km, or 94,500,000 mi). This seasonal difference in distance from the Sun causes a slight variation in the solar energy incoming to Earth but is not an immediate reason for seasonal change.

A plane touching all points of Earth's orbit is termed the **plane of the ecliptic**. Earth's tilted axis remains fixed relative to this plane as Earth revolves around the Sun. The plane of the ecliptic is important to our discussion of Earth's seasons. The structure of Earth's orbit is not a constant, but instead exhibits change over long periods. As shown in Chapter 17, Figure 17.31, Earth's distance from the Sun varies more than 17.7 million km (11 million mi) during a 100,000-year cycle, placing it closer or farther at different periods in the cycle. This variation is thought to be one of several factors that create Earth's cyclical pattern of glaciations (colder) and interglacial (warmer) periods.

Solar Energy: From Sun to Earth

Our Sun is unique to us but is a commonplace star in the Galaxy. It is only average in temperature, size, and color when compared with other stars, yet it is the ultimate energy source for most life processes in our biosphere.

The Sun captured about 99.9% of the matter from the original solar nebula. The remaining 0.1% of the matter formed all the planets, their satellites, asteroids, comets, and debris. Consequently, the dominant object in our region of space is the Sun. In the entire Solar System, it is the only object having the enormous mass needed to sustain a nuclear reaction and produce radiant energy.

The solar mass produces tremendous pressure and high temperatures deep in its dense interior. Under these conditions, the Sun's abundant hydrogen atoms, the lightest of all the natural elements, are forced together, as

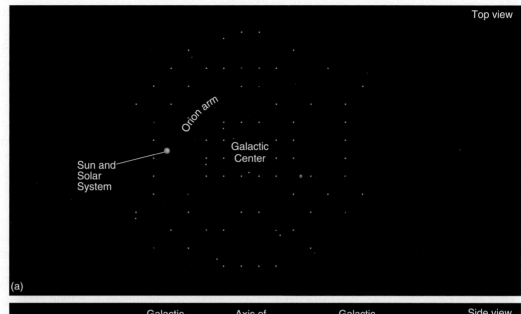

Top view

Orion arm

Galactic Center

Sun and Solar System

(a)

Side view

Galactic North Pole

Axis of rotation

Galactic plane

Center

Our Solar System

Spiral arms

—100,000 light-years—

(b)

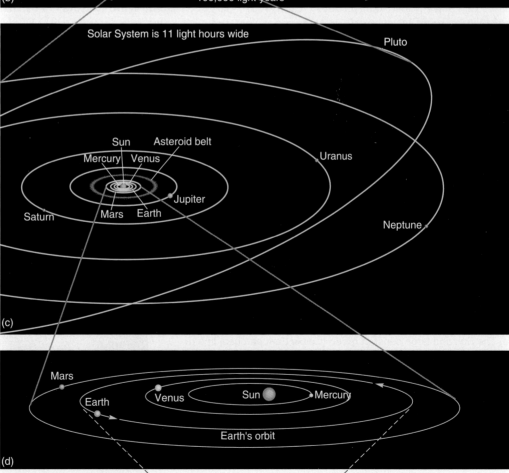

Solar System is 11 light hours wide

Pluto

Sun   Asteroid belt

Mercury   Venus

Uranus

Jupiter

Saturn   Mars   Earth

Neptune

(c)

Mars

Venus   Sun   Mercury

Earth

Earth's orbit

(d)

Farthest from Sun—aphelion, July 4
152,083,000 km
(94,500,000 mi)

Sun

Closest to Sun—perihelion, January 3
147,255,000 km
(91,500,000 mi)

Focus of Earth's elliptical orbit

ANIMATION Nebular Hypothesis

FIGURE 2.1 Milky Way Galaxy, Solar System, and Earth's orbit.
The Milky Way Galaxy viewed from above (a) and cross-section side view (b). Our Solar System of nine planets and asteroids is some 30,000 light-years from the center of the Galaxy. All of the planets except Pluto have orbits closely aligned to the plane of the ecliptic (c). The four inner terrestrial planets and the structure of Earth's elliptical orbit, illustrating perihelion (closest) and aphelion (farthest) positions during the year, are given in (d). Have you ever observed the Milky Way Galaxy in the night sky?

pairs of hydrogen nuclei are joined in a process called **fusion**. In the fusion reaction, hydrogen nuclei form helium, the second-lightest element in nature, and enormous quantities of energy are liberated—literally, disappearing solar mass becomes energy.

A sunny day can seem so peaceful, certainly belying the violence taking place on the Sun. The Sun's principal outputs consist of the solar wind and radiant energy in portions of the electromagnetic spectrum. Let us trace each of these emissions across space to Earth.

Solar Activity and Solar Wind

The Sun constantly emits clouds of electrically charged particles (principally, hydrogen nuclei and free electrons) that surge outward in all directions from the Sun's surface. This stream of energetic material travels much more slowly than light—only about 50 million km (31 million mi) a day—taking approximately 3 days to reach Earth. The term **solar wind** was first applied to this phenomenon in 1958. Solar wind extends from the Sun to a distance beyond Pluto's orbit. The *Voyager* and *Pioneer* spacecraft launched in the 1970s are now far beyond our Solar System and have yet to escape the solar wind.

The Sun's most conspicuous features are large **sunspots**, caused by magnetic storms on the Sun. Indi-

vidual sunspots may range in diameter from 10,000 to 50,000 km (6200 to 31,000 mi), with some growing as large as 160,000 km (100,000 mi), more than 12 times Earth's diameter (Figure 2.2a). These surface disturbances produce flares and prominences (Figure 2.2b). In addition, outbursts of charged material referred to as *coronal mass ejections* contribute to the flow of material to space in the solar wind.

A regular cycle exists for sunspot occurrences, averaging 11 years from maximum to maximum; however, the cycle may vary from 7 to 17 years. In recent cycles, a solar minimum occurred in 1976 and a solar maximum took place during 1979, with more than 100 sunspots visible at maximum. Another minimum was reached in 1986, and an extremely active solar maximum followed in 1990, with more than 200 sunspots visible at some time during the year—11 years after the previous maximum. A sunspot minimum in 1997, an intense maximum in 2001, and expected minimum in 2006–2007 maintain the average. (For more on the sunspot cycle, see http://www.ssl.msfc. nasa.gov/ssL/pad/solar/sunspots.htm.) News Report 2.1 discusses the order and patterns we see in nature, such as this sunspot cycle, as the result of chaotic interactions of many systems.

Solar Wind Effects   The charged particles of the solar wind first interact with Earth's magnetic field as they

Approx. size of Earth ⟶ •

(a)

(b)

FIGURE 2.2 Image of the Sun and sunspots.
(a) The Sun and large sunspot group (visible dark patches) in a recent active cycle, imaged March 30, 2001, by the MDI (Michelson Doppler Imager) instrument aboard satellite *SOHO*. This group was the source of numerous flares and coronal mass ejections, including the largest flare in 25 years on April 2. The area within the sunspot group is more than 13 times the entire surface area of Earth (13 times 500 million km²)—Earth is shown for scale. (b) A dramatic Sun captured by instruments aboard the *SOHO* satellite, February 12, 2001. A twirling prominence rises into the Sun's corona, and other prominences are also visible. [(a) Image courtesy of *SOHO/EIT* (Solar and Heliospheric Observatory/Extreme Ultraviolet Imaging Telescope) Consortium. (b) Image courtesy of *SOHO/MDI* Consortium—MDI is from Stanford-Lockheed/Martin Institute for Space Research. *SOHO* is an international project of cooperation between the European Space Agency and NASA. See http://sohowww.nascom.nasa.gov/.]

News Report 2.1

The Nature of Order Is Chaos

In 1960, Edward Lorenz, an MIT scientist, rocked the scientific world with the statement that a butterfly flapping its wings in Brazil might produce a tornado in Texas. He used this strange example to suggest that the interaction of orderly and deterministic systems may produce chaotic and unpredictable results.

For example, ice has a rigid internal structure, forced by bonding between water molecules. This structure dictates that all ice crystals are six-sided, yet no two ice crystals, despite this similarity, are identical. Beneath this chaos of design exists an order dictated by physical principles (see Figure 7.5).

A major shift in our real-world view occurred with the advent of *chaos theory*, a revolution in science that considers the nonlinear and unpredictable behavior of operational systems. This theory suggests that the scientific method must consider the coexistence of disorder and order, randomness and pattern, and symmetry and chaos in natural systems—the science of complexity and complex systems.

Consider the weather: Mathematical models and numerical equations describe the behavior of water vapor, temperature, and pressure patterns. Yet weather systems are sensitive to very small fluctuations in any of those ingredients. Therefore, it is difficult to exactly predict how a weather system will develop, what track it will follow, or how severe it might be (see Chapter 8). Two similar chaotic weather systems might produce a similar result, although it is not possible to say exactly what the output will be. This understanding of the role of chaos is helping scientists to improve forecasting of weather phenomena. In a dynamic weather system, chaos is the rule, just as it is in chemical and biological systems.

Chaos theory is useful in studying all of Earth's physical environments. For example, a river flows in branched channels over a floodplain in a pattern to conserve energy. The channels constantly shift in a randomness that is irregular and difficult to predict. See the *Terra* satellite image of the many mouths of the Ganges River in Figure 14.24 for an example of such *fractal branching*—irregular, curving channels that may or may not repeat their pattern.

This may seem strange, but the nature of order we observe is the result of chaos and the almost infinite interaction among physical elements! Chaos theory is a new dimension of the scientific method and physical geography.

---

approach Earth. The **magnetosphere** is a magnetic field surrounding Earth, generated by dynamo-like motions within our planet. The magnetosphere deflects the solar wind toward both of Earth's poles so that only a small portion of it enters the atmosphere.

Because the solar wind does not reach Earth's surface, research on this phenomenon must be conducted in space. In 1969 the *Apollo XI* astronauts exposed a piece of foil on the lunar surface as a solar wind experiment (Figure 2.3). When examined back on Earth, the exposed foil exhibited particle impacts that confirmed the presence and character of the solar wind.

Interaction of the solar wind and the upper layers of Earth's atmosphere produces the remarkable **auroras** that occur toward both poles. These lighting effects are the *aurora borealis* (northern lights) and *aurora australis* (southern lights) in the upper atmosphere, 80–500 km (50–300 mi) above Earth's surface (Figure 2.4). During the 2001 solar maximum, auroras were visible as far south as Jamaica, Texas, and California. Also, the solar wind disrupts certain radio broadcasts and some satellite transmissions, causes overloads on Earth-based electrical systems, and may affect weather patterns.

Our understanding of the solar wind is increasing dramatically as data are collected by a variety of satellites: the *SOHO* (*Solar and Heliospheric Observatory*), *FAST* (*Fast Auroral Snapshot*), *WIND*, *Ulysses*, the *Dynamics Explorer*, and

FIGURE 2.3 Astronaut and solar wind experiment.
Without a protective atmosphere, the lunar surface receives charged particles of the solar wind and all of the Sun's electromagnetic radiation. A sheet of foil is deployed by an *Apollo XI* astronaut in 1969 in the solar wind experiment. Earth-bound scientists analyzed the foil, which revealed the composition of the solar wind. Why wouldn't this experiment work if deployed on Earth's surface? [NASA photo.]

(a)

(b)

FIGURE 2.4 Auroras from an orbital perspective and from the ground in Alaska.
(a) *Polar* satellite false-color image of auroral halo over Earth's North Pole region from April 1996—the UVI sensor is able to capture the aurora on the day and night sides of Earth. (b) Surface view of an aurora borealis in the night sky over Alaska. [(a) Image from the Ultraviolet Imager (UVI) aboard *Polar* satellite courtesy of NASA. (b) Aurora photo by Johnny Johnson/Tony Stone Images, Inc.]

Genesis with solar wind collectors, among others. All satellite data are available on the Internet. (For auroral activity, see **http://www.sel.noaa.gov/pmap/** and for forecasts see **http://www.gi.alaska.edu**, or for auroral photographs see **http://northernlightsnome.homestead.com/**.)

Weather Effects   Another effect of the solar wind in the atmosphere is its possible influence on weather and climate cycles. Why do wetter periods in some midlatitude areas tend to coincide with every other solar maximum? Why do droughts often occur near the time of every other solar minimum? For example, sunspot cycles during more than 260 years from 1740 to 1998 coincided with periods of wetness and drought, as estimated by an analysis of tree-growth rings and weather records for that period throughout the western United States and elsewhere. These variations in weather tend to lag 2 or 3 years behind the solar maximum or minimum.

Regardless of the *cause* for the cyclical patterns of drought and wetness that do occur, a remarkable failure in current planning worldwide is the lack of *attention* given to these cycles. Preparing for such patterns could reduce property loss and casualties. As knowledge of the solar wind–weather relation improves, and as the half-decade-long western drought intensifies, the attention of policy makers and the public is mandatory.

Electromagnetic Spectrum of Radiant Energy

The key essential solar input to life is electromagnetic energy of various wavelengths. Solar radiation occupies a portion of the **electromagnetic spectrum** of radiant energy. This radiant energy travels at the speed of light to Earth. The total spectrum of this radiant energy is made up of different wavelengths. Figure 2.5 shows that a **wavelength** is the distance between corresponding

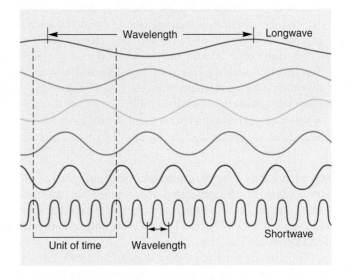

FIGURE 2.5 Wavelength and frequency.
Wavelength and frequency are two ways of describing the same phenomenon—electromagnetic wave motion. More short wavelengths pass a given point during a unit of time, so they are higher in frequency; whereas fewer long wavelengths pass a point in a unit of time, so they are lower in frequency.

points on any two successive waves. The number of waves passing a fixed point in 1 second is the *frequency*.

The Sun emits radiant energy composed of 8% ultraviolet, X-ray, and gamma-ray wavelengths; 47% visible light wavelengths; and 45% infrared wavelengths. A portion of the electromagnetic spectrum is illustrated in Figure 2.6, with wavelengths increasing from the top of the illustration to the bottom. Note the wavelengths at which various phenomena and human applications of energy occur.

An important physical law states that all objects radiate energy in wavelengths related to their individual surface temperatures: the hotter the object, the shorter the wavelengths emitted. This law holds true for the Sun and Earth.

FIGURE 2.6 A portion of the electromagnetic spectrum of radiant energy.
The spectrum is oriented with shorter wavelengths toward the top and longer wavelengths toward the bottom.

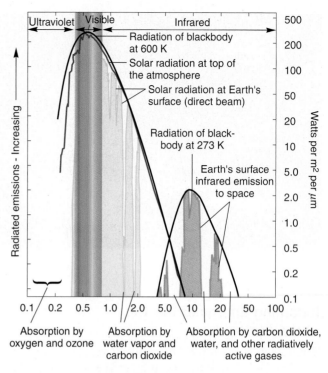

FIGURE 2.7 Solar and terrestrial energy distribution by wavelength.
The solar output peaks in shorter wavelengths of visible light in relation to its higher surface temperature, whereas Earth's emissions are concentrated in the infrared portion of the spectrum in relation to its lower surface temperature. Ideal blackbody curves for a radiating body like the Sun (hotter) and Earth (cooler) illustrate the concept. The dropouts in the plot lines of solar and terrestrial radiation represent absorption bands of water vapor, water, carbon dioxide, oxygen, ozone (O_3), and other gases. [Adapted from W.D. Sellers, **Physical Climatology** (Chicago: University of Chicago Press, 1965), p. 20. Used by permission.]

ANIMATION   Electromagnetic Spectrum and Plants

Figure 2.7 shows that the hot Sun radiates shorter wavelength energy, concentrated around 0.5 μm (micrometer).

The Sun's surface temperature is about 6000°C (11,000°F), and its emission curve shown in the figure is similar to that predicted for an idealized 6000°C surface, or *blackbody radiator* (shown in Figure 2.7). An ideal blackbody absorbs all the radiant energy that it receives and subsequently emits all that it receives. A hotter object like the Sun emits a much greater amount of energy per unit area of its surface than does a similar area of a cooler object like Earth. Shorter wavelength emissions are dominant at these higher temperatures.

Although cooler than the Sun, Earth radiates nearly all that it absorbs and acts as a blackbody (also shown in Figure 2.7). Earth is a cooler radiating body, so longer wavelengths are emitted as radiation mostly in the infrared portion of the spectrum. Some of the atmospheric gases vary in their response to radiation received, being transparent to some while absorbing others. The maximum energy output from Earth occurs in longer wavelengths, centered around 10.0 μm and entirely within the infrared portion of the spectrum.

Figure 2.8 illustrates the flows of energy into and out of Earth systems. To summarize, the Sun's radiated energy is *shortwave radiation* that peaks in the short visible wavelengths, whereas Earth's radiated energy is *longwave radiation* concentrated in infrared wavelengths. In Chapter 4, we see that Earth, clouds, sky, ground, and things that are terrestrial radiate longer wavelengths in contrast to the Sun.

FIGURE 2.8 Earth's energy budget simplified.
Inputs of shorter wavelengths arrive at Earth from the Sun. Outputs of longer wavelengths of infrared radiate to space from Earth. The data plotted in Figure 2.10 and the map in Figure 2.11 are derived from data gathered along the top of the atmosphere.

Incoming Energy at the Top of the Atmosphere

The region at the top of the atmosphere, approximately 480 km (300 mi) above Earth's surface, is termed the **thermopause**. It is the outer boundary of Earth's energy system and provides a useful point at which to assess the arriving solar radiation before it is diminished by scattering and absorption in passage through the atmosphere.

Earth's distance from the Sun results in its interception of only one two-billionth of the Sun's total energy output. Nevertheless, this tiny fraction of the Sun's overall output is an enormous amount of energy input to Earth's systems. Solar radiation that reaches a horizontal plane at Earth is called **insolation** (*in*coming *sol*ar radi*ation*), a term specifically applied to radiation arriving at Earth's atmosphere and surface. Insolation at the top of the atmosphere is expressed as the *solar constant*.

Solar Constant   Knowing the amount of insolation incoming to Earth is important to climatologists and other scientists. The **solar constant** is the average insolation received at the thermopause when Earth is at its average distance from the Sun. That value is 1372 W/m^2 (watts per square meter).* The constancy of the solar constant over time is important, for small variations of even 0.5% or 1.0% could prove dramatic for Earth's energy system.

As we follow insolation through the atmosphere to Earth's surface (Chapters 3 and 4), we see that the value of the solar constant is reduced by half or more through reflection, scattering, and absorption of shortwave radiation.

---

*A watt is equal to 1 joule (a unit of energy) per second and is the standard unit of power in the SI–metric system. (See Appendix C of this text for more information on measurement conversions.) In nonmetric calorie heat units, the solar constant is expressed as approximately 2 calories per square centimeter per minute, or 2 langleys per minute (a langley is 1 cal/cm^2). A calorie is the amount of energy required to raise the temperature of 1 g of water (at 15°C) 1 degree Celsius and is equal to 4.184 joules.

Uneven Distribution of Insolation   Earth's curved surface presents a continually varying angle to the incoming parallel rays of insolation (Figure 2.9). Differences in the angle of solar rays at each latitude result in an uneven distribution of insolation and heating. The place receiving maximum insolation is the point where insolation rays are perpendicular to the surface (radiating from directly overhead), called the **subsolar point**. All other places receive insolation at less than a 90° angle and thus experience more diffuse energy receipts. Solar beam angles become more pronounced at higher latitudes. As a result, during a year's time, the thermopause above the equatorial region receives 2.5 times more insolation than the thermopause above the poles. Lower-angle solar rays toward the poles must pass through a greater thickness of atmosphere, resulting in further losses of energy due to scattering, absorption, and reflection.

Figure 2.10 illustrates the daily variations throughout the year of energy at the top of the atmosphere for various latitudes in watts per square meter (W/m^2). The chart shows a decrease in insolation from the equatorial regions northward and southward toward the poles. However, in June, the North Pole receives more than 500 W/m^2 per day, which is more than is ever received at 40° N latitude or at the equator. Such high values result from long daylengths at the poles in summer: 24 hours a day, compared with only 15 hours of daylight at 40° N latitude and 12 hours at the equator. However, at the poles the summertime Sun at noon is low in the sky, so a daylength twice that of the equator yields only about a 100 W/m^2 difference.

In December, the pattern reverses. Note that the top of the atmosphere at the South Pole receives even more insolation than the North Pole does in June (more than 550 W/m^2). This is a function of Earth's closer location to the Sun at perihelion (January 3 on Figure 2.1d).

Along the equator, two maximum periods of approximately 430 W/m^2 occur at the spring and fall equinoxes, when the subsolar point is at the equator. Find your latitude on the graph and follow across the months to determine the seasonal variation of insolation where you live.

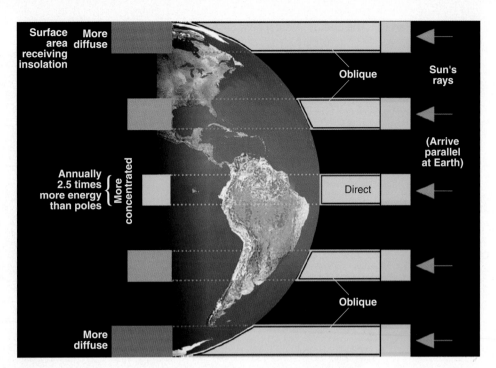

FIGURE 2.9 Insolation receipts and Earth's curved surface.
Solar insolation angles determine the concentration of energy receipts by latitude. Lower latitudes receive more concentrated energy from a more direct solar beam. Higher latitudes receive slanting (oblique) rays and more diffuse energy. Note the area covered by identical columns of solar energy arriving at Earth's surface at higher latitudes (more diffuse, larger area covered) and at lower latitudes (more concentrated, smaller area covered).

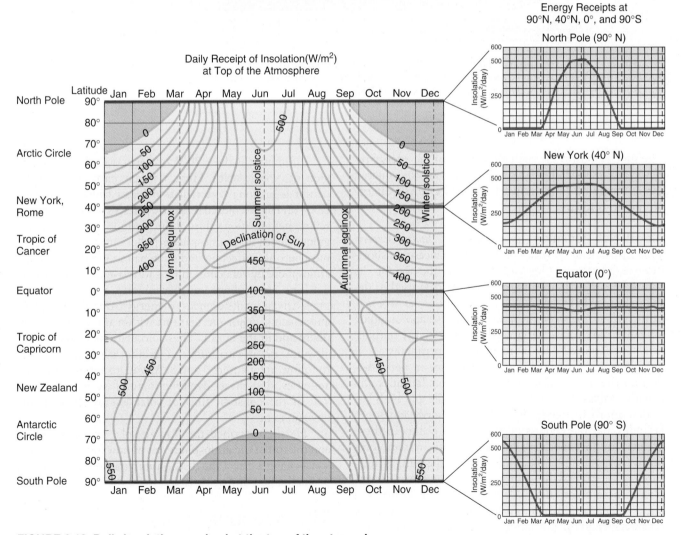

FIGURE 2.10 Daily insolation received at the top of the atmosphere.
The total daily insolation received at the top of the atmosphere is charted in watts per square meter per day by latitude and month (1 W/m²/day = 2.064 cal/cm²/day). A profile of annual energy receipts is graphed to the right for the North Pole, along 40° north latitude, along the equator, and for the South Pole. [Reproduced by permission of the Smithsonian Institution Press from *Smithsonian Miscellaneous Collections: Smithsonian Meteorological Tables*, vol. 114, 6th edition. Robert List, ed. (Washington, DC: Smithsonian Institution, 1984), p. 419, Table 134.]

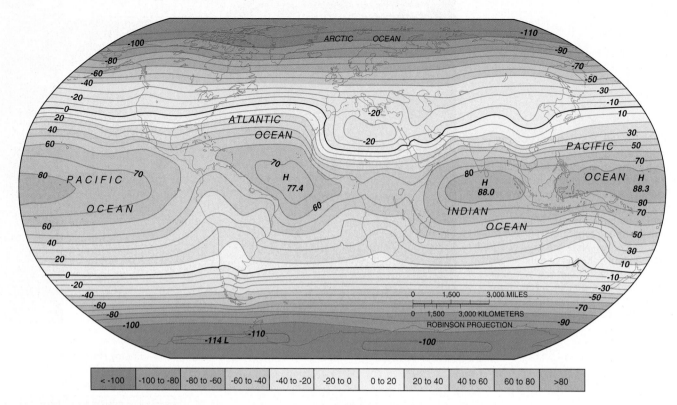

< -100	-100 to -80	-80 to -60	-60 to -40	-40 to -20	-20 to 0	0 to 20	20 to 40	40 to 60	60 to 80	>80

FIGURE 2.11 Daily net radiation patterns at the top of the atmosphere.
Averaged daily net radiation flows for a 9-year period (1979–1987), measured at the top of the atmosphere by the Earth Radiation Budget Experiment (ERBE) aboard the *Nimbus-7* satellite. Units are W/m². [Data for map courtesy of Dr. H. Lee Kyle, Goddard Space Flight Center, NASA.]

The four graphs to the right show plots of the energy received at the North Pole, along 40° N, along the equator, and at the South Pole during the year to give you energy profiles to compare.

Global Net Radiation *Earth Radiation Budget Experiment* instruments aboard several satellites measured shortwave and longwave flows of energy at the top of the atmosphere. ERBE sensors collected the data used to develop the map in Figure 2.11. This map shows *net radiation*, or the balance between incoming shortwave and outgoing longwave radiation—energy inputs minus energy outputs. See News Report 2.2 for more on these experiments.

Note the latitudinal energy imbalance in net radiation on the map—positive values in lower latitudes and negative values toward the poles. In middle and high latitudes, approximately poleward of 36° north and south latitudes, net radiation is negative. This occurs in these higher latitudes because Earth's climate system loses more energy to space than it gains from the Sun, as measured at the top of the atmosphere. In the lower atmosphere, these polar energy deficits are offset by flows of energy from tropical energy surpluses (as we will see in Chapters 4 and 6). The largest net radiation values, averaging 80 W/m², are above the tropical oceans along a narrow equatorial zone. Net radiation minimums are lowest over Antarctica.

Of interest is the −20 W/m² area over the Sahara region of North Africa. Here, typically clear skies—which permit great longwave radiation losses from Earth's surface—and light-colored reflective surfaces work together to reduce net radiation values at the thermopause. In other regions, clouds and atmospheric pollution in the lower atmosphere also affect net radiation patterns at the top of the atmosphere by reflecting more shortwave energy to space.

The atmosphere and ocean form a giant heat engine, driven by differences in energy from place to place and causing major circulations within the lower atmosphere and in the ocean. These circulations include global winds, ocean currents, and weather systems—subjects to follow in Chapters 6 and 8. As you go about your daily activities, let these dynamic natural systems remind you of the constant flow of solar energy through the environment.

Having examined the flow of solar energy to Earth and the top of the atmosphere, let us now look at how seasonal changes affect the distribution of insolation as Earth orbits the Sun during the year.

The Seasons

Earth's periodic rhythms of warmth and cold, dawn and daylight, twilight and night, have fascinated humans for centuries. In fact, many ancient societies demonstrated an

News Report 2.2

Monitoring Earth Radiation Budget

Earth's weather and climate are a direct result of the balance between sunlight received from space and energy reflected and radiated to space. Patterns of insolation absorption and energy reradiation by Earth systems produce an energy budget. Since 1978, the Earth–atmosphere energy budget has been monitored by the Earth Radiation Budget (ERB) package on board the *Nimbus-7* satellite, and later by the *ERB*, *NOAA-9*, and *NOAA-10* satellites. These satellites collected more than a decade of data. ERB sensors mapped the complex exchanges of energy among atmosphere, ocean, and land.

Measurements include monthly energy budget and variations, sea-sonal shifts in energy between equator and poles, and daily energy budgets at the regional scale. The map in Figure 2.11 is a direct result of ERBE, showing regions that absorb more energy than radiated (a heat-energy source, rose color on map in Figure 2.11) and that radiate more energy than received (a heat-energy sink, purple color).

A new tool in understanding Earth's radiation budget is the CERES (Clouds and Earth Radiant Energy System) sensor aboard satellite *Terra*. CERES monitors short-wave (light) and longwave (heat energy) and, thus, Earth's energy balance of incoming and outgoing radia-tion. A pair of CERES images from March 2000 are in Chapter 4.

In addition, ERB measurements determined that solar irradiance, or the luminous brightness of solar radiation, is directly correlated to the sunspot cycle discussed earlier. During the solar maximums of 1979 and 1991, the solar constant exceeded 1374 W/m^2; the 1986 minimum produced a constant of 1371 W/m^2 (remember the solar constant averages 1372 W/m^2). The ERBE findings help scientists better understand the systems described in Part 1 in this textbook (see **http://asd-www.larc. nasa.gov/erbe/ASDerbe.html**).

---

intense awareness of seasonal change and formally commemorated these natural energy rhythms with festivals, monuments, and calendars. Such ancient seasonal monuments and calendar markings occur worldwide, including thousands of sites in North America. Unfortunately, the 12-month Gregorian calendar in common use has no relation to natural rhythms of Sun, Earth, the Moon, and the march of the seasons.

Seasonality

Seasonality refers to both the seasonal variation of the Sun's position above the horizon and changing daylengths during the year. Seasonal variations are a response to changes in the Sun's **altitude**, or the angle between the horizon and the Sun. At sunrise or sunset, the Sun is at the horizon, so its altitude is 0°. During the day, if the Sun reaches halfway between the horizon and directly overhead, it is at 45° altitude. If the Sun reaches the point directly overhead, it is at 90° altitude. The Sun is directly overhead (90° altitude, or *zenith*) only at the *subsolar point*, where insolation is at a maximum. At all other surface points, the Sun is at a lower altitude angle, producing more-diffuse insolation.

The Sun's **declination** is the latitude of the subsolar point. Declination annually migrates through 47° of latitude, moving between the *Tropic of Cancer* at 23.5° N and the *Tropic of Capricorn* at 23.5° S latitude. Other than Hawai'i, which is between 19° N and 22° N, the subsolar point does not reach the continental United States or Canada; all other states and provinces are too far north.

In addition to changing Sun altitude and declination, seasonality means changing **daylength**, or duration of exposure to insolation. Daylength varies during the year, depending on latitude. The equator always receives equal hours of day and night: If you live in Ecuador, Kenya, or Singapore, every day and night is 12 hours long, year-round. People living along 40° N latitude (Philadelphia, Denver, Madrid, Beijing), or 40° S latitude (Buenos Aires, Capetown, Melbourne), experience about 6 hours' difference in daylight between winter and summer. Those at 50° N or S latitude (Winnipeg, Paris, Falkland Islands) experience almost 8 hours of annual daylength variation.

At the North and South Poles, the range of daylength is extreme and extends from a 6-month period of no insolation (ranging from twilight to darkness then to dawn) to a 6-month period of continuous 24-hour insolation (daylight)—literally the poles experience one long day and one long night each year! This is evident in Figure 2.14, if you note the illumination of the North Pole in June (to the left) and South Pole in December (to the right). Given this observation, turn to the back cover of this textbook and see whether you can determine the month during which the Apollo astronaut made the Earth photo. (The answer is on this book's copyright page.)

Reasons for Seasons

Seasons result from variations in the Sun's *altitude* above the horizon, the Sun's *declination* (latitude of the subsolar point), and *daylength* during the year. These in turn are

South Pole the Sun is setting—a dramatic moment for the people working there.

From March, the seasons move on to June 20 or 21 and the moment of the **summer solstice**, or **June solstice**. The subsolar point migrates from the equator to 23.5° N latitude, the Tropic of Cancer. Because the circle of illumination now includes the North Polar region, everything north of the Arctic Circle receives 24 hours of daylight—the "midnight Sun." Figure 2.16 is a multiple-image photo of the midnight Sun as seen north of the Arctic Circle. In contrast, the region from the Antarctic Circle to the South Pole (66.5°–90° S latitude) is in darkness the entire 24 hours. Those working in Antarctica call this *Midwinter's Day*.

September 22 or 23 is the moment in time of the **autumnal equinox**, or **September equinox**, when Earth's orientation is such that the circle of illumination again passes through both poles so that all parts of the globe experience a 12-hour day and a 12-hour night. The subsolar point returns to the equator, with days growing shorter to the north and longer to the south. Researchers stationed at the South Pole see the disk of the Sun just rising, ending their 6 months of darkness. In the Northern Hemisphere, autumn arrives, a time of many colorful changes in the landscape, whereas in the Southern Hemisphere it is spring.

Dawn and Twilight   *Dawn* is the period of diffused light that occurs before sunrise. The corresponding evening time after sunset is *twilight*. During both periods, light is scattered by molecules of atmospheric gases and reflected by dust and moisture illuminating the atmosphere. The duration of both is a function of latitude, because the angle of the Sun's path above the horizon determines the thickness of the atmosphere through which the Sun's rays must pass. This effect may be enhanced by the presence of pollution aerosols and suspended particles from volcanic eruptions or forest and grassland fires.

At the equator, where the Sun's rays are almost directly above the horizon throughout the year, dawn and twilight are limited to 30–45 minutes each. These times increase to 1–2 hours each at 40° latitude, and at 60° latitude they each range upward from 2.5 hours, with little true night in summer. The poles experience about 7 weeks of dawn and 7 weeks of twilight, leaving only 2.5 months of "night" during the 6 months when the Sun is completely below the horizon.

Seasonal Observations   In the midlatitudes of the Northern Hemisphere, the position of sunrise on the horizon migrates from day to day, from the southeast in December to the northeast in June. Over the same period, the point of sunset migrates from the southwest to the northwest. The Sun's altitude at local noon at 40° N latitude increases from a 26° angle above the horizon at the winter (December) solstice to a 73° angle above the horizon at the summer (June) solstice—a range of 47° (Figure 2.17).

Seasonal change is quite noticeable across the landscape away from the equator. Think back over the past year. What seasonal changes have you observed in vegetation, temperatures, and weather? Figure 2.18 presents two composite images of vegetation cover in winter and late summer as recorded by the AVHRR sensors aboard polar-orbiting satellites. Recently, the timing of the seasons is changing as global climates shift in middle and high latitudes. Spring and leafing out is occurring as much as 2 weeks earlier than expected from average conditions. Likewise, the timing of fall is happening later.

Midnight
Sun
↓

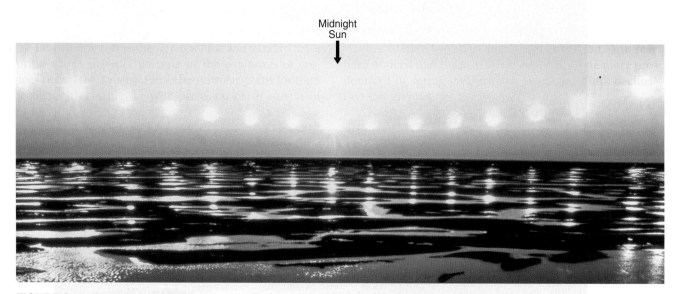

FIGURE 2.16 The midnight Sun.
The midnight Sun north of the Arctic Circle captured in a series of 18 exposures on the same piece of film. The camera is facing due north. Midnight is the exposure showing the Sun closest to the horizon. The photographer removed the lens cap at regular intervals to make the multiple exposures. [Photo by Gary Braasch/Tony Stone Images.]

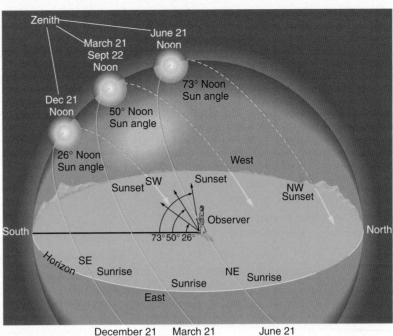

FIGURE 2.17 Seasonal observations—sunrise, noon, and sunset through the year. Seasonal observations are at 40° N latitude for the December solstice, March equinox, June solstice, and September equinox. The Sun's altitude increases from 26° in December to 73° above the horizon in June—a difference of 47°. Note the changing position of sunrise and sunset along the horizon during the year. For a useful sunrise and sunset calculator for any location, go to http://www.srrb.noaa.gov/highlights/sunrise/sunrise.html.

FIGURE 2.18 Seasonal change from orbit. Seasonal change is monitored and measured by sensors aboard polar-orbiting satellites. Compare and contrast similar regions for January and July: the American Midwest, China, Argentina, and Europe. Orbital remote sensing is tracking an earlier spring season worldwide. [Images courtesy of Garik Gutman, NOAA, National Environmental Satellite, Data, and Information Service.]

Summary and Review—Solar Energy to Earth and the Seasons

▶ *Distinguish* among galaxies, stars, and planets and *locate* Earth.

Our Solar System—the Sun and nine planets—is located on a remote, trailing edge of the **Milky Way Galaxy**, a flattened, disk-shaped mass estimated to contain up to 400 billion stars. **Gravity**, the mutual attracting force exerted by the mass of an object upon all other objects, is an organizing force in the Universe. The process of stars (like our Sun) condensing from nebular clouds with planetesimals (protoplanets) forming in orbits around their central masses is the **planetesimal hypothesis**.

Milky Way Galaxy (p. 42)
gravity (p. 42)
planetesimal hypothesis (p. 42)

1. Describe the Sun's status among stars in the Milky Way Galaxy. Describe the Sun's location, size, and relationship to its planets.
2. If you have seen the Milky Way at night, briefly describe it. Use specifics from the text in your description.
3. Briefly describe Earth's origin as part of the Solar System.
4. Compare the locations of the nine planets of the Solar System.

▶ *Overview* the origin, formation, and development of Earth and *construct* Earth's annual orbit about the Sun.

The Solar System, planets, and Earth began to condense from a nebular cloud of dust, gas, debris, and icy comets approximately 4.6 billion years ago. Distances in space are so vast that the **speed of light** (300,000 kmps, or 186,000 mps, which is about 9.5 trillion km, or nearly 6 trillion mi, per year) is used to express distance.

In its orbit, Earth is at **perihelion** (its closest position to the Sun) during our Northern Hemisphere winter (January 3 at 147,255,000 km, or 91,500,000 mi). It is at **aphelion** (its farthest position from the Sun) during our Northern Hemisphere summer (July 4 at 152,083,000 km, or 94,500,000 mi). Earth's average distance from the Sun is approximately 8 minutes and 20 seconds in terms of light speed. In the Solar System, an imaginary plane touching all points of Earth's orbit is termed the **plane of the ecliptic**.

speed of light (p. 42)   aphelion (p. 42)
perihelion (p. 42)   plane of the ecliptic (p. 42)

5. How far is Earth from the Sun in terms of light speed? In terms of kilometers and miles?
6. Briefly describe the relationship among these concepts: Universe, Milky Way Galaxy, Solar System, Sun, and Planet Earth.
7. Diagram in a simple sketch Earth's orbit about the Sun. How much does it vary during the course of a year?

▶ *Describe* the Sun's operation and *explain* the characteristics of the solar wind and the electromagnetic spectrum of radiant energy.

The fusion process—hydrogen atoms forced together under tremendous temperature and pressure in the Sun's interior—generates incredible quantities of energy. The Sun's most conspicuous features are large **sunspots**, caused by magnetic disturbances. Solar energy in the form of charged particles of **solar wind** travels out in all directions from disturbances on the Sun. Solar wind is deflected by Earth's **magnetosphere**, producing various effects in the upper atmosphere, including spectacular **auroras**, the northern and southern lights, which surge across the skies at higher latitudes. Another effect of the solar wind in the atmosphere is its possible influence on weather.

The **electromagnetic spectrum** of radiant energy travels outward in all directions from the Sun. The total spectrum of this radiant energy is made up of different **wavelengths**—the distance between corresponding points on any two successive waves of radiant energy. Eventually, some of this radiant energy reaches Earth's surface.

fusion (p. 44)   auroras (p. 45)
solar wind (p. 44)   electromagnetic spectrum (p. 46)
sunspots (p. 44)   wavelength (p. 46)
magnetosphere (p. 45)

8. How does the Sun produce such tremendous quantities of energy?
9. What is the sunspot cycle? At what stage was the cycle in the year 2001?
10. Describe Earth's magnetosphere and its effects on the solar wind and the electromagnetic spectrum.
11. Summarize the presently known effects of the solar wind relative to Earth's environment.
12. Describe the various segments of the electromagnetic spectrum, from shortest to longest wavelength. What are the main wavelengths produced by the Sun? Which are principally radiated by Earth to space?

▶ *Portray* the intercepted solar energy and its uneven distribution at the top of the atmosphere.

Electromagnetic radiation from the Sun passes through Earth's magnetic field to the top of the atmosphere—the **thermopause**, at approximately 500 km (300 mi) altitude. Incoming solar radiation that reaches a horizontal plane at Earth is called **insolation**, a term specifically applied to radiation arriving at Earth's surface and atmosphere. The term **solar constant** describes insolation at the top of the atmosphere: the average insolation received at the thermopause when Earth is at its average distance from the Sun. The solar constant is measured as 1372 W/m² (2.0 cal/cm²/min; 2 langleys/min). The place receiving maximum insolation is the **subsolar point**, where solar rays are perpendicular to the surface (radiating from directly overhead). All other locations away from the subsolar point receive slanting rays and more diffuse energy.

thermopause (p. 48)   solar constant (p. 48)
insolation (p. 48)   subsolar point (p. 48)

13. What is the solar constant? Why is it important to know?
14. Select 40° or 50° north latitude on Figure 2.10 and plot the amount of energy in watts per square meter (W/m²) per day on a graph for each month throughout the year. Compare this with the amount at the North Pole and at the equator.
15. If Earth were flat and oriented perpendicularly to incoming solar radiation (insolation), what would be the latitudinal distribution of solar energy at the top of the atmosphere?

▶ *Define* solar altitude, solar declination, and daylength and *describe* the annual variability of each—Earth's seasonality.

The angle between the Sun and the horizon is the Sun's **altitude**. The Sun's **declination** is the latitude of the subsolar point. Declination annually migrates through 47° of latitude, moving between the *Tropic of Cancer* at 23.5° N (June) and the *Tropic of Capricorn* at 23.5° S latitude (December). Seasonality means an annual change in the Sun's altitude and changing **daylength**, or duration of exposure.

Earth's distinct seasons are produced by interactions of **revolution** (annual orbit about the Sun), **rotation** (turning on the axis), **axial tilt** (about 23.5° from a perpendicular to the plane of the ecliptic), **axial parallelism** (the parallel alignment of the axis throughout the year), and sphericity. Earth rotates about its **axis**, an imaginary line extending through the planet from the geographic North Pole to the South Pole. As it rotates, the traveling boundary that divides daylight and darkness is called the **circle of illumination**. Earth's **axial tilt** is about 23.5° from a perpendicular to the plane of the ecliptic, and it remains oriented to the stars in the same direction throughout the year, or **axial parallelism**. Daylength is the interval between **sunrise**, the moment when the disk of the Sun first appears above the horizon in the east, and **sunset**, that moment when it totally disappears below the horizon in the west.

On December 21 or 22, at the moment of the **winter solstice** ("winter Sun stance"), or **December solstice**, the circle of illumination excludes the North Pole but includes the South Pole. The subsolar point is at 23.5° S latitude, the parallel called the Tropic of Capricorn. The moment of the **vernal equinox**, or **March equinox**, occurs on March 20 or 21. At that time, the circle of illumination passes through both poles so that all locations on Earth experience a 12-hour day and a 12-hour night.

June 20 or 21 is the moment of the **summer solstice**, or **June solstice**. The subsolar point migrates from the equator to 23.5° N latitude, the Tropic of Cancer. Because the circle of illumination now includes the North Polar region, everything north of the Arctic Circle receives 24 hours of daylight—the "midnight Sun." September 22 or 23 is the time of the **autumnal equinox**, or **September equinox**, when Earth's orientation is such that the circle of illumination again passes through both poles so that all parts of the globe experience a 12-hour day and a 12-hour night.

altitude (p. 51)
declination (p. 51)
daylength (p. 51)

16. Assess the 12-month Gregorian calendar, with its months of different lengths, and leap years, and its relation to the annual seasonal rhythms—the march of the seasons. What do you find?

17. The concept of seasonality refers to what specific phenomena? How do these two aspects of seasonality change during a year at 0° latitude? At 40°? At 90°?

18. Differentiate between the Sun's altitude and its declination at Earth's surface.

19. For the latitude at which you live, how does daylength vary during the year? How does the Sun's altitude vary? Does your local newspaper publish a weather calendar containing such information?

20. List the five physical factors that operate together to produce seasons.

21. Describe Earth's revolution and rotation and differentiate between them.

22. Define Earth's present tilt relative to its orbit about the Sun.

23. Describe seasonal conditions at each of the four key seasonal anniversary dates during the year. What are the solstices and equinoxes, and what is the Sun's declination at these times?

 NetWork

The *Geosystems* Home Page provides on-line resources for this chapter on the World Wide Web. To begin: Once on the Home Page, click on this textbook, scroll the Table of Contents menu, select this chapter, and click "Begin." You will find self-tests that are graded, review exercises, specific updates for items in the chapter, and many links to interesting related pathways on the Internet under "Destinations." *Geosystems* is at **http://www.prenhall.com/christopherson.**

Critical Thinking

A. Using the concepts in Figure 2.17, use a protractor and stick or ruler to measure the angle of the Sun's altitude at noon (or 1 P.M., if in daylight saving time). Do not look at the Sun; rather, with your back to the Sun, align the stick so that it casts no shadow as you measure its rays against the protractor. Place this measurement in your notebook and affix a Post-It® in the textbook to remind you to repeat the measurement near the end of the semester. Compare and analyze seasonal change using your different measurements of the Sun's altitude.

B. Also, in reference to the concepts in Figure 2.17, post a reminder in your notebook to check the position of sunrise and sunset at least twice during the semester. If you have a magnetic compass, note the degrees from north for sunrise and sunset (*azimuth* is read from north in a clockwise direction—0° and 360° being the same point, north). Find some place where you can see the horizon. If new to such observations, it will be a surprise as to the degree of location change over the span of months in a school term. Be sure and ask your teacher about the magnetic declination of your location to adjust your compass readings.

C. The variability of Earth's axial tilt, orbit about the Sun, and a wobble to the axis is described in Figure 17.31. Please refer to this figure and compare these changing conditions to the information in Table 2.1 and the related figures in this chapter. Speculate on the effects of each of these changes on the annual march of the seasons.

Rain clouds over the upper Midwest cast shadows toward a full-Moon rise. Some of the droplets trailing from the clouds are virga, which evaporate before reaching the ground. Such scenes are the continuous beauty of our atmosphere, yet the atmosphere provides important functions that protect and sustain our lives. [Photo by Bobbé Christopherson.]

Earth's Modern Atmosphere

▶ Key Learning Concepts

After reading the chapter, you should be able to:

▶ **Construct** a general model of the atmosphere based on the criteria composition, temperature, and function, and **diagram** this model in a simple sketch.

▶ **List** the stable components of the modern atmosphere and their relative percentage contributions by volume, and **describe** each.

▶ **Describe** conditions within the stratosphere; specifically, **review** the function and status of the ozonosphere (ozone layer).

▶ **Distinguish** between natural and anthropogenic variable gases and materials in the lower atmosphere.

▶ **Describe** the sources and effects of carbon monoxide, nitrogen dioxide, and sulfur dioxide, and **construct** a simple equation that illustrates photochemical reactions that produce ozone, peroxyacetyl nitrates, nitric acid, and sulfuric acid.

Earth's atmosphere is a unique reservoir of gases, the product of nearly 5 billion years of development. It protects us from hostile radiation and particles from the Sun and beyond. The atmosphere serves as an efficient filter. When astronauts work in space above the atmosphere, they must wear a bulky spacesuit that does everything to sustain them that the atmosphere does for us all the time.

In this chapter: We examine the modern atmosphere through its composition, temperature, and function. Our consideration of the atmosphere also includes the spatial aspects of both natural and human-produced air pollution. We all participate in the atmosphere with each breath we take, the energy we consume, the traveling we do, and the products we buy. Human activities caused the stratospheric ozone predicament and the blight of acid deposition on ecosystems. These topics are essential to physical geography, for we are influencing the atmospheric composition of the future.

Atmospheric Composition, Temperature, and Function

The *modern* atmosphere probably is the fourth general atmosphere in Earth's history. Therefore, this modern atmosphere is a gaseous mixture of ancient origin, the sum of all the exhalations and inhalations of life on Earth throughout time. The principal substance of this atmosphere is air, the medium of life as well as a major industrial and chemical raw material. *Air* is a simple mixture of gases that is naturally odorless, colorless, tasteless, and formless, blended so thoroughly that it behaves as if it were a single gas.

In his book *The Lives of a Cell*, the late physician and self-styled "biology watcher" Lewis Thomas compared the atmosphere of Earth to an enormous cell membrane. The membrane around a cell regulates the interactions between the cell's delicate inner workings and the potentially disruptive outer environment. Each cell membrane is selective as to what it will allow to pass through. The modern atmosphere acts as Earth's protective membrane, as Thomas described so vividly (Figure 3.1).

As a practical matter, we consider the top of our atmosphere to be around 480 km (300 mi) above Earth's surface, the same altitude we use in Chapter 2 for measuring the solar constant and insolation receipt. Beyond that altitude, the atmosphere is rarefied (nearly a vacuum) and is called the **exosphere**, which means "outer sphere." It contains scarce lightweight hydrogen and helium atoms, weakly bound by gravity as far as 32,000 km (20,000 mi) from Earth.

Atmospheric Profile

Earth's modern atmosphere is in a series of imperfectly shaped concentric "shells" or "spheres" that grade into one another, all bound to the planet by gravity. As critical as the atmosphere is to us, it represents only the thinnest envelope, amounting to less than one-millionth of Earth's total mass. We study the atmosphere by viewing it in layers that have distinctive properties and purposes. Figure 3.2 charts the atmosphere in a vertical cross section, or side view, and is a useful reference for the following discussion. We simplify by using three atmospheric criteria: *composition*, *temperature*, and *function*—noted along the left side of Figure 3.2a with the relations among the criteria detailed in Table 3.1.

Table 3.1	Atmospheric Criterion	
Composition	**Temperature**	**Function**
Heterosphere	Thermosphere	Ionosphere
	Mesosphere	
Homosphere {	Stratosphere	Ozonosphere
	Troposphere	

Viewed from the distance of the Moon, the astonishing thing about Earth, catching the breath, is that it is alive. The photographs show the dry, pounded surface of the Moon in the foreground, dead as an old bone. Aloft, floating free beneath the moist, gleaming membrane of a bright blue sky, is the rising Earth, the only exuberant thing in this part of the cosmos. . . . It has the organized, self-contained look of a live creature, full of information, marvelously skilled in handling the sun. . . . When the Earth came alive it began constructing its own membrane, for the general purpose of editing the sun. . . . Taken all in all, the sky is a miraculous achievement. It works, and for what it is designed to accomplish it is as infallible as anything in nature.

Lewis Thomas

FIGURE 3.1 Earthrise.
Earthrise over the stark and lifeless lunar surface. [Photo by NASA; quotation from "The World's Biggest Membrane" from *The Lives of a Cell* by Lewis Thomas. Copyright © 1973 by the Massachusetts Medical Society. Originally published in the *New England Journal of Medicine*. Reprinted by permission of Viking Penguin, a division of Penguin Books USA, Inc.]

(a)

(b)

FIGURE 3.2 Profile of the modern atmosphere.
(a) An integrated chart of our modern atmosphere. The chart spans the atmosphere from
Earth's surface to the top of the atmosphere at 480 km (300 mi). The columns along the left
side show the division of the atmosphere by composition, temperature, and function. The plot
of temperature and the scale along the bottom permit you to tell the temperature at any altitude. (The small balloon shows the height achieved by Kittinger, discussed in News Report 3.1.)
(b) Space Shuttle astronauts captured a dramatic sunset through various atmospheric
layers across the "edge" of our planet—called *Earth's limb.* A silhouetted cumulonimbus
thunderhead cloud is seen topping out at the tropopause. [Space Shuttle photo from NASA.]

Earth's atmosphere exerts its weight, pressing downward under the pull of gravity. Air molecules create air
pressure through their motion, size, and number. Pressure is exerted on all surfaces in contact with the air. The
weight (force over a unit area) of the atmosphere, or

air pressure, pushes in on all of us. Fortunately, that same
pressure also exists inside us, pushing outward; otherwise,
we would be crushed by the mass of air around us.

The atmosphere exerts an average force of ap-
proximately 1 kg/cm^2 (14 lb/in^2) at sea level. Gravity

FIGURE 3.3 Density decreases with altitude.
(a) The atmosphere, denser near Earth's surface, rapidly decreases in density with altitude. The difference in density is easily measured because air exerts its weight as pressure. Have you experienced pressure changes that you could feel on your eardrums? How high above sea level were you at the time? (b) Pressure profile of the atmosphere. Note the rapid decrease in atmospheric pressure with altitude and that about 90% of the atmospheric mass resides in the troposphere.

compresses air, making it denser near Earth's surface; it thins rapidly with increasing altitude (Figure 3.3a). Consequently, over half the total mass of the atmosphere is compressed below 5500 m (18,000 ft), 75% occurs below 10,700 m (35,100 ft), and 90% is below 16,000 m (52,500 ft). All but 0.1% of the atmosphere exists below an altitude of 50 km (31 mi), as shown in the pressure profile in Figure 3.3b (percentage column is farthest to the right). Worldwide, the U.N. estimates that approximately 140 million people live and work above 2440 m (8000 ft.) altitude and must adapt to reduced air pressure, which carries less oxygen in each breath.

Few people are aware that in routine air travel they are sitting above 80% of the total atmospheric volume! To better understand this pressure profile, imagine sky diving from a high-altitude balloon 33 km (20 mi) above Earth. What would you experience? How fast would you fall? What sounds would you hear? See News Report 3.1 to read about someone who actually did this.

At sea level, the atmosphere exerts a pressure of 1013.2 mb (*millibar*, or *mb*; a measure of force per square meter of surface area) or 29.92 in. of mercury (symbol, Hg), as measured by a barometer. In Canada and other countries, normal air pressure is expressed as 101.32 kPa (*kilopascal*; 1 kPa = 10 mb). More information on air pressure, the instruments that measure it, and the role air pressure plays in generating winds is in Chapter 6.

Atmospheric Composition Criterion

Using chemical *composition* as a criterion, the atmosphere divides into two broad regions, the *heterosphere* (80 to 480 km altitude) and the *homosphere* (Earth's surface to 80 km altitude), as shown along the left side of Figure 3.2a. As you read, note that the text follows the same path that incoming solar radiation travels through the atmosphere to Earth's surface.

Heterosphere   The **heterosphere** is the outer atmosphere in terms of composition. It begins at about 80 km (50 mi) altitude and extends outward to the transition to the exosphere and interplanetary space (see Figure 3.2). Less than 0.001% of the atmosphere's mass is in this rarified heterosphere. The International Space Station and most Space Shuttle missions orbit in the upper heterosphere.

As the prefix *hetero-* implies, this region is not uniform—its gases are *not evenly mixed*. This distribution is quite different from the blended gases we breathe near Earth's surface, in the homosphere. Gases in the heterosphere occur in distinct layers sorted by gravity according to their atomic weight, with the lightest elements (hydrogen and helium) at the margins of outer space and the heavier elements (oxygen and nitrogen) dominant in the lower heterosphere.

News Report 3.1

Falling Through the Atmosphere—The Highest Sky Dive

Imagine a small, unpressurized compartment large enough for only one person, floating at 31.3 km (19.5 mi) altitude, carried to such height by a helium-filled balloon. Even though it is daytime, the sky is almost black, for this is above all but 1.0% of the atmospheric mass. The air pressure at 31,300 m (102,800 ft) is barely measurable—this altitude is the beginning of space in experimental aircraft testing. The year is 1961. (See the position of this balloon on Figures 3.2a and 3.3b.)

Captain Joseph Kittinger, Jr., an Air Force officer, stood looking out of the opening in his capsule. He paused for the view of Earth's curved horizon, the New Mexico landscape over 31 km below, and the dark heavens above, and then leaped into the stratospheric void (Figure 3.1.1). He placed himself at tremendous personal risk for an experimental reentry into the atmosphere. What do you think he would initially hear and feel? At what speed would he fall? What would the temperature be? How long would he fall before opening his parachute?

Initially frightening to him was that he heard nothing, no rushing sound, for there was not enough air to produce any sound. The fabric of his pressure suit did not flutter, for there was not enough air to create friction against the fibers. He had no sensation

of movement until he glanced back at the balloon and capsule. They retreated rapidly from his motionless perception. This lack of a sense of speed as he fell was unnerving.

His speed was remarkable. Near the ground, objects fall at "terminal velocity," reaching about 200 kmph (125 mph) in a few seconds. However, in the rarefied stratosphere the lack of air resistance permits incredible speeds. Captain Kittinger quickly accelerated to 988 kmph (614 mph), slightly less than the speed of sound at sea level.

His free fall took him through the stratosphere and its ozone layer. As he encountered denser layers of atmospheric gases, frictional drag slowed his body. As he fell past 15,240 m (50,000 ft), he slowed to 400 kmph (250 mph). He dropped into the lower atmosphere, finally falling below airplane flying altitudes. His free fall lasted 4 minutes and 25 seconds, slowing to terminal velocity and the opening of his main chute at 5500 m (18,000 ft). He safely drifted to Earth's surface. This remarkable 13-minute 35-second voyage through 99% of the atmospheric mass remains a record to this day. To better understand the atmosphere, Kittinger placed himself in harm's way in this experiment at the early beginnings of human space exploration.

FIGURE 3.1.1 Stratospheric leap into history.
Moments into Captain Joseph Kittinger's historic exploration of the atmosphere, captured by a remotely triggered camera. The lanyard cord is attached to his parachute pack and snaps to set a timer that opens a small stabilization chute 16 seconds into the free fall. The clouds are more than 26,000 m (85,000 ft) below him. He carries an instrument pack on his seat, his main chute, and pure oxygen for his breathing mask. [Volkmar Wentzel/NGS Image Collection, used by permission from the National Geographic Society, *National Geographic Magazine*, Dec. 1960, p. 855. All rights reserved.]

Homosphere   Below the heterosphere is the other compositional shell of the atmosphere, the **homosphere**. This region extends from an altitude of 80 km (50 mi) to Earth's surface. Even though the atmosphere rapidly changes density in the homosphere—increasing toward Earth's surface—the blend of gases is nearly uniform throughout the homosphere. The only exceptions are the concentration of ozone (O_3) in the "ozone layer," from 19 to 50 km (12 to 31 mi), and the variations in water vapor, pollutants, and some trace chemicals in the lowest portion of the atmosphere.

The stable mixture of gases comprising air in the homosphere evolved slowly. The present proportion, which includes oxygen, evolved approximately 500 million years ago. Table 3.2 lists by volume the stable ingredients that constitute dry, clean air in the homosphere.

The homosphere is a vast reservoir of relatively inert *nitrogen*, originating principally from volcanic sources. Nitrogen is a key element of life, yet we exhale all the nitrogen we inhale. The explanation for this contradiction is that nitrogen integrates into our bodies not from the air we breathe, but through compounds in food. In the soil, nitrogen is bound to these compounds by nitrogen-fixing bacteria, and it returns to the atmosphere by denitrifying bacteria that remove nitrogen from organic materials. A complete discussion of the nitrogen cycle is in Chapter 19.

Oxygen, a by-product of photosynthesis, also is essential for life processes. Slight spatial variations occur in the percentage of oxygen in the atmosphere because of variations in photosynthetic rates with latitude, seasonal changes, and the lag time as atmospheric circulation

Table 3.2   Stable Components of the Modern Homosphere

Gas (Symbol)	Percentage by Volume	Parts per Million (ppm)
Nitrogen (N_2)	78.084	780,840
Oxygen (O_2)	20.946	209,460
Argon (Ar)	0.934	9,340
Carbon dioxide (CO_2)*	0.038	375.6
Neon (Ne)	0.001818	18
Helium (He)	0.000525	5
Methane (CH_4)	0.00014	1.4
Krypton (Kr)	0.00010	1.0
Ozone (O_3)	Variable	
Nitrous oxide (N_2O)	Trace	
Hydrogen (H)	Trace	
Xenon (Xe)	Trace	

*2003 average measured at Mauna Loa, Hawai'i (see: http://cdiac.esd.ornl.gov/ftp/maunaloa-co2/maunaloa.co2).

slowly mixes the air. Although it forms about one-fifth of the atmosphere, oxygen forms compounds that compose about half of Earth's crust. Oxygen readily reacts with many elements to form these materials. Both nitrogen and oxygen reserves in the atmosphere are so extensive that, at present, they far exceed human capabilities to disrupt or deplete them.

The gas *argon*, constituting about 1% of the homosphere, is completely inert (an unreactive "noble" gas) and therefore is unusable in life processes. Argon is a residue from the radioactive decay of an isotope, or chemical form, of potassium called potassium-40 (symbolized ^{40}K). Slow accumulation over millions of years accounts for all the argon present in the modern atmosphere. Because industry has found uses for inert argon (in lightbulbs, welding, and some lasers), it is extracted or "mined" from the atmosphere, in addition to nitrogen and oxygen, for commercial and industrial uses.

Carbon dioxide is a natural by-product of life processes. It is essentially a stable atmospheric component, qualifying it for inclusion in Table 3.2. Although its present percentage in the atmosphere is small at more than 0.038%, it is important in maintaining global temperatures. Over the past 200 years, its percentage increased as a result of human activities. Chapter 10 discusses the implications of this increase in global warming and climate change.

Atmospheric Temperature Criterion

Shifting to temperature as a criterion, the atmosphere has four distinct temperature zones—the *thermosphere*, *mesosphere*, *stratosphere*, and *troposphere* (labeled in Figure 3.2).

Thermosphere   We define the **thermosphere** ("heat sphere") as roughly corresponding to the heterosphere (80 km out to 480 km, or 50–300 mi). The upper limit of the thermosphere is the **thermopause** (the suffix *-pause* means "to change"). During periods of a less active Sun when there are fewer sunspots and coronal bursts, the thermopause may lower in altitude from the average 480 km (300 mi) to only 250 km altitude (155 mi). An active Sun causes the outer atmosphere to swell to an altitude of 550 km (340 mi), where it can create frictional drag on satellites in low orbit.

The temperature profile in Figure 3.2 (yellow curve) shows that temperatures rise sharply in the thermosphere, to 1200°C (2200°F) and higher. Despite such high temperatures, the thermosphere is not "hot" in the way you might expect. Temperature and heat are different concepts. The intense solar radiation in this portion of the atmosphere excites individual molecules (principally nitrogen and oxygen) to high levels of vibration. This **kinetic energy**, the energy of motion, is the vibrational energy that we measure as *temperature*.

However, the actual heat involved is very small. The reason is that the density of molecules is so low. There is little actual **heat** produced, or the flow of kinetic energy from one body to another because of a temperature difference between them. Heating in the atmosphere near Earth's surface is different because the greater number of

FIGURE 3.4 Noctilucent clouds.
In the mesosphere, fine ice crystals form on dust, principally at high latitudes. An observer may see these bands of ice crystals illuminated in unusual night clouds. The photo is made by a 747 airline pilot at 11,278 m (37,000 ft) at 56° N over the Atlantic Ocean, June 15, 2004. For more examples, please see www.BrianWhittaker.com. [Photo by Brian Whittaker.]

molecules in the denser atmosphere transmits their kinetic energy as **sensible heat**, meaning that we can measure it. (Density, temperature, and heat capacity determine the sensible heat of a substance.) There is more on this topic in Chapters 4 and 5.

Mesosphere The **mesosphere** is the area from 50 to 80 km (30 to 50 mi) above Earth and is the highest in altitude of the three temperature regions within the homosphere. As Figure 3.2 shows, the mesosphere's outer boundary, the *mesopause*, is the coldest portion of the atmosphere, averaging −90°C (−130°F), although that temperature may vary considerably (25–30 C°, or 45–54 F°). Note in Figure 3.3b the extremely low pressures (low density of molecules) in the mesosphere.

The mesosphere sometimes receives cosmic or meteoric dust, which acts as nuclei around which fine ice crystals form. At high latitudes, an observer may see these bands of ice crystals glow in rare and unusual night clouds called **noctilucent clouds** (Figure 3.4). For reasons not clearly understood, these unique clouds are on the increase. See, among several sites, http://lasp.colorado.edu/noctilucent_clouds/.

Stratosphere The **stratosphere** extends from 18 to 50 km (11 to 31 mi) from Earth's surface. Temperatures increase with altitude throughout the stratosphere, from −57°C (−70°F) at 18 km at the tropopause, warming to 0°C (32°F) at 50 km at the stratosphere's outer boundary, the *stratopause*.

Troposphere The **troposphere** is the final layer encountered by incoming solar radiation as it surges through the atmosphere to the surface. It is the home of the biosphere, the atmospheric layer that supports life, and the region of principal weather activity.

Approximately 90% of the total mass of the atmosphere and the bulk of all water vapor, clouds, air pollution, and life forms are within the troposphere. An average temperature of −57°C (−70°F) defines the **tropopause**, the troposphere's upper limit, but its exact elevation varies with the season, latitude, and surface temperatures and pressures. Near the equator, because of intense heating from the surface, the tropopause occurs at 18 km (11 mi); in the middle latitudes, it occurs at an average of 12 km (8 mi); and at the North and South Poles it only averages 8 km (5 mi) or less above Earth's surface.

Figure 3.5 illustrates the normal temperature profile within the troposphere during daytime. As the graph shows, temperatures decrease rapidly with increasing altitude at an average of 6.4 C° per kilometer (3.5 F° per 1000 ft), a rate known as the **normal lapse rate**. This temperature plot also appears in Figure 3.2.

The normal lapse rate is an average. The actual lapse rate at any particular time and place, which may deviate considerably because of local weather conditions, is the **environmental lapse rate**. This variation in temperature gradient in the lower troposphere is central to our discussion of weather processes in Chapter 7.

In the stratosphere, the marked warming with increasing altitude causes the tropopause to act like a lid,

FIGURE 3.5 The temperature profile of the troposphere.
During daytime, temperature decreases with increased altitude at a rate known as the *normal lapse rate*. Scientists use a concept called the *standard atmosphere* as an accepted description of air temperature and pressure changes with altitude. The values in this profile are part of the standard atmosphere. Note the approximate locations of the equatorial and midlatitude tropopauses.

essentially preventing whatever is in the cooler (denser) air below from mixing into the warmer (less dense) stratosphere. However, the tropopause may be disrupted above the midlatitudes wherever jet streams produce vertical turbulence and an interchange between the troposphere and the stratosphere, as we see in Chapter 6. Also, hurricanes occasionally inject moisture above this inverted temperature layer at the tropopause, and powerful volcanic eruptions may loft ash and sulfuric acid mists into the stratosphere, as did the Mount Pinatubo eruptions in 1991.

Atmospheric Function Criterion

Looking at our final atmospheric criterion of *function*, the atmosphere has two specific zones that remove most of the harmful wavelengths of incoming solar radiation and charged particles: the *ionosphere* and the *ozonosphere* (*ozone layer*). Figure 3.6 depicts in a general way the absorption of radiation by the various functional layers of the atmosphere.

Ionosphere   The outer functional layer, the **ionosphere**, extends throughout the thermosphere and into the

mesosphere below (Figure 3.2). The ionosphere absorbs cosmic rays, gamma rays, X-rays, and shorter wavelengths of ultraviolet radiation, changing atoms to positively charged ions and giving the ionosphere its name. The glowing auroral lights discussed in Chapter 2 occur principally within the ionosphere.

Figure 3.2 shows the average daytime altitudes of four regions within the ionosphere, known as the D, E, F_1, and F_2 layers. These are important to broadcast communications for they reflect certain radio wavelengths, including AM radio and other shortwave broadcasts, especially at night. However, during the day the ionosphere actively absorbs arriving radio signals, preventing distant reception. Normally unaffected are FM or television broadcast wavelengths, which pass through to space, thus necessitating use of communication satellites.

Ozonosphere   That portion of the stratosphere that contains an increased level of ozone is the **ozonosphere**, or **ozone layer**. Ozone is a highly reactive oxygen molecule made up of three oxygen atoms (O_3) instead of the usual two atoms (O_2) that make up oxygen gas. Ozone absorbs certain wavelengths of ultraviolet (principally, all the UVC, 100–290 nm, and about 90% of UVB, 290–320 nm).* This absorbed energy subsequently reradiates at longer wavelengths, as infrared radiation. This process converts most harmful ultraviolet radiation, effectively "filtering" it and safeguarding life at Earth's surface.

The ozone layer is presumed to have been relatively stable over the past several hundred million years (allowing for daily and seasonal fluctuations). Today, however, it is in a state of continuous change. Focus Study 3.1 presents an analysis of the crisis in this critical portion of our atmosphere. Fortunately, international treaties to prevent further ozone losses appear to be working.

Variable Atmospheric Components

The troposphere contains natural and human-caused variable gases, particles, and other chemicals. In fact, researchers in 2004 reported that the increase in aerosols, particularly soot, has produced a "dimming" of sunlight reaching the surface—globally, a measured 4% to 8% reduction in surface receipts. The spatial aspects of these variables are important applied topics in physical geography and appear in this and the next chapters.

Air pollution is not a new problem. Romans complained more than 2000 years ago about the foul air of their cities. Filling Roman air was the stench of open sewers, smoke from fires, and fumes from ceramic-making kilns and smelters (furnaces) that converted ores into metals. In human experience, cities are always the place where the environment's natural ability to process and recycle

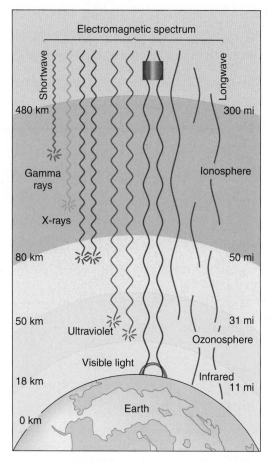

FIGURE 3.6 The atmosphere protects Earth's surface.
As solar energy passes through the atmosphere, the shortest wavelengths are absorbed. Only a fraction of the ultraviolet radiation, and most of the visible light and infrared, reaches Earth's surface. When above these protective layers, astronauts must wear a spacesuit that duplicates this filtering process in order to survive.

Ozone Breakdown, Ozone Hole
ANIMATION

*Nanometer (nm) = one-millionth of a millimeter; 1 nm = 10^{-9} m. For comparison, a micrometer, or micron, (μm) = one-millionth of a meter; 1μm = 10^{-6} m. A millimeter (mm) = one-thousandth of a meter; 1 mm = 10^{-3} m.

Focus Study 3.1

Stratospheric Ozone Losses: A Worldwide Health Hazard

Consider these news reports and conditions:

- The stratospheric ozone above Antarctica during the months of Antarctic spring (September to November) reached record losses in 2000, 2001, and 2002, covering an area three times larger than the United States. Each year, the "ozone hole" (actually a thinning) widens and deepens.

- As stratospheric ozone flows into the depleted region from lower latitudes, the protective layer thins over southern South America, southern Africa, Australia, and New Zealand. In Ushuaia, Argentina, surface UVB intensity increased 236% compared to normal levels.

- An international scientific consensus confirmed previous assessments of the anthropogenic, or human-caused, disruption of the ozone layer—chlorine atoms and chlorine monoxide molecules in the stratosphere are of human origin. (See *Scientific Assessment of Ozone Depletion*, by NASA, NOAA, United Nations Environment Programme, and World Meteorological Organization.)

- At Earth's opposite pole, a similar ozone depletion over the Arctic annually exceeds 30%, to a record 45% loss during spring 1997. The Canadian and U.S. governments report an "ultraviolet index" to help the public protect themselves (http://www.ec.gc.ca/ozone/).

- Overall ozone losses in the midlatitudes are continuing at 6% to 8% per decade. In North America, related skin cancers are increasing, totaling more than 1 million cases a year, of which some 41,000 are malignant melanomas, averaging 10,000 deaths. Skin cancer is increasing at 4.0% per year. Most affected across the globe are light-skinned persons who live at higher elevations and those who work principally outdoors (see http://www.cancer.org/).

- Increased ultraviolet radiation is affecting atmospheric chemistry, biological systems, oceanic phytoplankton (small photosynthetic organisms that form the basis of the ocean's primary food production), and fisheries, crop yields, and human skin, eye tissues, and immunity.

More ultraviolet radiation than ever before is breaking through Earth's protective ozone layer. What is going on in the stratosphere everywhere and in the polar regions specifically? Why is this happening, and how are people and their governments responding? What effects does this have on you personally?

Monitoring Earth's Fragile Safety Screen

A sample from the ozone layer's densest part (at 29 km, or 18 mi altitude) contains only 1 part ozone per 4 million parts of air—compressed to surface pressure, the ozone layer would be only 3 mm thick. Yet, this rarefied layer was in steady-state equilibrium for several hundred million years, absorbing intense ultraviolet radiation and permitting life to proceed safely on Earth.

The ozone layer has been monitored since the 1920s. Ground stations with instrumented balloons, aircraft, orbiting satellites, and a 30-station ozone-monitoring network (mostly in North America) observe stratospheric ozone. The total ozone mapping spectrometer (TOMS) began operations in 1978 aboard *Nimbus-7*, later on the *Upper Atmosphere Research Satellite* (*UARS*), *Adeos*, and *Meteor-3*, and, at the time of this writing, aboard the *Earth Probe* satellite. An October 2003 image is in Figure 3.1.1. (See http://jwocky.gsfc.nasa.gov/.)

Ozone Losses Explained

What is causing the decline in stratospheric ozone? In 1974, two atmospheric chemists, F. Sherwood Rowland and Mario Molina, hypothesized that some synthetic chemicals were releasing chlorine atoms that decompose ozone. These **chlorofluorocarbons**, or **CFCs**, are synthetic molecules of chlorine, fluorine, and carbon. (See Rowland and Molina's report: "Stratospheric sink for chlorofluoromethanes: Chlorine atom catalyzed destruction of ozone," *Nature*, 249 (1974): 810.)

CFCs are stable, or inert, under conditions at Earth's surface and they possess remarkable heat properties. Both qualities made them valuable as propellants in aerosol sprays and as refrigerants. Also, some 45% of CFCs were solvents in the electronics industry and used as foaming agents. Being inert, CFC molecules do not dissolve in water and do not break down in biological processes. (In contrast, chlorine compounds derived from volcanic eruptions and the ocean are water-soluble and rarely reach the stratosphere.)

Researchers Rowland and Molina hypothesized that stable CFC molecules slowly migrate into the stratosphere, where intense ultraviolet radiation splits them, freeing chlorine (Cl) atoms. This process produces a complex set of reactions that breaks up ozone molecules (O_3) and leaves oxygen gas molecules (O_2) in their place. The effect is severe, for a single chlorine atom decomposes more than 100,000 ozone molecules. The long residence time of chlorine atoms in the ozone layer (40 to 100 years) is likely to produce long-term consequences through this century from the chlorine already in place. More than 22 million metric tons (24 million tons) of CFCs were sold worldwide and subsequently released into the atmosphere.

Of many identified reactions, here are simplified equations that

(continued)

Focus Study 3.1 *(continued)*

EP/TOMS Total Ozone for Oct 1, 2003

Dobson Units
Dark Gray < 100, Red > 500 DU

FIGURE 3.1.1 The Antarctic ozone hole.
TOMS image for October 1, 2003, uses a revised color scale to represent ozone concentrations in Dobson units, with purples for amounts less than 180, black below 125. (One Dobson unit equals 2.69 × 1016 molecules of of O_3/cm^2) Measurements have dropped below 100 Dobson units. The ozone "hole" has grown larger since 1979, covering more than 25 million km^2 (9 million mi^2), or more than double the surface area of Antarctica—an area larger than Canada, the United States, and Mexico combined. [NASA.]

show chlorine monoxide-produced (ClO) and chlorine-produced ozone losses. Note that chlorine monoxide (ClO) reacts with ozone (O_3) to produce a free chlorine atom (Cl) and two oxygen gas molecules (O_3). The free chlorine atom then reacts with an ozone molecule (O_3) to produce more chlorine monoxide (ClO) and oxygen gas (O_2).

Ozone depletion by ClO and by Cl:

$$ClO + O_3 \longrightarrow Cl + 2O_2$$
$$Cl + O_3 \longrightarrow ClO + O_2$$

The chlorine monoxide is then available to go through the cycle all over again. The *net reaction*, with the chlorine catalyst removed, is a loss of ozone (O_3) and its conversion to oxygen gas molecules. The problem then is that oxygen gas (O_2), unlike

(O_3), is transparent to ultraviolet radiation. (See http://www.epa.gov/ozone/index.html.)

Political Realities—An International Response and the Future
Between 1976 and 1979, Canada, Sweden, Norway, and the state of Oregon banned CFC propellants in aerosols, and a U.S. federal ban began in 1978. However, more than half of U.S. production was exempted, including CFCs used as air-conditioning refrigerants and to make polyurethane foam. CFC sales initially dropped in the late 1970s, but they rose again under a 1981 presidential order that permitted the export and sales of banned products. Sales increased and hit a new peak in 1987 at 1.2 million metric tons (1.32 million tons).

Chemical manufacturers once claimed that no hard evidence existed to prove the ozone-depletion model, and they successfully delayed remedial action for 15 years. Today, with extensive scientific evidence and verification of losses, even the CFC manufacturers admit that the problem is serious. A few remaining critics are outside of the scientific community.

The *Montreal Protocol on Substances That Deplete the Ozone Layer*, as amended three times in 1990, 1992, and 1997, aims to reduce and eliminate CFC damage to the ozone layer. CFC sales are declining as many countries reduce demand and as worldwide industry phases in alternative chemicals. All production of harmful CFCs will cease by 2010. If the protocol is fully implemented, it is estimated that the stratosphere will return to nominal conditions by the end of this century. (See the United Nations Ozone Secretariat at http://www.unep.org/ozone/index-en.shtml and the Montreal Protocol at http://www.ec.gc.ca/international/multilat/ozone1_e.htm.)

The protocol also bans other chemicals, such as bromine, that are damaging to the atmosphere, using different target dates. Bromine sources include halon-charged fire extinguishers and methyl bromide pesticides. These bromine compounds, like CFCs, are inert in the lower atmosphere and reach the stratosphere by slow transport. They are about 50 times more effective than chlorine as catalysts causing reactions that destroy ozone, although they have only a residence time of 2 years in the atmosphere.

The world without this treaty might prove challenging. Imagine possible stratospheric ozone losses of 50% in the midlatitudes, producing 1.5 million more cases of malignant melanoma (a skin cancer). We owe much to doctors Rowland and Molina.

Mario Molina stated, "It was frustrating for many years, but it really paid off with the Protocol, which was a marvelous example of what the international community can do working together. We can see from atmospheric measurements that it is already working" (*Nature* 389, September 18, 1997: 219). For their work, Rowland, Molina, and another colleague, Paul Crutzen, received the 1995 Nobel Prize for Chemistry. In making the award, the Royal Swedish Academy of Sciences said, "By explaining the chemical mechanism that affects the thickness of the ozone layer, the three researchers have contributed to our salvation from a global environmental problem that could have catastrophic consequences. It has been possible to make far-reaching decisions on prohibiting the release of the gases that destroy ozone."

Ozone Losses over the Poles
How do Northern Hemisphere CFCs become concentrated over the South Pole? Evidently, chlorine freed in the Northern Hemisphere midlatitudes concentrates over Antarctica through the work of atmospheric winds. Persistent cold temperatures over the South Pole and the presence of thin, icy clouds in the stratosphere promote development of the ozone hole. Over the North Pole, conditions are more changeable, so the hole is smaller, although growing each year.

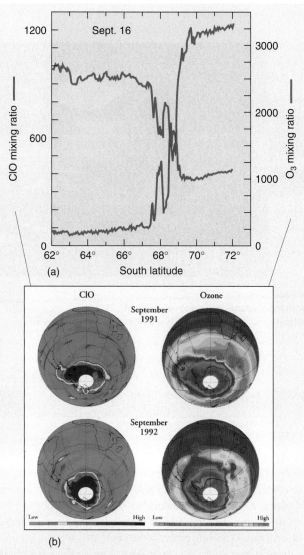

FIGURE 3.1.2 Chemical evidence of ozone damage by humans.
(a) The negative correlation between ClO and O_3 over the Antarctic
continent poleward of 68° S latitude. The data were collected from flights
during September 1987 at stratospheric altitudes. (b) Chlorine monoxide
(ClO) and ozone (O_3) concentrations above 20 km (12.5 mi) measured
during September 1991 and 1992–the beginning of the Antarctic
spring–confirm a causal relationship between the presence of chlorine
and the loss of ozone. [(a) Data from NASA; (b) Jet Propulsion Laboratory
and Goddard Space Flight Center.]

Polar stratospheric clouds (*PSCs*) are
thin clouds that are important catalysts in the release of chlorine for
ozone-depleting reactions. During
the long, cold winter months, a tight
circulation pattern forms over the
Antarctic continent—the polar vortex. Chlorine that is freed from
droplets in PSCs triggers the breakdown of the otherwise inert molecules
and frees the chlorine for catalytic reactions. Figure 3.1.2a graphs the negative correlation between ClO and O_3
over the Antarctic continent within
the polar vortex. You can see that concentrations of ClO increase toward
the pole as O_3 levels decline (Figure 3.1.2b). The ozone hole usually
fills by early December with stratospheric ozone from lower latitudes,
thus thinning ozone in middle and
high latitudes.

Atmospheric science is active
in Antarctica. The U.S. scientific
base, Palmer Station (64°; 46′ S) on
Anvers Island, Antarctic Peninsula,

experienced a 285% increase in surface UVB radiation above normal levels in fall 2000. Nearby at Ukraine's
Akademik Vernadsky Station, the
former British Faraday Base in the
Argentine Islands (65° 9′ S), one of
the original ozone spectrophotome-
ters instruments is actively measuring ultraviolet radiation (Figure 3.1.3).
Ukraine scientists are studying UVA
(320–400 nm) and UVB wavelengths
using a variety of instruments, including biodosimeters that use biological
material to measure radiation impact.

UV Index Helps Save Your Skin
Newspaper and television weather
reports regularly include the *UV Index*
(UVI) in forecasts, as reported by the
National Weather Service (NWS) and
the Environmental Protection Agency
(EPA) since 1994. After 10 years, the
index went through a revision in 2004
to align it with guidelines adopted by
the World Health Organization and
World Meteorological Organization.
Note the new index scale does not rely
on "burn times." People were misin-
terpreting time to burn estimates
thinking there was some safe level of
exposure, or safe tanning time—which
was the wrong message. Table 1
presents the new UVI. Fair-skinned
people (melano-compromised) must
exercise more caution than moderate to darker skin types (melano-
competent and melano-protected),
but all skin types should feel warned,
especially for babies and children.

As stratospheric ozone levels
continue to thin, surface exposure to
cancer-causing radiation climbs. The
public now is alerted to take extra
precautions in the form of sun-
screens, hats, protective clothing, and
sunglasses. Remember, damage is
cumulative and it may be decades before you experience the ill effects
triggered by this summer's sunburn.
(For more information, contact the
American Cancer Society, 800-227-
2345, http://www.cancer.org/; or
write the American Academy of
Dermatology, P.O. Box 681069,
Schaumburg, IL 60168-1069.)

(continued)

Focus Study 3.1 *(continued)*

FIGURE 3.1.3 Atmospheric science in Antarctica.
(a) Ukraine's Vernadsky Station operates year-round performing valuable scientific research. (b) This ozone spectrophotometer instrument measures the ultraviolet component of insolation arriving through the opening in the science station's roof. [Photos by Bobbé Christopherson.]

Table 1	UV Index (EPA, NWS, WHO, WMO)	
Exposure Risk Category	**UVI Range**	**Comments**
Low	less than 2	Low danger for average person. Wear sunglasses on bright days. Watch out for reflection off snow.
Moderate	3–5	Take covering precautions, such as sunglasses, sunscreen, hats, protective clothing, and stay in shade during midday hours.
High	6–7	Use sunscreens with SPF of 15 or higher. Reduce time in the Sun between 11 A.M. and 4 P.M. Use protections mentioned above.
Very High	8–10	Minimize Sun exposure 10 A.M. to 4 P.M. Use sunscreens with SPF ratings of over 15. Use protections mentioned above.
Extreme	11+	Unprotected skin is at risk of burn. Sunscreen application every two hours if out-of-doors. Avoid direct Sun exposure during midday hours. Use protections mentioned above.

See: http://www.epa.gov/sunwise/uvindex.html

waste is most taxed. Air pollution is closely linked to our production and consumption of energy and resources.

Solutions require regional, national, and international strategies, because the pollution sources often are distant from the observed impact—moving across political boundaries and even crossing oceans. Regulations to curb human-caused air pollution have met with great success, although much remains to be done. Before we discuss these topics, let's examine some natural pollution sources.

Natural Sources

Natural air pollution sources produce a greater quantity of pollutants—nitrogen oxides, carbon monoxide, hydrocarbons from plants and trees, and carbon dioxide—than do human-made sources. Table 3.3 lists some of these natural

sources and the substances they contribute to the air. However, any attempt to dismiss the impact of human-made air pollution through a comparison with natural sources is irrelevant, for we evolved with and adapted to the natural ingredients in the air. We did not evolve in relation to the comparatively recent concentrations of *anthropogenic* (human-caused) contaminants in our metropolitan regions.

A dramatic natural source of pollution was the 1991 eruption of Mount Pinatubo in the Philippines (15° N 120° E), perhaps the twentieth century's second largest eruption. This event injected between 15 and 20 million tons of sulfur dioxide (SO_2) into the stratosphere. The spread of these emissions is shown in a sequence of satellite images in Figure 6.1, Chapter 6.

Table 3.3	Sources of Natural Variable Gases and Materials
Source	**Contribution**
Volcanoes	Sulfur oxides, particulates
Forest fires	Carbon monoxide and dioxide, nitrogen oxides, particulates
Plants	Hydrocarbons, pollens
Decaying plants	Methane, hydrogen sulfides
Soil	Dust and viruses
Ocean	Salt spray and particulates

Devastating wildfires on several continents, including widespread fires across the western United States, produce natural air pollution as each year's fire season seems to be more severe. Soot, ash, and gases darkened skies and damaged health in affected regions. Wind patterns spread the pollution from the fires to nearby cities, closing airports and forcing evacuations to avoid the health-related dangers. Wildfire smoke contains particulate matter (dust, smoke, soot, ash), nitrogen oxides, carbon monoxide, and volatile organic compounds. During August 2002 in Alaska, record heat and dryness coupled with more than 12,000 lightning strikes combined to spark numerous wildfires (Figure 3.7a). In drought-plagued southern California, dozens of wildfires across the region burned thousands of acres in October 2003 (Figure 3.7b).

Natural Factors That Affect Air Pollution

The problems resulting from both natural and human-made atmospheric contaminants are made worse by several important natural factors. Among these, wind, local and regional landscape characteristics, and temperature inversions in the troposphere dominate.

Winds   Winds gather and move pollutants, sometimes reducing the concentration of pollution in one location while increasing it in another. Wind can produce dramatic episodes of dust movement. (Dust is defined as particles less than 62 μm,—microns—or 0.0025 in.) Traveling on prevailing winds, dust from Africa contributes to the soils of South America and Europe, and Texas dust ends up across the Atlantic. Such movement is confirmed by chemical analysis, frequently used by scientists to track dust to its source area (Figure 3.8). Imagine at any one moment a billion tons of dust aloft in the atmospheric circulation, carried by the winds!

Such air movements make the atmosphere's condition an international issue. For example, prevailing winds transport air pollution from the United States to Canada, causing much complaint and negotiation between the two

(a) Alaska, 2002

(b) Southern California, 2003

FIGURE 3.7 Alaskan and southern California wildfires fill the atmosphere with smoke.
(a) *Terra* images of wildfires in drought- and high-temperature-plagued portions of Alaska, August 4, 2002. More than 289,000 acres (117,000 h) were estimated to be burning in the image; stretching northeastward to the Brooks Range. Nearly 1.5 million acres (607,000 h) burned in wildfires in Alaska during 2002. The smoke contains particulate matter (dust, smoke, soot, ash), nitrogen oxides, carbon monoxide, and volatile organic compounds. (b) Southern California wildfires consume thousands of acres of drought-damaged forests; image made October 28, 2003. [Both images are *Terra* MODIS sensor image, courtesy of the MODIS Land Rapid Response Team, NASA/GSFC.]

(a)

(b)

(c)

FIGURE 3.8 Natural variable dust in the atmosphere.
(a) A dust storm in central Nevada. (b) A *Terra* orbital view of a sinuous wind-blown dust plume flowing northward from Africa, across the Mediterranean Sea, toward Turkey. The Nile Delta appears obscured by the dust storm. (c) Alkali dust, a serious air pollutant, rises from the exposed shorelines of Mono Lake, California. [(a) Photo by author; (b) *Terra* MODIS sensor image from 5/13/2001 courtesy of the MODIS Land Rapid Response Team NASA/GSFC; (c) photo by Bobbé Christopherson.]

governments. Pollution from North America adds to European air problems. In Europe, the cross-boundary drift of pollution is a major issue because of the proximity of countries. This issue has led in part to the formation of the European Union (EU), through which the European countries address pollution problems. See "Arctic Haze," High Latitude Connection 3.1, for another aspect of pollution drift.

Local and Regional Landscapes Local and regional landscapes are another important factor in air pollution. Surrounding mountains and hills can form barriers to air movement or can direct pollutants from one area to another. Some of the worst incidents result when local landscapes trap and concentrate air pollution.

Places with volcanic landscapes such as Iceland and Hawai'i have their own natural pollution with which to deal. During periods of sustained volcanic activity at Kīlauea, some 2000 metric tons (2200 tons) of sulfur dioxide are produced a day. Concentrations are sometimes high enough to merit broadcast warnings about health concerns. The resulting acid rain and volcanic smog, called *vog* by Hawaiians (for *vol*canic smo*g*), causes losses to agriculture as well as other economic impacts.

Temperature Inversion Vertical temperature and atmospheric density in the troposphere also can worsen pollution conditions. A **temperature inversion** occurs when the normal temperature decrease with altitude (normal lapse rate) begins to *increase* at some point. This can happen at any elevation from ground level to several thousand meters. Figure 3.9 compares a normal temperature profile with that of a temperature inversion. The normal profile (Figure 3.9a) permits warmer (less dense) air at the surface to rise, ventilating the valley and moderating surface pollution. But the warm-air inversion (Figure 3.9b) prevents the rise of cooler (denser) air beneath, halting the vertical mixing of pollutants with other atmospheric gases. Thus, instead of being carried away, pollutants are trapped under the *inversion layer*.

Inversions most often result from certain weather conditions, such as when the air near the ground is radiatively cooled on clear nights, or from topographic situations that produce cold-air drainage into valleys. In addition, the air above snow-covered surfaces or beneath subsiding air in a high-pressure system may cause a temperature inversion. (The concepts of high- and low-pressure systems are in Chapter 6.)

High Latitude Connection 3.1

Arctic Haze

Arctic haze is a term from the 1950s, when pilots noticed decreased visibility, either horizontally or at a slant angle from their aircraft. Since there is no heavy industry at these latitudes and sparse population, this seasonal haze (concentration of microscopic particles and air pollution that diminishes air clarity) was a mystery at first. When looking toward the Sun the haze has a reddish-brown tint, especially when the polar air mass is dominated by stable high-pressure systems and calm weather conditions.

A remarkable attribute of industrialization in the Northern Hemisphere is the haze across the Arctic region, especially from November to April. This winter haze is worse toward Alaska's North Slope. Although the haze can extend up to 8 km (5 mi) in altitude, it sometimes appears layered since sources are at varying distances from the Arctic and the warmer transporting winds are variously lifted over the cold air mass that hugs the surface.

Simply, winds of atmospheric circulation transport pollution to sites far distant from points of origin. There is no comparable haze over the Antarctic continent. Based on this analysis, can you think of why Antarctica lacks such a condition?

(a)

(b)

(c)

FIGURE 3.9 Normal and inverted temperature profiles.
A comparison of a normal temperature profile in the atmosphere (a) with a temperature inversion in the lower atmosphere (b). Note how the warmer air layer prevents mixing of the denser (cooler) air below the inversion, thereby trapping pollution. An inversion layer is visible in the morning hours over a valley (c). [Photo by Bobbé Christopherson.]

Anthropogenic Pollution

Anthropogenic, or human-caused, air pollution remains most prevalent in urbanized regions. Approximately 2% of annual deaths in the United States are attributable to air pollution. Comparable risks are identified in Canada, Europe, Mexico, Asia, and elsewhere.

The human population is moving to cities and is therefore coming in increasing contact with air pollution. By the year 2010, approximately 3.3 billion people (48% of world population) will live in metropolitan regions, some one-third with unhealthful levels of air pollution. This represents a potentially massive public health issue in this century.

Table 3.4 lists the names, chemical symbols, principal sources, and impacts of variable anthropogenic components in the air. The first seven pollutants in the table result from combustion of fossil fuels in transportation (specifically automobiles) and at stationary sources such as

Table 3.4	Anthropogenic Gases and Materials in the Lower Atmosphere		
Name	**Symbol**	**Source**	**Description and Effects of Criteria Pollutants**
Carbon monoxide	CO	Incomplete combustion of fuels	Odorless, colorless, tasteless gas Toxicity; affinity for hemoglobin Displaces O_2 in bloodstream 50 to 100 ppm causes headaches, vision and judgement losses
Nitrogen oxides	$NO_x(NO, NO_2)$	High temperature/pressure combustion	Reddish-brown choking gas Inflames respiratory system, destroys lung tissue Damages plants 3 to 5 ppm is dangerous
Volatile organic compounds	VOC	Incomplete combustion of fossil fuels such as gasoline; cleaning and paint solvents	Prime agents of ozone formation
Ozone	O_3	Photochemical reactions	Highly reactive, unstable gas Oxidizes surfaces, dries rubber and elastic Damages plants at 0.01 to 0.09 ppm Agricultural losses at 0.1 ppm 0.3 to 1.0 ppm irritates eyes, nose, throat
Peroxyacetyl nitrates	PAN	Photochemical reactions	Produced by NO + VOC photochemistry No human health effects Major damage to plants, forests, crops
Sulfur oxides	$SO_x(SO_2, SO_3)$	Combustion of sulfur-containing fuels	Colorless; irritating smell 0.1 to 1 ppm impairs breathing, taste threshold Human asthma, bronchitis, emphysema Leads to acid deposition
Particulate matter	PM	Dust, dirt, soot, salt, metals, organics; fugitive dust from agriculture, construction, roads, and wind erosion	Complex mixture of solid and aerosol particles Dust, smoke, and haze effect visibility Various health effects; bronchitis, pulmonary function PM_{10} negative health effects established by researchers
Carbon dioxide	CO_2	Complete combustion, mainly from fossil fuel consumption	Principal greenhouse gas Atmospheric concentration increasing 64% of greenhouse warming effect
Methane	CH_4	Organic processes	Secondary greenhouse gas Atmospheric concentration increasing 19% of greenhouse warming effect
Water vapor	H_2O vapor	Combustion processes, steam	See Chapter 7 for more on the role of water vapor in the atmosphere.

power plants and factories. Overall, automobiles contribute more than 60% of United States and 50% of Canadian human-caused air pollution. Figure 3.10 identifies major human-caused pollutants and their proportional sources in the United States in 1999 (still the latest report available). The proportion of sources for these pollutants may be considered representative for most developed countries (see http://www.epa.gov/).

In the United States in 1999, transportation produced 77% of the carbon monoxide, 47% of volatile organic compounds (VOC), 56% of the nitrogen oxides, and 25% of particulates. Environment Canada, an agency similar to the U.S. Environmental Protection Agency,

reported that with Canada's smaller transportation fleet, Canadian automobile emissions contribute 40% of the carbon monoxide, 21% of the VOCs, and 60% of the nitrogen oxides in Canada (see http://www.ec.gc.ca/). The United States and Canada are negotiating an Air Quality Agreement, a process formally begun in 1991 and coordinated by an International Joint Commission. Reduction in ozone, fine dust, and acid deposition is the main focus of the effort (see http://www.ijc.org/). "Border Air Quality Pilot Studies" began in 2004.

The apparent manageability of this transportation-pollution problem is interesting. In one California study, only 7% of the vehicles contributed half of the carbon

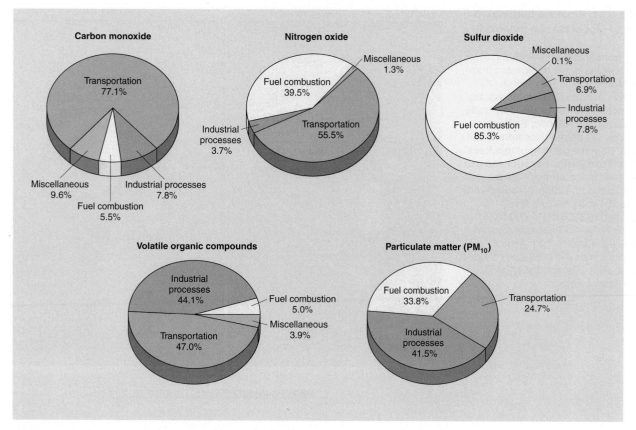

FIGURE 3.10 Human-caused air pollution and sources.
Major types of human-caused air pollution and their sources in the United States in 1999.
These proportions are typical of developed, industrialized countries. [Adapted from Office of Air
Quality, *National Air Quality and Emission Trends Report, 1999*, U.S. EPA (March 2001), Figures 2-4,
2-19, 2-33, 2-40, and 2-67, EPA 454/R-01-004.]

monoxide and only 10% contributed half of the VOC pollution. These "gross polluting" vehicles are not old cars—a common misconception—but include new cars! In random highway checks, 41% of vehicles had pollution equipment that was deliberately tampered with and 25% had defective or missing emission controls. Reducing air pollution from the transportation sector involves known technologies and strategies.

Stationary sources, such as electric power plants and industrial plants that use fossil fuels, contribute the most sulfur oxides and particulates. For this reason, concentrations are focused in the Northern Hemisphere and the industrial, developed countries.

The last three gases shown in Table 3.4 are discussed elsewhere in this text: Water vapor is examined with water and weather (Chapters 7 and 8); carbon dioxide and methane are discussed with greenhouse gases and climate (Chapters 4, 5, and 10).

Carbon Monoxide Pollution Carbon monoxide (CO) is a combination of one atom of carbon and one of oxygen. Incomplete combustion (burning with limited oxygen) of fuels or other carbon-containing substances produces carbon monoxide. A log decaying in the woods produces carbon monoxide, as does a forest fire or other

organic decomposition. Natural sources produce up to 90% of existing carbon monoxide, whereas anthropogenic sources, principally transportation, produce the other 10%. A dangerous point source of carbon monoxide for individuals is from primary and secondary tobacco smoke. Generally, the carbon monoxide from human sources is concentrated in urban areas, where it directly affects human health (see Table 3.4).

In the tropics of south-central Africa and the Amazon region of South America, a large natural source of carbon monoxide is the widespread burning of biomass (trees, grasses, brush) from August to October. This carbon monoxide spreads throughout the Southern Hemisphere. Figure 3.11 compares carbon monoxide emissions from both human-caused and wildfire biomass burning for March and September 2000 in South America.

Photochemical Smog Pollution Photochemical smog was not generally experienced in the past, but developed with the advent of the automobile. Today, it is the major component of anthropogenic air pollution (Figure 3.12). **Photochemical smog** results from the interaction of sunlight and the combustion products in automobile exhaust (nitrogen oxides and VOCs). Although the term *smog*—a combination of the words *smoke*

FIGURE 3.11 Biomass burning—a source of global carbon monoxide (CO).

Images from March and September, 2000, show the onset of biomass burning. In March, the slightly higher CO levels over the equatorial latitudes are from African fires, transported by trade winds across the Atlantic. By September, a large CO plume rises from Brazil, primarily from the burning of rain forest in the Amazon Basin. The soot and ash is lifted high in the atmosphere by strong cloud convection. Measurement is at the 700-mb pressure level, approximately 3650 m (12,000 ft).
[Images from MOPITT (Measurement of Pollution in the Troposphere) sensor aboard the *Terra* satellite courtesy of David Edwards, John Gille, and the MOPITT Science Team, UCAR.]

MOPITT Carbon Monoxide at 700 mb

March 2000

September 2000

low   high

and *fog*—is a misnomer, it is generally used to describe this phenomenon. Smog is responsible for the hazy sky and reduced sunlight in many of our cities, as shown in China and the U.S. Mid-Atlantic states in Figure 3.13—both regions under an air pollution siege.

The connection between automobile exhaust and smog was not determined until 1953 in Los Angeles, long after society had established its dependence upon individualized transportation. Despite this discovery, widespread mass transit declined, the railroads dwindled, and the polluting individual automobile remains America's preferred transportation. Strangely, the U.S. fleet of 2004 model cars and trucks had worse gas mileage than in 2001. This poor showing is principally due to gas-guzzling sport-utility vehicles and pickups that account for more than half of new sales. Efficiency standards for this class of vehicles have not changed since 1975; they are set lower than for cars.

Mexico City is notorious for poor air quality; its 22 million inhabitants and 3.5 million vehicles work, commute, and live in the world's second-largest metropolitan region. In 1999, World Resources Institute named it as the unhealthiest city in the world for children; this was based on 1995 data. In 2002, an air pollution emergency gripped the metropolitan region and 350,000 cars were ordered off the roads! Conditions are worsened by frequent subtropical high-pressure systems (descending, stable air) that act as effective air traps over the Valley of Mexico, in which the city lies.

Mexico enacted new laws in 1990 to reduce unhealthy photochemical smog conditions: more public rapid transportation, controls on automobiles, limitations on factory operations, and more pollution-absorbing park space and trees. Based on some success and an increase in clear days, a new 10-year plan began in 2000 directed at reducing ozone and particulates. The government has spent nearly $2 billion on pollution controls.

Figure 3.14 summarizes how car exhaust is converted into major air pollutants—ozone, **peroxyacetyl nitrates (PAN)**, and nitric acid. PAN produces no known health effect in humans, but it is particularly damaging to plants, including both agricultural crops and forests. Damage in California is estimated to exceed $1 billion a year, and it is

FIGURE 3.12 Photochemical smog near Denver.
Photochemical reactions in the skies over Denver, Colorado, produce the blanket of smog that moves over mountains and forest. What do you notice about air pollution conditions where you live or go to college? [Photo by Bobbé Christopherson.]

(a)

(b)

FIGURE 3.13 Serious air pollution in China and the United States.
(a) Smog and haze blanket most of eastern China. Beijing is about 150 km (93 mi) west of the coast where the heaviest pollution exists. Poor air pollution controls and the burning of coal contribute to the problem. (b) A serious air pollution alert with un-healthy conditions masks most of the Mid-Atlantic region of the U.S.—from northern Georgia to New York. Prevailing winds were blowing the mass offshore over the Atlantic Ocean. [(a) *Terra* MODIS sensor image from 3/12/2002 courtesy of MODIS Land Rapid Response Team NASA/GSFC; (b) *Terra* MODIS image from 6/11/2002 courtesy of Liam Gumley, Space Science and Engineering Center, University of Wisconsin–Madison and NASA.]

estimated to be several billion dollars nationwide in the farming and forestry sectors.

Worldwide, the problem with **nitrogen dioxide** production is its concentration in metropolitan regions. North American urban areas may have from 10 to 100 times higher nitrogen dioxide concentrations than nonurban areas. Nitrogen dioxide interacts with water vapor to form nitric acid (HNO_3) a contributor to acid deposition by precipitation, the subject of Focus Study 3.2.

Ozone is the primary ingredient in photochemical smog. (This is the same gas that is beneficial to us in the stratosphere in absorbing ultraviolet radiation.) The reactivity of ozone causes health concerns, for it damages biological tissues. For several reasons, children are at greatest risk from ozone pollution—one in four children in U.S. cities is at risk of developing health problems from ozone pollution. This ratio is significant; it means that more than 12 million children are vulnerable in those cities with the worst polluted air (Los Angeles, New York City, Atlanta, Houston, and Detroit).

The **volatile organic compounds (VOCs)**, including hydrocarbons from gasoline, surface coatings, and combustion at electric utilities, are important factors in

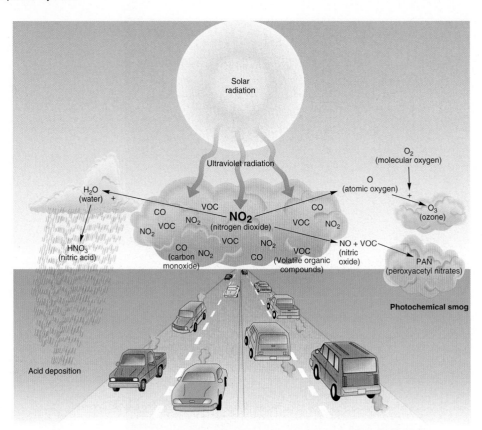

FIGURE 3.14 Photochemical reactions.
The interaction of automobile exhaust (NO$_2$, VOCs, CO) and ultraviolet radiation in sunlight causes photochemical reactions. The high temperatures in modern automobile engines produce reactions that form nitrogen dioxide (NO$_2$). This nitrogen dioxide, derived from automobiles and to a lesser extent from power plants, is highly reactive with ultraviolet light. The reaction liberates atomic oxygen (O) and a nitric oxide (NO) molecule from the NO$_2$. The free oxygen atom combines with an oxygen molecule, O$_2$, to form the oxidant ozone, O$_3$. In addition, the nitric oxide (NO) molecule reacts with VOCs to produce a family of chemicals called *peroxyacetyl nitrates* (*PAN*). To the left, note the formation of nitric acid and acid deposition.

ozone formation. States such as California base their standards for control of ozone pollution on VOC emission controls—a scientifically accurate emphasis.

Industrial Smog and Sulfur Oxides Over the past 300 years, except in some developing countries, coal slowly replaced wood as the basic fuel used by society. The Industrial Revolution required high-grade energy to run machines. The changes involved conversion from *animate* energy (energy from animal sources, such as animal-powered farm equipment) to *inanimate* energy (energy from nonliving sources, such as coal, steam, and water). The air pollution associated with coal-burning industries is known as **industrial smog** (Figure 3.15). The term *smog* was coined by a London physician at the turn of the twentieth century to describe the combination of fog and smoke containing sulfur gases (sulfur is an impurity in fossil fuels).

Once in the atmosphere, **sulfur dioxide** (**SO$_2$**) reacts with oxygen (O) to form sulfur trioxide (SO$_3$), which is highly reactive and, in the presence of water or water vapor, forms **sulfate aerosols**, tiny particles about 0.1 to 1 μm in diameter. Sulfuric acid (H$_2$SO$_4$) can form even in moderately polluted air at normal temperatures. Sulfur dioxide-laden air is dangerous to health, corrodes metals, and deteriorates stone building materials at accelerated rates. Focus Study 3.2 discusses this vital atmospheric issue.

In the United States, coal-burning electric utilities and steel manufacturing are the main sources of sulfur dioxide, principally in the East and Midwest. And,

because of the prevailing movement of air masses, they are the main sources of sulfur dioxide in adjacent Canadian regions. As much as 70% of Canadian sulfur dioxide is initiated within the United States.

Particulates Particulate matter (**PM**) is a diverse mixture of fine particles, both solid and aerosol, that impact human health. Haze, smoke, and dust are visible reminders of particulate material in the air we breathe.

FIGURE 3.15 Typical industrial smog.
Pollution generated by industry differs from that produced by transportation. Industrial pollution has high concentrations of sulfur oxides, particulates, and carbon dioxide.
[Photo by author.]

Focus Study 3.2

Acid Deposition: Damaging to Ecosystems

Acid deposition is a major environmental problem in some areas of the United States, Canada, Europe, and Asia. Such deposition is most familiar as "acid rain," but it also occurs as "acid snow" and in dry form as dust or aerosols. (Aerosols are tiny liquid droplets or solid particles.) In addition, winds can carry the acid-producing chemicals many kilometers from their sources before they settle on the landscape, where they enter streams and lakes as runoff and groundwater flows.

Acid deposition is causally linked to serious problems: declining fish populations and fish kills in the northeastern United States, southeastern Canada, Sweden, and Norway; widespread forest damage in these same places and Germany; widespread changes in soil chemistry; and damage to buildings, sculptures, and historic artifacts. In New Hampshire's Hubbard Brook Experimental Forest, a study covering 1960 to the present found half the nutrient calcium and magnesium base cations (see Chapter 18) were leached from the soil. Excess acids are the cause of the decline.

Despite scientific agreement about the problem, which the U.S. General Accounting Office calls a "combination of meteorological, chemical, and biological phenomena," corrective action was delayed by its complexity and politics.

The acidity of precipitation is measured on the pH scale, which expresses the relative abundance of free hydrogen ions (H^+) in a solution. Free hydrogen ions in a solution are what make an acid corrosive, for they easily combine with other ions. The pH scale is logarithmic: Each whole number represents a 10-fold change. A pH of 7.0 is neutral (neither acidic nor basic). Values less than 7.0 are increasingly *acidic*, and values greater than 7.0 are increasingly *basic*, or *alkaline*. (A pH scale for soil acidity and alkalinity is portrayed graphically in Chapter 18.)

Natural precipitation dissolves carbon dioxide from the atmosphere to form carbonic acid. This process releases hydrogen ions and produces an average pH reading for precipitation of 5.65. The normal range for precipitation is 5.3–6.0. Thus, normal precipitation is always slightly acidic.

Some anthropogenic gases are converted to acids in the atmosphere and then are removed by wet and dry deposition processes. Specifically, nitrogen and sulfur oxides released in the combustion of fossil fuels can produce nitric acid (HNO_3) and sulfuric acid (H_2SO_4) in the atmosphere.

Acid Precipitation Damage

Precipitation as acidic as pH 2.0 has fallen in the eastern United States, Scandinavia, and Europe. By comparison, vinegar and lemon juice register slightly less than 3.0. Aquatic plant and animal life perishes when lakes drop below pH 4.8.

(a)

(b)

FIGURE 3.2.1 The blight of acid deposition.
The harm done to forests and crops by acid deposition is well established, especially in Europe, such as in the Czech Republic (a), and in the forests of the Appalachian Mountains in the United States, such as in the forests on Mount Mitchell (b). [(a) Photo by Simon Fraser/Science Photo Library/Photo Researchers, Inc.; (b) Will and Deni McIntyre/Photo Researchers, Inc.]

(continued)

Focus Study 3.2 *(continued)*

More than 50,000 lakes and some 100,000 km (62,000 mi) of streams in the United States and Canada are at a pH level below normal (that is, below 5.3), with several hundred lakes incapable of supporting any aquatic life—15% of the lakes in New England and 41% in the Adirondack Mountains. Acid deposition causes the release of aluminum and magnesium from clay minerals in the soil, and both of these are harmful to fish and plant communities.

Also, relatively harmless mercury deposits in lake-bottom sediments convert in acidified lake waters into highly toxic *methylmercury*, which is deadly to aquatic life. Local health advisories in two provinces and 22 U.S. states are regularly issued to warn those who fish of the methylmercury problem. Mercury atoms rapidly bond with carbon and move through biological systems as an *organometallic compound*.

Damage to forests results from the rearrangement of soil nutrients, the death of soil microorganisms, and an aluminum-induced calcium deficiency

that is currently under investigation. The most advanced impact is seen in forests in Europe, especially in eastern Europe, principally because of its long history of burning coal and the density of industrial activity. In Germany and Poland, up to 50% of the forests are dead or damaged; in Switzerland, 30% are afflicted (Figure 3.2.1a).

In the United States, regional-scale decline in forest cover is significant, especially red spruce and sugar maples. In some maples, aluminum is collecting around rootlets; in spruce, acid fogs and rains leach calcium from needles directly. Affected trees are susceptible to winter cold, insects, and droughts. In New England, some stands of spruce are as much as 75% affected, as evidenced through analysis of tree-growth rings, which become narrower in adverse growing years. An indicator of forest damage is the reduction by almost half of the annual production of U.S. and Canadian maple sugar. Acid-laden cloud cover damages trees at higher elevations in the Appalachians (Figure 3.2.1b).

Government estimates of damage in the United States, Canada, and Europe exceed $50 billion annually. Because wind and weather patterns are international, efforts at reducing acidic deposition also must be international in scope. The decline of sulfur dioxide by more than 40% between 1973 and the present is a result of the U.S. Clean Air Act. Figure 3.2.2 maps the reduction in sulfate deposition between the 1983–1988 and 1995–1999 periods. However, this progress is only a beginning. According to a study in *BioScience*, power plants and other sources must cut emissions 80% beyond the Clean Air Act mandate.

Ten leading acid-deposition researchers reported in an extensive study in *BioScience*, March 2001:

> Model calculations suggest that the greater the reduction in atmospheric sulfur deposition, the greater the magnitude and rate of chemical recovery. Less aggressive proposals for controls of sulfur emissions will result in

FIGURE 3.2.2 Improvement in sulfate wet deposition rate.
Spatial portrayal of annual sulfate (principally SO$_4$) wet deposition on the landscape, 1983–1988 and 1995–1999 in kilograms per hectare. Clean Air Act regulations have lowered levels of emissions that add acid to the environment. According to a new study, more improvement is necessary to help ecosystems recover. [Maps courtesy of J. W. Lynch and J. A. Grimm, U.S. Forest Service, Northeast Forest Experimental Station, Northern Global Change Research Program.]

slower chemical and biological recovery and in delays in regaining the services of a fully functional ecosystem. . . . North America and Europe are in the midst of a large-scale experiment. Sulfuric and nitric acids have acidified soils, lakes, and streams, thereby stressing or killing terrestrial and aquatic biota.*

At best, acid deposition is an issue of global spatial significance for which science is providing strong incentives for action. Reductions in troublesome emissions are closely tied to energy conservation and therefore directly related to production of greenhouse gases and global warming concerns—thus, linking these environmental issues.

*C. T. Driscoll *et al.*, "Acidic deposition in the northeastern United States: Sources and Inputs, Ecosystems Effects, and Management Strategies," *BioScience* 51 (March 2001): 195.

PM_{10}, particulate smaller than 10 microns (10 μm or less) in diameter, was designated a matter for concern in 1987. $PM_{2.5}$ is currently being debated as an appropriate standard for human health. New studies are implicating even smaller particles, called ultrafines, that are $PM_{0.1}$. These are many times more potent than the larger $PM_{2.5}$ and PM_{10} size because they can get into smaller channels in lung tissue and cause scarring, abnormal thickening, and damage called *fibrosis*.

Studies in Provo and Orem, Utah (1989), Philadelphia (1992), and other cities and by the American Cancer Society (1995) established links between PM pollution and health. Nationally, a study of major cities disclosed a 26% greater risk of premature death due to respirable particulate pollution as compared with nonpolluted air—further driving up medical costs and related expenses.

In Utah County, Utah, researchers correlated PM_{10} concentrations with increased rates of hospitalization for bronchitis, asthma, pneumonia, and pleurisy (especially in children) and greatly increased medical costs. Similar studies of children affected by related illnesses in seven other cities revealed sickness rates twice as high for the city having the dirtiest air as for the city having the cleanest air. Pollution source owners dispute these findings and the concern generated by such studies. However, the health data show that the number of children with asthma doubled between 1980 and 1995 (3.6% to 7.5%) in the United States. By 2001, this percentage rose to 8.7%, or 6.3 million children.

Certainly, society would have difficulty halting two centuries of industrialization to slow air pollution, since the resulting economic chaos would likely be devastating. But if pollution production continues unabated, scientific evidence suggests that catastrophic environmental and human health damage, and continued climate changes, will result. Fortunately, economically viable solutions are available. We are now contributing significantly to the creation of the **anthropogenic atmosphere**, a tentative label for Earth's next atmosphere. The urban air we breathe today may be just a preview. What is the air quality like where you live, work, and go to college? Where could you go to find out its status?

Benefits of the Clean Air Act

The concentration of many air pollutants declined over the past several decades because of Clean Air Act legislation (1970, 1977, 1990), resulting in the avoidance of trillions of dollars in health, economic, and environmental losses. Despite this reality, air pollution controls are subject to a continuing political debate, a weakening of some standards, and proposals that will increase pollution emissions. According to a report prepared by the EPA, the *National Air Pollutant Emission Trends, 1999*, the 1999 emissions of five primary pollutants totaled 162 million metric tons (178 million tons), compared to 224 million metric tons (246 million tons) in 1970, the first year of the Clean Air Act (CAA). Figure 3.16 illustrates the trends in CO, NO_x, VOCs, SO_2, PM_{10}, and lead (Pb). Only nitrogen oxides increased between 1970 and 1999. In Canada, between 1980 and 1997, sulfur dioxide emissions decreased 44%.

Significant lead (Pb) reductions have global beneficial impact. Prior to the CAA, lead was added to gasoline, emitted in the exhaust, dispersed over great distances, and settled in living tissues, especially in children.

To be justified and worthwhile, abatement (mitigation and prevention) costs must not exceed the financial benefits derived from reducing pollution damage. Compliance with the CAA affected patterns of industrial production, employment, and capital investment. Although these expenditures must be viewed as investments that generated benefits and opportunities, the dislocation and job loss in some regions were severe: reductions in high-sulfur coal mining and cutbacks in polluting industries such as steel, for example.

In 1990, Congress requested the Environmental Protection Agency (EPA) to answer the question: How do the overall health, welfare, ecological, and economic benefits of CAA programs compare with the costs of these programs? In response, the EPA performed an exhaustive cost–benefit analysis and published a report in 1997. *The Benefits of the Clean Air Act, 1970 to 1990* (Office of Policy, Planning, and Evaluation, U.S. EPA) calculated the following:

- The *total direct cost* to implement all the federal, state, and local Clean Air Act rules from 1970 to 1990 was *$523 billion* (in 1990-value dollars). This cost was borne by businesses, consumers, and government entities.
- The estimate of *direct monetized benefits* from the Clean Air Act from 1970 to 1990 falls in a range from $5.6 to $49.4 trillion, with a central mean of *$22.2 trillion*. (The uncertainty of the assessment is indicated by the range of benefit estimates.)

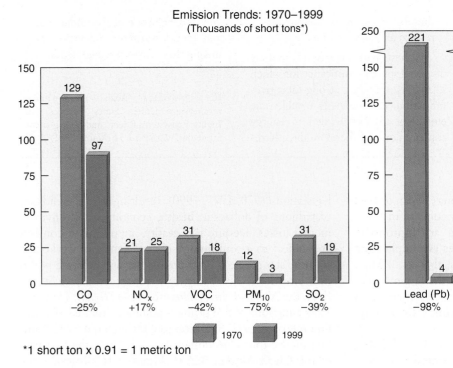

FIGURE 3.16 Trends in air pollutants.
Between 1970 and 1999, sulfur dioxide levels lowered due to scrubbers on smokestacks and emission controls. Nitrogen oxides increased, but volatile organic compounds and carbon monoxide decreased. All three involve exhaust emission controls. Particulate material declined. A huge lead reduction of −98%, or 234,000 tons, resulted from the use of unleaded gas and reduced industrial emissions. [Adapted from Office of Air Quality, *National Air Quality and Emission Trends Report* and *Latest Findings on National Air Quality*, U.S. EPA.]

- Therefore, the estimated *net financial benefit* of the Clean Air Act is *$21.7 trillion*! "The finding is overwhelming. The benefits far exceed the costs of the CAA in the first 20 years," said Richard Morgenstern, associate administrator for policy planning and evaluation at the EPA.

The benefits to society, directly and indirectly, were widespread across the entire population, including improved health and environment, less lead to harm children, lowered cancer rates, less acid deposition, and an estimated 206,000 fewer deaths related to air pollution in 1990 alone. These benefits took place during a period in which the U.S. population grew by 22% and the economy expanded by 70%. The benefits continued between 1990 and 2000 as air quality continued to improve.

As you reflect on this chapter and our modern atmosphere, the treaties to protect stratospheric ozone, and the EPA study of benefits from the CAA, you should feel encouraged. Scientists did the research and society knew what to do, took action, and reaped enormous economic and health benefits. An important role for physical geographers is to explain these global impacts through spatial analysis and to guide an informed citizenry toward a better understanding.

Summary and Review—Earth's Modern Atmosphere

▶ *Construct* a general model of the atmosphere based on the criteria composition, temperature, and function, and *diagram* this model in a simple sketch.

Our modern atmosphere is a gaseous mixture so evenly mixed it behaves as if it were a single gas. It is naturally odorless, colorless, tasteless, and formless. The principal substance of this atmosphere is air—the medium of life.

Above 480 km (300 mi) altitude, the atmosphere is rarefied (nearly a vacuum) and is the **exosphere**, which means "outer sphere." The weight (force over a unit area) of the atmosphere, exerted on all surfaces, is **air pressure**. It decreases rapidly with altitude.

By *composition*, we divide the atmosphere into the **heterosphere**, extending from 480 km (300 mi) to 80 km (50 mi), and the **homosphere**, extending from 80 km to Earth's surface. Within the heterosphere and using *temperature* as a criterion, we identify the **thermosphere**. Its upper limit, called the **thermopause**, is at approximately 480 km altitude. **Kinetic energy**, the energy of motion, is the vibrational energy that we measure and call temperature. The actual heat produced in the thermosphere is very small. The density of the molecules is so low that little actual **heat**, the flow of kinetic energy from one body to another because of a temperature difference between them, is produced. Nearer Earth's surface, the greater number of molecules in the denser atmosphere transmits their kinetic energy as **sensible heat**, meaning that we can feel it.

The homosphere includes the **mesosphere**, **stratosphere**, and **troposphere**, as defined by temperature criteria. Within the mesosphere, cosmic or meteoric dust acts as nuclei around which fine ice crystals form to produce rare and unusual night clouds called **noctilucent clouds**.

The normal temperature profile within the troposphere during the daytime decreases rapidly with increasing altitude at an average of 6.4 C° per kilometer (3.5 F° per 1000 ft), a rate known as the **normal lapse rate**. The top of the troposphere is wherever a temperature of $-57°C$ ($-70°F$) is recorded, a transition known as the **tropopause**. The actual lapse rate at any particular time and place may deviate considerably because of local weather conditions and is called the **environmental lapse rate**.

We distinguish a region in the heterosphere by its *function*. The **ionosphere** absorbs cosmic rays, gamma rays, X-rays, and shorter wavelengths of ultraviolet radiation and converts them into kinetic energy. A functional region within the stratosphere is the **ozonosphere**, or **ozone layer**, which absorbs life-threatening ultraviolet radiation, subsequently raising the temperature of the stratosphere.

> exosphere (p. 62)
> air pressure (p. 63)
> heterosphere (p. 64)
> homosphere (p. 65)
> thermosphere (p. 66)
> thermopause (p. 66)
> kinetic energy (p. 66)
> heat (p. 66)
> sensible heat (p. 67)
> mesosphere (p. 67)
> noctilucent clouds (p. 67)
> stratosphere (p. 67)
> troposphere (p. 67)
> tropopause (p. 67)
> normal lapse rate (p. 67)
> environmental lapse rate (p. 67)
> ionosphere (p. 68)
> ozonosphere, ozone layer (p. 68)

1. What is air? Where did the components in Earth's present atmosphere originate?

2. In view of the analogy by Lewis Thomas, characterize the various functions the atmosphere performs that protect the surface environment.

3. What three distinct criteria are employed in dividing the atmosphere for study?

4. Describe the overall temperature profile of the atmosphere, and list the four layers defined by temperature.

5. Describe the two divisions of the atmosphere on the basis of composition.

6. What are the two primary functional layers of the atmosphere and what does each do?

⬗ *List* the stable components of the modern atmosphere and their relative percentage contributions by volume, and *describe* each.

Even though the atmosphere's density decreases with increasing altitude in the homosphere, the blend (proportion) of gases is nearly uniform. This stable mixture of gases has evolved slowly.

The homosphere is a vast reservoir of relatively inert *nitrogen*, originating principally from volcanic sources and from bacterial action in the soil; *oxygen*, a by-product of photosynthesis; *argon*, constituting about 1% of the homosphere and completely inert; and *carbon dioxide*, a natural by-product of life processes and fuel combustion.

7. Name the four most prevalent stable gases in the homosphere. Where did each originate? Is the prevalence of any of these changing at this time?

⬗ *Describe* conditions within the stratosphere; specifically, *review* the function and status of the ozonosphere (ozone layer).

The overall reduction of the stratospheric ozonosphere, or ozone layer, during the past several decades represents a hazard for society and many natural systems and is caused by chemicals introduced into the atmosphere by humans. Since World War II, quantities of human-made **chlorofluorocarbons** (**CFCs**) and bromine-containing compounds have made their way into the stratosphere. The increased ultraviolet light at those altitudes breaks down these stable chemical compounds, thus freeing chlorine and bromine atoms. These atoms act as catalysts in reactions that destroy ozone molecules.

> chlorofluorocarbons (CFCs) (p. 69)

8. Why is stratospheric ozone (O_3) so important? Describe the effects created by increases in ultraviolet light reaching the surface.

9. Summarize the ozone predicament, and describe trends and any treaties that intend to protect the ozone layer.

10. Evaluate Crutzen, Rowland, and Molina's use of the scientific method in investigating stratospheric ozone depletion.

⬗ *Distinguish* between natural and anthropogenic variable gases and materials in the lower atmosphere.

Within the troposphere, both natural and human-caused variable gases, particles, and other chemicals are part of the atmosphere. We coevolved with natural "pollution" and thus are adapted to it. But we are not adapted to cope with our own anthropogenic pollution. It constitutes a major health threat, particularly where people are concentrated in cities.

Vertical temperature and atmospheric density distribution in the troposphere can worsen pollution conditions. A **temperature inversion** occurs when the normal temperature decrease with altitude (normal lapse rate) reverses. In other words, temperature begins to increase at some altitude.

> temperature inversion (p. 74)

11. Why are anthropogenic gases more significant to human health than are those produced from natural sources?

12. In what ways does a temperature inversion worsen an air pollution episode? Why?

⬗ *Describe* the sources and effects of carbon monoxide, nitrogen dioxide, and sulfur dioxide, and *construct* a simple equation that illustrates photochemical reactions that produce ozone, peroxyacetyl nitrates, nitric acid, and sulfuric acid.

Odorless, colorless, and tasteless, **carbon monoxide** (**CO**) is produced by incomplete combustion (burning with limited oxygen) of fuels or other carbon-containing substances. Transportation is the major human-caused source for carbon monoxide. The toxicity of carbon monoxide is due to its affinity for blood hemoglobin, which is the oxygen-carrying pigment in red blood cells. In the presence of carbon monoxide, the oxygen is displaced and the blood becomes deoxygenated (Table 3.4).

Photochemical smog results from the interaction of sunlight and the products of automobile exhaust, the single largest contributor of pollution that produces smog. Car exhaust, containing *nitrogen dioxide* and *volatile organic compounds* (*VOCs*), in the presence of ultraviolet light in sunlight converts into major air pollutants—*ozone*, *peroxyacetyl nitrates* (*PAN*), and *nitric acid*. The principal photochemical by-products include *ozone* (O_3), which causes negative health effects, oxidizes surfaces, and kills or damages plants; and **peroxyacetyl nitrates** (**PAN**), which produce no known health effects in humans but are particularly damaging to plants, including both agricultural crops and forests. **Nitrogen dioxide** (**NO_2**) damages and inflames human respiratory systems, destroys lung tissue, and damages plants. Nitric oxides participate in reactions that form nitric acid (HNO_3) in the atmosphere, forming both wet and dry acidic deposition. The **volatile organic compounds** (**VOCs**), including hydrocarbons from gasoline, surface coatings, and electric utility combustion, are important factors in ozone formation.

The distribution of human-produced **industrial smog** over North America, Europe, and Asia is related to transportation and electrical production. Such characteristic pollution contains **sulfur dioxide**. Sulfur dioxide in the atmosphere reacts to produce **sulfate aerosols**, which produce sulfuric acid (H_2SO_4) deposition, which can be dangerous to health, and affect Earth's energy budget by scattering and reflecting solar energy. **Particulate matter** (**PM**) consists of dirt, dust, soot, and ash from industrial and natural sources.

Energy conservation and efficiency and reducing emissions are key to reducing air pollution. Earth's next atmosphere most accurately may be described as the **anthropogenic atmosphere** (human-influenced atmosphere).

carbon monoxide (CO) (p. 77)
photochemical smog (p. 77)
peroxyacetyl nitrates (PAN) (p. 78)
nitrogen dioxide (p. 79)
volatile organic compounds (VOCs) (p. 79)
industrial smog (p. 80)
sulfur dioxide (p. 80)
sulfate aerosols (p. 80)
particulate matter (PM) (p. 80)
anthropogenic atmosphere (p. 83)

13. What is the difference between industrial smog and photochemical smog?

14. Describe the relationship between automobiles and the production of ozone and PAN in city air. What are the principal negative impacts of these gases?

15. How are sulfur impurities in fossil fuels related to the formation of acid in the atmosphere and acid deposition on the land?

16. In summary, what are the results from the first 20 years under Clean Air Act regulations?

NetWork

The *Geosystems* Home Page provides online resources for this chapter on the World Wide Web. To begin: Once on the Home Page, click on this textbook, scroll the Table of Contents menu, select this chapter, and click "Begin." You will find self-tests that are graded, review exercises, specific updates for items in the chapter, and many links to interesting related pathways on the Internet under "Destinations." *Geosystems* is at http://www.prenhall.com/christopherson.

Critical Thinking

A. A scientific study, *The Benefits of the Clean Air Act, 1970 to 1990* (Office of Policy, Planning, and Evaluation, U.S. EPA, October 1997), determined that the Clean Air Act provided the American people with benefits. Health, welfare, ecological, and economic benefits rated 42 to 1 over the costs of CAA implementation (estimated at $21.7 trillion in net benefit compared to $523 billion in costs). Do you think this is significant? Should this be part of the debate about future weakening or strengthening of the Clean Air Act? Do you think such benefits might result from other regulations such as the Clean Water Act or groundwater protection measures?

B. Relative to item A: In your opinion, why is the public generally unaware of these details? What are the difficulties in informing the public? Why is there such media attention given to antiscience and nonscientific opinions? Take a moment and brainstorm recommendations for action, education, and public awareness on these issues.

C. To determine the total ozone column at your present location, go to the TOMS Home Page at **http://toms. gsfc.nasa.gov/teacher/ozone_overhead.html**, "What was the total ozone column at your house?" Select a point on the map or enter your latitude and longitude and the date you want to check. The ozone column refers principally to stratospheric ozone and not to the photochemical pollutant in the lower troposphere. Note the instrument and satellite platform used. (Note also the limitations listed on the extent of data availability.) For several different dates, when do the lowest values occur? The highest values? Briefly explain what your results mean. How do you interpret the values found?

Jet condensation trails stimulate high-cirrus cloud formation. Inset photos and image show thin, newer contrails in contrast to older contrails widening over time as they develop. Studies found that such contrails affect daily temperature ranges between daytime maximum and nighttime minimum readings. This is another connection between transportation systems and Earth's atmospheric energy budget. [Photos by Bobbé Christopherson; inset image *Terra* MODIS satellite courtesy of MODIS Land Rapid Response Team, NASA/GSFC.]

Atmosphere and Surface Energy Balances

▶ **Key Learning Concepts**

After reading the chapter, you should be able to:

▶ *Identify* the pathways of solar energy through the troposphere to Earth's surface: transmission, scattering, diffuse radiation, refraction, albedo (reflectivity), conduction, convection, and advection.

▶ *Describe* what happens to insolation when clouds are in the atmosphere, and *analyze* the effect of clouds and air pollution on solar radiation received at ground level.

▶ *Review* the energy pathways in the Earth–atmosphere system, the greenhouse effect, and the patterns of global net radiation.

▶ *Plot* the daily radiation curves for Earth's surface, and *label* the key aspects of incoming radiation, air temperature, and the daily temperature lag.

▶ *Portray* typical urban heat island conditions, and *contrast* the microclimatology of urban areas with that of surrounding rural environments.

E arth's biosphere pulses with flows of solar energy that sustain our lives and empower natural systems. Changing seasons, daily weather conditions, and Earth's variety of climates remind us of this constant flow of energy cascading through the atmosphere. Earth's shifting seasonal rhythms are described in Chapter 2, and a profile of the atmosphere in Chapter 3. Energy and moisture exchanges between Earth's surface and the atmosphere are essential elements of weather and climate, discussed in later chapters.

In this chapter: We follow solar energy through the troposphere to Earth's surface. The *outputs* of reflected light and emitted infrared energy from the atmosphere and surface environment counter the *input* of insolation. Together, this input and output determines the net energy available to perform work. We examine surface energy budgets and analyze how net radiation is spent. The chapter concludes with a

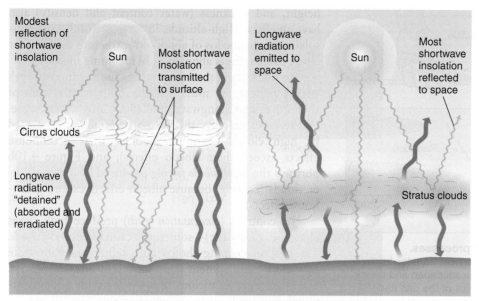

Modest reflection of shortwave insolation

Sun

Most shortwave insolation transmitted to surface

Cirrus clouds

Longwave radiation "detained" (absorbed and reradiated)

Longwave radiation emitted to space

Sun

Most shortwave insolation reflected to space

Stratus clouds

(a) High clouds: net greenhouse forcing and atmospheric warming

(b) Low clouds: net albedo forcing and atmospheric cooling

(c)

FIGURE 4.10 Energy effects of two cloud types.
Cloud effects vary depending on cloud type. (a) High, ice-crystal clouds (called *cirrus*) transmit most of the insolation. However, they absorb and delay losses of outgoing longwave infrared, producing a greater greenhouse forcing and a net warming of Earth. (b) Low, thick clouds (*stratus*) reflect most of the incoming insolation and radiate longwave infrared to space, producing a greater albedo forcing and a net cooling of Earth. (c) Jet aircraft exhaust triggers cirrus cloud development. Newer, thinner contrails spread to form cirrus clouds.
[(c) Astronaut photo May 15, 2002 courtesy of NASA/JSC.]

To better understand the role of natural clouds as well as jet contrails, NASA operates the CERES program, an acronym for Clouds and Earth's Radiant Energy System. CERES sensors aboard satellites began operating in 1997 (*TRMM*), the newest in 2000 (*Terra*). CERES data are an important part of the study of aerosols and the monsoons in Figure 4.8. (See http://asd-www.larc.nasa.gov/ceres/ASDceres.html.)

Data collected by CERES sensors cover both reflected and emitted radiation. Figure 4.11 shows the first global monthly images from satellite *Terra* for March 2000 in W/m² flux (meaning energy "flow"). In Figure 4.11a, lighter regions indicate where more sunlight is reflected into space than is absorbed—for example, by lighter-colored land surfaces such as deserts, or by cloud cover such as over tropical lands. Green and blue areas illustrate where less light was reflected.

In Figure 4.11b, orange and red pixels indicate regions where more heat was absorbed and emitted to space; less heat energy is escaping in the blue and purple regions.

Those blue regions of lower longwave emissions over tropical lands are due to tall, thick clouds along the equatorial region (the Amazon, equatorial Africa, and Indonesia); these clouds also caused higher shortwave reflection. Subtropical desert regions exhibit greater longwave radiation emissions owing to the presence of little cloud cover and greater radiative energy losses from surfaces that have absorbed a lot of energy, as we saw in Figure 4.2.

Earth–Atmosphere Radiation Balance

If Earth's surface and its atmosphere are considered separately, neither exhibits a balanced radiation budget. The average annual energy distribution is positive (an energy surplus) for Earth's surface and negative (an energy deficit) for the atmosphere as it radiates energy to space. Considered together, these two equal each other, making it possible for us to construct an overall energy balance. However, regionally and seasonally Earth absorbs more energy in the tropics and less in the polar regions, establishing the imbalance that drives global circulation patterns.

(a) Outgoing shortwave – Earth's albedo.

(b) Longwave energy flux to space.

FIGURE 4.11 Shortwave and longwave images show Earth's radiation budget components.
The CERES sensors aboard *Terra* made these portraits in March 2000, capturing (a) outgoing short-wave energy flux reflected from clouds, land, and water—Earth's albedo; and, (b) longwave energy flux emitted by these surfaces back to space. The scale beneath each image displays value gradations in watts per square meter. [Images courtesy of CERES Instrument Team, Langley Research Center, NASA.]

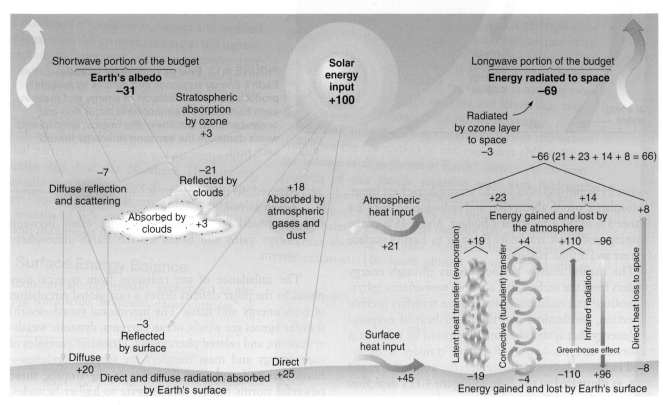

FIGURE 4.12 Detail of the Earth–atmosphere energy balance.
Solar energy cascades through the lower atmosphere (left-hand portion of the illustration), where it is absorbed, reflected, and scattered. Clouds, atmosphere, and the surface reflect 31% of this insolation back to space. Atmospheric gases and dust and Earth's surface absorb energy and radiate infrared radiation. Earth and atmosphere exchange energy through latent heat transfer in water vapor, convective transfer (moving air), and infrared radiation (right-hand portion of the illustration). Over time, Earth emits, on average, 69% of incoming energy to space. When added to Earth's average albedo (31%, reflected energy), this equals the total energy input from the Sun.

Global Warming, Climate Change, Earth-Atmosphere Energy Balance

ANIMATION

Figure 4.12 summarizes the Earth–atmosphere radiation balance. It brings together all the elements discussed to this point in the chapter by following 100% of arriving insolation through the troposphere. The shortwave portion of the budget is on the left in the illustration; the longwave part of the budget is on the right. Out of 100% of the solar energy arriving, Earth's average albedo is 31%. Figure 4.11a shows this shortwave radiation reflected from the Earth–atmosphere system during March 2000. Absorption by atmospheric clouds, dust, and gases involves

FIGURE 4.23 Heat island exposed.
Urban Heat Island Pilot Project (UHIPP) thermal infrared image of Sacramento, California, taken June 29, 1998, at 1 P.M. PDT from a NASA jet at 2000 m (6600 ft). The Sacramento River runs from north to south to the left of center of the frame, the American River is to the upper right, and the California state capitol building is the red dot in the green rectangle right of center from the river. False color tags white and red as relatively hot sites; greens and blues are relatively cool on this preliminary image. [Image courtesy of UHIPP, Marshall Space Flight Center, Global Hydrology and Climate Center, Huntsville, Alabama.]

Sacremento urban heat island in the distance

measurements in an outdoor mall parking lot found temperatures of 48°C (118°F), but a small planter with trees in the same lot was significantly cooler at 32°C (90°F)—16 C° (29 F°) lower in temperature! In Central Park in New York City, daytime temperatures average 5–10 C° (9–18 F°) cooler than urban areas outside the park.

Figure 4.23 is a thermal infrared image of a portion of Sacramento, California, taken at midday. The false colors of red and white are relatively hot areas (60°C, 140°F),

blues and greens are relatively cool areas (29°–36°C, 85°–96°F); clearly buildings are the hottest objects in the scene. As you look across this landscape, what strategies do you think would reduce the urban heat island effect? Possibilities include lighter-colored surfaces for buildings, streets, and parking lots; more reflective roofs; and more trees, parks, and open space. (The images from all the test cities are available at http://science.nasa.gov/newhome/headlines/essd01jul98%5F1.htm.)

Summary and Review—Atmosphere and Surface Energy Balances

▶ **Identify** the pathways of solar energy through the troposphere to Earth's surface: transmission, scattering, diffuse radiation, refraction, albedo (reflectivity), conduction, convection, and advection.

Radiant energy from the Sun that cascades through complex circuits to the surface powers Earth's biosphere. Our budget of atmospheric energy comprises shortwave radiation *inputs*

(ultraviolet light, visible light, and near-infrared wavelengths) and longwave radiation *outputs* (thermal infrared).

Transmission refers to the passage of shortwave and longwave energy through either the atmosphere or water. The gas molecules redirect radiation, changing the direction of the light's movement *without altering its wavelengths*. This phenomenon is known as **scattering** and represents 7% of

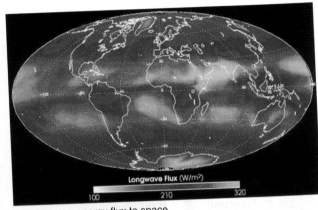

(a) Outgoing shortwave – Earth's albedo.

(b) Longwave energy flux to space.

FIGURE 4.11 Shortwave and longwave images show Earth's radiation budget components.
The CERES sensors aboard *Terra* made these portraits in March 2000, capturing (a) outgoing short-wave energy flux reflected from clouds, land, and water—Earth's albedo; and, (b) longwave energy flux emitted by these surfaces back to space. The scale beneath each image displays value gradations in watts per square meter. [Images courtesy of CERES Instrument Team, Langley Research Center, NASA.]

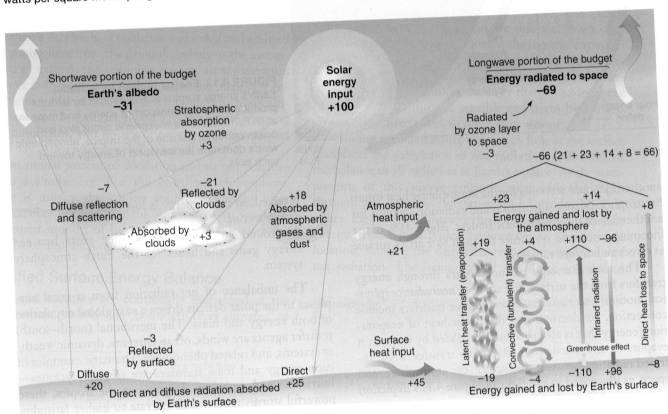

FIGURE 4.12 Detail of the Earth–atmosphere energy balance.
Solar energy cascades through the lower atmosphere (left-hand portion of the illus-tration), where it is absorbed, reflected, and scattered. Clouds, atmosphere, and the surface reflect 31% of this insolation back to space. Atmospheric gases and dust and Earth's surface absorb energy and radiate infrared radiation. Earth and atmos-phere exchange energy through latent heat transfer in water vapor, convective transfer (moving air), and infrared radiation (right-hand portion of the illustration). Over time, Earth emits, on average, 69% of incoming energy to space. When added to Earth's average albedo (31%, reflected energy), this equals the total energy input from the Sun.

ANIMATION

Global Warming, Climate Change, Earth-Atmosphere Energy Balance

Figure 4.12 summarizes the Earth–atmosphere radia-tion balance. It brings together all the elements discussed to this point in the chapter by following 100% of arriving insolation through the troposphere. The shortwave por-tion of the budget is on the left in the illustration; the longwave part of the budget is on the right. Out of 100% of the solar energy arriving, Earth's average albedo is 31%. Figure 4.11a shows this shortwave radiation reflected from the Earth–atmosphere system during March 2000. Absorption by atmospheric clouds, dust, and gases involves

FIGURE 4.23 Heat island exposed.
Urban Heat Island Pilot Project (UHIPP)
thermal infrared image of Sacramento,
California, taken June 29, 1998, at
1 P.M. PDT from a NASA jet at 2000 m
(6600 ft). The Sacramento River runs
from north to south to the left of center
of the frame, the American River is to
the upper right, and the California state
capitol building is the red dot in the
green rectangle right of center from the
river. False color tags white and red as
relatively hot sites; greens and blues
are relatively cool on this preliminary
image. [Image courtesy of UHIPP,
Marshall Space Flight Center, Global
Hydrology and Climate Center, Huntsville,
Alabama.]

Sacremento urban heat island in the distance

measurements in an outdoor mall parking lot found temperatures of 48°C (118°F), but a small planter with trees in the same lot was significantly cooler at 32°C (90°F)—16 C° (29 F°) lower in temperature! In Central Park in New York City, daytime temperatures average 5–10 C° (9–18 F°) cooler than urban areas outside the park.

Figure 4.23 is a thermal infrared image of a portion of Sacramento, California, taken at midday. The false colors of red and white are relatively hot areas (60°C, 140°F),

blues and greens are relatively cool areas (29°–36°C, 85°–96°F); clearly buildings are the hottest objects in the scene. As you look across this landscape, what strategies do you think would reduce the urban heat island effect? Possibilities include lighter-colored surfaces for buildings, streets, and parking lots; more reflective roofs; and more trees, parks, and open space. (The images from all the test cities are available at **http://science.nasa.gov/newhome/headlines/essd01jul98%5F1.htm**.)

Summary and Review—Atmosphere and Surface Energy Balances

▶ *Identify* the pathways of solar energy through the troposphere to Earth's surface: transmission, scattering, diffuse radiation, refraction, albedo (reflectivity), conduction, convection, and advection.

Radiant energy from the Sun that cascades through complex circuits to the surface powers Earth's biosphere. Our budget of atmospheric energy comprises shortwave radiation *inputs*

(ultraviolet light, visible light, and near-infrared wavelengths) and longwave radiation *outputs* (thermal infrared).

Transmission refers to the passage of shortwave and longwave energy through either the atmosphere or water. The gas molecules redirect radiation, changing the direction of the light's movement *without altering its wavelengths*. This phenomenon is known as **scattering** and represents 7% of

Earth's reflectivity, or albedo. Dust particles, pollutants, ice, cloud droplets, and water vapor produce further scattering. Some incoming insolation is diffused by clouds and atmosphere and is transmitted to Earth as **diffuse radiation**, the downward component of scattered light. The speed of insolation entering the atmosphere changes as it passes from one medium to another; the change of speed causes a bending action called **refraction**.

A portion of arriving energy bounces directly back into space without being converted into heat or performing any work. **Albedo** is the reflective quality (intrinsic brightness) of a surface; this returned energy is **reflection**. Albedo is an important control over the amount of insolation that is available for absorption by a surface. We state albedo as the percentage of insolation that is reflected. Earth and its atmosphere reflect 31% of all insolation when averaged over a year.

An increase in albedo and reflection of shortwave radiation caused by clouds is **cloud-albedo forcing**. Also, clouds can act as insulation, thus trapping longwave radiation and raising minimum temperatures. An increase in greenhouse warming caused by clouds is **cloud-greenhouse forcing**.

Absorption is the assimilation of radiation by molecules of a substance and its conversion from one form to another—for example, visible light to infrared radiation. **Conduction** is the molecule-to-molecule transfer of energy as it diffuses through a substance.

Energy also is transferred in gases and liquids by **convection** (when the physical mixing involves a strong vertical motion) or **advection** (when the dominant motion is horizontal). In the atmosphere or bodies of water, warmer portions tend to rise (they are less dense) and cooler portions tend to sink (they are more dense), establishing patterns of convection.

transmission (p. 90)
scattering (p. 90)
diffuse radiation (p. 91)
refraction (p. 91)
reflection (p. 92)
albedo (p. 92)
cloud-albedo forcing (p. 92)
cloud-greenhouse forcing (p. 92)
absorption (p. 95)
conduction (p. 95)
convection (p. 95)
advection (p. 95)

1. Diagram a simple energy balance for the troposphere. Label each shortwave and longwave component and the directional aspects of related flows.
2. What would you expect the sky color to be at 50 km (30 mi) altitude? Why? What factors explain the lower atmosphere's blue color?
3. Define refraction. How is it related to daylength? To a rainbow? To the beautiful colors of a sunset?
4. List several types of surfaces and their albedo values. Explain the differences among these surfaces. What determines the reflectivity of a surface?
5. Using Figure 4.6, explain the seasonal differences in albedo values for each hemisphere. Be specific, using the albedo values given in Figure 4.5 where appropriate.
6. Define the concepts transmission, absorption, diffuse radiation, conduction, and convection.

 Describe what happens to insolation when clouds are in the atmosphere, and *analyze* the effect of clouds and air pollution on solar radiation received at ground level.

Clouds reflect *insolation*, thus cooling Earth's surface—*cloud-albedo forcing*. Yet clouds also act as insulation, thus trapping longwave radiation and raising minimum temperatures—*cloud-greenhouse forcing*. Clouds affect the heating of the lower atmosphere, depending on cloud type, height, and thickness (water content and density). High-altitude, ice-crystal clouds reflect insolation with albedos of about 50% and redirect infrared to the surface, producing a net cloud-greenhouse forcing (warming); thick, lower cloud cover reflects about 90%, producing a net cloud-albedo forcing (cooling).

Emissions of sulfur dioxide and the subsequent chemical reactions in the atmosphere form *sulfate aerosols*, which act either as insolation-reflecting haze in clear-sky conditions or as a stimulus to condensation in clouds that increases reflectivity.

7. What role do clouds play in the Earth–atmosphere radiation balance? Is cloud type important? Compare high, thin cirrus clouds and lower, thick stratus clouds.
8. Jet contrails affect the Earth–atmosphere balance in what ways? Describe the recent scientific findings.
9. In what way does the presence of sulfate aerosols affect solar radiation received at ground level? How does it affect cloud formation?

 Review the energy pathways in the Earth–atmosphere system, the greenhouse effect, and the patterns of global net radiation.

The Earth–atmosphere energy system naturally balances itself in a steady-state equilibrium. It does so through energy transfers that are *nonradiative* (convection, conduction, and the latent heat of evaporation) and *radiative* (by infrared radiation between the surface, the atmosphere, and space).

Carbon dioxide, water vapor, methane, CFCs (chlorofluorocarbons), and other gases in the lower atmosphere absorb infrared radiation that is then emitted to Earth, thus delaying energy loss to space. This process is the **greenhouse effect**. In the atmosphere, infrared radiation is not actually trapped, as it would be in a greenhouse, but its passage to space is delayed (heat energy is detained in the atmosphere) through absorption and counterradiation.

Between the tropics, high insolation angle and consistent daylength cause more energy to be gained than lost (there are energy surpluses). In the polar regions, an extremely low insolation angle, highly reflective surfaces, and up to 6 months of no insolation annually cause more energy to be lost (there are energy deficits). This imbalance of net radiation from tropical surpluses to the polar deficits drives a vast global circulation of both energy and mass.

Surface energy balances are used to summarize the energy expenditure for any location. Surface energy measurements are used as an analytical tool of **microclimatology**. Adding and subtracting the energy flow at the surface lets us calculate **net radiation (NET R)**, or the balance of all radiation at Earth's surface available to do work—shortwave (SW) and longwave (LW).

greenhouse effect (p. 96)
microclimatology (p. 101)
net radiation (NET R) (p. 101)

10. What are the similarities and differences between an actual greenhouse and the gaseous atmospheric greenhouse? Why is Earth's greenhouse changing?

11. In terms of energy expenditures for latent heat of evaporation, describe the annual pattern as mapped in Figure 4.18.

12. Generalize the pattern of global net radiation. How might this pattern drive the atmospheric weather machine? (See Figures 4.13 and 4.17.)

13. In terms of surface energy balance, explain the term *net radiation* (NET R).

14. What are the expenditure pathways for surface net radiation? What kind of work is accomplished?

15. What is the role played by latent heat in surface energy budgets?

16. Compare the daily surface energy balances of El Mirage, California, and Pitt Meadows, British Columbia. Explain the differences.

▶ *Plot* the daily radiation curves for Earth's surface, and *label* the key aspects of incoming radiation, air temperature, and the daily temperature lag.

The greatest insolation input occurs at the time of the summer solstice in each hemisphere. Air temperature responds to seasons and variations in insolation input. Within a 24-hour day, air temperature peaks between 3:00 and 4:00 P.M. and dips to its lowest point right at or slightly after sunrise.

Air temperature lags behind each day's peak insolation. The warmest time of day occurs not at the moment of maximum insolation, but at that moment when a maximum of insolation is absorbed.

17. Why is there a temperature lag between the highest Sun altitude and the warmest time of day? Relate your answer to the insolation and temperature patterns during the day.

▶ *Portray* typical urban heat island conditions, and *contrast* the microclimatology of urban areas with that of surrounding rural environments.

A growing percentage of Earth's people lives in cities and experiences a unique set of altered microclimatic effects: increased conduction, lower albedos, higher NET R values, increased water runoff, complex radiation and reflection patterns, anthropogenic heating, and the gases, dusts, and aerosols of urban pollution. Urban surfaces of metal, glass, asphalt, concrete, and stone conduct up to 3 times more energy than wet, sandy soil and thus are warmed, producing an **urban heat island**. Air pollution, including gases and aerosols, is greater in urban areas than in rural ones. Every major city produces its own **dust dome** of airborne pollution.

urban heat island (p. 107)
dust dome (p. 109)

18. What is the basis for the urban heat island concept? Describe the climatic effects attributable to urban as compared with nonurban environments. What did NASA determine from the UHIPP overflight of Sacramento (Figure 4.23)?

19. Which of the items in Table 4.2 have you yourself experienced? Explain.

20. Assess the potential for solar energy applications in our society. What are some negatives? What are some positives?

NetWork

The *Geosystems* Home Page provides online resources for this chapter on the World Wide Web. To begin: Once on the Home Page, click on this textbook, scroll the Table of Contents menu, select this chapter, and click "Begin." You will find self-tests that are graded, review exercises, specific updates for items in the chapter, and many links to interesting related pathways on the Internet under "Destinations." *Geosystems* is at **http://www.prenhall.com/christopherson**.

Critical Thinking

A. Given what you now know about reflection, albedo, absorption, and net radiation expenditures, assess your wardrobe (fabrics and colors), house or apartment (colors of walls, especially south and west facing; or, in the Southern Hemisphere, north and west facing), and roof (orientation relative to the Sun), automobile (color, use of sun shades), bicycle seat (color), and other aspects of your environment to determine a personal "Energy IQ." What grade do you give yourself? Be cool!

B. On the *Geosystems* Home Page, Chapter 4, "Destinations," and in this chapter of the text, there are several listings of URLs relating to solar energy applications. Take some time and explore the Internet for a personal assessment of these necessary technologies (solar-thermal, solar-electric, photovoltaic cells, solar-box cookers, and the like). As we near the climatic limitations of the fossil fuel era, and the depletion of the resource itself, these available technologies are destined to become part of the fabric of our lives. Briefly describe your search results. Given these findings, determine if solar technology is available in your area.

C. Examine the photograph to the left. In Antarctica, on Petermann Island off the Graham Land Coast, we noticed this piece of kelp (a seaweed) dropped by a passing bird. The kelp was resting about 10 cm (4 in.) deep in the snow, in a hole about the shape of the kelp. In your opinion, what factors operated to make this scene? Now, expand your conclusion to the issue of mining in Antarctica. There are coal deposits and other minerals in Antarctica. The international Antarctic Treaty (1959, additional measures in 1972, 1982, 1991) blocks mining exploitation, although reconsideration is possible after 2048. Given your energy budget analysis of the kelp in the snow, information on surface energy budgets in this chapter, the dust and particulate output of mining, and the importance of a stable sea level, construct a case opposed to such mining activities based on these factors.

The fertile soils of western Iowa produce grains and feed livestock. These are the original grassland soils. Like a living history, soils bear climatic and geologic evidence of the last 15,000 years or more, long before humans began working the land. Agriculture depends on consistent temperature and precipitation, yet changes in each are producing shifting seasons—earlier spring and later fall—and negative changes in soil moisture. Our study of global temperatures and analysis of possible dynamic system changes is an important aspect of physical geography. [Photo by Bobbé Christopherson.]

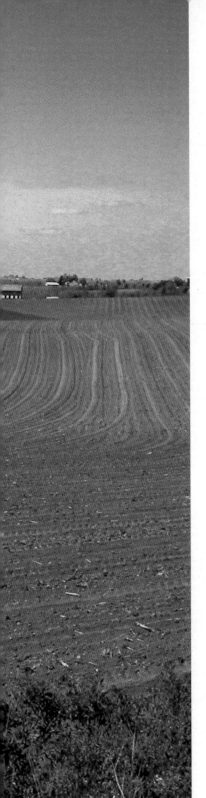

CHAPTER 5

Global Temperatures

> ## ▶ Key Learning Concepts

After reading the chapter, you should be able to:

▶ *Define* the concepts of temperature, kinetic energy, and sensible heat, and *distinguish* among Kelvin, Celsius, and Fahrenheit temperature scales and how they are measured.

▶ *List* and *review* the principal controls and influences that produce global temperature patterns.

▶ *Review* the factors that produce different marine effects and continental effects as they influence temperatures, and *utilize* several pairs of stations to illustrate these differences.

▶ *Interpret* the pattern of Earth's temperatures from their portrayal on January and July temperature maps and on a map of annual temperature ranges.

▶ *Contrast* wind chill and heat index, and *determine* human response to these apparent temperature effects.

What is the temperature now—both indoors and outdoors—as you read these words? How is it measured, and what does the value mean? How is air temperature influencing your plans for the day? Air temperature plays a remarkable role in our lives, both at the micro level and at the macro and global levels. Our bodies subjectively sense temperature and judge comfort and react to changing temperatures.

We read of heat waves and cold spells affecting people, crops, events, and energy consumption. For example, in 1995 devastating summer heat and high humidity killed more than 1000 people in the Midwest and East; and Europe lost more than 20,000 people in 2003 record heat-index conditions(see Focus Study 5.1). Record warmth is reaching into the Arctic. Tree-ring analysis, ice core data, and other proxy measures indicate that present temperatures are warmer than Earth has experienced during the last 1000 years. This warming trend in air temperature is the subject of much scientific, geographic, and political interest.

In this chapter: A variety of temperature regimes affect cultures, decision making, and resources consumed across the globe. Understanding some temperature concepts helps us begin our study of these Earth systems. We look at the principal temperature controls of latitude, altitude, cloud cover, and land-water heating differences as they interact to produce Earth's temperature patterns. We end by examining the effect of temperature on the human body and the resultant spatial aspects of hot and cold conditions on human experience.

Temperature Concepts and Measurement

Heat and temperature are not the same. *Heat* is a form of energy that flows from one system or object to another because the two are at different temperatures. **Temperature** is a measure of the average kinetic energy (motion) of individual molecules in matter. We feel the effect of temperature as the *sensible heat* transfer from warmer objects to cooler objects. For instance, when you jump into a cool lake, you can sense the heat transfer from your skin to the water as kinetic energy leaves your body and flows to the water; a chill develops.

Temperature and heat are related because changes in temperature are caused by the absorption or emission (gain or loss) of heat energy. The term *heat energy* is frequently used to describe energy that is added to or removed from a system or substance.

Temperature Scales

The temperature at which all atomic and molecular motion in matter completely stops is called *0° absolute temperature*, or "absolute zero." Its value on the different temperature-measuring scales is −273° Celsius (C), −459.4° Fahrenheit (F), and 0 Kelvin (K). Figure 5.1 compares these three scales. Formulas for converting between Celsius, SI (Système International), and English units are in Appendix C.

The Fahrenheit scale places the melting point of ice at 32°F, with 180 subdivisions to the boiling point of water at 212°F. The scale is named for its developer Daniel G. Fahrenheit, a German physicist (1686–1736). (Note that there is only one melting point for ice, but there are many freezing points for water, ranging from 32°F down to −40°F, depending on its purity, volume, and certain conditions in the atmosphere.)

About a year after the adoption of the Fahrenheit scale, Swedish astronomer Anders Celsius (1701–1744) developed the Celsius scale (formerly called centigrade). He placed the melting point of ice at 0°C and boiling temperature of water at sea level at 100°C, dividing his scale into 100 degrees using a decimal system.

FIGURE 5.1 Temperature scales.
Scales for expressing temperature in Kelvin (K) and degrees Celsius (°C) and Fahrenheit (°F).

British physicist Lord Kelvin (born William Thomson, 1824–1907) proposed the Kelvin scale in 1848. Science uses this scale because it starts at temperature absolute zero and thus readings are proportional to the actual kinetic energy in a material. The Kelvin scale's melting point for ice is 273 K, and its boiling point of water is 373 K.

Most countries use the Celsius scale to express temperature. The United States remains the only major country still using the Fahrenheit scale. The continuing pressure from the scientific community and corporate interests, not to mention the rest of the world, makes adoption of Celsius and SI units inevitable. This textbook presents Celsius (with Fahrenheit equivalents in parentheses) throughout to help bridge this transitional era in the United States.

Measuring Temperature

A *mercury thermometer* or *alcohol thermometer* is a sealed glass tube that measures outdoor temperatures. Fahrenheit invented both the alcohol and mercury thermometers. Cold climates demand alcohol thermometers because alcohol freezes at −112°C (−170°F), whereas mercury freezes at −39°C (−38.2°F). The principle of these thermometers is simple: When fluids are heated, they expand; upon cooling, they contract. A thermometer stores fluid in a small reservoir at one end and is marked with calibrations to measure the expansion or contraction of the fluid, which reflects the temperature of the thermometer's environment.

Figure 5.2 shows a mercury minimum–maximum thermometer. It preserves readings of the day's highest and lowest temperatures until reset (by moving the markers with a magnet). Another type, the recording thermometer, creates an inked record on a turning drum; usually set for one full rotation every 24 hours or every 7 days.

Thermometers for standardized official readings are placed outdoors in small shelters that are white (for high albedo) and louvered (for ventilation) to avoid overheating of the instruments. They are placed at least 1.2–1.8 m (4–6 ft) above a surface, usually on turf; in the United States 1.2 m (4 ft) is standard. Official temperature measurements are therefore in the shade to prevent the effect of direct insolation. If you install a thermometer for outdoor temperature reading, be sure to avoid direct sunlight on the instrument and place it in an area of good ventilation and at least 1.2 m off the ground.

FIGURE 5.3 Instrument shelter.
This standard thermometer shelter is white (for high albedo) and louvered (for ventilation), replacing the traditional cotton-region shelter. [Photo by Bobbé Christopherson.]

The instrument shelter in Figure 5.3 contains a *thermistor* that measures temperature by sensing the electrical resistance of a semiconducting material. Resistance changes at 4% per C°, which is reported electronically to the weather station.

Temperature readings are taken daily, sometimes hourly, at more than 16,000 weather stations worldwide. One of the goals of the Global Climate Observing System (GCOS) is to establish a reference network of one station per 250,000 km² (95,800 mi²). Some stations with recording equipment also report the duration of temperatures, rates of rise or fall, and temperature variation over time throughout the day and night. (See the World Meteorological Organization at http://www.wmo.ch/.)

The *daily mean temperature* is an average of daily minimum–maximum readings. The *monthly mean temperature* is the total of daily mean temperatures for the month divided by the number of days in the month. An *annual temperature range* expresses the difference between the lowest and highest monthly mean temperatures for a given year.

Principal Temperature Controls

The interaction of complex physical controls produces Earth's temperature patterns. These principal influences upon temperatures include latitude, altitude, cloud cover, and land–water heating differences.

FIGURE 5.2 A minimum–maximum thermometer.
[Photo by Bobbé Christopherson.]

Latitude

Insolation is the single most important influence on temperature variations. Figure 2.9 shows how insolation intensity decreases as one moves away from the subsolar point—a point that migrates annually between the Tropic of Cancer and Tropic of Capricorn (between 23.5° N and 23.5° S). In addition, daylength and Sun angle change throughout the year, increasing seasonal effect with increasing latitude. The five cities graphed in Figure 5.4 demonstrate the effects of latitudinal position on temperature. From equator to poles, Earth ranges from continually warm, to seasonally variable, to continually cold.

Altitude

Within the troposphere, temperatures decrease with increasing altitude above Earth's surface. (Recall that the *normal lapse rate* of temperature change with altitude is 6.4 C°/1000 m, or 3.5 F°/1000 ft; see Figure 3.5). Thus, worldwide, mountainous areas experience lower temperatures than do regions nearer sea level, even at similar latitudes. The density of the atmosphere also diminishes with increasing altitude. In fact, the density of the atmosphere at an elevation of 5500 m (18,000 ft) is about half of that at sea level. As the atmosphere thins, there is a loss in its ability to absorb and radiate sensible heat.

The consequences are that, at high elevations, average air temperatures are lower, nighttime cooling is greater, and the temperature range between day and night is greater than at low elevations. The temperature difference between areas of sunlight and shadow is greater than at sea level. You may have felt temperatures decrease

noticeably in the shadows and shortly after sunset when you were in the mountains. Surfaces both gain energy rapidly and lose energy rapidly to the thinner atmosphere.

Also, at higher elevations, the insolation received is more intense because of the reduced mass of atmospheric gases. As a result of this intensity, the ultraviolet energy component makes sunburn a distinct hazard.

The snowline seen in mountain areas indicates where winter snowfall exceeds the amount of snow lost through summer melting and evaporation. The snowline's location is a function both of elevation and latitude, so glaciers can exist even at equatorial latitudes if the elevation is great enough. Permanent ice fields and glaciers exist on equatorial mountain summits in the Andes and East Africa. In equatorial mountains, the snowline occurs at approximately 5000 m (16,400 ft) because of the latitude (more insolation in the tropics produces temperatures that place the snowline at higher altitude). With increasing latitude, snowlines gradually lower in elevation from 2700 m (8850 ft) in the midlatitudes to lower than 900 m (2950 ft) in southern Greenland.

Two cities in Bolivia illustrate the interaction of the two temperature controls, latitude and altitude. Figure 5.5 displays temperature data for the cities of Concepción and La Paz, which are near the same latitude (about 16°S). Note the elevation, average annual temperature, and precipitation for each location noted on the figure.

The hot, humid climate of Concepción at its much lower elevation stands in marked contrast to the cool, dry climate of highland La Paz. People living around La Paz actually grow wheat, barley, and potatoes—crops characteristic of the cooler midlatitudes—despite the fact that

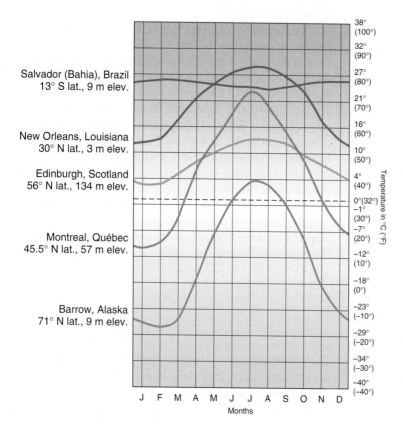

FIGURE 5.4 Latitude affects temperatures.
A comparison of five cities from near the equator to above the Arctic Circle demonstrates changing seasonality and increasing differences between average minimum and maximum temperatures.

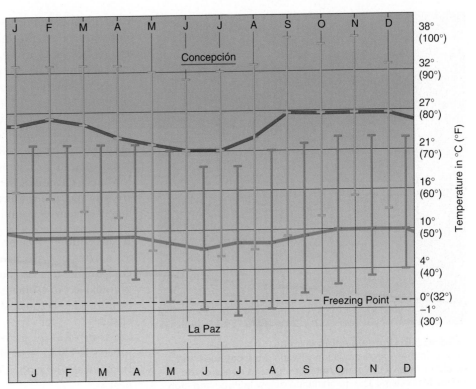

	Station	
	Concepción, Bolivia	**La Paz, Bolivia**
Latitude/longitude	16° 15′ S 62° 03′ W	16° 30′ S 68° 10′ W
Elevation	**490 m (1608 ft)**	**4103 m (13,461 ft)**
Avg. ann. temperature	24°C (75.2°F)	9°C (48.2°F)
Ann. temperature range	5 C° (9 F°)	3 C° (5.4 F°)
Ann. precipitation	121.2 cm (47.7 in.)	55.5 cm (21.9 in.)
Population	10,000	810,300 (Administrative division 1.6 million)

FIGURE 5.5 Effects of latitude and altitude.
Comparison of temperature patterns in two Bolivian cities: La Paz (4103 m; 13,461 ft) and Concepción (490 m; 1608 ft).

La Paz is 4103 m (13,461 ft) above sea level (Figure 5.6). (For comparison, the summit of Pikes Peak in Colorado is at 4301 m or 14,111 ft, and Mount Rainier, Washington, is at 4392 m or 14,410 ft.)

The combination of elevation and low-latitude location guarantees La Paz nearly constant daylength and moderate temperatures, averaging about 9°C (48°F) every month. Such moderate temperature and moisture conditions lead to the formation of more fertile soils than those found in the warmer, wetter climate of Concepción.

Cloud Cover

Orbiting satellites reveal that approximately 50% of Earth is cloud covered at any given moment. Clouds moderate temperature, and their effect varies with cloud type, height, and density, as discussed in Chapter 4. Their moisture reflects, absorbs, and liberates large amounts of energy (released upon condensation and cloud formation). Clouds reduce the insolation that reaches the surface. In general, they lower daily maximum temperatures

and raise nighttime minimum temperatures. Clouds also reduce latitudinal and seasonal temperature differences.

At night, clouds act as insulation and radiate longwave energy, preventing rapid energy loss. During the day, clouds reflect insolation as a result of their high albedo values. In the last chapter, Figures 4.7 and 4.10 portrayed cloud-albedo forcing and cloud-greenhouse forcing as they relate to the presence of clouds and cloud types.

Clouds are the most variable factor influencing Earth's radiation budget, making them the subject of much investigation and simulation in computer models of atmospheric behavior. The International Satellite Cloud Climatology Project, part of the World Climate Research Programme, is presently in the midst of such research. The Earth Radiation Budget experiments (ERB), including the Clouds and the Earth Radiant Energy System (CERES) sensors aboard the *TRMM* and *Terra* satellites, are assessing cloud effects on longwave, shortwave, and net radiation patterns as never before possible. (For more about clouds, see the ISCCP and

FIGURE 5.6 High-elevation farming.
People in these high-elevation villages grow potatoes and wheat in view of the permanent ice-covered peaks of the Andes Mountains. The combination of low latitude and high elevation creates consistent but moderate temperatures throughout the year, averaging about 9°C (48°F). [Photo by Mireille Vautier/Woodfin Camp & Associates.]

WCRP Home Pages at http://isccp.giss.nasa.gov/ and http://www.wmo.ch/web/wcrp/wcrp-home.html.)

Land–Water Heating Differences

Another major control over temperature is the different way in which land and water respond to insolation. Earth presents these two surfaces in an irregular arrangement of continents and oceans. Each absorbs and stores energy differently than the other and therefore contributes to the global pattern of temperature. Water bodies tend to produce moderate temperature patterns, whereas inland in continental interiors more extreme temperatures occur.

The physical nature of land (rock and soil) and water (oceans, seas, and lakes) is the reason for these **land–water heating differences**—land heats and cools faster than water. Figure 5.7 visually summarizes the following discussion of the five land–water temperature controls: evaporation, transparency, specific heat, movement and ocean currents, and sea-surface temperatures.

Evaporation   Evaporation consumes more of the energy arriving at the ocean's surface than is expended over a comparable area of land, simply because so much water is available. An estimated 84% of all evaporation on Earth is from the oceans. When water evaporates and thus changes to water vapor, *heat energy is absorbed in the process and is stored in the water vapor as latent heat.* We saw this process in Figure 4.18, the map of energy expended for the latent heat of evaporation. Chapter 7 fully covers latent heat.

You experience this evaporative heat loss (cooling) by wetting the back of your hand and then blowing on the moist skin. Sensible heat energy is drawn from your skin to supply some of the energy for evaporation, and you feel the cooling. Similarly, as surface water evaporates, it absorbs substantial energy from the immediate environment, resulting in a lowering of temperatures. Temperatures over land, with far less water, are not as moderated by evaporative cooling as are marine locations.

Transparency   The transmission of light obviously differs between soil and water: Solid ground is opaque, water is transparent. Consequently, light striking a soil surface

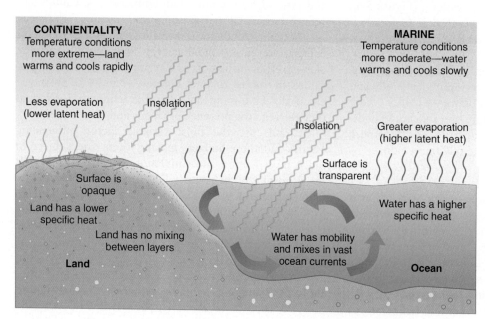

CONTINENTALITY
Temperature conditions more extreme—land warms and cools rapidly

Less evaporation (lower latent heat)

Insolation

Surface is opaque

Land has a lower specific heat

Land has no mixing between layers

Land

Insolation

MARINE
Temperature conditions more moderate—water warms and cools slowly

Greater evaporation (higher latent heat)

Surface is transparent

Water has a higher specific heat

Water has mobility and mixes in vast ocean currents

Ocean

FIGURE 5.7 Land–water heating differences.
The differential heating of land and water produces contrasting marine (more moderate) and continental (more extreme) temperature regimes. (*LE*, latent heat of evaporation, is the energy stored in water vapor.)

FIGURE 5.8 Land is opaque.
Profile of air and soil temperatures in Seabrook, New Jersey. Note that little temperature change occurs throughout the year at depths beyond a meter, whereas the daily extremes of temperature register along the surface, where insolation is absorbed. [Adapted from John R. Mather, *Climatology: Fundamentals and Applications* (New York: McGraw-Hill, 1974), p. 36. Adapted by permission.]

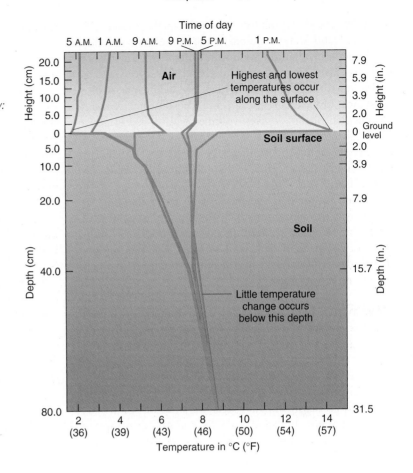

does not penetrate but is absorbed, heating the ground surface. That energy is accumulated during times of exposure and is rapidly lost at night or in shadows.

Figure 5.8 shows the profile of diurnal (daily) temperatures for a column of soil and the atmosphere above it at a midlatitude location. You can see that maximum and minimum temperatures generally are experienced right at ground level. Below the surface, even at shallow depths, temperatures remain about the same throughout the day. You encounter this at a beach, where surface sand may be painfully hot to your feet, but as you dig in your toes and feel the sand a few centimeters below the surface, it is cooler, offering relief.

In contrast, when light reaches a body of water, it penetrates the surface because of water's **transparency**—water is clear and light transmits through it to an average depth of 60 m (200 ft) in the ocean (Figure 5.9). This illuminated zone is known as the *photic layer* and has been recorded in some ocean waters to depths of 300 m (1000 ft). This characteristic of water results in the distribution of available heat energy over a much greater depth and volume, forming a larger energy reservoir than that of the surface layers of the land.

FIGURE 5.9 Ocean is transparent.
The transparency of the ocean at Taveuni Reef near Fiji permits insolation to penetrate to average depths of 60 m (200 ft), greatly increasing the volume of water that absorbs energy.
[Photo by Copr. F. Stuart Westmorland/Photo Researchers, Inc.]

Specific Heat When comparing equal volumes of water and land, water requires far more energy to increase its temperature than does land. In other words, *water can hold more heat than can soil or rock*, and therefore water is said to have a higher **specific heat**, the heat capacity of a substance. On average, the specific heat of water is about 4 times that of soil. A given volume of water represents a more substantial energy reservoir than an equal volume of land, so changing the temperature of the oceanic energy reservoir is a slower process than changing the temperature of land.

Likewise, for that oceanic heat reservoir to lose heat energy requires more time than would a similar volume of land. The temperature response of water bodies is "sluggish" in comparison with land surfaces. For this reason, day-to-day temperatures near a substantial body of water tend to be moderated.

Movement Land is a rigid, solid material, whereas water is a fluid and is capable of movement. Differing temperatures and currents result in a mixing of cooler and warmer waters, and that mixing spreads the available energy over an even greater volume than if the water were still. Surface water and deeper waters mix, redistributing energy. Both ocean and land surfaces radiate longwave radiation at night, but land loses its energy more rapidly than does the greater mass of the moving oceanic energy reservoir.

Ocean Currents and Sea-Surface Temperatures
Warm water adds energy to overlying air through high evaporation rates and transfers of latent heat. Thus, in an air mass, the amount of water vapor that ocean temperature affects forms an interesting *negative feedback* mechanism. Across the globe, ocean water is rarely found warmer than 31°C (88°F). Higher ocean temperatures produce higher evaporation rates and more energy is lost from the ocean as latent heat. As the water vapor content of the overlying air mass increases, the ability of the air to absorb longwave radiation also increases. Therefore, the air mass becomes warmer, enhancing the greenhouse effect. The warmer the air and the ocean become, the more evaporation will occur and the more water vapor will enter the air mass. More water vapor leads to cloud formation, which reflects insolation and produces lower temperatures. Lower temperatures of air and ocean reduce evaporation rates and the ability of the air mass to absorb water vapor.

As a specific example, the **Gulf Stream** (described in Chapter 6) moves northward off the east coast of North America, carrying warm water far into the North Atlantic (Figure 5.10). As a result, the southern third of Iceland experiences much milder temperatures than would be expected for a latitude of 65° N, just below the Arctic Circle (66.5°). In Reykjavík, on the southwestern coast of Iceland, monthly temperatures average above freezing during all months of the year. Similarly, the Gulf Stream moderates

FIGURE 5.10 The Gulf Stream. Satellite image of the warm Gulf Stream as it flows northward along the North American eastern coast. It is the streamlike flow in false-color red, orange, and yellow. Instruments sensitive to infrared wavelengths produced this remote-sensing image, which covers approximately 11.4 million km² (4.4 million mi²). Temperature differences are distinguished by computer-enhanced false coloration: reds/oranges = 25°–29°C (76°–84°F), yellows/greens = 17°–24°C (63°–75°F); blues = 10°–16°C (50°–61°F); and purples = 2°–9°C (36°–48°F). [Imagery by Rosenstiel School of Marine and Atmospheric Science, University of Miami.]

(a) February

FIGURE 5.11 Sea-surface temperatures.
Average annual sea-surface temperatures for February (a) and July (b) 1999 from satellites in the NOAA/NASA Pathfinder AVHRR data set, aboard *NOAA-7, -9, -11,* and *-14.* The Western Pacific Warm Pool, the warmest area of all oceans, is well defined. These remotely sensed data are closely correlated with actual measurements of the ocean's surface temperature. [Satellite image data from the NASA Physical Oceanography Distributed Active Archive Center at the Jet Propulsion Laboratory, California Institute of Technology.]

Global Sea-surface Temperatures

SATELLITE

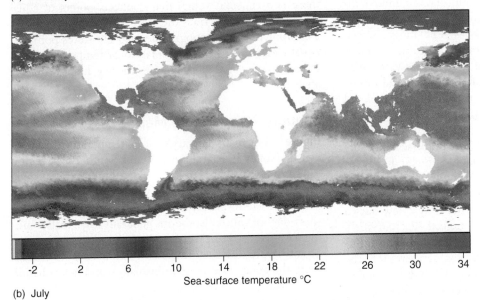

-2   2   6   10   14   18   22   26   30   34
Sea-surface temperature °C

(b) July

temperatures in coastal Scandinavia and northwestern Europe. In the western Pacific Ocean, the Kuroshio, or Japan Current, similar to the Gulf Stream, functions much the same in its warming effect on Japan, the Aleutians, and the northwestern margin of North America.

In contrast, along midlatitude and subtropical west coasts, cool ocean currents flowing toward the equator influence air temperatures. When conditions in these regions are warm and moist, fog frequently forms in the chilled air over the cooler currents. Chapter 6 discusses ocean currents.

Remote sensing from satellites is providing sea-surface temperature (SST) data that are well correlated with actual sea-surface measurements. This correlation permits a thorough global assessment of SSTs in programs such as Tropical Ocean Global Atmosphere (TOGA) and Coupled Ocean-Atmosphere Response Experiment (COARE). (See http://lwf.ncdc.noaa.gov/oa/coare/index.html)

The Physical Oceanography Distributed Active Archive Center (PODAAC, http://podaac.jpl.nasa.gov/), part of the Earth Observing System Data Information

System (EOSDIS, http://eosdismain.gsfc.nasa.gov/eosinfo/welcome/), is responsible for storage and distribution of data relevant to the physical state of the ocean. Sea-level height, currents, and ocean temperatures are measured as part of the effort to understand oceans and climate interactions.

Figure 5.11 displays SST data for February and July 1999 from satellites in the NOAA/NASA Pathfinder AVHRR data set, aboard *NOAA-7, -9, -11,* and *-14.* The Western Pacific Warm Pool is the red and maroon area in the southwestern Pacific Ocean (north of New Guinea) with temperatures above 30°C (86°F). This region has the highest average ocean temperatures in the world. Note the seasonal change in ocean temperatures; for example, compare the waters around Australia on the two images. Mean annual SSTs increased steadily from 1982 through 2004 to record levels for the past several hundred years. Increasing warmth is measured at depths to 1000 m (3200 ft) and even slight increases were found in the temperature of deep bottom water as reported in 2004. Scientists suggest that the ocean's ability to absorb excess heat energy from the atmosphere may be nearing its capacity.

Summary of Marine Effects vs. Continental Effects

As noted, Figure 5.7 summarizes the operation of the land–water temperature controls presented: evaporation, transparency, specific heat, movement, ocean currents, and sea-surface temperatures. The term **marine effect**, or *maritime*, describes locations that exhibit the moderating influences of the ocean, usually along coastlines or on islands. **Continental effect**, a condition of *continentality*, refers to areas less affected by the sea and therefore having a greater range between maximum and minimum temperatures on both a daily and yearly basis.

The Canadian cities of Vancouver, British Columbia, and Winnipeg, Manitoba, exemplify marine and continental conditions (Figure 5.12). Both cities are at approximately 49° N latitude. However, Vancouver has a more moderate pattern of average maximum and minimum temperatures. Vancouver's annual range of 16.0 C° (28.8 F°) is far less than Winnipeg's 38.0 C° (68.4 F°) range. In fact, Winnipeg's continental temperature pattern is more extreme in every aspect than that of maritime Vancouver.

A similar comparison of San Francisco, California, and Wichita, Kansas, provides us another example of marine and continental conditions (Figure 5.13). Both cities are at approximately 37° 40′ N latitude. The cooling waters of the Pacific Ocean and San Francisco Bay surround San Francisco on three sides. Summer fog helps delay the warmest summer month in San Francisco until September. In 100 years of weather records, marine effects moderated temperatures so that only a few days a year have summer maximums that exceed 32.2°C (90°F). Winter minimums rarely drop below freezing. This influence changes, however, as ocean temperatures off the coast rose to record levels during El Niño episodes, slightly reducing its moderating effects so coastal temperatures were a bit higher—20°C (68°F), some 7 C° (12 F°) above normal in the 1998 El Niño (more on El Niños in Chapter 10).

In contrast, Wichita is susceptible to freezes from late October to mid-April, with daily variations slightly increased by its elevation, experiencing −30°C (−22°F) as a record low. Wichita's temperature reaches 32.2°C (90°F) or higher at least 65 days each year, with 46°C (114°F) as a record high. In 1980, 1990, and again in 2000, temperatures exceeded 38°C (100°F) two to three weeks in a row—21 nonconsecutive days reached over 100°C in 2003. West of Wichita, winters increase in severity with increasing distance from the moderating influences of invading air masses from the Gulf of Mexico.

Earth's Temperature Patterns

Earth's temperature patterns result from the combined effect of the controlling factors in our discussion. Let us now look at temperatures portrayed on maps that show worldwide mean air temperatures and on maps focused on the polar regions for January (Figure 5.14) and July (Figure 5.16). To complete the analysis, Figure 5.17 presents the temperature range differences between the January and July maps, or the difference between averages of the coolest and warmest months.

We use maps for January and July instead of the solstice months of December and June because a lag occurs between insolation received and maximum or minimum temperatures experienced, as explained earlier in Chapter 4. The U.S. National Climate Data Center provided the data for this map preparation. Some ship reports go back to 1850 and land reports to 1890, although the bulk of the record is representative of conditions since 1950. An important consideration in using these maps is to remember that their small scale permits only generalizations about actual temperatures at specific locations.

The lines on temperature maps are known as *isotherms*. An **isotherm** is an isoline (a line along which there is a constant value) that connects points of equal temperature and portrays the temperature pattern, just as a contour line on a topographic map illustrates points of equal elevation. The spatial analysis of temperatures is a concern of geographers, and isotherms help with this analysis.

January Temperature Map

Figure 5.14a maps January's mean temperatures for the world, North Polar region (b), and South Polar region (c). In the Southern Hemisphere, the higher Sun altitude causes longer days and summer weather conditions; in the Northern Hemisphere, the lower Sun angle causes the short days of winter. Isotherms generally are zonal, trending east–west, parallel to the equator, and they appear to be interrupted by the presence of landmasses. Isotherms mark the general decrease in insolation and net radiation with distance from the equator.

The **thermal equator** (an isotherm connecting all points of highest mean temperature, roughly 27°C, 80°F) trends southward into the interior of South America and Africa, indicating higher temperatures over landmasses. In the Northern Hemisphere, isotherms shift equatorward as cold air chills the continental interiors. The oceans, in contrast, are more moderate, with warmer conditions extending farther north than over land at comparable latitudes.

As an example in Figure 5.14a, follow along 50° north latitude (the 50th parallel) and compare isotherms: 3° to 6°C in the North Pacific and 3° to 9°C in the North Atlantic, as contrasted to −18°C in the interior of North America and −24° to −30°C in central Asia. Also, note the orientation of isotherms over areas where there are mountain ranges and how they illustrate the cooling effects of altitude. Check the South American Andes as an example of the effects of altitude.

On the polar map in Figure 5.14b, the island of Greenland has a summit elevation of 3240 m (10,630 ft) on Earth's second-largest ice sheet, the highest elevation north of the Arctic Circle. Two-thirds of the island is north of the Arctic Circle. Its north shore is only 800 km (500 mi) from the North Pole. This combination of high latitude and interior high altitude on the ice sheet produces cold midwinter temperatures.

FIGURE 5.12 Marine and continental cities—Canada.
Comparison of temperatures in coastal Vancouver, British Columbia, and continental Winnipeg, Manitoba. Note that the freezing levels on the two graphs are positioned differently to accommodate the contrasting data.
[Vancouver waterfront photo by author; Winnipeg photo by John Eastcott/Yva Momatiak/The Image Works.]

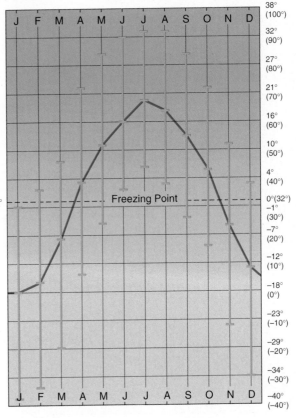

Station: Vancouver, British Columbia
Lat/long: 49° 11′ N 123° 10′ W
Avg. ann. temp.: 10°C (50°F)
Total ann. precip.:
  104.8 cm (41.3 in.)
Elevation: sea level
Population: 520,000
Ann. temp. range:
  16 C° (28.8 F°)

Station: Winnipeg, Manitoba
Lat/long: 49° 54′ N 97° 14′ W
Avg. ann. temp.: 2°C (35.6°F)
Total ann. precip.:
  51.7 cm (20.3 in.)
Elevation: 248 m (813.6 ft)
Population: 620,000
Ann. temp. range:
  38 C° (68.4 F°)

Station: San Francisco, California
Lat/long: 37° 37′ N 122° 23′ W
Avg. ann. temp.: 14°C (57.2°F)
Total ann. precip.:
    47.5 cm (18.7 in.)

Elevation: 5 m (16.4 ft)
Population: 777,000
Ann. temp. range:
    9 C° (16.2 F°)

Station: Wichita, Kansas
Lat/long: 37° 39′ N 97° 25′ W
Avg. ann. temp.: 13.7°C (56.6°F)
Total ann. precip.:
    72.2 cm (28.4 in.)

Elevation: 402.6 m (1321 ft)
Population: 327,000
Ann. temp. range:
    27 C° (48.6 F°)

FIGURE 5.13 Marine and continental cities—United States.
Comparison of temperatures in coastal San Francisco, California, and continental Wichita, Kansas.
[San Francisco photo by Bobbé Christopherson; Wichita region photo by Philip Gould/Corbis.]

FIGURE 5.14 Global mean temperatures for January.
Temperatures are in Celsius (convertible to Fahrenheit by means of the scale) as taken from separate air-temperature data bases for ocean and land. Note the inset map of North America and the equatorward-trending isotherms in the interior—(a) world map, (b) North Polar region, and (c) South Polar region. **Compare with Figure 5.16.** [Adapted and redrawn from National Climatic Data Center, *Monthly Climatic Data for the World,* 47 (January 1994). Prepared in cooperation with the World Meteorological Organization, Washington, DC, National Oceanic and Atmospheric Administration.]

Global Surface Temperatures,
Land and Ocean

(c)

(a)

(b)

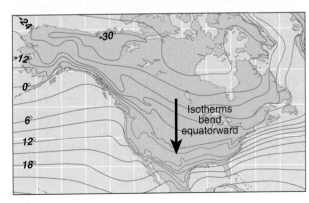

In Antarctica, "summer" is underway in December. This is Earth's coldest and highest average elevation land-mass (Figure 5.14c). January average temperatures for the three scientific bases noted on the map are: McMurdo Station, at the coast on Ross Island, averages −3°C; the Admundsen–Scott Station at the South Pole (elevation 2835 m, 9301 ft) averages −28°C; and the Russian Vostok Station at its more continental location (elevation 3420 m, 11,220 ft) averages −32°C; respectively, 26.6°F, −18.4°F, and −25.6°F.

For a continental region other than Antarctica, Russia is the coldest area on the map, specifically northeastern Siberia. The intense cold results from winter conditions of consistent clear, dry, calm air, small insolation input, and an inland location far from moderating maritime effects. Verkhoyansk, Russia (located within the −48°C isotherm on the map), actually recorded a minimum temperature of −68°C (−90°F) and experiences a daily average of −50.5°C (−58.9°F) for January. Verkhoyansk (Figure 5.15) has 7 months of temperatures below freezing, including at least 4 months below −34°C (−30°F). In contrast, this town has hit a maximum temperature of +37°C (+98°F) in July—an incredible 105°C (189 F°) min–max range! People do live and work in Verkhoyansk, which has a population of 1400; the town has been occupied continuously since 1638 and is today a secondary mining district.

Trondheim, Norway, is near the latitude of Verkhoyansk and at a similar elevation. But, Trondheim's coastal location moderates its annual temperature regime (Figure 5.15). January minimum and maximum temperatures range between −17° and +8°C (+1.4° and +46°F), and the minimum–maximum range for July is from +5° to +27°C (+41° to +81°F). The most extreme minimum and maximum temperatures ever recorded in Trondheim are −30° and +35°C (−22° and +95°F)—quite a difference from the continentality extremes at Verkhoyansk.

July Temperature Map

Average July temperatures are presented in Figure 5.16a for the world, North Polar region (b), and South Polar region (c). The longer days of summer and higher Sun altitude are in the Northern Hemisphere. Winter dominates the Southern Hemisphere, although it is milder than winters north of the equator because continental land-masses are smaller and the dominant oceans and seas store and release more energy. The thermal equator shifts northward with the high summer Sun and reaches the Persian Gulf–Pakistan–Iran area. The Persian Gulf is the site of the highest recorded sea-surface temperature of 36°C (96°F), difficult to imagine for a large water body.

During July in the Northern Hemisphere, isotherms shift poleward over land, as higher temperatures dominate continental interiors. July temperatures in Verkhoyansk average more than 13°C (56°F), which represents a 63 C° (113 F°) seasonal range between winter and summer averages. The Verkhoyansk region of Siberia is probably Earth's most dramatic example of continental effects on temperature.

The hottest places on Earth occur in Northern Hemisphere deserts during July. The reasons are simple: clear skies, strong surface heating, virtually no surface water, and few plants. Prime examples are portions of the Sonoran Desert of North America and the Sahara of Africa. Africa recorded a shade temperature higher than 58°C (136°F), a record set on September 13, 1922, at Al 'Azīzīyah, Libya (32° 32′N; 112 m, or 367 ft elevation).

The highest maximum and annual average temperatures in North America occurred in Death Valley, California, where the Greenland Ranch Station reached 57°C (134°F) in 1913. (The station is at 37° N and is −54.3 m, or −178 ft, below sea level.) Such hot, arid lands are discussed further in Chapter 15.

July is a time when nights in Antarctica are 24 hours long. This lack of insolation results in the lowest natural temperature reported on Earth, a frigid −89.2°C (−128.6°F), recorded on July 21, 1983, at the Russian Vostok Station, Antarctica (Figure 5.16c). Such a temperature is 11 C° (19.8 F°) colder than the freezing point of dry ice (solid carbon dioxide)! If the concentration of carbon dioxide were large enough, such a cold temperature would theoretically freeze tiny carbon dioxide dry-ice particles out of the sky.

The average July temperature around Vostok is −68°C (−90.4°F). For average temperature comparisons, the Admundsen–Scott Station at the South Pole experiences −60°C (−76°F) and McMurdo Station has a −26°C (−14.8°F) reading. Note that the coldest temperatures in Antarctica are usually in August, not July, just before the equinox sunrise in September at the end of the long polar night. High Latitude Connection 5.1 presents an overview of temperature trends in the polar regions.

Annual Temperature Range Map

The temperature range map helps identify areas that experience the greatest annual extremes (continental locations) and the most moderate temperature regimes (marine locations), as demonstrated in Figure 5.17. As you might expect, the largest temperature ranges occur in subpolar locations in North America and Asia, where average ranges of 64 C° (115 F°) are recorded (dark brown area on the map). The Southern Hemisphere, in contrast, has little seasonal variation in mean temperatures, owing to the lack of large landmasses and the vast expanses of water to moderate temperature extremes.

Southern Hemisphere temperature patterns are generally maritime and Northern Hemisphere patterns feature continentality. For example, in January (Figure 5.14), Australia is dominated by isotherms of 20°–30°C (68°–86°F), whereas in July (Figure 5.16), Australia is crossed by the 12°C (54°F) isotherm. The Northern Hemisphere, with greater land area overall, registers a slightly higher average surface temperature than does the Southern Hemisphere.

Imagine living in some of these regions and the degree to which you would need to adjust your personal wardrobe and make other comfort adaptations. See Focus Study 5.1 for more on air temperature and the human body. Meanwhile, the recent summer heat-wave deaths remain a powerful reminder of the role of temperatures in our lives. (Focus Study 5.1, "Air Temperature and the Human Body," begins on page 133.)

Station: Trondheim, Norway
Lat/long: 63° 25′ N 10° 27′ E
Avg. ann. temp.: 5°C (41°F)
Total ann. precip.:
  85.7 cm (33.7 in.)
Elevation: 115 m (377.3 ft)
Population: 139,000
Ann. temp. range:
  17 C° (30.6 F°)

Station: Verkhoyansk, Russia
Lat/long: 67° 35′ N 135° 23′ E
Avg. ann. temp.: −15°C (5°F)
Total ann. precip.:
  15.5 cm (6.1 in.)
Elevation: 137 m (449.5 ft)
Population: 1400
Ann. temp. range:
  63 C° (113.4 F°)

FIGURE 5.15 Marine and continental cities—Eurasia.
Comparison of temperatures in coastal Trondheim, Norway, and continental Siberian Russia. Note that the freezing levels on the two graphs are positioned differently to accommodate the contrasting data. [Trondheim photo by Norman Benton/Peter Arnold, Inc.; Verkhoyansk photo by TASS/Sovfoto/Eastfoto.]

FIGURE 5.16 Global mean temperatures for July.
Temperatures are in Celsius (convertible to Fahrenheit by means of the scale) as taken from separate air-temperature data bases for ocean and land. Note the inset map of North America and the poleward-trending isotherms in the interior—(a) world map, (b) North Polar region, and (c) South Polar region. Compare with Figure 5.14.
[Adapted and redrawn from National Climatic Data Center, *Monthly Climatic Data for the World*, 47 (July 1994). Prepared in cooperation with the World Meteorological Organization, Washington, DC, National Oceanic and Atmospheric Administration.]

Global Surface Temperatures, Land and Ocean

(c)

(a)

(b)

High Latitude Connection 5.1

Overview of Trends in the Polar Regions

Climate models used by scientists to predict and direct measurements show that the polar regions are experiencing some dramatic global climate change impacts. These changes lead some to say that the high latitudes are providing the global community an early warning signal. The question is, how do we read changing temperature signals, how do they relate to natural circulation cycles, and what are the spatial implications to the rest of the planet?

Arctic

Many proxy and direct measurements confirm that the twentieth century was a time of increasing temperatures at both middle and low latitudes and in the polar regions, especially during the last half century. Daily maximum and minimum temperatures show there is warming in the Arctic at a rate of about 5 C° (9 F°) per century. Since 1978, warming increased to a rate of 1.2 C° (2.2 F°) per decade, which means the last 20 years warmed at nearly 7 times the rate of the last 100 years. The trend is not uniform across the Arctic polar region, with areas of slight cooling in eastern Canada, the North Atlantic, and portions of Greenland, although an overall warming dominates.

In response to rising air and water temperatures, Arctic sea ice is on a significant decline. The sea ice is melting at a rate of about 9% per decade since 1978, especially in the eastern and central Arctic Ocean. Atmospheric pressure patterns cause sea ice to shift as well. Record losses marked 2002 and 2003, as sea ice reached its smallest spatial distribution in the satellite record. Figure 5.1.1 is a composite satellite image for 1979 and 2003 comparing sea-ice coverage in the Arctic. The sea ice thins as warmer air melts the surface and warmer waters attack the base of the ice. The summer melt period is increasing as spring arrives earlier and fall is delayed. Since 1970, about 43% of the Arctic sea ice has disappeared.

1979

2003

FIGURE 5.1.1 Loss of Arctic sea ice.
A composite of satellite data (DMSP, Special Sensor Microwave Imager) and surface data from 1979 and 2003 shows remarkable loss of Arctic sea ice. Each composite is a portrait of minimum sea-ice concentrations during summer months. For scale, the amount of sea-ice loss is approximately the size of Arizona, or 300,000 km^2 (116,000 mi^2)—approximately 43% of sea ice has disappeared since 1970. Not only did the atmosphere record increased warmth but the Arctic Ocean warmed as well. In addition to sea ice losses, changing pressure patterns cause it to shift. [Images courtesy of Scientific Visualization Studio, GSFC/NASA.]

More open water means darker surfaces that absorb more insolation that leads to more warming.

Chapters 1 and 17 discuss the 400% increase in meltponds on the surface of the Greenland ice sheet, ice shelves, and glaciers. This increase represents a significant positive feedback mechanism. In response to the increased discharge of freshwater in the Arctic, the northern oceans are freshening—actually, a lowering in salinity. There is valid concern that these meltwaters riding atop saline water can disrupt important circulation systems in the northern oceans.

In Alaska, temperatures rose over the past 30 years at a 3 C° (5 F°) per

(continued)

High Latitude Connection 5.1 *(continued)*

decade rate. All 11 mountain ranges experienced glacial ice losses. Glacial thinning reached 1.8 m per year by the late 1990s across Alaska. In response to this unloading of ice mass, Earth's crust is rebounding upward. A News Report in Chapter 17 explains the negative net mass balance of 67 Alaskan glaciers.

All of these changes are affecting living systems as specific high-latitude ecosystems shift across the landscape poleward at 6.1 km (3.8 mi) per decade and spring arrives earlier at a pace of 2.3 days per decade. Changes in tundra soils and the increased depth of active-layer melting in permafrost indicate regional warming, as discussed in Chapter 17. These thawing lands provide less nutrition to caribou and other grazing animals, worsen forest fire conditions, and are an additional source of carbon dioxide to the atmosphere.

By 2080, the Intergovernmental Panel on Climate Change (IPCC) forecasts increasing warmth in the Arctic ranging from 4.0° to 7.5 C° in summer and 2.5° to 14.0 C° in winter over land; and 0.5° to 4.5 C° in summer and 3.0° to 16.0 C° in winter over the Arctic Ocean. With these temperature changes precipitation will increase since warmer air can contain more moisture.

Antarctica

In Antarctica and the Southern Ocean, the temperature records of the permanently occupied research stations over the last 45 years demonstrate an overall warming. This record includes periods of slight cooling within this time frame. The present warming rate is less than that of the Arctic, at about a 0.9° to 1.2 C° increase per century.

However, along the Antarctic Peninsula, the location of numerous ice-shelf collapses and disintegrations, the rate of warming is more than twice this rate, as the January 0° isotherm continues to shift southward. The exposed land is open to

exotic plants and animals that move into these altered ecosystems. In terms of sea ice, the Antarctic trend is not as pronounced as that of the Arctic, with sea-ice extent reduced by 0.4°–1.8° of latitude over the last century. Continued changes will affect marine life and birds.

By 2080, the IPCC forecasts increasing warmth in the Antarctic ranging from 1.0° to 4.8 C° in summer and 1.0° to 5.0 C° in winter over the continent; and 0.0° to 2.8 C° in summer and 0.5° to 5.0 C° in winter over the Southern Ocean.

For more information on both polar regions, please see: Chapter 16, "Polar Regions (Arctic and Antarctic)," in *Climate Change 2001, Impacts, Adaptation, and Vulnerability, Contribution of Working Group II to the Third Assessment Report of the Intergovernmental Panel on Climate Change* (New York: Cambridge University Press, © 2001), pp. 801–41. The web site is at http://www.ipcc.ch/.

F°	5	9	18	27	36	45	54	63	72	81	90	99	108	F°
C°	3	5	10	15	20	25	30	35	40	45	50	55	60	C°

FIGURE 5.17 Global annual temperature ranges.
A generalized portrait of the annual range of global temperatures in Celsius degrees (conversions to Fahrenheit degrees shown on scale). The mapped data show the difference between January and July temperature means.

Air Temperature and the Human Body

We describe our perception of temperature with the term *apparent temperature*, or *sensible temperature*. This perception varies among individuals and cultures. Through complex mechanisms, our bodies maintain an average internal temperature within a degree of 36.8°C (98.2°F), slightly lower in the morning or in cold weather and slightly higher at emotional times or during exercise and work.*

Taken together, the water vapor content of air, wind speed, and air temperature affect each individual's sense of comfort. The most discomfort comes with high temperatures, high humidity, and low winds. More comfort comes with low humidity, moderate temperatures, and light winds.

Although modern heating and cooling systems, where available, reduce the impact of extreme temperatures indoors, the danger to human life from excessive heat or cold persists. Deaths still occur, especially among groups of people who can't afford to alter temperatures in their living spaces.

When changes occur in the surrounding air, the human body reacts to maintain its core temperature and to protect the brain at all costs. Table 1 summarizes the human body's response to stress induced by exposure to *hypothermia* (low temperature) and *hyperthermia* (high temperature). Carefully review these two lists of responses for future reference.

Wind Chill

The wind-chill index is important to people who experience winters with freezing temperatures and was first proposed by Antarctic explorer Paul

*The traditional value for "normal" body temperature, 37°C (98.6°F), was set in 1868 using old methods of measurement. According to Dr. Philip Mackowiak of the University of Maryland School of Medicine, a more accurate modern assessment places the normal value at 36.8°C (98.2°F), with a range of 2.7 C°(4.8 F°) for the human population overall (*Journal of the American Medical Association*, September 23–30, 1992).

Table 1	The Human Body's Response to Temperature Stress
At Low Temperatures (Heat-gaining mechanism)	**At High Temperatures (Heat-dissipation mechanism)**

Temperature-Regulation Methods

At Low Temperatures (Heat-gaining mechanism)	At High Temperatures (Heat-dissipation mechanism)
Constriction of surface blood vessels	Dilation of surface blood vessels
Concentration of blood	Dilution of blood
Flexing to reduce surface exposure	Extending to increase exposure
Increased muscle tone	Decreased muscle tone
Decrease in sweating	Sweating
Inclination to increase activity	Inclination to decrease activity
Shivering	
Increased cell metabolism	

Consequent Disturbances

At Low Temperatures	At High Temperatures
Increased urine volume	Decreased urine volume
Danger of inadequate blood supply to exposed parts; frostbite	Reduced blood supply to brain; dizziness, nausea, fainting
Discomfort leading to neuroses	Discomfort leading to neuroses
Increased appetite	Decreased appetite
	Mobilization of tissue fluid
	Thirst and dehydration
	Reduced chloride balance; heat cramps

Failure of Regulation

At Low Temperatures	At High Temperatures
Falling body temperature	Rising body temperature
Drowsiness	Impaired heat-regulating center
Cessation of heartbeat and respiration	Failure of nervous regulation; cessation of breathing

Source: Climatology—Arid Zone Research X. © UNESCO 1958. Reproduced by permission of UNESCO.

Siple in 1939. The *wind-chill factor* indicates the enhanced rate at which body heat is lost to the air and at best represents an estimate of heat-energy loss. As wind speeds increase, heat loss from the skin increases. A formula for calculating these relationships was developed in 1945, and the National Weather Service (NWS) began reporting wind-chill temperatures in 1970. This older index tended to overestimate heat loss from skin.

The NWS and the Meteorological Services of Canada (MSC, http://www.msc-smc.ec.gc.ca/) revised the wind-chill formula and standard assumptions for the 2001–2002 winter season. The new Wind Chill

Temperature (WCT) Index is an effort to improve the accuracy of heat-loss calculations. Computer modeling, clinical trials and testing, and advances in technology make this revision possible. Figure 5.1.1 is the new version that went into service in November 2001.

For example, using the new chart, if the air temperature is −7°C (20°F) and the wind is blowing at 32 kmph (20 mph), skin temperatures will be at −16°C (4°F). The lower wind-chill values present a serious freezing hazard to exposed flesh. Imagine the wind chill experienced by a downhill

(continued)

Focus Study 5.1 *(continued)*

	Actual Air Temperature in °C (°F)								
Calm	4° (40°)	−1° (30°)	−7° (20°)	−12° (10°)	−18° (0°)	−23° (−10°)	−29° (−20°)	−34° (−30°)	−40° (−40°)
8 (5)	2° (36°)	−4° (25°)	−11° (13°)	−17° (1°)	−24° (−11°)	−30° (−22°)	−37° (−34°)	−43° (−46°)	−49° (−57°)
16 (10)	1° (34°)	−6° (21°)	−13° (9°)	−20° (−4°)	−27° (−16°)	−33° (−28°)	−41° (−41°)	−47° (−53°)	−54° (−66°)
24 (15)	0° (32°)	−7° (19°)	−14° (6°)	−22° (−7°)	−28° (−19°)	−36° (−32°)	−43° (−45°)	−50° (−58°)	−57° (−71°)
32 (20)	−1° (30°)	−8° (17°)	−16° (4°)	−23° (−9°)	−30° (−22°)	−37° (−35°)	−44° (−48°)	−52° (−61°)	−59° (−74°)
40 (25)	−2° (29°)	−9° (16°)	−16° (3°)	−24° (−11°)	−31° (−24°)	−38° (−37°)	−46° (−51°)	−53° (−64°)	−61° (−78°)
48 (30)	−2° (28°)	−9° (15°)	−17° (−1°)	−24° (−12°)	−32° (−26°)	−39° (−39°)	−47° (−53°)	−55° (−67°)	−62° (−80°)
56 (35)	−2° (28°)	−10° (14°)	−18° (0°)	−26° (−14°)	−33° (−27°)	−41° (−41°)	−48° (−55°)	−56° (−69°)	−63° (−82°)
64 (40)	−3° (27°)	−11° (13°)	−18° (−1°)	−26° (−15°)	−34° (−29°)	−42° (−43°)	−49° (−57°)	−57° (−71°)	−64° (−84°)
72 (45)	−3° (26°)	−11° (12°)	−19° (−2°)	−27° (−16°)	−34° (−30°)	−42° (−44°)	−50° (−58°)	−58° (−72°)	−66° (−86°)
80 (50)	−3° (26°)	−11° (12°)	−19° (−3°)	−27° (−17°)	−35° (−31°)	−43° (−45°)	−51° (−60°)	−59° (−74°)	−67° (−88°)

Wind speed, kmph (mph)

Frostbite times: 30 min.   ▨ 10 min.   ▨ 5 min.

FIGURE 5.1.1 Wind Chill Temperature Index.
WCT Index factor for various temperatures and wind speeds. A new version of the wind-chill chart—effective November 2001. For quick calculations see http://205.156.54.206/om/windchill/index.shtml#calculator (in English units the new formula is: Wind chill (F°) $= 35.74 + 0.6125T - 35.75(V^{0.16}) + 0.4275T(V^{0.16})$, where T = air temperature in °F, V = wind speed in mph. In Canada, heat loss is expressed in watts per square meter.) [Adapted from the National Weather Service and Meteorological Services of Canada, version 11/01/01.]

ski racer going 130 kmph (80 mph)—frostbite is a definite possibility during a 2-minute run. The wind chill does omit consideration of sunlight intensity, a person's physical activity, and the use of protective clothing, such as a windbreaker, that prevents the wind access to your skin.

Heat Index

The *heat index* (HI) indicates the human body's reaction to the combination of air temperature and water vapor. The heat index, combining the effects of temperature and humidity, indicates how hot the air feels to an average person. Water vapor in air is expressed as relative humidity, a concept presented in Chapter 7. For now, it is sufficient to say that the amount of water vapor in the air affects the

evaporation rate of perspiration on the skin. The reason is simply available space: The more water vapor (higher humidity), the less space the air has to absorb water vapor evaporated from perspiration.

Figure 5.1.2 is an abbreviated version of the heat index that the NWS now includes in its daily weather summaries during summer months. The table beneath the graph describes the effects of heat-index categories on higher-risk groups. A combination of high temperature and high humidity can severely reduce the body's natural ability to regulate internal temperature (see http://weather.noaa.gov/weather/hwave.html).

Summer Heat-Index Deaths   For
nearly a week during July 1995,

Chicago's heat-index values went to category I in dwellings that lacked air-conditioning. Chicago had never experienced a 48-hour period during which temperatures did not go below 31.5°C (89°F).

With high pressure (hot, stable air) dominating the Midwest to the Atlantic and moist air from the Gulf of Mexico, all the ingredients for this disaster were present. Afternoon temperatures were above 32°C (90°F) for a week in Chicago, 38°C (100°F) in South Dakota, and 40°C (104°F) in Toledo, Ohio, and New York City; the temperature hit 43°C (109°F) in Omaha, Nebraska. Cities in New England that had reached 37.7°C

(continued)

Focus Study 5.1 *(continued)*

FIGURE 5.1.2 Heat index.
Heat-index graph for various
temperatures and relative humidity levels.
[Courtesy of the National Weather Service.]

(100°F) only twice in their history exceeded that for 4 or 5 days during July 1995.

This combination of high temperatures and water vapor in the air produced stifling conditions for everyone, the sick and elderly in particular. On July 13, some apartments without air-conditioning exceeded indoor heat-index temperatures of 54°C (130°F) when the official temperature at Midway Airport reached a record 41°C (106°F). The heat-index death toll totaled 700 people in Chicago. Nearly 1000 people died overall in the Midwest and East from these conditions.

Europe was paralyzed in the summer of 2003 with high heat-index conditions when temperatures topped 40°C (104°F) in June, July, and August. Official estimates across the continent placed the death count at more than 20,000, but this triggered a great undercount controversy, as governments were ill prepared to adequately deal with such a catastrophe.

France was hardest hit with more than 14,000 deaths, some 11,000 in

Level of concern	Category	Heat Index Apparent Temperature	General Effect of Heat Index on People in High-Risk Groups
Extreme danger	I	54°C (130°F) or higher	Heat/sunstroke highly likely with continued exposure
Danger	II	41° – 54°C (105° – 130°F)	Sunstroke, heat cramps, or heat exhaustion likely and heatstroke possible with prolonged exposure and/or physical activity
Extreme caution	III	32° – 41°C (90° – 105°F)	Sunstroke, heat cramps, and heat exhaustion possible with prolonged exposure and/or physical activity
Caution	IV	27° – 32°C (80° – 90°F)	Fatigue possible with prolonged exposure and/or physical activity

FIGURE 5.1.3 Chicago heat wave.
The Midwest and eastern United States were hit with a devastating heat wave during July 1995. City emergency, medical, and coroner services were overwhelmed by the tragedy. [Photo by Dave Weaver/Gamma Liaison Agency, Inc.]

the first two weeks of August alone. Although the heat wave began August 1, a full emergency response was not begun until August 13—the day temperatures began to fall.

The rush for medical help in the Midwest and in Europe quickly swamped hospitals, which had to turn away hundreds of patients. An added problem in France was that many doctors, nurses, and fellow citizens were out of town as this was the beginning of the typical holiday season.

In Chicago, the rate at which the heat claimed victims overwhelmed the coroner, so the dead were stored in many refrigerated trucks parked outside the morgue (Figure 5.1.3). We can only wonder how many lives would have been saved if refrigeration for the living were somehow available.

Global Temperatures Suggest a Greenhouse Warming

There is a distinct possibility that future humans may experience greater temperature-related challenges owing to complex changes now under way in the lower atmosphere. Scientists agree that human activities are enhancing Earth's natural greenhouse. Certain gases that absorb and emit radiation are increasing and are absorbing more longwave radiation and delaying losses of heat energy to space, thus changing the natural Earth–atmosphere energy equilibrium.

The Intergovernmental Panel on Climate Change (IPCC), through major reports in 1990, 1992, 1995, its *Third Assessment Report*, and summarizing *Synthesis Report* in 2001, confirms that global warming is occurring. Its latest report states, "... there is new and stronger evidence that most of the warming observed over the past 50 years is attributable to human activities. ... Both temperature and sea level are projected to continue to rise throughout the twenty-first century for all scenarios studied." Uncertainty remains in forecasting the pace and implications of this warming for Earth systems.

The Arctic Climate Impact Assessment (ACIA) reported confirmation of conditions in the Arctic in their science report released late 2004. ACIA stated, "The Arctic is extremely vulnerable to observed and projected climate change and its impacts. The Arctic is now experiencing some of the most rapid and severe climate change on Earth."

The period from 1970 to 2003 registered the warmest years in the history of instrumental measurements. Effects of this warming are the subject of many scientific studies: melting and retreat of mountain glaciers worldwide, disintegration of coastal ice shelves around Antarctica, sea level rising at a faster rate than previously observed, more intense thunderstorms and weather extremes, changes in vegetation on land and the distribution of marine organisms in the oceans, and heat-wave effects on grain production. The snowline in portions of the European Alps rose 100 m (330 ft) in elevation since 1980.

An array of sophisticated satellites, remote-sensing capabilities, and powerful computers equip scientists to run global circulation models and decipher the climate trends. Actual field measurements and observations confirm these remotely sensed conclusions. With the blame resting on the exhaust from our fossil fuel consumption and our modern technological society, the political fallout and debate is intense. Chapter 10 discusses global warming and climate change. There we address what is known and unknown, how global models work, the radiatively active gases that are forcing the warming, the consequences to natural systems, the uncertainties, and what international action is underway to slow the effects of such change.

Summary and Review—Global Temperatures

▶ *Define* the concepts of temperature, kinetic energy, and sensible heat, and *distinguish* among Kelvin, Celsius, and Fahrenheit temperature scales and how they are measured.

Temperature is a measure of the average kinetic energy (motion) of individual molecules in matter. We feel the effect of temperature as the sensible heat transfer from warmer objects to cooler objects when these objects are touching. Temperature scales include

- Kelvin scale: 100 units between the melting point of ice (273 K) and the boiling point of water (373 K).
- Celsius scale: 100 degrees between the melting point of ice (0°C) and the boiling point of water (100°C).
- Fahrenheit scale: 180 degrees between melting point of ice (32°F) and the boiling point of water (212°F).

The Kelvin scale is used in scientific research because temperature readings start at absolute zero and thus are proportional to the actual kinetic energy in a material.

temperature (p. 116)

1. Distinguish between sensible heat and sensible temperature.
2. What does air temperature indicate about energy in the atmosphere?
3. Compare the three scales that express temperature. Find "normal" human body temperature on each scale and record the three values in your notes.

4. What is your source of daily temperature information? Describe the highest temperature you have experienced and the lowest temperature. From what we have discussed in this chapter, can you identify the factors that may have contributed to these temperatures?

▶ *List* and *review* the principal controls and influences that produce global temperature patterns.

Principal controls and influences upon temperature patterns include *latitude* (the distance north or south of the equator), *altitude* (location above sea level), *cloud cover* (reflection, absorption, and reradiation of energy), and *land–water heating differences* (the nature of evaporation, transparency, specific heat, movement, and ocean currents and sea-surface temperatures).

5. Explain the effect of altitude on air temperature. Why is air at higher altitudes lower in temperature? Why does it feel cooler standing in shadows at higher altitudes than at lower altitudes?
6. What noticeable effect does air density have on the absorption and radiation of energy? What role does altitude play in that process?
7. How is it possible to grow moderate-climate-type crops such as wheat, barley, and potatoes at an elevation of 4103 m (13,460 ft) near La Paz, Bolivia, so near the equator?
8. Describe the effect of cloud cover with regard to Earth's temperature patterns. From the last chapter, review the

cloud-albedo forcing and cloud-greenhouse forcing of different cloud types and relate the concepts with a simple sketch.

▶ *Review* the factors that produce different marine effects and continental effects as they influence temperatures, and *utilize* several pairs of stations to illustrate these differences.

The physical nature of land (rock and soil) and water (oceans, seas, and lakes) is the reason for **land–water heating differences**, the fact that land heats and cools faster than water. Moderate temperature patterns are associated with water bodies, and extreme temperatures occur inland. The controls that cause differences between land and water surfaces are *evaporation, transparency, specific heat, movement,* and *ocean currents and sea-surface temperatures.*

Light penetrates water because of its **transparency**. Water is clear and light transmits to an average depth of 60 m (200 ft) in the ocean. This penetration distributes available heat energy over a much greater volume than could occur on opaque land; thus, a larger energy reservoir is formed. When equal volumes of water and land are compared, water requires far more energy to increase its temperature than does land. In other words, water can hold more energy than can soil or rock, so water has a higher **specific heat**, the heat capacity of a substance, averaging about 4 times that of soil.

Ocean currents affect temperature. An example of the effect of ocean currents is the **Gulf Stream**, which moves northward off the east coast of North America, carrying warm water far into the North Atlantic. As a result, the southern third of Iceland experiences much milder temperatures than would be expected for a latitude of 65° N, just below the Arctic Circle (66.5°).

Marine effect, or maritime, describes locations that exhibit the moderating influences of the ocean, usually along coastlines or on islands. **Continental effect** refers to the condition of areas that are less affected by the sea and therefore have a greater range between maximum and minimum temperatures diurnally and yearly.

land–water heating differences (p. 120)
transparency (p. 121)
specific heat (p. 122)
Gulf Stream (p. 122)
marine effect (p. 124)
continental effect (p. 124)

9. List the physical aspects of land and water that produce their different responses to heating from absorption of insolation. What is the specific effect of transparency in a medium?

10. What is specific heat? Compare the specific heat of water and soil.

11. Describe the pattern of sea-surface temperatures (SSTs) as determined by satellite remote sensing. Where is the warmest ocean region on Earth?

12. What effect does sea-surface temperature have on air temperature? Describe the negative feedback mechanism created by higher sea-surface temperatures and evaporation rates.

13. Differentiate between marine and continental temperatures. Give geographic examples of each from the text: Canada, the United States, Norway, and Russia.

▶ *Interpret* the pattern of Earth's temperatures from their portrayal on January and July temperature maps and on a map of annual temperature ranges.

Maps for January and July instead of the solstice months of December and June are used for temperature comparison because of the natural lag that occurs between insolation received and maximum or minimum temperatures experienced. Each line on these temperature maps is an **isotherm**, an isoline that connects points of equal temperature. Isotherms portray temperature patterns.

Isotherms generally are zonal, trending east–west, parallel to the equator. They mark the general decrease in insolation and net radiation with distance from the equator. The **thermal equator** (isoline connecting all points of highest mean temperature) trends southward in January and shifts northward with the high summer Sun in July. In January it extends farther south into the interior of South America and Africa, indicating higher temperatures over landmasses.

In the Northern Hemisphere in January, isotherms shift equatorward as cold air chills the continental interiors. The coldest area on the map is in Russia, specifically northeastern Siberia. The intense cold experienced there results from winter conditions of consistent clear, dry, calm air, small insolation input, and an inland location far from any moderating maritime effects.

isotherm (p. 124)
thermal equator (p. 124)

14. What is the thermal equator? Describe its location in January and in July. Explain why it shifts position annually.

15. Observe trends in the pattern of isolines over North America and compare the January average temperature map with the July map. Why do the patterns shift locations?

16. Describe and explain the extreme temperature range experienced in north-central Siberia between January and July.

17. Where are the hottest places on Earth? Are they near the equator or elsewhere? Explain. Where is the coldest place on Earth?

18. From the maps in Figures 5.14, 5.16, and 5.17, determine the average temperature values and annual range of temperatures for your present location. Compare the maps in Figures 5.14c and 5.16c of the South polar region, and describe seasonal changes between the two maps. Characterize conditions along the Antarctic Peninsula between the two maps (around 60° west longitude).

▶ *Contrast* wind chill and the heat index, and *determine* human response to these apparent temperature effects.

The wind-chill factor indicates the enhanced rate at which body heat is lost to the air. As wind speeds increase, heat loss from the skin increases. The *heat index* (HI) indicates the human body's reaction to air temperature and water vapor. The level of humidity in the air affects our natural ability to cool through evaporation from skin.

19. What is the wind-chill temperature on a day with an air temperature of −12°C (10°F) and a wind speed of 32 kmph (20 mph)?

20. On a day when temperature reaches 37.8°C (100°F), how does a relative humidity reading of 50% affect apparent temperature?

NetWork

The *Geosystems* Home Page provides online resources for this chapter on the World Wide Web. To begin: Once on the Home Page, click on this textbook, scroll the Table of Contents menu, select this chapter, and click "Begin." You will find self-tests that are graded, review exercises, specific updates for items in the chapter, and many links to interesting related pathways on the Internet under "Destinations." *Geosystems* is at http://www.prenhall.com/christopherson.

Critical Thinking

A. With each temperature map (Figures 5.14, 5.16, and 5.17), begin by finding your own city or town and noting the temperatures indicated by the isotherms for January and July and the annual temperature range. Record the information from these maps in your notebook. As you work through the different maps throughout this text, note atmospheric pressure and winds, annual precipitation, climate type, landforms, soil orders, vegetation, and terrestrial biomes. By the end of the course you will have recorded a complete physical geography profile for your regional environment. The scale of all these maps allows you to make only a rough approximation for your location.

B. Have you ever experienced any of the different responses of the human body to the low-temperature or high-temperature stress noted in Focus Study 5.1, Table 1? Where were you at the time of each experience? Make copies of the wind-chill and heat-index charts given in the Focus Study for later reference. On an appropriate day, check air temperature and wind speed and determine their combined effect on skin chilling. In contrast, on an appropriate day, check air temperature and relative humidity values and determine their combined effect on your personal comfort. The *Geosystems* Home Page presents several interesting temperature links under "Destinations."

Career Link 5.1

Dr. Louwrens Hacquebord, Professor of Arctic and Antarctic Studies

The Hornsund fjord system is in southwestern Spitsbergen, the largest island of the Svalbard Archipelago in the Arctic Ocean, 965 km (600 mi) north of Norway. The surrounding mountains have a light covering of fresh snow, steep slopes sweep to the water's edge. The air is 4°C (39°F) with a brisk wind chill near −1°C and a cloudy, moody sky overhead. A weary eye for polar bears adds to the feel of the place. We are at 77° N latitude.

We walk inland along a curving, rocky shoreline toward a mound. Whalebones are stacked against the windward side of the ruins of a small dwelling. Here a cooper (a barrel maker) lived in A.D. 1618, and English whalers worked the waters for Greenland right whales. Many of these sites, and ones operated by the Dutch in the north, were abandoned by the 19th century.

We come upon four Dutch researchers excavating several sites in the area. Here we meet project leader

Dr. Louwrens Hacquebord standing in a meter-deep excavation pit in a 400-year-old midden deposit (garbage dump) where the cooper threw his trash. Louwrens describes environmental conditions in 1618 and the changes in sea ice and climate that were just ahead in this cooper's life, as he carefully brushes soil away from a dark object emerging from the side of the pit.

You hear Louwrens' interdisciplinary perspective as he weaves physical geography, history, biology, and archeology together in recreating life at this site (Figure CL 5.1.1). He explains that with the eradication of the whales there was a population explosion of zooplankton. He wants to know how this affected bird life and impacted ecosystems—synthesizing the evidence he finds from the middens with past temperatures, climate, and ecosystems, to recreate the environment in which these whalers worked.

The moment arrives when he gently removes a leather shoe and

places it on the ground, continuing to remove its earthen covering. The shoe's design is clearly visible. For 386 years, from the day the cooper discarded it, this shoe sat awaiting Louwrens' discovering hands. The heavy boots we wear are in stark contrast to this slipper of a shoe. Life was rough at these whaling stations. Louwrens explains that, "Some 80% of the men died of scurvy (lack of Vitamin C)."

Louwrens Hacquebord was born and raised in The Netherlands, a landscape partially reclaimed from the sea. He recalls, "I was Boy Scout between ages 12 and 16 and loved using maps and compasses. I loved to make maps, and when I was a boy, I created maps of imaginary countries, naming features, and cities, capitals. I noticed the artificial 'dwellings hills,' like bumps on the landscape on which houses were built, and wanted to know more; wanted to understand what they were for and who built them, and when they were built. I was interested in many aspects of my country, with its open delta and extensive ocean front."

By the time he went to high school he knew he was interested in landscape studies and maps. "When we went to outdoor camps, I realized I loved being out in the field, out in nature. I started my studies with physical geography."

At the University of Utrecht he majored in physical geography because it was an allied field for graduate-level studies in archeology, a field he wanted to pursue. But he explains, "As I studied more physical geography and completed more field work, I found that physical geography was simply too interesting to leave behind for archeology. I decided to stay in physical geography and do my Masters program in it." Louwrens was shaping a multidisciplinary program and career with geography at the core.

After earning his Masters' degree, he taught high school for six years before returning to graduate school for a Ph.D. at the University of Amsterdam. For his doctorate, Louwrens studied a 17th century Dutch whaling settlement at Smeerenburg, on Spitsbergen. He was able to bring in past environmental conditions as he reconstructed this relict whaling station. His Ph.D. was awarded with *cum laude* (highest honors) distinction for the nature of his multidisciplinary approach and synthesis of varied subject areas—he forged a new methodology with his work.

Dr. Hacquebord took a position at the University of Groningen in 1984 and was appointed head of the Arctic Centre. He achieved full professor status and assumed a newly created position as Chair in Arctic

Studies in 1994. Describing his work at Groningen, he says, "I teach Arctic Studies, and physical geography is at the core of the discipline, along with history, geography, archeology, and biology."

His approach spread to the international stage: "I became the The Netherlands representative to the International Arctic Science Committee (IASC) and serve on the IASC Executive Committee. I asked all the human and social scientists to join in and meld with the physical and life scientists, for studies circum-Arctic in scope. IASC is an advisory board to the Arctic Council. I was there at the beginning of the Arctic Climate Impact Assessment (ACIA). I am now working on plans for the International Polar Year, which begins March 2007."

To a question about climate change, he replies, "The impact of climate change is much bigger in the polar region than the rest of the globe. I saw the latest scientific maps with temperature trends over the last 30 years in the Arctic and it was scary to see the magnitude of change! Especially in the Eurasian and American continents; the shifting of the tree line and the permafrost distribution is remarkable. This is why it is a key interest at IASC."

About students and teaching, "I encourage my students to use physical geography as a core and then move outward to bring in many approaches so they can get a complete view of the environment. You will be successful when you go over the borders of the discipline. However, it is important to never forget your starting point in physical geography."

As an example, Louwrens describes one of his Ph.D. students. She has taken old Antarctic maps from

FIGURE 5.1.1 Dr. Louwrens Hacquebord in the field.
Professor Hacquebord teaching us about an early 17th century English whaling station, which he is excavating on Spitsbergen Island, Arctic Ocean.
[Photo by Bobbé Christopherson.]

1800 to 1957—the last International Geophysical Year—and digitized them to reconstruct ice shelf movement. She uses GIS to synthesize dynamic maps, linking past conditions with climate and other variables, to understand the patterns. Louwrens explains, "This is a first for this kind of mapping."

Louwrens concludes, "I want to write about how the Arctic makes us feel, how nature makes us feel, and how we respond. I want to see young people experiencing the Arctic, and deciphering what is occurring so dramatically in these polar regions." Just as Louwrens taught us on that cold morning in Spitsbergen, as he lifted the leather shoe from its four-century rest and described climatic conditions in 1618, we hear a calling for geographers to be the synthesizers. Louwrens is truly a scientist pushing the edges of the discipline.

Almost constant trade winds naturally prune this tree near South Point, Hawai'i, along the southern coast of the big island. A real potential exists for harvesting wind to produce electrical power. At present, wind power is the fastest growing form of electrical generation on Earth. [Photo by Bobbé Christopherson.]

CHAPTER 6

Atmospheric and Oceanic Circulations

▶ **Key Learning Concepts**

After reading the chapter, you should be able to:

▶ *Define* the concept of air pressure, and *describe* instruments used to measure air pressure.

▶ *Define* wind, and *describe* how wind is measured, how wind direction is determined, and how winds are named.

▶ *Explain* the four driving forces within the atmosphere—gravity, pressure gradient force, Coriolis force, and friction force—and *describe* the primary high- and low-pressure areas and principal winds.

▶ *Describe* upper-air circulation and its support role for surface systems, and *define* the jet streams.

▶ *Overview* several multiyear oscillations of air temperature, air pressure, and circulation in the Arctic, Atlantic, and Pacific oceans.

▶ *Explain* several types of local winds: land–sea breezes, mountain–valley breezes, katabatic winds, and the regional monsoons.

▶ *Discern* the basic pattern of Earth's major surface ocean currents and deep thermohaline circulation.

Early in April 1815, on an island named Sumbawa in present-day Indonesia, the volcano Tambora erupted violently. It spewed an estimated 150 km³ (36 mi³) of material, 25 times the volume produced by the 1980 Mount St. Helens eruption in Washington State. Global atmospheric circulation carried material from Tambora worldwide, creating a stratospheric veil of dust and acid mist. Remarkable optical and meteorological effects lasted in Earth's atmosphere years after the eruption—beautiful sunrises and sunsets and a temporary lowering of temperatures worldwide. Scientists in 1815 lacked the remote-sensing capability of satellite technology, and they had no way of knowing the global impact of Tambora's eruption.

141

(a)

Luzon Island, Philippines

6/15 – 6/19/91

(b)

7/4 – 7/10/91

(c)

7/25 – 8/1/91

(d)

8/15 – 8/21/91

Effects cover 42% of globe in just 60 days

(e)

FIGURE 6.1 Volcanic eruption effects spread worldwide by winds.
Satellite image shows Mount Pinatubo as it erupted (a). False-color images (b–e) show aerosols from Mount Pinatubo, smoke from fires, and dust storms, all swept about the globe by the general atmospheric circulation. The dramatic increase in aerosols over the oceans from mid-June 1991 (b) to mid-August (e) is due to the Mount Pinatubo eruption on June 15. The false color shows aerosol concentration, measured by the atmosphere's aerosol optical thickness (AOT): White is densest, dull yellow indicates medium values, and brown areas have the lowest aerosol concentration. In (b) note the dust moving westward from Africa, smoke from Kuwaiti oil well fires set during the Persian Gulf War, smoke from forest fires in Siberia, haze off the east coast of the United States, and the Pinatubo aerosol layer beginning to emerge north of Indonesia and expanding globally in (c) through (e).
[(a) *AVHRR* satellite eruption image courtesy of U.S. Geological Survey, EROS Data Center. *AVHRR* satellite images of aerosols courtesy of Dr. Larry L. Stowe, NESDIS/NOAA. Used by permission.]

After 635 years of dormancy, Mount Pinatubo in the Philippines erupted in 1991 (Figure 6.1a). This event had tremendous atmospheric impact, lofting 15–20 million tons of ash, dust, and sulfur dioxide (SO_2) into the atmosphere. As the sulfur dioxide rose into the stratosphere, it quickly formed sulfuric acid (H_2SO_4) aerosols, which became concentrated at 16–25 km (10–15.5 mi) altitude. This debris increased atmospheric albedo of about 1.5%, giving scientists an estimate of the aerosol volume generated by the eruption (see Figure 1.6).

The AVHRR instrument aboard *NOAA-11* monitored the reflected solar radiation from Mount Pinatubo's aerosols as global winds swept them around Earth. Figure 6.1b–d shows images made at three-week intervals that clearly track the spread of the debris worldwide. Some 60 days after the eruption (the last satellite image in the sequence), the aerosol cloud spanned about 42% of the globe, from 20° S to 30° N. Again, for almost two years colorful sunrises and sunsets and a small lowering of average temperatures followed. The eruption provided a unique insight into the dynamics of atmospheric circulation.

Today, technology permits a depth of analysis unknown in the past—satellites now track the atmospheric effects from dust storms, forest fires, industrial haze, warfare, and the dispersal of volcanic explosions, among many other things. As an example, millions of tons of dust from African soils cross the Atlantic each year, borne by the winds of atmospheric circulation. This dust sometimes increases the iron content of the waters off Florida, promoting the toxic algal blooms (*Karenia brevis*) known as "Red Tides."

Global winds are certainly an important reason why the United States, the former Soviet Union, and Great Britain signed the 1963 Limited Test Ban Treaty. That treaty banned above-ground testing of nuclear weapons because atmospheric circulation spread radioactive contamination worldwide. Such agreements illustrate how the fluid movement of the atmosphere socializes humanity more than perhaps any other natural or cultural factor. Our atmosphere makes the world a spatially linked society—one person's or country's exhalation is another's inhalation.

In this chapter: We begin with a discussion of wind essentials, consisting of air pressure and its measurement and a description of wind. The driving forces that produce surface winds are pressure gradient, Coriolis, and friction. We examine the circulation of Earth's atmosphere and the patterns of global winds, including principal pressure systems and winds. We also consider Earth's wind-driven oceanic currents. The chapter explains multiyear oscillations in atmospheric flows, local winds, and the powerful monsoons. The energy driving all this movement comes from one source: the Sun.

Wind Essentials

Earth's atmospheric circulation transfers both energy and mass on a grand scale. In the process, the imbalance between equatorial energy surpluses and polar energy deficits is partly resolved, Earth's weather patterns formed, and ocean currents produced. Air pollutants, whether natural or human-caused, are spread worldwide by atmospheric circulation, far from their point of origin.

Atmospheric circulation is generally categorized at three levels: *primary circulation* consisting of general worldwide circulation, *secondary circulation* of migratory high-pressure and low-pressure systems, and *tertiary circulation* that includes local winds and temporal weather patterns. Winds that move principally north or south along meridians are known as *meridional flows*. Winds moving east or west along parallels of latitude are called *zonal flows*.

Air Pressure and Its Measurement

Important to an understanding of wind is the concept of air pressure, its measurement and expression. The molecules that constitute air create *air pressure* through their motion, size, and number—the factors that determine the temperature and density of the air. Air pressure, then, is a product of the temperature and density of a mass of air. Pressure is exerted on all surfaces in contact with the air.

FIGURE 6.2 Developing the barometer.
Evangelista Torricelli developed the barometer to measure air pressure as a by-product of trying to solve a mine-drainage problem (a). Two types of instruments are used to measure atmospheric pressure: (b) a mercury barometer (idealized sketch), and (c) an aneroid barometer. Have you used a barometer? If so, what type is it? Have you tried to reset it using a local weather information source? [(c) Courtesy of Qualimetrics, Inc., Sacramento, California.]

In 1643, Galileo's pupil Evangelista Torricelli was working on a mine-drainage problem. His work led him to discover a method for measuring air pressure (Figure 6.2a). He knew that pumps in the mine were able to "pull" water upward about 10 m (33 ft), but no higher, and he did not know why. Careful observation led him to discover that this limitation was caused not by weak pumps but by the atmosphere itself. Torricelli noted that the water level in the vertical pipe fluctuated from day to day. He figured out that air pressure varies with weather conditions.

To simulate the problem at the mine, Torricelli devised an instrument at Galileo's suggestion, using a much denser fluid than water—mercury (Hg)—in a glass laboratory tube only 1 m high. Torricelli sealed the glass tube at one end, filled it with mercury, and inverted it into

a dish of mercury (Figure 6.2b). He determined that the average height of the column of mercury in the tube was 760 mm (29.92 in.) and that it did vary day-to-day as the weather changed. He concluded that the mass of surrounding air was exerting pressure on the mercury in the dish that counterbalanced the column of mercury.

Using similar instruments, scientists set a standard of *normal sea-level pressure* at 1013.2 mb (*millibar*, or *mb*, which expresses force per square meter of surface area), or 29.92 in. mercury (Hg). In Canada and other countries, normal air pressure is expressed as 101.32 kPa (*kilopascal*; 1 kPa = 10 mb).

Any instrument that measures air pressure is called a *barometer* (from the Greek *baros*, meaning "weight"). Torricelli developed a **mercury barometer**. A more compact barometer design, which works without a meter-long tube of mercury, is the aneroid barometer, shown in Figure 6.2c. *Aneroid* means "using no liquid." The **aneroid barometer** principle is simple: Imagine a small chamber, partially emptied of air, which is sealed and connected to a mechanism attached to a needle on a dial. As air pressure increases, it presses on the chamber; as air pressure decreases, it relieves pressure from the chamber. The chamber responds to these changes in air pressure and moves the needle. An aircraft altimeter is a type of aneroid barometer. It accurately measures altitude because air pressure diminishes with altitude above sea level. For accuracy the altimeter must be adjusted for temperature changes.

Figure 6.3 illustrates comparative scales in millibars and inches used to express air pressure and its relative force. The normal range of Earth's atmospheric pressure from strong high pressure to deep low pressure is

about 1050 to 980 mb (31.00 to 29.00 in.). The figure also indicates the extreme highest and lowest pressures ever recorded in the United States, Canada, and on Earth.

Wind: Description and Measurement

Simply stated, **wind** is generally the horizontal motion of air across Earth's surface. Turbulence adds wind updrafts and downdrafts and a vertical component to this definition. *Differences in air pressure (density) between one location and another produce wind.* Wind's two principal properties are speed and direction, and instruments measure each. An **anemometer** measures wind speed in kilometers per hour (kmph), miles per hour (mph), meters per second (mps), or knots. (A *knot* is a nautical mile per hour, covering 1 minute of Earth's arc in an hour, equivalent to 1.85 kmph,

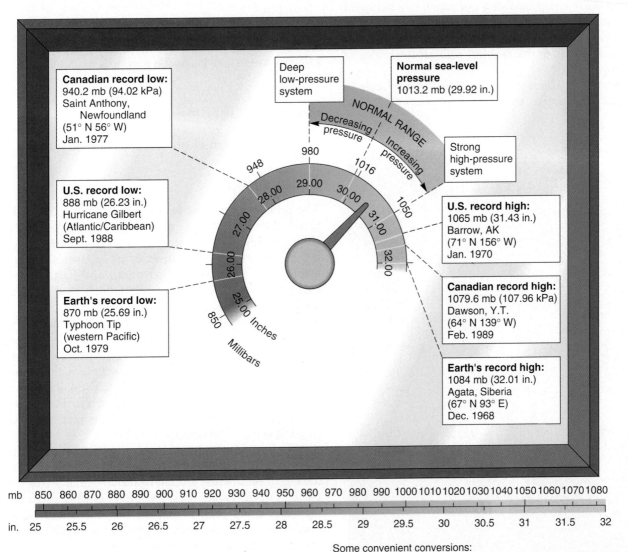

FIGURE 6.3 Air pressure readings and conversions.
Scales for expressing barometric air pressure in millibars and inches, with average air-pressure values and recorded pressure extremes. Canadian values include kilopascal equivalents, 10 mb = 1 kPa.

or 1.15 mph.) A **wind vane** determines wind direction; the standard measurement is taken 10 m (33 ft) above the ground to reduce the effects of local topography on wind direction (Figure 6.4).

Winds are named for the direction *from which they originate*. For example, a wind from the west is a *westerly* wind (it blows eastward); a wind out of the south is a *southerly* wind (it blows northward). Figure 6.5 illustrates a simple wind compass, naming 16 principal wind directions used by meteorologists.

The traditional *Beaufort wind scale* is a descriptive scale useful in visually estimating wind speed (Table 6.1). In 1806, Admiral Beaufort of the British Navy introduced his wind scale. In 1926, G. C. Simpson expanded Beaufort's scale to include wind speeds on land. The National Weather Service (old Weather Bureau) standardized the scale in 1955 (see **http://www.crh.noaa. gov/lot/webpage/beaufort/**). The scale is still referenced on ocean charts, enabling estimation of wind speed without instruments. Although most ships use sophisticated equipment to perform such measurements, you will find a small poster of the Beaufort scale on the bridge of many ships.

Using no instruments, try out the Beaufort scale in Table 6.1 to estimate wind speed as you walk across

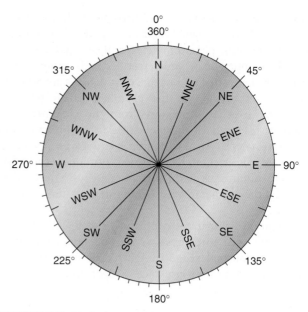

FIGURE 6.5 A wind compass.
Sixteen wind directions identified on a wind compass. Winds are named for the direction from which they originate. For example, a wind from the west is a westerly wind.

campus today. Then, moisten your finger, hold it in the air, and sense evaporative cooling on one side of your finger to indicate from which direction the wind is blowing.

Global Winds

The primary circulation of winds across Earth has fascinated travelers, sailors, and scientists for centuries, although only in the modern era is a true picture emerging of the pattern and causes of global winds. Breakthroughs in space-based observations and Earth-bound computer technology are refining models that simulate total atmospheric and oceanic circulation.

Scientists at the National Center for Atmospheric Research determined that atmospheric flows of global winds redistribute much of the energy between the tropics and the polar regions. The atmosphere is the dominant medium from about 35° latitude to the poles in each hemisphere, whereas ocean currents redistribute more heat in a zone straddling the equator between the 17th parallels in each hemisphere.

Scientists at the Jet Propulsion Laboratory and the University of California, Los Angeles, painstakingly assembled a remarkable portrait of surface winds across the Pacific Ocean (Figure 6.6). The *Seasat* satellite produced the image using radar to measure the motion and direction of ocean waves. Because wind drives waves on the ocean surface, wave patterns indicate winds.

The patterns in the figure are the result of specific forces at work in the atmosphere: *pressure gradient force, Coriolis force, friction force,* and *gravity*. These forces are our next topic. As we progress through this chapter, you may want to refer to this *Seasat* image to identify the winds, eddies, and vortexes it portrays.

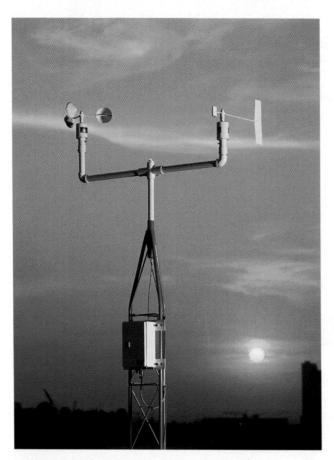

FIGURE 6.4 Wind vane and anemometer.
Instruments used to measure wind direction (wind vane, right) and wind speed (anemometer, left) at a weather station installation. [Photo by Belfort Instruments.]

Table 6.1   Beaufort Wind Scale

Wind Speed			Beaufort Wind Scale			
kmph	mph	knots	Beaufort Number	Wind Description	Observed Effects at Sea	Observed Effects on Land
<1	<1	<1	0	Calm	Glassy calm, like a mirror	Calm, no movement of leaves
1–5	1–3	1–3	1	Light Air	Small ripples; wavelet scales; no foam on crests	Slight leaf movement; smoke drifts; wind vanes still
6–11	4–7	4–6	2	Light breeze	Small wavelets; glassy look to crests, which do not break	Leaves rustling; wind felt; wind vanes moving
12–19	8–12	7–10	3	Gentle breeze	Large wavelets; dispersed whitecaps as crests break	Leaves and twigs in motion; small flags and banners extended
20–29	13–18	11–16	4	Moderate breeze	Small, longer waves; numerous whitecaps	Small branches moving; raising dust, paper, litter, and dry leaves
30–38	19–24	17–21	5	Fresh breeze	Moderate, pronounced waves; many whitecaps; some spray	Small trees and branches swaying; wavelets forming on inland waterways
39–49	25–31	22–27	6	Strong breeze	Large waves, white foam crests everywhere; some spray	Large branches swaying; overhead wires whistling; difficult to control an umbrella
50–61	32–38	28–33	7	Moderate (near) gale	Sea mounding up; foam and sea spray blown in streaks in the direction of the wind	Entire trees moving; difficult to walk into wind
62–74	39–46	34–40	8	Fresh gale (or gale)	Moderately high waves of greater length; breaking crests forming sea spray; well-marked foam streaks	Small branches breaking; difficult to walk; moving automobiles drifting and veering
75–87	47–54	41–47	9	Strong gale	High waves; wave crests tumbling and the sea beginning to roll; visibility reduced by blowing spray	Roof shingles blown away; slight damage to structures; broken branches littering the ground
88–101	55–63	48–55	10	Whole gale (or storm)	Very high waves and heavy, rolling seas; white appearance to foam-covered sea; overhanging waves; visibility reduced	Uprooted and broken trees; structural damage; considerable destruction; seldom occurring
102–116	64–73	56–63	11	Storm (or violent storm)	White foam covering a breaking sea of exceptionally high waves; small- and medium-sized ships lost from view in wave troughs; wave crests frothy	Widespread damage to structures and trees, a rare occurrence
>117	>74	>64	12–17	Hurricane	Driving foam and spray filling the air; white sea; visibility poor to nonexistent	Severe to catastrophic damage; devastation to affected society

(a)

(b)

FIGURE 6.6 Wind portrait of the Pacific Ocean.
Surface wind measured by radar scatterometer aboard the *Seasat* satellite on a September day. Scientists analyzed 150,000 measurements to produce this image (a). Colors are correlated with wind speeds, and the white arrows denote wind direction. Try comparing the wind patterns with a visible-light image of the same region (b). Can you identify the pattern of trade winds, westerlies, high-pressure cells, and low-pressure cells from the cloud patterns on the image? [Wind portrait courtesy of Dr. Peter Woiceshyn, Jet Propulsion Laboratory, Pasadena, California. Satellite image inset from Laboratory of Planetary Studies, Cornell University. Used by permission.]

Driving Forces Within the Atmosphere

Four forces determine both speed and direction of winds:

- Earth's *gravitational force* on the atmosphere is virtually uniform. Gravity equally compresses the atmosphere worldwide, with the density decreasing as altitude increases. Without gravity, there would be no atmospheric pressure—or atmosphere, for that matter.

- **Pressure gradient force** drives air from areas of higher barometric pressure (more dense air) to areas of lower barometric pressure (less dense air), thereby causing winds. Without a pressure gradient force, there would be no wind.

- The **Coriolis force**, a deflective force, makes wind that travels in a straight path appear to be deflected in relation to Earth's rotating surface. The Coriolis force deflects wind to the right in the Northern Hemisphere and to the left in the Southern Hemisphere. Without Coriolis force, winds would move along straight paths between high- and low-pressure areas.

- The **friction force** drags on the wind as it moves across surfaces; it decreases with height above the surface. Without friction, winds would simply move in paths parallel to isobars and at high rates of speed.

All four of these forces operate on moving air and ocean currents at Earth's surface and influence global wind circulation patterns. The following sections describe the actions of the pressure gradient, Coriolis, and friction forces. (The gravitational force operates uniformly worldwide.)

Pressure Gradient Force

High- and low-pressure areas exist in the atmosphere principally because Earth's surface is unequally heated. For example, cold, dense air at the poles exerts greater pressure than warm, less dense air along the equator. These pressure differences establish a *pressure gradient force*.

An **isobar** is an *isoline* (a line along which there is a constant value) plotted on a weather map to connect points of equal pressure. A pattern of isobars on a weather map provides a portrait of the pressure gradient between an area of higher pressure and one of lower pressure. The spacing between isobars indicates the intensity of the pressure difference, or *pressure gradient*.

Just as closer contour lines on a topographic map indicate a steeper slope on land, so closer isobars denote steepness in the pressure gradient. In Figure 6.7a, note the spacing of the isobars (green lines). A steep gradient causes faster air movement from a high-pressure area to a low-pressure area. Isobars spaced wider apart from one another mark a more gradual pressure gradient, one that

creates a slower airflow. Along a horizontal surface, the pressure gradient force alone acts at right angles to the isobars, so wind blows across them at right angles. Note the location of steep ("strong winds") and gradual ("light winds") pressure gradients and their relationship to wind intensity on the map in Figure 6.7b.

Figure 6.8 illustrates the forces that direct winds. Figure 6.8a shows the pressure gradient force acting alone. A field of subsiding, or sinking, air develops in a high-pressure area. Air diverges out of the high-pressure area at the surface, moving outward in all directions. On the other hand, in a low-pressure area, as air rises, it converges from all directions into the area of lower pressure at the surface. For instance, on a warm day, the temperature of the air increases and the air is less dense and more buoyant, so it rises. In contrast, on a cold day, the temperature of the air is lower, the air is denser and less buoyant, so it descends. Chapter 7 discusses this behavior of a parcel of air.

Coriolis Force

You might expect surface winds to move in a straight line from areas of higher pressure to areas of lower pressure. On a nonrotating Earth, they would. But on our rotating planet, the Coriolis force deflects anything that flies or flows across Earth's surface—wind, an airplane, or ocean currents—from a straight path. This

(a)

(b)

FIGURE 6.7 Pressure gradient determines wind speed. Pressure gradient (a). On a weather map (b), the closer spacing of isobars represents a steeper pressure gradient that produces stronger winds; wider spacing of isobars denotes a gradual pressure gradient that leads to lighter winds. Here we see surface winds spiraling clockwise out of a high-pressure system and spiraling counterclockwise into a low-pressure system.

(a)

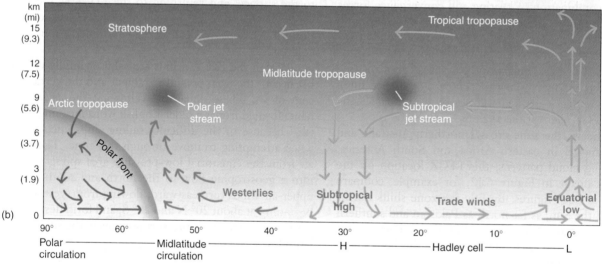

(b)

FIGURE 6.13 General atmospheric circulation model.
Two views of the general atmospheric circulation: (a) general circulation schematic;
(b) equator-to-pole cross section of the Northern Hemisphere. Both views show
Hadley cells, subtropical highs, the polar front, the subpolar low-pressure cells,
and approximate locations of the subtropical and polar jet streams.

Global Wind Circulation,
ANIMATION Hadley Cells

Global Infrared
SATELLITE

and heats by compression on its descent to the surface, as
in Figure 6.13. Warmer air has a greater capacity to
absorb evaporation than does cooler air, making this
descending warm air relatively dry (large water-vapor
holding capacity, low water-vapor content). The air is also

dry because heavy precipitation along the equatorial portion of the circulation removes moisture.

Surface air diverging from the subtropical high-
pressure cells generates Earth's principal surface winds:
the westerlies and the trade winds. The **westerlies** are the

dominant surface winds from the subtropics to high latitudes. They diminish somewhat in summer and are stronger in winter in both hemispheres.

As you examine the global pressure maps in Figure 6.11, you find several high-pressure areas. In the Northern Hemisphere, the Atlantic subtropical high-pressure cell is called the **Bermuda high** (in the western Atlantic) or the **Azores high** (when it migrates to the eastern Atlantic in winter). The area in the Atlantic under this subtropical high features clear, warm waters and large quantities of *Sargassum* (a seaweed), which gives the area its name—the Sargasso Sea. The **Pacific high**, or *Hawaiian high*, dominates the Pacific in July, retreating southward in January. In the Southern Hemisphere, three large high-pressure centers dominate the Pacific, Atlantic, and Indian Oceans, especially in January, and tend to move along parallels of latitude in shifting zonal positions.

The entire high-pressure system migrates with the summer high Sun, fluctuating about 5°–10° in latitude. The *eastern sides* (right-hand side) of these anticyclonic systems are drier and more stable (less convective activity) and feature cooler ocean currents than do the *western sides* (left-hand side). The drier eastern side of these systems and dry-summer conditions influence climate along subtropical and midlatitude west coasts (Figure 6.14). In fact, Earth's major deserts generally occur within the subtropical belt and extend to the west coast of each continent except Antarctica. In Figures 6.14 and 6.22, note the desert regions of Africa come right to the shore in both hemispheres, with the cool, southward-flowing *Canaries Current* offshore in the north, and the cool, northward-flowing *Benguela* current offshore in the south.

The western sides of subtropical high-pressure cells tend to be moist and unstable. These conditions cause warm, moist weather in Hawai'i, Japan, southeastern China, and the southeastern United States, depicted in Figure 6.14.

Because the subtropical belts are near 25° N and 25° S latitudes, these areas sometimes are known as the *calms of Cancer* and the *calms of Capricorn*. These zones of windless, hot, dry desert air, so deadly in the era of sailing ships, earned the name *horse latitudes*. The origin of this term is popularly attributed to becalmed and stranded sailing crews of past centuries, who destroyed the horses on board, not wanting to share food or water with the livestock. The term's true origin may never be known; the *Oxford English Dictionary* calls it "uncertain."

Subpolar Low-Pressure Cells: Cool and Moist Air

In January, two low-pressure cyclonic cells exist over the oceans around 60° N latitude, near their namesakes: the North Pacific **Aleutian low** and the North Atlantic **Icelandic low** (see Figure 6.11a). Both cells are dominant in winter and weaken or disappear in summer with the strengthening of high-pressure systems in the subtropics. The area of contrast between cold (from higher latitudes) and warm (from lower latitudes) air forms the **polar front**, where masses of air with different characteristics

FIGURE 6.14 Subtropical high-pressure system in the Atlantic.
Characteristic circulation and climate conditions related to the Atlantic subtropical high-pressure anticyclone in the Northern Hemisphere. Note deserts extending to the shores of Africa with offshore cool currents, whereas the southeastern United States is moist and humid, with offshore warm currents.

battle. This front encircles Earth and is focused in these low-pressure areas.

Figure 6.13 illustrates this confrontation between warm, moist air from the westerlies and cold, dry air from the polar and Arctic regions. Cooling and condensation in the lifted air occurs because of upward displacement of the warm air. Low-pressure cyclonic storms migrate out of the Aleutian and Icelandic frontal areas and may produce precipitation in North America and Europe, respectively. Northwestern sections of North America and Europe generally are cool and moist as a result of the passage of these cyclonic systems onshore—consider the weather in British Columbia, Washington, Oregon, Ireland, and the United Kingdom.

In the Southern Hemisphere, a discontinuous belt of subpolar low-pressure systems surrounds Antarctica. The spiraling cloud patterns produced by these cyclonic systems are visible on the spacecraft image in Figure 6.15. Severe cyclonic storms can cross Antarctica, producing strong winds and new snowfall. How many cyclonic systems can you identify around Antarctica on the image?

Polar High-Pressure Cells: Frigid, Dry Deserts

Polar high-pressure cells are weak. The polar atmospheric mass is small, receiving little energy from the Sun to put it into motion. Variable winds, cold and dry, move away from the polar region in an anticyclonic direction. They descend and diverge clockwise in the Northern Hemisphere (counterclockwise in the Southern Hemisphere) and form weak, variable winds named **polar easterlies**.

Daytime valley-breeze conditions

Nighttime mountain-breeze conditions

FIGURE 6.20 Pattern of mountain–valley breezes during day and night.

coastal areas. Compression heats the air as it flows from higher to lower elevations, and with increasing speed it moves through constricting valleys to the southwest. These winds bring dust, dryness, and heat to populated areas near the coast and create dangerous wildfire conditions.

Katabatic winds, or gravity drainage winds, are of larger regional scale and usually stronger than mountain–valley breezes, under certain conditions. An elevated plateau or highland is essential to their formation, where layers of air at the surface cool, become denser, and flow downslope. The ferocious winds that can blow off the ice sheets of Antarctica and Greenland are katabatic in nature.

Worldwide, a variety of terrain produces such winds, which bear many local names. The *mistral* of the Rhône Valley in southern France can cause frost damage to vineyards as the cold north winds move over the region to the Gulf of Lions and the Mediterranean Sea. The frequently stronger *bora*, driven by the cold air of winter high-pressure systems inland, flows across the Adriatic Coast

to the west and south. In Alaska such winds are called the *taku*.

Regionally, wind represents a significant source of renewable energy of great promise. Focus Study 6.1 briefly explores the potential for development of wind resources.

Monsoonal Winds

Some regional wind systems seasonally change direction. Intense, seasonally shifting wind systems occur in the tropics over Southeast Asia, Indonesia, India, northern Australia, and equatorial Africa. A milder version of such monsoonal-type flow affects the extreme southwestern United States. These winds involve an annual cycle of returning precipitation with the summer Sun. The Arabic word for season, *mausim*, or **monsoon**, names them. (Specific monsoonal weather, associated climate types, and vegetation regions are discussed in Chapters 8, 10, and 20.) The location and size of the Asian landmass and

Wind Power: An Energy Resource for the Present and Future

The principles of wind power are ancient, but the technology is modern and the benefits are substantial. Relative to wind energy applications, rough estimates place the global wind resource at 500% more than present global energy use.

In more-developed countries, energy sources are dominated by the use of nonrenewable fuels—coal, gas, oil, nuclear power—and centralized energy production. In less-developed countries, however, many people rely on renewable energy—small hydroelectric plants, wind-power systems, diminishing wood supplies for burn-ing, solar energy—for cooking, heating, and pumping water. These resources are considered renewable because they are not depleted in the span of a human lifetime.

The Nature of Wind Energy

Power generation from wind is site-specific, because only certain regions have adequate winds. Electricity is generated at wind farms, on groups of wind machines. Figure 6.1.1 shows such farms in California and Germany. Three basic settings favor wind resources: (1) along coastlines influenced by trade winds and west-erly winds; (2) where mountain passes constrict airflow and interior valleys develop thermal low-pressure areas, thus drawing air across the landscape; or (3) where localized winds occur, such as katabatic or monsoonal flows. Cash-short developing countries are generally located in areas blessed by such steady winds, such as the trade winds in the tropics. Where wind is reliable less than 25%–30% of the time, only small-scale uses are economically feasible.

The potential of wind power in the United States is enormous. For example, the winds of North and South Dakota and Texas could meet all U.S. electrical needs and the investment would be made in rural regions of the country. In the California Coast Ranges, because of land–sea breezes between the Pacific and Central Valley, peak winds blow from April to October, matching peak electrical demands for residential, commercial, and agricultural water pumping.

Wind-power development is enhanced by financial incentives. Farmers in Iowa and Minnesota receive $2000 in annual income from electrical production by a leased turbine and $20,000 a year from an owned turbine—requiring only one-quarter acre to site the wind turbine. The corn harvested annually from that same site is worth about $100. Farmers are exploring this income source with assistance from the American Corn Growers Association (ACGA) and are forming wind-energy cooperatives. The limitation seems to be a shortage of transmission capacity from rural sites to cities (see the ACGA report at http://www.acga. org/programs/2001WindCCS/).

FIGURE 6.1.1 Wind farms in California and Germany.
(a) Wind farm in the Cajon Pass of southern California, along Interstate 15.
(b) Wind farm near Harle, East Friesland, Germany. [Photos by (a) Bobbé Christopherson; (b) © Uwe Walz/CORBIS.]

Wind-Power Assessment

Wind-generated energy resources are the fastest-growing energy technology,

(continued)

Focus Study 6.1 (continued)

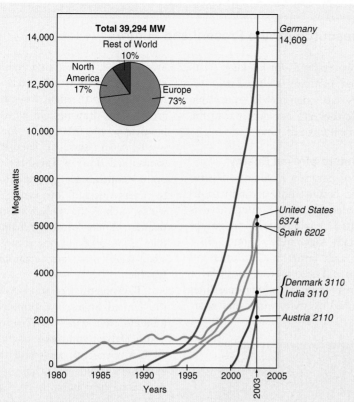

FIGURE 6.1.2 Installed wind-generating capacity.
In 1980, only 10 MW of wind-generated capacity was in operation in the world. By 1997 it had risen to 7630 MW, and by 2003, it surged to 39,294 MW worldwide. New capacity is growing at record rates in Germany, Denmark, Spain, India, and to a lesser extent in the United States. [*Source:* the American Wind Energy Association (AWEA) and the European Wind Energy Association (EWEA) annual reports.]

in terms of new capacity, outside the United States—there was more than a 1000% increase in capacity between 1990 and 2003, a tripling just since 1998. Total world capacity reached 39,294 MW (megawatts; 39,294 million watts) by the end of 2003—an increase of 8133 MW over the previous year. Globally, the wind-power industry employs more than 100,000 people with investments passing the U.S.$9 billion mark in 2003. Most of the jobs and about 80% of the industry are focused in Europe. Germany, Spain, Denmark, India, and Austria are surging ahead in annual additions to capacity (Figure 2). Germany now produces almost 5% of its electricity from wind power. The European Wind Energy Association announced an installed capacity goal of 60,000 MW by 2010, more than the current worldwide total.

The United States is second in installed wind-power capacity, on a pace with Spain. In the United States, the industry suffers from political hindrances, such as congressional failure to renew the federal Production Tax Credit that expired in December 2003, a credit that assists wind development. Only modest installations occurred in 2004, despite the fact that the U.S. wind-energy potential exceeds present electricity consumption by 300%.

To put numbers in meaningful perspective, every 10,000 MW of wind-generation capacity reduces carbon dioxide emissions by 33 million metric tons if it replaces coal, or 21 million metric tons if it replaces mixed fossil fuels. As an example, if countries rally and create a proposed $400-billion-dollar industry by installing 500,000 MW of worldwide wind capacity by 2020, as much as a

fourth of carbon dioxide emissions would be avoided. For the United States, 225,000 wind turbines would reduce coal emissions of carbon dioxide by 59%. In the 1997 Kyoto Protocol, the United States proposed reducing greenhouse gas emissions by just 7% below 1990 levels.

The Benefits of the Wind Resource

Economic reality should override further delaying actions because of the economic benefits from using wind resources, especially where peak winds are in concert with peak electrical demand for air cooling, space heating, or agricultural water pumping. Natural gas prices in California reached 15 to 20 cents per kilowatt-hour (kWh) in the early 2000 price spike [electricity on the spot (short-term, daily) market hit 33 cents/kWh], whereas wind-power contracts came in at 3 cents/kWh in the same year.

In Europe, smaller turbines produce at 4 cents per kWh. Two Stanford University engineers calculated that the contracted price for wind power of 3 cents/kWh in the United States is cheaper than coal (estimated at 5.5 to 8.3 cents per kWh), when considering all costs. Lower industry estimates for coal costs exclude consideration of the health impacts of coal-mine dust, acid deposition, smog, and global warming. (See M. Z. Jacobsen and G. M. Masters, "Exploiting Wind vs. Coal," *Science* 293 (August 24, 2001): 1438; also see American Wind Energy Association, "Global Wind Energy Market Report," May 2001.)

Long-term costs favor wind-energy deployment. With *all* costs considered, wind energy is cost-competitive and actually cheaper than oil, coal, and nuclear power. The two Stanford engineers, in the study cited previously, stated,

> Much of the recent energy debate in the United States has focused on increasing coal use. However, the

(continued)

Focus Study 6.1 *(continued)*

cost of wind energy is now less than coal. Shifting from coal to wind would address health, environmental, and energy problems. (p. 1438)

Whether or not governments and transnational energy corpora-

tions support full-scale implementation of renewable resources such as wind, the energy realities of the near future will leave no alternative, and global political realities affirm this urgency. By the middle of this cen-

tury, wind-generated electricity could be routine, along with other renewable energy sources, conservation, and energy efficiency. Germany, Spain, and Denmark are leading the way for now.

---

its proximity to the Indian Ocean drive the monsoons of southern and eastern Asia (Figure 6.21). Also important to the generation of monsoonal flows are wind and pressure patterns in the upper-air circulation.

The extreme temperature range from summer to winter over the Asian landmass is due to its continentality (isolation from the modifying effects of the ocean). An intense high-pressure anticyclone dominates this continental landmass in winter (see Figures 6.11a and 6.21a), whereas the equatorial low-pressure trough (ITCZ) dominates the central area of the Indian Ocean. This pressure gradient produces cold, dry winds from the Asian interior over the Himalayas and across India. Average temperatures range between 15° and 20°C (60°–68°F) at lower elevations.

These winds desiccate, or dry out, the landscape and then give way to hot weather from March through May. The photo of the GPS unit in News Report 1.1 (Chapter 1) was made 18 m (60 ft) below the summit of Mount Everest on May 20, 1998—the monsoons dictate climbing schedules.

During the June–September wet period, the subsolar point (direct overhead rays of sunlight) shifts northward to the Tropic of Cancer, near the mouths of the Indus and Ganges Rivers. The ITCZ shifts northward over southern Asia, and the Asian continental interior develops a thermal low pressure, associated with high average temperatures (remember the summer warmth in Verkhoyansk, Siberia, from Chapter 5). Meanwhile, subtropical high pressure dominates the Indian Ocean, with a surface temperature of

FIGURE 6.21 The Asian monsoons.
Asian monsoon pressure and wind patterns during (a) winter and (b) summer. Note the shifting location of the ITCZ, the changing pressures over the Indian Ocean, and the different conditions over the Asian landmass. The inset climograph for Nagpur shows the severe contrast in seasonal precipitation. [Adapted from Joseph E. Van Riper, *Man's Physical World*, p. 215. Copyright 1971 by McGraw-Hill. Adapted by permission.]

(a) Winter, dry monsoon

(b) Summer, wet monsoon

(c) Precipitation at Nagpur, India

Station: Nagpur, India   Elevation: 310 m (1016 ft)
Lat/long: 21° 1′ N 79° 1′ E   Total ann. precip.: 124.2 cm (48.9 in.)

30°C (86°F). As a result of this reversed pressure gradient, hot subtropical air sweeps over the warm ocean, producing extremely high evaporation rates (Figure 6.21b).

By the time this air mass reaches India and the convergence zone, it is laden with moisture in thunderous, dark clouds. When the monsoonal rains arrive from June to September, they are welcome relief from the dust, heat, and parched land of Asia's springtime. Likewise, the annual monsoon is an integral part of Indian music, poetry, and life. World-record rainfalls of the wet monsoon drench India. Cherrapunji, India, received both the second-highest average annual rainfall (1143 cm, or 450 in.) and the highest single-year rainfall (2647 cm, or 1042 in.) on Earth.

The fact that the monsoons of southern Asia involve vast global pressure systems leads us to ask whether future climate change might affect present monsoonal patterns. Researchers are conducting a field experiment in the Indian Ocean, the Indian Ocean Experiment (INDOEX), to examine this question (**http://www-indoex.ucsd.edu/ ProjDescription.html**). Their model projections forecast higher rainfall if consideration is given to increases in carbon dioxide and its greenhouse warming alone. However, when consideration is given to increases in aerosols—principally sulfur compounds and black carbon—then a drop in precipitation of 7%–14% appears in the model. Air pollution reduces surface heating and therefore decreases the pressure differences at the heart of monsoonal flows. Considering that 70% of the annual precipitation for the entire region comes during the wet monsoon, the actual occurrence of such changes would force difficult societal adjustments to changing water resources. (See the satellite images in Figure 4.8 and review the text discussion.)

Oceanic Currents

The driving force for ocean currents is the frictional drag of the winds, thus linking the atmospheric and oceanic systems. Also important in shaping these currents is the interplay of the Coriolis force, density differences caused by temperature and salinity, the configuration of the continents and ocean floor, and astronomical forces (the tides).

Surface Currents

Figure 6.22 portrays the general patterns of major ocean currents. Because ocean currents flow over distance and through time, the Coriolis force deflects them. However, their pattern of deflection is not as tightly circular as that of the atmosphere. Compare this ocean-current map with the map showing Earth's pressure systems (see Figure 6.11) and you can see that ocean currents are driven by the circulation around subtropical high-pressure cells in both hemispheres. These circulation systems are known as *gyres* and generally appear to be offset toward the western side of each ocean basin. (Remember, in the Northern Hemisphere, winds and ocean currents move *clockwise* about high-pressure cells—note the currents in the North Pacific and North Atlantic on the map (see Figure 6.14). In the Southern Hemisphere, circulation is *counterclockwise* around high-pressure cells, evident on the map.)

In sailing days, the Spanish galleons (the *Manila Galleons*) would leave San Blas and Acapulco, Mexico (16.5° N), and sail southwest, catching the northeast trade winds across the Pacific to the Philippines and Manila (14° N). Goods would be traded and new commodities obtained, and the ships would move northward to catch the westerlies in the midlatitudes to be blown across the ocean to the shores of present-day Alaska or British

FIGURE 6.22 Major ocean currents.
[After the U. S. Navy Hydrographic Office.]

Ocean Circulation

News Report 6.3

A Message in a Bottle and Rubber Duckies

To give you an idea of the dynamic circulation of the ocean, consider the following examples. A 9-year-old child at Dana Point, California (33.5° N), a small seaside community south of Los Angeles, placed a letter in a glass juice bottle in July 1992 and tossed it into the waves. Thoughts of distant lands and fabled characters filled the child's imagination as the bottle disappeared. The vast circulation around the Pacific high, clockwise-circulating gyre, now took command (Figure 1).

Three years passed before ocean currents carried the message in a bottle to the coral reefs and white sands of Mogmog, a small island in Micronesia (7° N). A 7-year-old child there had found a pen pal from afar and immediately sent a photo and card to the sender of the message.

Imagine the journey of that note from California—traveling through storms and calms, clear moonlit nights and typhoons, as it floated on ancient currents as the galleons once had.

In January 1994, a large container ship from Hong Kong loaded with toys and other goods was ravaged by a powerful storm. One of the containers on board split apart in the wind off the coast of Japan, dumping nearly 30,000 rubber ducks, turtles, and frogs into the North Pacific. Westerly winds and the North Pacific current swept this floating cargo at up to 29 km (18 mi) a day across the ocean to the coast of Alaska, Canada, Oregon, and California. Other toys, still adrift, went through the Bering Sea and into the Arctic Ocean (this route is indicated by the dashed line in

Figure 1). Their drifting migration took a decade but some objects made it around the Arctic Ocean to the Labrador Sea and will end up in the Atlantic Ocean.

The high-floating message bottle and rubber ducks offered a much better opportunity to study winds than did an earlier spill of 60,000 athletic shoes near Japan. The low-floating shoes tracked across the Pacific Ocean until they landed in the Pacific Northwest. By the way, the shoes that did not make landfall headed (or footed) back around the Pacific gyre into the tropics and westward, back toward Japan! Scientists are using incidents such as the message in a bottle and rubber duckies to learn more about wind systems and ocean currents.

FIGURE 6.3.1 Pacific Ocean currents transport human artifacts.
The approximate route of a message in a bottle from Dana Point, California, to Mogmog, Micronesia, assumes a partial circuit around the Pacific gyre. Given the 3-year travel time, we do not know if the message circumnavigated the Pacific Ocean more than once. The rubber duckies voyaged eastward across the Pacific and beyond.

Columbia and sail along the coast to Northern California. The galleons would sail south, fighting the rain, frequent fog, and the right-hand Coriolis deflection pushing them away from the coast. The journey would end back in Mexico with cargo from the Orient. Thus the history of the *Manilla Galleons* is a lesson in oceanic currents around the Pacific gyre. News Report 6.3 dramatically portrays this clockwise circulation around the Pacific Ocean.

Equatorial Currents In Figure 6.22, you can see that trade winds drive the ocean surface waters westward in a

concentrated channel along the equator. These currents, called *equatorial currents*, are kept near the equator by the Coriolis force, which diminishes to zero at the equator. As these surface currents approach the western margins of the oceans, the water actually piles up against the eastern shores of the continents. The average height of this pileup is 15 cm (6 in.). This phenomenon is the **western intensification**.

The piled-up ocean water then goes where it can, spilling northward and southward in strong currents, flowing in tight channels along the eastern shorelines. In

FIGURE 6.23 Deep-ocean thermohaline circulation.
Scientists are deciphering centuries-long deep circulation in the oceans, sluggishly driven by temperature- and salinity-caused density differences. This global circulation is like a vast conveyor belt of water, drawing heat energy from surface environs and transporting it for release in the depths of the ocean basins.

 Ocean Circulation

the Northern Hemisphere, the *Gulf Stream* and the *Kuroshio* (current east of Japan) move forcefully northward as a result of western intensification. Their speed and depth are increased by the constriction of the area they occupy. The warm, deep, clear water of the ribbon-like Gulf Stream (Figure 5.10) usually is 50–80 km (30–50 mi) wide and 1.5–2.0 km (0.9–1.2 mi) deep, moving at 3–10 kmph (1.8–6.2 mph). In 24 hours, ocean water can move 70–240 km (40–150 mi) in the Gulf Stream.

Upwelling and Downwelling Flows Where surface water is swept away from a coast, either by surface divergence (induced by the Coriolis force) or by offshore winds, an **upwelling current** occurs. This cool water generally is nutrient-rich and rises from great depths to replace the vacating water. Such cold upwelling currents exist off the Pacific coasts of North and South America and the subtropical and midlatitude west coast of Africa. These areas are some of Earth's prime fishing regions.

In other regions where there is an accumulation of water—such as the western end of an equatorial current, or the Labrador Sea, or along the margins of Antarctica—the excess water gravitates downward in a **downwelling current**. These are the deep currents that flow vertically and along the ocean floor and travel the full extent of the ocean basins, carrying heat energy and salinity.

Thermohaline Circulation—The Deep Currents

Differences in temperatures and salinity produce density differences important to the flow of deep currents. This is Earth's **thermohaline circulation**, and it is different from wind-driven surface currents. Traveling at slower speeds

than surface currents, the thermohaline circulation hauls larger volumes of water. Figure 16.3 illustrates the ocean's physical structure and profiles of temperature, salinity, and dissolved gases; note the temperature and salinity differences with depth on this illustration.

To picture such a deep current, imagine a continuous channel of water beginning with cold water downwelling in the North Atlantic and upwelling in the Indian Ocean and North Pacific (Figure 6.23). Here it warms and then is carried in surface currents back to the North Atlantic. A complete circuit of the current may require 1000 years from downwelling in the Labrador Sea off Greenland to its reemergence in the southern Indian Ocean and to its return. Even deeper Antarctic bottom water flows northward in the Atlantic Basin beneath these currents.

These current systems appear to play a profound role in global climate; in turn, global warming has the potential to disrupt the downwelling in the North Atlantic and the thermohaline circulation. If this sinking of surface waters slowed or stopped, the ocean's ability to redistribute absorbed heat energy would be disrupted and surface accumulation of heat energy would add to the warming already underway. There is a freshening of ocean surface waters in both polar regions, contrasted with large increases in salinity in the surface waters at lower latitudes. Increased rates of glacier, sea-ice, and ice-sheet melting produce these fresh lower-density surface waters that ride on top of the denser saline water. The concern is that such changes in ocean temperature and salinity could dampen the rate of the North Atlantic Deep Water downwelling in the North Atlantic. One study claims to have detected a slowing of 10% to 15% in this downwelling from what is thought to be the long-term average.

This is a vital research frontier because Earth's hydrologic cycle is still little understood. Research groups such as Scripps Institution of Oceanography (http://sio.ucsd.edu/), Woods Hole Oceanographic Institution (http://www.whoi.edu/), Institute for the Study of Earth, Oceans, and Space (http://www.eos.sr.unh.edu/), Potsdam Institute for Climate Impact Research (http://www.pik-potsdam.de/), Bedford Institute of Oceanography (http://www.bio.gc.ca/welcome-e.html), to name only a few of many, are places to periodically check for updated information.

Summary and Review—Atmospheric and Oceanic Circulations

▸ *Define* the concept of air pressure, and *describe* instruments used to measure air pressure.

The weight (created by motion, size, and number of molecules) of the atmosphere is *air pressure*, which exerts an average force of approximately 1 kg/cm² (14.7 lb/in.²). A **mercury barometer** measures air pressure at the surface (mercury in a tube—closed at one end and open at the other, with the open end placed in a vessel of mercury—that changes level in response to pressure changes) or an **aneroid barometer** (a closed cell, partially evacuated of air, that detects changes in pressure).

> mercury barometer (p. 144)
> aneroid barometer (p. 145)

1. How does air exert pressure? Describe the basic instrument used to measure air pressure. Compare the operation of two different types of instruments discussed.

2. What is normal sea-level pressure in millimeters? Millibars? Inches? Kilopascals?

▸ *Define* wind, and *describe* how wind is measured, how wind direction is determined, and how winds are named.

Volcanic eruptions such as those of Tambora in 1815 and Mount Pinatubo in 1991 dramatically demonstrate the power of global winds to disperse aerosols and pollution worldwide in a matter of weeks. Atmospheric circulation facilitates important transfers of energy and mass on Earth, thus maintaining Earth's natural balances. Earth's atmospheric and oceanic circulations represent a vast heat engine powered by the Sun.

Wind is the horizontal movement of air across Earth's surface. Its speed is measured with an **anemometer** (a device with cups that are pushed by the wind) and its direction with a **wind vane** (a flat blade or surface that is directed by the wind). A descriptive scale useful in visually estimating wind speed is the traditional *Beaufort wind scale*.

> wind (p. 145)
> anemometer (p. 145)
> wind vane (p. 146)

3. What is a possible explanation for the beautiful sunrises and sunsets during the summer of 1992 in North America? Relate your answer to global circulation.

4. Explain this statement: "The atmosphere socializes humanity, making all the world a spatially linked society." Illustrate your answer with some examples.

5. Define wind. How is it measured? How is its direction determined?

6. Distinguish among primary, secondary, and tertiary general classifications of global atmospheric circulation.

7. What is the purpose of the Beaufort wind scale? Characterize winds given Beaufort numbers of 4, 8, and 12, and describe specific effects over both water and land.

▸ *Explain* the four driving forces within the atmosphere—gravity, pressure gradient force, Coriolis force, and friction force—and *describe* the primary high- and low-pressure areas and principal winds.

Earth's gravitational force on the atmosphere operates uniformly worldwide. Winds are driven by the **pressure gradient force** (air moves from areas of high pressure to areas of low pressure), deflected by the **Coriolis force** (an apparent deflection in the path of winds or ocean currents caused by the rotation of Earth; deflecting objects to the right in the Northern Hemisphere and to the left in the Southern Hemisphere), and dragged by the **friction force** (Earth's varied surfaces exert a drag on wind movements in opposition to the pressure gradient). Maps portray air pressure patterns using the **isobar**—an isoline that connects points of equal pressure. A combination of the pressure gradient and Coriolis forces alone produces **geostrophic winds**, which move parallel to isobars, characteristic of winds above the surface frictional layer.

Winds descend and diverge, spiraling outward to form an **anticyclone** (clockwise in the Northern Hemisphere), and they converge and ascend, spiraling upward to form a **cyclone** (counterclockwise in the Northern Hemisphere). The pattern of high and low pressures on Earth in generalized belts in each hemisphere produces the distribution of specific wind systems. These primary pressure regions are the **equatorial low-pressure trough**, the weak **polar high-pressure cells** (at both the North and South Poles), and the **subtropical high-pressure cells** and **subpolar low-pressure cells**.

All along the equator winds converge into the equatorial low creating the **intertropical convergence zone (ITCZ)**. Air rises along the equator and descends in the subtropics in each hemisphere. The winds returning to the ITCZ from the northeast in the Northern Hemisphere and from the southeast in the Southern Hemisphere produce the **trade winds**.

Winds flowing out of the subtropics to higher latitudes produce the **westerlies** in either hemisphere. The subtropical high-pressure cells on Earth, generally between 20° and 35° in either hemisphere, are variously named the **Bermuda high**, **Azores high**, and **Pacific high**.

Along the polar front and the series of low-pressure cells, the **Aleutian low** and **Icelandic low** dominate the North Pacific and Atlantic, respectively. This region of contrast between colder air toward the poles and warmer air toward the equator is the **polar front**. The weak and variable **polar easterlies** diverge from the polar high-pressure cells, particularly the **Antarctic high**.

> pressure gradient force (p. 148)
> Coriolis force (p. 148)
> friction force (p. 148)
> isobar (p. 149)
> geostrophic winds (p. 152)
> anticyclone (p. 153)

8. What does an isobaric map of surface air pressure portray? Contrast pressures over North America for January and July.

9. Describe the effect of the Coriolis force. Explain how it apparently deflects atmospheric and oceanic circulations.

10. What are geostrophic winds, and where are they encountered in the atmosphere?

11. Describe the horizontal and vertical air motions in a high-pressure anticyclone and in a low-pressure cyclone.

12. Construct a simple diagram of Earth's general circulation, including the four principal pressure belts or zones and the three principal wind systems.

13. How is the intertropical convergence zone (ITCZ) related to the equatorial low-pressure trough? How does it appear on a satellite image?

14. Characterize the belt of subtropical high pressure on Earth: Name the specific cells. Describe the generation of westerlies and trade winds. Discuss sailing conditions.

15. What is the relation among the Aleutian low, the Icelandic low, and migratory low-pressure cyclonic storms in North America? In Europe?

▶ *Describe* upper-air circulation and its support role for surface systems, and *define* the jet streams.

Air pressure in the middle and upper troposphere is described using a **constant isobaric surface**, or surface along which the same pressure, such as 500 mb, is recorded regardless of altitude. The height of this surface above the ground is described in ridges and troughs that support the development of and help sustain surface pressure systems; lows are sustained by divergence aloft, and highs are sustained by convergence aloft.

Vast, flowing, longwave undulations in these upper-air westerlies form wave motions called **Rossby waves**. Prominent streams of high-speed westerly winds in the upper-level troposphere are called the **jet streams**. Depending on their latitudinal position in either hemisphere, they are termed the *polar jet stream* or the *subtropical jet stream*.

constant isobaric surface (p. 158)
Rossby waves (p. 158)
jet streams (p. 160)

16. What is the relation between wind speed and the spacing of isobars?

17. How is the constant isobaric surface (ridges and troughs) related to surface pressure systems? To divergence aloft and surface lows? To convergence aloft and surface highs?

18. Relate the jet-stream phenomenon to general upper-air circulation. How is the presence of this circulation related to airline schedules from New York to San Francisco and the return trip to New York?

▶ *Overview* several multiyear oscillations of air temperature, air pressure, and circulation in the Arctic, Atlantic, and Pacific oceans.

Several system fluctuations that occur in multiyear or shorter periods are important in the global circulation picture. The most famous of these is the El Niño–Southern Oscillation (ENSO) phenomenon. A north–south fluctuation of atmospheric variability marks the *North Atlantic Oscillation (NAO)* as pressure differences between the Icelandic low and the Azores high in the Atlantic alternate in strength. The *Atlantic Oscillation (AO)* is the variable fluctuation between middle- and high-latitude air mass conditions over the Northern Hemisphere. The AO is associated with the NAO especially in winter, and the two phases of the *AO Index* correlate to the two-phased *NAO Index*.

Across the Pacific Ocean the *Pacific Decadal Oscillation (PDO)* involves variability between two regions of sea-surface temperatures and related air pressure: the northern and tropical western Pacific (region #1) and the area of the eastern tropical Pacific, along the West Coast (region #2). The PDO switches between positive and negative phases in 20- to 30-year cycles.

19. What phases are identified for the AO and NAO indices? What winter weather conditions generally affect the eastern United States during each phase?

20. What is the apparent relation between the PDO and the strength of El Niño events? Between PDO phases and the intensity of drought in the southwestern United States?

▶ *Explain* several types of local winds: land–sea breezes, mountain–valley breezes, katabatic winds, and the regional monsoons.

Different heating characteristics of land and water surfaces create **land–sea breezes**. **Mountain–valley breezes** are caused by temperature differences during the day and evening between valleys and mountain summits. **Katabatic winds**, or gravity drainage winds, are of larger regional scale and are usually stronger than mountain–valley breezes, under certain conditions. An elevated plateau or highland is essential, where layers of air at the surface cool, become denser, and flow downslope.

Intense, seasonally shifting wind systems occur in the tropics over Southeast Asia, Indonesia, India, northern Australia, equatorial Africa, and southern Arizona. These winds involve an annual cycle of returning precipitation with the summer Sun and are named after the Arabic word for season, *mausim*, or **monsoon**. The monsoons of southern and eastern Asia are driven by the location and size of the Asian landmass and its proximity to the Indian Ocean.

land–sea breezes (p. 163)
mountain–valley breezes (p. 163)
katabatic winds (p. 164)
monsoon (p. 164)

21. People living along coastlines generally experience variations in winds from day to night. Explain the factors that produce these changing wind patterns.

22. The arrangement of mountains and nearby valleys produces local wind patterns. Explain the day and night winds that might develop.

23. Describe the seasonal pressure patterns that produce the Asian monsoonal wind and precipitation patterns. Contrast January and July conditions.

⏺ *Discern* the basic pattern of Earth's major surface ocean currents and deep thermohaline circulation.

Ocean currents are primarily caused by the frictional drag of wind and occur worldwide at varying intensities, temperatures, and speeds, both along the surface and at great depths in the oceanic basins. The circulation around subtropical high-pressure cells in both hemispheres is notable on the ocean circulation map—these *gyres* are usually offset toward the western side of each ocean basin.

The trade winds converge along the ITCZ and push enormous quantities of water that pile up along the eastern shore of continents in a process known as the **western intensification**. Where surface water is swept away from a coast, either by surface divergence (induced by the Coriolis force) or by offshore winds, an **upwelling current** occurs. This cool water generally is nutrient-rich and rises from great depths to replace the vacating water. In other portions of the sea where there is an accumulation of water, the excess water gravitates downward in a **downwelling current**. These currents generate important mixing currents that flow along the ocean floor and travel the full extent of the ocean basins, carrying heat energy and salinity.

Differences in temperatures and salinity produce density differences important to the flow of deep, sometimes vertical, currents; this is Earth's **thermohaline circulation**. Traveling at slower speeds than wind-driven surface currents, the thermohaline circulation hauls larger volumes of water. There is scientific concern that increased surface temperatures in the ocean and atmosphere, coupled with climate-related changes in salinity, can alter the rate of thermohaline circulation in the oceans.

western intensification (p. 169)
upwelling current (p. 170)
downwelling current (p. 170)
thermohaline circulation (p. 170)

24. What is the relationship between global atmospheric circulation and ocean currents? Relate oceanic gyres to patterns of subtropical high pressure.

25. Define the western intensification. How is it related to the Gulf Stream and Kuroshio currents?

26. Where on Earth are upwelling currents experienced? What is the nature of these currents?

27. What is meant by deep-ocean thermohaline circulation? At what rates do these currents flow? How might this circulation be related to the Gulf Stream in the western Atlantic Ocean?

28. Relative to Question 27, what relation do these deep currents have to global warming and possible climate change?

NetWork

The *Geosystems* Home Page provides online resources for this chapter on the World Wide Web. To begin: Once on the Home Page, click on this textbook, scroll the Table of Contents menu, select this chapter, and click "Begin." You will find self-tests that are graded, review exercises, specific updates for items in the chapter, and many links to interesting related pathways on the Internet under "Destinations." *Geosystems* is at **http://www. prenhall.com/christopherson**

Critical Thinking

A. Using Table 6.1 and Figure 6.5, on a day with wind, estimate wind speed and wind direction at least twice during the day, and record them in your notebook. If possible, check these against the reports from some local source of weather information. What changes in winds do you notice over several days? How did these changes relate to the weather you experienced?

B. Refer to the "Short Answer" section in Chapter 6 of the *Geosystems* Home Page. Questions 1 through 5 deal with satellite images of pressure systems, wind, and cloud patterns. Using information in this chapter, complete these five items. Can you identify any causes for these circulation patterns? Note the other nine short-answer items for further thought.

C. Go to **http://www.awea.org/**, the web site of the American Wind Energy Association. Sample the materials presented as you assess Focus Study 6.1 and the potential for wind-generated electricity. What are your thoughts concerning this resource—its potential, reasons for delays, and the competitive economics presented? What countries are leading the way? Propose a brief action plan for more rapid progress, or a justification for more delays, depending on your point of view.

PART II

The Water, Weather, and Climate Systems

Earth is the water planet. Surface water in such quantity is unique in the Solar System. Chapter 7 explains why water covers more than two-thirds of the globe and describes the remarkable qualities and properties it possesses. We see the daily dynamics of the atmosphere—the powerful interaction of moisture and energy, the resulting stability and instability, and the variety of cloud forms—as a preamble to understanding weather. Chapter 8 examines weather and its causes. Topics include the interaction of air masses, the daily weather map, and analysis of violent phenomena in thunderstorms, tornadoes, and hurricanes, and the recent trends for each of these.

Chapter 9 explains water circulation on and over Earth in the hydrologic cycle—an output of the water–weather system. We examine the water-budget concept, which is useful in understanding soil-moisture and water-resource relationships on global, regional, and local scales. Potable (drinking) water could be the critical global political issue in this century. In Chapter 10, we see the spatial implications over time of the energy–atmosphere and water–weather systems and the generation of Earth's climatic patterns. In this way, Chapter 10 interconnects all the system elements from Chapters 2 through 9. Part Two closes with a discussion of global climate change, current conditions, and future climate trends.

Ice, water, and water vapor in the Arctic region comprise features of Earth's water, weather, and climate system. These huge icebergs are about 86% more massive beneath the ocean's surface than what you see. The mirrored reflections with cloud-diffused evening light create a scene of great beauty. The polar regions are the scientific focus of the International Polar Year 2007–2008 that will involve topics from PART TWO. [Photo by Bobbé Christopherson.]

Cumulonimbus and cumulus (congestus) clouds over the upper Midwest in spring 2004. These towering cloud tops reach above 9000 m (29,500 ft) altitude. Another cumulonimbus thunderhead is visible in the distance. Chapter 7 explains the processes that produce such cloud formations and leads us to the next chapter on weather. [Photo by Bobbé Christopherson.]

Water and Atmospheric Moisture

▶ **Key Learning Concepts**

After reading the chapter, you should be able to:

▶ *Describe* the origin of Earth's waters, *define* the quantity of water that exists today, and *list* the locations of Earth's freshwater supply.

▶ *Describe* the heat properties of water, and *identify* the traits of its three phases: solid, liquid, and gas.

▶ *Define* humidity and the expressions of the relative humidity concept, and *explain* dew-point temperature and saturated conditions in the atmosphere.

▶ *Define* atmospheric stability, and *relate* it to a parcel of air that is ascending or descending.

▶ *Illustrate* three atmospheric conditions—unstable, conditionally unstable, and stable—with a simple graph that relates the environmental lapse rate to the dry adiabatic rate (DAR) and moist adiabatic rate (MAR).

▶ *Identify* the requirements for cloud formation, and *explain* the major cloud classes and types, including fog.

Walden is blue at one time and green at another, even from the same point of view. Lying between the earth and the heavens, it partakes of the color of both. . . . A lake is the landscape's most beautiful and expressive feature. It is earth's eye; looking into which the beholder measures the depth of his own nature. . . . Sky water. It needs no fence. Nations come and go without defiling it. It is a mirror which no stone can crack. . . . Nature continually repairs. . . . a mirror in which all impurity presented to it sinks, swept and dusted by the sun's hazy brush. A field of water. . . . is continually receiving new life and motion from above. It is intermediate in its nature between land and sky.*

*Reprinted by permission of Merrill, an imprint of Macmillan Publishing Company, from *Walden* by Henry David Thoreau, pp. 192, 202, 204. Copyright © 1969 by Merrill Publishing. Originally published 1854.

Thus did Thoreau speak of the water so dear to him—Walden Pond in Massachusetts, where he lived along the shore.

Water is critical to our daily lives and is an extraordinary compound in nature. Several satellite probes detected ice beneath the poles of the Moon. At Mars in 2004, three orbiting spacecraft and two robot landers imaged and detected polar water ice and erosional features formed by flowing water early in Martian history and found evidence of subsurface water. Farther from the Sun, icy expanses can be seen on two of Jupiter's moons, Europa and Calisto—exciting discoveries about water in our Solar System. Yet, in the Solar System, water occurs in significant quantities only on our planet. Water covers 71% of Earth.

Pure water is colorless, odorless, and tasteless; yet, because it is a solvent (dissolves solids), pure water rarely occurs in nature. Water weighs 1 g/cm^3 (gram per cubic centimeter), or 1 kg/L (kilogram per liter). In the English system, water weighs 62.3 lb/ft^3, or 8.337 lb/gal.

Water constitutes nearly 70% of our bodies by weight and is the major ingredient in plants, animals, and our food. A human can survive 50 to 60 days without food but only 2 or 3 days without water. The water we use must be adequate, both in quantity and quality, for its many tasks—everything from personal hygiene to vast national water projects. Water indeed occupies that place between land and sky, mediating energy and shaping both the lithosphere and atmosphere, as Thoreau revealed. Water is the medium of life.

In this chapter: We examine water on Earth and the dynamics of atmospheric moisture and stability—the essentials of weather. The key questions answered include:

What were its origins? How much is there? And where is water located? An irony exists with this most common of compounds that possesses the most uncommon physical characteristics. Water's unique heat properties and existence in all three states in nature are critical in powering Earth's weather systems. Condensation of water vapor and atmospheric conditions of stability and instability are key to cloud formation. We end with clouds, our beautiful indicators of atmospheric conditions.

Water on Earth

Earth's hydrosphere contains about 1.36 billion cubic kilometers of water (specifically, 1,359,208,000 km^3, or 326,074,000 mi^3). According to scientific evidence, much of Earth's water originated from icy comets and hydrogen- and oxygen-laden debris that formed part of the planetesimals that coalesced to form the planet. The water within the planet reaches the surface by outgassing.

Outgassing is a continuing process by which water and water vapor emerge from layers deep within and below the crust, 25 km (15.5 mi) or more below Earth's surface. Figure 7.1 presents two such areas, among many sites worldwide, in New Zealand and Iceland. The Haukadalur Geysir geothermal area is in southeast Iceland, about 110 km from its capital, Reykjavik (Figure 7.1b). Written accounts of the Geysir activity date back to A.D. 1294 and it is the namesake of the term *geyser* we use today.

In the early atmosphere, massive quantities of outgassed water vapor condensed and then fell to Earth in torrential rains. For water to remain on Earth's surface, land temperatures had to drop below the boiling point of

(a)

(b)

FIGURE 7.1 Water outgassing from the crust.
Outgassing of water from Earth's crust in geothermal areas: (a) near Wairakei on the North Island of New Zealand; and (b) the Smidur boiling spring in the Geysir area in Haukadalur, Iceland. [Photos by (a) Bill Bachman/Photo Researchers, Inc.; (b) Bobbé Christopherson.]

100°C (212°F), something that occurred about 3.8 billion years ago. The lowest places across the face of Earth then began to fill with water—first ponds, then lakes and seas, and eventually ocean-sized bodies of water. Massive flows of water washed over the landscape, carrying both dissolved and solid materials to these early seas and oceans. Outgassing of water has continued ever since and is visible in volcanic eruptions, geysers, and seepage to the surface.

Worldwide Equilibrium

Today, water is the most common compound on the surface of Earth, having attained the present volume of 1.36 billion cubic kilometers approximately 2 billion years ago. This quantity has remained relatively constant, even though water is continuously being lost from the system. Water is lost when it dissociates into hydrogen and oxygen and the hydrogen escapes Earth's gravity to space, or when it breaks down and forms new compounds with other elements. Pristine water not previously at the surface, which emerges from within Earth's crust, replaces lost water in the system. The net result of these water inputs and outputs is that Earth's hydrosphere quantity is in a steady-state equilibrium where inputs and outputs are about equal.

Despite this overall net balance in water quantity, worldwide changes in sea level, called **eustasy** do occur. Eustatic changes relate to the water volume in the oceans and not changes in the overall quantity of planetary water. Some of these changes result when the amount of water stored in glaciers and ice sheets varies; these are called **glacio-eustatic** factors (see Chapter 17). In cooler times, as more water is bound up in glaciers (on mountains worldwide and in high latitudes) and in ice sheets (Greenland and Antarctica), sea level lowers. In warmer times, less water is stored as ice, so sea level rises.

Some 18,000 years ago, during the most recent ice-age pulse, sea level was more than 100 m (330 ft) lower than it is today; 40,000 years ago it was 150 m (about 500 ft) lower. Over the past 100 years, mean sea level has risen by 20–40 cm (8–16 in.) and is still rising worldwide as higher temperatures melt more ice.

Isostasy refers to actual physical changes in landmasses, such as continental uplift or subsidence, that cause apparent changes in sea level relative to coastal environments. Chapter 11 discusses such land elevation changes.

Distribution of Earth's Water Today

From a geographic point of view, ocean and land surfaces are distributed unevenly. If you examine a globe, it is obvious that most of Earth's continental land is in the Northern Hemisphere, whereas water dominates the Southern Hemisphere. In fact, when you look at Earth from certain angles, it appears to have an *oceanic hemisphere* and a *land hemisphere* (Figure 7.2).

An illustration of the present location of all of Earth's liquid and frozen water—whether fresh or saline, surface or underground—is in Figure 7.3. The oceans contain 97.22% of all water, about 1.321 billion cubic kilometers, or 317 million cubic miles, of saltwater (Figure 7.3a).

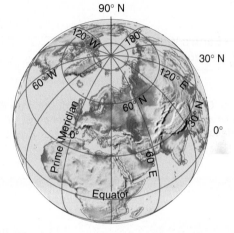

FIGURE 7.2 Land and water hemispheres.
Two perspectives that roughly illustrate how Earth's surface features an ocean hemisphere and a land hemisphere.

A table in the figure lists and details the four oceans—Pacific, Atlantic, Indian, and Arctic. The extreme southern portions of the Pacific, Atlantic, and Indian Oceans that surround the Antarctic continent are sometimes collectively called the "Southern Ocean."

Only 2.78% of all of Earth's water is freshwater (nonsaline and nonoceanic). The middle pie chart, along with Table 7.1, details this freshwater portion—surface water and subsurface water (Figure 7.3b). Ice sheets and glaciers are the greatest single repository of surface freshwater; they contain 77.14% of all of Earth's freshwater. Adding subsurface groundwater to frozen surface water accounts for 99.36% of all freshwater.

The remaining freshwater, which resides in lakes, rivers, and streams so familiar to us, actually represents less than 1% of all water (Figure 7.3c). All the world's freshwater lakes total only 125,000 km³ (30,000 mi³), with 80% of this volume in just 40 of the largest lakes and about 50% contained in just 7 lakes (listed with Table 7.1).

The greatest single volume of lakewater resides in 25-million-year-old Lake Baykal in Siberian Russia. It contains almost as much water as all five U.S. Great Lakes combined. Africa's Lake Tanganyika contains the next-largest volume, followed by the five Great Lakes. Overall, 70% of lakewater is in North America, Africa, and Asia,

Ocean	Earth's Ocean Area (%)	*Area (km² [mi²])	*Volume (km³ [mi³])	Mean Depth of Main Basin (m [ft])
Pacific	48	179,670 (69,370)	724,330 (173,700)	4280 (14,040)
Atlantic	28	106,450 (41,100)	355,280 (85,200)	3930 (12,890)
Indian	20	74,930 (28,930)	292,310 (70,100)	3960 (12,900)
Arctic	4	14,090 (5440)	17,100 (4100)	1205 (3950)

*Data in thousands (000): includes all marginal seas.

Earth's Water and the Hydrologic Cycle

ANIMATION

FIGURE 7.3 Ocean and freshwater distribution on Earth. Pie diagrams show (a) all-water, (b) freshwater only, and (c) surface-water percentages.

Table 7.1   Distribution of Freshwater on Earth

Location	Amount [km³ (mi³)]		Percentage of Freshwater	Percentage of Total Water
Surface Water				
Ice sheets and glaciers	29,180,000	(7,000,000)	77.14	2.146
Freshwater lakes*	125,000	(30,000)	0.33	0.009
Saline lakes and inland seas	104,000	(25,000)	0.28	0.008
Atmosphere	13,000	(3,100)	0.03	0.001
Rivers and streams	1,250	(300)	0.003	0.0001
Total surface water	29,423,250	(7,058,400)	77.78	2.164
Subsurface Water				
Groundwater—surface to 762 m (2500 ft) depth	4,170,000	(1,000,000)	11.02	0.306
Groundwater—762 to 3962 m (2500 to 13,000 ft) depth	4,170,000	(1,000,000)	11.02	0.306
Soil moisture storage	67,000	(16,000)	0.18	0.005
Total subsurface water	8,407,000	(2,016,000)	22.22	0.617
Total Freshwater (rounded)	**37,800,000**	**(9,070,000)**	**100.00%**	2.78%

*Major Freshwater Lakes	Volume [km³ (mi³)]		Surface Area [km² (mi²)]		Depth (m [ft])	
Baykal (Russia)	22,000	(5280)	31,500	(12,160)	1620	(5315)
Tanganyika (Africa)	18,750	(4500)	39,900	(15,405)	1470	(4923)
Superior (U.S./Canada)	12,500	(3000)	83,290	(32,150)	397	(1301)
Michigan (U.S.)	4,920	(1180)	58,030	(22,400)	281	(922)
Huron (U.S./Canada)	3,545	(850)	60,620	(23,400)	229	(751)
Ontario (U.S./Canada)	1,640	(395)	19,570	(7,550)	237	(777)
Erie (U.S./Canada)	485	(115)	25,670	(9,910)	64	(210)

FIGURE 7.4 Mono Lake, California, a saline lake.
Mono Lake is in the western Great Basin, with the Sierra Nevada range as a backdrop (aerial view is to the southwest in February). Formed approximately 3 million years ago, this lake is saline because it has no outlets other than evaporation. At its maximum, some 18,000 years ago, Lake Russell was about five times larger and six times deeper than the present Mono Lake in the photo, easily filling the basin marked by its ancient shorelines. [Photo by Bobbé Christopherson.]

with about a fourth of lakewater worldwide in innumerable small lakes. More than 3 million lakes exist in Alaska alone, and Canada has over 750 km² of lake surface.

Not connected to the ocean are saline lakes and salty inland seas. They usually exist in regions of interior river drainage (no outlet to the ocean), which allows salts to become concentrated. They contain 104,000 km³ (25,000 mi³) of water. Examples of such lakes include Utah's Great Salt Lake, California's Mono Lake, Southwest Asia's Caspian and Aral Seas, and the Dead Sea between Israel and Jordan (Figure 7.4).

Think of all the moisture in the atmosphere and Earth's thousands of flowing rivers and streams. Combined, they amount to only 14,250 km³ (3400 mi³), or only 0.033% of freshwater, or 0.0011% of all water! Yet, this small amount is very dynamic. A water molecule traveling through atmospheric and surface-water paths moves through the entire hydrologic cycle (ocean–atmosphere–precipitation–runoff) in less than two weeks. Contrast this to a water molecule in deep-ocean circulation, groundwater, or a glacier; moving slowly, it takes thousands of years to migrate through the system. Chapter 9 explores the hydrologic system and water resources in more detail.

Unique Properties of Water

Earth's distance from the Sun places it within a most remarkable temperate zone when compared with the locations of the other planets. This temperate location allows all three states of water—ice, liquid, and vapor—to occur naturally on Earth.

Even though water is the most common compound on Earth's surface, it exhibits most uncommon properties. Two atoms of hydrogen and one of oxygen, which readily bond, comprise each water molecule (suggested in Figure 7.5, upper left). Once hydrogen and oxygen atoms combine (in a covalent, or double, bond), they are difficult to separate, thereby producing a water molecule that remains stable in Earth's environment. This water molecule

Water Phase Changes

FIGURE 7.5 Three states of water and water's phase changes.
The three physical states of water: (a) gas, or water vapor, (b) water, and (c) ice. Note the molecular arrangement in each state and the terms that describe the changes from one phase to another. Also note how the polarity of water molecules bonds them to one another, loosely in the liquid state and firmly in the solid state. The plus and minus symbols denote whether heat energy is absorbed (+) or liberated (released) (−) during the phase change.

is a versatile solvent and possesses extraordinary heat characteristics.

The nature of the hydrogen–oxygen bond gives the hydrogen side of a water molecule a positive charge and the oxygen side a negative charge (see Figure 7.5, upper right). As a result of this *polarity*, water molecules attract each other: The positive (hydrogen) side of a water molecule attracts the negative (oxygen) side of another. This bonding between water molecules is *hydrogen bonding*.

The polarity of water molecules also explains why water "acts wet," sticking to things and dissolving many substances. Because of this solvent ability, pure water is rare in nature because something is usually dissolved in it.

The effects of hydrogen bonding in water are observable in everyday life. Hydrogen bonding creates the *surface tension* that allows you to float a steel needle on the surface of water, even though steel is much denser than water. This *surface tension* allows you to slightly overfill a glass with water; the water, actually above the rim of the glass, is held by a web of millions of hydrogen bonds.

Hydrogen bonding is the cause of *capillarity*, which you observe when you "dry" something with a paper towel. The towel draws water through its fibers because hydrogen bonds make each molecule pull on its neighbor. In chemistry laboratory classes, students observe the curved *meniscus*, or surface of the water, which forms in a cylinder or a test tube because hydrogen bonding allows the water to slightly "climb" the glass test-tube walls. *Capillary action* is an important component of soil–moisture processes, discussed in Chapter 9 and 18. Without hydrogen bonding to hold molecules together in water and ice, water would be a gas at normal surface temperatures.

Heat Properties

For water to change from one state to another (solid, liquid, or vapor), *heat energy must be absorbed or liberated (released)*. To cause a change of state, the amount of heat energy must be sufficient to affect the hydrogen bonds between molecules. This relation between water and heat energy is important to atmospheric processes. In fact, the heat exchanged between physical states of water provides more than 30% of the energy that powers the general circulation of the atmosphere.

Figure 7.5 presents the three states of water and the terms describing a change from one state to another, a **phase change**. Along the bottom of the illustration, *melting* and *freezing* describe the familiar phase change between solid and liquid. At the right, the terms *condensation* and *evaporation* (or *vaporization* at boiling temperature) apply to the change between liquid and vapor. At the left, the term **sublimation** refers to the direct change of ice to water vapor or water vapor to ice; although the latter is usually referred to as *deposition*, when water vapor attaches directly to an ice crystal. The deposition of water vapor to ice may form *frost* on surfaces.

Ice, the Solid Phase As water cools, it behaves like most compounds and contracts in volume. However, it reaches its greatest density not as ice, but as water at 4°C (39°F). Below that temperature, water behaves differently from other compounds. It begins to expand as more hydrogen bonds form among the slower-moving molecules, creating the hexagonal (six-sided) structures shown in Figure 7.5c and 7.6a. This six-sided preference applies to ice crystals of all shapes: plates, columns, needles, and dendrites (branching or treelike forms).

This expansion of water and ice that begins at 4°C continues to a temperature of −29°C (−20°F)—up to a 9% increase in volume is possible. This expansion is important in the weathering of rocks and in highway and pavement damage. News Report 7.1 discusses some effects of ice (for more on ice crystals and snowflakes, see **http://www.its.caltech.edu/~atomic/snowcrystals/** or see **http://www.lpsi.barc.usda.gov/emusnow/**).

News Report

News Report 7.1

Breaking Roads and Pipes and Sinking Ships

Road crews are busy in the summer repairing winter damage to streets and freeways in regions where winters are cold. A major contributor to this damage is the expansive phase change from water to ice. Rainwater seeps into roadway cracks and then expands as it freezes, thus breaking up the pavement. Perhaps you have noticed that bridges suffer the greatest damage. The reason is that cold air can circulate beneath a bridge and produces more freeze–thaw cycles on the bridge than in the roadbed on rock and soil.

The expansion of freezing water exerts a tremendous force—enough to crack plumbing or an automobile radiator or engine block. Wrapping water pipes with insulation to avoid damage is a common winter task in many places. People living in very cold climates use antifreeze and engine-block and battery heaters to avoid damage to vehicles. Historically, this physical property of water was put to useful work in quarrying rock for building materials. Holes were drilled and filled with water before winter so that when cold weather arrived, the water would freeze and expand, cracking the rock into manageable shapes.

A major hazard to ships in higher latitudes is posed by floating ice. Since ice has approximately 0.86 the density of water, an iceberg sits with approximately 6/7 of its mass below water level. The irregular edges of submarine ice can impact the side of a passing ship. Hitting an iceberg buckled plates and substandard rivets along the side of the RMS *Titanic* on its maiden voyage in 1912, causing its sinking and perhaps a momentary lowering of society's faith in technology.

(a)

(b)

(c)

(d)

FIGURE 7.6 The uniqueness of ice forms.
(a) Computer-enhanced photo reveals ice-crystal patterns, which are dictated by the internal structure between water molecules. This open structure also explains the lower density of ice and why it floats in water. Ice crystals demonstrate a unique interaction of chaos (all ice crystals are different) and the determinism of physical principles (all have a six-sided structure). (b) Icebergs off the Greenland coast melt faster below the ocean's surface, making them inherently unstable. Formerly underwater portions of the iceberg sculpted by currents and melt channels are visible. (c) Ice needles form in the freezing process of a meltwater stream in Greenland. (d) A bergy bit (small iceberg) off the coast of Antarctica illustrates the density–buoyancy state of floating ice. [(a) Photo enhancement © Scott Camazine/Photo Researchers, Inc., after W. A. Bentley; (b) (c) and (d) photos by Bobbé Christopherson.]

The expansion in volume that accompanies the freezing process results in a decrease in density (the same number of molecules occupy greater space). Specifically, pure ice has 0.91 times the density of water, so it floats. Without this change in density, much of Earth's freshwater would be bound in masses of ice on the ocean floor.

In nature, this density varies slightly given the age of the ice and air content. Therefore, icebergs float with slightly varying displacements that average about 1/7 (14%) of their mass exposed and about 6/7 (86%) for the submerged portion hidden beneath the ocean's surface (Figure 7.6d). With underwater portions melting faster than those above water, icebergs are inherently unstable and will overturn. You see in Figure 7.6b ruffled and fluted ice that was previously underwater. Look at the ice tower on the cover of this book, exceeding 60 m in height (200 ft), and imagine the mass of ice that is below the sea surface. No doubt it has run aground.

Chapter 13 discusses the freezing action of ice as an important physical weathering process. Chapter 17 discusses the freeze-and-thaw action of surface and subsurface water affecting approximately 30% of Earth's surface in periglacial landscapes, producing a variety of processes and landforms. An iceberg photo gallery is in Figure 17.36.

Water, the Liquid Phase   As a liquid, water assumes the shape of its container and is a noncompressible fluid. For ice to change to water, heat energy must increase the

FIGURE 7.7 Water's heat-energy characteristics.
The phase changes of water absorb or release a lot of latent heat energy. To transform 1 g of ice at 0°C to 1 g of water vapor at 100°C requires 720 calories: 80 + 100 + 540. The landscape illustrates phase changes between water (lake at 20°C) and water vapor under typical conditions in the environment.

motion of the water molecules to break some of the hydrogen bonds (Figure 7.5b). Despite the fact that there is no change in sensible temperature between ice at 0°C (32°F) and water at 0°C, 80 calories* of heat energy must be absorbed for the phase change of 1 g of ice to melt to 1 g of water (Figure 7.7, upper left). Heat energy involved in the phase change is **latent heat** and is hidden within the structure of water. It becomes liberated whenever the phase reverses and a gram of water freezes. The *latent heat of freezing* or the *latent heat of melting* involves 80 calories.

To raise the temperature of 1 g of water at 0°C (32°F) to boiling at 100°C (212°F), we must add 100 calories, gaining an increase of 1 C° (1.8 F°) for each calorie added. No phase change is involved in this temperature gain.

Water Vapor, the Gas Phase Water vapor is an invisible and compressible gas in which each molecule moves independently of the others (Figure 7.5a). The phase change from liquid to vapor at boiling temperature, under normal sea-level pressure, requires the addition of a much greater amount of heat energy than does the phase change from solid to liquid: Each gram of water changed to water vapor requires 540 calories (Figure 7.7). Those calories are the **latent heat of vaporization**. When water vapor condenses to a liquid, each gram gives up its hidden 540 calories as the **latent heat of condensation**. Perhaps you have felt the liberation of the latent heat of condensation on your skin from steam, as when you drain steamed vegetables or pasta or fill a hot teakettle.

In summary, taking 1 g of ice at 0°C and changing its phase to water, then to water vapor at 100°C—from a solid, to a liquid, to a gas—*absorbs* 720 calories (80 cal + 100 cal + 540 cal). Reversing the process, or

*Remember from Chapter 2: A calorie is the amount of energy required to raise the temperature of 1 gram of water (at 15°C), 1 degree Celsius and is equal to 4.184 joules.

changing the phase of 1 g of water vapor at 100°C to water then to ice at 0°C, *liberates* 720 calories into the surrounding environment. Your *Student Animation CD* that accompanies this book has an excellent animation illustrating these concepts.

Heat Properties of Water in Nature

In a lake or stream or in soil water, at 20°C (68°F), every gram of water that breaks away from the surface through evaporation must absorb from the environment approximately 585 cal as the *latent heat of evaporation* (see the natural scene in Figure 7.7). This is slightly more energy than would be required if the water were at a higher temperature such as boiling (540 cal). You can feel this absorption of latent heat as evaporative cooling on your skin when it is wet. This latent heat exchange is the dominant cooling process in Earth's energy budget.

The process reverses when air cools and water vapor condenses back into the liquid state, forming moisture droplets and thus liberating 585 cal for every gram of water as the *latent heat of condensation*. When you realize that a small, puffy, fair-weather cumulus cloud holds 500–1000 tons of moisture droplets, think of the tremendous latent heat released when water vapor condensed to droplets. Government meteorologists estimated that the moisture in Hurricane Andrew (1992) weighed nearly 30 trillion metric tons at its maximum power and mass. With 585 cal released for every gram as the latent heat of condensation, you can see that a weather event such as a hurricane involves a staggering amount of energy.

The **latent heat of sublimation** absorbs 680 cal as a gram of ice transforms into vapor. Water vapor freezing directly to ice releases a comparable amount of energy.

Humidity

Humidity refers to water vapor in the air. The capacity of air for water vapor is primarily a function of temperature—the temperatures of both the air and the water vapor, which are usually the same. There are several ways of expressing humidity, as discussed in this section.

We are all aware of humidity in the air, for its relationship to air temperature determines our sense of comfort. North Americans spend billions of dollars a year to adjust humidity, either with air-conditioning (extracting water vapor and cooling) or with air-humidifying (adding water vapor). We discussed the relation between humidity and temperature and the heat index in Chapter 5. To determine the energy available for powering weather, one has to know the water-vapor *content* of air and relate this in a ratio to the air's *saturation equilibrium* at a given temperature.

Relative Humidity

After air temperature and barometric pressure, the most common piece of information in local weather broadcasts is relative humidity. **Relative humidity** is a ratio (expressed as a percentage) of the amount of water vapor that is actually in the air compared to the maximum water vapor possible in the air at a given temperature. Warmer air increases the evaporation rate from water surfaces, whereas cooler air tends to increase the condensation rate of water vapor to water surfaces. Because there is a maximum amount of water vapor that can exist in a volume of air at a given temperature, the rates of evaporation and condensation can reach equilibrium at some point; the air is then saturated with humidity. Relative humidity tells us how near the air is to saturation and is an expression of an ongoing process of water molecules moving between air and moist surfaces, for condensation and evaporation operate continuously.

If air is relatively dry, the relative humidity percentage is low; if the air is relatively moist, the relative humidity percentage is higher. If the air is saturated with maximum water vapor for its temperature, the relative humidity percentage is 100% (Figure 7.8). Figure 7.8 shows that at 5 P.M. the evaporation rate exceeds condensation at these higher temperatures. At 11 A.M. the evaporation rate still exceeds condensation, though not by as much since daytime temperatures at 5 P.M. are not as high. At 5 A.M., in the cooler morning air, saturation equilibrium exists and any further cooling or addition of water vapor will produce net condensation.

FIGURE 7.8 Water vapor, temperature, and relative humidity.
The maximum water vapor possible in warm air is greater (net evaporation) than that of cold air (net condensation), so relative humidity changes with temperature, even though in this example the actual water vapor in the air stays the same during the day.

Cold glass chills the surrounding air layer to the dew-point temperature

Dew (active condensation)

Somebody forgot to put a coaster under the glass!

(a)

FIGURE 7.9 Dew-point temperature examples.
(a) The low temperature of the glass chills the surrounding air layer to the dew-point temperature and saturation. Thus, water vapor condenses out of the air and onto the glass as dew. (b) Cold air above the rain-soaked rocks is at the dew point and is saturated. Water evaporates from the rock into the air and condenses in the cold air in a changing veil of clouds. (c) The cold ocean surface chills the moist air layer to the dew point and saturation. As water vapor condenses, a dense fog forms. In the evening when temperatures drop over coastal lands, fog forms there too, giving the appearance that it is moving inland. [Photos by author.]

(b)

(c)

Relative humidity varies because of evaporation, condensation, or temperature changes. All three affect both the moisture in the air (the numerator) and the water vapor possible (the denominator) in the air. The formula to calculate the relative humidity ratio and express it as a percentage is simply:

$$\text{Relative humidity} = \frac{\text{Actual water vapor in the air}}{\begin{array}{c}\text{Maximum water vapor}\\\text{possible in the air at}\\\text{that temperature}\end{array}} \times 100$$

Saturation As stated, air is **saturated** when the rate of evaporation and the rate of condensation reach equilibrium—that is, 100% relative humidity. In saturated air, the net transfer of water molecules between a moist surface and the air reaches equilibrium. Saturation indicates that any further addition of water vapor or any decrease in temperature that reduces the evaporation rate will result in active condensation (clouds, fog, or precipitation). Therefore, relative humidity indicates the nearness of air to a saturated condition and when active condensation will begin.

The temperature at which a given mass of air becomes saturated and net condensation begins to form

water droplets is the **dew-point temperature**. *The air is saturated when the dew-point temperature and the air temperature are the same.* When temperatures are below freezing, the term *frost point* is sometimes used.

A cold drink in a glass provides a common example of these conditions. The water droplets that form on the outside of the glass condense from the air because the air layer next to the glass is chilled to below its dew-point temperature and thus becomes saturated (Figure 7.9). Figure 7.9 shows two additional examples of saturated air and active condensation above a rock surface and in a fog formation overlying a cool ocean surface.

Satellites now routinely sense water vapor in the lower atmosphere. Water vapor absorbs infrared wavelengths, making it possible to distinguish areas of relatively high water vapor from areas of low water vapor, using infrared sensors. Figure 7.10 includes images of Hurricane Michelle and the Western Hemisphere showing water-vapor content recorded by sensors of the "water-vapor channel." This knowledge is important to forecasting because it shows the moisture available to weather systems and therefore the available energy (latent heat) and precipitation potential of those systems.

(a)

(b)

FIGURE 7.10 Images of water vapor in the atmosphere.
GOES-8 infrared images portray water vapor in the atmosphere. (a) On the color scale used for this image, the lighter-gray tones denote higher water-vapor content, and color denotes high-altitude cloud tops that are cooler. Over the Gulf of Mexico and Caribbean, Hurricane Michelle and its strong vertical development is clearly visible heading toward western Cuba and near Florida, November 4, 2001. (b) Water vapor over the full Western Hemisphere; note subpolar low-pressure circulation.
[*GOES* images courtesy of NESDIS Satellite Services Division NOAA.]

Daily and Seasonal Relative Humidity Patterns An inverse relation occurs during a typical day between air temperature and relative humidity—as temperature rises, relative humidity falls (Figure 7.11a). Relative humidity is highest at dawn, when air temperature is lowest. If you park outdoors, you know about the wetness of the dew that condenses on the car overnight. In your own experience, you probably have noticed this pattern—the morning dew on windows, cars, and lawns evaporates by late morning as net evaporation increases with air temperature.

Relative humidity is lowest in the late afternoon, when higher temperatures increase the rate of evaporation. As shown in Figure 7.8, the actual water vapor present in the air may remain the same throughout the day. However, relative humidity changes because the temper-ature, and therefore the rate of evaporation, varies from morning to afternoon.

The seasonal variation in relative humidity by time of day confirms the relationship of temperature and relative humidity (Figure 7.11b). January readings are higher than July readings because air temperatures are lower overall in winter. Similar relative humidity records at most weather stations demonstrate the same relation among season, temperature, and relative humidity.

Expressions of Relative Humidity

There are several ways to express humidity and relative humidity. Each has its own utility and application. Two measures involve vapor pressure and specific humidity.

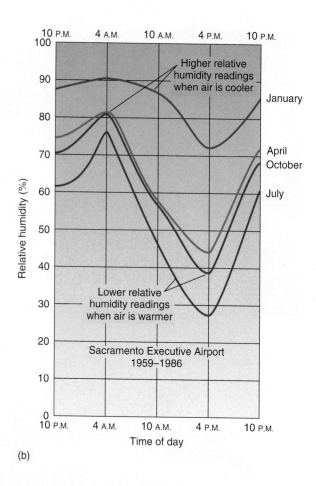

FIGURE 7.11 Daily and seasonal relative humidity patterns.
(a) Typical daily variations demonstrate temperature and relative humidity relations;
(b) seasonal variations in daily relative humidity.

Vapor Pressure As free water molecules evaporate from surfaces into the atmosphere, they become water vapor. Now part of the air, water-vapor molecules become a part of the air pressure measurement, for they exert a portion of the air pressure along with nitrogen and oxygen molecules. The share of air pressure that is made up of water-vapor molecules is **vapor pressure**. Millibars (mb) express vapor pressure, the same as air pressure.

As explained earlier, *saturation* is reached when the movement of water molecules between surface and air is in equilibrium. Air that contains as much water vapor as possible at a given temperature is at the *saturation vapor pressure*, which indicates how much water vapor is necessary to make the air saturated. Any temperature increase or decrease will change the saturation vapor pressure.

Figure 7.12 graphs the saturation vapor pressure at various air temperatures. The graph illustrates that, for every temperature increase of 10 C° (18 F°), the saturation water vapor pressure in air nearly doubles. This relation explains why warm tropical air over the ocean can hold so much water vapor, thus providing great latent heat to power tropical storms. It also explains why cold air is "dry" and why cold air toward the poles does not

produce a lot of precipitation (it contains too little water vapor, even though it is near the dew-point temperature).

As the graph shows, air at 20°C (68°F) has a saturation vapor pressure of 24 mb; that is, the air is saturated if the water-vapor portion of the air pressure is at 24 mb. Thus, if the water vapor actually present is exerting a vapor pressure of only 12 mb in 20°C air, the relative humidity is 50% (12 mb ÷ 24 mb = 0.50 × 100 = 50%). The inset in Figure 7.12 compares saturation vapor pressure over water and over ice surfaces at subfreezing temperatures. You can see that saturation vapor pressure is greater above a water surface than over an ice surface—that is, it takes more water-vapor molecules to saturate air above water than it does above ice. This fact is important to condensation processes and rain-droplet formation, both of which are described later in this chapter.

Specific Humidity A useful humidity measure is one that remains constant as temperature and pressure change. **Specific humidity** is the mass of water vapor (in grams) per mass of air (in kilograms) at any specified temperature. Because it is measured in mass, specific humidity is not affected by changes in temperature or pressure, such as when an air parcel rises to higher

FIGURE 7.12 Saturation vapor pressure.
Saturation vapor pressure of air at various temperatures—the maximum possible water vapor, as measured by the pressure it exerts. Inset compares saturation vapor pressures over water surfaces with those over surfaces at subfreezing temperatures. Note the 24 mb label for the text discussion.

FIGURE 7.13 Maximum specific humidity.
Maximum specific humidity for a mass of air at various temperatures—the maximum possible water vapor in mass of water vapor per unit mass of air (g/kg). Note the 47 g, 15 g, and 4 g labels for the text discussion.

elevations. Specific humidity stays constant despite volume changes.*

The maximum mass of water vapor possible in a kilogram of air at any specified temperature is the *maximum specific humidity*, plotted in Figure 7.13. The graph shows that a kilogram of air could hold a maximum specific humidity of 47 g of water vapor at 40°C (104°F), 15 g at 20°C (68°F), and about 4 g at 0°C (32°F). Therefore, if a kilogram of air at 40°C has a specific humidity of 12 g, its relative humidity is 25.5% (12 g ÷ 47 g = 0.255 × 100 = 25.5%). Specific humidity is useful in describing the moisture content of large air masses that are interacting in a weather system and provides necessary information for weather forecasting.

Instruments for Measuring Humidity Various instruments measure relative humidity. The **hair hygrometer** uses the principle that human hair changes as much as 4%

*Another similar measure used in relative humidity that approximates specific humidity is the *mixing ratio*, that is, the ratio of the mass of water vapor (grams) per mass of dry air (kilograms), as in g/kg.

in length between 0% and 100% relative humidity. The instrument connects a standardized bundle of human hair through a mechanism to a gauge. As the hair absorbs or loses water in the air, it changes length, indicating relative humidity (Figure 7.14a).

Another instrument used to measure relative humidity is a **sling psychrometer**. Figure 7.14b shows this device, which has two thermometers mounted side-by-side on a metal holder. One is the *dry-bulb thermometer*; it simply records the ambient (surrounding) air temperature. The other thermometer is the *wet-bulb thermometer*; it is set lower in the holder and its bulb is covered by a moistened cloth wick. The psychrometer is then spun ("slung") by its handle or placed where a fan forces air over the wet bulb.

The rate at which water evaporates from the wick depends on the relative saturation of the surrounding air. If the air is dry, water evaporates quickly, *absorbing the latent heat of evaporation from the wet-bulb thermometer and its wick*, cooling the thermometer and causing its temperature to drop (the *wet-bulb depression*). In conditions of high humidity, little water evaporates from the wick; in low humidity, more water evaporates. After being spun a minute or two, the temperature on each bulb is compared on a relative humidity (psychrometric) chart, from which relative humidity can be determined.

Now that you know something about atmospheric moisture, dew point, and relative humidity, let us examine the concept of stability in the atmosphere.

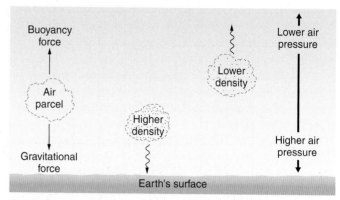

Rotating drum and graph paper on which relative humidity readings are traced

Bundled strands of human hair are held tightly between two brackets

Mechanism to sense tension from hair strands

Inked needle

(a)

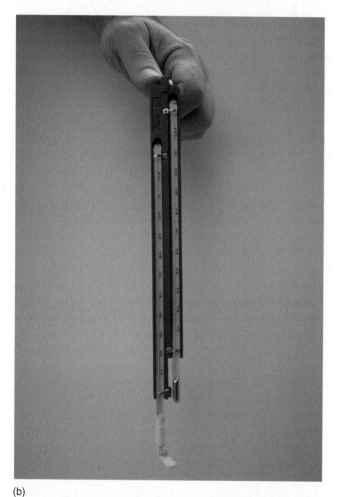

(b)

FIGURE 7.14 Instruments that measure relative humidity.
(a) The principle of a hair hygrometer. (b) Sling psychrometer with wet and dry bulbs. [Photo by Bobbé Christopherson.]

Atmospheric Stability

Meteorologists use the term *parcel* to describe a body of air that has specific temperature and humidity characteristics. Think of an air parcel as a volume of air, perhaps 300 m (1000 ft) in diameter, or more. The temperature of the volume of air determines the density of the air parcel. Warm air produces a lower density in a given volume of air; cold air produces a higher density. Two opposing forces work on a parcel of air: an upward *buoyant force* and

Buoyancy force

Air parcel

Gravitational force

Lower density

Higher density

Earth's surface

Lower air pressure

Higher air pressure

FIGURE 7.15 The forces acting on an air parcel.
Buoyancy and gravitational forces work on an air parcel. Different densities produce rising or falling parcels in response to imbalance in these forces.

ANIMATION Atmospheric Stability

a downward *gravitational force*. A parcel of lower density than the surrounding air rises (is more buoyant); a rising parcel expands as external pressure decreases. In contrast, a parcel of higher density descends (is less buoyant); a falling parcel compresses as external pressure increases. Figure 7.15 shows air parcels and illustrates these relationships. (In Chapter 8 we discuss air masses, which are larger parcels and regional in extent.)

An indication of weather conditions is the relative stability of air parcels and air masses in the atmosphere. **Stability** refers to the tendency of an air parcel, with its water-vapor cargo, either to remain in place or to change vertical position by ascending (rising) or descending (falling). An air parcel is *stable* if it resists displacement upward or, when disturbed, tends to return to its starting place. An air parcel is *unstable* if it continues to rise until it reaches an altitude where the surrounding air has a density and temperature similar to its own.

To visualize this, imagine a hot-air balloon launch. The air-filled balloon sits on the ground with the same air temperature inside as in the surrounding environment, like a stable air parcel. As the burner ignites, the balloon fills with hot (less-dense) air and rises buoyantly (Figure 7.16), like an unstable parcel of air. The concepts of stable and unstable air allow us to examine the specific temperature characteristics that produce these conditions.

Adiabatic Processes

Determining the degree of stability or instability requires measuring two temperatures: the temperature inside an air parcel and the temperature in the air surrounding the parcel. The contrast of these two temperatures determines stability. Such temperature measurements are made daily with instrument packages called *radiosondes* carried aloft by helium-filled balloons at thousands of weather stations.

The *normal lapse rate*, introduced in Chapter 3, is the average decrease in temperature with increasing altitude,

FIGURE 7.16 Principles of air stability and balloon launches.
Hot-air balloons being launched in the Swiss Alps illustrate the principles of stability. As the temperature inside a balloon increases, the air in the balloon becomes less dense than the surrounding air and the buoyancy force causes the balloon to rise, acting like a warm-air parcel. [Photo by Bap Vandystadt/Photo Researchers, Inc.]

a value of 6.4 C°/1000 m (3.5 F°/1000 ft). This rate of temperature change is for still, calm air, and it can vary greatly under different weather conditions. Consequently, the *environmental lapse rate* is the actual lapse rate at a particular place and time. It can vary by several degrees per thousand meters.

An ascending parcel of air cools by expansion, an expansion resulting from the reduced air pressure surrounding the parcel at higher altitudes (Figure 7.17a). A descending parcel heats by compression (Figure 7.17b). Physical laws that govern the behavior of gases explain these temperature changes internal to a moving air parcel.

When pressure on a parcel decreases as it rises to higher altitudes, the parcel expands and its temperature and density decrease (the expansion process consumes

sensible heat). Conversely, as an air parcel sinks toward the surface, air pressure increases, causing the parcel to compress. Compression increases the temperature and density in the parcel (compression produces sensible heat).

Temperature changes in both ascending and descending air occur *without any significant heat exchange between the surrounding environment and the vertically moving parcel of air*. The warming and cooling rates for a parcel of expanding or compressing air are termed **adiabatic**. (*Diabatic* means occurring with an exchange of heat; *adiabatic* means occurring *without a loss or gain* of heat energy to or from the environment.) There are two adiabatic rates, depending on moisture conditions in the vertically moving air parcel: a dry adiabatic rate (DAR) and a moist adiabatic rate (MAR).

Dry Adiabatic Rate The **dry adiabatic rate (DAR)** is the rate at which "dry" air cools by expansion (if ascending) or heats by compression (if descending). "Dry" refers to air that is less than saturated (relative humidity is less than 100%). The average DAR is 10 C°/1000 m (5.5 F°/1000 ft).

The rising parcel at the left in Figure 7.18a illustrates the principle. To see how a specific example of dry air behaves, consider an unsaturated parcel of air at the surface with a temperature of 27°C (81°F). It rises, expands, and cools adiabatically at the DAR as it rises to 2500 m (approximately 8000 ft). What happens to the temperature of the parcel? Calculate the temperature change in the parcel, using the dry adiabatic rate:

$$(10 \text{ C°}/1000 \text{ m}) \times 2500 \text{ m} = 25 \text{ C° of total cooling}$$

$$(5.5 \text{ F°}/1000 \text{ ft}) \times 8000 \text{ ft} = 44 \text{ F° of total cooling}$$

Subtracting the 25 C° (44 F°) of adiabatic cooling from the starting temperature of 27°C (81°F) gives the temperature in the air parcel at 2500 m as 2°C (36°F).

Air parcel cools internally as it expands under lower air pressure

Decreasing air pressure

(a) Cooling by expansion

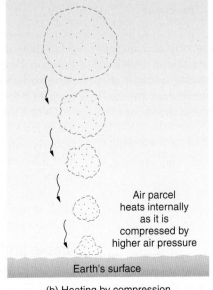

Air parcel heats internally as it is compressed by higher air pressure

(b) Heating by compression

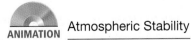

ANIMATION Atmospheric Stability

FIGURE 7.17 Vertically moving air experiences temperature changes.
(a) A rising air parcel cools by expansion. (b) A falling air parcel heats by compression.

(a) Air parcel *cools* adiabatically at the DAR

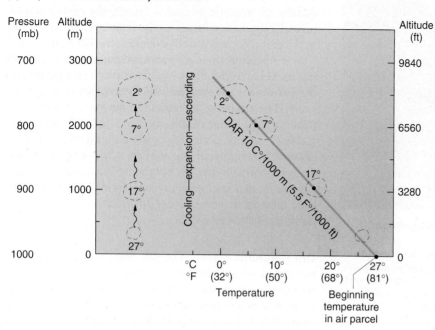

FIGURE 7.18 Adiabatic cooling and heating.

Vertically moving air parcels expand when they rise (because air pressure is less with increasing altitude) and are compressed when they descend. (a) An air parcel that is less than saturated cools adiabatically at the dry adiabatic rate (DAR). (b) A descending air parcel that is less than saturated heats adiabatically by compression at the DAR.

 ANIMATION Atmospheric Stability

(b) Air parcel *heats* adiabatically at the DAR

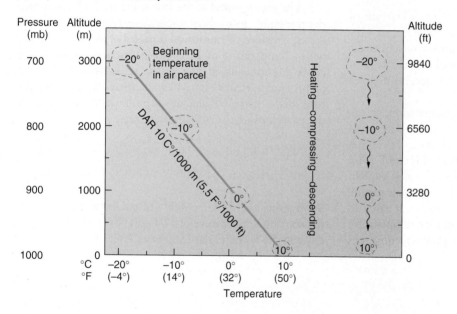

In Figure 7.18b, assume that an unsaturated air parcel with a temperature of −20°C is at 3000 m (−4°F at 9800 ft) descends to the surface, heating adiabatically. Using the dry adiabatic lapse rate, we determine the temperature of the air parcel when it arrives at the surface:

$$(10 \text{ C°}/1000 \text{ m}) \times 3000 \text{ m} = 30 \text{ C° of total warming}$$
$$(5.5 \text{ F°}/1000 \text{ ft}) \times 9800 \text{ ft} = 54 \text{ F° of total warming}$$

Adding the 30 C° of adiabatic warming from the starting temperature of −20°C gives the temperature in the air parcel at the surface as 10°C (50°F).

Moist Adiabatic Rate The **moist adiabatic rate (MAR)** is the rate at which an ascending air parcel that is moist, or saturated, cools by expansion or that a descending parcel warms by compression. The average MAR is

6 C°/1000 m (3.3 F°/1000 ft). This is roughly 4 C° (2 F°) less than the dry adiabatic rate. From this average, the MAR varies with moisture content and temperature and can range from 4 C° to 10 C° per 1000 m (2 F° to 5.5 F° per 1000 ft).

The cause of this variability, and the reason that the rate is lower than the DAR, is the latent heat of condensation. As water vapor condenses in the saturated air, latent heat is liberated, becoming sensible heat, and the adiabatic rate of cooling lowers. The release of latent heat may vary with temperature and water-vapor content. The MAR is much lower than the DAR in warm air, whereas the two rates are more similar in cold air.

Stable and Unstable Atmospheric Conditions

Now we bring this discussion together to determine atmospheric stability. The relation among the dry adiabatic

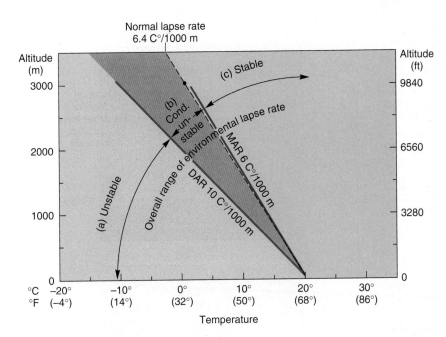

FIGURE 7.19 Temperature relationships and atmospheric stability.
The relationship between dry and moist adiabatic rates and environmental lapse rates produces three atmospheric conditions: (a) unstable (environmental lapse rate exceeds the DAR), (b) conditionally unstable (environmental lapse rate is between the DAR and MAR), and (c) stable (environmental lapse rate is less than DAR and MAR).

rate (DAR), moist adiabatic rate (MAR), and the environmental (actual) lapse rate at a given time and place determines the stability of the atmosphere over an area. You see examples of possible stability relationships in Figure 7.19. Let's examine what produces this variation in atmospheric stability.

Temperature relationships in the atmosphere produce the three different conditions: unstable, conditionally unstable, and stable. For the sake of illustration, the three examples in Figure 7.20 begin with an air parcel at the surface at 25°C (77°F). In each example, compare the temperatures of the air parcel and the surrounding environment. Assume that there is a lifting mechanism present, such as surface heating, a mountain range, or weather fronts, to get the parcel started (we examine lifting mechanisms in Chapter 8).

Given unstable conditions in Figure 7.20a, the air parcel continues to rise through the atmosphere because it is warmer (less dense and more buoyant) than the surrounding environment. Note that the environmental lapse rate on this occasion is at 12 C°/1000 m (6.6 F°/ 1000 ft). That is, the air surrounding the air parcel is cooler by 12 C° for every 1000-m increase in altitude. By 1000 m (about 3300 ft), the rising air parcel adiabatically cooled by expansion at the DAR from 25° to 15°C, while the surrounding air cooled from 25°C at the surface to 13°C. By comparing the temperature in the air parcel and the surrounding environment, you see that the temperature in the parcel is 2 C° (3.6 F°) warmer than the surrounding air at 1000 m. *Unstable* describes this condition because the less-dense air parcel will continue to lift.

Eventually, as the air parcel continues rising and cooling, it may achieve the dew-point temperature, saturation, and active condensation. This point of saturation forms the *lifting condensation level* that you see in the sky as the flat bottoms of clouds.

The example in Figure 7.20c shows stable conditions resulting when the environmental lapse rate is only 5 C°/1000 m (3 F°/1000 ft). An environmental lapse rate

of 5 C°/1000 m is less than both the DAR and the MAR, a condition in which the air parcel has a lower temperature (is more dense and less buoyant) than the surrounding environment. The relatively cooler air parcel tends to settle back to its original position—it is *stable*. The denser air parcel resists lifting, unless forced by updrafts or a barrier, and the sky remains generally cloud-free. If clouds form, they tend to be stratiform (flat clouds) or cirroform (wispy), lacking vertical development. In regions experiencing air pollution, stable conditions in the atmosphere worsen the pollution by slowing exchanges in the surface air.

You may be wondering what stability condition exists if the environmental lapse rate is somewhere between the DAR and MAR, causing conditions to be neither unstable nor stable. In Figure 7.20b, the environmental lapse rate is measured at 7 C°/1000 m. Under these conditions, the air parcel resists upward movement, unless forced, if it is less than saturated. But, if the air parcel becomes saturated and cools at the MAR, it acts unstable and continues to rise.

One example of such conditionally unstable air occurs when stable air is forced to lift as it passes over a mountain range. As the air parcel lifts and cools to the dew point, the air becomes saturated and condensation begins. Now the MAR is in effect and the air parcel behaves in an unstable manner. The sky may be clear and without a cloud, yet huge clouds may develop over a nearby mountain range.

With these stability relationships in mind, let us look at the most visible expressions of stability and humidity in the atmosphere: clouds and fog.

Clouds and Fog

Clouds are more than whimsical, beautiful decorations in the sky; they are fundamental indicators of overall conditions, including stability, moisture content, and weather. They form as air becomes saturated with water. Clouds are the subject of much scientific inquiry, especially regarding

FIGURE 7.20 Stability—three examples.
Specific examples of (a) unstable, (b) conditionally unstable, and (c) stable conditions in the lower atmosphere. Note the response to these three conditions in the air parcel on the right side of each diagram.

their effect on net radiation patterns, as discussed in Chapters 4 and 5. With a little knowledge and practice, you can learn to "read" the atmosphere from its signature clouds.

A **cloud** is an aggregation of tiny moisture droplets and ice crystals that are suspended in air, great enough in volume and concentration to be visible. Fog is simply a cloud in contact with the ground. Cloud types are too numerous to fully describe here, so we present the most common examples in a simple classification scheme.

Cloud Formation Processes

Clouds may contain raindrops, but not initially. At the outset, clouds are a great mass of moisture droplets, each

invisible without magnification. A **moisture droplet** is approximately 20 μm (micrometers) in diameter (0.002 cm, or 0.0008 in.). It takes a million or more such droplets to form an average raindrop of 2000 μm in diameter (0.2 cm, or 0.078 in.), as shown in Figure 7.21.

As an air parcel rises it may cool to the dew-point temperature where relative humidity is 100%. (Under certain conditions, condensation may occur at slightly less or more than 100% relative humidity.) More lifting of the air parcel cools it further, producing condensation of water vapor into water. Water does not just condense among the air molecules. Condensation requires **cloud-condensation nuclei**, microscopic particles that always are present in the atmosphere.

FIGURE 7.21 Moisture droplets and raindrops.
Cloud-condensation nuclei, moisture droplets, and a raindrop enlarged many times—compared at roughly the same scale.

Continental air masses average 10 billion cloud-condensation nuclei per cubic meter. Ordinary dust, soot, and ash from volcanoes and forest fires and particles from burned fuel, such as sulfate aerosols, typically provide these nuclei. The air over cities contains great concentrations of such nuclei. In maritime air masses, a high con-

centration of sea salts derived from ocean sprays, which average 1 billion nuclei per cubic meter, supply the needed nuclei. The lower atmosphere never lacks cloud-condensation nuclei.

Given the presence of saturated air, the availability of cloud-condensation nuclei, and the presence of cooling (lifting) mechanisms in the atmosphere, condensation occurs. Two principal processes account for the majority of the world's raindrops and snowflakes: the *collision-coalescence process* and the *Bergeron ice-crystal process*. Figure 7.22 summarizes raindrop and snowflake formation.

Cloud Types and Identification

As noted, a cloud is a collection of moisture droplets and ice crystals suspended in air in sufficient volume and concentration to be visible. In 1803, English biologist and amateur meteorologist Luke Howard, in his article "On the Modification of Clouds," established a classification system for clouds and coined Latin names for them that we still use. About Howard's accomplishment his biographer stated,

Clouds were no longer exempt from human comprehension, and Howard, in contributing both a system of

FIGURE 7.22 Raindrop and snowflake formation.
Principal processes for raindrop and snowflake formation: the *collision–coalescence process* and the *ice-crystal process*. (a) The collision–coalescence process predominates in clouds that form at above-freezing temperatures, principally in the warm clouds of the tropics. Initially, simple condensation takes place on small nuclei, some of which are larger and produce larger water droplets. As those larger droplets respond to gravity and fall through a cloud, they combine with smaller droplets, gradually coalescing into a raindrop. (b) The ice-crystal process occurs as supercooled water droplets (minute droplets of water that are below freezing and still in liquid form) evaporate rapidly near ice crystals, which then absorb the vapor. The ice crystals feed on the supercooled cloud droplets, grow in size, and eventually fall as snow or rain. Precipitation in middle and high latitudes can begin as ice and snow high in the clouds, then melt and gather moisture as it falls through the warmer portions of the cloud. [Art adapted from Frederick K. Lutgens and Edward J. Tarbuck, *The Atmosphere: An Introduction to Meteorology*, 3d ed., copyright © 1986, p. 127. Reprinted by permission of Prentice Hall, Inc., Englewood Cliffs, NJ.]

Table 7.2   Cloud Classes and Types

Class	Altitude/Composition at Midlatitudes	Type	Symbol	Description
Low clouds (C_L)	Up to 2000 m (6500 ft) Water	Stratus (St)		Uniform, featureless, gray, like high fog
		Stratocumulus (Sc)		Soft, gray, globular masses in lines, groups, or waves, heavy rolls, irregular overcast patterns
		Nimbostratus (Ns)		Gray, dark, low, with drizzling rain
Middle clouds (C_M)	2000–6000 m (6500–20,000 ft) Ice and water	Altostratus (As)		Thin to thick, no halos, Sun's outline just visible, gray day
		Altocumulus (Ac)		Patches of cotton balls, dappled, arranged in lines or groups, rippling waves, the lenticular clouds associated with mountains
High clouds (C_H)	6000–13,000 m (20,000–43,000 ft) Ice	Cirrus (Ci)		Mares'-tails, wispy, feathery, hairlike, delicate fibers, streaks, or plumes
		Cirrostratus (Cs)		Veil of fused sheets of ice crystals, milky, with Sun and Moon halos
		Cirrocumulus (Cc)		Dappled, "mackerel sky," small white flakes, tufts, in lines or groups, sometimes in ripples
Vertically developed clouds	Near surface to 13,000 m (43,000 ft) Water below, ice above	Cumulus (Cu)		Sharply outlined, puffy, billowy, flat-based, swelling tops, fair weather
		Cumulonimbus (Cb)		Dense, heavy, massive, dark thunderstorms, hard showers, explosive top, great vertical development, towering, cirrus-topped plume blown into anvil-shaped head

analysis and a full Latin nomenclature covering their families and genera, had contributed more than anyone to easing the path of understanding . . . But the naming of clouds was a different kind of gesture for the hand of classification to have made. Here was the naming not of a solid, stable thing but of a series of self-canceling evanescences [disappearing entities]. Here was the naming of a fugitive presence that hastened to its onward dissolution. Here was the naming of clouds.*

Altitude and *shape* are key to cloud classification. Clouds occur in three basic forms—flat, puffy, and wispy—and in four primary altitude classes and ten basic cloud types. Horizontally developed clouds—flat and layered—are *stratiform* clouds. Vertically developed clouds—puffy and globular—are *cumuliform* clouds. Wispy clouds usually are quite high in altitude and are made of ice crystals; these are *cirroform*.

These three basic forms occur in four altitudinal classes: low, middle, high, and those vertically developed through the troposphere. Table 7.2 presents the basic cloud classes and types. The cloud symbols used today were of Luke Howard's invention. Figure 7.23 illustrates the general appearance of each type and includes representative photographs.

*R. Hamblyn, *The Invention of Clouds, How An Amateur Meteorologist Forged the Language of the Skies* (New York: Farrar, Straus, and Giroux, 2001), pp. 165, 171.

Low clouds, ranging from the surface up to 2000 m (6500 ft) in the middle latitudes, are simply called *stratus* or *cumulus* (Latin for "layer" and "heap," respectively). **Stratus** clouds appear dull, gray, and featureless. When they yield precipitation, they become **nimbostratus** (*nimbo-* denotes "stormy" or "rainy"), and their showers typically fall as drizzling rain (Figure 7.23e).

Cumulus clouds appear bright and puffy, like cotton balls. When they do not cover the sky, they float by in infinitely varied shapes. Vertically developed cumulus clouds are in a separate class in Table 7.2 because further vertical development can produce cumulus clouds that extend beyond low altitudes into middle and high altitudes (illustrated at the far right in Figure 7.23 and 7.23d).

Sometimes near the end of the day **stratocumulus** may fill the sky in patches of lumpy, grayish, low-level clouds. Near sunset, these spreading puffy stratiform remnants may catch and filter the Sun's rays, sometimes indicating clearing weather.

The prefix *alto-* (meaning "high") denotes middle-level clouds. They are made of water droplets and, when cold enough, can be mixed with ice crystals. **Altocumulus** clouds, in particular, represent a broad category that occurs in many different styles: patchy rows, wave patterns, a "mackerel sky," or lens-shaped (lenticular) clouds.

Ice crystals in thin concentrations compose clouds occurring above 6000 m (20,000 ft). These wispy filaments, usually white except when colored by sunrise or sunset, are **cirrus** clouds (Latin for "curl of hair"), sometimes dubbed mares' tails. Cirrus clouds look as though an artist took a brush and added delicate feathery strokes high in the sky. Cirrus clouds indicate an oncoming storm, especially if they thicken and lower in elevation. The prefix *cirro-*, as in *cirrostratus* and *cirrocumulus*, indicates other high clouds that form a thin veil or puffy appearance, respectively.

A cumulus cloud can develop into a towering giant called **cumulonimbus** (again, *-nimbus* in Latin denotes "rain storm" or "thundercloud"; Figure 7.24). Such clouds are *thunderheads* because of their shape and associated lightning and thunder. Note the surface wind gusts, updrafts and downdrafts, heavy rain, and the presence of ice crystals at the top of the rising cloud column. High-altitude winds may then shear the top of the cloud into the characteristic anvil shape of the mature thunderhead.

Fog

By international definition, **fog** is a cloud layer on the ground, with visibility restricted to less than 1 km (3300 ft). The presence of fog tells us that the air temperature and the dew-point temperature at ground level are nearly identical, indicating saturated conditions. A temperature-inversion layer generally caps a fog layer (warmer temperatures above and cooler temperatures below the inversion altitude), with as much as 22 C° (40 F°) difference in air temperature between the cool ground under the fog and the warmer, sunny skies above.

Almost all fog is warm—that is, its moisture droplets are above freezing. Supercooled fog, which occurs when the moisture droplets are below freezing, is special because it can be dispersed by means of artificial seeding with ice crystals or other crystals that mimic ice, following the principles of the ice-crystal formation process described earlier. Let's briefly look at several types of fog.

Advection Fog   When air in one place *migrates* to another place where conditions are right for saturation, an **advection fog** forms. For example, when warm, moist air overlays cooler ocean currents, lake surfaces, or snow masses, the layer of migrating air directly above the surface becomes chilled to the dew point, and fog develops. Off all subtropical west coasts in the world, summer fog forms in the manner just described (Figure 7.25). Some coastal desert communities actually extract usable water from such fog formations, as described in News Report 7.2.

News Report 7.2

Harvesting Fog

Desert organisms have adapted remarkably to the presence of coastal fog along western coastlines in subtropical latitudes. For example, sand beetles in the Namib Desert in extreme southwestern Africa harvest water from the fog. They hold up their wings so condensation collects and runs down to their mouths. As the day's heat arrives, they burrow into the sand, only to emerge the next night or morning when the advection fog brings in more water for harvesting. Throughout history, people harvested water from fog. For centuries, coastal villages in the deserts of Oman collected water drips deposited on trees by coastal fogs.

In the Atacama Desert of Chile and Peru, residents stretch large nets to intercept the fog; moisture condenses on the netting and drips into trays, flowing through pipes to a 100,000-liter (26,000-gallon) reservoir. Large sheets of plastic mesh along a ridge of the El Tofo Mountains harvest water from advection fog (Figure 7.2.1). Chungungo, Chile, receives 10,000 liters (2,600 gallons) of water from 80 fog-harvesting collectors in a project developed by Canadian (International Development Research Center) and Chilean interests and made operational in 1993.

At least 30 countries across the globe experience conditions suitable for this water resource technology. (See **http://www.idrc.ca/nayudamma/fogcatc_72e.html**, click on "Health," and "Tapping into Fog.")

FIGURE 7.2.1 Fog harvesting.
In the mountains inland from Chungungo, Chile, polypropylene mesh stretched between two posts captures advection fog for local drinking-water supplies. [Photo by Robert S. Schemenauer.]

FIGURE 7.23 Principal cloud types.

Principal cloud types, classified by form (cirroform, stratiform, and cumuliform) and altitude (low, middle, high, and vertically developed across altitude): (a) altocumulus, (b) cirrus, (c) cirrostratus, (d) cumulonimbus, (e) nimbostratus, (f) stratus, (g) altostratus, and (h) cumulus.
[Photos (a), (b), (d), (g), and (h) by Bobbé Christopherson; (c), (e), and (f), by author.]

Cirrus

Cirrostratus

Anvil-shaped head

Cumulonimbus

Clouds with vertical development

Cumulus (fair weather)

(c)

(d)

(g)

(h)

(a)

(b)

(c)

FIGURE 7.24 Cumulonimbus thunderhead.
(a) Structure and form of a cumulonimbus cloud. Violent updrafts and downdrafts mark the circulation within the cloud. Blustery wind gusts occur along the ground.
(b) Space Shuttle astronauts capture a dramatic cumulonimbus thunderhead as it moves over Galveston Bay, Texas.
(c) Few acts of nature can match the sheer power released by an intense thunderstorm and related turbulence for air travelers. [(b) Space Shuttle photo from NASA; (c) photo by Bobbé Christopherson.]

Another type of advection fog forms when cold air lies over the warm water of a lake, ocean surface, or even a swimming pool. This wispy **evaporation fog,** or *steam fog,* may form as water molecules evaporate from the water surface into the cold overlying air, effectively humidifying the air to saturation followed by condensation to form fog (Figure 7.26). When evaporation fog happens at sea, it is a shipping hazard called *sea smoke.*

Advection fog also forms when moist air flows to higher elevations along a hill or mountain. This upslope lifting leads to adiabatic cooling by expansion as the air rises. The resulting **upslope fog** forms a stratus cloud at the condensation level of saturation. Along the Appalachians and the eastern slopes of the Rockies such fog is common in winter and spring. Another advection fog associated with topography is **valley fog.** Because cool air is denser than warm air, it settles in low-lying areas,

producing a fog in the chilled, saturated layer near the ground in the valley (Figure 7.27).

Radiation Fog When radiative cooling of a surface chills the air layer directly above that surface to the dewpoint temperature, creating saturated conditions, a **radiation fog** forms. This fog occurs over moist ground, especially on clear nights; it does not occur over water because water does not cool appreciably overnight. Slight movements of air deliver even more moisture to the cooled area for more fog formation of greater depth (Figure 7.28).

In Figure 7.29, can you tell what two kinds of fog are pictured? The river water is warmer than the cold overlying air, producing an evaporation fog, especially beyond the bend in the river. The moist farmlands have radiatively cooled overnight, chilling the air along the surface to

FIGURE 7.25 Advection fog.
San Francisco's Golden Gate Bridge shrouded by an invading advection fog characteristic of summer conditions along a western coast. [Photo by author.]

the dew point and active condensation. You see wisps of radiation fog as light air movements flow from right-to-left in the photo.

Every year the media carry stories of multicar pileups on stretches of highway where vehicles drive at high speed in foggy conditions. These crash scenes can involve

FIGURE 7.26 Evaporation fog
Evaporation fog (sea smoke) highlights dawn on a cold morning at Donner Lake, California. Later that morning as air temperatures rose, what do you think happened to the evaporation fog? [Photo by Bobbé Christopherson.]

FIGURE 7.27 Valley fog.
Cold air settles in the valleys of the Appalachian Mountains, chilling the air to the dew point and forming a valley fog. [Photo by author.]

dozens of cars and trucks. Fog is a hazard to drivers, pilots, sailors, pedestrians, and cyclists, and its conditions of formation are quite predictable. The spatial aspects of fog occurrence should be a planning element for any proposed airport, harbor facility, or highway. The prevalence of fog throughout the United States and Canada is shown in Figure 7.30.

FIGURE 7.28 Radiation fog.
Satellite image of a radiation fog in the southern Great Valley of California, November 20, 2002. This fog is locally known as a tule fog (pronounced "toolee") because of its association with the tule (bulrush) plants that line the low-elevation islands and marshes of the Sacramento River and San Joaquin River delta regions. [*Terra* MODIS image courtesy of MODIS Rapid Response Team, GSFC/NASA.]

humidity (p. 185)
relative humidity (p. 185)
saturated (p. 186)
dew-point temperature (p. 186)
vapor pressure (p. 188)
specific humidity (p. 188)
hair hygrometer (p. 189)
sling psychrometer (p. 189)

11. What is humidity? How is it related to the energy present in the atmosphere? To our personal comfort and how we perceive apparent temperatures?

12. Define relative humidity. What does the concept represent? What is meant by the terms *saturation* and *dew-point temperature*?

13. Using different measures of humidity in the air than those used in the chapter, derive relative humidity values (vapor pressure/saturation vapor pressure; specific humidity/maximum specific humidity) in Figures 7.12 and 7.13.

14. How do the two instruments described in this chapter measure relative humidity?

15. How does the daily distribution of relative humidity compare with the daily distribution of air temperature?

▶ *Define* atmospheric stability, and *relate* it to a parcel of air that is ascending or descending.

Meteorologists use the term *parcel* to describe a body of air that has specific temperature and humidity characteristics. Think of a parcel of air as a volume of air, perhaps 300 m (1000 ft) in diameter. The temperature of the volume of air determines the density of the air parcel. Warm air has a lower density in a given volume of air; cold air has a higher density.

Stability refers to the tendency of an air parcel, with its water-vapor cargo, either to remain in place or to change vertical position by ascending (rising) or descending (falling). An air parcel is *stable* if it resists displacement upward or, when disturbed, it tends to return to its starting place. An air parcel is *unstable* if it continues to rise until it reaches an altitude where the surrounding air has a density (air temperature) similar to its own.

stability (p. 190)

16. Differentiate between stability and instability relative to a parcel of air rising vertically in the atmosphere.

17. What are the forces acting on a vertically moving parcel of air? How are they affected by the density of the air parcel?

▶ *Illustrate* three atmospheric conditions—unstable, conditionally unstable, and stable—with a simple graph that relates the environmental lapse rate to the dry adiabatic rate (DAR) and moist adiabatic rate (MAR).

An ascending (rising) parcel of air cools by expansion, responding to the reduced air pressure at higher altitudes. A descending (falling) parcel heats by compression. These temperature changes internal to a moving air parcel are explained by physical laws that govern the behavior of gases. Temperature changes in both ascending and descending air parcels occur without any significant heat exchange between the surrounding environment and the vertically moving parcel of air. The warming and cooling rates for a parcel of expanding or compressing air are termed **adiabatic**.

The **dry adiabatic rate (DAR)** is the rate at which "dry" air cools by expansion (if ascending) or heats by compression (if descending). The term *dry* is used when air is less than saturated (relative humidity less than 100%). The DAR is 10 C°/1000 m (5.5 F°/1000 ft). The **moist adiabatic rate (MAR)** is the average rate at which ascending air that is moist (saturated) cools by expansion, or descending air warms by compression. The average MAR is 6 C°/1000 m (3.3 F°/1000 ft). This is roughly 4 C° (2 F°) less than the dry rate. The MAR, however, varies with moisture content and temperature and can range from 4 to 10 C° per 1000 m (2 to 5.5 F° per 1000 ft).

A simple comparison of the dry adiabatic rate (DAR) and moist adiabatic rate (MAR) in a vertically moving parcel of air with that of the environmental lapse rate in the surrounding air determines the atmosphere's stability—whether it is unstable (lifting of air parcel continues), stable (air parcel resists vertical displacement), or conditionally unstable (air parcel behaves as though unstable if the MAR is in operation, and stable otherwise).

adiabatic (p. 191)
dry adiabatic rate (DAR) (p. 191)
moist adiabatic rate (MAR) (p. 192)

18. How do the adiabatic rates of heating or cooling in a vertically displaced air parcel differ from the normal lapse rate and environmental lapse rate?

19. Why is there a difference between the dry adiabatic rate (DAR) and the moist adiabatic rate (MAR)?

20. What atmospheric temperature and moisture conditions would you expect on a day when the weather is unstable? When it is stable? Relate in your answer what you would experience if you were outside watching.

21. Use the "Atmospheric Stability" animation on your *Student Animation CD*. Try different temperature settings on the sliders to produce stable and unstable conditions.

▶ *Identify* the requirements for cloud formation, and *explain* the major cloud classes and types, including fog.

A **cloud** is an aggregation of tiny moisture droplets and ice crystals suspended in the air. Clouds are a constant reminder of the powerful heat-exchange system in the environment. **Moisture droplets** in a cloud form when saturated air and the presence of **cloud-condensation nuclei** lead to *condensation*. Raindrops are formed from moisture droplets through either the *collision–coalescence process* or the *Bergeron ice-crystal process*.

Low clouds, ranging from the surface up to 2000 m (6500 ft) in the middle latitudes, are **stratus** (flat clouds, in layers) or **cumulus** (puffy clouds, in heaps). When stratus clouds yield precipitation, they are **nimbostratus**. Sometimes near the end of the day, lumpy, grayish, low-level clouds called **stratocumulus** may fill the sky in patches. Middle-level clouds are denoted by the prefix *alto-*. **Altocumulus** clouds, in particular, represent a broad category that occurs in many different styles. Clouds at high altitude, principally composed of ice crystals, are called **cirrus**. A cumulus cloud can develop into a towering giant **cumulonimbus** cloud (-*nimbus* in Latin denotes "rain storm" or "thundercloud"). Such clouds are called *thunderheads* because of their shape and their associated lightning, thunder, surface wind gusts, updrafts and downdrafts, heavy rain, and hail.

Fog is a cloud that occurs at ground level. **Advection fog** forms when air in one place migrates to another place where conditions exist that can cause saturation—for example, when warm, moist air moves over cooler ocean currents. Another type of advection fog forms when cold air flows over the warm water of a lake, ocean surface, or swimming pool. This **evaporation fog**, or steam fog, may form as the water molecules evaporate from the water surface into the cold overlying air. **Upslope fog** is produced when moist air is forced to higher elevations along a hill or mountain. Another fog caused by topography is **valley fog**, formed because cool, denser air settles in low-lying areas, producing fog in the chilled, saturated layer near the ground. Radiative cooling of a surface that chills the air layer directly above the surface to the dew-point temperature creates saturated conditions and a **radiation fog**.

cloud (p. 194)
moisture droplet (p. 194)
cloud-condensation nuclei (p. 194)
stratus (p. 196)
nimbostratus (p. 196)
cumulus (p. 196)
stratocumulus (p. 196)
altocumulus (p. 196)
cirrus (p. 197)

cumulonimbus (p. 197)
fog (p. 197)
advection fog (p. 197)
evaporation fog (p. 200)
upslope fog (p. 200)
valley fog (p. 200)
radiation fog (p. 200)

22. Specifically, what is a cloud? Describe the droplets that form a cloud.

23. Explain the condensation process: What are the requirements? What two principal processes are discussed in this chapter?

24. What are the basic forms of clouds? Using Table 7.2, describe how the basic cloud forms vary with altitude.

25. Explain how clouds might be used as indicators of the conditions of the atmosphere and of expected weather.

26. What type of cloud is fog? List and define the principal types of fog.

27. Describe the occurrence of fog in the United States and Canada. Where are the regions of highest incidence?

NetWork

The *Geosystems* Home Page provides online resources for this chapter on the World Wide Web. To begin: Once on the Home Page, click on this textbook, scroll the Table of Contents menu, select this chapter, and click "Begin." You will find self-tests that are graded, short essay and review exercises, specific updates for items in the chapter, and many links to interesting related pathways on the Internet under "Destinations." *Geosystems* is at **http://www.prenhall.com/christopherson**

Critical Thinking

A. Examine the iceberg photographs in Figure 7.6b and 7.6d and on this book's cover. These photos show Greenland and Antarctic waters. Determine what caused the "shoreline" thermal notch above the ocean surface around the iceberg in 7.6b. Why do you think the iceberg appears to be riding higher in the water than several weeks earlier? In rough terms, how much ice (in percent) do you think is beneath the surface in comparison to the amount you see above sea level in the iceberg cover photo? Explain why this physical trait can be a hazard to shipping.

B. Using Figure 7.23, begin to observe clouds on a regular basis. See if you can relate the cloud type to particular weather conditions at the time of observation. You may want to keep a log in your notebook during this physical geography course. Seeing clouds in this way will help you to understand weather and make studying Chapter 8 easier.

C. Given our discussion of relative humidity, air temperature, dew-point temperature, and saturation, please refer to the "Short Answer" section of Chapter 7, items 1 and 2, on the *Geosystems* Home Page. View the dew-point temperature map in relation to the map of air temperatures. As you compare and contrast the maps, describe in general terms what relative humidity conditions you find in the Northwest and along the Gulf Coast in the Southeast.

February 1, 2000: Parts of Washington and Oregon receive heavy snow in the mountains and rain elsewhere from a jet-stream-guided storm system. Wind gusts reach 110 kmph (70 mph) along the Oregon Coast as air rushes into the low-pressure center. The next system is far to the west over the Pacific Ocean. [*GOES-10* image courtesy of NESDIS/NOAA; weather map inset courtesy of "Daily Weather Maps," National Weather Service, NOAA.]

CHAPTER 8

Weather

▶ **Key Learning Concepts**

After reading the chapter, you should be able to:

▶ **Describe** air masses that affect North America, and **relate** their qualities to source regions.

▶ **Identify** four types of atmospheric lifting mechanisms, and **describe** four principal examples.

▶ **Analyze** the pattern of orographic precipitation, and **describe** the link between this pattern and global topography.

▶ **Describe** the life cycle of a midlatitude cyclonic storm system, and **relate** this to its portrayal on weather maps.

▶ **List** the measurable elements that contribute to modern weather forecasting, and **describe** the technology and methods employed.

▶ **Analyze** various forms of violent weather and the characteristics of each, and **review** several examples of each from the text.

Water has a leading role in the vast drama played out daily on Earth's stage. It affects the stability of air masses and their interactions and produces powerful and beautiful special effects in the lower atmosphere. Air masses come into conflict; they move and shift, dominating now one region then another, varying in strength and characteristics. Think of the weather as a play, North America the stage, and the air masses as actors of varying ability.

In this chapter: We follow huge air masses across North America, observe powerful lifting mechanisms in the atmosphere, revisit the concepts of stable and unstable conditions, examine migrating cyclonic systems with attendant cold and warm fronts, and conclude with a portrait of violent and dramatic weather. Water, with its ability to absorb and release vast quantities of heat energy, drives this daily drama in the atmosphere. Also, modern weather forecasting is introduced. The spatial implications of these weather phenomena and their relationship to human activities strongly link meteorology to physical geography and this chapter. We begin with a look at some weather essentials.

Weather Essentials

Weather is the short-term, day-to-day condition of the atmosphere, contrasted with *climate*, which is the long-term average (over decades) of weather conditions and extremes in a region. Weather is, at the same time, both a "snapshot" of atmospheric conditions and a technical status report of the Earth–atmosphere heat-energy budget. Important elements that contribute to weather are temperature, air pressure, relative humidity, wind speed and direction, and seasonal factors such as insolation receipt related to daylength and Sun angle.

We turn for weather forecasts to the National Weather Service (NWS) in the United States (**http://www.nws .noaa.gov**) or the Canadian Meteorological Centre, a branch of the Meteorological Service of Canada (MSC) (**http://www.msc-smc.ec.gc.ca/cmc/index_e.html**) to see current satellite images and to hear weather analysis. Internationally, the World Meteorological Organization coordinates weather information (see **http://www. wmo.ch/**). Many sources of weather information and related topics are found in the "Destinations" section for this chapter on the *Geosystems* Home Page.

Meteorology is the scientific study of the atmosphere. (*Meteor* means "heavenly" or "of the atmosphere.") Meteorologists study the atmosphere's physical character-istics and motions, related chemical, physical, and geologic processes, the complex linkages of atmospheric systems, and weather forecasting. Computers handle volumes of data for accurate forecasting of near-term weather and for studying trends in long-term weather, climatology, and climatic change. New developments in supercomputing, an Earth-bound instrument network in the Automated Surface Observing System (ASOS) arrays, orbiting observation systems, and Doppler radar installations are rapidly advancing the science of the atmosphere.

An important tool in each NWS Weather Forecast Office is the Advanced Weather Interactive Processing System (AWIPS, see Figure 8.17). AWIPS allows forecasters to integrate meteorological, hydrological, satellite, and radar data at a computer workstation. The forecaster and scientist have an interactive ability to observe, analyze, and manipulate geospatial weather data. The network of 135 AWIPS stations (including 13 River Forecast Centers) eventually will increase to 148 stations.

An essential part of understanding weather is Doppler radar. Using backscatter from two radar pulses, it detects the direction of moisture droplets toward or away from the radar, indicating wind direction and speed. This information is critical to weather forecasting and severe storm warnings. Through June 2004, 158 WSR-88D (*Weather Surveillance Radar*) Doppler radar systems as part of the NEXRAD (Next Generation Weather Radar) program were operational through the National Weather Service (125 systems) and in conjunction with the Federal Aviation Administration (12 systems) and the Department of Defense (21 systems) (Figure 8.1). In Canada through 2003, the National

(a)

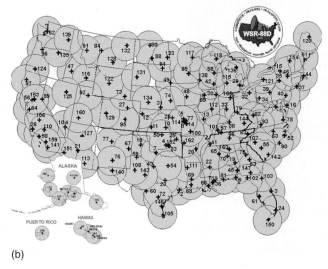

(b)

FIGURE 8.1 Weather installation.
(a) Doppler radar installation at the Indianapolis International Airport operated by the National Weather Service. The radar antenna is sheltered within the dome structure. (b) WSR-88D coverage in the United States. Other installations are maintained in Okinawa, Guam, South Korea, and the Azores.
[(a) Photo by Bobbé Christopherson; (b) adapted map courtesy of Radar Operations Center, Norman, Oklahoma.]

Radar Project had placed 25 CWSR-98 radars in service out of an eventual 30 units.

Weather data collection is expanding—China installed a remotely operated weather station at 5200 m (17,060 ft) altitude on the Tibetan side of Mount Everest. Now, measures of wind speed and direction, air pressure, temperature, humidity, and precipitation will assist forecasters and mountain climbers alike. One hundred seventy-five people died climbing Earth's highest mountain above sea level over the past 80 years; it is hoped that this satellite-linked system will reduce some of the risk.

The cost of weather-related destruction has risen more than 500% over the past two decades, from an average of $2 billion to $10 billion annually. For instance, 1998 weather damage alone topped $90 billion worldwide, which exceeds the total for all the 1980s, even when adjusted for inflation! Floods, droughts, ice storms, tropical cyclones and hurricanes, tornadoes, coastal storm surges, and heat waves add to the need that we improve our understanding of weather.

Air Masses

Each area of Earth's surface imparts its temperature and moisture characteristics to the air it touches. The effect of the surface on the air creates regional masses of air having specific conditions of temperature, humidity, and stability. These masses of air interact to produce weather patterns—in essence, they are the actors in our weather drama. Such a distinctive body of air is called an **air mass**, and it initially reflects the characteristics of its *source region*. For example, weather forecasters speak of a "cold Canadian air mass" and "moist tropical air mass."

The longer an air mass remains stationary over a region, the more definite its physical attributes become. Within each air mass there is a homogenous mix of temperature and humidity that sometimes extends through the lower half of the troposphere. Such masses of air possess all the physical characteristics of the atmosphere discussed in earlier chapters and thus link the Earth–atmosphere energy budget and water–weather systems.

Air Masses Affecting North America

Air masses generally are classified according to the moisture and temperature characteristics of their source regions:

1. *Moisture*—designated **m** for maritime (wet) and **c** for continental (dry).
2. *Temperature* (latitude factor)—designated **A** (arctic), **P** (polar), **T** (tropical), **E** (equatorial), and **AA** (Antarctic).

The principal air masses that affect North America in winter and summer are mapped in Figure 8.2.

Continental polar (cP) air masses form only in the Northern Hemisphere and are most developed in winter when they dominate cold weather conditions. These cP air masses are major players in middle- and high-latitude weather. The cold, dense cP air lifts moist, warm air in its

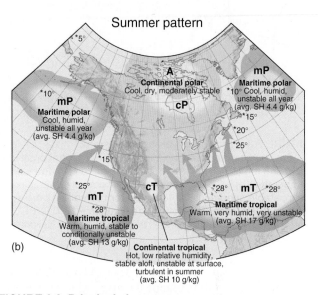

FIGURE 8.2 Principal air masses.
These are the air masses that influence North America in winter (a) and summer (b). (*Sea-surface temperature in °C; SH = specific humidity.)

path, producing lifting, cooling, and condensation. As we will see, the greater the temperature difference between this cold air mass and warmer air masses, the more dramatic is the weather produced. An area covered by cP air in winter experiences cold, stable air, clear skies, high pressure, and anticyclonic wind flow, all visible on the weather map in Figure 8.3. The Southern Hemisphere lacks the necessary landmasses to produce continentality at high latitudes, to create continental polar characteristics.

Maritime polar (mP) air masses in the Northern Hemisphere exist northwest and northeast of the North American continent over the northern oceans. Within them, cool, moist, unstable conditions prevail throughout the year. The Aleutian and Icelandic subpolar low-pressure cells reside within these mP air masses, especially in their well-developed winter pattern (see the chapter-opening satellite image).

Two maritime tropical (mT) air masses—the *mT Gulf/Atlantic* and the *mT Pacific*—influence North America.

FIGURE 8.3 Winter high-pressure system.
A cP air mass dominates the Midwest with a central pressure of 1042.8 mb (30.76 in.), air temperature of −17°C, dew-point temperature of −21°C (2° and −5°F, respectively), with clear, calm, stable conditions. Note the pattern of isobars portraying the cP air mass. The dotted lines are the −18°C (0°F) and 0°C (32°F) isotherms. [Weather map courtesy of National Weather Service, NOAA.]

The humidity experienced in the East and Midwest is created by the mT Gulf/Atlantic air mass, which is particularly unstable and active from late spring to early fall (Figure 8.4). In contrast, the mT Pacific is stable to conditionally unstable and generally lower in moisture content and available energy. As a result, the western United States, influenced by this weaker Pacific air mass, receives lower average precipitation than the rest of the country.

Please review Figure 6.14 and the discussion of subtropical high-pressure cells off the coast of North America—the moist, unstable conditions on the western edge of the Atlantic (east coast), and the drier, stable conditions on the eastern edge of the Pacific (west coast). These conditions, coupled respectively with ocean currents that are warmer (Gulf Stream) and cooler (California Current), produce the characteristics of each source region for these different maritime air masses.

Air Mass Modification

As air masses migrate from their source regions, their temperature and moisture characteristics modify and slowly take on the characteristics of the land over which they pass. For example, an mT Gulf/Atlantic air mass may carry humidity to Chicago and on to Winnipeg but gradually will lose its initial characteristics of high humidity and warmth with each day's passage northward.

Similarly, below-freezing temperatures occasionally reach into southern Texas and Florida, brought by an invading winter cP air mass from the north. However, that air mass warms from the −50°C (−58°F) of its source region in central Canada. In winter, as a cP air mass moves southward over warmer land, it moderates, warming especially after it leaves areas covered by snow.

Modification of cP air as it moves south and east produces snowbelts that lie to the east of each of the Great Lakes. As below-freezing cP air passes over the warmer Great Lakes, it absorbs heat energy and moisture from the lake surfaces and becomes *humidified*. This enhancement produces heavy lake-effect snowfall downwind into Ontario, Québec, Michigan, Pennsylvania, and New York—some areas receiving in excess of 250 cm (100 in.) in average snowfall a year (Figure 8.5).

Atmospheric Lifting Mechanisms

For air masses to cool adiabatically (by expansion) and to reach the dew-point temperature and saturate, condense, form clouds, and perhaps precipitate, they must lift and rise in altitude. Four principal lifting mechanisms operate in the atmosphere:

- Convergent lifting—air flows toward an area of low pressure

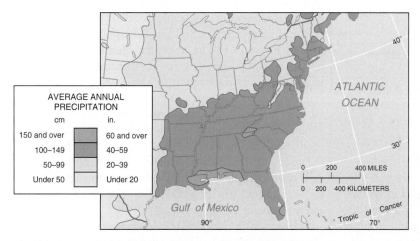

FIGURE 8.4 Influence of mT Gulf/Atlantic air mass.
The pattern of precipitation over the southeastern United States shows the influence of the warm, moist, and generally unstable mT Gulf/Atlantic air mass. Precipitation decreases with distance inland from the source region over the Gulf of Mexico and Atlantic Ocean.

FIGURE 8.5 Lake-effect snowbelts of the Great Lakes.
(a) Locally, heavy areas of snowfall are associated with the lee side of each of the Great Lakes. In winter, cold cP and cA air masses pass from colder land surfaces across the relatively warmer water of these lakes. (b) The air masses are warmed and humidified (water vapor is added) from the lake water. The humid, now unstable air yields heavy snowfall as it moves onshore and becomes chilled. The strongest effect is generally limited to about 50 km (30 mi) inland up to 100 km (60 mi). (c) Satellite image shows the lake-effect weather. [(a) Snowfall data from the *Climatic Atlas of the United States* (Washington, DC: Department of Commerce, NOAA, 1983), p. 53. (c) MODIS sensor *Terra* image, December 7, 2002, courtesy of the Rapid Response Team, NASA/GSFC]

- Convectional lifting—stimulated by local surface heating
- Orographic lifting—air is forced over a barrier such as a mountain range
- Frontal lifting—along the leading edges of contrasting air masses

Descriptions of all four mechanisms follow and are in Figure 8.6.

Convergent Lifting

Air flowing from different directions into the same low-pressure area is converging, displacing air upward in **convergent lifting**. In the tropics, convergent lifting of warm, moist air produces disturbances that can lead to the development of a tropical storm. All along the equatorial region, the southeast and northeast trade winds converge, forming the intertropical convergence zone (ITCZ) and

(a) Convergent

(b) Convectional (local heating)

(c) Orographic (barrier)

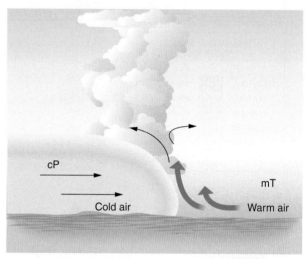

(d) Frontal (e.g. cold front)

FIGURE 8.6 Four atmospheric lifting mechanisms.
(a) Convergent lifting. (b) Convectional lifting. (c) Orographic lifting. (d) Frontal lifting.

areas of extensive uplift, towering cumulonimbus cloud development, and high average annual precipitation (see Figures 6.12 and 6.13b).

Convectional Lifting

When an air mass passes from a maritime source region to a warmer continental region, heating from the warmer land causes lifting and convection in the air mass. Other sources of surface heating might include an urbanized area, which causes a heat island, or an area of dark soil in a plowed field—the warmer surfaces producing **convectional lifting.** If conditions are unstable, initial lifting continues and clouds develop. Figure 8.7 illustrates convectional action stimulated by local heating, with unstable conditions present in the atmosphere. The rising parcel of air continues its ascent because it is warmer and therefore less dense than the surrounding environment.

Florida's precipitation generally illustrates both these lifting mechanisms: convergence and convection. Heating

of the land produces convergence of onshore winds from the Atlantic and the Gulf of Mexico. As an example of local heating and convectional lifting, Figure 8.8 depicts a day on which the landmass of Florida was warmer than the surrounding Gulf of Mexico and Atlantic Ocean. Because the Sun's radiation gradually heats the land throughout the day and warms the air above it, convectional showers tend to form in the afternoon and early evening, causing the highest frequency of days with thunderstorms in the United States. Florida appears highlighted and painted with clouds.

Towering cumulonimbus clouds are summertime features in the regions of North America that experience the mT Gulf/Atlantic air mass and, to a lesser extent, the weaker mT Pacific air mass. Convectional precipitation also dominates along the ITCZ, over tropical islands, and anywhere that moist, unstable air is heated from below or where the inflowing trade winds produce a dynamic convergence.

FIGURE 8.7 Local heating and convection.
Local heating and convection occur under unstable atmospheric conditions. In the air parcel, note that specific humidity is 8 g/kg and beginning temperature is 25°C. If you look back to Figure 7.13, you find that air with a specific humidity of 8 g/kg must be cooled to 11°C to achieve the dew-point temperature. In the example, this dew point is reached after 14 C° of adiabatic cooling at 1400 m (4600 ft). Note that the DAR (dry adiabatic rate) is used when the air parcel is less than saturated, changing to the MAR (moist adiabatic rate) above the lifting-condensation level at 1400 m.

ANIMATION Atmospheric Stability

FIGURE 8.8 Convectional activity over the Florida peninsula.
Cumulus clouds cover the land, with several cells developing into cumulonimbus thunderheads. Local heating lifts warm, moist air from the Gulf of Mexico as it passes over the relatively warmer land. [MODIS sensor *Terra* image courtesy of the Rapid Response Team, NASA/GSFC.]

Orographic Lifting

The physical presence of a mountain acts as a topographic barrier to migrating air masses. **Orographic lifting** (*oro* means "mountain") occurs when air is forcibly lifted upslope as it is pushed against a mountain. The lifting air cools adiabatically. Stable air forced upward in this manner may produce stratiform clouds, whereas unstable or conditionally unstable air usually forms a line of cumulus and cumulonimbus clouds. An orographic barrier enhances convectional activity and causes additional lifting during the passage of weather fronts and cyclonic systems, thereby extracting more moisture from passing air masses.

Figure 8.9a illustrates the operation of orographic lifting under unstable conditions. The wetter intercepting slope is the *windward slope*, as opposed to the drier far-side

slope, known as the *leeward slope*. Moisture condenses from the lifting air mass on the windward side of the mountain; on the leeward side, the descending air mass heats by compression, and any remaining water in the air evaporates. Thus, air beginning its ascent up a mountain can be warm and moist, but finishing its descent on the leeward slope it becomes hot and dry.

In North America, **chinook winds** (called *föhn* or *foehn* winds in Europe) are the warm, downslope airflows characteristic of the leeward side of mountains. Such winds can bring a 20 C° (36 F°) jump in temperature and greatly reduced relative humidity on the lee side of the mountains.

The term **rain shadow** is applied to dry regions leeward of mountains. East of the Cascade Range, Sierra

(a)

(b)

FIGURE 8.9 Orographic precipitation.
(a) Orographic barrier and precipitation patterns—unstable conditions assumed. Prevailing winds force warm, moist air upward against a mountain range, producing adiabatic cooling, eventual saturation and net condensation, cloud formation, and precipitation. On the leeward slope, as the "dried" air descends, compressional heating warms it and net evaporation dominates, creating the hot, relatively dry rain shadow of the mountain.
(b) Rain shadow produced by descending, warming air contrasts with the clouds of the windward side. Dust is stirred up by leeward downslope winds. [Photo by author.]

Nevada, and Rocky Mountains, such rain-shadow patterns predominate. In fact, the precipitation pattern of windward and leeward slopes persists worldwide, as confirmed by the precipitation maps for North America (Figure 9.6) and the world (Figure 10.2). See News Report 8.1 for more about the role of orographic barriers in setting precipitation records.

The state of Washington provides an excellent example of this concept, as shown in Figure 8.10. The Olympic Mountains and Cascade Mountains orographically lift invading mP air masses from the North Pacific Ocean, squeezing out annual precipitation of more than 500 cm and 400 cm, respectively (200 in. and 160 in.).

The Quinault Ranger and Rainier Paradise weather stations demonstrate windward-slope rainfall. The cities of Sequim, in the Puget Trough, and Yakima, in the

Columbia Basin, are in the rain shadow on the leeward side of these mountain ranges and are characteristically low in annual rainfall. Find these stations on the landscape profile and on the precipitation map.

FIGURE 8.10 Orographic patterns in Washington State.
(a) Four stations in Washington provide examples of orographic effects: windward precipitation and leeward rain shadows. Isohyets (isolines of equal precipitation amounts) on the map indicate rainfall (in inches). (b) Vegetation and landscapes in this true-color MODIS image from *Terra* demonstrate these precipitation patterns. [(a) Data from J. W. Scott and others, *Washington: A Centennial Atlas* (Bellingham, WA: Center for Pacific Northwest Studies, Western Washington University, 1989), p. 3; (b) MODIS sensor *Terra* image courtesy of Land Response Team, NASA/GSFC.]

Quinault Ranger Station—Olympic National Forest (windward)

310 cm/yr

Precipitation in centimeters (inches)

Temperature, °C (°F)

Month

Sequim (leeward)

41 cm/yr

ANNUAL PRECIPITATION

cm		in.
500 and above		200 and above
405–499		160–199
250–404		100–159
150–249		60–99
50–149		20–59
25–49		10–19
Under 25		Under 10

Olympic Mountains   Puget Trough   Cascade Mountains   Columbia Basin

W   Quinault   Sequim   Rainier   Yakima   E

(a)

Rainier Paradise Station (windward)

263 cm/yr

Yakima (leeward)

21 cm/yr

(b)

News Report 8.1

Mountains Set Precipitation Records

Because orographic lifting is limited in areal extent to locations where a topographic barrier exists, it is the least dominant lifting mechanism worldwide. But mountain ranges are the most consistent of all the precipitation-inducing mechanisms. Both the greatest average annual precipitation and the greatest maximum annual precipitation on Earth occur on the windward slopes of mountains that intercept moist tropical trade winds.

The world's greatest average annual precipitation occurs in the United States on Mount Waialeale, on the island of Kaua'i, Hawai'i. This mountain rises 1569 m (5147 ft) above sea level. On its windward slope, rainfall averaged 1234 cm (486 in., or 40.5 ft) a year for the years 1941–1992 (topping the mountain's previous average of 460 in.). In contrast, the rain-shadow side of Kaua'i receives only 50 cm (20 in.) of rain annually. If no islands existed at this location, this portion of the Pacific Ocean would receive only an average 63.5 cm (25 in.) of precipitation a year. (Note the importance of using established weather stations with a consistent record of weather data reporting when making such assessments. There are several stations that claim higher values but at this time fall short of these dependable criteria.)

Another place receiving world-record precipitation is Cherrapunji, India, 1313 m (4309 ft) above sea level at 25° N latitude, in the Assam Hills south of the Himalayas. Because of the summer monsoons that pour in from the Indian Ocean and the Bay of Bengal, Cherrapunji has received 930 cm (366 in., or 30.5 ft) of rainfall in one month and a total of 2647 cm (1042 in., or 86.8 ft) in one year! Not surprisingly, Cherrapunji is the all-time precipitation record holder for a single year and for every other time interval from 15 days to 2 years. The average annual precipitation there is 1143 cm (450 in., 37.5 ft), placing it second only to Mount Waialeale.

Frontal Lifting (Cold and Warm Fronts)

The leading edge of an advancing air mass is called its *front*. Vilhelm Bjerknes (1862–1951) first applied the term while working with a team of meteorologists in Norway during World War I. Weather systems seemed to them to be migrating air-mass "armies," doing battle along fronts. A front is a place of atmospheric discontinuity, a narrow zone forming a line of conflict between two air masses of different temperature, pressure, humidity, wind direction and speed, and cloud development. The leading edge of a cold air mass is a **cold front**, whereas the leading edge of a warm air mass is a **warm front**.

Cold Front   On weather maps, such as those shown in the chapter opener, in Figure 8.14, and in News Report 8.2, a cold front is a line with triangular spikes that point in the direction of frontal movement along an advancing cP or mP air mass. The steep face of the cold air mass suggests its ground-hugging nature, caused by its greater density and uniform physical character compared to the warmer air mass it displaces (Figure 8.11).

Warm, moist air in advance of the cold front lifts upward abruptly and experiences the same adiabatic rates of cooling and factors of stability or instability that pertain to all lifting air parcels. A day or two ahead of the cold

FIGURE 8.11 A typical cold front.
Denser, advancing cold air forces warm, moist air to lift abruptly. As the air is lifted, it cools by expansion at the DAR, cooling to the dew-point temperature as it rises to a level of condensation and cloud formation. Cumulonimbus clouds may produce large raindrops, heavy showers, lightning and thunder, and hail.

ANIMATION   Cold and Warm Fronts

FIGURE 8.12 Cold front and squall line.
A sharp line of cumulonimbus clouds near the Texas coast and
Gulf of Mexico mark a cold front and squall line. The cloud
formation rises to 17,000 m (56,000 ft). The passage of such a
frontal system over land often produces strong winds and
possibly tornadoes. [Space Shuttle astronaut photo from NASA.]

front's passage, high cirrus clouds appear, telling observers that a lifting mechanism is on the way.

A wind shift, temperature drop, and lowering barometric pressure mark a cold front's advance due to lifting
along the front. As the line of most intense lifting passes,
usually just ahead of the front itself, air pressure drops to a
local low. Clouds may build along the cold front into
characteristic cumulonimbus form and may appear as an
advancing wall of clouds. Precipitation usually is heavy,

containing large droplets, and can be accompanied by
hail, lightning, and thunder.

The aftermath of a cold front passage usually brings
northerly winds in the Northern Hemisphere and
southerly winds in the Southern Hemisphere as anticyclonic high-pressure advances; lower temperatures;
increasing air pressure produced by the cooler, more
dense air; and broken cloud cover.

The particular shape and size of the North American
landmass and its latitudinal position present conditions
where cP and mT air masses are best developed and have
the most direct access to each other. The resulting contrast can lead to dramatic weather, particularly in late
spring, with sizable temperature differences from one side
of a cold front to the other.

A fast-advancing cold front can cause violent lifting
and create a zone right along or slightly ahead of the
front called a **squall line**. Along a squall line, such as the
one in the Gulf of Mexico shown in Figure 8.12, wind
patterns are turbulent and wildly changing, and precipitation is intense. The well-defined frontal clouds in the
photograph rise abruptly to almost 17,000 m (56,000 ft),
with new thunderstorms forming along the front.
Tornadoes also may develop along such a squall line.

Warm Front   A line with semicircles facing in the direction of frontal movement denotes a warm front on weather maps (see Figure 8.14 map). The leading edge of an
advancing warm air mass is unable to displace cooler, passive air, which is denser along the surface. Instead, the
warm air tends to push the cooler, underlying air into a
characteristic wedge shape, with the warmer air sliding up
over the cooler air. Thus, in the cooler-air region a temperature inversion is present, sometimes causing poor air
drainage and stagnation.

Figure 8.13 illustrates a typical warm front in which
mT air is gently lifted, leading to stratiform cloud development and characteristic nimbostratus clouds and

FIGURE 8.13 A typical warm front.
Note the sequence of cloud development as the warm front approaches. Warm air slides
upward over a wedge of cooler, passive air near the ground. Gentle lifting of the warm, moist air
produces nimbostratus and stratus clouds and drizzly rain showers, in contrast to the more
dramatic precipitation associated with the passage of a cold front.

ANIMATION   Cold and Warm
Fronts

drizzly precipitation. A warm front presents a progression of cloud development to an observer: High cirrus and cirrostratus clouds announce the advancing frontal system; then clouds lower and thicken to altostratus; and finally the clouds lower and thicken to stratus within several hundred kilometers of the front.

Midlatitude Cyclonic Systems

The conflict between contrasting air masses can develop a **midlatitude cyclone,** or **wave cyclone**. This migrating low-pressure center with converging, ascending air spirals inward counterclockwise in the Northern Hemisphere, or inward clockwise in the Southern Hemisphere. Because of the undulating nature of frontal boundaries and the steering flow of the jet streams, the term *wave* is appropriate. The combination of the *pressure gradient force, Coriolis force*, and *surface friction* generates this cyclonic motion (see discussion in Chapter 6).

Before World War I, weather maps displayed only pressure and wind patterns. V. Bjerknes added the concept of fronts, and his son Jacob Bjerknes contributed the concept of migrating centers of cyclonic low-pressure systems.

Wave cyclones dominate weather patterns in the middle and higher latitudes of both the Northern and Southern Hemispheres and act as a catalyst for air-mass conflict. Such a midlatitude cyclone can initiate along the polar front, particularly in the region of the Icelandic and Aleutian subpolar low-pressure cells in the Northern Hemisphere. The intense high-speed winds of the jet streams guide cyclonic systems along their tracks (see Figures 6.17 and 6.18).

Life Cycle of a Midlatitude Cyclone

Figure 8.14 shows the birth, maturity, and death of a typical midlatitude cyclone in several stages, along with an idealized weather map. On the average, a midlatitude cyclonic system takes 3–10 days to progress through these stages from the area where it develops to the area where it finally dissolves. However, chaos rules and every day's weather map departs from the ideal in some manner.

Cyclogenesis   The atmospheric process in which low-pressure wave cyclones develop and strengthen is **cyclogenesis**. This process usually begins along the polar front, where cold and warm air masses converge and are drawn into conflict.

The polar front is a discontinuity of temperature, moisture, and winds that establishes potentially unstable conditions. For a wave cyclone to form along the polar front, a surface point of air *convergence* is matched by a compensating area of *divergence* aloft. Even a slight disturbance along the polar front, perhaps a small change in the path of the jet stream, can initiate the converging, ascending flow of air and thus a surface low-pressure system (illustrated in Figure ure 8.14a).

In addition to the polar front, certain other areas are associated with wave cyclone development and intensifi-

cation: the eastern slope of the Rockies and other north–south mountain barriers, the Gulf Coast, and along the east coasts of North America and Asia.

Open Stage   To the east of the developing low-pressure center, warm air begins to move northward along an advancing front, while cold air advances southward to the west of the center. See this organization on Figure 8.14b and around the center of low pressure, located over western Nebraska. The growing circulation system then vents into upper-level winds. As the midlatitude cyclone matures, the counterclockwise flow draws the cold air mass from the north and west and the warm air mass from the south. In the cross section, you can see the profiles of both a cold front and a warm front and each air-mass segment. Compare with the illustrations in Figures 8.11 and 8.13.

On the map, Denver has just experienced the passage of a cold front. Before the front passed, you see on the map that winds were from the southwest, but now the winds have shifted and are from the northwest. Temperature and humidity went from the warm, moist mT air mass to colder conditions as the cP air mass moves over Denver. Meanwhile, Wichita, Kansas, experienced the passage of a warm front and now is in the midst of the warm-air segment of the cyclone.

Such an open stage of a midlatitude cyclone occurred April 20, 2000, with low pressure centered over the Iowa–Missouri border. Figure 8.15 shows you a portion of the daily weather map for that day and an image from *SeaWiFS* sensors aboard satellite *OrbView-2*. Note how the isobars portray the cyclone and the low of 997.6 mb (29.45 in.) on the map. Note the cloud pattern stretching along the cold front and the swirling clouds around the low.

Occluded Stage   Remember the relation between air temperature and the density of an air mass. The colder cP air mass is denser than the warmer mT air mass. This cooler, more unified, air mass acts like a bulldozer blade and therefore moves faster than the warm front. Cold fronts can travel at an average 40 kmph (25 mph), whereas warm fronts average roughly half that at 16–24 kmph (10–15 mph). Thus, a cold front often overtakes the cyclonic warm front, wedging beneath it, producing an **occluded front** (*occlude* means "to close") in Figure 8.14c. On the idealized weather map an occluded front stretches south from the center of low pressure in Virginia to the border between the Carolinas (Figure 8.14c). Precipitation may be moderate to heavy initially and then taper off as the warmer air wedge is lifted higher by the advancing cold air mass. Note the still active warm front in the extreme Southeast and the flow of mT air. What conditions do you observe in Tallahassee, Florida? What might it be like there in the next 12 hours if the cold front passes south of the city?

Also note the designation of a **stationary front** on the weather map (in Arkansas and Mississippi). This frontal symbol tells you that there is a stalemate between cooler and warmer air masses where air flow on either side is almost parallel to the front, although in opposite

FIGURE 8.14 Idealized stages of a midlatitude wave cyclone.
A system may take from 3 to 10 days to cross the continent along storm tracks that vary
seasonally. Standard weather station symbols and current conditions for six cities appear
below the map. The block diagrams show (a) cyclogenesis where surface convergence and
lifting begin, (b) the open stage, (c) the occluded stage, (d) the dissolving stage, reached at
the end of the storm track as the cyclone spins down, no longer energized by the latent heat
from condensing moisture. After studying the text, can you describe conditions in Boise,
Denver, Wichita, Columbus, Atlanta, and Tallahassee as depicted on this weather map?

Midlatitude
Cyclones

directions. Some gentle lifting is producing light to moderate precipitation. Eventually the stationary front will
begin to move, as one of the air masses assumes dominance, evolving into a warm or a cold front.

Dissolving Stage The final, dissolving, stage of the
midlatitude cyclone occurs when its lifting mechanism is
completely cut off from the warm air mass, which was its
source of energy and moisture. Remnants of the cyclonic

system then dissipate in the atmosphere, perhaps after
passage across the country (Figure 8.14d).

Storm Tracks Cyclonic storms—1600 km (1000 mi)
wide—and their attendant air masses move across the
continent along **storm tracks,** which shift in latitude with
the Sun and the seasons. Typical storm tracks that cross
North America are farther northward in summer and
farther southward in winter (Figure 8.16). As the storm

(a) 3:00 P.M. EST

(b) 7:00 A.M. EST

FIGURE 8.15 Open stage of a midlatitude cyclone.
(a) *SeaWiFS* image of a cyclonic system over the Midwest. The cloud patterns are areas of precipitation; clear skies are in a wedge behind the cold front as a cP air mass moves over the landscape.
(b) A segment from the April 20, 2000, daily weather map, about 8 hours earlier than the image, showing the low-pressure system centered on 997.6 mb (29.45 in.). Counterclockwise winds circulate around the low. [(a) *SeaWiFS* image used with permission of ORBIMAGE. All rights reserved. (b) Segment of "Daily Weather Map," courtesy of National Weather Service, NOAA.]

tracks begin to shift northward in the spring, cP and mT air masses are in their clearest conflict. This is the time of strongest frontal activity, featuring thunderstorms and tornadoes. Storm tracks follow the path of upper-air winds, which direct storm systems across the continent.

A map of actual storm tracks for March 1991 in Figure 8.16b demonstrates several areas of cyclogenesis: the northwest over the Pacific Ocean, the Gulf of Mexico, the eastern seaboard, and the Arctic. Cyclonic circulation also frequently develops on the lee side of mountain ranges, as along the Rockies from Alberta south to Colorado. By crossing the mountains, such systems gain access to the moisture-laden, energy-rich mT air masses from the Gulf of Mexico and are strengthened.

Analysis of Daily Weather Maps— Forecasting

Synoptic analysis is the evaluation of weather data collected at a specific time. Building a database of wind, pressure, temperature, and moisture conditions is key to *numerical* (computer-based) *weather prediction* and the development of weather-forecasting models. Development of numerical models is a great challenge because the atmosphere operates as a nonlinear system, tending toward chaotic

behavior. Slight variations in input data or slight changes in the basic assumptions of the model's behavior can produce widely varying forecasts. As our knowledge of the interactions that produce weather and our instruments and software improve, so too will the accuracy of our forecasts.

Weather data necessary for the preparation of a synoptic map and forecast include:

- Barometric pressure (sea level and altimeter setting)
- Pressure tendency (steady, rising, falling)
- Surface-air temperature
- Dew-point temperature
- Wind speed, direction, and character (gusts, squalls)
- Type and movement of clouds
- Current weather
- State of the sky (current sky conditions)
- Visibility; vision obstruction (fog, haze)
- Precipitation since last observation

For links to weather maps, current forecasts, satellite images, and the latest radar, please go to the *Geosystems* Home Page, to Chapter 8, "Destinations," and you will find many related links to the Internet. With the Internet, there is no need to wait for television or the newspaper to

(a) Average storm tracks

(b) Actual storm tracks in March 1991

FIGURE 8.16 Typical and actual storm tracks.
(a) Cyclonic storm tracks over North America vary seasonally. The tracks indicate several locations of cyclogenesis.
(b) Actual cyclonic tracks during March 1991 over North America. [(b) From *Storm Data* vol. 33, no. 3 (March 1991); Asheville, NC: NOAA/NESDIS, National Climatic Data Center.]

bring you the latest satellite images, for they are at your fingertips.

NOAA's Forecast Systems Laboratory (FSL), Boulder, Colorado, is developing new forecasting tools (see http://www.fsl.noaa.gov). A few innovations include: wind profilers using radar to profile winds from the surface to high altitudes; a High Performance Computing System (277 CPUs networked to act as one) to model software for 3-D weather models and other computations; improved international cooperation and weather data sharing; and a standard Advanced Weather Interactive Processing System (AWIPS) enhanced by the installation of 883 Automated Surface Observing System, or ASOS (Figuree 8.17). An FSL research meteorologist demonstrates an AWIPS in the Career Link at the end of this chapter.

ASOS sensor instrument arrays are a primary surface weather-observing network (Figure 8.17a). An ASOS installation includes: rain gauge (tipping bucket type), temperature/dew-point sensor, barometer, present weather identifier, wind-speed indicator, direction sensor,

cloud-height indicator, freezing-rain sensor, thunderstorm sensor, and visibility sensor, among other items.

Preparing a weather report and forecast requires analysis of daily weather maps and the use of standard weather symbols, such as those in Figure 8.14. Although the actual pattern of cyclonic passage over North America is widely varied in shape and duration, you can apply this general model, along with your understanding of warm and cold fronts, to reading the daily weather map. A weather map is a useful tool, especially when weather turns dramatic in violent episodes.

Violent Weather

Weather provides a continuous reminder of the flow of energy across the latitudes that at times can set into motion destructive, violent weather conditions. In dollar value, weather-related damage is increasing each year as population increases in areas prone to violent weather. We focus on thunderstorms, tornadoes, and hurricanes.

However, we must include mention of such things as *ice storms* of **sleet** (freezing rain, ice glaze, and ice pellets), snow blizzards, and low temperatures. Sleet is caused when precipitation falls through a below-freezing layer of air near the ground. During the ice storm of 1998, a large region in Canada and the United States with 700,000 residents was without power for weeks (Figure 8.18). Imagine a coating of ice on everything, weighing on power lines and tree limbs. More than 80 hours of freezing rain and drizzle hit the region; this is more than double typical ice-storm duration. Hypothermia claimed 25 human lives.

Weather-related losses exceeded $10 billion per year through the 1990s, eclipsing the previous average of less than $2 billion a year. As mentioned earlier, 1998 broke all records for weather-related damage worldwide. We now look at specific types of violent weather: thunderstorms, derechos, tornadoes, and tropical cyclones. Government research and monitoring of violent weather is centered at NOAA's National Severe Storms Laboratory and Storm Prediction Center (see http://www.nssl.noaa.gov, http://www.spc.noaa.gov/, and http://www.fsl.noaa.gov/; consult these sites for each of the topics that follow).

Thunderstorms

The condensation of large quantities of water vapor in clouds liberates tremendous amounts of energy. This process locally heats the air, rapidly changing its density and buoyancy and causing violent updrafts. As rising air parcels pull surrounding air into the cumulonimbus cloud system, updrafts become stronger. Raindrops form and descend through the cloud, and their frictional drag pulls air toward the ground, causing violent downdrafts. The resulting giant cumulonimbus clouds signal dramatic weather moments—heavy precipitation, lightning, thunder, hail, blustery winds, and maybe tornadoes. See these conditions illustrated in Figure 7.24.

Such thunderstorms may develop under three conditions: within an air mass (particularly in warm, moist air),

(a)

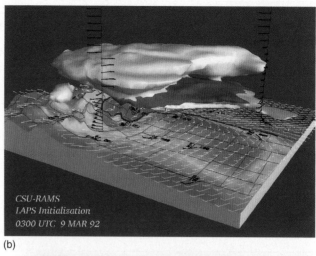

CSU-RAMS
LAPS Initialization
0300 UTC 9 MAR 92

(b)

(c)

(d)

(e)

FIGURE 8.17 ASOS weather instruments and AWIPS workstation display.
(a) Automated Surface Observing System (ASOS) installation, one of 883 ASOS stations in use for data gathering as part of the U.S. primary surface weather observing network. (b) Forecast Systems Laboratory (FSL) scientists develop 3-D modeling programs to portray weather and simulate future weather for more accurate forecasting. On the image note the gray area that depicts significant cloud water content and the red layer within the cloud where aircraft icing is forecast. The black wind barbs off the two vertical axes denote wind profiles through the atmosphere. The view is from the southeast. (c) Powerful computers are needed to process the data and produce these 3-D virtual images on the Advanced Weather Interactive Processing System (AWIPS), employing screens with multiple frames. Of the many items that can be displayed we see in Figure 8.17c a simulated 3-D pressure analysis, water vapor, and visible satellite image—including lightning strikes—and humidity. (d) An AWIPS workstation. (e) A portion of the High Performance Computing System at the FSL. [Photo (a) and (e) by Bobbé Christopherson; (b–d) images courtesy of the FSL, NOAA, Boulder, Colorado.]

FIGURE 8.18 Aftermath of record ice storm.
For more than a week in January 1998, record blizzards and ice storms from a nor'easter storm blanketed New England and eastern Canada, causing almost a billion dollars in damage. These broken birch trees in Maine, snapped under the weight of an ice coating as thick as 8 cm (3 in.), remain as reminders of this weather disaster. Power was out for weeks.
[Photo by Bobbé Christopherson.]

in a line along a cold front or other convergent boundary, or where mountain slopes cause orographic lifting. With a slowly rotating updraft, a dangerous convective supercell thunderstorm might develop, featuring strong winds, hail, and persistent tornadoes of great strength.

Thousands of thunderstorms occur on Earth at any given moment. Equatorial regions and the ITCZ experience many of them, exemplified by the city of Kampala in Uganda, East Africa (north of Lake Victoria), which sits virtually on the equator and averages a record 242 days a year of thunderstorms. Figure 8.19 shows the annual distribution of days with thunderstorms across the United States and Canada. You can see that, in North America, most thunderstorms occur in areas dominated by mT air masses.

Atmospheric Turbulence   Most airplane flights experience at least some turbulence—the encountering of air of different densities or air layers moving at different speeds and directions. This is a natural state of the atmosphere and passengers are asked to keep seat belts fastened when in their seats to avoid injury, even if the seat-belt light is turned off.

Thunderstorms can produce severe turbulence in the form of downbursts, which are exceptionally strong downdrafts. Downbursts are classified by size: a *macroburst* is at least 4.0 km (2.5 mi) wide and in excess of 210 kmph (130 mph); a *microburst* is smaller in size and speed. A microburst causes rapid changes in wind speed and direction and it causes the dreaded *wind shear* that can bring down aircraft. Such turbulence events are short-lived and hard to detect, although the Forecast Systems Laboratory, among others, is making progress in developing forecasting methods.

Lightning and Thunder   An estimated 8 million lightning strikes occur each day on Earth. **Lightning** refers to flashes of light caused by enormous electrical discharges—tens of millions to hundreds of millions of volts—that briefly superheat the air to temperatures of 15,000°–30,000°C (27,000°–54,000°F). A buildup of electrical energy polarity between areas within a cumulonimbus cloud or between the cloud and the ground creates lightning (Figure 8.20a). The violent expansion of this abruptly heated air sends shock waves through the atmosphere as the sonic bang of **thunder**.

NASA's Lightning Imaging Sensor (LIS) aboard the *Tropical Rainfall Measuring Mission* (*TRMM*) satellite monitors lightning and other phenomena. The LIS can image lightning strikes day or night, within clouds or between cloud and ground. The sensor's data show that about 90% of all strikes occur over land in response to increased convection over relatively warmer continental surfaces, with expected seasonal shifts with the high Sun, as shown in Figure 8.20b and c (see **http://thunder.msfc.nasa.gov/lis/**).

Lightning poses a hazard to aircraft, people, animals, trees, and structures. Certain precautions are mandatory when a lightning discharge threatens, because lightning causes nearly 200 deaths and thousands of injuries each year in the United States and Canada. When lightning is imminent, the National Weather Service issues *severe storm warnings* and cautions people to remain indoors. If the hair on your head or neck begins to stand on end, get indoors, or if in the open get low to the ground, crouched on both feet and not laying down, in a low spot, if possible. Your hair is telling you that a charge is building in that area. The place *not* to seek shelter is beneath a tree, for trees are good conductors of electricity and often are hit by lightning.

Hail   Ice pellets called **hail** generally form within a cumulonimbus cloud. Raindrops circulate repeatedly above and below the freezing level in the cloud, adding layers of ice until the circulation in the cloud can no longer support their weight (Figure 8.21a). Hail may also grow from the addition of moisture on a snow pellet.

Pea-sized and marble-sized hail is common, although hail the size of golf balls and baseballs happens at least once or twice a year somewhere in North America. For larger hail to form, the frozen pellets must stay aloft for longer periods. The largest authenticated hailstone in the world landed in Kansas in 1970; it measured 14 cm (5.6 in.) in diameter and weighed 758 g (1.67 lb)!

Hail is common in the United States and Canada, although somewhat infrequent at any given place. Hail occurs perhaps every one or two years in the highest-frequency areas. Annual hail damage in the United States tops $800 million (Figure 8.21b). The pattern of hail occurrence across the United States and Canada is similar to that of thunderstorms shown in Figure 8.19.

Derechos

Although tornadoes and hurricanes grab headlines, straight-line winds associated with thunderstorms and bands of showers cause significant damage and crop

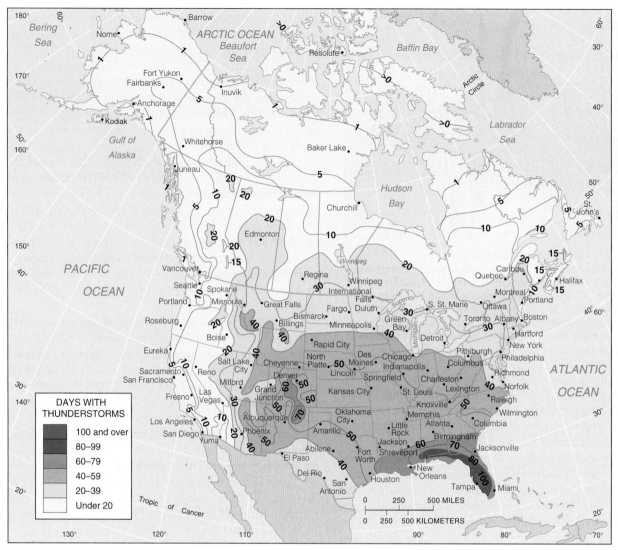

FIGURE 8.19 Thunderstorm occurrence.
Average annual number of days experiencing thunderstorms. [Data courtesy of U.S. National
Weather Service; Map Series 3; *Climatic Atlas of Canada*, Atmospheric Environment Service, Canada.]

losses. These strong linear winds in excess of 26 m/s (58 mph) are known as **derechos,** or in Canada sometimes called "plow winds." A derecho in 1998 in eastern Wisconsin exceeded 57 m/s (128 mph). Researchers identified 377 events between 1986 and 2003, an average of about 21 per year. The name, coined by physicist G. Hinrichs in 1888, derives from a Spanish word meaning "direct" or "straight ahead." Wind outbursts from convective storms can produce groups of downburst clusters from a thunderstorm system. These derechos tend to blast in linear paths fanning out along curved-wind fronts over a wide swath of land.

Derechos pose distinct hazards to summer outdoor activities by overturning boats, hurling flying objects, and causing broken trees and limbs. Their highest frequency (about 70%) is from May to August in the region stretching from Iowa, across Illinois, and into the Ohio River Valley in the upper Midwest. By September and on through to April, areas of activity, although less than in summer, migrate southward to eastern Texas through Alabama. For more information and a description of

numerous derechos, see **http://www.spc.noaa.gov/misc/ AbtDerechos/derechofacts#definition**. Derecho conditions are similar to the strong migrating cyclonic systems and rotating thunderstorms that give birth to tornadoes, although they can occur without any tornado activity.

Tornadoes

The updrafts associated with a cumulonimbus cloud appear on satellite images as pulsing bubbles of clouds, sometimes as supercells of rotating updrafts (Figure 8.22). Friction with the ground slows surface winds, but higher in the troposphere, winds blow faster. Thus, a body of air moves faster at higher altitude than at the surface, creating a rotation in the air along a horizontal axis parallel to the ground (picture a rolling pin; see Figure 8.23a). When that rotating air encounters the strong updrafts associated with frontal activity, the axis of rotation shifts to a vertical alignment perpendicular to the ground (Figure 8.23b).

According to one hypothesis for *supercell tornadoes*, this spinning, rising column of mid-troposphere-level air forms

(a)

FIGURE 8.20 Seasonal images show global lightning.
(a) Multiple cloud-to-ground lightning strikes captured in a time-lapse photo. (b) A composite of three months' data derived from NASA's Lightning Imaging Sensor (LIS) records all lightning strikes between 35° N and 35° S latitudes during (b) winter (December 1999–February 2000) and (c) summer (June, July, August 2000). The LIS is aboard the *Tropical Rainfall Measuring Mission* satellite launched in 1997. [(a) Photo from C. Clark, NOAA Photo Library, NOAA Central Library, NSSL; (b) and (c) *TRMM* images courtesy of NASA's Global Hydrology and Climate Center, Marshall Spaceflight Center, Huntsville, Alabama.]

(b)   Winter (Dec. 1999, Jan. and Feb. 2000)

(c)   Summer (June, July, August 2000)

a mesocyclone. A **mesocyclone** can range up to 10 km (6 mi) in diameter and rotate vertically within a *supercell cloud* (the parent cloud) to a height of thousands of meters (Figure 8.23c). As a mesocyclone extends vertically and contracts horizontally, wind speeds accelerate in an inward vortex (as ice skaters draw their arms in to accelerate while spinning or as water speeds up when it spirals down a sink drain). A well-developed mesocyclone will produce heavy rain, large hail, blustery winds, and lightning; some mature mesocyclones generate tornado activity.

(a)

(b)

FIGURE 8.21 Hailstones and hailstone damage.
(a) Irregular layers give evidence of numerous convective trips through a cumulonimbus cloud.
(b) Hail can severely damage windows, automobiles, crops, livestock, and people. [(a) © Arthur
C. Smith III/Grant Heilman Photography, Inc.; (b) Photo by *The Denver Post*, 1990.]

As more moisture-laden air is drawn up into the circulation of a mesocyclone, more energy is liberated and the rotation of air becomes more rapid as a *wall cloud* forms (Figure 8.24). The swirl of the mesocyclone itself is visible, as are dark gray **funnel clouds** that pulse from the bottom side of the parent cloud. The terror of this stage of development is the lowering of such a funnel to Earth.

This tightly rotating vortex is the dreaded **tornado,** from the Latin *tornare*, "to make round by turning" (Figures 8.23f and g and 8.24). When tornado circulation occurs over water, a **waterspout** forms, and the sea or lake surface water is drawn up into the funnel.

A tornado's scoured path is visible on the ground in a series of semicircular "suction" prints and, of course, in the trail of damage, such as in the image of a tornado's path through Maryland (Figure 8.23d). Figure 8.23e shows F5 destruction from a tornado in Oklahoma City, a few blocks east of I-35. At its peak, this extreme tornado was 1610 m (1 mi) wide at its base and lasted an hour, killing 36 people. Resulting property damage exceeded $1 billion.

Tornado Measurement and Science A tornado's diameter can range from a few meters to a few hundred meters, or larger. It can last from a few seconds to tens of

FIGURE 8.22 Thunderstorms on the Great Plains.
These boiling cloud tops signify severe supercell thunderstorms on the Great Plains along a cold front with upper-level support on October 9, 2001. This storm system produced nearly 100 reports of hail and strong winds and 22 tornado strikes in Nebraska and Oklahoma. [*NOAA-12* AVHRR image courtesy of NOAA.]

FIGURE 8.23 Mesocyclone and tornado formation.
(a) Strong wind aloft establishes spinning along a horizontal axis. (b) Updraft from thunderstorm development tilts the rotating air along a vertical axis. (c) Mesocyclone forms as a rotating updraft within the thunderstorm. If one forms, a tornado will descend from the lower portion of the mesocyclone. (d) A tornado leaves its mark as this *Landsat-7* image captures. The damage path tracks for 39 km (24 mi) where an F4 tornado tore through La Plata, Maryland; more than 900 homes and 200 businesses were destroyed. (e) F5 tornado damage in May 1999 in Oklahoma. The church visible in the background was used for an emergency shelter. (f) A fully developed tornado swirls dust and debris as it locks onto the ground. (g) A tornado emerges from the base of a wall cloud near Oakfield, Wisconsin. [(d) *Landsat-7* image courtesy of Satellite Systems Branch, USGS EROS Data Center. Photos by (e) Todd Lindlay; (f) C. Lavoie/First Light; (g) Don Lloyd/AP/Wide World Photos.]

Stronger winds

Spinning along
horizontal axis

Weaker winds

(a)

Thunderstorm
forming

Updraft

(b)

Clouds overshoot
top of thunderstorm

Mesocyclone
(3 to 10 km
diameter)

Anvil

Tornado

Air inflows

(c)

(e)

ANIMATION   Tornado Wind Patterns

(g)

FIGURE 8.24 Supercell tornado.
A supercell tornado descends from the cloud base near Spearman, Texas. Strong hail is falling to the left of the tornado. [Photo by Howard Bluestein, all rights reserved.]

phy and the nature of the landscape, and shape permit contrasting air masses to confront one another. Canada experiences an average of 80 observed tornadoes per year, although unpopulated rural areas go unreported. University of Leeds researchers report that in the United Kingdom as many as 100 tornadoes may hit in a year, although observers report 60 to 80 a year. Other continents report a small number of tornadoes each year.

Using satellites, aircraft, and surface measurements, the National Center for Atmospheric Research and the University of Oklahoma completed a two-year project in 1995 called VORTEX (Verification of the Origin of Rotation in Tornadoes). VORTEX observed 10 tornadoes for detailed study. The VORTEX Project paved the way for Subvortex, the latest cooperative research effort. In 1997, a Subvortex team equipped with a twin-Doppler radar unit on a van obtained high-resolution images of an F1 tornado southwest of Tulsa. (See **http://www.stormresearch.com/vortex/** for a summary of these efforts.)

The Storm Prediction Center in Kansas City, Missouri, provides short-term forecasting for thunderstorms and tornadoes to the public and to NWS field offices. Warning times of about 12 minutes are possible with current technology (see the Storm Prediction Center at **http://www.spc.noaa.gov/**).

minutes. Air pressure inside a tornado funnel usually is about 10% less than in the surrounding air. This disparity is similar to the pressure difference between sea level and an altitude of 1000 m (3300 ft). Such a sharp horizontal-pressure gradient causes an in-rushing convergence and severe wind speeds at the funnel. Winds can exceed 485 kmph (300 mph), with at least a third of all tornadoes exceeding 181 kmph (113 mph).

The late Theodore Fujita, a noted meteorologist from the University of Chicago, designed the Fujita Scale, which classifies tornadoes according to wind speed as indicated by related property damage (Table 8.1). Of all tornadoes, the frequency distribution on the Fujita Scale is F0 − F1 = 74%, F2 − F3 = 25%, and F4 − F5 = 1%. Theoretically, the scale continues from F6 to F12, although no tornado is expected to ever reach F6.

Tornadoes have struck all 50 states and all the Canadian provinces and territories. May and June are the peak months, as you see in Figure 8.25, based on 54 years of records. News Report 8.2 details a tornado outbreak in May 2003 right through the heart of the infamous "Tornado Alley"—a corridor that experiences the highest average number of tornadoes on Earth. Looking at the spatial distribution map, can you identify this region?

North America receives more tornadoes than anywhere on Earth because its latitudinal position, topogra-

Table 8.1	The Fujita Scale
F-number	**Wind Speed; Damage Specs**
F0	18–32 m/s (64–116 kmph; 40–72 mph); *light damage*: break branches, damage chimneys.
F1	33–49 m/s (117–180 kmph; 73–112 mph); *moderate damage*: beginning of hurricane wind-speed designation, roof coverings peel off, mobile homes pushed off foundations.
F2	50–69 m/s (181–253 kmph; 113–157 mph); *considerable damage*: roofs torn off frame houses, large trees uprooted or snapped, box cars pushed over, small missiles generated.
F3	70–92 m/s (254–332 kmph; 158–206 mph); *severe damage*: roofs torn off well-constructed houses, trains overturned, trees uprooted, cars thrown.
F4	93–116 m/s (333–419 kmph; 207–260 mph); *devastating damage*: well-built houses leveled, cars thrown, large missiles generated.
F5	117–142 m/s (420–512 kmph; 261–318 mph); *incredible damage*: houses lifted and carried distance to disintegration, car-sized missiles fly farther than 100 m, bark removed from trees.

See **http://www.tornadoproject.com/fscale/fscale.htm**

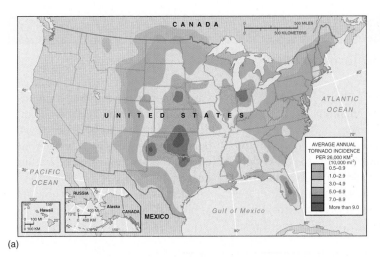

FIGURE 8.25 Tornado occurrence in the United States, 1950–2003.
(a) Average number of tornadoes per 26,000 km² (10,000 mi²) for 1950 to 2003. (b) Average number of tornadoes per month. Remarkably, the May average was exceeded by 366 tornadoes that hit in May 2004. [Data courtesy of the Storm Prediction Center, National Weather Service, Kansas City, Missouri.]

(a)

(b)

Recent Tornado Records   In the United States, 44,633 tornadoes were recorded in the 54 years from 1950 through 2003, causing 4652 deaths, or about 86 deaths per year, although tornadoes took 130 lives in the United States in 1998 alone. Of note is the fact that these tornadoes resulted in more than 78,000 injuries and property damage of over $25 billion for this half century. The yearly average of $500 million in damage is rising at about $50 million a year.

The annual average number of tornadoes before 1990 was 787. Interestingly, since 1990 the average per year has risen to over 1000, with 2003 experiencing 1376 tornadoes. For the year 2004 through the month of November, more than 1600 tornadoes made the preliminary count. Importantly, the number of F4 and F5 tornadoes is on the increase, reaching more than a dozen each season.

Reasons for the annual increase in tornado frequency in North America range from global climate change and more intense thunderstorm activity to better reporting by a larger population armed with videotape recorders. Researchers are pursuing many questions with an enthu-

siasm for the science and an awe of the subject, as summarized by Howard Bluestein, foremost tornado researcher and meteorologist:

> The source of rotation in some tornadoes appears to be preexisting vortices near the ground above where convective storms are growing; in others, the source seems to be linked to the mesocyclone. Does the mesocyclone itself descend to the ground and intensify to become the tornado, or does it trigger events near the ground that create tornadoes from other sources of rotation? Can it do both? We hope the answers to these questions are soon forthcoming. Our quest for discovery has not taken away from the respect we have for the awesome power the tornado harbors, nor the thrill for viewing the violent motions in the tornado or the beauty of the storm. We eagerly await the next act in the atmospheric play starring the tornado.*

*H. Bluestein, *Tornado Alley, Monster Storms of the Great Plains* (New York: Oxford University Press, 1999), p. 162.

News Report 8.2

May 2003 Tornado Outbreak in Tornado Alley

During May 2003, 543 tornadoes struck the United States. The first 12 days of the month featured 354 tornadoes. Figure 8.2.1 maps the tracks for these record-setting May tornadoes across some 20 states. This is 366 beyond the average for May, and more tornadoes for any single month since 1950. The results caused more than 40 deaths and hundreds of millions of dollars in property damage.

Eighty-four tornadoes hit on Sunday, May 4, in a path across eight states. Tornado damage ratings of four F4s and eight F3s struck in the Midwest. Five tornadoes struck in or near Kansas City. One of the Kansas City F4 tornadoes achieved a funnel-base width of 450 m (1500 ft). Some damage appeared to be from down-bursts and derecho linear-wind effects. (For an animation of satellite images, see http://www.osei.noaa. gov/Events/Severe/US_Midwest/ 2003/SVRusMW125_G12.avi)

FIGURE 8.2.1 May 2003 tornado outbreak.
Map of 543 tornado tracks that hit during May 2003 as listed in *Storm Data* by NWS Forecast Offices. Tornado paths may blend, or cross, or fall on top of one another on a map of this small-scale. This was a record for the month of May. [NWS Forecast Offices in *Storm Data*, Storm Prediction Center/NOAA.]

FIGURE 8.2.2 May 5, 2003, 7:00 A.M. weather map segment.
The cold front, now partially occluded and still potent, that swept across Oklahoma, Kansas, and Missouri May 4. The conflict between cold-dry and warm-moist air masses sets up perfect frontal conditions for the formation of supercells, spawning many mesocyclones and tornadoes. [Weather map segment courtesy of "Daily Weather Maps," NWS, NOAA.]

The weather map for May 5, 7:00 A.M., shows the system that produced these storms the day before (Figure 8.2.2). Local NWS forecasters broadcast in advance that conditions were perfect for a severe weather outbreak. Several supercell storms resulted from the passage of a cold front, where cold, dry air from the Rockies collided with warm, moist air from the Gulf of Mexico. Upper-air flows with divergence aloft helped trigger giant supercells along the frontal boundary. Following numerous hits, the entire state of Oklahoma was declared a federal disaster area on May 10 (Figure 8.2.3).

For those of you who live in this region of the country, tornado alerts and tornado warnings are something with which you are familiar. Unfortunately, the first six months of 2004 didn't show any letup, with 962 tornadoes on the books. Compare this with the annual average of only 787 tornadoes for pre-1990.

FIGURE 8.2.3 Tornado damage near Oklahoma City. A chaplain inspects what is left of his church in Bethany, Oklahoma, following a May 10, 2003 outbreak. The entire state was declared a disaster area on this day following a week of severe weather. [Photo by Richard Michael Pruitt/Dallas Morning News.]

Tropical Cyclones

A powerful manifestation of the Earth–atmosphere energy budget is the **tropical cyclone**, which originates entirely within tropical air masses. The tropics extend from the Tropic of Cancer at 23.5° N to the Tropic of Capricorn at 23.5° S, containing the equatorial zone between 10° N and 10° S. Approximately 80 tropical cyclones occur annually worldwide. Some 45 per year are powerful enough to be classified as hurricanes, typhoons, and cyclones (different regional names for the same type of tropical storm)—30% of these occur in the western North Pacific. Of note, the lowest sea-level pressure recorded on Earth was 870 mb (25.69 in.) in the center of Typhoon Tip, northwest of Guam, in October 1979.

Cyclonic systems forming in the tropics are quite different from midlatitude cyclones because the air of the tropics is essentially homogeneous, with no fronts or conflicting air masses of differing temperatures. In addition, the warm air and warm seas ensure abundant water vapor and thus the necessary latent heat to fuel these storms. Tropical cyclones convert heat energy from the ocean into mechanical energy in the wind.

What mechanism triggers the start of a tropical cyclone? Meteorologists now think that cyclonic motion begins with slow-moving easterly waves of low pressure in the trade-wind belt of the tropics, such as the Caribbean area (Figure 8.26). For these processes, sea-surface temperatures must exceed approximately 26°C (79°F). Tropical cyclones form along the eastern (leeward) side of these migrating troughs of low pressure, a place of convergence and rainfall. Surface airflow then converges into the low-pressure area, ascends, and flows outward as it diverges aloft. This important divergence aloft acts as a chimney, pulling more moisture-laden air into the developing system. To maintain and strengthen this vertical convective circulation, there must be little or no wind shear to interrupt or block the airflow.

The map in Figure 8.27 shows areas in which tropical cyclones form and some of the characteristic storm tracks traveled. The map indicates the range of months during which tropical cyclones are most likely to appear. They tend to occur when the equatorial low-pressure trough, or the ITCZ, is farthest from the equator during the months following the summer solstice in each hemisphere. For example, storms that strike the southeastern United States do so mostly between August and October, following the June 21 solstice. Officially, tropical cyclone season is from June 1 to November 30 each year. (For comparison, the average midlatitude cyclonic storm tracks are in blue on the map.)

Relative to North and Central America, tropical depressions (low-pressure areas) intensify into tropical storms as they cross the Atlantic. A rule of thumb has emerged: If tropical storms mature early along their track, they tend to curve northward toward the North Atlantic and miss the United States. The critical position is approximately 40° W longitude. If a tropical storm matures after it reaches the longitude of the Dominican Republic (70° W), then it has a higher probability of hitting the United States.

Figure 8.28 shows two hurricanes with unusual tracks, one in Canada (2003) and one in Brazil (2004). Hurricane Juan, a category 2 storm, was one of the most powerful to hit Canada, making landfall in Nova Scotia in September 2003. Coastal and port facilities were swamped by heavy rain and storm surges of up to 1.5 m (5 ft). Wave heights coming into Halifax Harbor ranged from 9 to 19 m (30–62 ft). Warmer sea-surface temperature might fuel more such storms into the midlatitudes in the years to come.

In the Southern Hemisphere, no hurricane was ever observed turning from the equator into the south Atlantic. However, an unnamed March 2004 hurricane made landfall in Brazil south of the resorts at Laguna,

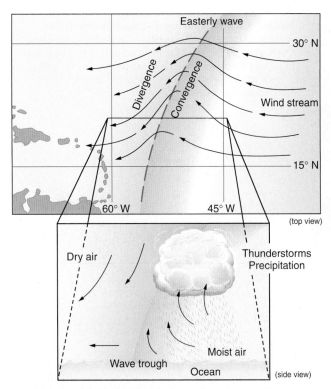

FIGURE 8.26 Easterly wave in the tropics.
A low-pressure center develops along an easterly (westward-moving) wave. Moist air rises in an area of convergence at the surface to the east of the wave trough. Wind flows bend and converge before the trough and diverge downwind from the trough.

state of Santa Catarina. Experts first described it as an extratropical depression, a type of "cutoff" low-pressure system, and debated whether to label it a hurricane. However, NASA's *Terra* satellite in Figure 8.28b clearly shows an organized pattern around a central eye and rain bands characteristic of a hurricane, thus resolving the question. Compare the circulation patterns between Hurricane Juan and Hurricane "Brazil 1," what do you see? Note the Coriolis force in action, counterclockwise in the Northern Hemisphere (8.28a) and clockwise in the Southern Hemisphere (8.28b).

Hurricanes, Typhoons, and Cyclones   Tropical cyclones are potentially the most destructive storms experienced by humans, claiming thousands of lives each year worldwide. This is especially true when they attain wind speeds and low-pressure readings that upgrade their status to a full-fledged **hurricane, typhoon,** or *cyclone* (>65 knots, 74 mph, 119 kmph). Across the globe, such storms bear these three different names, among others: *hurricane* around North America, *typhoon* in the western Pacific (Japan, Philippines), and *cyclone* in Indonesia, Bangladesh, and India. By any name they can be destructive killers. Worldwide, about 10% of all tropical disturbances have the right ingredients to become hurricanes or typhoons (see the National Hurricane Center at http://www.nhc.noaa.gov/ or the Joint Typhoon Warning Center at http://www.npmoc.navy.mil/jtwc.html).

FIGURE 8.27 Worldwide pattern of the most intense tropical cyclones.
Typical tropical storm tracks with principal months of occurrence and regional names. For the North Atlantic region from 1871 to 2000, 1084 tropical cyclones (storms and hurricanes) developed. The peak day in this Atlantic region (assuming a 9-day moving average calculation) is September 10, with 64 cyclones for the 129-year period. Also shown for comparison are characteristic storm tracks of midlatitude cyclones.

(a)

(b)

FIGURE 8.28 Hurricanes in Canada and Brazil.
(a) Hurricane Juan approaches Halifax Harbor, Nova Scotia, Canada. (b) Unnamed Hurricane "Brazil 1" approaches Brazil's southeastern coast, March 27, 2003. A hurricane in the south Atlantic is a unique occurrence in recorded history. The storm is passing over water with sea-surface temperatures of 22C° (72°F). The yellow and red colored rain bands are producing 30 to 50 mm (1.2 to 2.0 in.) of rainfall; green, 10 to 20 mm (0.4 to 0.8 in.); blue, less than 0.4 mm.
[(a) and (b) MODIS sensor *Terra* images courtesy of the Rapid Response Team, NASA/GSFC.]

Table 8.2 presents the criteria for classification of tropical cyclones on the basis of wind speed and lists features of each designation. When you hear meteorologists speak of a "category 4" hurricane, they are using the Saffir–Simpson Hurricane Damage Potential Scale to estimate possible damage from hurricane-force winds (see Table 8.2). This scale, using wind speeds and central pressure criteria, ranks hurricanes and typhoons in five categories, from smaller, category 1 storms to extremely dangerous (and rare) category 5. Damage depends on the degree of property development at a storm's landfall site, how prepared citizens are for the blow, the storm surge, the presence of embedded tornadoes, and the overall intensity of the storm.

The cyclone that struck Bangladesh in 1970 killed an estimated 300,000 people, and the one in 1991 claimed more than 200,000 lives. The Galveston, Texas, hurricane of 1900 killed 6000; Hurricane Audry (1957), 400; Hurricane Gilbert (1988), 318; Hurricane Camille (1969), 256; and Hurricane Mitch (1998), more than 12,000 dead with many still missing, making it the deadliest Atlantic hurricane in 220 years. The record holder for property damage is 1992's Hurricane Andrew, which tallied $20 billion in damage. In 2003, Hurricane Isabel struck the Outer Banks and Cape Hatteras of North Carolina in September 2003, causing extensive flooding and wind damage to these fragile barrier islands as well as 36 deaths and about $2 billion in damages (Figure 8.29). Hurricane Charley brought category 4 winds to Florida's Gulf Coast in 2004, causing approximately $12 billion in damage and taking about two-dozen lives. As happened with Hurricane Andrew and Florida's Atlantic coast, the

Table 8.2   Tropical Cyclone Classification

Designation	Winds	Features
Tropical disturbance	Variable, low	Definite area of surface low pressure; patches of clouds
Tropical depression	Up to 34 knots (63 kmph, 39 mph)	Gale force, organizing circulation; light to moderate rain
Tropical storm	35–63 knots (63–118 kmph, 39–73 mph)	Closed isobars; definite circular organization; heavy rain; assigned a name
Hurricane (Atlantic and E. Pacific) Typhoon (W. Pacific) Cyclone (Indian Ocean, Australia)	Greater than 65 knots (119 kmph, 74 mph)	Circular, closed isobars; heavy rain, storm surges; tornadoes in right-front quadrant

Saffir-Simpson Hurricane Damage Potential Scale

Category	Wind Speed	Notable Atlantic Examples
1	65–82 knots (74–95 mph)	—
2	83–95 knots (96–110 mph)	—
3	96–113 knots (111–130 mph)	1985 Elena; 1991 Bob; 1995 Roxanne, Marilyn; 1998 Bonnie
4	114–135 knots (131–155 mph)	1979 Frederic; 1985 Gloria; 1995 Felix, Luis, Opal; 2004 Charley 1998 Georges
5	>135 knots (>155 mph)	1935 No. 2; 1938 No. 4; 1960 Donna; 1961 Carla; 1969 Camille; 1979 David; 1988 Gilbert; 1989 Hugo; 1992 Andrew; 1998 Mitch

Ft. Meyers region will take years to recover from Charley.

We appear to be in a more intense hurricane cycle similar to the 1941 to 1970 pattern when there were 24 major hurricanes, as compared to 14 between 1971 and 1994. Warmer Atlantic waters along the 20th parallel between 20° W and 90° W characterize this more intense cycle.

Since record keeping began in 1870, the second-greatest year for number of Atlantic hurricanes was 1995. Atmospheric scientists at Colorado State University, led by William Gray, successfully predicted the near-record 1995 Atlantic hurricane season and developed a forecasting model that accurately predicts Atlantic hurricanes. The 1995 season and this forecasting approach are discussed in Focus Study 8.1.

Physical Structure   Fully organized tropical cyclones have an intriguing physical appearance (Figure 8.30). They range in diameter from a compact 160 km (100 mi), to 1000 km (600 mi), to some western Pacific typhoons that attain 1300–1600 km (800–1000 mi). Vertically, these storms dominate the full height of the troposphere.

The inward-spiraling clouds form dense rain bands, with a central area designated the *eye*. Around the eye swirls a thunderstorm cloud called the *eyewall*, which is the area of most intense precipitation. The eye remains an enigma, for in the midst of devastating winds and torrential rains the eye is quiet and warm with even a glimpse of blue sky or stars possible. The structure of the rain bands, central eye, and eyewall is clearly visible for Hurricane Gilbert in Figure 8.30a (top view), 8.30b (oblique view from an artist's perspective), and 8.30c (side-radar view).

The entire storm moves along at 16–40 kmph (10–25 mph). When it makes **landfall**, or moves ashore, the storm pushes seawater inland, causing dangerous **storm surges** often several meters in depth. Storm surges often catch people by surprise and cause the majority of hurricane drownings. In 1998, Hurricanes Bonnie, Georges, and Mitch moved slowly onshore, producing damaging storm surges along North Carolina, Mississippi, and Central America, respectively. The slow progress of these storms produced extensive flooding from the sustained rains.

The strongest winds of a tropical cyclone are usually recorded in its right-front quadrant (relative to the storm's directional path), where dozens of fully developed tornadoes may be embedded at the time of landfall. For example, Hurricane Camille in 1969 had up to 100 tornadoes embedded in that quadrant.

Hurricane Andrew–1992   The period between August 24 and September 11, 1992 displayed nature's power and

FIGURE 8.29 Cape Hatteras and the Outer Banks, hit again.
At one point in its path when far out at sea, Hurricane Isabel reached a category 5 rating. At the time of landfall during the afternoon of September 18, 2003, it had slowed to a category 2—still a damaging storm. Rainfall exceeded 12.7 cm (5 in.) over eastern North Carolina. Its remnants reached Lake Erie 24 hours later. [MODIS sensor *Terra* image, September 18, 2003, courtesy of Rapid Response Team, NASA/GSFC.]

capacity to cause personal, societal, and economic tragedy. Those 19 days saw Hurricane Andrew (Florida and Louisiana), Hurricane Iniki (Kaua'i, Hawai'i), and Typhoon Omar (Guam) strike with record-breaking fury. Figure 8.31 presents an AVHRR image from August 24, showing Andrew after it moved across Florida into the Gulf of Mexico. The locator map shows the entire track. The central eye is clearly visible, as are the tight rain bands marking the rapid inward-spiraling, counterclockwise wind flow. The tragedy of Andrew is that the storm destroyed or seriously damaged 70,000 homes and left 200,000 people homeless between Miami and the Florida Keys.

Hurricane Andrew's sustained winds were 225 kmph (140 mph), with gusts to 282+ kmph (175+ mph).

Studies completed by meteorologist Theodore Fujita estimated that winds in the eyewall reached 320 kmph (200 mph) in small vortexes. The remnants of Hurricane Andrew continued northward, reaching Québec and eastern Canada by August 28. The storm produced heavy rains as the remnants of Andrew interacted with weather fronts moving through the area, breaking many 24-hour precipitation records in Québec.

The Everglades, the coral reefs north of Key Largo, some 10,000 acres of mangrove wetlands, and coastal southern pine forests all had significant damage from Andrew. Natural recovery will take years, which offers scientists a research opportunity. Scientific assessments judge the Everglades to be quite resilient because the region's ecosystems evolved naturally with periodic hurricanes over the millennia. An important fact is that storms damage human structures more than they damage natural systems. Remember this perspective: *Urbanization, agriculture, pollution, and water diversion pose a greater ongoing threat to the Everglades than do hurricanes!*

Statistically, the damage caused by tropical cyclones is increasing substantially as more and more development occurs along susceptible coastlines, whereas loss of life is decreasing in most parts of the world owing to better forecasting of these storms. *Science* offered this "Perspective":

> The risk of human losses is likely to remain low, however, because of a well-established warning and rescue system and ongoing improvements in hurricane prediction. A main concern is the risks of high damage costs (up to $100 billion in a single event) because of ongoing population increases in coastal areas and increasing investment in buildings and extensive infrastructure in general.*

Ironically, the relatively mild hurricane seasons from 1971 to 1994 encouraged weak zoning and rapid development of vulnerable coastal lowlands. New buildings, apartments, and government offices are opening right next to still visible rubble and bare foundation pads. The consequence is that our entire society bears the financial cost of planning failure, whether in coastal regions or other areas with natural hazards, not to mention those victims who directly shoulder the physical, emotional, and economic hardship of the event. Remember, had Hurricane Andrew made landfall a few kilometers farther north, it would have blasted Miami; damage estimates for that missed catastrophe are around $85 billion. Hurricane Charley's destruction of coastal properties in Florida in 2004 further underscores the problem.

This recurrent, yet avoidable, cycle—construction, devastation, reconstruction, devastation—was reinforced ironically in *Fortune* magazine more than 30 years ago

(text continued on page 239)

---

*L. Bengtsson, "Hurricane Threats," *Science* 293 (July 20, 2001): p. 441.

(a)

(b)

(c)

FIGURE 8.30 Profile of a hurricane.
For the Western Hemisphere, Hurricane Gilbert attained the record size (1600 km, or 1000 mi, in diameter) and record low barometric pressure (888 mb, or 26.22 in.). Gilbert's sustained winds reached 298 kmph (185 mph) with peaks exceeding 320 kmph (200 mph). Hurricane Gilbert, September 13, 1988: (a) *GOES-7* satellite image; (b) a stylized portrait of a mature hurricane, drawn from an oblique perspective (cutaway view shows the eye, rain bands, and wind flow patterns); (c) *SLAR* (side-looking airborne radar) image from an aircraft flying through the center of the storm. Rain bands of greater cloud density are false-colored in yellows and reds. The clear sky in the central eye is dramatically portrayed. [(a), (b), (c) NOAA and the National Hurricane Center, Miami.]

Hurricane Wind
Patterns

Hurricane
Georges

Focus Study 8.1

Forecasting Atlantic Hurricanes

Forecasters at the National Hurricane Center (NHC) in Miami were busy throughout the 1995–2003 hurricane seasons. This was the most active nine-year period in the history of the NHC, with 117 named tropical storms, including 68 hurricanes (26 intense, meaning category 3 or higher). This was a record level of activity despite the reduced number of tropical cyclones during the 1997

El Niño season of reduced tropical storm intensity. In 2003, the Atlantic tropical storm season was 168% of expected net tropical cyclone activity, which translates to 14 named storms (9.6 is average), including 7 hurricanes (5.9 is average), or 118% of normal. For comparison, tropical storm activity for the 1995 season, discussed shortly, was 235% of an average season.

An important innovation to help the NHC and state and local governments assess these storms is an accurate forecasting method developed by a team at Colorado State University led by William M. Gray (see http://typhoon.atmos.colostate.edu/forecasts/).

Their analysis of weather records for the period 1900–2003 disclosed a significant causal relationship between

Atlantic tropical storms, including hurricanes that make landfall along the U.S. Gulf, and East Coasts, and several meteorological variables. Of 16 potential storm predictors used by the Colorado scientists, five key forecast variables emerge:

1. *The presence or absence of warm water off the coast of Peru in the eastern Pacific.* Sea-surface temperatures are part of the El Niño and La Niña phenomena (see Chapter 10). A strong El Niño featuring warm surface water in the eastern Pacific Ocean tends to dampen the development of Atlantic hurricanes. Associated with El Niño events are strong westerly winds in the upper troposphere (12,000 m, 40,000 ft) that tend to shear off the tops of developing tropical depressions and storms. The opposite effect occurs during La Niña episodes, when cool water occurs in the eastern Pacific Ocean. These periods are associated with weak westerly surface winds, thus encouraging storm intensification in the Atlantic. During 1997, the Pacific El Niño was the strongest of the century, more than double the intensity of the 1982–1983 El Niño. Such an episode reduces the Atlantic storm season.

2. *Rainfall, temperature, and pressure patterns in West Africa,* specifically the Sahel along the southern margins of the Sahara (10°–20° N). Temperature and pressure regimes from February to May can set the stage for heavy rainfall during the normal June to September rainy season. A wet February to May stimulates storm formation off the African coast. A drought in this same region tends to retard Atlantic storm development.

3. *Directional flows of equatorial stratospheric winds (20,000–23,000 m altitude, 68,000–75,000 ft), which tend to shift and reverse roughly every 12 to 16 months.* These are the quasi-biennial oscillation (QBO) winds. Westerly winds double Atlantic tropical storm probability and intensity, whereas when the QBO

easterly flow intensifies, storm tops shear off and storm development is discouraged.

4. *Sea-surface temperatures in the Atlantic Ocean between Africa and the Cape Verde Islands and the Caribbean and along the North American East Coast.* Warmer surface water favors tropical storm development. Important is the functioning of the Atlantic *thermohaline circulation,* or *Atlantic conveyor belt.* Imagine a river of water flowing northward from the Caribbean. Near Greenland, the current sinks deep beneath the surface and flows southward to the southern Atlantic Ocean. Higher-than-average sea-surface temperature and salinity level are related to a more intense tropical storm/hurricane season—the trend since 1994. Lower temperatures and salinity tend to dampen storm activity—characteristic of the 1900–1925 and 1970–1993 periods.

5. *The status of the Azores high-pressure anticyclone* (see Figure 6.14 and text discussion). If the circulation system is weak and below average in strength, then the east-Atlantic trade winds are weakened and tropical storm development is favored. A strong Azores high dampens development.

Dr. Gray's team has issued forecasts for 21 years and scored better than climatology and the long-term averages, especially when it comes to predicting named storms. The Colorado State team issues its forecast each December (for the next year) and issues updates in April, June, and August. (To check the accuracy of the present forecast, see the URL listed previously.)

The 1995 Atlantic Hurricane Season—An Example

As Atlantic hurricane seasons go (June 1 to November 30), 1995 was the second-most active since record keeping began in 1870. Of 19 tropical storms, 11 became hurricanes, and 5 of these achieved category 3 status or higher. Only 1933 (21 tropical storms) and

1969 (12 hurricanes) exceeded this total. During the Atlantic hurricane season, the year 1995 experienced tropical storms 52% of the season and hurricanes 27% of the time—both records. Hurricanes Allison, Erin, and Opal together killed 66 and damaged $3.7 billion in U.S. property—the season totaled $8 billion in overall damage.

Figure 8.1.1 maps all 19 tropical storms, including 11 hurricanes (colored tracks). The season began with Allison on June 3–11 and ended with Tanya from October 27 through November 1. Figure 8.1.2 is a satellite image from August 30 that shows tropical storms Karen and Luis, Hurricanes Humberto and Iris, and the remnants of tropical storm Jerry off the image to the far left.

Each storm went through stages of tropical depression, tropical storm, hurricane, and extratropical depression, based on wind speed (depression, storm, hurricane) and location of the low-pressure center (tropical or extratropical). Top wind speed and lowest central pressure are noted on the map for each hurricane.

These storms hit several areas in particular. Three hurricanes crossed the Florida panhandle within 160 km (100 mi) of each other in June, July, and September. Also, the eastern Caribbean was savaged, especially the islands of Dominica, Saint Thomas, and Saint Maarten, where Luis and Marilyn hit within a few days of each other in September. As one researcher noted, "Maybe lightning does strike the same place twice!"

No matter how accurate storm forecasts become, coastal and lowland property damage will continue to increase until better hazard zoning and development restrictions are in place. The property insurance industry appears to be taking action to promote these improvements. They are requiring tougher building standards to obtain coverage—or, in some cases, they are refusing to insure property along vulnerable coastal lowlands.

Focus Study 8.1 (continued)

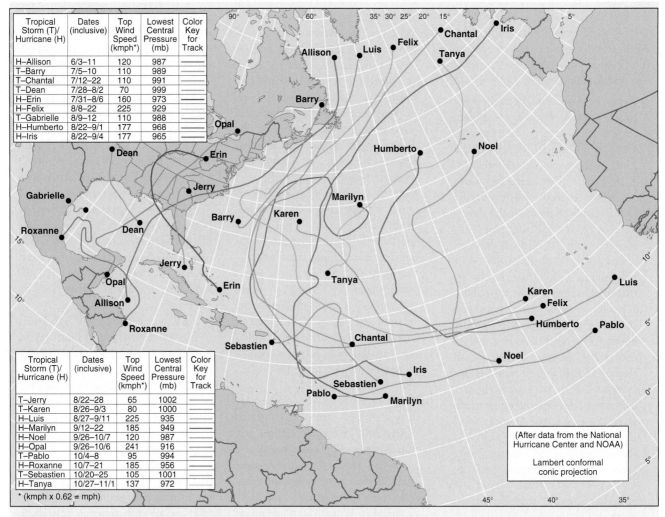

Tropical Storm (T)/ Hurricane (H)	Dates (inclusive)	Top Wind Speed (kmph*)	Lowest Central Pressure (mb)	Color Key for Track
H–Allison	6/3–11	120	987	
T–Barry	7/5–10	110	989	
T–Chantal	7/12–22	110	991	
T–Dean	7/28–8/2	70	999	
H–Erin	7/31–8/6	160	973	
H–Felix	8/8–22	225	929	
T–Gabrielle	8/9–12	110	988	
H–Humberto	8/22–9/1	177	968	
H–Iris	8/22–9/4	177	965	

Tropical Storm (T)/ Hurricane (H)	Dates (inclusive)	Top Wind Speed (kmph*)	Lowest Central Pressure (mb)	Color Key for Track
T–Jerry	8/22–28	65	1002	
T–Karen	8/26–9/3	80	1000	
H–Luis	8/27–9/11	225	935	
H–Marilyn	9/12–22	185	949	
H–Noel	9/26–10/7	120	987	
H–Opal	9/26–10/6	241	916	
T–Pablo	10/4–8	95	994	
H–Roxanne	10/7–21	185	956	
T–Sebastien	10/20–25	105	1001	
H–Tanya	10/27–11/1	137	972	

* (kmph x 0.62 = mph)

(After data from the National Hurricane Center and NOAA)

Lambert conformal conic projection

FIGURE 8.1.1 1995 Atlantic storm season.
This is the Atlantic, Caribbean, and Gulf of Mexico hurricane season for 1995. Noted on the map are inclusive dates from each storm's birth as a tropical depression to dissipation, highest wind speed in kmph (kmph × 0.62 = mph), and lowest pressure (mb). [Data courtesy of NOAA and the National Hurricane Center, Miami.]

FIGURE 8.1.2 Satellite observes four storms at one time.
Tropical storms Karen and Luis and Hurricanes Humberto and Iris march westward across the Atlantic Ocean in fairly tight formation during late August 1995. The remnant of tropical storm Jerry is seen at the far left. [Image from National Weather Service and NOAA, (NESDIS), National Climatic Data Center.]

(a)

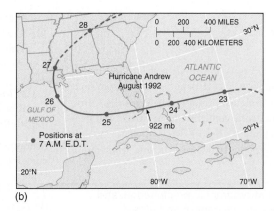
(b)

FIGURE 8.31 Hurricane Andrew.
(a) *NOAA-10* AVHRR image of Hurricane Andrew, August 24, 1992 (4:41 P.M. EDT) right after passage across southern Florida. (b) The track of Hurricane Andrew from August 23 through 28, 1992. [Image from NOAA as supplied by the EROS Data Center.]

Hurricanes Charley, Frances, and Ivan

after the destruction by Hurricane Camille:

> Before long the beachfront is expected to bristle with new motels, apartments, houses, condominiums, and office buildings. Gulf Coast businessmen, incurably optimistic, doubt there will ever be another hurricane like Camille, and even if there is, they vow, the Gulf Coast will rebuild bigger and better after that.*

*Fortune, October 1969, p. 62.

Summary and Review—Weather

Weather is the short-term condition of the atmosphere; **meteorology** is the scientific study of the atmosphere. The spatial implications of atmospheric phenomena and their relationship to human activities strongly link meteorology to physical geography. Analyzing and understanding patterns of wind, air pressure, temperature, and moisture conditions portrayed on daily weather maps is key to numerical (computer) weather forecasting.

weather (p. 208)
meteorology (p.208)

▶ *Describe* air masses that affect North America, and *relate* their qualities to source regions.

Specific conditions of humidity, stability, and cloud coverage occur in a regional, homogenous **air mass**. The longer an air mass remains stationary over a region, the more definite its physical attributes become. The homogeneity of temperature and humidity in an air mass sometimes extends through the lower half of the troposphere. Air masses are categorized by their moisture content—**m** for maritime (wetter) and **c** for continental (drier)—and their temperature, a function of latitude—designated **A** (arctic), **P** (polar), **T** (tropical), **E** (equatorial), and **AA** (Antarctic).

air mass (p. 209)

1. How does a source region influence the type of air mass that forms over it? Give specific examples of each basic classification.

2. Of all the air masses, which are of greatest significance to the United States and Canada? What happens to them as they migrate to locations different from their source regions? Give an example of air-mass modification.

▶ *Identify* four types of atmospheric lifting mechanisms, and *describe* four principal examples.

Air masses can be lifted by **convergent lifting** (airflows conflict, forcing some of the air to lift); **convectional lifting** (air passing over warm surfaces gains buoyancy); **orographic lifting** (passage over a topographic barrier); and *frontal lifting*. In North America, **chinook winds** (called *föhn* or *foehn* winds in Europe) are the warm, downslope airflows characteristic of the leeward side of mountains. Orographic lifting creates wetter windward slopes and drier leeward slopes situated in the **rain shadow** of the mountain. Conflicting air masses at a front produce a **cold front** (and sometimes a zone of strong wind and rain) or a **warm front**. A zone right along or slightly ahead of the front, called a

squall line, is characterized by turbulent and wildly changing wind patterns and intense precipitation.

convergent lifting (p. 211)
convectional lifting (p. 212)
orographic lifting (p. 213)
chinook winds (p. 213)
rain shadow (p. 213)
cold front (p. 216)
warm front (p. 216)
squall line (p. 217)

3. Explain why it is necessary for an air mass to be lifted if there is to be saturation, condensation, and precipitation.

4. What are the four principal lifting mechanisms that cause air masses to ascend, cool, condense, form clouds, and perhaps produce precipitation? Briefly describe each.

5. Differentiate between the structure of a cold front and a warm front.

▶ *Analyze* the pattern of orographic precipitation, and *describe* the link between this pattern and global topography.

The physical presence of a mountain acts as a topographic barrier to migrating air masses. *Orographic lifting* (*oro-* means "mountain") occurs when air is forcibly lifted upslope as it is pushed against a mountain. It cools adiabatically. An orographic barrier enhances convectional activity and causes additional lifting during the passage of weather fronts and cyclonic systems, thereby extracting more moisture from passing air masses. The precipitation pattern of windward and leeward slopes persists worldwide. The state of Washington is presented in the chapter as an excellent example of this concept.

6. When an air mass passes across a mountain range, many things happen to it. Describe each aspect of a mountain crossing by a moist air mass. What is the pattern of precipitation that results?

7. Explain how the distribution of precipitation in the state of Washington is influenced by the principles of orographic lifting.

▶ *Describe* the life cycle of a midlatitude cyclonic storm system, and *relate* this to its portrayal on weather maps.

A **midlatitude cyclone**, or **wave cyclone**, is a vast low-pressure system that migrates across the continent, pulling air masses into conflict along fronts. **Cyclogenesis**, the birth of the low-pressure circulation, can occur off the west coast of North America, along the polar front, along the lee slopes of the Rockies, in the Gulf of Mexico, and along the East Coast. A midlatitude cyclone can be thought of as having a life cycle of birth, maturity, old age, and dissolution. An **occluded front** is produced when a cold front overtakes a warm front in the maturing cyclone. Sometimes a **stationary front** develops between conflicting air masses, where airflow is parallel to the front on both sides. These systems are guided by the jet streams of the upper troposphere along seasonally shifting **storm tracks**.

midlatitude cyclone (p. 218)
wave cyclone (p. 218)
cyclogenesis (p. 218)
occluded front (p. 218)
stationary front (p. 218)
storm tracks (p. 219)

8. Differentiate between a cold front and a warm front as types of frontal lifting and what you would experience with each one.

9. How does a midlatitude cyclone act as a catalyst for conflict between air masses?

10. What is meant by cyclogenesis? In what areas does it occur and why? What is the role of upper-tropospheric circulation in the formation of a surface low?

11. Diagram a midlatitude cyclonic storm during its open stage. Label each of the components in your illustration, and add arrows to indicate wind patterns in the system.

▶ *List* the measurable elements that contribute to modern weather forecasting, and *describe* the technology and methods employed.

Synoptic analysis involves the collection of weather data at a specific time, as shown on the synoptic weather maps inserted in the chapter opener, in Figures 8.3, 8.15b, and in News Report 8.2. The standard weather station symbols used on these maps are in the legend to Figure 8.14. Building a database of wind, pressure, temperature, and moisture conditions is key to computer-based weather prediction and the development of weather forecasting models. Preparing a weather report and forecast requires analysis of a daily weather map and satellite images that depict atmospheric conditions. Weather data from these sources include the following:

- Barometric pressure (sea level and altimeter setting)
- Pressure tendency (steady, rising, falling)
- Surface air temperature
- Dew-point temperature
- Wind speed, direction, and character (gusts, squalls)
- Type and movement of clouds
- Current weather
- State of the sky (current sky conditions)
- Visibility; vision obstruction (fog, haze)
- Precipitation since last observation

Technology employed includes Dopler radar, wind profilers, AWIPS consoles, ASOS instrument arrays, various satellite platforms, and high-speed computers.

12. What is your principal source of weather data, information, and forecasts? Where does your source obtain its data? Have you used the Internet and World Wide Web to obtain weather information? In what ways will you personally apply this knowledge in the future? What benefits do you see?

▶ *Analyze* various forms of violent weather and the characteristics of each, and *review* several examples of each from the text.

The violent power of some weather phenomena poses a hazard to society. Severe ice storms involve **sleet** (freezing rain, ice glaze, and ice pellets), snow blizzards, and crippling ice coatings on roads, power lines, and crops. Thunderstorms produce **lightning** (electrical discharges in the atmosphere), **thunder** (sonic bangs produced by the rapid expansion of air after intense heating by lightning), and **hail** (ice pellets formed within cumulonimbus clouds).

Strong linear winds in excess of 26 m/s (58 mph) are known as **derechos**, associated with thunderstorms and bands of showers crossing a region. Straight-line winds can cause significant damage and crop losses. A spinning, cyclonic column rising to midtroposphere level—a **mesocyclone**—is sometimes visible as the swirling mass of a cumulonimbus cloud, especially in a *supercell* system. Dark gray **funnel clouds** pulse from the bottom side of the parent cloud. A **tornado** is formed when the funnel connects with Earth's surface. A **waterspout** forms when a tornado circulation occurs over water.

Within tropical air masses, large low-pressure centers can form along easterly wave troughs. Under the right conditions, a tropical cyclone is produced. Depending on wind speeds and central pressure, a **tropical cyclone** can become a **hurricane**, **typhoon**, or cyclone, when winds exceed 65 knots (74 mph, 119 kmph). As forecasting and the public's perception of weather-related hazards improve, loss of life decreases, although property damage continues to increase. Great damage occurs to occupied coastal lands when hurricanes make **landfall** and when winds drive ocean water inland in **storm surges**.

sleet (p. 221)
lightning (p. 223)
thunder (p. 223)
hail (p. 223)
derechos (p. 224)
mesocyclone (p. 225)
funnel clouds (p. 226)
tornado (p. 226)
waterspout (p. 226)
tropical cyclone (p. 231)

hurricane (p. 232)
typhoon (p. 232)
landfall (p. 234)
storm surges (p. 234)

13. What constitutes a thunderstorm? What type of cloud is involved? What type of air mass would you expect in an area of thunderstorms in North America?

14. Lightning and thunder are powerful phenomena in nature. Briefly describe how they develop.

15. Describe the formation process of a mesocyclone. How is this development associated with that of a tornado?

16. Evaluate the pattern of tornado activity in the United States. Where is Tornado Alley? What generalizations can you make about the distribution and timing of tornadoes? What happened in 2003?

17. What are the different classifications for tropical cyclones? List the various names used worldwide for hurricanes. Have any hurricanes ever occurred in the south Atlantic?

18. What factors contributed to the incredible damage cost of Hurricane Andrew? Why have such damage figures increased even though loss of life has decreased over the past 30 years?

19. What forecast factors did scientists use to accurately predict the 1995 Atlantic hurricane season?

20. Relative to improving weather forecasting, what are some of the technological innovations discussed in this chapter?

NetWork

The *Geosystems* Home Page provides online resources for this chapter on the World Wide Web. To begin: Once on the Home Page, click on this textbook, scroll the Table of Contents menu, select this chapter, and click "Begin." You will find self-tests that are graded, short essay and review exercises, specific updates for items in the chapter, and many links to interesting related pathways on the Internet under "Destinations." *Geosystems* is at **http://www.prenhall.com/christopherson**

Critical Thinking

A. Select "Destinations" from the left-hand frame of Chapter 8 on the *Geosystems* Home Page. Be prepared—this is a fantastic listing of many links on the Internet. Any item in color and underlined will open and take you directly to a website. Sample at least five of these weather links and describe what you find. You may want to mark some of these as links with your own browser for ready reference when you need weather information.

B. Relative to coastal devastation from tropical weather, the following line appears in the text: "This recurrent, yet avoidable cycle—construction, devastation, reconstruction, devastation. . . . " This describes the ever-increasing dollar losses to property from tropical storms and hurricanes, yet at a time when improved forecasts have resulted in a significant reduction in loss of life. (These issues are discussed further in Chapter 16's Focus Study 16.1.) In your opinion, what is the solution to halt these increasing losses, this cycle of destruction? How would you implement your plan?

C. From your local newspaper (or an Internet source), collect daily weather maps for five successive days. Compare your map sequence and identify principal weather elements on the maps. Were published forecasts for your region accurate? What source is listed for these forecast maps? Government or private contractor?

Tracy Smith, Research Meteorologist

After almost two decades at the Forecast Systems Laboratory (FSL), Tracy Smith is still enthusiastic about her work. Tracy asserts, "Weather is exciting! There is such satisfaction in developing a model and forecasting something and then having it happen—predict it and then find it. We are challenged to improve weather forecasting. If we can add a few minutes to severe weather warning times and narrow the focus of the warning area, or improve aviation safety, lives can be saved."

FSL developed the Advanced Weather Interactive Processing System (AWIPS) to assist forecasters across the country (pictured in Figure 8.17). Tracy works on computer models to improve the "data ingest" of all the information acquired from a growing list of new technologies. As part of her Rapid Update Cycle (RUC) effort, she is working to increase the update frequency, accuracy, and resolution of forecasting models. Tracy says, "This is one of the reasons atmospheric science is such a wide-open field as we attack these challenges."

When I asked about her interest in weather, she responded with a spirited laugh, "I didn't really decide to be a meteorologist until I was 15 or 16 years old!" I reminded her that this was an interesting thing for a teenager to say. She said, "When I was four or five my brother and sister brought a schoolbook home called *Hurricanes and Twisters*. I read it over and over." She was fascinated by the sky and, living in northeastern Ohio, she remembers those "big booming nocturnal thunderstorms in the summer."

When she was seven years old, the great Palm Sunday tornado outbreak of 1965 hit Ohio at Pittsfield (with an F5) and nearby Grafton (with an F4), where she visited her grandparents. She said, "I remember that night, waking up and hearing this great storm outside." Tracy's grandparents' house was slightly damaged but the homes around them were leveled—she walked along the street and saw the devastation. Her childhood fascination and experiences blossomed into a great career working with the atmospheric sciences she loves. Tracy admits to sometimes using vacation time to storm chase!

Tracy graduated from Bowling Green University, Ohio, with a degree in geography and a minor in math. She then earned her Master's degree in atmospheric sciences at Colorado State University, Fort Collins. For her thesis, she studied the radiation budget at the top of the atmosphere over the Indian monsoon region, using satellite data. Her committee was made up of both atmospheric science and geography professors. "Geography is important to atmospheric science because we are dealing with spatial elements of the atmosphere, hydrologic cycle, and topography. And, because of the impact on people, geography brings in the human aspect," she states.

Tracy participates in the daily weather briefings at FSL and once a month conducts the briefing. She says it is important to practice communication skills by using these weather briefings in order to be able to explain the forecast to others.

As to the future: "We want to refine the system, gain higher resolution, integrate all new data sources as they develop, and improve accuracy. We are just scratching the surface to get all this operational. The speed and capability of these computers are increasing so fast that this will be the next revolution in forecasting for us. Improving scale and accuracy are our ongoing challenges." And with her infectious laugh she asserts, "I like my job! I am paid to learn and study things and follow my instincts to solve big problems. What better way to spend your day than trying to understand the world around you while helping society!" Tracy's enthusiasm is obvious as she demonstrates the AWIPS workstation to me.

FIGURE 8.1.1 Tracy Smith, Research Meteorologist. Tracy works at the Forecast Systems Laboratory, Boulder, Colorado, National Weather Service, NOAA. Here she shows this author the workings of an AWIPS workstation. [Photo by Bobbé Christopherson.]

July 1983

July 2003

CHAPTER 9

Water Resources

▶ Key Learning Concepts

After reading the chapter, you should be able to:

▶ *Illustrate* the hydrologic cycle with a simple sketch, and *label* it with definitions for each water pathway.

▶ *Relate* the importance of the water-budget concept to your understanding of the hydrologic cycle, water resources, and soil moisture for a specific location.

▶ *Construct* the water-balance equation as a way of accounting for the expenditures of water supply, and *define* each of the components in the equation and their specific operation.

▶ *Describe* the nature of groundwater, and *define* the elements of the groundwater environment.

▶ *Identify* critical aspects of freshwater supplies for the future, and *cite* specific issues related to sectors of use, regions and countries, and potential remedies for any shortfalls.

The western United States, particularly the Southwest, is experiencing a prolonged drought. The Colorado River is a major water resource in this region. Hoover Dam on the Arizona–Nevada border backs up Lake Mead. The maximum reservoir elevation of 375 m (1229 ft) is shown in the July 1983 photo when Lake Mead was at capacity. The rarely used emergency spillways (foreground) were put into service to prevent the lake from topping the roadway and dam. Twenty years later, in July 2003, Lake Mead was at 348 m (1142 ft), or about 27 m (88.6 ft) lower than 1983. The Colorado River average annual flow (1931–2002) was 14.8 million acre-feet (maf); however, in 2001 it was 10.75 maf, 2002 was only 3.1 maf, and 2003 was 6.4 maf—all far below average. By July 2004, Lake Mead dropped to 343 m (1126.7 ft), or another 6 m lower than the 2003 photo. More on this in Chapter 15 Focus Study 15.1, "The Colorado River: A System Out of Balance." [Photos by (1983) author; (2003) Bobbé Christopherson.]

The physical reality of life is defined by water. Our lives are bathed and infused with water. Our own bodies are about 70% water, as are plants and animals. We use water to cook, bathe, wash clothing, and dilute our wastes. We water small gardens and vast agricultural tracts. Most industrial processes would be impossible without water. It is the essence of our existence and is therefore the most critical resource supplied by Earth systems—the essential resource for life.

In the last two chapters, we saw how the exchanges of energy between water and the atmosphere drive Earth's weather systems. The flow of water links the atmosphere, ocean, land, and living things through exchanges of energy and matter. In particular, the energy and moisture exchange between plants and the atmosphere is important to the status of the climate system and the range of responses from plant communities to climate change.

Aerial and surface views of eastern Nevada's Humboldt National Forest. The midlatitude cold desert climate produces a landscape dotted with a Pinyon–Juniper woodland (*Pinus monophylla* and *Juniperus osteosperma*, respectively). A light snow highlights the trees and mountains of the Great Basin. The inset photo shows this desert forest in summer. Total average rainfall in this region is about 20 cm (8 in.) with warm summers and cold winters. [Photos by Bobbé Christopherson.]

CHAPTER 10

Global Climate Systems

▶ Key Learning Concepts

After reading the chapter, you should be able to:

▶ **Define** climate and climatology, and **explain** the difference between climate and weather.

▶ **Review** the role of temperature, precipitation, air pressure, and air mass patterns used to establish climatic regions.

▶ **Review** the development of climate classification systems, and **compare** genetic and empirical systems as ways of classifying climate.

▶ **Describe** the principal climate classification categories other than deserts, and **locate** these regions on a world map.

▶ **Explain** the precipitation and moisture efficiency criteria used to determine the arid and semiarid climates, and **locate** them on a world map.

▶ **Outline** future climate patterns from forecasts presented, and **explain** the causes and potential consequences of climate change.

E arth experiences an almost infinite variety of *weather*—conditions of the atmosphere at any given time and place. But if we consider the weather over many years, including its variability and extremes, a pattern emerges that constitutes **climate**. Think of climatic patterns as dynamic rather than static. Climate is more than a consideration of simple averages of temperature and precipitation.

Today, climatologists know that intriguing global-scale linkages exist in the Earth–atmosphere–ocean system. For instance, strong monsoonal rains in West Africa are correlated with the development of intense Atlantic hurricanes; yet, an El Niño in the Pacific is tied to drought-breaking rains in the American West, floods in Louisiana and Northern Europe, and a weak Atlantic hurricane season. Lacking such a condition, the West is locked into drought that has persisted more than six years. The El Niño phenomenon is the subject of Focus Study 10.1.

Climatologists, among other scientists, are analyzing global climate change—record-breaking global average temperatures, glacial ice melt, drying soil-moisture conditions, changing crop yields, spreading of infectious disease, changing distributions of plants and animals, declining coral reef health and fisheries, and the thawing of high-latitude lands and seas.

In this chapter: Climates are so diverse that no two places on Earth's surface experience exactly the same climatic conditions; in fact, Earth is a vast collection of microclimates. However, broad similarities among local climates permit their grouping into climatic regions.

Many of the physical elements of the environment, studied in the first nine chapters of this text, link together to explain climates. New to this edition of *Geosystems* is a simplified classification system based on physical factors that help uncover the "why" question—why climates are in certain locations. Though imperfect, this method is easily understood and is based on a widely used classification system devised by climatologist Wladimir Köppen (pronounced KUR-pen). For reference, Appendix B details the Köppen climate classification system and all its criteria.

Climatologists use powerful computer models to simulate changing complex interactions in the atmosphere,

hydrosphere, lithosphere, and biosphere. This chapter concludes with a discussion of climate change and its vital implications for society. Climate patterns are changing at a pace not evidenced in the records of the past millennia—changes in climate and natural vegetation during the next 50 years could exceed the total of all changes since the peak of the last ice-age episode, some 18,000 years ago. Especially significant are changes occurring in the polar regions of the Arctic and Antarctic.

Earth's Climate System and Its Classification

Climatology, the study of climate, is the analysis of long-term weather patterns over time and space, including extreme weather events, to find areas of similar weather statistics. These areas of similar conditions are **climatic regions**. Observed patterns grouped into regions are at the core of climate classification.

The climate where you live may be humid with distinct seasons, or dry with consistent warmth, or moist and cool—almost any combination is possible. There are places where it rains more than 20 cm (8 in.) each month, with monthly average temperatures remaining above

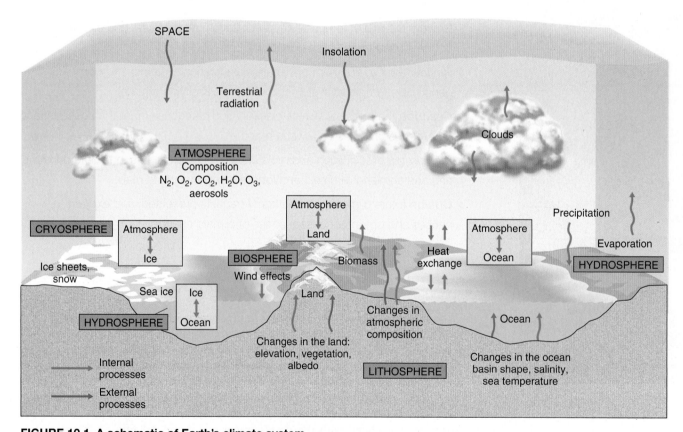

FIGURE 10.1 A schematic of Earth's climate system.
Imagine you are hired to write a computer program that simulates Earth's climates. To do this, you must consider including all the material covered in this text. *Internal processes* that influence climate involve the atmosphere, hydrosphere (streams and oceans), cryosphere (polar ice masses and glaciers), biosphere, and lithosphere (land)—all energized by insolation. All interact to produce climatic patterns. *External processes*, principally from human activity, affect this climatic balance and force climate change. [After J. Houghton, *The Global Climate* (Cambridge, UK: Cambridge University Press, 1984), and the Global Atmospheric Research Program.]

27°C (80°F) year-round. Other places may be rainless for a decade at a time. A climate may have temperatures that average above freezing every month yet still threaten severe frost problems for agriculture. Students reading *Geosystems* in Singapore experience precipitation every month, ranging from 13.1 to 30.6 cm (5.1 to 12.0 in.), or 228.1 cm (89.8 in.) during an average year, whereas students at the university in Karachi, Pakistan, measure only 20.4 cm (8 in.) of rain over an entire year.

Climates greatly influence *ecosystems*, the natural, self-regulating communities formed by plants and animals in their nonliving environment. On land, the basic climatic regions determine to a large extent the location of the world's major ecosystems. These regions, called *biomes*, include forest, grassland, savanna, tundra, and desert. Plant, soil, and animal communities are associated with these biomes. Because climate cycles through periodic change—it is never really stable—ecosystems should be thought of as being in a constant state of adaptation and response. Figure 10.1 presents a schematic view of Earth's climate system, showing both internal and external processes and linkages that influence climate and thus regulate such changes.

Climate Components: Insolation, Temperature, Pressure, Air Masses, and Precipitation

The principal elements of climate are insolation, temperature, pressure, air masses, and precipitation. The first nine chapters discussed each of these elements. We review them briefly here. Insolation is the energy input for the climate system, but it varies widely over Earth's surface by latitude (see Chapter 2 and Figures 2.9, 2.10, and 2.11). Daylength

and temperature patterns vary diurnally (daily) and seasonally. The principal controls of temperature are latitude, altitude, land–water heating differences, and cloud cover. The pattern of world temperatures and their annual ranges are in Chapter 5 (see Figures 5.14, 5.16, and 5.17).

Temperature variations result from a coupling of dynamic forces in the atmosphere to Earth's pattern of atmospheric pressure and resulting global wind systems (see Figures 6.11 and 6.13). Important, too, are the location and physical characteristics of air masses, those vast bodies of homogeneous air that form over oceanic and continental source regions.

Moisture is the remaining input to climate. The hydrologic cycle transfers moisture, with its tremendous latent heat energy, through Earth's climate system (see Figure 9.1). The moisture input to climate is precipitation in all its forms. Figure 10.2 shows the worldwide distribution of precipitation, our moisture supply. Its patterns are important, for it is a key climate control factor. Average temperatures and daylength help us approximate POTET (potential evapotranspiration), a measure of natural moisture demand.

Most of Earth's desert regions, areas of permanent water deficit, are in lands dominated by subtropical high-pressure cells, with bordering lands grading to grasslands and to forests as precipitation increases. The most consistently wet climates on Earth straddle the equator in the Amazon region of South America, the Congo region of Africa, and Indonesia and Southeast Asia, all of which are influenced by equatorial low pressure and the intertropical convergence zone (ITCZ, see Figure 6.11).

Simply relating the two principal climatic components—temperature and precipitation—reveals general climate types (Figure 10.3). Temperature and

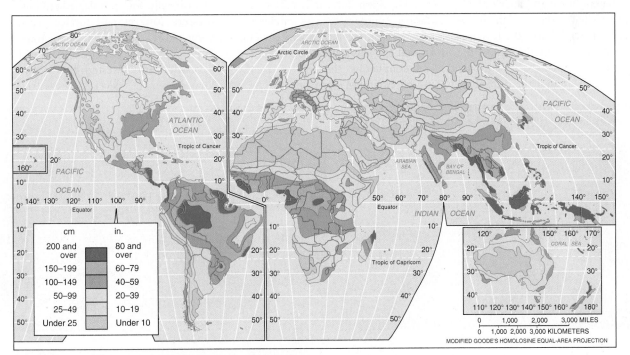

FIGURE 10.2 Worldwide average annual precipitation.
The causes that produce these patterns should be recognizable to you: temperature and pressure patterns; air mass types; convergent, convectional, orographic, and frontal lifting mechanisms; and the general energy availability that decreases toward the poles.

FIGURE 10.3 Climatic relationships. Temperature and precipitation schematic reveals climatic relationships. Based on general knowledge of your college location, can you identify its approximate location on the schematic diagram? Now locate the region of your birthplace.

Focus Study 10.1

The El Niño Phenomenon—Record Intensity, Global Linkages

Climate is the consistent behavior of weather over time, but average weather conditions also include extremes that depart from normal. The El Niño–Southern Oscillation (ENSO) in the Pacific Ocean forces the greatest interannual variability of temperature and precipitation on a global scale. The two strongest ENSO events in 120 years hit in 1997–1998 and 1982–1983. The spring wildflower bloom in Death Valley in 1998 provides visible evidence of the resultant heavy rains (Figure 10.1.1). Peruvians coined the name El Niño ("the boy child") because these episodes seem to occur around the traditional

December celebration time of Christ's birth. Actually, El Niños can occur as early as spring and summer and persist through the year.

Revisit Figure 6.22 and see that the northward-flowing Peru current dominates the region off South America's west coast. These cold waters move toward the equator and join the westward movement of the south equatorial current.

The Peru current is part of the normal counterclockwise circulation of winds and surface ocean currents around the subtropical high-pressure cell dominating the eastern Pacific in the Southern Hemisphere. As a

result, a location such as Guayaquil, Ecuador, normally receives 91.4 cm (36 in.) of precipitation each year under dominant high pressure, whereas islands in the Indonesian archipelago receive more than 254 cm (100 in.) under dominant low pressure. This normal alignment of pressure is shown in Figure 10.1.2a.

What Is ENSO?

Occasionally, for unexplained reasons, pressure patterns and surface ocean temperatures shift from their usual locations. Higher pressure than normal develops over the western Pacific, and lower pressure develops over the

(a) 1998

(b) 2002

FIGURE 10.1.1 El Niño's impact on the desert.
Death Valley, southeastern California, in (a) full spring bloom following record rains triggered by the 1997–1998 El Niño and (b) the same scene in spring 2002 in its stark desert grandeur. A dramatic effect caused by changes in the distant tropics of the Pacific Ocean. [Photos by Bobbé Christopherson.]

FIGURE 10.1.2 Normal, El Niño, and La Niña changes in the Pacific.
(a) Normal patterns in the Pacific; (b) El Niño wind and weather patterns across the Pacific Ocean and *TOPEX/Poseidon* satellite image for November 1997 (white and red colors indicate warmer surface water—a warm pool). (c) *TOPEX/Poseidon* image of La Niña conditions in transition in the Pacific on October 1998 (purple and blue colors for cooler surface water—a cool pool). (d) A persistent La Niña in March 2000 satellite image. (e) Image from June 2001, showing no El Niño as equatorial waters slowly warm with sea-surface temperature near normal. (f) The warm pool of waters retreated westward leaving neutral or a mild La Niña off South America, July 2004 image. [(a) and (b) Adapted and author corrected from C. S. Ramage, "El Niño." © 1986 by *Scientific American*, Inc.; (b)–(e) *TOPEX/Poseidon* and (f) *Jason–1* images courtesy of Jet Propulsion Laboratory, NASA.]

ANIMATION El Niño/El Niña

eastern Pacific. Trade winds normally moving from east to west weaken and can be reduced or even replaced by an eastward (west-to-east) flow. The shifting of atmospheric pressure and wind patterns across the Pacific is the *Southern Oscillation*. Chapter 6 discussed the Pacific Decadal Oscillation (PDO) and its interrelation with ENSO.

Sea-surface temperatures increase, sometimes more than 8 C° (14 F°) above normal in the central and eastern Pacific during an ENSO,

(continued)

News Report 10.1

What's in a Boundary?

The boundary between mesothermal and microthermal climates is sometimes placed along the isotherm where the coldest month is −3°C (26.6°F) or lower. That might be an accurate criterion for Europe, but for conditions in North America, the 0°C (32°F) isotherm is considered more appropriate. The difference between the 0 and −3°C isotherms covers an area about the width of the state of Ohio. Remember, these isotherm lines are really transition zones and do not mean abrupt change from one temperature to another.

A line denoting at least one month below freezing runs from New York City roughly along the Ohio River, trending westward until it meets the dry climates in the southeastern corner of Colorado. In Figure 10.5 you see this boundary used. A line marking −3°C as the coldest month would run farther north along Lake Erie and the southern tip of Lake Michigan. In addition, remember that from year to year, the position of the 0°C isotherm for January can shift several hundred kilometers as weather conditions vary.

Climate change adds another dimension to this question of accurate placement of statistical boundaries—for they are shifting. The Inter-governmental Panel on Climate Change (IPCC, discussed later in this chapter) predicts a 150- to 550-km (90- to 350-mi) range of possible poleward shift of climatic patterns in the midlatitudes during this century. Such change in climate would place Ohio, Indiana, and Illinois within climate regimes now experienced in Arkansas and Oklahoma. As you examine North America in Figure 10.5, use the graphic scale to get an idea of the magnitude of these potential shifts. Boundaries are indeed dynamic.

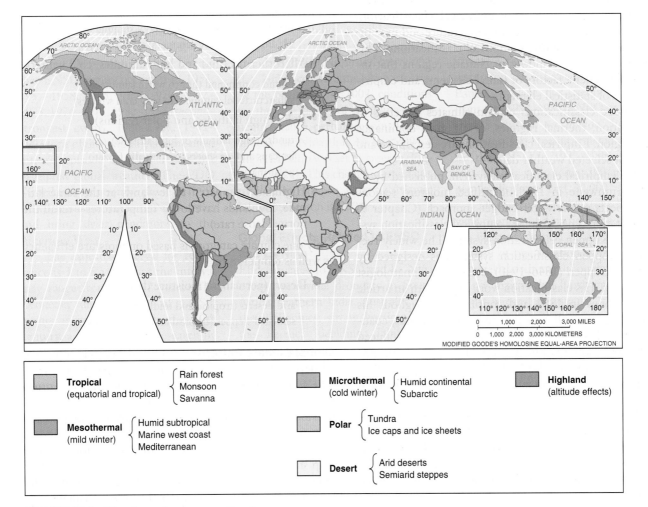

FIGURE 10.4 Climate regions generalized.
Six general climate categories. Relative to the question asked about your campus and birthplace in the caption to Figure 10.3, locate these two places on this climate map.

average annual temperature, total annual precipitation, elevation, the local population, annual temperature range, annual hours of sunshine (if available, as an indication of cloudiness), and a location map. Along the top of each climograph, find the dominant weather features that are influential in that climate.

Discussions of soils, vegetation, and major terrestrial biomes that fully integrate these global climate patterns are in Chapters 18 and 19. Table 20.1 synthesizes all this information and enhances your understanding of this chapter, so please place a tab on that page and refer to it as you read.

Tropical Climates (equatorial and tropical latitudes)

Tropical climates occupy about 36% of Earth's surface, including both ocean and land areas—Earth's most extensive climate category. The tropical climates straddle the equator from about 20° N to 20° S, roughly between the Tropics of Cancer and Capricorn, thus the designation *tropical*. Tropical climates stretch northward to the tip of Florida and south-central Mexico, central India, and Southeast Asia. These climates are truly winterless. Consistent daylength and almost perpendicular Sun angle throughout the year generate this warmth. Important causal elements include:

- Consistent daylength and insolation input, which produce consistently warm temperatures;
- Intertropical convergence zone (ITCZ), which brings rain as it shifts seasonally with the high Sun;
- Warm ocean temperatures and unstable maritime air masses.

Tropical climates have three distinct regimes: *tropical rain forest* (ITCZ present all year), *tropical monsoon* (ITCZ present for

6 to 12 months), and *tropical savanna* (ITCZ present for less than 6 months).

| Tropical rain forest Tropical monsoon | Tropical savanna |

Tropical Rain Forest Climates

The *tropical rain forest* climate is constantly moist and warm. Convectional thunderstorms, triggered by local heating and trade-wind convergence, peak each day from midafternoon to late evening inland and earlier in the day where marine influence is strong along coastlines. Precipitation follows the migrating intertropical convergence zone (ITCZ, Chapter 6). The ITCZ shifts northward and southward with the summer Sun throughout the year, but it influences *tropical rain forest* regions all year long. Not surprisingly, water surpluses are enormous, creating the world's greatest stream discharges in the Amazon and Congo Rivers.

High rainfall sustains lush evergreen broadleaf tree growth, producing Earth's equatorial and tropical rain forests (Figure 10.6). The leaf canopy is so dense that little light diffuses to the forest floor, leaving the ground surface dim and sparse in plant cover. Dense surface vegetation occurs along riverbanks, where light is abundant. (Widespread deforestation of Earth's rain forest is detailed in Chapter 20.)

High temperature promotes energetic bacterial action in the soil so that organic material is quickly consumed. Heavy precipitation washes away certain minerals and nutrients. The resulting soils are somewhat sterile and can support intensive agriculture only if supplemented by fertilizer.

Uaupés, Brazil (Figure 10.7), is characteristic of *tropical rain forest*. On the climograph you can see that the lowest-precipitation month receives nearly 15 cm

(6 in.), and the annual temperature range is barely 2 C° (3.6 F°). In all such climates, the diurnal (day-to-night) temperature range exceeds the annual average minimum–maximum (coolest to warmest) range: Day–night temperatures can range more than 11 C° (20 F°), more than 5 times the annual monthly average range.

The only interruption of *tropical rain forest* climates across the equatorial region is in the highlands of the South American Andes and in East Africa (see Figure 10.5). There, higher elevations produce lower temperatures; Mount Kilimanjaro is less than 4° south of the equator, but at 5895 m (19,340 ft) it has permanent glacial ice on its summit (although this ice is shrinking due to higher air temperatures). Such mountainous sites fall within the highland climate category.

Tropical Monsoon Climates

The *tropical monsoon* climates feature a dry season that lasts one or more months. Rainfall brought by the ITCZ affects these areas from 6 to 12 months of the year. (Remember, the ITCZ affects the *tropical rain forest* climate region throughout the year.) The dry season occurs when the convergence zone is not overhead. Yangon, Myanmar (formerly Rangoon, Burma), is an example of this climate type, as illustrated by the climograph and photograph in Figure 10.8.

Tropical monsoon climates lie principally along coastal areas within the tropical rain forest climatic realm and

(text continued on page 287)

ESOTHERM

Humid M

Humid W

Marine M

Medite S

ml

mP

SH

ATLANTIC OCEAN

Continental air mass

Station: Churchill, Manitoba
Lat/long: 58°45' N 94°04' W
Avg. Ann. Temp.: −7°C (19.4°F)
Total Ann. Precip.:
          44.3 cm (17.4 in.)

Elevation: 35 m (114.8 ft)
Population: 1400
Ann. Temp. Range:
          40 C° (72 F°)
Ann. Hr of Sunshine:
          1732

(a)

(b)

(c)

FIGURE 10.19 Subarctic cool summer climate.
(a) Climograph for Churchill, Manitoba (*subarctic,* cool summer). (b) Winter scene with foraging polar bear in the Churchill area near Hudson Bay. (c) Summer scene as forests give way to tundra landscapes in Churchill area. [(b) Photo by Art Wolfe/Tony Stone Images. (c) Photo by Norbert Rosing/Animals Animals/Earth Scenes.]

Continental air mass

Station: Verkhoyansk, Russia
Lat/long: 67°35' N 133°27' E
Avg. Ann. Temp.: −15°C (5°F)
Total Ann. Precip.:
　　　15.5 cm (6.1 in.)

Elevation: 137 m (449.5 ft)
Population: 1500
Ann. Temp. Range:
　　63 C° (113.4 F°)

FIGURE 10.20 Extreme subarctic cold winter climate.
(a) Climograph for Verkhoyansk, Russia (*subarctic,* very
cold winter). (b) Scene in the town of Verkhoyansk during
the short summer. [Photo by Dean Conger/National
Geographic Society.]

Polar and Highland Climates

The polar climates cover about 19% of Earth's total surface
and about 17% of its land area. These climates have no true
summer like that in lower latitudes. Poleward of the Arctic and
Antarctic Circles, daylength in summer becomes continuous,
yet average monthly temperatures never rise above 10°C
(50°F). These temperature conditions are intolerant to tree
growth. Daylength, which in part determines the amount of insolation received, and low Sun altitude in the sky are the principal climatic factors in these frozen and barren regions. Yet,
in winter, the Sun drops below the horizon poleward of 66.5°
latitude, producing continuous night. An extended dawn and
twilight period eases the seasonal shock and reduces the time
of true night.

Principal climatic factors in these frozen and barren regions are the following:

- Extremes of daylength between winter and summer determine the amount of insolation received.
- Low Sun altitude even during the long summer days is
the principal climatic factor.
- Extremely low humidity produces low precipitation
amounts—these regions are Earth's frozen deserts.
- Light-colored surfaces of ice and snow reflect substantial
energy away from the ground surface, thus reducing net
radiation.

Polar climates have three regimes: *tundra* (high latitude, or altitude); *ice caps* and *ice sheet* (perpetually frozen); and *polar marine* (oceanic association, slight moderation of extreme cold).

Also in this climate category we include highland
climates. For even at low latitudes the effects of altitude can
produce tundra and polar conditions. Glaciers on tropical
mountain summits attest to the cooling effects of altitude.
Highland climates on the map follow the pattern of Earth's
mountain ranges.

(a)

(b)

FIGURE 10.23 Desert landscape.
Desert plants are particularly well adapted to the harsh environment. (a) The silt-laden Colorado River,
banks lined with moisture-demanding cottonwood and tamarisk trees, flows north of Moab, Utah, cutting
through beautiful red sandstone. (b) Joshua Tree National Park is part of the Mojave desert in southeastern California, here with yucca and desert lupine, among other plants. [Photos by Bobbé Christopherson.]

extend to western continental margins, where cool, stabilizing ocean currents operate offshore and summer
advection fog forms. The Atacama Desert of Chile, the
Namib Desert of Namibia, the Western Sahara of
Morocco, and the Australian Desert each lie adjacent to
such a coastline.

Orographic lifting intercepts moisture-bearing
weather systems to create rain shadows along mountain
ranges that extend these dry regions into higher latitudes.
Note these rain shadows in North and South America on
the climate map. The isolated interior Asia, far distant
from any moisture-bearing air masses, falls within the *dry
arid and semiarid* climates as well.

Major subdivisions include: *deserts* (precipitation
supply roughly less than one-half of the natural moisture
demand) and *semiarid steppes* (precipitation supply roughly
more than one-half of natural moisture demand). Important is whether precipitation falls principally in the winter
with a dry summer, in the summer with a dry winter, or is
evenly distributed. Winter rains are most effective because they fall at a time of lower moisture demand. Relative to temperature, the lower-latitude deserts and steppes
tend to be hotter with less seasonal change than the midlatitude deserts and steppes, where mean annual temperatures are below 18°C (64.4°F) and freezing winter
temperatures are possible.

Low-Latitude Hot Desert Climates

Low-latitude hot desert climates are Earth's true tropical
and subtropical deserts and feature annual average temperatures above 18°C (64.4°F). They generally reside on
the western sides of continents, although Egypt, Somalia,
and Saudi Arabia also fall within this classification.
Rainfall is from local summer convectional showers.

Some regions receive almost no rainfall, whereas others
may receive up to 35 cm (14 in.) of precipitation a year.
A representative *low-latitude hot desert* city is Ar Riyāḍ
(Riyadh), Saudi Arabia (Figure 10.24).

Along the Sahara's southern margin is a drought-
tortured region. Human populations suffered great hardship as desert conditions gradually expanded over their
homelands. The sparse environment sets the stage for a
rugged lifestyle and subsistence economies, pictured here
near Timbuktu, Mali (Figure 10.24c). Chapter 15 presents
the process of desertification (expanding desert conditions).

Midlatitude Cold Desert Climates

Midlatitude cold desert climates cover only a small area: the
countries along the southern border of Russia, the Gobi
Desert, and Mongolia in Asia; the central third of Nevada
and areas of the American Southwest, particularly at high
elevations; and Patagonia in Argentina. Because of lower
temperatures and lower moisture-demand values, rainfall
must be low for a station to qualify as a *midlatitude cold
desert* climate; consequently, total annual average rainfall
is only about 15 cm (6 in.).

A representative station is Albuquerque, New Mexico,
with 20.7 cm (8.1 in.) of precipitation and an annual
average temperature of 14°C (57.2°F) (Figure 10.25).
Across central Nevada stretches a characteristic expanse of
midlatitude cold desert, greatly modified by a century of
livestock grazing (Figure 10.25b). Review the chapter-
opening photos of this region. Comparing *midlatitude cold
desert* and *low-latitude hot desert* regions reveals an interesting similarity in the annual temperature range and a distinct difference in precipitation patterns. The role of
summer convectional thunderstorms is evident on the
climograph in Figure 10.25.

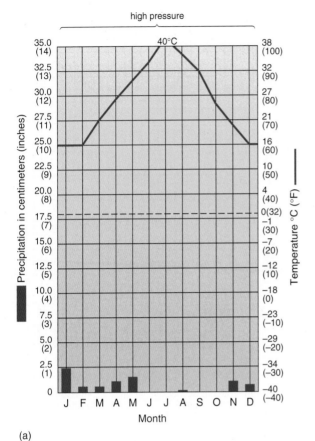

Station: Ar Riyāḍ (Riyadh), Saudi Arabia
Lat/long: 24°42'N 46°43' E
Avg. Ann. Temp.: 26°C (78.8°F)
Total Ann. Precip.: 8.2 cm (3.2 in.)
Elevation: 609 m (1998 ft)
Population: 5,024,000
Ann. Temp. Range: 24 C° (43.2 F°)

(a)

FIGURE 10.24 Low-latitude hot desert climate.
(a) Climograph for Ar Riyāḍ (Riyadh), Saudi Arabia (*low-latitude hot desert*). (b) The Arabian desert sand dunes near Ar Riyāḍ. (c) Herders bring a few cattle to market near Timbuktu, Mali, in west Africa. Precipitation has been below normal in the region since 1966. [(b) Photo by Ray Ellis/Photo Researchers, Inc. (c) Photo by Betty Press/Woodfin Camp & Associates.]

Low-Latitude Hot Steppe Climates

Low-latitude hot steppe climates generally exist around the periphery of hot deserts, where shifting subtropical high-pressure cells create a distinct summer-dry and winter-wet pattern. Average annual precipitation in *low-latitude hot steppe* areas is usually below 60 cm (23.6 in.). Walgett, in interior New South Wales, Australia, provides a Southern Hemisphere example of this climate (Figure 10.26). This climate is seen around the Sahara's periphery and in the Iran, Afghanistan, Turkmenistan, and Kazakstan region.

Midlatitude Cold Steppe Climates

The *midlatitude cold steppe* climates occur poleward of about 30° latitude and the *midlatitude cold desert* climates. Such midlatitude steppes are not generally found in the Southern Hemisphere. As with other dry climate regions, rainfall in the steppes is widely variable and undependable, ranging from 20 to 40 cm (7.9 to 15.7 in.). Not all rainfall is convectional, for cyclonic storm tracks penetrate the continents; however, most storms produce little precipitation.

Figure 10.27 presents a comparison between Asian and North American *midlatitude cold steppe*. Consider

(text continued on page 306)

(b)

(c)

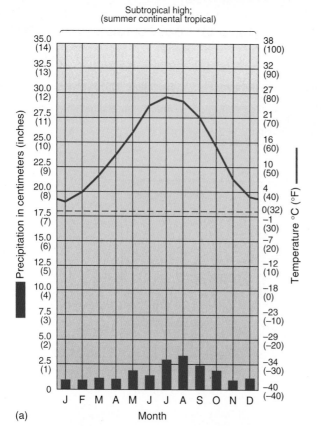

Subtropical high;
(summer continental tropical)

(a)

Month

Station: Albuquerque,
New Mexico
Lat/long: 35°03′ N 106°37′ W
Avg. Ann. Temp.: 14°C (57.2°F)
Total Ann. Precip.:
20.7 cm (8.1 in.)

Elevation: 1620 m (5315 ft)
Population: 479,000
Ann. Temp. Range:
24 C° (43.2 F°)
Ann. Hr of Sunshine:
3420

(b)

FIGURE 10.25 Midlatitude cold desert climate.
(a) Climograph for Albuquerque, New Mexico (*midlatitude
cold desert*). (b) Cold, high desert landscape of the Basin and
Range Province in Nevada, east of Ely along highway U.S. 50.
This area is in the region shown in the chapter-opening
photos. [Photo by Bobbé Christopherson.]

Subtropical high

(a)

Month

Station: Walgett, New South
Wales, Australia
Lat/long: 30°0′ S 148°07′ E
Avg. Ann. Temp.: 20°C (68°F)
Total Ann. Precip.:
45.0 cm (17.7 in.)

Elevation: 133 m (436 ft)
Population: 8200
Ann. Temp. Range:
17 C° (31 F°)

(b)

FIGURE 10.26 Low-latitude hot steppe climate.
(a) Climograph for Walgett, New South Wales, Australia
(*low-latitude hot steppe*). (b) Vast plains characteristic
of north-central New South Wales. [Photo by Otto
Rogge/Stock Market.]

(a)

Station: Semey (Semipalatinsk), Kazakstan
Lat/long: 50°21' N 80°15' E
Avg. Ann. Temp.: 3°C (37.4°F)
Total Ann. Precip.: 26.4 cm (10.4 in.)

Elevation: 206 m (675.9 ft)
Population: 270,500
Ann. Temp. Range: 39 C° (70.2 F°)

(b)

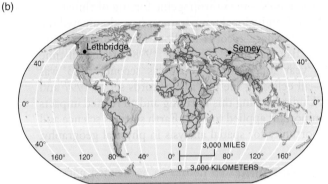

Station: Lethbridge, Alberta
Lat/long: 49°42' N 110°50' W
Avg. Ann. Temp.: 2.9°C (37.3°F)
Total Ann. Precip.: 25.8 cm (10.2 in.)

Elevation: 910 m (2985 ft)
Population: 73,000
Ann. Temp. Range: 24.3 C° (43.7 F°)

(c)

(d)

FIGURE 10.27 Midlatitude cold steppe climates, Kazakstan and Canada.
Climograph for (a) Semey (Semipalatinsk) in Kazakstan; (b) herders in the region near Semey. (c) Climograph for Lethbridge in Alberta. (d) Canadian prairies and grain elevators of southern Alberta. [(b) Photo by Sovfoto/Eastfoto; (d) photo by author.]

FIGURE 10.31 Disintegrating ice shelves along the Antarctic coast.
(a) 2003 image of a broken B-15 iceberg which calved off the Ross Ice Shelf near Roosevelt Island, Antarctica, in 2000. This iceberg was originally twice the size of Delaware. (b) Map of the Antarctic Peninsula showing the status of ice shelves. Note the location of the Larsen ice shelves. (c) Disintegration and retreat of Larsen-B ice shelf between January 31 and March 7, 2002. Note the meltponds in the January image (see the High-Latitude Connection in Chapter 1 for more on meltponds; also see News Report 17.3). [(a) and (c) Satellite *Terra* courtesy Rapid Response Team, NASA. (b) Map provided by the British Antarctic Survey. All rights reserved.]

(a)

(b)

(c) January 31, 2002

March 7, 2002

that "no regrets" opportunities to reduce carbon dioxide emissions are available in most countries. The IPCC Working Group III defines this as follows:

> No regrets options are by definition greenhouse gas emissions reduction options that have negative net costs, because they generate direct and indirect benefits that are large enough to offset the costs of implementing the options.

Benefits that equal or exceed their cost to society include reduced energy cost, improved air quality and health, reduction in tanker spills and oil imports, and deployment of renewable and sustainable energy sources, among others. This holds true without even considering the benefits of slowing the rate of climate change. One key to "no regrets" is the untapped energy-efficiency potential. For Europe, scientists determined that carbon emissions could be reduced to less than half the 1990 level by 2030, at a negative cost. In the United States, five Department of Energy national laboratories (Oak Ridge, Lawrence Berkeley, Pacific Northwest, National Renewable Energy, and Argonne) reported that the United States can meet the Kyoto carbon emission reduction targets with negative overall costs (cash benefit savings) ranging from −$7 to −$34 billion. (For more, see Working Group III, *Climate Change 2001*, *Mitigation*, London: Cambridge University Press, 2001, pp. 21, 474–76, and 506–507.)

Summary and Review—Global Climate Systems

▶ *Define* climate and climatology, and *explain* the difference between climate and weather.

Climate is dynamic, not static. **Climate** is a synthesis of weather phenomena at many scales, from planetary to local, in contrast to weather, which is the condition of the atmosphere at any given time and place. Earth experiences a wide variety of climatic conditions that can be grouped by general similarities into climatic regions. **Climatology** is the study of climate and attempts to discern similar weather statistics and identify **climatic regions**.

climate (p. 275)
climatology (p. 276)
climatic regions (p. 276)

1. Define climate and compare it with weather. What is climatology?
2. Explain how a climatic region synthesizes climate statistics.
3. How does the El Niño phenomenon produce the largest interannual variability in climate? What are some of the changes and effects that occur worldwide?

▶ *Review* the role of temperature, precipitation, air pressure, and air mass patterns used to establish climatic regions.

Climatic inputs include insolation (pattern of solar energy in the Earth–atmosphere environment), temperature (sensible heat energy content of the air), precipitation (rain, sleet, snow, and hail; the supply of moisture), air pressure (varying patterns of atmospheric density), and air masses (regional-sized homogeneous units of air). Climate is the basic element in ecosystems, the natural, self-regulating communities of plants and animals that thrive in specific environments.

4. How do radiation receipts, temperature, air-pressure inputs, and precipitation patterns interact to produce climate types? Give an example from a humid environment and one from an arid environment.
5. Evaluate the relationships among a climatic region, ecosystem, and biome.

▶ *Review* the development of climate classification systems, and *compare* genetic and empirical systems as ways of classifying climate.

Classification is the process of ordering or grouping data in related categories. A **genetic classification** is based on causative factors, such as the interaction of air masses. An **empirical classification** is one based on statistical data, such as temperature or precipitation. This text analyzes climate using aspects of both approaches, with a map based on climatological elements.

classification (p. 281)
genetic classification (p. 281)
empirical classification (p. 281)

6. What are the differences between a genetic and an empirical classification system?
7. What are some of the climatological elements used in classifying climates? Why each of these?

▶ *Describe* the principal climate classification categories other than deserts, and *locate* these regions on a world map.

Here we focus on temperature and precipitation measures. Keep in mind these are measurable results produced by interacting elements of weather and climate. These data are plotted on a **climograph** to display the characteristics of the climate.

There are six basic climate categories. Temperature and precipitation considerations form the basis of five climate categories and their regional types:

- Tropical (equatorial and tropical latitudes)
  rain forest (rainy all year)
  monsoon (6 to 12 months rainy)
  savanna (less than 6 months rainy)
- Mesothermal (midlatitudes, mild winters)
  humid subtropical (hot summers)
  marine west coast (warm to cool summers)
  Mediterranean (dry summers)
- Microthermal (mid- and high latitudes, cold winters)
  humid continental (hot to warm summers)
  subarctic (cool summers to very cold winters)
- Polar (high latitudes and polar regions)
  tundra (high latitude or high altitude)
  ice caps and ice sheets (perpetually frozen)
  polar marine
- Highland (compared to lowlands at the same latitude, highlands have lower temperatures—recall the normal lapse rate)

Only one climate category is based on moisture efficiency as well as temperature:

- Desert (permanent moisture deficits)

    arid deserts (tropical and midlatitudes)
    semiarid steppes (tropical and midlatitudes)

climograph (p. 281)

8. List and discuss each of the principal climate categories. In which one of these general types do you live? Which category is the only type associated with the annual distribution and amount of precipitation?

9. What is a climograph, and how is it used to display climatic information?

10. Which of the major climate types occupies the most land and ocean area on Earth?

11. Characterize the tropical climates in terms of temperature, moisture, and location.

12. Using Africa's tropical climates as an example, characterize the climates produced by the seasonal shifting of the ITCZ with the high Sun.

13. Mesothermal (subtropical and midlatitude, mild winter) climates occupy the second-largest portion of Earth's entire surface. Describe their temperature, moisture, and precipitation characteristics.

14. Explain the distribution of the *humid subtropical hot-summer* and *Mediterranean dry-summer* climates at similar latitudes and the difference in precipitation patterns between the two types. Describe the difference in vegetation associated with these two climate types.

15. Which climates are characteristic of the Asian monsoon region?

16. Explain how a *marine west coast* climate type can occur in the Appalachian region of the eastern United States.

17. What role do offshore ocean currents play in the distribution of the *marine west coast* climates? What type of fog is formed in these regions?

18. Discuss the climatic conditions for the coldest places on Earth outside the poles.

▶ *Explain* the precipitation and moisture efficiency criteria used to determine the arid and semiarid climates, and *locate* them on a world map.

The dry and semiarid climates are described by precipitation rather than temperature. Dry climates are the world's arid deserts and semiarid regions, with their unique plants, animals, and physical features. The arid and semiarid climates occupy more than 35% of Earth's land area, clearly the most extensive climate over land.

Major subdivisions are *arid deserts* in tropical and midlatitude areas (precipitation—natural water supply—less than one-half of natural water demand) and *semiarid steppes* in tropical and midlatitude areas (precipitation more than one-half of natural water demand).

19. In general terms, what are the differences among the four desert classifications? How are moisture and temperature distributions used to differentiate these subtypes?

20. Relative to the distribution of arid and semiarid climates, describe at least three locations where they occur across the globe and the reasons for their presence in these locations.

▶ *Outline* future climate patterns from forecasts presented, and *explain* the causes and potential consequences of climate change.

Various activities of present-day society are producing climatic changes, particularly a global warming trend. The highest average annual temperatures experienced since the advent of instrumental measurements dominated the last 25 years. There is a scientific consensus building that global warming is related to the anthropogenic impacts on the natural greenhouse effect.

These conditions were further supported by the 2001 *Third Assessment Report* from the Intergovernmental Panel on Climate Change. IPCC predicted surface-temperature response to a doubling of carbon dioxide with a range of increase from 1.4 C° (2.5 F°) to 5.8 C° (10.4 F°) between the present and 2100. Natural climatic variability over the span of Earth's history is the subject of **paleoclimatology**. A **general circulation model (GCM)** is used to forecast climate patterns and is evolving to greater capability and accuracy than in the past. People and their political institutions can use GCM forecasts to form policies aimed at reducing unwanted climate change.

paleoclimatology (p. 306)
general circulation model (GCM) (p. 309)

21. Explain climate forecasts. How do general circulation models (GCMs) produce such forecasts?

22. Describe the potential climatic effects of global warming on polar and high-latitude regions. What are the implications of these climatic changes for persons living at lower latitudes?

23. How is climatic change affecting agricultural and food production? Natural environments? Forests? The possible spread of disease?

24. What are the present actions being taken to delay the effects of global climate change? What is the Kyoto Protocol? The operation of the Conference of the Parties? What is the current status of U.S. and Canadian government action on the Protocol?

NetWork

The *Geosystems* Home Page provides online resources for this chapter on the World Wide Web. To begin: Once on the Home Page, click on this textbook, scroll the Table of Contents menu, select this chapter, and click "Begin." You will find self-tests that are graded, review exercises, specific updates for items in the chapter, and many links to interesting related pathways on the Internet under "Destinations." *Geosystems* is at http://www. prenhall.com/christopherson.

Critical Thinking

A. The text asked that you find the climate conditions for your campus and your birthplace and locate these two places on Figures 10.3, 10.4, and 10.5. Briefly describe the information sources you used: library, Internet, teacher, and phone calls to state and provincial climatologists. Now, refer to Appendix B to refine your assessment of climate for the two locations. Briefly show how you worked through the Köppen climate criteria given in the appendix that established the climate classification for your two cities.

B. El Niño conditions are forecast to return to the Pacific during the life of this edition of *Geosystems*. On the *Geosystems* Home Page and in Focus Study 10.1 there are URL references for El Niño and La Niña. Sample three or four sources to determine the up-to-date status of the El Niño phenomenon at the present time. How are current conditions different from the record El Niño event in 1997–1998? Or La Niña in 1998–2000? What Internet links were most helpful to you in completing this status report?

C. Many external factors force climate. The chart "Global and annual mean radiative forcing for the year 2000, relative to 1750" is presented here (from *Climate Change 2001, The*

Scientific Basis, Washington: Cambridge University Press, 2001, Figure 3, p. 8, and Figure 6.6, p. 392).

The estimates of radiative forcing in Watts per square meter units are given on the *y*-axis (vertical axis). The level of scientific understanding is noted along the *x*-axis (horizontal axis), arranged from "high" to "very low." Those columns above the "0" value (in red) indicate *positive forcing*, such as the greenhouse gases grouped in the far-left column. Columns that fall below the "0" value (in blue) indicate *negative forcing*, such as the haze from sulfate aerosols, fourth column from the left. The vertical line between the markers on each column is an estimate of the uncertainty range. Where no column appears but there is instead a line denoting a range, there is no central estimate given present uncertainties, such as for mineral dust.

Assume you are a policy maker with a goal of reducing the rate of global warming, that is, reducing positive radiative forcing of the climate system. What strategies do you suggest to alter the height of the columns and adjust the mix of elements that cause warming? Assign priorities to each suggested strategy to denote most-to-least effective in moderating climate change. Brainstorm and discuss your strategies with others.

The Global Mean Radiative Forcing of the Climate System for the Year 2000, Relative to 1750

PART III

The Earth–Atmosphere Interface

Earth is a dynamic planet whose surface is shaped by active physical agents of change. Two broad systems organize these agents in Part 3—endogenic and exogenic. The **endogenic system** (Chapters 11 and 12) encompasses internal processes that produce flows of heat and material from deep below Earth's crust. Radioactive decay principally powers these processes. The materials involved constitute the *solid realm* of Earth. Earth's surface responds by moving, warping, and breaking, sometimes in dramatic episodes of earthquakes and volcanic eruptions, constructing the crust.

At the same time, the **exogenic system** (Chapters 13 through 17) involves external processes that set into motion air, water, and ice, all powered by solar energy—this is the *fluid realm* of Earth's environment. These media carve, shape, and reduce the landscape. One such process, weathering, breaks up and dissolves the crust. Erosion picks up these materials; transports them in rivers, winds, coastal waves, and flowing glaciers; and deposits them along the way. Thus, Earth's surface is the interface between two vast open systems: one that builds the landscape and topographic relief and one that tears it down into sedimentary plains.

Dramatic Hawaiian landscape where active lava flows form the Kamoamoa bench and represent some of the newest land surface on Earth. The glowing lava is 2200°C (3990F°) as it sluggishly makes its way to the sea, setting the surf to boil. The Kilauea eruptions are now the longest continuous period of active eruptions in history. The present episode is active since January 1983. [Photo by Bobbé Christopherson.]

Only in Iceland does a portion of Earth's 64,000 km (40,000 mi) mid-ocean ridge system surface. Here at Thingvellir, the rift between the North American and European plates of Earth's crust divide the landscape of lichen- and moss-covered basalt. This is about 45 km (28 mi) from the capital city of Reykjavík in southwestern Iceland. (Locate Iceland and the sea-floor spreading center that causes this rift in the upper-left of the map that opens Chapter 12.)
[Photo by Bobbé Christopherson.]

11

The Dynamic Planet

▶ Key Learning Concepts

After reading the chapter, you should be able to:

▶ *Distinguish* between the endogenic and exogenic systems, *determine* the driving force for each, and *explain* the pace at which these systems operate.

▶ *Diagram* Earth's interior in cross section, and *describe* each distinct layer.

▶ *Illustrate* the geologic cycle, and *relate* the rock cycle and rock types to endogenic and exogenic processes.

▶ *Describe* Pangaea and its breakup, and *relate* several physical proofs that crustal drifting is continuing today.

▶ *Portray* the pattern of Earth's major plates, and *relate* this pattern to the occurrence of earthquakes, volcanic activity, and hot spots.

The twentieth century was a time of great discovery about Earth's internal structure and dynamic crust, yet much remains on the scientific frontier of Earth systems science. Discoveries have revolutionized our understanding of how continents and oceans came to be arranged as they are. A new era of understanding is emerging, combining various sciences within the study of physical geography. One task of physical geography is to explain the *spatial implications* of all this new information and its effect on the landscape—the pattern of geomagnetic reversals, the coincidence of earthquakes and volcanic activity and plate boundaries, and the tracks and trends of hot-spot movement.

In this chapter: Earth's interior is organized, with a core surrounded by roughly concentric shells of material. It is unevenly heated by the radioactive decay of unstable elements. A rock cycle produces three classes of rocks through igneous, sedimentary, and metamorphic processes. Internal processes coupled with the rock cycle, hydrologic cycle, and tectonic cycle result in a varied crustal surface, featuring irregular fractures, extensive mountain ranges both on land and the ocean floor, drifting continental and oceanic crust, and frequent earthquakes and volcanic events. All of this movement of materials results from **endogenic** forces within Earth—the subject of this chapter.

The Pace of Change

The **geologic time scale** is a summary timeline of all Earth history, shown in Figure 11.1. It reflects currently accepted names of time intervals for each segment of Earth's history, from vast *eons* through briefer *eras*, *periods*, and *epochs*. Present thinking places Earth's age at 4.567 billion years, with the Moon about 30 million years younger, formed when a Mars-sized object struck the early Earth.

The time scale depicts two important kinds of time: *relative* (what happened in what order) and *absolute* (actual number of years before the present). Also on the geologic time scale, see the labels denoting the six major extinctions of life forms in Earth history. Rather than being complete extinctions, think of some of these spasms in planetary life as significant *depletions*. These range from 440 million years ago (m.y.a.) to the ongoing present-day episode caused by modern civilization.

Relative time is the *sequence* of events, based on the relative positions of rock strata above or below each other. Relative time is based on the important general principle of *superposition*, which states that *rock and sediment always are arranged with the youngest beds "superposed" toward the top of a rock formation and the oldest at the base, if they have not been disturbed*. The study of these sequences is called *stratigraphy*. Thus, relative time places the Precambrian at the bottom (beginning) of the time scale and the Holocene (today) at the top. Important time clues—namely, *fossils*, the remains of ancient plants and animals—lie embedded within these strata. Since approximately 3.6 billion years ago, life has left its evolving imprint in the rocks.

Scientific methods such as radiometric dating determine *absolute time*, the actual "millions of years ago" (m.y.a.) shown on the time scale. These absolute ages permit scientists to actively update geologic time, refining the time-scale sequence and lending greater accuracy to relative dating sequences. See News Report 11.1 for more information on radioactivity and Earth's time clock. For more on the geologic time scale, see **http://www.ucmp.berkeley.edu/exhibit/geology.html**.

Holocene is the name given to the youngest epoch in the geologic time scale, characteristic of postglacial conditions since the retreat of the continental glaciers, approximately the last 11,500 years. Because of the impact of human society on planetary systems, discussion is underway about designating a new name for the human epoch. A proposal is to call this time the *Anthropocene*.

However, when would such a time begin? Some think the beginning of the Industrial Revolution in the 18th century is the starting point. Whereas others think

---

News Report 11.1

Radioactivity: Earth's Time Clock

The age of Earth and the age of the earliest known crustal rock are astounding, for we think in terms of Earth's trips around the Sun and the pace of our own lives. We need something greater than human time to measure the vastness of geologic time. Nature has provided a way: *radiometric dating*. It is based on the steady decay of certain atoms.

An atom contains protons and neutrons in its nucleus. Certain forms of atoms called isotopes have unstable nuclei; that is, the protons and neutrons do not remain together indefinitely. As particles break away and the nucleus disintegrates, radiation is emitted and the atom decays into a different element—this process is *radioactivity*.

Radioactivity provides the steady time clock needed to measure the age of ancient rocks. It works because the decay rates for different isotopes are determined precisely, and they do not vary beyond established uncertainties. Further refining of these geochronologies, the chronologic sequence of geologic time, is an ongoing process.

The decay rate is expressed as *half-life*, the time required for one-half of the unstable atoms in a sample to decay into "daughter" isotopes. Some examples of unstable elements that become stable elements and their half-lives include uranium-238 to lead-206 (4.5 billion years); thorium-232 to lead-208 (14.1 billion years); potassium-40 to argon-40 (1.3 billion years); and, in organic materials, carbon-14 to nitrogen-14 (5730 years), although with correlation from ice core data, carbon-dating accuracy is being pushed back to more than 24,000 years.

The presence of these decaying elements and stable end products in sediment or rock allows scientists to read the radiometric "clock." They compare the amount of original isotope in the sample with the amount of decayed end product in the sample. If the two are in a ratio of 1:1 (equal parts), one half-life has passed. Errors can occur if the sample was disturbed or subjected to natural weathering processes that might alter its radioactivity.

To increase accuracy, investigators may check a sample using more than one radiometric measurement. Calibration of the past 10,000 years correlates through tree-ring analysis; the past 45,000 years is calibrated using fossils found in lake sediments, among several other cross checks available. The dates for Earth's oldest known rocks from Canada and Greenland were verified using several radiometric methods.

FIGURE 11.1 Geologic time scale.
Both relative and absolute dating methods calibrate the geologic time scale. In the column at the left, note that 88% of geologic time occurred during the Precambrian Era. Highlights of Earth's history also are shown in the figure as bulleted items, including the six major extinctions or depletions of life forms, printed in red (the sixth extinction episode is underway at the present time), and dates appear in m.y.a., million years ago. Note certain revisions from previous editions of *Geosystems* that reflect new radiometric measurements. [Data and update from Geological Society of America and *Nature* 429 (May 13, 2004): 124–25.]

ANIMATION

Applying Relative Dating Principles

an appropriate time is some 5000 years ago when Asians began flooding rice fields that released methane gas into the atmosphere. Or perhaps 8000 years ago is a better chronology with the spread of domesticated crops, widespread clearing of forests, and the human use of fire to modify environments. Human impacts on the atmosphere began at the roots of civilization.

A fundamental principle of Earth science is **uniformitarianism**, which assumes that *the same physical processes active in the environment today have been operating throughout geologic time*. For example, if streams carve valleys now, they must have done so 500 m.y.a. The phrase "the present is the key to the past" describes this principle. Evidence from exploration and from the landscape record of volcanic eruptions, earthquakes, and exogenic processes support uniformitarianism. The concept was first proposed by James Hutton in his *Theory of the Earth* (1795) and was later amplified by Charles Lyell in *Principles of Geology* (1830). Today, the earlier Hutton version guides scientific thinking because of his view that rates and some processes do change, whereas essential underlying principles of biology, physics, geology, and chemistry remain uniform.

In contrast, the principle of *catastrophism* places the vastness of Earth's age and the complexity of its rocks into a shortened time span. Catastrophism holds that Earth is young and that mountains, canyons, and plains formed through catastrophic events that did not require eons of time. Because there is little physical evidence to support catastrophism, it is more appropriately considered a belief than a serious scientific hypothesis. Ancient landscapes, such as the Appalachian Mountains or the heartland–shield regions of the continents, or the vast drifting of the continents and other fragments of crust, represent a much broader time scale of events, beyond the severe limitations of catastrophism.

Within the principle of uniformitarianism, catastrophic events such as massive landslides, earthquakes, volcanic eruptions, and sometime cyclic episodes of mountain building punctuate geologic time. Vast super floods occurred as ice-dammed lakes tore through their frozen dams in the later days of the last ice age, such as glacial Lake Missoula that left evidence across a broad region. These episodes occur as interruptions in the generally uniform processes that shape the slowly evolving landscape. Here, the *punctuated equilibrium* concept studied in the life sciences and paleontology might apply to aspects of Earth's long developmental history.

To understand the phenomena seen at Earth's surface, we must have knowledge of our planet's internal structure and energy. Let us now journey deep within Earth to see its inner workings.

Earth's Structure and Internal Energy

Along with the other planets and the Sun, Earth is thought to have condensed and congealed from a nebula of dust, gas, and icy comets about 4.6 billion years ago (review this in Chapter 2). Scientists are observing this same formation process underway elsewhere in our Milky Way Galaxy and the Universe. Previously, the oldest surface rock discovered on Earth (known as the Acasta Gneiss) was found in northwestern Canada; it was radiometrically dated to an age of 3.96 billion years. Scientists from Missouri's Saint Louis University, the Canadian Geological Survey, and Australian National University confirmed the age.

Recent research found detrital zircons (particles of preexisting zirconium silica oxides transported and deposited, forming rock) in Western Australia dating to about 4.3 billion years old; these are possibly the oldest materials in Earth's crust. These discoveries tell us something significant: Earth was forming continental crust at least 4 billion years ago, during the Archean Eon. Although not an official designation, note the Hadean Eon in Figure 11.1, a proposal to describe the period before the Archean.

As Earth solidified, gravity sorted materials by density. Heavier substances such as iron gravitated slowly to its center, and lighter elements such as silica slowly welled upward to the surface and became concentrated in the crust. Consequently, Earth's interior is sorted into roughly concentric layers, each one distinct in either chemical composition or temperature. Heat energy migrates outward from the center by conduction and by physical convection in the more fluid or plastic layers in the mantle and nearer the surface.

Our knowledge of Earth's *internal differentiation* into these layers is acquired entirely through indirect evidence, because we are unable to drill more than a few kilometers into Earth's crust. There are several physical properties of Earth materials that enable us to know the nature of the interior. For example, when an earthquake or underground nuclear test sends shock waves through the planet, the cooler areas, which generally are more rigid, transmit these **seismic waves** at a higher velocity than do the hotter areas, where seismic waves are slowed to lower velocity. This is the science of *seismic tomography*, as if Earth is subjected to a kind of CAT scan with every earthquake.

Density also affects seismic-wave velocities. Plastic zones simply do not transmit some seismic waves; they absorb them. Some seismic waves are reflected as densities change, whereas others are refracted, or bent, as they travel through Earth. Thus, the distinctive ways in which seismic waves pass through Earth and the time they take to travel between two surface points help seismologists deduce the structure of Earth. Research is active and the model of Earth's interior is being refined through each new scientific paper. Figure 11.2 is the present model of Earth's interior.

Figure 11.3 illustrates the dimensions of Earth's interior compared with surface distances in North America to give you a sense of size and scale. An airplane flying from Halifax, Nova Scotia, to San Francisco would travel the same distance as that from Earth's center to its surface. Note the thinness of the crust, extending only from Vallejo to San Francisco on the far west coast.

FIGURE 11.2 Earth in cross section.
(a) Cutaway showing Earth's interior. (b) Earth's interior in cross section, from the inner core to the crust. (c) Detail of the structure of the lithosphere and its relation to the asthenosphere. (For comparison to the densities noted, the density of water is 1.0g/cm³, and mercury, a liquid metal, is 13.0g/cm³.) A revised estimate of Earth's mass (or weight) made in 2000 set it at 5.972 sextillion metric tons (5972 followed by 18 zeros). [Photo from NASA.]

Earth's Core and Magnetism

A third of Earth's entire mass, but only a sixth of its volume, lies in its dense core. The **core** is differentiated into two regions—*inner core* and *outer core*—divided by a transition zone several hundred kilometers wide (see Figure 11.2b). The inner core is thought to be solid iron that is well above the melting temperature of iron at the surface but remains solid because of tremendous pressure. The inner core is thought to have formed first shortly after Earth condensed. The iron in the core is impure, probably combined with silicon and possibly oxygen and sulfur. Recent research points to the conclusion that the inner core may be a single, enormous crystal of iron. The outer core is molten, metallic iron with a lighter density than the inner core.

Earth's Magnetism The fluid outer core generates at least 90% of Earth's magnetic field and the magnetosphere that surrounds and protects Earth from the solar wind and cosmic radiation. One hypothesis explains that circulation in the outer core converts thermal and gravitational energy into magnetic energy, producing Earth's magnetic field. The north magnetic pole (NMP) is near 83° N by 114° W (in the year 2005) in the Canadian Arctic, and the south magnetic pole is off the coast of Wilkes Land, Antarctica. These surface expressions of Earth's magnetic field migrate; for example, the NMP has moved 1100 km (685 mi) just this past century.

An intriguing feature of Earth's magnetic field is that its polarity sometimes fades to zero and then returns to full strength, with north and south magnetic poles reversed. In the process, the field does not blink on and off but instead diminishes slowly to low intensity, perhaps 25% strength, and then rapidly regains to full power. This **geomagnetic reversal** has taken place 9 times during the past 4 million years and hundreds of times over Earth's history. During the transition interval of low strength, Earth's surface receives higher levels of cosmic radiation and solar particles, but note that past reversals do not correlate with species extinctions. The evolution of life has weathered many of these transitions.

The average period of a magnetic reversal is about 500,000 years; several hundred documented occurrences varied from as short as 20 to 30 thousand years to 50 million years. Transition periods of low intensity last from 1000 to 10,000 years. The last reversal was 790,000 years ago, preceded by a transition ranging 2000 years in length along the equator to 10,000 years in the midlatitudes. The obvious question for us is, Why does this happen?

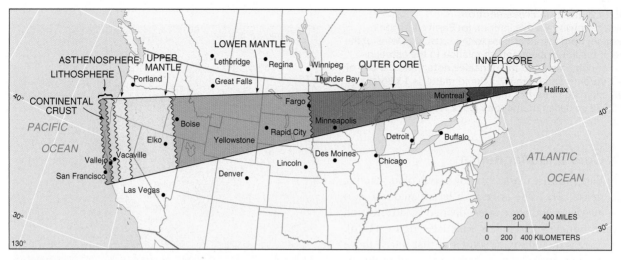

FIGURE 11.3 Distances from core to crust.
Surface map, using the distance from Halifax, Nova Scotia, to San Francisco, to compare the distance from Earth's center to the surface (6370 km, or 3963 mi). The thickness of the continental crust is the same as the distance between Vallejo (in the eastern portion of San Francisco Bay) and the city of San Francisco (roughly 20 mi, 32 km).

The reasons for these magnetic reversals are unknown. However, the spatial patterns they create at Earth's surface are a key tool in understanding the evolution of landmasses and the movements of the continents. When new iron-bearing rocks solidify from molten material (lava) at Earth's surface, the small magnetic particles in the rocks align according to the orientation of the magnetic poles at that time. As the rocks cool and solidify, this alignment locks in place.

All across Earth, rocks of the same age bear an identical record of magnetic reversals in the form of measurable magnetic "stripes" in any magnetic material they contain, such as iron particles. These stripes illustrate global patterns of changing magnetism. Matching segments allow scientists to reassemble past continental arrangements. Later in this chapter we see the importance of these magnetic reversals.

An apparent trend in recent geologic time is toward more frequent reversals. When Earth is without polarity in its magnetic field, a random pattern of magnetism in crustal rock results. Given present rates of magnetic field decay over the last 150 years, we are perhaps 1000 years away from entering the next phase of field changes, although there is no expected pattern to forecast the timing.

Earth's Mantle

Figure 11.2b shows a transition zone several hundred kilometers wide, at an average depth of about 2900 km (1800 mi), dividing Earth's outer core from its mantle. Scientists at the California Institute of Technology analyzed more than 25,000 earthquakes and determined that this transition area is uneven, with ragged peak-and-valley-like formations. New scientific discoveries determined that some of the motions in the mantle are created by this zone of rough texture and low velocity, caused by slower speeds of seismic waves in hot material, at what is called the *Gutenberg discontinuity*. A *discontinuity* is a place where physical differences occur between adjoining regions in Earth's interior, such as between the outer core and lower mantle.

The **mantle** (lower and upper together) represents about 80% of Earth's total volume. The mantle is rich in oxides of iron and magnesium and silicates (FeO, MgO, and SiO_2). They are dense and tightly packed at depth, grading to lesser densities toward the surface. A broad transition zone of several hundred kilometers, centered between 410 and 660 km (255–410 mi) below the surface, separates the upper mantle from the lower mantle. The entire mantle experiences a gradual temperature increase with depth and a stiffening due to increased pressures. The denser lower mantle is thought to contain a mixture of iron, magnesium, and silicates, with some calcium and aluminum. An exciting new hypothesis states that a tremendous amount of water is bound up with minerals in the mantle.

The upper mantle divides into three fairly distinct layers: upper mantle, asthenosphere, and *uppermost mantle*, shown in Figure 11.2c. Next to the crust is the uppermost mantle, a high-velocity zone just below the crust, where seismic waves transmit through a rigid, cooler layer. This uppermost mantle, along with the crust, makes up the *lithosphere*, approximately 45–70 km thick.

Below the lithosphere, from about 70 km down to 250 km, is the **asthenosphere**, or plastic layer (from the Greek *asthenos*, meaning "weak"). It contains pockets of increased heat from radioactive decay and is susceptible to slow convective currents in these hotter, and therefore less dense, materials.

Because of its dynamic condition, the asthenosphere is the least rigid region of the mantle, with densities averaging 3.3 g/cm^3 (or 3300 kg/m^3). About 10% of the asthenosphere is molten in asymmetrical patterns and hot spots. The resulting slow movement in this zone disturbs the overlying crust and creates tectonic activity—the folding, faulting, and general deformation of surface rocks. In return, movement of the crust apparently influences currents throughout the mantle.

The depth affected by convection currents is the subject of much scientific research. One body of evidence states that mixing occurs throughout the entire mantle, upwelling from great depths at the core–mantle boundary, sometimes in small blobs of material; other times "megablobs" of mantle convect toward and away from the crust. Another view states that mixing in the mantle is layered, segregated above and below the 660-km boundary. Presently, evidence indicates some truth in both positions.

As an example, there are hot spots on Earth, such as those under Hawai'i, Easter Island, and Tahiti, among others, that appear to be at the top of tall plumes of rising mantle rock that are anchored deep in the lower mantle, where the upward flow of warmer, less-dense material begins. The plume beneath Iceland appears to begin at the 660-km transition. Yet, there are other surface hot spots that do not sit atop such tall plumes, but over smaller ones going down some 200 km (124 mi). In other surface regions, slabs of crust descend and penetrate to the lower mantle. For basics on Earth's interior and plate tectonics, see http://www.solarviews.com/eng/earthint.htm.

Earth's Lithosphere and Crust

The lithosphere includes the **crust** and uppermost mantle to about 70 km (43 mi) in depth (Figure 11.2c). An important internal boundary between the crust and the high-velocity portion of the uppermost mantle is another discontinuity, called the **Mohorovičić discontinuity**, or **Moho** for short. It is named for the Yugoslavian seismologist who determined that seismic waves change at this depth owing to sharp contrasts of materials and densities.

Figure 11.2c illustrates the relation of the crust to the rest of the lithosphere and the asthenosphere below. Crustal areas beneath mountain masses extend deep, perhaps to 50–60 km (31–37 mi), whereas the crust beneath continental interiors averages about 30 km (19 mi) in thickness. Oceanic crust averages only 5 km (3 mi). The crust is only a fraction of Earth's overall mass. Drilling through the crust into the uppermost mantle remains an elusive scientific goal (see News Report 11.2).

News Report 11.2

Drilling the Crust to Record Depths

Scientists wanting to sample mantle material directly have unsuccessfully tried for decades to penetrate Earth's crust to the Moho discontinuity (the crust–mantle boundary). The longest-lasting deep-drilling attempt is on the northern Kola Peninsula near Zapolyarny, Russia, 250 km north of the Arctic Circle—the *Kola Borehole (KSDB)*. Twenty years of high-technology drilling (1970–1989) produced a hole 12,262 km deep (7.6 mi, or 40,230 ft), purely for exploration and science. Crystalline rock 1.4 billion years old at 180°C (356°F) was reached. The site has other active boreholes, and the fifth is underway. (See an analysis log at http://www.icdp-online.de/.) A record holder for depth for a gas well is in Oklahoma; it was stopped at 9750 m (32,000 ft) when the drill bit ran into molten sulfur.

Oceanic crust is thinner than continental crust and is the object of several drilling attempts. The Integrated Ocean Drilling Program (IODP), a cooperative effort between the governments of Japan and the United States, replaced the former ODP effort directed by Texas A & M University in 2003. However, the university program is still the Science Operator of the research ship *JOIDES Resolution* and lead coordinator.

FIGURE 11.2.1 Ocean drilling ship. A modern ocean-floor drilling ship, the *JOIDES Resolution*. The International Ocean Drilling Program operates the research ship. Since it began operations in 1984, through 2003, *JOIDES* has spent more than 5000 days at sea, traveled more than half a million kilometers, drilled 1555 holes into the seafloor, and recovered hundreds of kilometers of core for analysis. [Photo by ODP/Texas A & M University.]

Previously, ODP drilled a 2.5-km-deep hole in an oceanic rift near the Galápagos Islands from 1975 to 1993, still the record. The new IODP plans further drilling at the site (Figure 11.2.1). An ocean floor of distinct layers of lava and deeper magma chambers was found. But the Moho and Earth's mantle remain untapped.

Over the past 40 years, about 1700 bore holes were completed, yielding 160 km of sediment and rock cores and more than 35,000 samples for researchers. Much of this science windfall helped put together the puzzle of geologic time and plate tectonics. See http://www-odp.tamu.edu/ for background or http://www.oceandrilling.org/ for the IODP home page information.

Construction of the largest deep-ocean drilling ship is ongoing in Okayama, Japan, with completion scheduled for 2006. Japan's *Chikyu* will be the largest research ship for this purpose and the first capable of high-latitude drilling in the Arctic. The new IODP ship will be able to drill to 7 km, more than three times the *JOIDES Resolution* limits. This *JOIDES*, workhorse of the ODP, is being refurbished for the new IODP program.

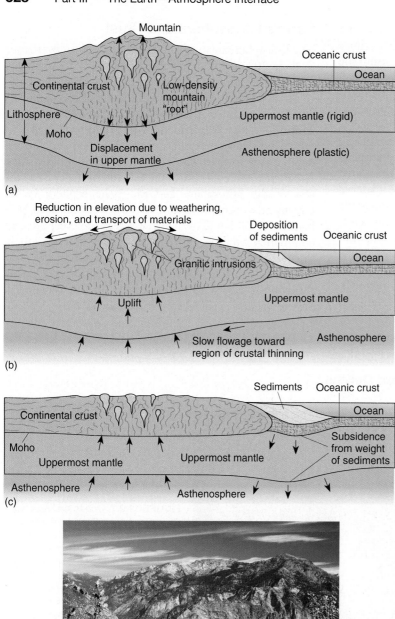

(a)

(b)

(c)

(d)

FIGURE 11.4 Isostatic adjustment of the crust.
Earth's entire crust is in a constant state of compensating adjustment, as suggested by these three sequential stages. In (a), the mountain mass slowly sinks, displacing mantle material. In (b), because of the loss of mass through erosion and transportation, the crust isostatically adjusts upward and sediments accumulate in the ocean. As the continental crust thins (c), the heavy sediment load offshore begins to deform the lithosphere beneath the ocean. (d) The melting of ice from the last ice age and losses of overlying sediments are thought to produce an ongoing isostatic uplift of portions of the Sierra Nevada batholith. [Photo by author.]

The composition and texture of continental and oceanic crusts are quite different, and this difference is a key to the concept of drifting continents. The oceanic crust is denser than continental crust. In collisions, the denser oceanic material plunges beneath the lighter, more buoyant continental crust.

- *Continental crust* is essentially **granite**; it is crystalline and high in silica, aluminum, potassium, calcium, and sodium. (Sometimes continental crust is called *sial*, shorthand for *si*lica and *al*uminum.) Most important, continental crust is relatively low in density, averaging 2.7 g/cm³ (or 2700 kg/m³). Compare this with other densities given in Figure 11.2.

- *Oceanic crust* is **basalt**; it is granular and high in silica, magnesium, and iron. (Sometimes oceanic crust is called *sima*, shorthand for *si*lica and *ma*gnesium.) It is denser than continental crust, averaging 3.0 g/cm³ (or 3000 kg/m³).

Buoyancy is the principle that something less dense, such as wood, floats in something denser, such as water. The principles of buoyancy and balance were combined in the 1800s into the important principle of **isostasy**, which explains certain vertical movements of Earth's crust.

Think of Earth's crust as floating on the denser layers beneath, much as a boat floats on water. Where the load is greater, owing to glaciers, sediment, or mountains, the

High Latitude Connection 11.1

Isostatic Rebound in Alaska

The retreat of glacial ice following the last ice-age cycle unloaded much weight off the crust in Alaska, among those regions affected. Researchers from the Geophysical Institute at the University of Alaska–Fairbanks, using an array of global positioning systems (GPS) receivers, measured isostatic rebound of the crust following glacial ice losses. They expected to find a slowed rate of crustal rebound in southeastern Alaska as compared to the more rapid response in the distant past when the ice first retreated.

Instead, they detected some of the most rapid vertical motion on Earth, averaging about 36 mm (1.42 in.) per year. Scientists attribute this isostatic rebound to the loss of glaciers in the region, especially the areas from Yakutat Bay and the Saint Elias Mountains in the north, through Glacier Bay, Juneau, and on south through the inland passage. Only about 1 mm of this uplift is from post-ice age response. This rapid rebound is attributable to glacial melt and retreat over the past 150 years and

correlates with accelerating ice losses and record warmth across Alaska (see News Report 17.1).

For instance, Glacier Bay experienced a retreat of ice of more than 97 km (60 mi) since 1794. The land that was beneath this ice has isostatically rebounded 5.5 m (18 ft). The vertical action appears to be affecting some of the fault systems in the region, although the extent of this and its consequences are unknown at this writing.

---

crust tends to sink, or ride lower in the asthenosphere. Without that load (for example, when a glacier melts), the crust rides higher, in a recovery uplift known as *isostatic rebound*. Thus, the entire crust is in a constant state of compensating adjustment, or isostasy, slowly rising and sinking in response to its own burdens, as it is pushed, and dragged, and pulled over the asthenosphere (Figure 11.4). An interesting development in Alaska relates this concept of isostatic rebound and climate change, as described in High-Latitude Connection 11.1.

Earth's crust is the outermost shell: an irregular, brittle layer that resides restlessly on a dynamic and diverse interior. Let us examine the processes at work on this crust and the variety of rock types that compose the landscape.

The Geologic Cycle

Earth's crust is in an ongoing state of change, being formed, deformed, moved, and broken down by physical, chemical, and biological processes. While the endogenic (internal) system is at work building landforms, the exogenic (external) system is busily wearing them down. This vast give-and-take at the Earth–atmosphere–ocean interface is called the **geologic cycle**. It is fueled from two sources—Earth's internal heat and solar energy from space—influenced by the ever-present leveling force of Earth's gravity.

Figure 11.5 illustrates the geologic cycle, combining many of the elements presented in this text. The geologic cycle is composed of three subsystems:

- The *hydrologic cycle* is the vast system that circulates water, water vapor, ice, and energy throughout the Earth–atmosphere–ocean environment. This cycle rearranges Earth materials through erosion, transportation, and deposition, and it circulates water as

the critical medium that sustains life. (Chapters 7, 8, and 9 discussed water's properties, weather, and the hydrologic cycle.)
- The *rock cycle*, through processes in the atmosphere, crust, and mantle, produces three basic rock types—igneous, sedimentary, and metamorphic. We examine this next.
- The *tectonic cycle* brings heat energy and new materials to the surface and recycles old materials to mantle depths, creating movement and deformation of the crust. We discuss the tectonic cycle later in this chapter.

The Rock Cycle

To begin our look at the rock cycle, we see that only eight natural elements compose 99% of Earth's crust. Just two of these—oxygen and silicon—account for 74.3% of the crust (Table 11.1). Oxygen, the most reactive gas in the lower atmosphere, readily combines with other elements. For this reason, the percentage of oxygen is greater in the crust (about 47%) than in the atmosphere (about 21%). The internal differentiation process explains the relatively large percentages of lightweight elements such as silicon and aluminum in the crust. These less-dense elements migrate toward the surface, as discussed earlier.

Minerals and Rocks   Earth's elements combine to form minerals. A **mineral** is an inorganic, or nonliving, natural compound having a specific chemical formula and possessing a crystalline structure. The combination of elements and the crystal structure give each mineral its characteristic hardness, color, density, and other properties. For example, the common mineral *quartz* is silicon dioxide, SiO_2, and has a distinctive six-sided crystal (see photo inset in Table 11.1).

Of the more than 4200 minerals, about 30 minerals comprise the *rock-forming minerals* most commonly

(a)

(b)

FIGURE 11.5 The geologic cycle.
(a) The geologic cycle is a model showing the interactive relation among the hydrologic cycle, rock cycle, and tectonic cycle. (b) Earth's surface is where two dynamic systems interact—the endogenic (internal) and exogenic (external).

encountered. *Mineralogy* is the study of the composition, properties, and classification of minerals. (Note that mercury, a liquid metal, is an exception to this definition for it has no crystalline structure and that water is not a mineral but ice does fit the definition of a mineral.) For more on minerals see **http://webmineral.com/** or **http://www.minsocam.org/MSA/Research_Links.html**.

One of the most widespread mineral families on Earth is the *silicates*, because silicon and oxygen are so common and because they readily combine with each other and with other elements. Roughly 95% of Earth's crust comprises silicates. This mineral family includes quartz, feldspar, clay minerals, and numerous gemstones. *Oxides* are another group of minerals in which oxygen combines with metallic elements, such as iron, to form hematite,

Table 11.1   Common Elements in Earth's Crust

Element	Percentage of Earth's Crust by Weight
Oxygen (O)	46.6
Silicon (Si)	27.7
Aluminum (Al)	8.1
Iron (Fe)	5.0
Calcium (Ca)	3.6
Sodium (Na)	2.8
Potassium (K)	2.6
Magnesium (Mg)	2.1
All others	1.5
Total	100.00

Note. A quartz crystal (SiO_2) consists of Earth's two most abundant elements, silicon (Si) and oxygen (O). Inset photo from *Laboratory Manual in Physical Geology*, 3rd ed., R. M. Busch, ed. © 1993 by Macmillan Publishing Co.

Fe_2O_3. Also, there are the *sulfides* and *sulfates* groups in which sulfur compounds combine with metallic elements to form pyrite (FeS_2) and anhydrite ($CaSO_4$), respectively.

Another important mineral family is the *carbonate* group, which features carbon in combination with oxygen and other elements such as calcium, magnesium, and potassium. An example is the mineral calcite ($CaCO_3$), a form of calcium carbonate.

A **rock** is an assemblage of minerals bound together (such as granite, a rock containing three minerals), or a mass of a single mineral (such as rock salt), or undifferentiated material (such as the noncrystalline glassy obsidian, or volcanic glass), or even solid organic material (such as coal). Thousands of different rocks have been identified. All can be sorted into one of three kinds, depending on the processes that formed them: *igneous* (melted), *sedimentary* (from settling out), and *metamorphic* (altered). Figure 11.6 illustrates these three processes and the interrelations among them that constitute the **rock cycle**. Let us examine each rock-forming process.

Igneous Processes

An **igneous rock** is one that solidifies and crystallizes from a molten state. Familiar examples are granite, basalt, and rhyolite. Igneous rocks form from **magma**, which is molten rock beneath the surface (hence the name *igneous*, which means "fire-formed" in Latin). Magma is fluid, highly gaseous, and under tremendous pressure. It either *intrudes* into crustal rocks, cools, and hardens, or it *extrudes* onto the surface as **lava**.

The cooling history of an igneous rock—how fast it cooled and how steadily its temperature dropped—determines its crystalline physical characteristics, or crystallization. Igneous rocks range from coarse-grained (slower cooling, with more time for larger crystals to form) to fine-grained or glassy (faster cooling).

Igneous rocks comprise approximately 90% of Earth's crust, although sedimentary rocks (sandstone, shale, limestone), soil, or oceans frequently cover them. Figure 11.7 illustrates the variety of occurrences of igneous rocks, both on and beneath Earth's surface.

Intrusive and Extrusive Igneous Rocks   Intrusive igneous rock that cools slowly in the crust forms a **pluton**, a general term for any intrusive igneous rock body, regardless of size or shape, that invaded layers of crustal rocks. The Roman god of the underworld, Pluto, is the namesake. The largest pluton form is a **batholith**, defined as an irregular-shaped mass with a surface greater than 100 km², 40 mi² (Figure 11.8a). Batholiths form the mass of many large mountain ranges—for example, the Sierra Nevada batholith in California, the Idaho batholith, and the Coast Range batholith of British Columbia and Washington State.

Smaller plutons include the magma conduits of ancient volcanoes that have cooled and hardened. Those that form parallel to layers of sedimentary rock are *sills*; those that cross layers of the rock they invade are *dikes*. You see these two forms in Figure 11.7 at Petermann Island, Antarctica. Magma also can bulge between rock strata and produce a lens-shaped body called a *laccolith*, a type of sill. In addition, magma conduits themselves may solidify in roughly cylindrical forms that stand starkly above the landscape when finally exposed by weathering and erosion. Shiprock *volcanic neck* in New Mexico is such a feature, rising 518 m (1700 ft) above the surrounding plain, as suggested by the art in Figure 11.7 and shown in the inset photos; note the radiating dikes in the aerial photo. Weathering action of air, water, and ice can expose all of these intrusive forms.

Volcanic eruptions and flows produce extrusive igneous rock, such as lava that cools and forms basalt

FIGURE 11.6 The rock cycle.
A rock-cycle schematic demonstrating the relation among igneous, sedimentary, and metamorphic processes. The arrows indicate that each rock type can enter the cycle at various points and be transformed into other rock types. [Art adapted by permission from R. M. Busch, ed., *Laboratory Manual in Physical Geology*, 3rd ed., © 1993 by Macmillan Publishing Co. Photos by Bobbé Christopherson.]

NOTEBOOK   The Rock Cycle

(Figure 11.8b). Chapter 12 presents volcanism and treats this in detail.

Classifying Igneous Rocks Mineral composition and texture usually classify igneous rocks (Table 11.2). The two broad categories are:

1. *Felsic* igneous rocks—derived both in composition and name from *fel*dspar and *sil*i*ca*. Felsic minerals are generally high in silica, aluminum, potassium, and sodium and have low melting points. Rocks formed from felsic minerals generally are lighter in color and are less dense than mafic mineral rocks.

2. *Mafic* igneous rocks—derived both in composition and name from *ma*gnesium and *fer*ric (Latin for iron). Mafic minerals are low in silica, high in magnesium and iron, and have high melting points.

Aerial, Shiprock

Volcanic neck
(Shiprock, NM)

FIGURE 11.7 Igneous rock types.
The variety of occurrences of igneous rocks, both intrusive (below the surface) and extrusive
(on the surface). Inset photographs show samples of granite (intrusive) and basalt (extrusive)
and other igneous forms. [Photos of rock samples by Bobbé Christopherson; photos of Shiprock
volcanic neck by author.]

Formation of
Intrusive Igneous
Features

Rocks formed from mafic minerals are darker in
color and of greater density than felsic mineral
rocks.

Table 11.2 shows that the same magma that produces
coarse-grained granite (when it slowly cools beneath the
surface) can form fine-grained *rhyolite* (when it cools
above the surface—see inset photo). If it cools quickly,
magma having silica content comparable to granite and
rhyolite can form the dark, smoky, glassy-textured rock
called *obsidian*, or volcanic glass (see inset photo). Another

glassy rock, called *pumice*, forms when escaping gases bub-
ble a frothy texture into the lava. Pumice is full of small
holes, is light in weight, and is low enough in density to
float in water (see inset photo).

On the mafic side, basalt is the most common fine-
grained extrusive igneous rock. It makes up the bulk
of the ocean floor, accounting for 71% of Earth's surface.
It appears in lava flows such as those on the big island
of Hawai'i (Figure 11.8b). An intrusive counterpart to
basalt, formed by slow cooling of the parent magma, is
gabbro.

FIGURE 11.8 Intrusive and extrusive rocks.
(a) Exposed granites of the Sierra Nevada batholith. The boulders sitting on the granite are erratics left behind by glacial ice that melted and retreated thousands of years ago. (b) Basaltic lava flows on Hawai'i. The glowing opening is from a skylight into an active lava tube where molten lava is visible; the shiny surface is where lava recently flowed out of the skylight. (Do you see a suggestion of a dragon head and glowing red eye in the lava sculpture?). [Photos by Bobbé Christopherson.]

Sedimentary Processes

Solar energy and gravity drive the process of *sedimentation*, with water as the principal transporting medium. Existing rock is disintegrated and dissolved by weathering, picked up and moved by erosion and transportation, and deposited along river, beach, and ocean sites, where burial initiates the rock-forming process. The formation of **sedimentary rock** involves **lithification** processes of cementation, compaction, and a hardening of sediments.

Most sedimentary rocks derive from fragments of existing rock or organic materials. Bits and pieces of former rocks—principally quartz, feldspar, and clay minerals—erode and then are mechanically transported by water (lake, stream, ocean, overland flow), ice (glacial action), wind, and gravity. They are transported from "higher-energy" sites, where the carrying medium has the energy to pick up and move them, to "lower-energy" sites, where the material is dumped.

The common sedimentary rocks are *sandstone* (sand that became cemented together), *shale* (mud that became compacted into rock), *limestone* (bones and shells that became cemented or calcium carbonate that precipitated in ocean and lake waters), and *coal* (ancient plant remains that became compacted into rock). Some minerals, such as calcium carbonate, dissolve into solution and form sedimentary deposits by precipitating from those solutions to form rock. This is an important process in the oceanic and karst (weathered limestone; Chapter 13) environments.

Characteristically, sedimentary rocks are laid down by wind, water, or ice in horizontally layered beds. Different environmental conditions produce a variety of sedimentary forms. Various cements, depending on availability, fuse rock particles together. Lime, or calcium carbonate ($CaCO_3$), is the most common, followed by iron oxides (Fe_2O_3) and silica (SiO_2). Drying (dehydration), heating, or chemical reactions can also unite particles.

The layered strata of sedimentary rocks form an important record of past ages. **Stratigraphy** is the study of the sequence (superposition), thickness, and spatial distribution of strata. These sequences yield clues to the age and origin of the rocks. Figure 11.9a shows a sandstone sedimentary rock in a desert landscape. Note the multiple layers in the formation and how differently they resist weathering processes. The climatic history of this rock's formation environment is disclosed in the stratigraphy—drier periods of former dunes near the top and wetter periods of horizontal deposits in the middle and lower portions. Figure 11.9c shows sedimentary rock strata on Mars that imply water deposition of sediments sometime in the distant Martian past.

The two primary sources of sedimentary rocks are the mechanically transported fragments of older rock, called *clastic sediments*, and the dissolved minerals in solution, termed *chemical sediments*.

Clastic Sedimentary Rocks Weathered and fragmented rocks that are further worn in transport provide

Table 11.2   Igneous Rock Minerals

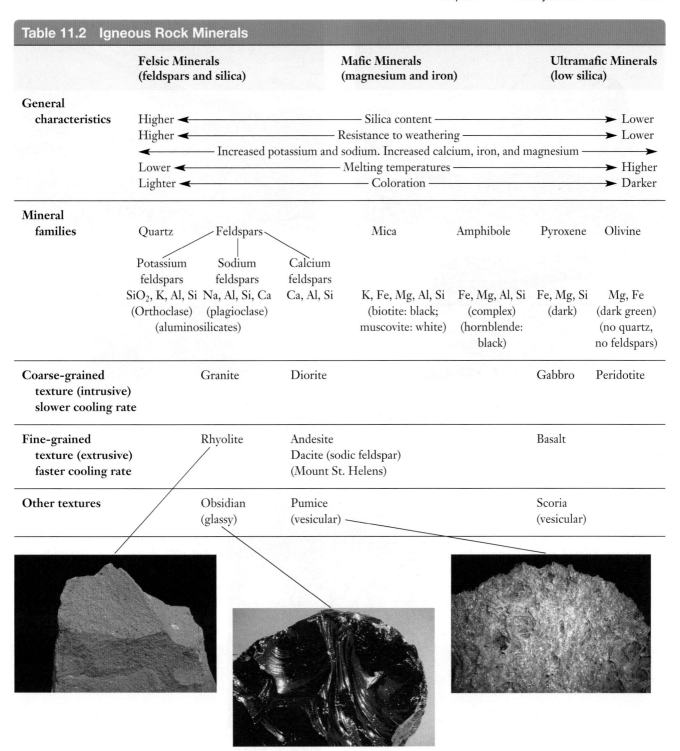

	Felsic Minerals (feldspars and silica)			**Mafic Minerals** (magnesium and iron)			**Ultramafic Minerals** (low silica)
General characteristics	Higher ← Silica content → Lower						
	Higher ← Resistance to weathering → Lower						
	← Increased potassium and sodium. Increased calcium, iron, and magnesium →						
	Lower ← Melting temperatures → Higher						
	Lighter ← Coloration → Darker						
Mineral families	Quartz	Feldspars		Mica	Amphibole	Pyroxene	Olivine
	Potassium feldspars SiO₂, K, Al, Si (Orthoclase) (aluminosilicates)	Sodium feldspars Na, Al, Si, Ca (plagioclase)	Calcium feldspars Ca, Al, Si	K, Fe, Mg, Al, Si (biotite: black; muscovite: white)	Fe, Mg, Al, Si (complex) (hornblende: black)	Fe, Mg, Si (dark)	Mg, Fe (dark green) (no quartz, no feldspars)
Coarse-grained texture (intrusive) slower cooling rate	Granite	Diorite				Gabbro	Peridotite
Fine-grained texture (extrusive) faster cooling rate	Rhyolite	Andesite Dacite (sodic feldspar) (Mount St. Helens)				Basalt	
Other textures	Obsidian (glassy)	Pumice (vesicular)				Scoria (vesicular)	

Source: Photos from R. M. Busch, ed., *Laboratory Manual in Physical Geology,* 3rd ed., © 1993 by Macmillan Publishing Co.

clastic sediments. Table 11.3 lists the range of *clast* sizes—everything from boulders to microscopic clay particles—and the form they take as *lithified rock.* Common sedimentary rocks from clastic sediments include siltstone or mudstone (silt-sized particles), shale (clay to silt-sized particles), and sandstone (sand-sized particles ranging from 0.06 to 2 mm in diameter).

Chemical Sedimentary Rocks   *Chemical sedimentary* rocks are not formed from physical pieces of broken rock but instead from dissolved minerals, transported in solution and chemically precipitated from solution (they are essentially nonclastic). The most common chemical sedimentary rock is **limestone**, which is lithified calcium carbonate, derived from inorganic and organic sources. A similar form

(a)

(a)

(b)

FIGURE 11.11 Hydrothermal deposits.
(a) Mammoth Hot Springs, Yellowstone National Park, is an example of a hydrothermal deposit principally composed of travertine ($CaCO_3$), deposited as a chemical precipitate from heated spring water as it evaporated. (b) Hydrothermal vents with black smokers along the Mid-Atlantic ridge.
[(a) Photo by author; (b) image courtesy of Scripps Institution of Oceanography.]

(b)

FIGURE 11.12 Metamorphic rocks.
(a) A metamorphic rock outcrop in Greenland, the Amitsoq Gneiss, at 3.8 billion years old, one of the oldest rock formations on Earth. (b) Precambrian Vishnu Schist composes these rock cliffs in the inner gorge of the Grand Canyon in northern Arizona. These metamorphic rocks lie beneath many layers of sedimentary rocks deep in the continent. Despite the fact that the schist is harder than steel, the Colorado River has cut down into the uplifted Colorado Plateau, exposing and eroding these ancient rocks. [(a) Photo by Kevin Schafer/Peter Arnold, Inc.; (b) photo by author.]

ANIMATION Foliati

Also, molten magma rising within the crust may "cook" adjacent rock, a process called *contact metamorphism*.

Metamorphic rocks may be changed both physically and chemically from the original rocks. Table 11.4 lists some metamorphic rock types, their parent rocks, and resultant textures. If the mineral structure demonstrates a particular alignment after metamorphism, the rock is *foliated*, and some minerals may appear as wavy striations (streaks or lines) in the new rock (Table 11.4, left photo). On the other hand, parent rock with a more homogeneous (evenly mixed) makeup may produce a *nonfoliated* rock (Table 11.4, middle photo). Which form do you see in the two inset photos of metamorphic rocks in Figure 11.6 (center right), foliated or nonfoliated?

Also in the table, note that sedimentary shale rock becomes metamorphic slate, which readily splits into slabs and thin plates. One piece of slate is placed on-edge in the photo to show this physical property. Commercially, slate is made into roofing shingles, monuments, and countertops.

In this section, you have seen how three rock-forming processes yield the igneous, sedimentary, and metamorphic materials of Earth's crust. Next we look at how heat energy and motions originating within Earth move huge slabs of lithosphere, causing the continents to *drift*. The fact that the continents migrated thousands of kilometers, and continue to migrate, is a revolutionary discovery in the *tectonic cycle* that came to light only in the last century.

Table 11.4   Metamorphic Rocks

Parent Rock	Metamorphic Equivalent	Texture
Shale (clay minerals)	Slate	Foliated
Granite, slate, shale	Gneiss	Foliated
Basalt, shale, peridotite	Schist	Foliated
Limestone, dolomite	Marble	Nonfoliated
Sandstone	Quartzite	Nonfoliated

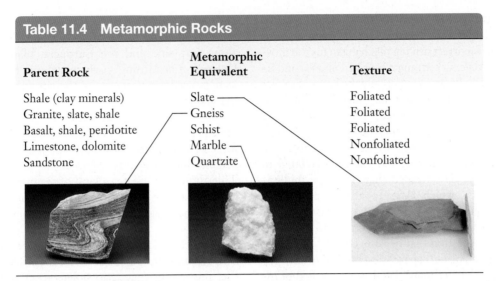

Plate Tectonics

Have you ever looked at a world map and noticed that a few of the continental landmasses—particularly South America and Africa—appear to have matching shapes, like pieces of a jigsaw puzzle, as along the African–Arabian Peninsula coastlines of the Red Sea (Figure 11.13)? The incredible reality is that the continental pieces indeed once were fitted together! Continental landmasses not only migrated to their present locations but continue to move at speeds up to 6 cm (2.4 in.) per year.

FIGURE 11.13 Jigsaw puzzle coastlines.
The Arabian Peninsula to the right and Africa to the left are rifted apart, forming the Red Sea. Winds are transporting a regionwide dust storm across the sea from Egypt and Sudan. As maps became more accurate several centuries ago, people's curiosity grew over the visual fit between some coastlines in the world. [May 31, 2003, image courtesy of satellite *Terra*, MODIS Rapid Response Team, NASA/GSFC.]

We say that the continents are *adrift* because the less-dense lithosphere is migrating around on the denser upper mantle, the asthenosphere, and the rest of the mantle that is active and in motion. The key point is that the arrangement of continents and oceans we see today is not permanent, but is in a continuing state of change.

A continent, such as North America, is actually a collage of crustal pieces that migrated from elsewhere to form the landscape we know. Through a historical chronology, let us trace the discoveries that led to the plate tectonics theory—a major revolution in Earth systems science.

A Brief History

As early mapping gained accuracy, some observers noticed the symmetry among continents, particularly the fit between South America and Africa. Abraham Ortelius (1527–1598), a geographer, noted the apparent fit of some continental coastlines in his *Thesaurus Geographicus* (1596). In 1620, English philosopher Sir Francis Bacon noted gross similarities between the edges of Africa and South America (although he did not suggest that they had drifted apart). Benjamin Franklin wrote in 1780 that the crust must be a shell that can break and shift by movements of fluid below. Others wrote—unscientifically—about such apparent relationships, but it was not until much later that a valid explanation came forward.

In 1912, German geophysicist and meteorologist Alfred Wegener publicly presented in a lecture his idea that Earth's landmasses migrate. His book, *Origin of the Continents and Oceans*, appeared in 1915. Today, Wegener is regarded as the father of this concept, which he first called **continental drift**. Scientists at the time were unreceptive to Wegener's revolutionary proposal. Thus, a great debate began, lasting almost 50 years until a scientific consensus developed that Wegener was right after all.

Wegener thought that all landmasses formed one supercontinent approximately 225 m.y.a., during the Triassic Period. This one landmass he called **Pangaea**, meaning "all Earth" (see Figure 11.17b). Although his initial model kept the landmasses together too long and his proposal included an incorrect driving mechanism for the moving continents, Wegener's arrangement of Pangaea and its breakup was correct, Pangaea being only the latest of earlier supercontinent arrangements over the span of Earth history.

To come up with his Pangaea fit, he studied the geologic record represented in the rock strata, the fossil record, and the climatic record for the continents. He concluded that South America and Africa correlated in many complex ways. He decided that the large midlatitude coal deposits, which date to the Permian and Carboniferous Periods (251–354 m.y.a.), exist because these regions once were nearer the equator and therefore covered by lush vegetation that became coal.

As modern scientific capabilities built the case for continental drift, the 1950s and 1960s saw a revival of interest in Wegener's concepts and, finally, confirmation. Aided by an avalanche of discoveries, the plate tectonics theory today is universally accepted as an accurate model of the way Earth's surface evolves. *Tectonic*, from the Greek *tektonikùs*, meaning "building" or "construction," refers to changes in the configuration of Earth's crust as a result of internal forces. **Plate tectonic** processes include upwelling of magma; lithospheric plate movements; seafloor spreading and lithospheric subduction; earthquakes; volcanic activity; and lithospheric deformation such as warping, folding, and faulting.

Sea-Floor Spreading and Production of New Crust

The key to establishing the theory of continental drift was a better understanding of the seafloor. The seafloor has a remarkable feature: an interconnected worldwide mountain chain, forming a ridge some 64,000 km (40,000 mi) in extent and averaging more than 1000 km (600 mi) in width. A striking view of this great undersea mountain chain opens Chapter 12. How did this global mountain chain get there?

In the early 1960s, geophysicists Harry H. Hess and Robert S. Dietz proposed **sea-floor spreading** as the mechanism that builds this mountain chain and drives continental movement. Hess said that these submarine mountain ranges, called the **mid-ocean ridges**, were the direct result of upwelling flows of magma from hot areas in the upper mantle and asthenosphere and perhaps from the deeper lower mantle. These are sites of intense hydrothermal activity, as mentioned earlier.

When mantle convection brings magma up to the crust, the crust fractures, and the magma extrudes onto the seafloor and cools to form new seafloor. This process builds the mid-ocean ridges and spreads the seafloor laterally, rifting as the two sides move apart. The concept is illustrated in Figure 11.14, which shows how the ocean floor is rifted (moves apart) and scarred along mid-ocean ridges. Figure 11.14b is a remote-sensing image of the mid-ocean ridge and Figure 11.14c is a map of the system in the Atlantic. Figure 11.14d shows a portion of the

mid-ocean ridge system that surfaces at Thingvellir, Iceland, forming these rifts—the North American plate along one side and the European plate on the other. Iceland's first Althing (parliament) met here in A.D. 930; leaders used the acoustics of the rift walls to be heard by the gathering.

As new crust generates and the seafloor spreads, magnetic particles in the lava orient with the magnetic field in force at the time the lava cools and hardens. The particles become locked in this alignment as new seafloor forms. This alignment creates a kind of magnetic tape recording in the seafloor. The continually forming oceanic crust records each magnetic reversal and reorientation of Earth's polarity.

Figure 11.15 illustrates just such a recording from the Mid-Atlantic Ridge south of Iceland. The colors denote the alternating magnetic polarization preserved in the minerals of the oceanic crust. Note the mirror images that develop on either side of the sea-floor rift as a result of the nearly symmetrical spreading of the seafloor; the relative ages of the rocks increase with distance from the ridge. These periodic reversals of Earth's magnetic field are a valuable clue to understanding sea-floor spreading, helping scientists fit together pieces of Earth's crust.

These sea-floor recordings of Earth's magnetic-field reversals and other measurements allowed the age of the seafloor to be determined. The complex harmony of the two concepts of plate tectonics and sea-floor spreading became clearer. The youngest crust anywhere on Earth is at the spreading centers of the mid-ocean ridges, and, with increasing distance from these centers, the crust gets steadily older (Figure 11.16). The oldest seafloor is in the western Pacific near Japan, dating to the Jurassic Period. Note on the map in the figure the distance between this basin and its spreading center in the South Pacific west of South America.

Overall, the seafloor is relatively young—nowhere does it exceed 208 m.y.a., remarkable when you remember that Earth's age is 4.6 billion years. The reason is that oceanic crust is short-lived—the oldest sections, farthest from the mid-ocean ridges, are slowly plunging beneath continental crust along Earth's deep oceanic trenches. The discovery that the seafloor is young demolished earlier thinking that the oldest rocks would be found there.

Subduction of the Lithosphere

In contrast to the upwelling zones along the mid-ocean ridges are the areas of descending lithosphere elsewhere. On the left side of Figure 11.14a, note how one plate of the crust is diving beneath another, into the mantle. Recall that the basaltic ocean crust has a density of 3.0 g/cm^3, whereas continental crust averages a lighter 2.7 g/cm^3. As a result, when continental crust and oceanic crust slowly collide, the denser ocean floor will grind beneath the lighter continental crust, thus forming a **subduction zone**, as shown in the figure. The subducting slab of crust exerts a gravitational pull on the rest of the plate—an important driving force in plate motion. This plate motion can trigger through shear traction along the base of the

(b)

Depth (meters)

4 km

(c)

(d)

Mid-Atlantic Ridge

Atlantic Ocean floor

Sea-floor spreading

Andes Mountains

South America

Oceanic crust

Lithosphere

Asthenosphere

Ocean trench

Rising magma

Subduction zone

Pacific Ocean floor

Diving plate

Upwelling of basaltic magma along mid-ocean ridge

Lithosphere

Asthenosphere

Mantle

Africa

Continental crust

Indian Ocean floor

(a)

FIGURE 11.14 Crustal movements.

(a) Sea-floor spreading, upwelling currents, subduction, and plate movements, shown in cross section. Arrows indicate the direction of the spreading. (b) A 4-km-wide image of the Mid-Atlantic Ridge showing linear faults, a volcanic crater, a rift valley, and ridges. The image was taken by the TOBI (towed ocean-bottom instrument) at approximately 29° N latitude. (c) Detail from Tharp's ocean-floor map. (d) The mid-Atlantic rift surfaces through Iceland. This scene is at Thingvellir in southeastern Iceland, where the North American and European plates of Earth's crust meet. [(a) After P. J. Wyllie, *The Way the Earth Works*, © 1976, by John Wiley & Sons, adapted by permission; (b) image courtesy of D. K. Smith, Woods Hole Oceanographic Institute, Woods Hole, Massachusetts. All rights reserved; (c) Bruce C. Heezen and Marie Tharp, courtesy Marie Tharp; (d) photo by Bobbé Christopherson.]

ANIMATION

Sea-floor Spreading, Subduction

FIGURE 11.17 Continents adrift, from 465 m.y.a. to the present.
Observe the formation and breakup of Pangaea and the types of motions occurring at plate boundaries. [(a) From R. K. Bambach, "Before Pangaea: The Geography of the Paleozoic World," *American Scientist* 68 (1980): 26–38, reprinted by permission; (b–e) from R. S. Dietz and J. C. Holden, *Journal of Geophysical Research* 75, no. 26 (September 10, 1970): 4939–4956, © The American Geophysical Union; (f) remote-sensing image courtesy of S. Tighe, University of Rhode Island, and R. Detrick, Woods Hole Oceanographic Institute, Woods Hole, Massachusetts.]

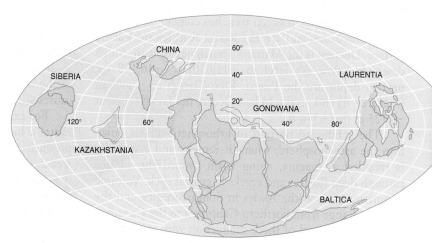

(a) 465 million years ago

Pangaea—all Earth. Panthalassa ("all seas") became the Pacific Ocean, and the Tethys Sea (partly enclosed by the African and the Eurasian plates) became the Mediterranean Sea; trapped portions of former ocean became the present-day Caspian Sea. The Atlantic Ocean did not exist.

Africa shared a common connection with both North and South America. Today, the Appalachian Mountains in the eastern United States and the Atlas Mountains of northwestern Africa reflect this common ancestry; they are, in fact, portions of the same mountain range, torn thousands of kilometers apart.

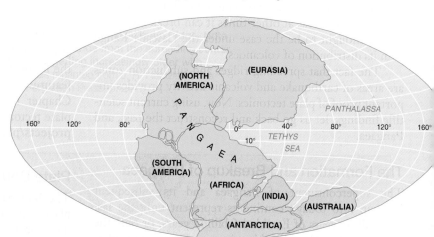

(b) 225 million years ago

New seafloor that formed since the last map is highlighted (gray tint). An active spreading center rifted North America away from landmasses to the east, shaping the coast of Labrador. India was farther along in its journey, with a spreading center to the south and a subduction zone to the north; the leading edge of the India plate was diving beneath Eurasia.

The outlines of South America, Africa, India, Australia, and southern Europe appear within the continent called *Gondwana* in the Southern Hemisphere. North America, Europe, and Asia, comprise *Laurasia*, the northern portion.

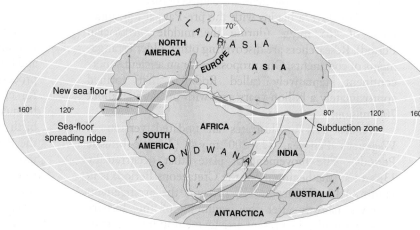

(c) 135 million years ago

ANIMATION   Pangaea Breakup, Plate Movements

(d) 65 million years ago

Sea-floor spreading along the Mid-Atlantic Ridge grew some 3000 km (almost 1900 mi) in 70 million years. Africa moved northward about 10° in latitude, leaving Madagascar split from the mainland and opening up the Gulf of Aden. The rifting along what would be the Red Sea began. The India plate moved three-fourths of the way to Asia, as Asia continued to rotate clockwise. Of all the major plates, India traveled the farthest—almost 10,000 km.

Convergent plate boundary—plates converge, producing a subduction zone. Coastal area features mountains, volcanoes, and earthquakes

ent plate boundary—diverge at mid-ocean

(e) Today

Transform fault—plates move laterally past each other between sea-floor spreading centers

Fracture zone
Transform fault

Plate Boundaries

NOTEBOOK

(f) East-Pacific rise, sea-floor spreading ridge (center of image at 9° N)

From 65 m.y.a. until the present, more than half of the ocean floor was renewed. The northern reaches of the India plate underthrust the southern mass of Asia through subduction, forming the Himalayas in the upheaval created by the collision. Plate motions continue to this day.

FIGURE 11.18 Earth's major lithospheric plates and their movements.
Each arrow represents 20 million years of movement. The longer arrows indicate that the Pacific and Nazca plates are moving more rapidly than the Atlantic plates. Compare the length of these arrows with those light-tan areas on Figure 11.17. [Adapted from U.S. Geodynamics Committee.]

Pangaea Breakup, Plate Movements, India Collision with Asia

Plate Boundaries

FIGURE 11.19 The ocean floor revealed.
A global gravity anomaly map derived from *Geosat* and *ERS-1* altimeter data. The radar altimeters measured sea-surface heights. Variation in sea-surface elevation is a direct indication of the topography of the ocean floor. [Global gravity anomaly map image courtesy of D. T. Sandwell, Scripps Institution of Oceanography. All rights reserved, 1995.]

FIGURE 11.20 Transform faults.
Appearing along fracture zones, a transform fault is only that section between spreading centers where adjacent plates move in opposite directions—between **C and D.** [Adapted from B. Isacks, J. Oliver, and L. R. Sykes, *Journal of Geophysical Research* 73 (1968): 5855–5899, © The American Geophysical Union. Detail insert from Tharp's ocean-floor map, Bruce C. Heezen and Marie Tharp, courtesy Marie Tharp.]

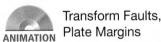

Transform Faults, Plate Margins

Andes Mountains chain and related volcanoes. The collision of India and Asia is another example of a convergent boundary, mentioned with Figure 11.17.

- *Transform boundaries* (lower right) occur where plates slide laterally past one another at right angles to a sea-floor spreading center, neither diverging nor converging, and usually with no volcanic eruptions. These are the right-angle fractures stretching across the mid-ocean ridge system worldwide (visible in Figures 11.18 and 11.19 and Chapter 12 opening illustration).

Related to plate boundaries, another piece of the tectonic puzzle fell into place in 1965 when University of

Toronto geophysicist Tuzo Wilson first described the nature of transform boundaries and their relation to earthquake activity. All the spreading centers on Earth's crust feature these perpendicular scars (Figure 11.20). Some transform faults are a few hundred kilometers long; others, such as those along the East Pacific rise, stretch out 1000 km or more (over 600 mi). Across the entire ocean floor, spreading-center mid-ocean ridges are the location of **transform faults.** The faults generally are parallel to the direction in which the plate is moving. How do they occur?

You can see in Figure 11.20 that mid-ocean ridges are not simple straight lines. When a mid-ocean rift begins, it opens at points of weakness in the crust. The fractures you see in the figure began as a series of offset breaks in the crust in each portion of the spreading center. As new material rises to the surface, building the mid-ocean ridges and spreading the plates, these offset areas slide past each other, in horizontal faulting motions. The resulting fracture zone, which follows these breaks in Earth's crust, is active only along the fault section between ridges of spreading centers, as shown in Figure 11.20.

Along transform faults (the fault section between C and D in the figure) the motion is one of horizontal displacement—no new crust is formed or old crust subducted. In contrast, beyond the spreading centers, the two sides of the fracture zones join and are inactive. In fact, the plate pieces on either side of the fracture zone are moving in the same direction, away from the spreading center (the fracture zones between A and B and between E and F in the figure).

The name *transform* was assigned because of this apparent transformation in the direction of the fault movement. The famous San Andreas fault system in California, where continental crust has overridden a transform system, relates to this type of motion. The fault that triggered the 1995 Kobe earthquake in Japan also has this type of horizontal motion, as does the North Anatolian fault in Turkey and its latest 1999 quake.

Earthquake and Volcanic Activity

Plate boundaries are the primary location of earthquake and volcanic activity, and the correlation of these phenomena is an important aspect of plate tectonics. The next chapter discusses earthquakes and volcanic activity in more detail.

Figure 11.21 maps earthquake zones, volcanic sites, hot spots, and plate motion. The "ring of fire" surrounding the Pacific Basin, named for the frequent incidence of volcanoes, is evident. The subducting edge of the Pacific plate thrusts deep into the crust and mantle, producing molten material that makes its way back toward the surface, causing active volcanoes along the Pacific Rim. Such processes occur similarly at plate boundaries throughout the world.

Hot Spots

A dramatic aspect of Earth's internal dynamics is the estimated 50 to 100 **hot spots** across Earth's surface (shown in Figure 11.21). These are individual sites of upwelling material arriving at the surface in tall plumes from the mantle or rising from near the surface that produce thermal effects in groundwater and crust. Some of these sites can be developed for geothermal power (Focus Study 11.1). Hot spots occur beneath both oceanic and continental crust. Some hot spots are anchored deep in

(text continued on page 351)

FIGURE 11.21 Earthquake and volcanic activity locations.
Earthquake and volcanic activity in relation to major tectonic plate boundaries and principal hot spots. [Earthquake and volcano data from *Earthquakes*, B. A. Bolt, © 1988 W. H. Freeman and Company; reprinted with permission. Hot spots adapted from USGS data.]

Focus Study 11.1

Heat from Earth—Geothermal Energy and Power

A tremendous amount of endogenic energy flows from Earth's interior toward the surface. Temperatures at the base of Earth's crust range from 200°–1000°C (392°–1800°F). Convection and conduction transport this geothermal energy in enormous quantities from the mantle to the crust, yet, geothermal power development is limited in extent to certain locations (Figure 11.1.1). Despite such site-specific limitations, geothermal energy for power production and hot water is a viable energy source for the present and future.

What Is Geothermal Energy?
Geothermal energy is produced when pockets of magma and hot portions of the crust heat groundwater. This energy is transmitted to the surface by heated water or steam accessed through the drilling of wells. For an effective *underground thermal reservoir* to form, an aquifer must have high porosity and high permeability that

allows heated water to move freely through connecting pore spaces. (See Figure 9.17 and the accompanying text for illustration and definition of these terms.)

Ideally, this aquifer should heat groundwater from 180° to 350°C (355°–600°F) and be accessible to drilling within 3 km (1.9 mi) of the surface, although 6–7 km depths are workable (3.7–4.3 mi). Lower temperatures will still produce steam. With drilling costs as a major overhead expense, the shallower the well needed, the better. Impermeable rock strata above the thermal reservoir aquifer helps preserve the resource and describes a *hydrothermal reservoir*, the most common form of thermal repository.

Geothermal energy literally refers to heat from Earth's interior, whereas *geothermal power* relates to specific applied strategies of geothermal electric or geothermal direct applications. *Geothermal electric* uses

steam, or hot water that flashes to steam, to drive a turbine-generator. In *geothermal direct*, hot water is used to heat buildings and to cool buildings using a heat-exchange system, to heat greenhouses, to heat soil, to heat swimming pools, and for manufacturing processes, and aquaculture, among many other uses.

Geothermal Power Production
The first geothermal electrical generating station was built in Larderello, Italy, in 1904, and it has been in continuous operation since 1913 with an installed capacity of 360 MWe (megawatts electric). Today, geothermal applications are in use in 30 countries, through some 200 plants, producing an output of 8000 MWe and direct heat of 12,000 MWt (megawatts thermal). In Reykjavik, Iceland, the majority of space heating (87%) and some of the electrical production (18%) is geothermal. The Svartsengi power plant (40 MWe) on

(a)

(b)

FIGURE 11.1.1 Surface geothermal activity in Yellowstone and Iceland.
(a) Yellowstone is the most famous national park that contains thermal features. Approximately 10,000 geysers, mudpots, and hot springs are evidence of molten pockets of heat in an unstable crust. This geothermal resource area crosses park boundaries to the north into Montana. A controversy involving development of those thermal aquifers and the possible changes that development might produce inside the national park is being debated. (b) This is the original Geysir, in Haukadalur, Iceland, from which we derive the name for all other geysers on Earth; historical accounts of Geysir's activity date back to A.D. 1294. [Photos (a) by author; (b) by Bobbé Christopherson.]

(continued)

Floor of the Oceans, 1975, by Bruce C. Heezen and Marie Tharp.
The scarred ocean floor is clearly visible: sea-floor spreading centers marked by oceanic ridges that stretch over 64,000 km (40,000 mi), subduction zones indicated by deep oceanic trenches, and transform faults slicing across oceanic ridges.

Follow the East Pacific rise (an ocean ridge and spreading center) northward as it trends beneath the west coast of the North American plate, disappearing under earthquake-prone California. The continents, offshore submerged continental shelves, and the expanse of the sediment-covered abyssal plain are all identifiable in this illustration.

On the floor of the Indian Ocean, you can see the wide track along which the India plate traveled northward to its collision with the Eurasian plate. Vast deposits of sediment cover the Indian Ocean floor, south of the Ganges River to the east of India. Sediments derived from the Himalayan Range blanket the floor of the Bay of Bengal (south of Bangladesh) to a depth of 20 km (12.4 mi). These sediments result from centuries of soil erosion in the land of the monsoons.

Isostasy can be examined in action in central and west-central Greenland, where the weight of the ice sheet has depressed portions of the land far below sea level—now visible with the ice artificially removed by the cartographer (see the upper-left corner of the map). In contrast, the region around Canada's Hudson Bay became ice-free about 8000 years ago and has isostatically rebounded 300 m (1100 ft).

In the area of the Hawaiian Islands, the hot-spot track is marked by a chain of islands and seamounts that you can follow along the Pacific plate from Hawai'i (near the page gutter) to the Aleutians. Subduction zones south and east of Alaska and Japan, and along the western coast of South and Central America, are visible as dark trenches. Along the left margin of the map, you can find Iceland's position on the Mid-Atlantic Ridge. [© 1980 by Marie Tharp. Reproduced by permission of Marie Tharp.]

Tectonics, Earthquakes, and Volcanism

▶ Key Learning Concepts

After reading the chapter, you should be able to:

▶ *Describe* first, second, and third orders of relief, and *relate* examples of each from Earth's major topographic regions.

▶ *Describe* the several origins of continental crust, and *define* displaced terranes.

▶ *Explain* compressional processes and folding, and *describe* four principal types of faults and their characteristic landforms.

▶ *Relate* the three types of plate collisions associated with orogenesis, and *identify* specific examples of each.

▶ *Explain* the nature of earthquakes, their measurement, and the nature of faulting.

▶ *Distinguish* between an effusive and an explosive volcanic eruption, and *describe* related landforms, using specific examples.

E arth's endogenic systems produce flows of heat and material toward the surface to form crust. Ongoing processes produce continental landscapes and oceanic sea-floor crust, sometimes in dramatic episodes. Earth's physical systems move to the front pages whenever an earthquake strikes or a volcanic eruption threatens a city. In 1999, more than 30,000 people perished in quakes in Turkey, Greece, Taiwan, and Mexico; in 2003, in one earthquake alone in Iran 30,000 were killed (Figure 12.1a); in other years, however, the death toll might fall below 1000.

A magnitude 7.9 near Denali National Park, Alaska, was notable as the largest earthquake on Earth in 2002 (Figure 12.1b). The remote location prevented loss of life but dramatic effects took place: cracked roads and bridges, huge boulders and rocks that fell onto glacial surfaces, offsets in glacial crevasses, many avalanches, and 150 km (92 mi) of broken supports along the Alaska pipeline. The fault

(a)

(b)

(c)

FIGURE 12.1 Earthquakes strike Iran, Alaska, and Taiwan.
(a) Captured in 1-m resolution, the mud-brick and straw buildings in a section of Bam, Iran, lay in ruins, leveled by a magnitude 6.9 quake. The buildings look as if they were smudged and blurred by some giant thumb. (b) A magnitude 7.9 earthquake struck central Alaska, 66 km (41 mi) southeast of Denali National Park. A portion of the Trans-Alaska oil pipeline was knocked off its supports. (c) A 1999 earthquake caused damage in Chi-Chi, Nantou County, Taiwan, and killed over 2000 people, leaving 100,000 homeless. Because of extensive instrumentation in the region, a wealth of data on this quake was acquired for analysis. Damage to this dam and Shihkang reservoir and other building failures were severe. [Image by (a) *IKONOS* satellite, used by permission Space Imaging, Inc.; photos by (b) Earthquake Engineering Research Institute; photos (b) and (c) by Earthquake Engineering Research Institute.]

populations as never before. For instance, scientists are closely monitoring the growing bulge on the western flanks of the South Sister volcano in central Oregon. And, now in its third decade of eruption, the outpouring of lava from Kīlauea volcano, Hawai'i, continues to amaze both scientists and tourists. Note the Internet links in this chapter and the wealth of information that is readily available so that the public is kept up-to-date on these and other hazards.

Such tectonic activity has repeatedly deformed, recycled, and reshaped Earth's crust during its 4.6-billion-year existence, sometimes in dramatic episodes. The principal tectonic and volcanic zones lie along plate boundaries. The arrangement of continents and oceans, the origin of mountain ranges, topography, and the locations of earthquake and volcanic activity are all the result of dynamic Earth processes.

In this chapter: We examine processes that construct Earth's surface and create world structural regions. Tectonic processes deform, recycle, and reshape Earth's crust. These processes occur sometimes in dramatic episodes but most often in slow, deliberate motions that build the landscape. Continental crust has been forming throughout most of Earth's 4.6-billion-year existence.

The arrangement of continents and oceans, the origin of mountain ranges, the topography of the land and the seafloor, and the locations of earthquake and volcanic activity are all evidence of our dynamic Earth. Principal seismic and volcanic zones occur along plate boundaries and hot-spot locations, thus linking plate tectonics to the potential devastation of earthquakes and local threats of major volcanic eruptions and their attendant global climatic impact.

We begin our look at tectonics and volcanism on the ocean floor, hidden from direct view. The illustration that opens this chapter is a striking representation of Earth with its blanket of water removed, as revealed to us through decades of direct and indirect observation. Careful examination of this portrait gives us a helpful review of the concepts learned in the previous chapter, laying the foundation for this and subsequent chapters. Be sure to take the quick tour presented in the caption for this chapter-opening map. Try to correlate this sea-floor

rupture extended eastward from the epicenter for over 340 km (210 mi), with right-lateral horizontal offsets of nearly 9 m (26 ft). This quake triggered more than 200 small earthquakes in far-distant Yellowstone, produced seismic waves that were felt in Texas, and altered water tables and well water in Pennsylvania.

Volcanic eruptions pose threats to life and property to those who live in proximity to the hazard. Earth systems science is providing analysis and warnings to affected illustration with the maps of crustal plates and plate boundaries shown in Figures 11.18 and the gravity anomaly image in Figure 11.19.

Earth's Surface Relief Features

Relief refers to vertical elevation differences in the landscape. Examples include the low relief of Nebraska and Saskatchewan, medium relief in foothills along mountain ranges, and high relief in the Rockies and Himalayas. The undulating form of Earth's surface, including its relief, is called **topography**, portrayed so effectively on topographic maps—the lay of the land.

The relief and topography of Earth's landforms played a vital role in human history: High mountain passes both protected and isolated societies, ridges and valleys dictated transportation routes, and vast plains necessitated developing faster methods of communication and travel. Earth's topography has stimulated human invention and spurred adaptation.

Crustal Orders of Relief

Our understanding of Earth's relief and topography is enhanced by modern computer capabilities and tools such as the global positioning system (GPS) for determining location and elevation. Scientists put elevation data in digital form; the data then are available for computer manipulation and display. The resulting *digital elevation models* (*DEMs*) aid in the scientific analysis of topography, area–altitude distributions, slopes, and local stream-drainage characteristics, among many benefits.

The U.S. Geological Survey has prepared a digitized shaded-relief map of the United States. The entire map is made up of 12 million spot elevations, with less than a kilometer between any two. A web site prepared by the USGS, "A Tapestry of Time and Terrain," uses these data, colored geologic regions, and detailed topography in a useful virtual map: **http://tapestry.usgs.gov**.

For convenience of description, geographers group the landscape's topography into three *orders of relief*. These orders classify landscapes by scale, from vast ocean basins and continents down to local hills and valleys.

First Order of Relief   The first order of relief is the coarsest level of landforms, consisting of the continents and oceans. **Continental landmasses** are those portions of crust that reside above or near sea level, including the undersea continental shelves along the coastlines. The **ocean basins**, featured in the chapter-opening illustration, are entirely below sea level. Approximately 71% of Earth is covered by water.

Second Order of Relief   The second order of relief is the intermediate level of landforms, for both continental and ocean-basin features. Continental features in the second order include mountain masses, plains,

and lowlands. A few examples are the Alps, Canadian and American Rockies, west Siberian lowland, and Tibetan Plateau. The great rock "shields" that form the heart of each continental mass are of this second order. In the ocean basins, second order of relief includes continental rises, slopes, abyssal plains, mid-ocean ridges, submarine canyons, and oceanic trenches (subduction zones)—all visible in the sea-floor illustration that opens this chapter.

Third Order of Relief   The third and most-detailed order of relief includes individual mountains, cliffs, valleys, hills, and other landforms of smaller scale. These features are identifiable as local landscapes.

Hypsometry   Figure 12.2 is a *hypsographic curve* (from the Greek *hypsos* meaning "height") that shows the distribution of Earth's surface by area and elevation in relation to sea level. Relative to Earth's diameter of 12,756 km (7926 mi), the surface is of low relief, only about 20 km (12.5 mi) from highest peak to lowest oceanic trench. For perspective, Mount Everest is 8.8 km (5.5 mi) above sea level and the Mariana Trench is 11 km (6.8 mi) below sea level. Note the GPS measurement for the revised summit elevation of Mount Everest announced in November 1999—8850 m (29,035 ft). See News Report 12.1 for more on this geographic milestone.

The average elevation of Earth's solid surface is actually under water: −2070 m (−6790 ft) below mean sea level. The average elevation for exposed land is only +875 m (+2870 ft). For the ocean depths, average elevation is −3800 m (−12,470 ft). From this description you can see that, on the average, the oceans are much deeper than continental regions are high. Overall, the underwater ocean basins, ocean floor, and submarine mountain ranges form Earth's largest "landscape."

Earth's Topographic Regions

The three orders of relief can be further generalized into six topographic regions: plains, high tablelands, hills and low tablelands, mountains, widely spaced mountains, and depressions (Figure 12.3). An arbitrary elevation or descriptive limit that is in common use defines each type of topography (see the map legend).

Four of the continents possess extensive *plains*, areas with local relief of less than 100 m (325 ft) and slope angles of 5° or less. Some plains have high elevations of more than 600 m (2000 ft); in the United States, the high plains attain elevations above 1220 m (4000 ft). The Colorado Plateau, Greenland, and Antarctica are notable *high tablelands* (the latter two of ice), with elevations exceeding 1520 m (5000 ft). *Hills* and *low tablelands* dominate Africa.

Mountain ranges, characterized by local relief exceeding 600 m (2000 ft), occur on each continent. Earth's relief and topography are undergoing constant change as a result of processes that form crust.

The announcement came November 11, 1999, in Washington, DC, at the opening reception of the 87th annual meeting of the American Alpine Club. Mount Everest had a new revised elevation, determined by direct Global Positioning System (GPS) placement on the mountain's icy summit!

In May 1999, mountaineers Pete Athans and Bill Crouse reached the summit with five Sherpas. The climbers operated Global Positioning System satellite equipment on the top of Mount Everest and determined the precise height of the world's tallest mountain. In Chapter 1, the photograph in News Report 1.1 shows a GPS unit placed by climber Wally Berg 18 m below the summit in 1998. This site is still Earth's highest benchmark, with a permanent metal marker installed in rock. Other GPS instruments in the region permit differential calibration for exact measurement (Figure 12.1.1). This is a continuation of a measurement effort begun in 1995 by Bradford Washburn, renowned mountain photographer/explorer and honorary director of Boston's Museum of Science.

Astronaut Thomas D. Jones made the photo of Mount Everest in Figure 12.1.2 during Space Shuttle flight STS-80 in 1996 aboard Space Shuttle *Columbia*. This startling view caught the mountain's triangular East Face bathed in morning light (east is to the

FIGURE 12.1.1 GPS installation measures Mount Everest.
A global positioning system (GPS) installation near Namche Bazar in the Khumbu (Everest) region of Nepal. The station is at latitude 27.8° N, longitude 86.7° E, at 3523 m (11,558 ft). The summit of Mount Everest is 30 km (18.6 mi) distant—visible above the right side of the weather dome. A network of such GPS installations correlated with the unit the climbers took to Everest's summit. [Photo by Charles Corfield, science manager to the 1998 and 1999 expeditions.]

lower left, north is to the lower right, south at the upper left; see locator map direction arrow). The North Face is in shadow. Several glaciers are visible flowing outward from the mountains, darkened by rock and debris.

Washburn announced the new measurement of 8850 m (29,035 ft). Everest's new elevation is close to the previous official measure of 8848 m (29,028 ft) set in 1954 by the Survey of India after picking the average measurements from 12 different survey points around the mountain.

Washburn stated, "The reading of 29,035 feet (8,850 meters) showed no measurable change in the height of Everest calculated since GPS observations began four years ago. But from these GPS readings it appears that the horizontal position of Everest seems to be moving steadily and slightly northeastward; between 3 and 6 mm a year (up to 0.25 in. a year)." The mountain range is being driven further into Asia by plate tectonics—the continuing collision of Indian and Asian landmasses.

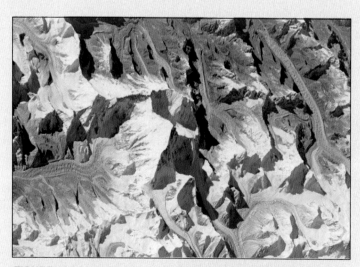

FIGURE 12.1.2 Mount Everest in morning light from orbit aboard Space Shuttle *Columbia*. [Photo courtesy of NASA, made by Dr. Thomas Jones, astronaut, 1996 (see Career Link 1.1).]

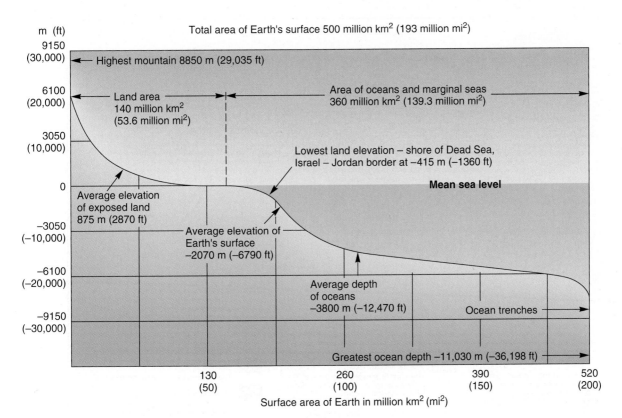

FIGURE 12.2 Earth's hypsometry.
Hypsographic curve of Earth's surface, charting area and elevation as related to mean sea level. From the highest point above sea level (Mount Everest) to the deepest oceanic trench (Mariana Trench), Earth's overall relief is almost 20 km (12.5 mi). The new height given for Mount Everest was announced in November 1999.

Crustal Formation Processes

How did Earth's continental crust form? What gave rise to the three orders of relief just discussed? Ultimately, the answers to both questions are the combined effects of tectonic activity, which is driven by our planet's internal energy, and the exogenic processes of weathering and erosion, powered by the Sun through the actions of air, water, waves, and ice.

Tectonic activity generally is slow, requiring millions of years. Endogenic (internal) processes result in gradual uplift and new landforms, with major mountain building occurring along plate boundaries. These uplifted crustal regions are quite varied, but we think of them in three general categories, all discussed in this chapter:

- Residual mountains and stable continental cratons, formed from inactive remnants of ancient tectonic activity;
- Tectonic mountains and landforms, produced by active folding, faulting, and crustal movements;
- Volcanic features, formed by the surface accumulation of molten rock from eruptions of subsurface materials.

Thus, several distinct processes operate in concert to produce the continental crust we see around us.

Continental Shields

All continents have a nucleus of ancient crystalline rock on which the continent "grows" with the addition of crustal fragments and sediments. This nucleus is the *craton,* or heartland region, of the continental crust. Cratons generally have been eroded to a low elevation and relief. Most date to the Precambrian and can exceed 2 billion years of age. The lack of basaltic components in these cratons offers a clue to their stability. The lithosphere in cratonic regions, which consists of crust and lithospheric uppermost mantle, is thicker than the lithosphere beneath younger portions of continents and oceanic crust.

A **continental shield** is a region where a craton is exposed at the surface. Figure 12.4 shows the principal areas of exposed shields and a photo of the Canadian shield. Layers of younger sedimentary rock surround these shields and appear quite stable over time. Examples of such a stable *platform* are the region that stretches from east of the Rockies to the Appalachians and northward into central and eastern Canada, a large portion of China, eastern Europe to the Ural Mountains, and across portions of Siberia.

Building Continental Crust and Terranes

The formation of continental crust is complex and takes hundreds of millions of years. It involves the entire

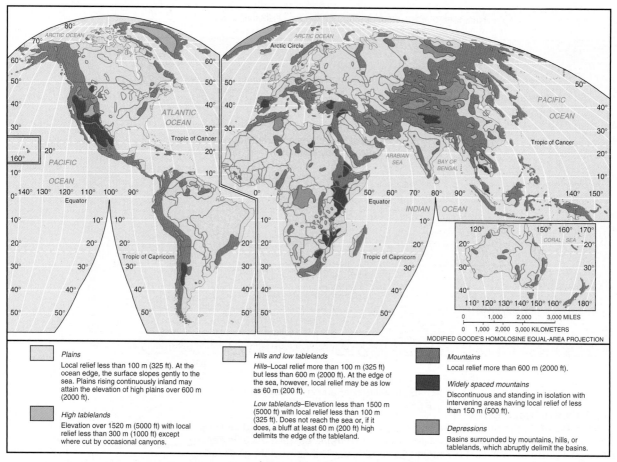

FIGURE 12.3 Earth's topographic regions.
Earth's topography is characterized as plains, high and low tablelands, hills, mountains, and depressions. [After R. E. Murphy, "Landforms of the World," *Annals of the Association of American Geographers* 58, no. 1 (March 1968). Adapted by permission.]

---

sequence of sea-floor spreading and formation of oceanic crust, its later subduction and remelting, and its subsequent rise as new magma, all summarized in Figure 12.5.

To understand this process, study Figure 12.5 (and Figure 11.14). Begin with the magma that originates in the asthenosphere and wells up along the mid-ocean ridges. Basaltic magma is formed from minerals in the upper mantle that are rich in iron and magnesium. Such magma has less than 50% silica and has a low-viscosity (thin) texture—it tends to flow. This mafic material rises to erupt at spreading centers and cools to form new basaltic seafloor, which spreads outward to collide with continental crust along its far edges. This denser oceanic crust plunges beneath the lighter continental crust, into the mantle, where it remelts. The new magma then rises and cools, forming more continental crust, in the form of intrusive granitic igneous rock.

As the subducting oceanic plate works its way under a continental plate, it takes with it trapped seawater and sediment from eroded continental crust. The remelting incorporates the seawater, sediments, and surrounding crust into the mixture. As a result, the magma, generally called a *melt*, that migrates upward from a subducted plate contains 50%–75% silica and aluminum (called andesitic

or silicic, depending on silica content). The melt has a high-viscosity (thick) texture—it tends to block and plug conduits to the surface.

Bodies of such silica-rich magma may reach the surface in explosive volcanic eruptions, or they may stop short and become subsurface intrusive bodies in the crust, cooling slowly to form granitic crystalline plutons such as batholiths (see Figures 11.7 and 11.8a). Note that this composition is quite different from the magma that rises directly from the asthenosphere at sea-floor spreading centers. In these processes of crustal formation, you can literally follow the cycling of materials through the tectonic cycle.

Each of Earth's major lithospheric plates actually is a collage of many crustal pieces acquired from a variety of sources. Crustal fragments of ocean floor, curving chains (or arcs) of volcanic islands, and other pieces of continental crust all have been forced against the edges of continental shields and platforms. These slowly migrating crustal pieces, which have become attached or accreted to the plates, are called **terranes** (not to be confused with "terrain," which refers to the topography of a tract of land).

These displaced terranes, sometimes called *microplate* or *exotic terranes*, have histories different from those of the continents that capture them. They are usually

(b)

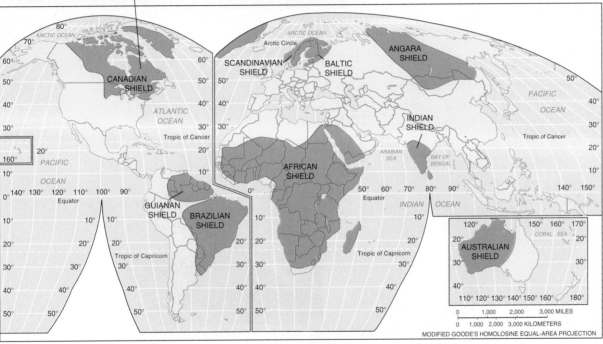

(a)

FIGURE 12.4 Continental shields.
(a) Portions of major continental shields that have been exposed by erosion. Adjacent portions
of these shields remain covered by younger sedimentary layers. (b) Canadian shield landscape
in central Labrador interior, stable for hundreds of millions of years, stripped by past glaciations, and marked by intrusive igneous dikes (magmatic intrusions). [(a) After R. E. Murphy,
"Landforms of the World," *Annals of the Association of American Geographers* 58, no. 1 (March
1968), adapted by permission. (b) Photo by John Eastcott/Yva Momatiuk/The Image Works.]

framed by fault-zone fractures and differ in rock composition and structure from their new continental homes.

In the region surrounding the Pacific, accreted terranes are particularly prevalent. At least 25% of the
growth of western North America can be attributed to
the accretion of terranes since the early Jurassic Period
(190 million years ago). A good example is the Wrangell
Mountains, which lie just east of Prince William Sound
and the city of Valdez, Alaska. The *Wrangellia terranes*—a
former volcanic island arc and associated marine

sediments from near the equator—migrated approximately 10,000 km (6200 mi) to form the Wrangell
Mountains and three other distinct formations along the
western margin of the continent (Figure 12.6).

The Appalachian Mountains, extending from Alabama
to the Maritime Provinces of Canada, possess bits of land
once attached to ancient Europe, Africa, South America,
Antarctica, and various oceanic islands. The discovery of
terranes, made only in the 1980s, demonstrates one of the
ways continents are assembled.

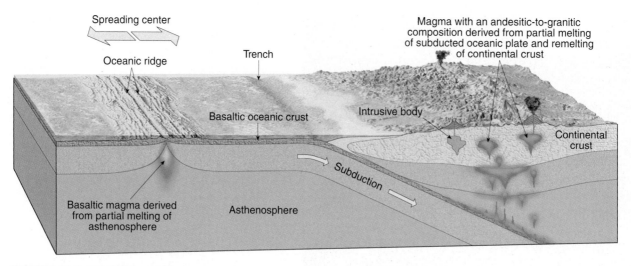

FIGURE 12.5 Crustal formation.
Material from the asthenosphere upwells along sea-floor spreading centers. Basaltic ocean floor is subducted beneath lighter continental crust, where it melts, along with its cargo of sediments, water, and minerals. This melting generates magma, which makes its way up through the crust to form igneous intrusions and extrusive eruptions. [After E. J. Tarbuck and F. K. Lutgens, *Earth, An Introduction to Physical Geology,* 5th ed., © Prentice Hall, 1996, Figure 20.20, p. 502.]

(b) Wrangell Mountains, Alaska

(a)

FIGURE 12.6 North American terranes.
(a) Wrangellia terranes, highlighted among the other terranes along the western margin of North America, occur in four segments. (b) Snow-covered Wrangell Mountains north of the Chugach Mountains in this November 7, 2001, image in east-central Alaska and across the Canadian border. Note Mount McKinley (Denali) at 6194 m (20,322 ft) in the upper-left part of the image, north of Cook Inlet, highest elevation in North America. [(a) Based on data from U.S. Geological Survey. (b) *Terra* image from MODIS sensor, courtesy of MODIS Land Rapid Response Team, NASA/GSFC.]

 ANIMATION   Terrane Formation

Crustal Deformation Processes

Rocks, whether igneous, sedimentary, or metamorphic, are subjected to powerful stress by tectonic forces, gravity, and the weight of overlying rocks. There are three types of stress: *tension* (stretching), *compression* (shortening), and *shear* (twisting or tearing), as shown in Figure 12.7.

Strain is how rocks respond to stress. Strain is expressed in rocks by *folding* (bending) or *faulting* (breaking). Think of stress as a force and resulting strain as the deformation in the rock. Whether a rock bends or breaks depends on several factors, including composition and how much pressure is on the rock. An important quality is whether the rock is *brittle* or *ductile*. The patterns created by these processes are clearly visible in the landforms we see today, especially in mountain areas. Figure 12.7 illustrates each type of stress and its resulting strain and the surface expressions that develop.

Folding and Broad Warping

When rock strata that are layered and flat are subjected to compressional forces, they become deformed (Figure 12.8). Convergent plate boundaries intensely compress rocks, deforming them in a process known as **folding**. As an analogy, if we take sections of thick fabric, stack them flat on a table, and then slowly push on opposite ends of the stack, the cloth layers will bend and rumple into folds similar to those shown in Figure 12.8a.

If we then draw a line down the center axis of a resulting ridge, and a line down the center of a resulting trough, we are able to see how the names of the folds are assigned. Along the *ridge* of a fold, layers *slope downward away from the axis*, which is called an **anticline**. In the *trough* of a fold, however, layers *slope downward toward the axis*, called a **syncline**.

If the axis of either type of fold is not "level" (horizontal, or parallel to Earth's surface), the fold–axis layers then *plunge* (dip down) at an angle. Knowledge of how folds are angled to Earth's surface and where they are located is important for the petroleum industry. For example, petroleum geologists know that oil and natural gas collect in the upper portions of anticlinal folds in permeable rock layers such as sandstone. This is the "anticlinal theory of petroleum accumulation" attributed to I. C. White, founder of the West Virginia Geological Survey.

FIGURE 12.7 Three kinds of stress, strain, and resulting surface expressions.
The bulldozers represent the stress, or force, on the rock strata. (a) Tension stress produces a stretching and thinning of the crust and a *normal fault.* (b) Compression produces shortening and folding and *reverse faulting* of the crust. In a back-and-forth horizontal motion, (c) shear stress produces a bending of the crust and, on breaking, a *strike-slip fault.*

(a)

(b)

(c)

FIGURE 12.8 Folded landscapes.
(a) Folded landscape and the basic types of fold structures. (b) A roadcut exposes a synclinal ridge in western Maryland. This syncline is a natural outdoor classroom, and the state of Maryland built an interpretive center with a walkway above the highway to permit study of the syncline. (c) Compressional folding and faulting in a roadcut along the San Andreas fault zone in Southern California. [Photos by (b) Mike Boroff/Photri-Microstock; (c) Bobbé Christopherson.]

ANIMATION Folds, Anticlines, and Synclines

Figure 12.8a further illustrates various folds that have been weathered and reduced by exogenic processes:

- A residual "synclinal ridge" may form within a syncline because different rock strata offer greater resistance to weathering processes. An actual synclinal ridge is exposed dramatically in an interstate highway roadcut in Figure 12.8b.
- Compressional forces often push folds far enough that they actually overturn upon their own strata ("overturned anticline").
- Further stress eventually fractures the rock strata along distinct lines, and some overturned folds are thrust upward, causing a considerable shortening of

the original strata ("thrust fault"). Areas of intense stress, compressional folding, and faulting are visible in a roadcut where the San Andreas fault in Southern California passes beneath the Antelope Valley Freeway (Figure 12.8c).

The Canadian Rocky Mountains, the Appalachian Mountains, and areas of the Middle East illustrate the complexity of folded landscapes. Satellites allow us to view many of these structures from an orbital perspective, as in Figure 12.9, north of the Persian Gulf and the Zagros Mountains of Iran. This area was a dispersed terrane that separated from the Eurasian plate. However, the collision produced by the northward push of the Arabian

FIGURE 12.9 Folding in the Zagros crush zone, Iran.
The Zagros Mountains are a product of the Zagros crush zone between the Arabian and Eurasian plates, where the northward push of the Arabian block is causing the folded mountains shown.
[NASA image.]

block is now shoving this terrane back into Eurasia and forming an active margin known as the Zagros crush zone, a zone of continuing collision more than 400 km (250 mi) wide. In the satellite image, anticlines form the parallel ridges; active weathering and erosion processes are exposing the underlying strata.

In addition to the rumpling of rock strata just discussed, broad warping actions also affect Earth's continental crust. These actions produce similar up-and-down bending of strata, but the bends are far greater in extent than those produced by folding. Warping forces include mantle convection, isostatic adjustment such as that caused by the weight of previous ice loads across northern Canada, or crustal swelling above an underlying hot spot. Warping features can be small, individual, foldlike structures called *basins* and *domes* (Figure 12.10a, b). They can also range up to regional features the size of the Ozark Mountain complex in Arkansas and Missouri, the Colorado Plateau in the West, the Richat dome in Mauritania, or the Black Hills of South Dakota (Figure 12.10 c, d).

Faulting

A freshly poured concrete sidewalk is smooth and strong. But stress the sidewalk by driving heavy equipment over it, and the resulting strain might cause a fracture. Pieces on either side of the fracture may move up, down, or horizontally, depending on the direction of stress. Similarly, when rock strata are stressed beyond their ability to remain a solid unit, they express the strain as a fracture.

Rocks on either side of the fracture displace relative to the other side in a process known as **faulting**. Thus, *fault zones* are areas where fractures in the rock demonstrate crustal movement. At the moment of fracture, a sharp release of energy occurs, called an **earthquake** or *quake*.

The fracture surface along which the two sides of a fault move is the *fault plane*. The names of the three basic types of faults, illustrated in Figure 12.11, are based on the tilt and orientation of the fault plane. A normal fault forms when rocks are pulled apart by tensional stress. A thrust or reverse fault results when rocks are forced together by compressional stress. And, a strike-slip fault forms when rocks are torn by lateral-shearing stress.

Normal Fault   When forces pull rocks apart, the tension causes a **normal fault**, or *tension fault*. When the break occurs, rock on one side moves vertically along an inclined fault plane (Figure 12.11a). The downward-shifting side is the *hanging wall*; it drops relative to the *footwall block*. The exposed fault plane sometimes is visible along the base of faulted mountains, where individual ridges are truncated by the movements of the fault and appear as triangular facets at the ends of the ridges. A cliff formed by faulting is commonly called a *fault scarp*, or *escarpment*.

Reverse (Thrust) Fault   Compressional forces associated with converging plates force rocks to move *upward* along the fault plane. This is a **reverse fault**, or *compression fault* (Figure 12.11b). On the surface, it appears similar to a normal fault, although more collapse and landslides may occur from the hanging wall component. In

FIGURE 12.10 Domes and basins.
(a) An upwarped dome. (b) A structural basin. (c) The Richat dome structure of Mauritania.
(d) The Black Hills of South Dakota is a dome structure, shown here on a digitized relief map.
[(c) Image from Terra sensor, October 7, 2000, Courtesy of NASA/GSFC/MITI and U.S./Japan ASTER team; (d) from USGS digital terrain map I-2206, 1992.]

England, when miners worked along a reverse fault, they would stand on the lower side (footwall) and hang their lanterns on the upper side (hanging wall), giving rise to these terms.

If the fault plane forms a low angle relative to the horizontal, the fault is termed a **thrust fault**, or *overthrust fault*, indicating that the overlying block has shifted far over the underlying block (see Figure 12.8, "thrust fault"). Place your hands palms-down on your desk, with fingertips together, and slide one hand up over the other—this is the motion of a low-angle thrust fault, with one side pushing over the other.

In the Alps, several such overthrusts result from compressional forces of the ongoing collision between the African and Eurasian plates. Beneath the Los Angeles Basin, overthrust faults produce a high risk of earthquakes and caused many quakes in the 20th century, including the $30 billion 1994 Northridge earthquake. These blind (unknown until they rupture) thrust faults beneath the Los Angeles region are a major earthquake threat in the future.

Strike-Slip Fault If movement along a fault plane is horizontal, such as produced along a transform fault, it forms a **strike-slip fault** (Figure 12.11c; refer to Figure 11.20 for review). The movement is *right-lateral* or *left-lateral*, depending on the motion perceived when you observe movement on one side of the fault relative to the other side.

Although strike-slip faults do not produce cliffs (scarps), as do the other types of faults, they can create linear rift valleys. This is the case with the San Andreas fault system of California. The rift valley is clearly visible in Figure 12.11c, where the edges of the North American and Pacific plates are grinding past one another as a result of transform-fault movement. The evolution of this fault system is shown in Figure 12.12.

Note in the figure how the East Pacific rise developed as a spreading center with associated transform faults (1), while the North American plate was progressing westward after the breakup of Pangaea. Forces then shifted the transform faults toward a northwest–southeast alignment along a weaving axis (2). Finally, the western margin of North America overrode those shifting transform faults (3). Note the convergence rate is a rapid 4 cm (1.6 in.) per year.

In *relative* terms, the motion along this series of transform faults is right-lateral, whereas in *absolute* terms the North American plate is still moving westward. Consequently, the San Andreas system is a series of faults that are *transform* (associated with a former spreading center), *strike-slip* (horizontal in motion), and *right-lateral* (one side is moving to the right relative to the other side).

The North Anatolian fault system in Turkey is a strike-slip fault and has a right-lateral motion similar to

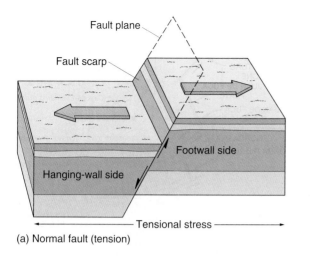

Fault plane

Fault scarp

Footwall side

Hanging-wall side

Tensional stress

(a) Normal fault (tension)

(a)

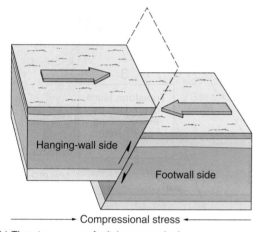

Hanging-wall side

Footwall side

Compressional stress

(b) Thrust or reverse fault (compression)

(b)

Right-lateral*   Left-lateral**

(c) Strike-slip fault (lateral shearing)

* Viewed from either dot on each road,
movement to opposite side is *to the right*.
** Viewed from either dot on each road,
movement to opposite side is *to the left*.

Pacific
plate

(c)

North
American
plate

(d)

FIGURE 12.11 Types of faults.

(a) A normal fault produced by tension in the crust, visible along the edge in Tashkent, Uzbekistan. (b) A thrust, or reverse fault, produced by compression in the crust, visible in these offset strata in coal seams and volcanic ash in British Columbia. (c) A strike-slip fault produced by lateral shearing, clearly seen looking north along the San Andreas fault rift zone on the eastern edge of the Coast Ranges in California—Pacific plate to the left, North American plate to the right. (d) Can you find the San Andreas fault on the satellite image? Look for a linear rift stretching from southeast to northwest, at the edge of the coastal mountains and San Joaquin Valley. Use the arrow as a guide. [(a) Photo by Fred McConnaughey/Photo Researchers, Inc.; (b) photo by Fletcher and Baylis/Photo Researchers, Inc.; (c) Kevin Schafer/Peter Arnold, Inc. (d) *Terra* MISR image courtesy of MISR Team, NASA/GSFC/JPL.]

ANIMATION Fault Types, Transform Faults, Plate Margins

Area of photos
(b) and (c)

(a)

(b)

(c)

FIGURE 12.18 The Appalachian Mountains.
(a) Appalachian Mountain region of Pennsylvania south through Maryland, Virginia, and West Virginia in a *Terra* MODIS sensor image from October 11, 2000, showing true fall colors. The highly folded nature of this entire region is clearly visible in the satellite image. (b) The folded Little and Line Mountains end at the Susquehana River, along the left edge of this aerial photo. (c) The distinctive Ridge and Valley Province is shown following a light March snowfall. The effects of topography on economic activity, transportation routes, and habitation is readily apparent. [(a) *Terra* image courtesy of MODIS Land Science Team, NASA/GSFC. Photos (b) and (c) by Bobbé Christopherson.]

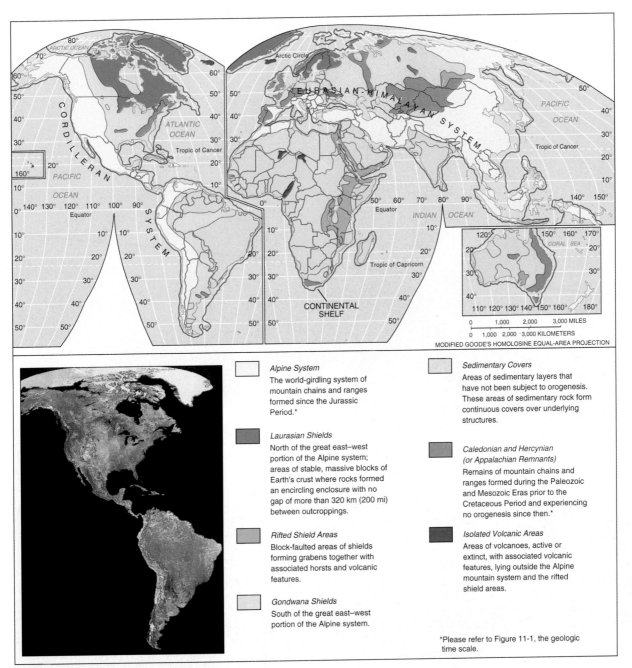

FIGURE 12.19 World structural regions and major mountain systems.
Some of the regions appear larger than the structures themselves because each region includes related landforms adjacent to the central feature. (Correlate aspects of this map with Figures 12.3 and 12.4.) Structural regions in the Western Hemisphere are visible on the composite false-color *Landsat* image inset (vegetation is portrayed in red). [After R. E. Murphy, "Landforms of the World," *Annals of the Association of American Geographers* 58, no. 1 (March 1968). Adapted by permission. Inset image courtesy of EROS Data Center and the National Geographic Society.]

The quake was caused by one of the many low-angle thrust faults that underlie the Los Angeles region. The January 17, 1994, Northridge (Reseda) earthquake, a magnitude 6.8, and more than 10,000 aftershocks, caused approximately $350 million damage to campus buildings two weeks before the beginning of spring classes. The semester began just three weeks late in 450 temporary trailers. Geography professors taught some class sessions outdoors for a brief time. Amazingly, graduation was still held in May.

What are the mechanisms that produce such different tectonic events? Why are some expected, whereas most strike in total surprise?

Focus, Epicenter, Foreshock, and Aftershock

P- and S- Waves, Seismology

ANIMATION

The subsurface area along a fault plane, where the motion of seismic waves is initiated, is the *focus*, or

FIGURE 12.20 Anatomy of an earthquake.
(a) The fault-plane solution for the 1989 Loma Prieta, California, earthquake shows the lateral and vertical (thrust) movements occurring at depth. There was no surface expression of this fault plane. Damage totaled $8 billion, 14,000 people were displaced from their homes, 4000 were injured, and 67 killed. (b) In only 15 seconds, more than 2 km (1.2 mi) of the Route 880 Cypress Freeway collapsed. (c) Section failure in the San Francisco–Oakland Bay Bridge, actually this is the way the bridge was designed to handle strain without completely failing. [(a) After P. J. Ward and R. A. Page, *The Loma Prieta Earthquake of October 17, 1989* (Washington, DC: U.S. Geological Survey, November 1989), p. 1; (b) and (c) courtesy of California Department of Transportation.]

hypocenter, of an earthquake (see labels in Figure 12.20). The area at the surface directly above the focus is the *epicenter*. Shock waves produced by an earthquake radiate outward through the crust from the focus and epicenter. Some of the seismic waves are conducted throughout the planet to distant instruments. Scientists explore Earth's interior using these seismic wave patterns and

the nature of their transmission through the layers of the planet.

An *aftershock* may occur after the main shock, sharing the same general area of the epicenter; some aftershocks rival the main tremor in magnitude. A *foreshock* also is possible, preceding the main shock. The pattern of foreshocks is now regarded as an important consideration in the forecasting effort. Before the 1992 Landers earthquake in southern California, at least two dozen foreshocks occurred along the soon-to-erupt portion of the fault.

Earthquake Intensity and Magnitude

Tectonic earthquakes are those quakes associated with faulting. A worldwide network of more than 4000 **seismograph** instruments record vibrations transmitted as waves of energy throughout Earth's interior and in the crust. Using this device and actual observations, scientists rate earthquakes on two kinds of scales: a *qualitative* damage intensity scale and a *quantitative* magnitude-of-energy-released scale.

A damage *intensity scale* is useful in classifying and describing damage to terrain and structures after an earthquake. Earthquake damage intensity is rated on the arbitrary *Mercalli scale*, a Roman numeral scale from I to XII representing "barely felt" to "catastrophic total destruction." It was designed in 1902 and modified in 1931. Table 12.1 shows this scale and the number of quakes in each category that are expected each year. The strongest earthquake thought to have struck was the magnitude 9.6 in Chile in 1960, producing Mercalli XII damage. The fault-line break in the crust extended over 1000 km (629 mi).

Moment Magnitude Scale Revises the Richter Scale
In 1935, Charles Richter designed a system to estimate earthquake magnitude. In this method, a seismograph located at least 100 km (62 mi) from the epicenter of the quake records the amplitude of seismic waves. That measurement is then charted on the **Richter scale**. The relation of magnitude to energy released is still a useful feature of his scale.

The Richter scale is logarithmic: Each whole number on it represents a 10-fold increase in the measured wave amplitude. Translated into energy, each whole number signifies a 31.5-fold increase in energy released. Thus, a magnitude of 3.0 on the Richter scale represents 31.5 times more energy than a 2.0, and 992 times more energy than a 1.0. It is difficult to imagine the power

released by the Tangshan quake, which was rated a 7.6 on the Richter scale.

Today, the Richter scale is improved and more quantitative. Revision was needed because the scale did not properly measure or differentiate between quakes of high intensity. Seismologists wanted to know more about what they call the *seismic moment* to understand a broader range of possible motions during an earthquake.

The **moment magnitude scale**, in use since 1993, is more accurate for large earthquakes than Richter's *amplitude magnitude* scale. Moment magnitude considers the amount of fault slippage produced by the earthquake, the size of the surface (or subsurface) area that ruptured, and the nature of the materials that faulted, including how resistant they were to failure. The new scale rates the 1994 Northridge quake, mentioned earlier, at magnitude 6.8 and considers extreme ground acceleration (movement upward), which the Richter amplitude magnitude method underestimated.

A reassessment of past quakes has increased the rating of some and decreased that of others. As an example, the 1964 earthquake at Prince William Sound in Alaska had an amplitude magnitude of 8.6, but on the moment magnitude scale it increases to a magnitude 9.2. Please note that Table 12.2 includes more earthquakes for the period following 1960 than for the years before 1960. This is not because earthquake frequency has increased; rather, it reflects an effort to include recent events that affected increased population densities in vulnerable areas.

Note the last event in the table: When a major quake hits an ancient city of mud brick, straw, and date-palm logs, enormous damage and casualties result. The city of Bam in southeastern Iran, near the frontier with Pakistan, was hit in the early dawn with a magnitude 6.6 quake in December 2003, damaging or destroying 85% of the buildings. Loss of life approached 30,000 people, with injuries totaling a comparable number (see Figure 12.1a).

The National Earthquake Information Center in Golden, Colorado, reports earthquake epicenters. For references, see http://wwwneic.cr.usgs.gov/ and the National Geophysical Data Center at http://www.ngdc.noaa.gov/seg/. As part of the "Learning from Earthquakes

---

Table 12.1 Earthquake Characteristics and Frequency Expected Each Year

Effects in Populated Areas	Approximate Intensity (modified Mercalli scale)	Approximate Magnitude (Moment Magnitude scale)	Number per Year
Nearly total damage	XII	>8.0	1 every few years
Great damage	X–XI	7–7.9	18
Considerable-to-serious damage to buildings; railroad tracks bent	VIII–IX	6–6.9	120
Felt-by-all, with slight damage to buildings	V–VII	5–5.9	800
Felt-by-some to felt-by-many	III–IV	4–4.9	6,200
Not felt, but recorded	I–II	2–3.9	500,000

Source: USGS, Earthquake Information Center.

Seismograph, How it Works

ANIMATION

Table 12.2 A Sampling of Significant Earthquakes**

Year	Date	Location	Number of Deaths	Mercalli Intensity	Moment Magnitude (Richter)
1556	Jan. 23	Shaanxi Province, China	830,000	*	*
1737	Oct. 11	Calcutta, India	300,000	*	*
1812	Feb. 7	New Madrid, Missouri	Several	XI–XII	*
1857	Jan. 9	Fort Tejon, California	*	X–XI	*
1870	Oct. 21	Montreal to Québec, Canada	*	IX	*
1886	Aug. 31	Charleston, South Carolina	*	IX	6.7
1906	Apr. 18	San Francisco, California	3,000	XI	7.7 (8.25)
1923	Sept. 1	Kwanto, Japan	143,000	XII	7.9 (8.2)
1939	Dec. 27	Erzincan, Turkey	40,000	XII	7.6 (8.0)
1960	May 22	Southern Chile	5,700	XII	9.6 (8.6)
1964	Mar. 28	Southern Alaska	131	X–XII	9.2 (8.6)
1970	May 31	Northern Peru	66,000	*	7.9 (7.8)
1971	Feb. 9	San Fernando, California	65	VII–IX	6.7 (6.5)
1972	Dec. 23	Managua, Nicaragua	5,000	X–XII	6.2 (6.2)
1976	Jul. 28	Tangshan, China	250,000	XI–XII	7.4 (7.6)
1978	Sept. 16	Iran	25,000	X–XII	7.8 (7.7)
1985	Sept. 19	Mexico City, Mexico	7,000	IX–XII	8.1 (8.1)
1988	Dec. 7	Armenia–Turkey border	30,000	XII	6.8 (6.9)
1989	Oct. 17	Loma Prieta (near Santa Cruz, California)	66	VII–IX	7.0 (7.1)
1991	Oct. 20	Uttar Pradesh, India	1,700	IX–XI	6.2 (6.1)
1994	Jan. 17	Northridge (Reseda), California	66	VII–IX	6.8
1995	Jan. 17	Kobe, Japan	5,500	XII	6.9
1996	Feb. 17	Indonesia	110	X	8.1
1997	Feb. 28	Armenia–Azerbaijan	1,100	XII	6.1
1997	May 10	Northern Iran	1,600	XII	7.3
1998	May 30	Afghanistan–Tajikistan	4,000	XII	6.9
1998	Jul. 17	Papua, New Guinea	2,200	X	7.1
1999	Jan. 26	Armenia, Colombia	1,000	VIII–IX	6.0
1999	Aug. 17	Izmit, Turkey	17,100	VIII–XI	7.4
1999	Sept. 7	Athens, Greece	150	VI–VIII	5.9
1999	Sept. 20	Chi-Chi, Taiwan	2,500	VI–X	7.6
1999	Sept. 30	Oaxaca, Mexico	33	VI	7.5
1999	Oct. 16	Hector Mine, California	0	*	7.1
1999	Nov. 12	Düzce, Turkey	700	VI–X	7.2
2001	Jan. 26	Gujarat state, India	19,998	X–XII	7.7
2002	Nov. 3	Near Denali National Park, Alaska	1	X	7.9
2003	Dec. 26	Bam, Iran	30,000	X–XII	6.9
2004	Dec. 26	Northern Sumatra (west coast)	150,000+***	X–XII	9.0

*Data not available.

**There is not a recent increase in earthquakes; this table merely reflects more detail on the recent record.

***Estimated deaths from quake and resulting tsunami that traveled across the Indian Ocean.

 ANIMATION Seismograph, How it Works

Project," reconnaissance teams examine the sites of major quakes; see http://www.eqnet.org. For a real-time listing of earthquakes worldwide, go to http://wwwneic.cr.usgs.gov/neis/bulletin/bulletin.html. In Canada, a listing of significant Canadian earthquakes is at http://www.seismo.nrcan.gc.ca/major_eq/majoreq_e.php.

The Nature of Faulting

We earlier described types of faults and faulting motions. The specific mechanics of how a fault breaks remain under study, but **elastic-rebound theory** describes the basic process. Generally, two sides along a fault appear to be locked by friction, resisting any movement despite the powerful forces acting on the adjoining pieces of crust. Stress continues to build strain along the *fault plane* surfaces, storing elastic energy like a wound-up spring. When the strain buildup finally exceeds the frictional lock, both sides of the fault abruptly move to a condition of less strain, releasing a burst of mechanical energy. This type of sudden movement rocked the region around

News Report 12.2

A Tragedy in Kobe, Japan—The Hyogo-ken Nanbu Earthquake

Early on the morning of January 17, 1995, a year to the day after the Northridge quake, a section of the Nojima fault zone, approximately 16 km (10 mi) below the surface, became the focus of a devastating earthquake. On the moment magnitude scale the quake registered 6.9; it killed 5500 people, injured 26,000, and caused more than $100 billion in damage. About 200,000 homes, or 10% of all the housing in the metropolitan area, were damaged; some 80,000 buildings completely collapsed; major transportation arteries were severed; and the busy port was brought to a standstill (Figure 12.2.1).

For Japan, this was the worst quake loss since the 1923 temblor that hit Tokyo. The Kobe region had large earthquakes in A.D. 868, 1596, and 1916, but after almost 80 years of relative calm, the people did not expect this surprise—despite the fact the fault and the earthquake mechanism are well understood. The fault moved in a right-lateral, strike-slip motion, similar to the San Andreas fault in California. This event moved the land along three segments, where displacement of 1.0 to 1.5 m occurred in a rupture zone that stretched over 30 km (18 mi) in length.

The lessons are many, especially the occurrence of liquefaction and the failure of filled land. Reclaimed areas of Osaka Bay taken from the sea by landfill liquefied quickly in the shaking. Soil surfaces failed, collapsing the structures they supported (Figure 12.2.1b). Importantly, these human-made landfills and islands are characteristic of the San Francisco Bay region, where, since the early 1900s, about half the original surface area of the bay is reclaimed with fill and occupied with buildings.

A positive note in Kobe and Osaka was the good performance of landfill sites where drains were installed and where deeply anchored pilings were in place to support structures. The cost of such retrofits in the United States could easily exceed several hundred billion dollars. In the meantime, the resemblance between the geology of the Kobe region and the San Francisco Bay region, specifically the menacing Hayward fault system, is troubling. The Kobe lesson raises great concern and a call to action. In Japan, this earthquake triggered an increase in spending by the Japanese government for earthquake-monitoring research.

(a)

(b)

FIGURE 12.2.1 1995 earthquake in Kobe, Japan.
(a) Catastrophic failure to an elevated freeway; note the failure of the supporting pillars.
(b) Ground failure, mainly liquefaction, tilts a building on the failing foundation soils to near collapse; note the space between the building and the ground on the right side.
[Photos by Haruyoshi Yamaguchi/Sygma.]

Kobe, Japan, in 1995, with a magnitude 6.9 quake. See News Report 12.2 for an account of this significant event.

Think of the fault plane as a surface with irregularities that act as sticking points, preventing movement, similar to two pieces of wood held together by drops of glue of different sizes rather than an even coating of glue. Research scientists at the USGS and the University of California identify these small areas of high strain as *asperities*. They are the points that break and release the sides of the fault.

If the fracture along the fault line is isolated to a small asperity break, the quake will be small in magnitude. Clearly, as some asperities break (perhaps recorded as small foreshocks), the strain increases on surrounding asperities that remain intact. Thus, small earthquakes in an area may be precursors to a major quake. However, if the break involves the release of strain along several asperities, the quake will be greater in extent and will involve the shifting of massive amounts of crust. The latest evidence points to a wavelike pattern, as rupturing spreads along the fault plane, rather than the entire fault surface giving way at once.

A study of the buildup and release of stress and strain along the North Anatolian fault system in Turkey (Figure 12.13) illustrates the process (Figure 12.21a). Strain accumulates as the two sides of the fault attempt to move

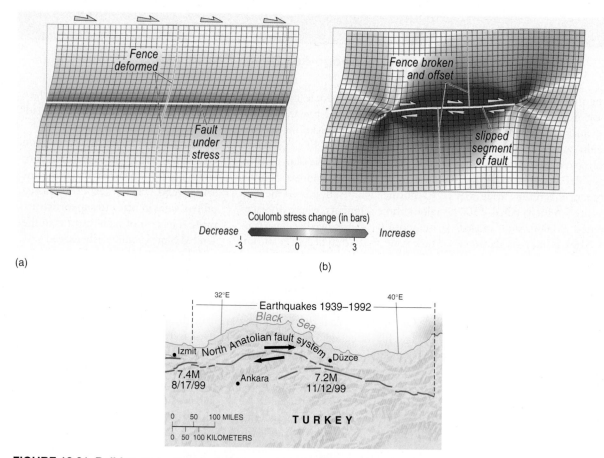

FIGURE 12.21 Buildup and release of stress and strain along a fault system.
The elastic-rebound mechanism illustrated here applies to strike-slip faults such as the North Anatolian in Turkey and the San Andreas system of faults in California. (a) As two sides of this right-lateral, strike-slip fault move, they remain locked by friction, creating stress. Such stress (a force) produces strain (a deformation) in the rock. Note the fence that crosses the fault line. This stress continues to build to the break point. (b) In an earthquake, such as the 7.4 magnitude shock that hit Izmit, Turkey, the two sides snap into new positions, strain is released, and the fence no longer forms a continuous property line. The segments of the fault that remain locked now have added strain. A Coulomb is a unit of shear-stress failure overcoming internal resisting friction and resistance to movement; the color scale ranges from blue (decreased stress) to red (increased stress). [Illustrations (a) and (b) courtesy of Serkan Bozkurt, USGS.]

laterally past one another, yet remain locked. Stress forces continue to build strain until it overwhelms the frictional sticking points and the two sides snap violently and at great speed into new positions. An earthquake hits. Note in Figure 12.21b that strain deformation is relieved in the area that broke, whereas on either side of the faulted region the sides remain locked, stresses are stronger, and resulting strain is increased.

Earthquakes and the San Andreas Fault

In 1906, San Francisco was a city of 400,000 people (there are several million in the metropolitan region today). In that year, a magnitude 7.7 (8.25 Richter) tremor rocked the city, felling buildings and initiating a firestorm fed by broken gas pipes. Movement along a fault was evident over 435 km (270 mi), prompting intensive research to discover the nature of such faulting. The elastic-rebound theory developed as a result of research done on this rupture. Realize that this quake occurred 6 years before Wegener proposed his continental drift hypothesis.

The fault system that devastated San Francisco in 1906, and coastal and inland California throughout recorded history, is the San Andreas. The San Andreas fault system provides a good example of a spreading center overridden by an advancing continental plate (refer to Figure 12.12 for its evolution).

The tectonic history of the San Andreas was brought home to hundreds of millions of television viewers as they watched the 1989 baseball World Series played in Candlestick Park near San Francisco. Less than half an hour before the call to "play ball," a powerful earthquake rocked the region, turning sportscasters into newscasters and sports fans into disaster witnesses. This magnitude 7.0 earthquake involved a portion of the San Andreas fault, near Loma Prieta, approximately 16 km (9.9 mi) east of Santa Cruz and 95 km (59 mi) south of San Francisco (see Figure 12.20). A fault had ruptured at a focus unusually deep for the San Andreas system, more than 18 km (11.5 mi) below the surface.

Unlike previous earthquakes—such as the one in 1906, when the plates shifted a maximum of 6.4 m (21 ft) relative to each other—there was no evidence of a fault plane or rifting at the surface in the Loma Prieta area. Instead, the fault plane suggested in Figure 12.20 shows the two plates moving horizontally approximately 2 m (6 ft) past each other deep below the surface, with the Pacific plate thrusting 1.3 m (4.3 ft) upward. This vertical motion is unusual for the San Andreas fault and indicates that this portion of the San Andreas system is more complex than previously thought.

In the San Francisco Bay Area the present forecast places a 62% chance of a magnitude 6.7, or higher, striking in the next 30 years (2000–2030) along the San Andreas system of faults. Added to this is a 14% chance of a significant earthquake occurring along some yet unknown fault.

Los Angeles Region

Since the mid-1980s, an area east of Los Angeles has experienced seven earthquakes greater than 6.0 magnitude. After a magnitude 6.1 quake in 1992, a magnitude 7.4 tremor rocked the lightly populated area near Landers, California (see the enlargement map of southern California in Figure 12.12). Although it was the single largest quake in California in 30 years, the remote location kept injuries and damage slight. Involved in this series were four different faults and some unknown segments. The main faulting caused a displacement of 6.1 m (20 ft).

For yet unexplained reasons, related earthquakes over the next few weeks struck in Mammoth Lakes, about 645 km (400 mi) to the north; at Mount Shasta in northern California; in southern Nevada and Utah; and 1810 km (1125 mi) distant in Yellowstone National Park, Wyoming. Evidence is growing that in complex fault

systems, such as in the Los Angeles Region, active faults can interact, transferring stress to adjoining systems, with one fault perhaps triggering activity along another fault. This complicates analysis and forecasting.

The 1971 San Fernando, 1987 Whittier, 1988 Pasadena, 1991 Sierra Madre, and 1994 Northridge (Reseda) earthquakes are a few of the many quakes associated with deeply buried thrust faults. Although not directly aligned with the San Andreas, scientists think that southern California will be affected by more of these thrust-fault actions as strain continues to build along the nearby San Andreas system of faults. The fault line that marks the contact between the Pacific and the North American plates is indeed restless.

Earthquake Forecasting and Planning

The map in Figure 12.22 plots the epicenters of earthquakes in the United States and southern Canada between 1899 and 1990. These occurrences give some indication of relative risk by region. The challenge is to discover how to predict the *specific time and place* for a quake in the short term. See News Report 12.3 for more on forecasting.

Actual implementation of an action plan to reduce death, injury, and property damage from earthquakes is difficult. The political environment adds complexity since studies show that an accurate earthquake prediction may threaten a region's economy (Figure 12.23). Research examining the potential socioeconomic impact of earthquake prediction on an urban community shows surprising negative economic impacts in the period before the quake hits. Imagine a chamber of commerce, bank, real estate agent, tax assessor, or politician who would welcome an earthquake prediction and such negative

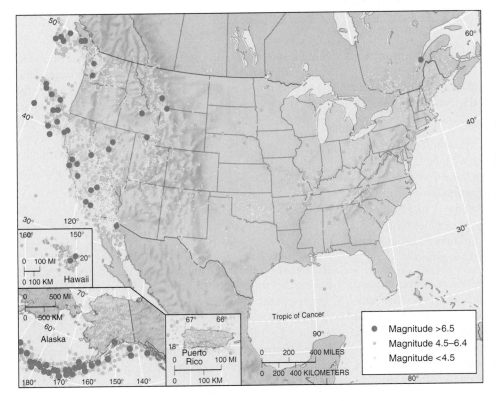

FIGURE 12.22 Seismicity of the United States and southern Canada, 1899–1990. Earthquake occurrences of magnitude 4.5, or greater, indicate areas of greatest seismic risk. Note the contrast between stable and unstable regions, related to tectonic activity. Active seismic regions include the West Coast, the Wasatch Front of Utah northward into Canada, the central Mississippi Valley, southern Appalachians, and portions of South Carolina, upstate New York, and Ontario. [From the USGS, National Earthquake Information Center. For more detailed checking on your locale, see http://eqhazmaps.usgs.gov/.]

News Report 12.3

Seismic Gaps, Nervous Animals, Dilitancy, and Radon Gas

How do you forecast an earthquake? One approach is to examine the history of each plate boundary and determine the frequency of past earthquakes, a study called *paleoseismology*. Paleoseismologists construct maps that provide an estimate of expected earthquake activity based on past performance. An area that is quiet and overdue for an earthquake is a *seismic gap*; such an area forms a gap in the earthquake occurrence record and is therefore a place that possesses accumulated strain. The area along the Aleutian Trench subduction zone had three such gaps until the great 1964 Alaskan earthquake filled one of them.

The areas around San Francisco and northeast of Los Angeles represent other such gaps where the fault system appears to be locked by friction and stress is producing accumulating strain. The U.S. Geological Survey in 1988 made a prediction that there was a 30% chance of an earthquake occurring with a magnitude 6.5 within 30 years in the Loma Prieta portion of the fault system. The actual 1989 quake dramatically filled a

portion of the seismic gap in that region. The Nojima fault in the area of Kobe occurred in another gap.

One potentially positive discovery from the Loma Prieta disaster was that Stanford University scientists found unusually great changes in Earth's magnetic field—about 30 times more than normal—3 hours before the main shock. These measurements were made 7 km (4.3 mi) from the epicenter, raising hopes that short-term warnings might be possible in the future. A valid research question is whether animals have the ability to detect these minute changes in magnetic fields, an ability that was perhaps lost in humans. If so, strange prequake animal behavior, often reported, might provide forecasting clues.

Dilitancy refers to the slight increase in volume of rock produced by small cracks that form under stress and accumulated strain. The affected region may tilt and swell in response to strain. Tiltmeters measure these changes in suspect areas. Another indicator of dilitancy is an increase in radon (a naturally occurring, slightly

radioactive gas) dissolved in groundwater. At present, earthquake hazard zones have thousands of radon monitors taking samples in test wells.

A seismic network in operation in Mexico City is providing 70-second warnings of arriving seismic-wave energy from distant epicenters. Following the Loma Prieta earthquake, freeway-repair workers were linked to a broadcast system that gave them 20-second alerts of aftershocks from distant epicenters. In southern California, the USGS is coordinating a Southern California Seismographic Network (SCSN at **http://www. trinet.org/scsn/scsn.html**) to correlate 350 instruments for immediate earthquake location analysis and disaster coordination. Also, for the southern California region, see **http:// www-socal.wr.usgs.gov/**.

Someday, accurate earthquake forecasting may be a reality, but several questions remain. How will humans respond to a forecast? Can a major metropolitan region be evacuated for short periods of time? Can cities relocate after a disaster to areas of lesser risk?

---

publicity for their city. These factors work against effective planning and preparation.

Long-range planning is a complex subject. After the Loma Prieta earthquake in 1989, the National Research Council concluded in a report:

> One of the most jarring lessons from these quakes may be that earthquake professionals have long known many of the things that could have been done to reduce devastation. . . . The cost-effectiveness of mitigation and the importance of closing the knowledge gap among researchers, building professionals, government officials, and the public are only two of the many lessons from the Loma Prieta that need immediate action.

A valid and applicable generalization is that *humans and their institutions are unable or unwilling to perceive hazards in a familiar environment*. In other words, we tend to feel secure in our homes and communities, even if they are sitting on a quiet fault zone. Such an axiom of human behavior certainly helps explain why large populations continue to live and work in earthquake-prone settings.

Similar questions also can be raised about populations in areas vulnerable to floods, droughts, hurricanes and coastal storm surge, and settlements on barrier islands. (See the Natural Hazards Observer at **http://www. colorado.edu/hazards/o/** as part of the Natural Hazards Center at the University of Colorado at **http://www. colorado.edu/hazards/index.html**. For coping with natural hazards in Canada, see **http://www.utoronto.ca/ env/nh/toc.htm**.)

Volcanism

More than 1300 identifiable volcanic cones and mountains exist on Earth, although fewer than 600 are active—having at least one eruption in recorded history. An average of 60 volcanic eruptions a year occur as reminders of Earth's internal energy, plate tectonics, and hot-spot activity. Eruptions in remote locations and at depths on the seafloor go largely unnoticed, but the occasional eruption of great magnitude near a population center makes headlines.

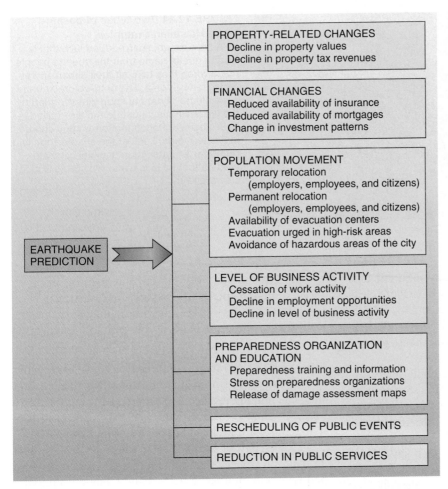

FIGURE 12.23 Socioeconomic impacts of an earthquake prediction. Business, political, and monetary interests are at stake as scientists get closer to being able to predict earthquakes. [After J. E. Haas and D. S. Mileti, *Socioeconomic Impact of Earthquake Prediction on Government, Business, and Community* (Boulder: University of Colorado, Institute of Behavioral Sciences). Used by permission.]

North America has about 70 volcanoes (mostly inactive) along the western margin of the continent. Mount St. Helens in Washington State is a famous active example, and more than 1 million visitors a year travel to Mount St. Helens Volcanic National Monument to see volcanism for themselves.

The map in Figure 11.21 located a sampling of volcanoes across the planet. Some of the many volcanoes that are currently active include the Soufriere Hills on Montserrat (Lesser Antilles/Caribbean Sea), Mount Etna (Italy), Mount Ruapehu (New Zealand), Popocatèpetl, (Mexico), Axial Summit (sea-floor spreading center off the coast of Oregon), Villarrica (Chile), Nyiragongo (Congo, Africa), Piton de la Fournaise (Island of Reunion), and Mayon (Philippines), and the ongoing Kīlauea eruptions in Hawai'i. For a listing of currently active volcanoes, with links, see http://volcano.und.nodak.edu/vwdocs/current_volcs/current.html.

Other useful Internet resources include an index to the world's volcanoes at http://vulcan.wr.usgs.gov/Volcanoes/framework.html and the Smithsonian's Global Volcanism Program at http://www.volcano.si.edu/gvp/. A source of information about volcanoes on the web is at the University of North Dakota's "Volcano World" http://volcano.und.nodak.edu/, or check http://www.geo.mtu.edu/volcanoes/links/observatories.html for links to volcano observatories across the globe.

Volcanic Features

A **volcano** forms at the end of a central vent or pipe that rises from the asthenosphere and upper mantle through the crust into a volcanic mountain. A **crater**, or circular surface depression, usually forms at or near the summit. Magma rises and collects in a magma chamber deep below the volcano until conditions are right for an eruption. This subsurface magma emits tremendous heat; in some areas it boils groundwater, producing *geothermal energy*, as seen in the thermal springs and geysers of Yellowstone National Park, or elsewhere in the world where it is harnessed for geothermal power production.

Lava (molten rock), gases, and **pyroclastics**, or *tephra* (pulverized rock and clastic materials of various sizes ejected violently during an eruption), pass through the vent to the surface and build a volcanic landform. There are two principal forms of flowing basaltic lava, both named in Hawaiian terms (Figure 12.24). **Aa** is a basaltic lava that is rough and jagged with sharp edges. This texture happens because the lava loses trapped gases, flows slowly, and develops a thick skin that cracks into the jagged surface. Another principal lava texture that is more fluid than aa is **pahoehoe**. As this lava flows it forms a thin crust that develops folds and appears "ropy," like coiled, twisted rope. Both forms can come from the same eruption, and sometimes pahoehoe will become aa as the flow progresses. Other types of basaltic magma are described later in this section.

(a) Aa

(b) Pahoehoe

FIGURE 12.24 Two types of basaltic lava—Hawaiian examples.
(a) Aa is a rough, sharp-edged lava that is said to get its name from the sounds people make when they take off their shoes and attempt to walk on it. The gold-colored strands are a clump of natural spun glass formed by the lava as it blew out into the air—called Pele's hair. (b) Pahoehoe forms ropy chords in twisted folds. Look at the chapter-opening photo for Chapter 13 to see highly weathered pahoehoe in Iceland. [Photos by Bobbé Christopherson.]

ANIMATION   Tectonic Settings and Volcanic Activity

The fact that lava can occur in many different textures and forms accounts for the varied behavior of volcanoes and the different landforms they build. In this section, we look at five volcanic landforms and their origins: cinder cones, calderas, shield volcanoes, plateau basalts, and composite volcanoes.

A **cinder cone** is a small, cone-shaped hill usually less than 450 m (1500 ft) high, with a truncated top formed from cinders that accumulate during moderately explosive eruptions. Cinder cones are made of pyroclastic material and *scoria* (cindery rock, full of air bubbles).

Another distinctive landform is a large, basin-shaped depression called a **caldera** (Spanish for "kettle"). It forms when summit material on a volcanic mountain collapses inward after an eruption or other loss of magma. In the Cape Verde Islands (15° N, 24.5° W), Fogo Island has such a caldera in the midst of a volcanic cone, opening to the sea on the eastern side. The best farmland is in the caldera, so the people choose to live and work on the floor of an active volcano. Unfortunately, eruptions began in April 1995, destroying farms and forcing the evacuation of 5000 people. A cinder cone 600 m (1800 ft) high formed in the former farmlands. The area experienced eruptions between A.D. 1500 and 1750, followed by six more in the next century, each destroying houses and crops. The previous eruption before the current activity was in 1951.

An example of a caldera in North America is beautiful Crater Lake in southern Oregon. News Report 12.4 discusses another, the Long Valley Caldera in California.

ANIMATION   Formation of Crater Lake

Location and Types of Volcanic Activity

The location of volcanic mountains on Earth is a function of plate tectonics and hot-spot activity. Volcanic activity occurs in three settings (with representative examples):

1. Along subduction boundaries at continental plate–oceanic plate convergence (such as Mount St. Helens and Kliuchevskoi, Siberia) or oceanic plate–oceanic plate convergence (Philippines and Japan).
2. Along sea-floor spreading centers on the ocean floor (Iceland on the Mid-Atlantic ridge or off the coast of Oregon and Washington) and areas of rifting on continental plates (the rift zone in east Africa).
3. At hot spots, where individual plumes of magma rise to the crust (such as Hawai'i and Yellowstone National Park).

Figure 12.25 illustrates these three types of volcanic activity, which you can compare with the active volcano sites and plate boundaries shown in Figure 11.21. Figures 12.26 to 12.31 illustrate the various aspects of volcanism presented in Figure 12.25 and in the following discussion.

The variety of forms among volcanoes makes them difficult to classify; most fall in a middle ground between one type and another. Even during a single eruption, a volcano may behave in several different ways. The primary factors in determining an eruption type are (1) the magma's chemistry, which is related to its source; and (2) the magma's viscosity. Viscosity is the magma's resistance to flow ("thickness"), ranging from low viscosity (very fluid) to high viscosity (thick and flowing slowly). We consider two types of eruptions—effusive and explosive—and the characteristic landforms they build.

Effusive Eruptions

Effusive eruptions are the relatively gentle ones that produce enormous volumes of lava annually on the seafloor and in places such as Hawai'i and Iceland. These direct eruptions from the asthenosphere and upper mantle produce a low-viscosity magma that is very fluid and cools to form a dark, basaltic rock low in silica (less than 50% silica and rich in iron and magnesium). Gases readily escape from this magma because of its low viscosity, adding to a gentle **effusive eruption** that pours out on the surface, with relatively small explosions and few pyroclastics. However, dramatic fountains of basaltic lava sometimes shoot upward, powered by jets of rapidly expanding gases.

An effusive eruption may come from a single vent or from the flank of a volcano, through a side vent. If such vents form a linear opening, they are called *fissures;* these sometimes erupt in a dramatic *curtain of fire* (sheets of molten rock spraying into the air). Rift zones capable of erupting tend to converge on the central crater, or vent, as they do in Hawai'i. The interior of such a crater, often a

News Report 12.4

Is the Long Valley Caldera Next?

Trees are dying in the forests on Mammoth Mountain, in a portion of the Long Valley Caldera, near the California–Nevada border. The oval caldera is about 15 by 30 km (9 by 19 mi) long in a north–south direction and sits at 2000 m (6500 ft) elevation, rising to 2600 m (8500 ft) along the western side.

About 1200 tons of carbon dioxide is coming up through the soil in the old caldera each day. Carbon dioxide levels reach 30% to 96% of gases in some soil samples. The source: active and moving magma at some 3-km depth. Elsewhere in the world, this and other gas emissions assist forecasts of potential volcanic activity. At Mammoth Mountain, these gases signify magmatic activity and may be a portent of things to come (Figure 12.4.1). There are now seven areas of dead trees in the area.

A powerful volcanic eruption 760,000 years ago formed the Long Valley Caldera. This ancient eruption exceeded the volume output of the 1980 Mount St. Helens event by more than 2000 times! Volcanic activity of a lesser extent has occurred in the area over the past several hundred thousand years, most recently between A.D. 1720 and 1850 in the Mono Lake area north of Long Valley.

In the late 1970s and again in 1996, continuing through 1998, swarms of earthquakes shook Long Valley—thousands of quakes overall, some three dozen greater than magnitude 3.0. The subsurface foci for these quakes are at about 11 km beneath the region. The combination of volcanic gas production and earthquakes is drawing scientific and public attention to this area because of the potential for a future massive eruption somewhere along the Mono-Inyo Craters stretching north of Long Valley. The region is a popular tourist mecca and recreational center, with second-home residential development and a growing year-round population. For updates, see **http://quake.wr.usgs. gov/VOLCANOES/LongValley/**.

FIGURE 12.4.1 Trees are dying. Forest killed by carbon dioxide and a warning sign about the danger. The hazard has killed one person to date. The CO_2 is gassing up through soils from subsurface magma and related volcanic activity. [Photo by Bobbé Christopherson.]

---

sunken caldera, may fill with low-viscosity magma during an eruption, forming a molten lake, which then may overflow lava downslope in dramatic rivers and falls of molten rock.

On the island of Hawai'i, the continuing Kīlauea eruption is the longest in recorded history, active since January 3, 1983—the 55th episode of this series began in February 1997. More than 60 separate eruptions occurred at Kīlauea's summit between 1778 (the arrival of westerners) and 1982. During 1989–1990, lava flows from Kīlauea actually consumed several visitor buildings in the Hawai'i Volcanoes National Park and two housing subdivisions in Kalāpana, Kapa'ahu, and Kaimū. Kīlauea has produced more lava than any other vent on Earth in recorded history (Part 3 opening photo, Figures 12.28 and 12.29). Pu'u O'o is the active crater on Kīlauea and is presently continuing to send streams of lava to the ocean (Figure 12.32).

A typical mountain landform built from effusive eruptions is gently sloped, gradually rising from the surrounding landscape to a summit crater. The shape is similar in outline to a shield of armor lying face up on the ground and therefore is called a **shield volcano**. The shield shape and size of Mauna Loa in Hawai'i is distinctive when compared with Mount Rainier in Washington, which is a different type of volcano (explained shortly) and the largest in the Cascade Range (Figure 12.33).

The height of the Mauna Loa shield is the result of successive eruptions, flowing one on top of another. Mauna Loa is one of five shield volcanoes that make up the island of Hawai'i. At least 1 million years were needed to accumulate its mass. Mauna Kea is slightly taller, but Mauna Loa is the most massive single mountain on Earth.

In another form, effusive eruptions send material out through hot spots or elongated fissures, such as a continental rift valley, forming extensive sheets of basaltic lava on the surface (see Figure 12.25). The Columbian Plateau of the northwestern United States, some 2–3 km thick, is the result of the eruption of **plateau basalts**, or *flood basalts* (foreground in Figure 12.27).

More than double the size of the Columbian Plateau is the Deccan Traps, which dominates west-central India. *Trap* is Dutch for "staircase," referring to the typical step-like eroded lavas. The Siberian Traps is more than twice

FIGURE 12.25 Tectonic settings of volcanic activity.
Magma rises and lava erupts from rifts, through crust above subduction zones, and where thermal plumes at hot spots break through the crust.
[After U.S. Geological Survey, *The Dynamic Planet* (Washington, DC: Government Printing Office, 1989).]

Continental rift valley

Plateau basalt

Lithosphere

Asthenosphere

FIGURE 12.26 Rift valley.
This rift is in Iceland, at the dramatic Gullfoss (waterfalls). Iceland is the only place where the Mid-Atlantic ridge system is seen above sea level. Iceland is growing as new material increases the size of this island country. [Photo by Bobbé Christopherson.]

(c)

FIGURE 12.27 Two expressions of volcanic activity.
In the foreground are plateau (flood) basalts characteristic of the Columbia Plateau in Oregon. Mount Hood, in the background to the west of the plateau, is a composite volcano, part of the Cascade Range of active and dormant volcanoes that result from subduction of the Juan de Fuca plate beneath the North American plate.
[Photo by author.]

(a)   (b)

FIGURE 12.28 Volcanic fountaining, boiling surf, and lava falls in Hawai'i.
From Earth's most active volcano: (a) Kīlauea erupts a fountain of lava from its East Rift spatter cone. (b) Lava flows reach the sea, adding new land to the big island of Hawai'i. (c) Continuing flows form spectacular lava falls over the pali in 2002.
[(a) Photo by Kepa Maly/U.S. Geological Survey; (b) and (c) photos by Bobbé Christopherson.]

FIGURE 12.30 Extrusive igneous rocks compared.
A dacite volcanic bomb (lighter rock on left, higher in silica) from Mount St. Helens and a basalt lava rock (darker rock on right, lower in silica) from Hawai'i. What does this coloration tell you about the rocks' history and composition? Find these rock types in Table 11.2. [Photos by Bobbé Christopherson.]

FIGURE 12.29 Kīlauea landscape.
erial photo of the newest land on the planet produced by the massive ows of basaltic lavas from the Kīlauea volcano, Hawai'i Volcano ational Park. In the distance, ocean water quenches the intense :00°C (2200°F) lava on contact, producing steam and hydrochloric :id mist. [Photo by Bobbé Christopherson.]

FIGURE 12.31 Volcanic island off the coast of New Zealand.
White's Isle, New Zealand (37.5° S, 177° E) actively erupted through 2001 without warning, sometimes emitting gas and steam, other times bursts of pyroclastics and ash. [Photo by Harvey Lloyd/Stock Market.]

Focus Study 12.1 *(continued)*

(a)

(b)

(c)

FIGURE 12.1.1 Mount St. Helens before and after the eruption—days and years later.
(a) Mount St. Helens prior to the 1980 eruption. (b) The devastated and scorched land shortly after the eruption in 1980. The scorched earth and tree blowdown area covered some 38,950 hectares (95,000 acres). (c) A landscape in recovery as life moves back in and takes hold to establish new ecosystems in 1999 along the Toutle River and debris flow. [(a) Photo by Pat and Tom Lesson/Photo Researchers; (b) photo by Krafft-Explorer/Photo Researchers, Inc.; (c) photo by Bobbé Christopherson.]

The mountain had been quiet since 1857. New activity began in March 1980, with a sharp earthquake registering a magnitude 4.1. The first eruptive outburst occurred one week later, beginning with a magnitude 4.5 quake and continuing with a thick black plume of ash and the development of a small summit crater. Ten days later, the first volcanic earthquake, called a *harmonic tremor*, registered on the many instruments that had been hurriedly placed around the volcano. Harmonic tremors are slow, steady vibrations, unlike the sharp releases of energy associated with tectonic earthquakes and faulting. Harmonic tremors told scientists that magma was on the move within the mountain.

Also developing was a massive bulge on the north side of the mountain. This bulge indicated the direction of the magma flow within the volcano. A bulge represents the greatest risk from a composite volcano, for it could signal a potential lateral burst through the bulge and across the landscape.

Early on Sunday, May 18, the area north of the mountain was rocked by a magnitude 5.0 quake, the strongest to date. The mountain, with its distended 245-m (800-ft) bulge, was shaken, but nothing happened. Then a second quake (magnitude 5.1) hit a few min-

utes later, loosening the bulge and launching the eruption. David Johnston, a volcanologist with the U.S. Geological Survey, was only 8 km (5 mi) from the mountain, servicing instruments, when he saw the eruption begin. He radioed headquarters in Vancouver, Washington, saying, "Vancouver, Vancouver, this is it!" He perished in the eruption. For a continuously updated live picture of Mount St. Helens from Johnston's approximate observation point and other specifics, go to **http://www.fs.fed. us/gpnf/mshnvm/volcanocam**. The camera is mounted below the roofline at the Johnston Ridge Observatory. The observatory is open to the public.

As the contents of the mountain exploded, a surge of hot gas (about 300°C, or 570°F), steam-filled ash, pyroclastics, and a *nuée ardente* (rapidly moving, very hot, explosive ash and incandescent gases) moved northward, hugging the ground and traveling at speeds up to 400 kmph (250 mph) for a distance of 28 km (17 mi).

The slumping north face of the mountain produced the greatest landslide witnessed in recorded history; about 2.75 km³ (0.67 mi³) of rock, ice, and trapped air, all fluidized with steam, surged at speeds approaching 250 kmph (155 mph). Landslide materials traveled for 21 km (13 mi) into the

valley, blanketing the forest, covering a lake, and filling the rivers below. A series of photographs, taken at 10-second intervals from the east looking west, records this sequence (Figure 12.1.2). The eruption continued with intensity for 9 hours, first clearing out old rock from the throat of the volcano and then blasting new material.

As destructive as such eruptions are, they also are constructive, for this is the way in which a volcano eventually builds its height. Before the eruption, Mount St. Helens was 2950 m (9677 ft) tall; the eruption blew away 418 m (1370 ft). Today, Mount St. Helens is building a lava dome within its crater. The thick lava rapidly and repeatedly plugs and breaks in a series of lesser dome eruptions that may continue for several decades. The dome already is more than 300 m (1000 ft) high, so a new mountain is being born from the eruption of the old. The dome is covered with a dozen survey benchmarks and several tiltmeters to monitor the status of the volcano. Swarms of minor earthquakes along the north flank of the mountain began in November 2001, marking this mountain as still unstable. Dome eruptions of varying intensities occurred during 2004 and 2005. Intensive scientific research and

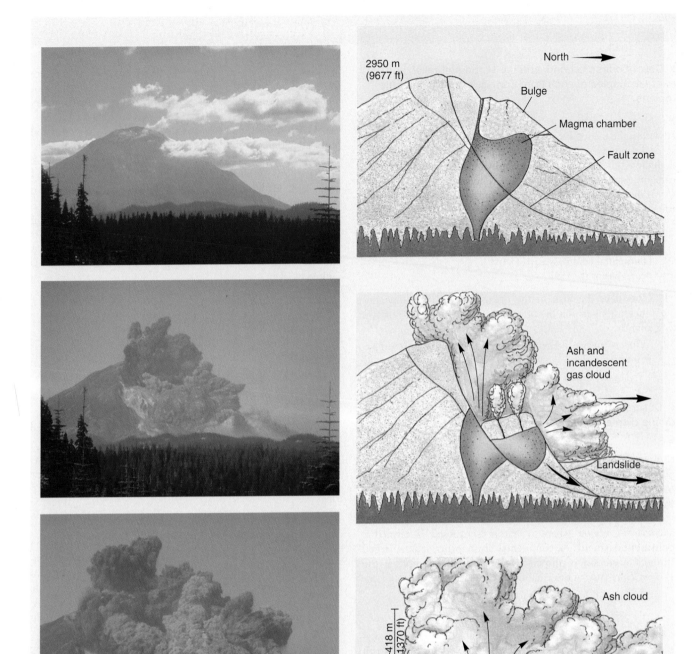

FIGURE 12.1.2 The Mount St. Helens eruption sequence and corresponding schematics. [Photo sequence by Keith Ronnholm. All rights reserved.]

monitoring has paid off, as every eruption since 1980 was successfully forecasted from days to as long as 3 weeks in advance, with the exception of one small eruption in 1984.

An ever-resilient ecology is recovering as plants and animals reclaim the devastated landscape. Please refer to the dramatic comparison photos from 1983 and 1999 in Figure 19.27. More than two decades have passed, yet strong interest in the area continues; scientists and more than 1 million tourists visit the Mount St. Helens Volcanic National Monument each year. For more information, see the Cascades Volcano Observatory at http://vulcan.wr.usgs.gov/ and specifically Mount St. Helens at http://vulcan.wr.usgs.gov/Volcanoes/MS/framework.html.

Summary and Review—Tectonics, Earthquakes, and Volcanism

▶ *Describe* first, second, and third orders of relief, and *relate* examples of each from Earth's major topographic regions.

Earth's surface is dramatically shaped by tectonic forces generated within the planet. **Relief** is the vertical elevation difference in a local landscape. The undulating physical surface of Earth, including relief, is called **topography**. Convenient descriptive categories are termed *orders of relief*. The coarsest level of landforms includes the **continental landmasses** and **ocean basins**; the finest comprises local hills and valleys.

relief (p. 359)
topography (p. 359)
continental landmasses (p. 359)
ocean basins (p. 359)

1. How does the map of the ocean floor (chapter-opening illustration) exhibit the principles of plate tectonics? Briefly analyze.
2. What is meant by an "order of relief"? Give an example from each order.
3. Explain the difference between relief and topography.

▶ *Describe* the several origins of continental crust, and *define* displaced terranes.

The continents are formed as a result of several processes, including upwelling material from below the crust and migrating portions of crust. The source of material generated by plate tectonic processes determines the behavior and the composition of the crust that results.

A continent has a nucleus of ancient crystalline rock called a *craton*. A region where a craton is exposed is termed a **continental shield**. As continental crust forms, it is enlarged through accretion of dispersed **terranes**. An example is the *Wrangellia terrane* of the Pacific Northwest and Alaska.

continental shield (p. 361)
terranes (p. 362)

4. What is a craton? Relate this structure to continental shields and platforms, and describe these regions in North America.
5. What is a migrating terrane, and how does it add to the formation of continental masses?
6. Briefly describe the journey and destination of the Wrangellia Terrane.

▶ *Explain* compressional processes and folding; and *describe* four principal types of faults and their characteristic landforms.

Folding, broad warping, and faulting deform the crust and produce characteristic landforms. Compression causes rocks to deform in a process known as **folding**, during which rock strata bend and may overturn. Along the *ridge* of a fold, layers *slope downward away from the axis*, which is an **anticline**. In the *trough* of a fold, however, layers *slope downward toward the axis*, called a **syncline**.

When rock strata are stressed beyond their ability to remain a solid unit, they express the strain as a fracture. Rocks on either side of the fracture are displaced relative to the other side

in a process known as **faulting**. Thus, *fault zones* are areas where fractures in the rock demonstrate crustal movement. At the moment of fracture, a sharp release of energy, called an **earthquake**, or *quake*, occurs.

When forces pull rocks apart, the tension causes a **normal fault**, sometimes visible on the landscape as a scarp, or escarpment. Compressional forces associated with converging plates force rocks to move upward, producing a **reverse fault**. A low-angle fault plane is referred to as a **thrust fault**. Horizontal movement along a fault plane that produces a linear rift valley is a **strike-slip fault**. In the U.S. interior west, the Basin and Range Province is an example of aligned pairs of normal faults and a distinctive horst-and-graben landscape. The term **horst** is applied to upward-faulted blocks; **graben** refers to downward-faulted blocks.

folding (p. 365)   reverse fault (p. 367)
anticline (p. 365)   thrust fault (p. 368)
syncline (p. 365)   strike-slip fault (p. 368)
faulting (p. 367)   horst (p. 371)
earthquake (p. 367)   graben (p. 371)
normal fault (p. 367)

7. Diagram a simple folded landscape in cross section, and identify the features created by the folded strata.
8. Define the four basic types of faults. How are faults related to earthquakes and seismic activity?
9. How did the Basin and Range Province evolve in the western United States? What other examples exist of this type of landscape?

▶ *Relate* the three types of plate collisions associated with orogenesis, and *identify* specific examples of each.

Orogenesis is the birth of mountains. An orogeny is a mountain-building episode, occurring over millions of years, that thickens continental crust. It can occur through large-scale deformation and uplift of the crust. It also may include the capture of migrating terranes and cementation of them to the continental margins and the intrusion of granitic magmas to form plutons.

Oceanic plate–continental plate collision orogenesis is now occurring along the Pacific coast of the Americas and has formed the Andes, the Sierra of Central America, the Rockies, and other western mountains. *Oceanic plate–oceanic plate* collision orogenesis produces either simple volcanic island arcs or more complex arcs such as Japan, the Philippines, the Kurils, and portions of the Aleutians. The region around the Pacific contains expressions of each type of collision in the **circum-Pacific belt**, or the **ring of fire**.

Continental plate–continental plate collision orogenesis is quite mechanical; large masses of continental crust, such as the Himalayan Range, are subjected to intense folding, overthrusting, faulting, and uplifting.

orogenesis (p. 371)
circum-Pacific belt (p. 372)
ring of fire (p. 372)

10. Define *orogenesis*. What is meant by the birth of mountain chains?
11. Name some significant orogenies.

12. Identify on a map Earth's two large mountain chains. What processes contributed to their development?

13. How are plate boundaries related to episodes of mountain building? Explain how different types of plate boundaries produce differing orogenic episodes and different landscapes.

14. Relate tectonic processes to the formation of the Appalachians and the Alleghany orogeny.

▶ *Explain* the nature of earthquakes, their measurement, and the nature of faulting.

Earthquakes generally occur along plate boundaries; major ones can be disastrous. Earthquakes result from faults, which are under continuing study to learn the nature of faulting, stress and the buildup of strain, irregularities along fault-plane surfaces, the way faults rupture, and the relationship among active faults. Earthquake prediction and improved planning are active concerns of *seismology*, the study of earthquake waves and Earth's interior. Seismic motions are measured with a **seismograph.**

Charles Richter developed the **Richter scale**, a measure of earthquake magnitude. A more precise and quantitative scale that assesses the *seismic moment* is now used, especially for larger quakes; this is the **moment magnitude scale.** The specific mechanics of how a fault breaks are under study, but the **elastic-rebound theory** describes the basic process. In general, two sides along a fault appear to be locked by friction, resisting any movement. This stress continues to build strain along the fault surfaces, storing elastic energy like a wound-up spring. When energy is released abruptly as the rock breaks, both sides of the fault return to a condition of less strain.

seismograph (p. 379)
Richter scale (p. 379)
moment magnitude scale (p. 379)
elastic-rebound theory (p. 380)

15. Describe the differences in human response between the two earthquakes that occurred in China in 1975 and 1976.

16. Differentiate between the Mercalli and moment magnitude and amplitude scales. How are these used to describe an earthquake? Why has the Richter scale been updated and modified?

17. What is the relationship between an epicenter and the focus of an earthquake? Give examples from the Loma Prieta, California, and Kobe, Japan, earthquakes.

18. What local soil and surface conditions in San Francisco severely magnify the energy felt in earthquakes?

19. How do the elastic-rebound theory and asperities help explain the nature of faulting? In your explanation, relate the concepts of stress (force) and strain (deformation) along a fault. How does this lead to rupture and earthquake?

20. Describe the San Andreas fault and its relationship to ancient sea-floor spreading movements along transform faults.

21. How is the seismic gap concept related to expected earthquake occurrences as described in News Report 12.3? Are any gaps correlated with earthquake events in the recent past? Explain.

22. What do you see as the biggest barrier to effective earthquake prediction?

▶ *Distinguish* between an effusive and an explosive volcanic eruption, and *describe* related landforms, using specific examples.

Volcanoes offer direct evidence of the makeup of the asthenosphere and uppermost mantle. A **volcano** forms at the end of a central vent or pipe that rises from the asthenosphere through the crust into a volcanic mountain. A **crater**, or circular surface depression, usually forms at the summit. Areas where magma is near the surface may heat groundwater, producing **geothermal energy.**

Eruptions produce **lava** (molten rock), gases, and **pyroclastics** (pulverized rock and clastic materials ejected violently during an eruption) that pass through the vent to openings and fissures at the surface and build volcanic landforms. Basaltic lava flows occur in two principal textures: **aa**, rough and sharp-edged lava, and **pahoehoe**, smooth, ropy folds of lava. Volcanic activity produces landforms such as a **cinder cone**, a small hill, or a large basin-shaped depression sometimes caused by collapse of a volcano's summit and called a **caldera.**

Volcanoes are of two general types, based on the chemistry and the viscosity of the magma involved. **Effusive eruption** produces a **shield volcano** (such as Kīlauea in Hawai'i) and extensive deposits of **plateau basalts**, or flood basalts. Magma of higher viscosity leads to an **explosive eruption** (such as Mount Pinatubo in the Philippines), producing a **composite volcano.** Volcanic activity has produced some destructive moments in history but constantly creates new seafloor, land, and soils.

volcano (p. 385)   cinder cone (p. 386)
crater (p. 385)   caldera (p. 386)
geothermal energy (p. 385)   effusive eruption (p. 386)
lava (p. 385)   shield volcano (p. 387)
pyroclastics (p. 385)   plateau basalts (p. 387)
aa (p. 385)   explosive eruption (p. 390)
pahoehoe (p. 385)   composite volcano (p. 390)

23. What is a volcano? In general terms, describe some related features.

24. Where do you expect to find volcanic activity in the world? Why?

25. Compare effusive and explosive eruptions. Why are they different? What distinct landforms are produced by each type? Give examples of each.

26. Describe several recent volcanic eruptions.

NetWork

The *Geosystems* Home Page provides online resources for this chapter on the World Wide Web. To begin: Once on the Home Page, click on this textbook, scroll the Table of Contents menu, select this chapter, and click "Begin." You will find self-tests that are graded, review exercises, specific updates for items in the chapter, and many links to interesting related pathways on the Internet under "Destinations." *Geosystems* is at http://www. prenhall.com/christopherson.

Critical Thinking

A. Using the topographic regions map in Figure 12.3, assess the region within 100 km (62 mi) and 1000 km (620 mi) of your campus. Describe the topographic character and the variety of relief within these two regional scales. You may want to consult local maps and atlases in your analysis. Do you perceive that this type of topographic region influences lifestyles? Economic activities? Transportation? History of the region?

B. Determine from library, Internet, and local agencies the seismic potential of the region in which your campus is sited. If a hazard exists, does it influence regional planning? Availability of property insurance? If appropriate, are there existing disaster or emergency plans for your campus? For your community, state, or province? Explain any other significant issues you found in working on this critical thinking challenge.

Career Link 12.1

Travis Heggie, Geographer/Park Ranger/Ph.D.

My wife and I arrived at the end of Chain of Craters Road in Hawai'i Volcanoes National Park (HAVO) on the Big Island to witness heightened lava flow activity, which by July 2002 had overrun the end of the road. This is part of the ongoing 1983 eruption of the Pu'u O'o Vent of the Kīlauea volcano. Our excitement built as we scrambled across the uneven wreckage of basaltic lava from earlier flows—glassy, crunchy sounds beneath our shoes. We could feel the intense heat of the lava a kilometer distant, as it flowed over the pali (steep slope) and made its way along and beneath the surface toward the ocean.

There was Travis, helping, guiding, instructing anyone in need or with questions. His enthusiasm glowed like the lava in the distance. Travis is a park ranger on the volcano. And this man is a geographer with an excitement and joy that is contagious (Figure 12.1.1).

Travis was born in Alberta, Canada, about 60 mi (96 km) from Waterton-Glacier International Peace Park. Travis and his family soon moved to Cedar City, Utah, where his father was attending Southern Utah University. His dad began his teaching career there. "My first six years were in Utah. I have fond memories of the canyons and red rock landscapes of southern Utah." The family moved back to Alberta when Travis was seven. "My dad was a teacher. I come from a family of teachers. I want to teach."

After high school, Travis went to South Korea on a two-year volunteer mission for his church, which required him to learn the Korean language. Travis said, "My interest in geography came during my mission. I was a small-town kid, thrown out into the world. I wanted to know where places were, why they were there, and how to communicate with local people." Following his mission, he met and married his wife Tracy, and they returned to South Korea to work at a training institute. His facility with the Korean language was important in helping diplomats prepare for overseas assignments.

The couple returned to the states in 1994 and enrolled at Rick's College, Idaho. A world regional geography class changed the course of his life. "When I took that class, I knew geography was my calling; that

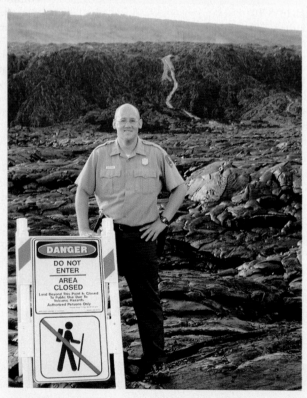

FIGURE 12.1.1 Travis Heggie, Park Ranger, Geographer. Park Ranger Travis standing at a warning barricade, placed to keep the public at a safe distance from active lava flows in Hawai'i Volcanoes National Park. [Photo by Bobbé Christopherson.]

class got me! This was what I wanted to learn, what I wanted to be. I realized I could combine my interests in world events, climate studies, biology, ecology, and geology into one discipline called 'geography,' for here I could examine the 'big picture,' because geography seemed to have it all, through the lens of spatial analysis. One course was an Earth science course for teachers. The teacher had spent time at the Hawai'i Volcanoes National Park. He showed a lot of slides and got me interested in Earth processes—this set up my next move after Rick's College. We went to Hawai'i. I think the geography department at the University of Hawai'i at Hilo is one of the best places in the world to do geography! Which I did for the next four years."

Travis said, "The professors at Hilo were so dynamic, such inspiring teachers and leaders. The work was rigorous but also rewarding." Travis started working with the National Park Service (NPS) while at the University of Hawai'i. He answered an ad for a student-training program and was hired. He wants to always keep this connection in his career.

In graduate school at Texas A&M University, Travis majored in physical geography, emphasizing biogeogra-

phy. His Master's thesis was "Rainforest Composition and Succession on a South Pacific Island (Island of Tutuila, American Samoa)," completed in 2001. Travis explained, "I have permanent research plots on the island and continue to gather data on several forest communities from sea level up to 245 m (800 feet) elevation." Travis analyzed abandoned agricultural land and old plantations, looking at what species were regenerating. He analyzed 18 years of land-use data and put together a broad picture of succession and regeneration.

Travis finished his Ph.D. at Texas A&M in late 2004, although he still finds time to work on the lava flows for the NPS in Hawai'i. His wife Tracy completed a Master's in geography at Texas A&M as well, researching water quality in Colorado.

Travis's dissertation topic is "Natural Hazards, Risk, and Emergency Management: A Case Study of Hawai'i Volcanoes National Park." Travis had a risk/hazard analysis advisor on his Ph.D. committee, a geographer who specializes in environmental impacts, and a professor from the Urban Planning Department. For his dissertation Travis obtained accident and injury data back to 1992 and he interviewed more than 800 tourists in the HAVO. He found that people not

familiar with this environment are most susceptible to injury; not adjusting hazard perception is common in human behavior. Injuries range from abrasions and blisters, to sprains, sunburns, cuts, and fractures, with 5% suffering thermal burns where the lava enters the ocean. The holiday mindset is powerful: "They are in Hawai'i and have to see the 'red glowing stuff,' no matter what! And local business and politics encourage this, because it represents tourism dollars. I found that only 24% of people actually stop and read at least one sign. If there is a sign that will work psychologically, I recommend a big STOP sign, with a warning message and posted accident reports."

Travis observed, "People tend to feel safe in a national park. Whether its lava in Hawai'i or grizzly bears in Yellowstone, people don't readily perceive hazards in a national park—a prevailing attitude according to my perception surveys. The reality is that national parks are inherently dangerous places, because they tend to be in areas of dramatic expressions of nature and Earth systems, yet we had difficulty getting everyone to respond to warnings."

With his Ph.D. in hand Travis, his wife, and new child are in Washington, DC. He accepted a new post for the National Park Service to institute and operate the first visitor injury program for the NPS/Department of Interior. He will be able to apply his research and experience to help the public as they experience physical geography in the national parks.

Travis's own words say it best as to what he wants readers of this Career Link to remember. Without hesitation, he implored, "Do Geography! I literally mean 'do it all.' You need to have an understanding of both human and physical geography, as well as climatology, economic geography, and regional courses, and an ability with analytical tools, such as statistics, field methods, and GIS."

This man is an enthusiastic geographer and skilled lifelong student of the world. Do geography!

FIGURE 12.1.2 Lava flows run over highway, July 2002.
The Chain of Craters Road is engulfed in 1200°C (2200°F) lava, incinerating the asphalt, releasing toxic smoke. The former ranger station hut was moved a kilometer from the end of the road to avoid the inferno. [Photo by Travis Heggie.]

Pahoehoe lava flows (more fluid than aa lava, with a ropy, folded texture) in Iceland. Physical and chemical weathering processes at work for thousands of years are breaking down the basalt. Lichens participate in the disintegration of the rock as soils slowly form. Recall from Table 11.2 that basalt and other mafic minerals are less resistant to weathering than felsic minerals such as granite and gabbro. The inset photo shows the landscape near this lava flow. [Photos by Bobbé Christopherson.]

13

Weathering, Karst Landscapes, and Mass Movement

▶ Key Learning Concepts

After reading the chapter, you should be able to:

▶ *Define* the science of geomorphology.

▶ *Illustrate* the forces at work on materials residing on a slope.

▶ *Define* weathering, and *explain* the importance of the parent rock and joints and fractures in rock.

▶ *Describe* frost action, crystallization, pressure-release jointing, and the role of freezing water as physical weathering processes.

▶ *Describe* the susceptibility of different minerals to the chemical weathering processes called hydration, hydrolysis, oxidation, carbonation, and solution.

▶ *Review* the processes and features associated with karst topography.

▶ *Portray* the various types of mass movements, and *identify* examples of each in relation to moisture content and speed of movement.

A benefit you receive from a physical geography course is a new appreciation of the scenery. Whether you go by foot, bicycle, car, train, or plane, travel is an opportunity to experience Earth's varied landscapes and to witness the active processes that produce them.

In the last two chapters, we discussed the endogenic (internal) processes of our planet and how they produce landforms. However, as the landscape is formed, several exogenic (external) processes simultaneously wear and waste it. This ongoing struggle between tectonics and climate shapes the landscape upon which the biosphere drama is played.

Here we begin a five-chapter examination of exogenic processes at work on the landscape. This chapter examines weathering and mass movement of the

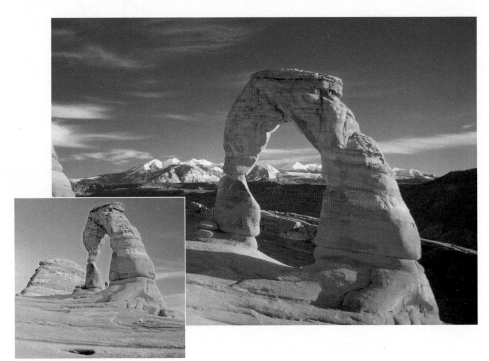

FIGURE 13.1 Weathered landform.
Delicate Arch—a dramatic example of differential weathering in Arches National Park, Utah. Resistant rock strata at the top of the structure have helped preserve the arch beneath as surrounding rock was eroded away. Note the person standing at the base in the inset photo for a sense of scale. [Photos by author.]

lithosphere. The next four chapters look at specific exogenic agents and their handiwork—river systems, wind-influenced landscapes, deserts in water-deficit regions, coastal processes and landforms, and regions worked by ice and glaciers. Whether you enjoy time along a river, love the desert or the waves along a coastline, or live in a place where glaciers once carved the land, you will find something of interest in these chapters.

In this chapter: We look at physical (mechanical) and chemical weathering processes that break up, dissolve, and generally reduce the landscape. Such weathering releases essential minerals from bedrock for soil formation or enrichment. Perhaps you live in a region that has caves and caverns. Water has dissolved enormous underground worlds of mystery and darkness, yet many caves remain undiscovered. In addition, we examine mass movement processes that continually operate in and upon the landscape. A news broadcast may describe an avalanche in Colombia; mudflows in Los Angeles or Peru; tragic landslides in China, Turkey, Indonesia, or Colombia; massive rockfall in Yosemite; or prediction of an imminent mudflow down the slopes of Mount Ruapehu volcano in New Zealand. We seek to understand the physical processes leading to these events.

Landmass Denudation

Geomorphology is the science of landforms—their origin, evolution, form, and spatial distribution. This science is an important aspect of physical geography. **Denudation** is any process that wears away or rearranges landforms. The principal denudation processes affecting surface materials include *weathering*, *mass movement*,

erosion, *transportation*, and *deposition*, as produced by the agents of moving water, air, waves, and ice—all influenced by the pull of gravity.

Interactions between the structural elements of the land and denudation processes are complex. They represent a continuing struggle between Earth's internal and external processes, between the resistance of materials and weathering and erosional processes. The 15-story-tall Delicate Arch in Utah is dramatic evidence of this struggle (Figure 13.1). Differing resistances of the rocks, coupled with variations in the processes at work on the rock, are carving this delicate sculpture—an example of **differential weathering**, where a more resistant cap rock protects supporting strata below.

Ideally, endogenic processes build *initial landscapes*, whereas exogenic processes develop *sequential landscapes* of low relief, gradual change, and stability. Several hypotheses were proposed to model denudation processes and to account for the appearance of the landscape in its developmental stages.

Geomorphic Models of Landform Development

Famed American geographer and geomorphologist William Morris Davis (1850–1934) proposed the erosion cycle, or *geomorphic cycle model*. Davis theorized that a landscape undergoes initial uplift that is accompanied by erosion or removal of materials. From the uplifted landscape, streams begin flowing more rapidly, cutting more energetically, both headward (upstream) and deeper, forming valleys. Davis believed that slope angle is gradually reduced and that ridges and divides become rounded and lowered over time. According to Davis's cyclic model, the landscape eventually evolves into an old erosional surface (discussed further in Chapter 14).

Davis's theory, which helped launch the science of geomorphology and was innovative at the time, did not account for the processes being observed in modern geomorphology studies. Academic support for his cyclic and evolutionary landform development model declined by 1960. Although not generally accepted today, his thinking about the evolution of landscapes is influential and some of his terminology is still used.

Dynamic Equilibrium View of Landforms

Most geomorphologists prefer the **dynamic equilibrium model**, which emphasizes a balance among force, form, and process. The dynamic equilibrium model summarizes this balancing act between tectonic uplift and reduction by weathering and erosion, between the resistance of rocks and the ceaseless attack of weathering and erosion. A dynamic equilibrium demonstrates a trend over time. According to current thinking, landscapes in a dynamic equilibrium feature ongoing adaptations to the ever-changing conditions of rock structure, climate, local relief, and elevation. In a sense, every landscape is unique unto itself.

A landscape is an open system, with highly variable inputs of energy and materials: Uplift creates the *potential energy of position* above sea level and therefore disequilibrium, an imbalance, between relief and energy. The Sun provides radiant energy that converts into *heat energy*. The hydrologic cycle imparts *kinetic energy* through mechanical motion. *Chemical energy* is made available from the atmosphere and various reactions within the crust. In response to this input of energy and materials, landforms constantly adjust toward *equilibrium*.

Endogenic events (such as earthquakes and volcanic eruptions) or exogenic events (such as heavy rainfall or forest fire) may provide new sets of relationships for the landscape. Following such destabilizing events, a landform system arrives at a **geomorphic threshold**—the point at which there is enough energy to overcome resistance against movement. At this threshold, the system breaks through to a new equilibrium as the landform adjusts. The pattern over time follows a sequence: (1) equilibrium stability (fluctuating around some average), (2) a destabilizing event, (3) a period of adjustment, and (4) development of a new and different condition of equilibrium stability.

The disturbed hillslope in Figure 13.2 is in the midst of compensating adjustment. The failure of saturated slopes caused a landslide into the river and set a disequilibrium condition. As a consequence, the new dam of material threw the stream into disequilibrium between its flow and sediment load.

Slow, continuous-change events, such as soil development and erosion, tend to maintain an approximate

FIGURE 13.2 A slope in disequilibrium.
Unstable, saturated soils gave way, leaving a debris dam partially blocking the river.
[Photo by author.]

equilibrium condition. Dramatic events such as a major landslide or dam collapse require longer recovery times before equilibrium is reestablished. (Figure 1.5 graphically illustrates both the distinction between a steady-state equilibrium and a dynamic equilibrium and the occurrence of a geomorphic threshold.)

Slopes Material loosened by weathering is susceptible to erosion and transportation. However, if gravity is to move loosened material downslope, the agents of erosion must overcome the forces of friction, inertia (the tendency of objects at rest to remain at rest), and the cohesion of particles to each other (Figure 13.3a). If the slope angle is steep enough for gravity to overcome frictional forces, or if material is dislodged by the impact of raindrops, hail, falling branches, moving animals, wind, trail bikes, off-road vehicles, logging, or construction equipment, then particle erosion and transport downslope occur.

Slopes or *hillslopes* are curved, inclined surfaces that form the boundaries of landforms. Figure 13.3b illustrates basic slope components that vary among slopes with conditions of rock structure and climate. Slopes generally feature an upper *waxing slope* near the top (*waxing* means "increasing"). This convex surface curves downward and grades into the *free face* below. The presence of a free face indicates an outcrop of resistant rock that forms a steep scarp or cliff.

Downslope from the free face is a *debris slope*, which receives rock fragments and materials from above. The condition of a debris slope reflects the local climate. In humid climates, continually moving water carries material away, lowering the debris slope. But in arid climates, debris slopes accumulate. A debris slope transitions into a *waning slope*, a concave surface along the base of the slope.

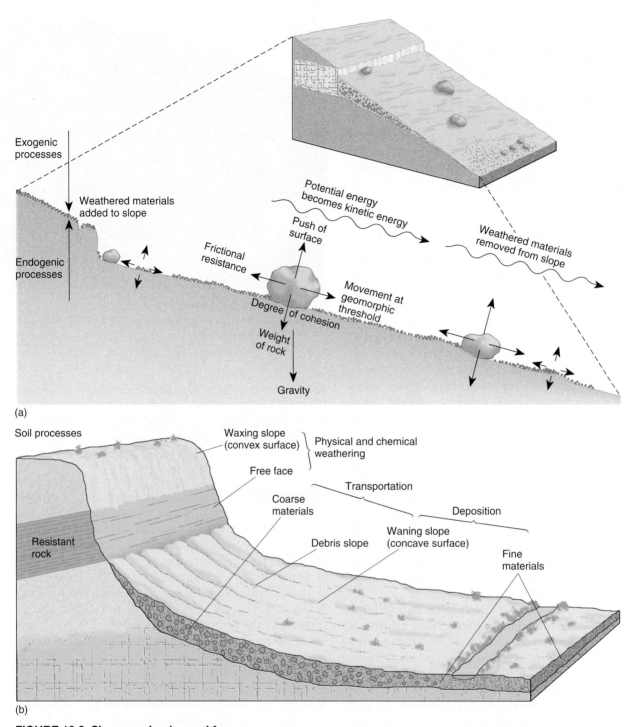

FIGURE 13.3 Slope mechanics and form.
(a) Directional forces (noted by arrows) act on materials along an inclined slope. (b) The principal elements of a slope.

This waning surface of erosional, coarser materials gently slopes at a continuously decreasing angle to the valley floor, where finer materials are deposited.

A slope is an open system seeking an *angle of equilibrium*. Conflicting forces work simultaneously on slopes to establish an optimum compromise incline that balances these forces. You can identify these slope components and conditions on the actual hillslope shown in Figure 13.4. When any condition in the balance is altered, all forces on the slope compensate by adjusting to a new dynamic equilibrium.

The relation between rates of weathering and breakup of slope materials, coupled with the rates of mass movement and material erosion, forms the shape of the slope. A slope is *stable* if its strength exceeds these denudation processes and *unstable* if materials are weaker than these processes. Why are hillslopes shaped in certain ways? How do slope elements evolve? How do hillslopes behave during rapid, moderate, or slow uplift? These are topics of active scientific study and research.

Now, with the concepts of landmass denudation, dynamic equilibrium, and slope development in mind, let

FIGURE 13.4 A hillslope example.
Compare this slope with the components mentioned in Figure 13.3. Note the role of the rock outcrop in interrupting the slope and the rock fragments that are loosened from the outcrop by frost action. [Photo by author.]

us examine specific processes that operate to wear away landforms.

Weathering Processes

Weathering processes attack rocks at Earth's surface and to some depth below the surface. **Weathering** processes either disintegrate rock into mineral particles or dissolve them in water. Weathering processes are both physical (mechanical) and chemical and the interplay of the two is complex. There is a synergy, a combined action, between physical and chemical weathering as the suite of processes work the rock, such as the interplay between hydration and hydrolysis now combined under chemical weathering in this edition.

Weathering does not transport the materials; it simply generates them for erosion and transport by the agents of water, wind, waves, and ice—all influenced by gravity. In most areas, the upper surface of bedrock undergoes continual weathering, creating broken-up rock called **regolith**. Loose surface material comes from further weathering of regolith and from transported and deposited regolith (Figure 13.5a). In some areas, regolith may be missing or undeveloped, thus exposing an outcrop of unweathered bedrock. The thickness of the soil cover depends on a balance between soil production rates and erosion (removal of soil particles)—an equilibrium between competing processes.

Bedrock is the *parent rock* from which weathered regolith and soils develop. Wherever a soil is relatively youthful, its parent rock is traceable through similarities in composition. For example, the sand in Figure 13.5c derives its color and character from the parent rock in the background, just as the sediments on Mars derive their char-

acteristics from the weathered parent material present (Figure 13.5d). Such sandy, unconsolidated fragmental material, known as **sediment**, combines with weathered rock to form the **parent material** from which soil evolves.

Factors Influencing Weathering Processes

Weathering is greatly influenced by the character of the bedrock: hard or soft, soluble or insoluble, broken or unbroken. *Jointing* in rock is important for weathering processes. **Joints** are fractures or separations in rock that occur without displacement of the sides (as would be the case in faulting). The presence of these usually flat spaces increases the surface area of rock exposed to both physical and chemical weathering.

Important controls on weathering rates are climatic elements—precipitation, temperature, and freeze–thaw cycles. There is a relation among climate (annual precipitation and temperature), physical weathering, and chemical weathering processes. In general, physical weathering dominates in drier, cooler climates, whereas chemical weathering dominates in wetter, warmer climates. Extreme dryness reduces weathering rates, as is experienced in desert climates (*low- and mid-latitude latitude hot desert*). In the hot, wet, tropical and equatorial rain forest climates (*tropical rain forest*), most rocks weather rapidly, and the weathering extends deep below the surface. Also related to climate and significant weathering is the position of the water table and water movement on and within rock structures and the subsurface environment.

Another control over weathering rates is the *geographic orientation* of a slope—whether it faces north, south, east, or west. Orientation controls the slope's exposure to Sun, wind, and precipitation. Slopes facing away from the Sun's rays tend to be cooler, moister, and more vegetated than slopes in direct sunlight. This effect of orientation is especially noticeable in the middle and higher latitudes.

Vegetation is also a factor in weathering. Although vegetative cover can protect rock by shielding it from raindrop impact and providing roots to stabilize soil, it also produces organic acids from the partial decay of organic matter; these acids contribute to chemical weathering. Plant roots can enter crevices and break up a rock, exerting enough pressure to drive rock segments apart, thereby exposing greater surface area to other weathering processes (Figure 13.6). You may have observed how tree roots can heave the sections of a sidewalk or driveway sufficiently to raise and crack the concrete.

The scale at which we analyze weathering processes is important. Research at *microscale* levels reveals greater complexity in the relation of climate and weathering. At the small scale of actual reaction sites on the rock surface, both physical and chemical weathering processes can occur across varied climate types. Hygroscopic water (a molecule-thin water layer on soil particles) and capillary water (soil water) activate chemical weathering processes, even in the driest landscape. (Review sections in Chapter 9 for these water types.)

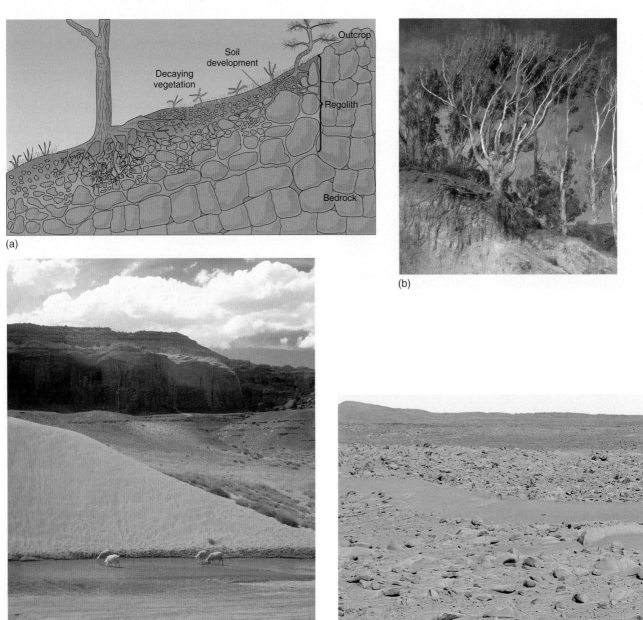

FIGURE 13.5 Regolith, soil, and parent materials.
(a) A cross section of a typical hillside. (b) A cliff exposes hillside components. (c) These reddish-colored dunes in the Navajo Tribal Park, near the Utah–Arizona border, derive their color from the red-sandstone parent materials in the background. (d) This image, from a larger panoramic scene, was made by the Mars Exploration Rover *Spirit* on March 12, 2004. Looking toward the Columbia Hills, you can see weathered rocks and windblown sand in approximate true colors. [Photos (b) and (c) by author; Mars image (d) courtesy of NASA/JPL and Cornell University.]

Imagine all the factors that influence weathering rates as operating in concert: rock composition and structure (jointing), climate influence (precipitation and temperature), groundwater and water movement, slope orientation, vegetation, and microscopic boundary-layer conditions at reaction sites. We separate weathering processes here for convenience of study, but they operate in a complex synergy. Of course in all of this, *time* is the crucial factor, for these processes require long periods of time to operate.

Physical Weathering Processes

When rock is broken and disintegrated without any chemical alteration, the process is called **physical weathering** or *mechanical weathering*. By breaking up rock,

FIGURE 13.6 Organic physical weathering.
(a) Can you find the tree root growing in and on the jointing fracture in the rock? (b) Roots exert a force on the sides of this joint in the rock. [Photos by author.]

(b)

physical weathering produces more surface area on which chemical weathering may operate. A single rock that becomes broken into eight pieces has doubled its surface area susceptible to weathering processes. We look briefly at three physical weathering processes: frost action, crystallization, and pressure-release jointing.

Frost Action When water freezes, its volume expands as much as 9% (see Chapter 7). Such expansion creates a powerful mechanical force called **frost action**, or *freeze–thaw action*, which can exceed the tensional strength of rock. Repeated freezing (expanding) and thawing (contracting) of water breaks rocks apart (Figure 13.7). Freezing actions are important in the humid microthermal climates (*humid continental* and *subarctic*) and polar climates, and they occur at higher elevations in mountains worldwide in the highland climates. In arctic and subarctic climates, frost action dominates soil conditions (discussed in further detail in Chapter 17).

The work of ice begins in small openings, gradually expanding until rocks are cleaved, or split. Figure 13.8

FIGURE 13.7 Climate and weathering.
Frost action shattered this granite rock; ice expansion forced the rock segments apart. [Photo by Bobbé Christopherson.]

shows blocks of rock on which this *joint-block separation* occurs along existing joints and fractures. This weathering action, called *frost-wedging*, pushes portions of the rock apart. Cracking and breaking create varied shapes in the rocks, depending on the rock structure. In Figure 13.8a, the softer supporting rock underneath the slabs is weathering at a faster pace in *differential weathering*.

In Figure 13.8c, frost wedging loosened rock from these cliffs. The angular pieces of rock fragments cascade down and form a **talus slope**, which is a poorly sorted, cone-shaped deposit of debris at the base of a steep slope. Several talus cones constitute such a talus slope in the photograph.

Frost action was important to various cultures as a force to quarry rock. Pioneers in the early American West drilled holes in rock, poured water in the holes, and then plugged them. During the cold winter months, expanding ice broke off large blocks along lines determined by the drill-hole patterns. In spring, the pioneers hauled the blocks to town for construction material. Frost action also produces unwanted fractures that damage road pavement and burst water pipes. Perhaps you have noticed rough and broken highways in areas that experience freezing temperatures. The pavement breaks into chunks each winter and develops potholes.

Spring can be a risky time to venture into mountainous terrain. As rising temperatures melt the winter's ice, newly fractured rock pieces fall without warning and may even start rock slides. The falling rock pieces may physically shatter on impact—another form of physical weathering (Figure 13.9). One such rockfall in Yosemite National Park in 1996 involved a 670-m (2200-ft) crashing drop of a 162,000-ton granite slab at 260 kmph (160 mph). The impact shattered the rock, covering 50 acres in powdered rock and felling 500 trees.

Crystallization Especially in arid climates, dry weather draws moisture to the surface of rocks. As the water evaporates, dissolved minerals in the water grow crystals. Over time, as the crystals grow and enlarge, they exert a force

(a)

(b)

(c)

FIGURE 13.9 Rockfall.
Shattered rock debris from a large rockfall in Yosemite National Park. A larger rockfall involving 162,000 tons of granite shocked Yosemite in July 1996, and another in 1999. Freshly exposed rock shows where the rockfall originated. [Photo by author.]

FIGURE 13.8 Physical weathering examples.
(a) Physical weathering along joints in rock produces discrete blocks in the back country of Canyonlands National Park, Utah. (b) Joint-block separation in slate at Alkehornet, Isfjord, on Spitsbergen Island in the Arctic Ocean where freezing is intense. (c) Frost action loosens and frees rock that falls to the base of the cliff, accumulating as a talus slope, along Road 258 in southwestern Norway. [Photos by (a) author; (b) and (c) Bobbé Christopherson.]

great enough to spread apart individual mineral grains and begin breaking up the rock. Such *crystallization*, or *salt-crystal growth*, is a form of physical weathering.

In the Colorado Plateau of the Southwest, salty water slowly flows from rock strata. As this salty water evapo-

rates, crystallization loosens the sand grains. Subsequent erosion and transportation by water and wind complete the sculpturing process. Deep indentations develop in sandstone cliffs, especially where the sandstone lies above an impervious layer such as shale. More than 1000 years ago, Native Americans built entire villages in these weathered niches at several locations, including Mesa Verde in Colorado and Arizona's Canyon de Chelly (pronounced "canyon duh shay," Figure 13.10).

Pressure-Release Jointing Recall from Chapter 11 how rising magma that is deeply buried and subjected to high pressure forms intrusive igneous rocks called plutons. These plutons cool slowly and produce coarse-grained, crystalline granitic rocks. As the landscape is subjected to uplift, the regolith overburden is weathered, eroded, and transported away, eventually exposing the pluton as a mountainous batholith (a plutonic batholith is illustrated in Figure 11.7).

As the tremendous weight of overburden is removed from the granite, the pressure of deep burial is relieved. Over millions of years, the granite slowly responds with an

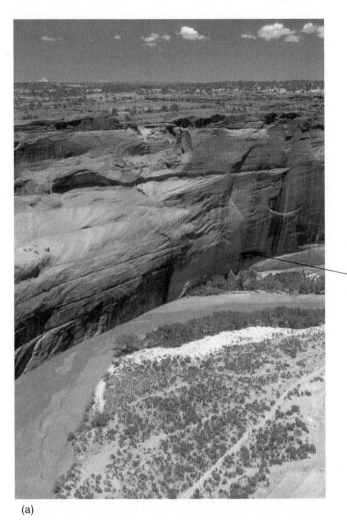

(a)

FIGURE 13.10 Physical weathering in sandstone.
(a) Cliff dwelling site in Canyon de Chelly, Arizona, occupied by the Anasazi people until about 900 years ago. The niche in the rock was formed partially by crystallization, which forced apart mineral grains and broke up the rock. The dark streaks on the rock are thin coatings of desert varnish, composed of iron oxides with traces of manganese and silica. (b) Water and an impervious rock layer helped concentrate weathering processes in a niche in the overlying sandstone. [Photo by author.]

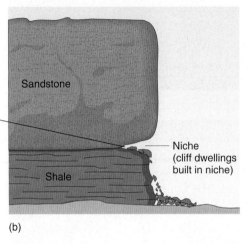

(b)

enormous physical heave. In a process known as *pressure-release jointing*, layer after layer of rock peels off in curved slabs or plates, thinner at the top of the rock structure and thicker at the sides. As these slabs weather, they slip off in a process called **sheeting**. This *exfoliation process* creates arch-shaped and dome-shaped features on the exposed landscape, sometimes forming an **exfoliation dome** (Figure 13.11). Such domes are probably the largest weathering features on Earth (in areal extent).

Chemical Weathering Processes

Chemical weathering is the actual decomposition, or chemical change, of minerals in rock. Chemical weathering involves reactions between air and water and minerals in rocks. Minerals may combine with oxygen or carbon dioxide from the atmosphere, or they may dissolve and combine with water. Water is important because of its general availability and its tremendous ability to dissolve. Chemical breakdown hastens as both temperature and precipitation increase.

Although individual minerals vary in susceptibility, all rock-forming minerals are at least slightly responsive to chemical weathering. A familiar example is the eating away of cathedral façades and the etching of tombstones by acidic precipitation. Figure 13.12a illustrates chemical weathering along joints in granite as chemical processes

dissolve its component minerals. This dissolution in turn opens spaces for frost action.

As an example of the way chemical weathering attacks rock, consider **spheroidal weathering**. The sharp edges and corners of rocks are rounded as the alteration of minerals progresses through the rock. Joints in the rock offer more surfaces of opportunity for more weathering. Water penetrates joints and fractures and dissolves the rock's weaker minerals or cementing materials. A boulder can be attacked from all sides, shedding spherical shells of decayed rock, like the layers of an onion. The resulting rounded edges are the basis for the name *spheroidal*. Spheroidal weathering of rock resembles exfoliation, but it does not result from pressure-release jointing.

We now look at chemical weathering processes: hydration and hydrolysis, oxidation, and carbonation and solution.

Hydration and Hydrolysis   Two processes involving water, one of simple combination with a mineral and the other of chemical reaction with a mineral, work to decompose rock. In this edition these are grouped under chemical weathering because both involve chemical actions.

Hydration, meaning "combination with water," involves little chemical change. Water becomes part of the chemical composition of the mineral (such a hydrate is gypsum, which is hydrous calcium sulfate: $CaSO_4 \cdot 2H_2O$).

(a)

(b)

(c)

(d)

FIGURE 13.11 Exfoliation in granite.
Exfoliation processes loosen slabs of granite, freeing them for further weathering and downslope movement. (a) Great arches form in the White Mountains of New Hampshire. (b) Exfoliated layers of rock are visible in characteristic dome formations in granites. The loosened slabs of rock are susceptible to further weathering and downslope movement. This view is from the east side of Half Dome in Yosemite National Park, California. (c) Half Dome perspective from the west; relief is approximately 1500 m (5000 ft) from the top of the dome to the glaciated valley below. (d) Rock sheeting exposed along Beverly Sund, Nordaustlandet Island in the Arctic Ocean. [Photos by (a) and (d) Bobbé Christopherson, (b) and (c) by author.]

When some minerals hydrate, they expand, creating a strong mechanical effect that stresses the rock, forcing grains apart.

A cycle of hydration and dehydration can lead to granular disintegration and further susceptibility of the rock to chemical weathering. Hydration works together with carbonation and oxidation to convert feldspar, a common mineral in many rocks, to clay minerals and silica. The hydration process is also at work on the sandstone niches shown in the cliff-dwelling photo.

When minerals chemically react with water, the process is **hydrolysis**. Hydrolysis is a decomposition process that breaks down silicate minerals in rocks. Compared with hydration, in which water combines with minerals in the rock, the hydrolysis process involves water in chemical reactions to produce different compounds.

For example, the weathering of feldspar minerals in granite can be caused by a reaction to the normal mild acids dissolved in precipitation:

feldspar (K, Al, Si, O) + carbonic acid and water $\rightarrow$
residual clays + dissolved minerals + silica

So the by-products of chemical weathering of feldspar in granite include clay (such as kaolinite) and silica. As clay forms from some minerals in the granite, quartz (SiO_2) particles are left behind. The resistant quartz may wash downstream, eventually becoming sand on some distant beach. Clay minerals become a major component in soil and in shale, a common sedimentary rock.

When weaker minerals in rock are changed by hydrolysis, the interlocking crystal network breaks down, so

(a)

(b)

(c)

FIGURE 13.12 Chemical weathering and spheroidal weathering.
(a) Chemical weathering processes act on the joints in granite to dissolve weaker minerals, leading to a rounding of the edges of the cracks. (b) Rounded granite outcrop demonstrates spheroidal weathering and the disintegration of rock. (c) The rugged, weathered Alabama Hills, with Mount Whitney, California, in the background, provides scenic backdrops for many commercials and movies. You can imagine a movie chase scene raising clouds of dust! [Photos by Bobbé Christopherson.]

the rock fails and *granular disintegration* takes place. Such disintegration in granite may make the rock appear etched, corroded, and softened (Figure 13.12b).

In Table 11.2, the second line shows resistance to chemical weathering in igneous rocks. On the ultramafic end of the table (right side), the low-silica minerals olivine and peridotite are most susceptible to chemical weathering. Stability gradually increases toward the high-silica minerals such as feldspar. On the far left side of the table, quartz is resistant to chemical weathering. You can see that, because of the nature of the constituent minerals, basalt weathers faster chemically than does granite.

Oxidation   Another example of chemical weathering occurs when certain metallic elements combine with oxygen to form oxides. This is a chemical weathering process known as **oxidation**. Perhaps the most familiar oxidation form is the "rusting" of iron in rocks or soil that produces a reddish brown stain of iron oxide (Fe_2O_3). We have all left a tool or nails outside only to find them weeks later, coated with iron oxide. The rusty color is visible on the surfaces of rock and in heavily oxidized soils such as those in the southeastern United States, southwestern deserts, or the tropics (Figure 13.13). Here is a simple oxidation reaction in iron:

iron (Fe) + oxygen (O_2) → iron oxide (hematite; Fe_2O_3)

As iron is removed from the minerals in a rock, the disruption of the crystal structures in the rock's minerals makes the rock more susceptible to further chemical weathering and disintegration.

Carbonation and Solution   The third form of chemical weathering occurs when a mineral dissolves into *solution*—for example, when sodium chloride (common table salt) dissolves in water. Water is the universal solvent because it is capable of dissolving at least 57 of the natural elements and many of their compounds.

Water vapor readily dissolves carbon dioxide, thereby yielding precipitation containing carbonic acid (H_2CO_3). This acid is strong enough to react with many minerals, especially limestone, in a process called **carbonation**. *Carbonation* simply means reactions whereby carbon combines with minerals.

Such carbonation chemical weathering transforms minerals that contain calcium, magnesium, potassium, and sodium. When rainwater attacks formations of limestone (which is calcium carbonate, $CaCO_3$), the constituent minerals dissolve and wash away with the mildly acidic rainwater:

calcium carbonate + carbonic acid and water →
calcium bicarbonate ($Ca_2^{2+}CO_2H_2O$)

Walk through an old cemetery and you can observe the carbonation of marble, a metamorphic form of limestone. Weathered limestone and marble, in tombstones or in rock

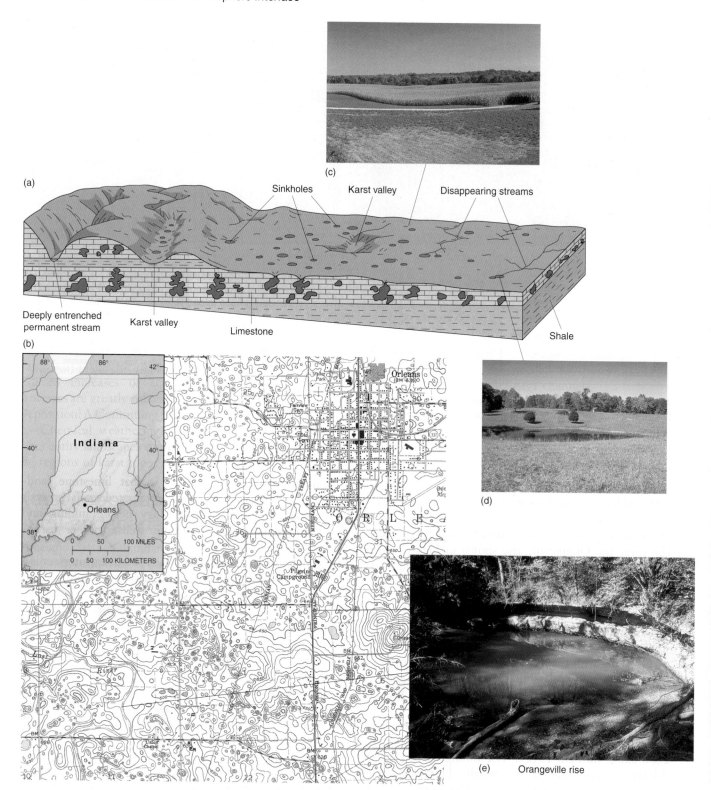

FIGURE 13.15 **Features of karst topography in Indiana.**
(a) Idealized features of karst topography in southern Indiana. (b) Karst topography southwest of the town of Orleans, Indiana. On average, 1022 sinkholes occur per 1 mi^2 in this area. On the map, note the contour lines: Depressions are indicated with small hachures (tick marks) on the downslope side of contour lines. (c) Gently rolling karst landscape and cornfields near Orleans, Indiana. (d) This pond is in a sinkhole depression near Palmyra, Indiana. (e) The Orangeville rise, near Orangeville, Indiana is just north of the Lost River rise. During periods of high rainfall, this rise is almost filled with water. [(a) Adapted from W. D. Thornbury, *Principles of Geomorphology,* illustration by W. J. Wayne, p. 326; © 1954 by John Wiley & Sons. (b) Mitchell, Indiana quadrangle, USGS. Photos (c), (d), and (e) by Bobbé Christopherson.]

Karst Farm Park, Indiana

(a)

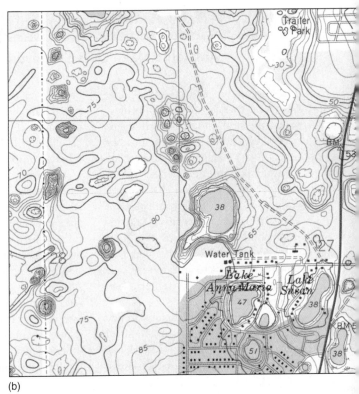

(b)

FIGURE 13.16 Sinkholes.
(a) Florida sinkhole, formed in 1981 in Winter Park, a suburb of Orlando. (b) Karst area 25 km (15.5 mi) north of Winter Park depicted on a topographic map. Note that the depressions are marked by small hachures (tick marks). [(a) Photo by Jim Tuten/Black Star; (b) Orange City quadrangle, USGS.]

FIGURE 13.17 Sinkhole swallows a bridge.
Downstream from the Hercules Quarry several dozen sinkholes have collapsed since 2000, taking down one of the bridges from State Route 33 and part of this small bridge in Stockertown, Pennsylvania. [Photo by Bobbé Christopherson.]

FIGURE 13.18 Deep-space research from a sinkhole.
Cockpit karst topography near Arecibo, Puerto Rico, provides a natural depression for the dish antenna of a giant radio telescope. The Arecibo Observatory is part of the National Astronomy and Ionosphere Center, which is operated by Cornell University under contract with the National Science Foundation. [Photo courtesy of Cornell University.]

News Report 13.1

Amateurs Make Cave Discoveries

The exploration and scientific study of caves is *speleology*. Although professional physical and biological scientists carry on investigations, amateur cavers, or "spelunkers," have made many important discoveries. As an example, in the early 1940s George Colglazier, a farmer southwest of Bedford, Indiana, awoke to find his farm pond at the bottom of a deep collapsed sinkhole. This sinkhole is now the entrance to an extensive cave system that includes a subterranean navigable stream.

Cave habitats are unique. They are nearly closed, self-contained ecosystems with simple food chains and great stability. In total darkness, bacteria synthesize inorganic elements and produce organic compounds that sustain many types of cave life, including algae, small invertebrates, amphibians, and fish.

In a cave discovered in 1986, near Movile in southeastern Romania, cave-adapted invertebrates were discovered after millions of years of sunless isolation. Thirty-one of these organisms were previously unknown. Without sunlight, the ecosystem in Movile is sustained on sulfur-metabolizing bacteria that synthesize organic matter using energy from oxidation processes. These chemosynthetic bacteria feed other bacteria and fungi that in turn support cave animals. The sulfur bacteria produce sulfuric acid compounds that may prove to be important in the chemical weathering of some caves.

The mystery, intrigue, and excitement of cave exploration lie in the variety of dark passageways, enormous chambers that narrow to tiny crawl spaces, strange formations, and underwater worlds that can be accessed only by cave diving. Private-property owners and amateur adventurers discovered many of the major caves, a fact that keeps this popular science/sport very much alive. (For nearly a thousand worldwide links and information, see the http://hum.amu.edu.pl/?sgp/spec/links.html web site.)

FIGURE 13.19 Tower karst of the Guangxi (Kwangsi) Province, China.
Resistant strata protect each tower as weathering removes the surrounding limestone. [Photo by Wolfgang Kaehler/ Wolfgang Kaehler Photography.]

Caves generally form just beneath the water table, where later lowering of the water level exposes them to further development. *Dripstones* form as water containing dissolved minerals slowly drips from the cave ceiling. Calcium carbonate precipitates out of the evaporating solution, literally one molecular layer at a time, and accumulates at a point below on the cave floor. Forming depositional features, *stalactites* grow from the ceiling and *stalagmites* build from the floor; sometimes the two grow until they connect and form a continuous *column* (Figure 13.20b). A dramatic subterranean world is thus created.

Cave science is an aspect of geomorphology where amateur cavers make important discoveries about these unique habitats and new life forms—*biospeleology*. More than 90% of known caves are not biologically surveyed, and some 90% of possible caves worldwide still lie undiscovered, making this a major research frontier (see News Report 13.1). (For more on caves and related formations, see http://www.goodearthgraphics.com/virtcave/virtcave.html.)

FIGURE 13.20 Cavern features.
(a) An underground cavern and related forms in limestone. (b) A column in Carlsbad Caverns, New Mexico, where a series of underground caverns includes rooms more than 1200 m (4000 ft) long and 190 m (625 ft) wide. Although a national park since 1930, unexplored portions remain. (c) A sinkhole in a Florida pasture. From Marengo Caves, Marengo, Indiana: (d) a flowstone and pool of water; (e) dripstone drapery formations; (f) soda straws hanging from ceiling cracks, forming one molecular layer at a time. [(b) Photo by author; (c) photo by Thomas M. Scott, Florida Geological Survey; photos (d), (e), (f) by Bobbé Christopherson.]

(a)
(b)
(c)
(d)
(e)
(f)

Surface stream

Sinkhole

Lost River

Collapsed sinkhole

Soluble rock (limestone)

Waterfall

Stalactite

Drip curtain

Collapse blocks

Column

Drip stones

Collapse blocks

Chamber

Stalagmite

Cave pool

Underground stream

Mass Movement Processes

Nevado del Ruiz, northernmost of two dozen dormant (not extinct, sometimes active) volcanic peaks in the Cordilleran Central of Colombia, had erupted six times during the past 3000 years, killing 1000 people during its last eruption in 1845. On November 13, 1985, at 11 P.M. after a year of earthquakes and harmonic tremors, a growing bulge on its northeast flank, and months of small summit eruptions, Nevado del Ruiz violently erupted in a lateral explosion. The mountain was back in action.

On this night, the familiar pyroclastics, lava, and blast were not the worst problem. The hot eruption quickly melted ice on the mountain's snowy peak, liquefying mud and volcanic ash, sending a hot mudflow downslope. Such a flow is a *lahar*, an Indonesian word referring to mudflows of volcanic origin. This lahar moved rapidly down the Lagunilla River toward the villages below. The wall of mud was at least 40 m (130 ft) high as it approached Armero, a regional center with a population of 25,000. The city slept as the lahar buried its homes: 23,000 people were killed there and in other afflicted river valleys; thousands were injured; 60,000 were left homeless. The volcanic debris flow is now a permanent grave for its victims. Not all mass movements are this destructive, but such processes play a major role in the denudation of the landscape.

For more on mass movement hazards, including landslides, see the web site of the Natural Hazards Center at the University of Colorado, Boulder, at **http://www. Colorado.edu/hazards/** or the USGS Geologic Hazards page at **http://landslides.usgs.gov/index.html**.

Mass Movement Mechanics

Physical and chemical weathering processes create an overall weakening of surface rock, which makes it more susceptible to the pull of gravity. The term **mass movement** applies to any unit movement of a body of material, propelled and controlled by gravity, such as the lahar just described. Mass movements can be surface processes or they can be submarine landslides beneath the ocean. Mass movement content can range from dry to wet, slow to fast, or small to large, and from free-falling to gradual or intermittent (see Figure 13.22).

The term *mass movement* is sometimes used interchangeably with **mass wasting**, which is the general process involved in mass movements and erosion of the landscape. To combine the concepts, we can say that the mass movement of material works to waste slopes and provide raw material for erosion, transportation, and deposition.

The Role of Slopes All mass movements occur on slopes under the influence of gravitational stress. If we pile dry sand on a beach, the grains will flow downslope until equilibrium is achieved. The steepness of the resulting slope depends on the size and texture of the grains; this steepness is called the **angle of repose**. This angle represents a balance of the driving force (gravity) and resisting force (friction and shearing). The angle of repose for various materials commonly ranges between 33° and 37° (from horizontal), and 30° to 50° for snow avalanche slopes.

The *driving force* in mass movement is gravity. It works in conjunction with the weight, size, and shape of the surface material; the degree to which the slope is oversteepened (how far it exceeds the angle of repose); and the amount and form of moisture available (frozen or fluid). The greater the slope angle, the more susceptible the surface material is to mass wasting processes.

The *resisting force* is the shearing strength of slope material, that is, its cohesiveness and internal friction, which work against gravity and mass wasting. To reduce shearing strength is to increase shearing stress, which eventually reaches the point at which gravity overcomes friction, initiating slope failure.

Clays, shales, and mudstones are highly susceptible to hydration (physical swelling in response to the presence of water). If such materials underlay rock strata in a slope, the strata will move with less driving-force energy. When clay surfaces are wet, they deform slowly in the direction of movement, and when saturated they form a viscous fluid with little shearing strength (resistance to movement) to hold back the slope. However, if the rock strata are such that material is held back from slipping, then more driving-force energy may be required, such as that generated by an earthquake.

Madison River Canyon Landslide In the Madison River Canyon near West Yellowstone, Montana, a blockade of dolomite (a magnesium-rich carbonate rock) held back a deeply weathered and *oversteepened slope* (40° to 60° slope angle) for untold centuries (white area in Figure 13.21). Then, shortly after midnight on August 17, 1959, a magnitude 7.5 earthquake broke the dolomite structure along the foot of the slope. The break released 32 million m^3 (1.13 billion ft^3) of mountainside, which moved downslope at 95 kmph (60 mph), causing gale force winds through the canyon. Momentum carried the material more than 120 m (about 400 ft) up the opposite canyon slope, trapping several hundred campers with about 80 m (260 ft) of rock, killing 28 people.

The mass of material also effectively dammed the Madison River and thus created a new lake, dubbed Quake Lake. The landslide debris dam established a new temporary equilibrium for the river and canyon. A channel was quickly excavated by the U.S. Army Corps of Engineers to prevent a disaster below the dam, for if Quake Lake overflowed the landslide dam, the water would quickly erode a channel and thereby release the entire contents of the new lake onto farmland downstream. This event conveys a dramatic example of the role of slopes and tectonic forces in creating massive land movements.

Classes of Mass Movements

In any mass movement, gravity pulls on a mass until the critical shear-failure point is reached—*a geomorphic*

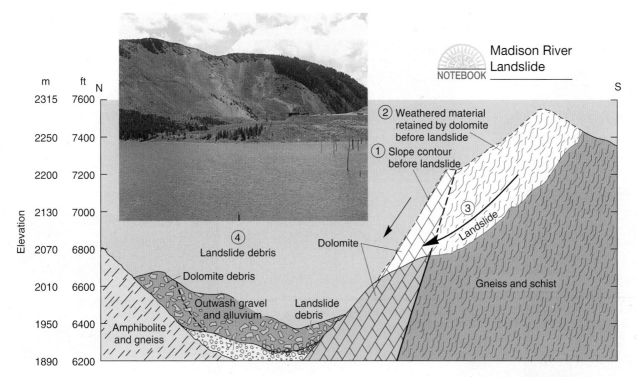

FIGURE 13.21 Madison River landslide.
Cross section showing geologic structure of the Madison River Canyon in Montana, where an earthquake triggered a landslide in 1959. (1) Prequake slope contour. (2) Weathered rock that failed. (3) Direction of landslide. (4) Landslide debris blocking the canyon and damming the Madison River. [After J. B. Hadley, *Landslides and Related Phenomena Accompanying the Hebgen Lake Earthquake of 17 August 1959*, U.S. Geological Survey Professional Paper 435-K, p. 115. Inset photo by Bobbé Christopherson.]

threshold. The material then can *fall, slide, flow,* or *creep*—the four classes of mass movement. Figure 13.22 summarizes these classes. Note the temperature and moisture gradients in the margins of the illustration, which show the relation between water content and movement rate. The Madison River Canyon event was a type of slide, whereas the Nevado del Ruiz lahar mentioned earlier was a flow. We now look at specific mass movement classes.

Falls and Avalanches   This class of mass movement includes rockfalls and debris avalanches. A **rockfall** is simply a volume of rock that falls through the air and hits a surface. During a rockfall, individual pieces fall independently and characteristically form a cone-shaped pile of irregular broken rocks called a *talus slope* at the base of a steep incline, where several *talus cones* coalesce (Figure 13.23).

A **debris avalanche** is a mass of falling and tumbling rock, debris, and soil. It is differentiated from a slower debris slide or landslide by the velocity of onrushing material. This speed often results from ice and water that fluidize the debris. The extreme danger of a debris avalanche results from its tremendous speed and consequent lack of warning.

In 1962 and again in 1970, debris avalanches roared down the west face of Nevado Huascarán, the highest peak in the Peruvian Andes. The 1962 debris avalanche

contained an estimated 13 million m³ (460 million ft³) of material, burying the city of Ranrahirca and eight other towns, killing 4000 people.

An earthquake initiated the 1970 event. Upward of 100 million m³ (3.53 billion ft³) of debris buried the city of Yungay, where 18,000 people perished (Figure 13.24). This avalanche attained velocities of 300 kmph (185 mph), which is especially incredible when you consider the quantity of material involved and the fact that some boulders weighed thousands of tons. The avalanche covered a vertical drop of 4144 m (13,600 ft) and a horizontal distance of 16 km (10 mi) in just a few minutes.

Another dramatic example of a debris avalanche, but one that did not kill, occurred west of the Saint Elias Range, north of Yakutat Bay in Alaska. A magnitude 7.1 earthquake triggered several dozen rockfalls, large debris avalanches, and snow avalanches. The largest single avalanche covered approximately 4.7 km² (1.8 mi²) of the Cascade Glacier (Figure 13.25). The photo reveals characteristic grooves, lobes, and large rocks associated with these fluid avalanches.

Landslides   A sudden rapid movement of a cohesive mass of regolith or bedrock that is not saturated with moisture is a **landslide**—a large amount of material failing simultaneously. Surprise creates the danger, for the downward pull of gravity wins the struggle for equilibrium

FIGURE 13.22 Mass movement classes.
Principal types of mass movement and mass wasting events. Variations in water content and
rates of movement produce a variety of forms. (a) A 1995 slide in La Conchita, California;
(b) saturated hillsides fail; (c) mudflows 2 m deep, both in Santa Cruz County, California.
[Photos by (a) Robert L. Schuster/USGS; (b) Alexander Lowry/Photo Researchers, Inc.;
(c) James A. Sugar.]

FIGURE 13.23 Talus slope.
Rockfall and talus deposits at the base of a steep slope along
Duve Fjord, Nordaustlandet Island. Can you see the lighter
rock strata that are the source for the three talus cones?
[Photo by Bobbé Christopherson.]

in an instant. Focus Study 13.1 describes one such surprise event that struck near Longarone, Italy, in 1963.

To eliminate the surprise element, scientists are using
the global positioning system (GPS) to monitor landslide
movement. With GPS, scientists measure slight land shifts
in suspect areas for clues to possible mass wasting. GPS was
applied in two cases in Japan and effectively identified
prelandslide movements of 2–5 cm per year, providing information to expand the area of hazard concern and warning.

Slides occur in one of two basic forms: translational
or rotational (see Figure 13.22 for an idealized view of
each). *Translational slides* involve movement along a planar
(flat) surface roughly parallel to the angle of the slope,
with no rotation. The Madison Canyon landslide
described earlier was a translational slide. Flow and creep
patterns also are considered translational in nature.

Rotational slides occur when surface material moves
along a concave surface. Frequently, underlying clay
presents an impervious surface to percolating water. As a

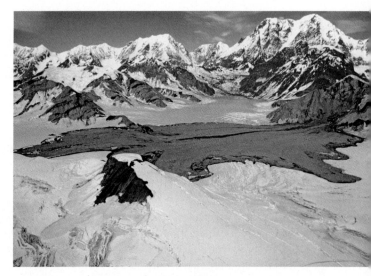

FIGURE 13.25 Debris avalanche, Alaska.
A debris avalanche covers portions of the Cascade Glacier, west of the Saint Elias Range in Alaska. The one pictured was the largest of several dozen triggered by a 1979 earthquake. [Photo by George Plafker.]

FIGURE 13.24 Debris avalanche, Peru.
A 1970 debris avalanche falls more than 4100 m (2.5 mi) down the west face of Nevado Huascarán, burying the city of Yungay, Peru. The same area was devastated by a similar avalanche in 1962 and by others in pre-Columbian times. A great danger remains for the towns in the valley from possible future mass movements. [Photo by George Plafker.]

result, water flows along the clay surface, undermining the overlying block and lubricating the contact, thereby reducing the friction force. The simplest form of rotational slide is a rotational slump, in which a small block of land shifts downward. The upper surface of the slide appears to rotate backward and often remains intact. The surface may rotate as a single unit, or it may present a stepped appearance.

A landslide slide show and solution diagrams may be viewed at **http://www.kingston.ac.uk/~ce_s011/slides. htm**. For a comprehensive look at an ongoing landslide situation in southern California, see **http://geology.wr. usgs.gov/wgmt/elnino/scampen/examples.html**. This site includes general background on mass movements, including many diagrams, and before and after photos.

Flows   Flows include *earthflows* and more fluid **mudflows**. When the moisture content of moving material is high, the suffix *-flow* is used (see Figure 13.22). Heavy rains can saturate barren mountain slopes and set them moving, as was the case east of Jackson Hole, Wyoming,

in the spring of 1925. Material above the Gros Ventre River (pronounced "grow vaunt") broke loose and slid downslope as a unit. The slide occurred because sandstone formations rested on weak shale and siltstone, which became moistened and soft, offering little resistance to the overlying strata. The slide is still visible after 75 years, as you can see in Figure 13.26.

Because of melted snow and rain, the water content of the Gros Ventre landslide was great enough to classify it as an *earthflow*. About 37 million m^3 (1.3 billion ft^3) of wet soil and rock moved down one side of the canyon and surged 30 m (100 ft) up the other side. The earthflow dammed the river and formed a lake, as did the landslide across the Madison River in the Hebgen Lake area in 1959. However, in 1925 equipment was not available to excavate a channel, so the new lake filled. Two years later, the lake water broke through the temporary earthflow dam, transporting a tremendous quantity of debris over the region downstream.

Creep   A persistent, gradual mass movement of surface soil is called **soil creep**. In creep, individual soil particles are lifted and disturbed by the expansion of soil moisture as it freezes, by cycles of moistness and dryness, diurnal temperature variations, or grazing livestock or digging animals.

In the freeze–thaw cycle, particles lift at right angles to the slope by freezing soil moisture, as shown in Figure 13.27. When the ice melts, however, the particles fall straight downward in response to gravity. As the process repeats, the surface soil gradually creeps its way downslope.

The overall wasting of a slope may cover a wide area and may cause fence posts, utility poles, and even trees to lean downslope. Various strategies are used to arrest the mass movement of slope material—grading the terrain, building terraces and retaining walls, planting ground cover—but the persistence of creep nearly always wins.

Focus Study 13.1

Vaiont Reservoir Landslide Disaster

Place: Northeastern Italy, Vaiont Canyon in the Italian Alps, rugged scenery, 680 m (2200 ft) above sea level.

   Location: 46.3° N, 12.3° E, near the border of the Veneto and Friuli-Venezia Giulia regions.

   Situation: Centrally located in the region, an ideal situation for hydroelectric power production.

   Site: Steep-sided, narrow glaciated canyon, opening to populated lowlands to the west; ideal site for a narrow-crested, high dam and deep reservoir.

   The Plan: Build the second-highest dam in the world, using a new thin-arch design, 262 m (860 ft) high, 190 m (623 ft) crest length, impounding a reservoir capacity of 150 million m³ (5.3 billion ft³) of water (third-largest reservoir in the world),

and generate hydroelectric power for distribution.

Geological Analysis

1. Steep canyon walls composed of interbedded limestone and shale; badly cracked and deformed structures; open fractures in the shale inclined toward the future reservoir body. The steepness of the canyon walls enhances the strong driving forces (gravitational) at work on rock structures.
2. High potential for bank storage (water absorption by canyon walls) into the groundwater system that will increase water pressure on all rocks in contact with the reservoir.
3. The nature of the shale beds is such that cohesion will be reduced as their clay minerals become saturated.

4. Evidence of ancient rockfalls and landslides on the north side of the canyon.
5. Evidence of creep activity along south side of the canyon.

Political/Engineering Decision
Begin design and construction of a thin-arch dam at this site.

Events During Construction and Filling Large volumes of concrete were injected into the bedrock as "dental work" in an attempt to strengthen the fractured rock. During reservoir filling in 1960, 700,000 m³ (2.5 million ft³) of rock and soil slid into the reservoir from the south side. A slow creep of slope materials along the entire south side began shortly thereafter, increasing to about 1 cm

(a)

(b)

Limit of landslide, October 9, 1963

Area filled by slide, 1963

Limit of flood, downstream from slide

Limit of 1960 slide

• Cities and towns

FIGURE 13.1.1 Vaiont Reservoir disaster.
(a) Cross section of the Vaiont River Valley.
(b) Map of the disaster site noted with evidence of previous activity. [After G. A. Kiersch, "The Vaiont Reservoir Disaster," California Division of Mines and Geology, Mineral Information Service, vol. 18, no. 7, 1964, pp. 129–138.]

per week by January 1963. Creep rate increased as the level of the reservoir rose. By mid-September 1963, the creep rate exceeded 40 cm (16 in.) a day.

This sequence of events set the stage for one of the worst dam disasters in history. Heavy rains began on September 28, 1963. Alarmed at the runoff from the rain into the reservoir and the increase in creep rate along the south wall, engineers opened the outlet tunnels on October 8 in an attempt to bring down the reservoir level—but it was too late.

The next evening, it took only 30 seconds for a landslide of 240 million m³ (8.5 billion ft³) to crash into the reservoir, shaking seismographs across Europe. A 150-m-thick (500-ft) slab of mountainside gave way (2 km by 1.6 km in area; 1.2 mi by 1 mi), sending shock waves of wind and water up the canyon 2 km (1.2 mi) and splashing a 100-m (330-ft) wave over the dam. The former reservoir was effectively filled with bedrock, regolith, and soil that almost entirely displaced its water content (Figure 13.1.1).

Amazingly, the experimental dam design held.

Downstream, in the unsuspecting town of Longarone, near the mouth of Vaiont Canyon on the Piave River, people heard the distant rumble and were quickly drowned by the 69-m (226-ft) wave of water that came out of the canyon. That night 3000 people perished. As for a lesson or recommendation, we need only to reread the preconstruction geologic analysis that began this Focus Study. The Italian courts eventually prosecuted the persons responsible.

Human-Induced Mass Movements (Scarification)

Every human disturbance of a slope—highway roadcut, surface mining, or building of a shopping mall, housing development, or home—can hasten mass wasting. The newly destabilized and oversteepened surfaces are thrust into a search for a new equilibrium. Imagine the disequilibrium in slope relations created by the highway roadcut pictured in Figure 12.8b.

Large open-pit surface mines—such as the Bingham Copper Mine west of Salt Lake City, the abandoned Berkeley Pit in Butte, Montana, and numerous large coal surface mines in the U.S. East and West, such as Black

Mesa (Figure 13.28a)—are examples of human-induced mass movements, generally called **scarification**. At the Berkeley Pit, toxic drainage water is threatening the regional aquifers, the Clark Fork River, and the local freshwater supply. Residues of copper, zinc, lead, and arsenic lace one creek, now devoid of life, and are in the soil on Butte Hill, where children play. Wind dispersal is a particular problem with radioactive uranium tailings in the West.

At the Bingham Copper Mine, a mountain literally was removed since mining began in 1906, forming a pit 4 km wide and 1 km deep (Figure 13.28b). This is easily the largest human-made excavation on Earth, with ores

This forested mass is the main portion of the earthflow.

FIGURE 13.26 The Gros Ventre earthflow near Jackson, Wyoming.
Evidence of the 1925 earthflow is still visible after more than 80 years. [Photo by Steven K. Huhtala.]

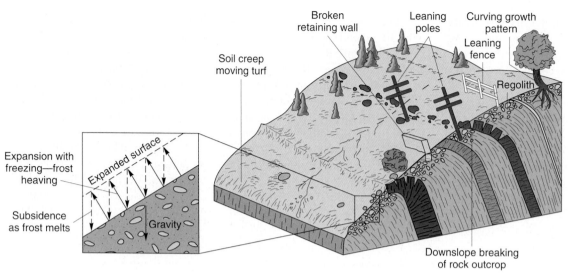

FIGURE 13.27 Soil creep and its effects.

FIGURE 13.28 Scarification.
(a) Black Mesa, Arizona, strip-mining for coal. (b) Bingham Canyon, Utah, west of Salt Lake City, strip-mining for copper and other minerals. (c) Spoil banks of tailings in a West Virginia coal-mining area. (d) The 55-m-deep Dally slate quarry in Pen Argyl, Pennsylvania. [Photos (a) and (c) by author; (b) and (d) by Bobbé Christopherson.]

containing copper, gold, silver, and molybdenum. The disposal of tailings (mined ore of little value) and waste material is a significant problem at any surface mine. Such large excavations produce tailing piles that are unstable and susceptible to further weathering, mass wasting, or wind dispersal. The leaching of toxic materials from tailings and waste piles poses an ever-increasing problem to streams, aquifers, and public health across the country.

Where underground mining is common, particularly for coal in the Appalachians, land subsidence and collapse produce further mass movements. Homes, highways, streams, wells, and property values are severely affected. Another controversial form of mining called *mountaintop removal* is done by removing ridges and summits and dumping the debris into stream valleys, thereby exposing the coal seams for mining and burying the stream channels. A change in the Clean Water Act in 2002 removed a 20-year-old restriction that barred dumping within 100 feet of a stream, to protect flows. Some 1930 km (1200 mi) of streams are now filled with tailings (Figure 13.28c).

In eastern Pennsylvania commercial slate deposits formed over millions of years. The Dally slate quarry in Pen Argyl is now working a second pit that is 55 m (180 ft) deep (Figure 13.28d). Removal of slate is guided by the grain of the metamorphic rock so as to produce the flattest pieces. Slate is mined for shingles and countertops. The local county turned the first quarry into a landfill—a practical use for the pit.

Scientists can informally quantify the scale of human-induced scarification for comparison with natural denudation processes. R. L. Hooke, Department of Geology and Geophysics, University of Minnesota, used estimates of U.S. excavations for new housing, mineral production (including the three largest—stone, sand and gravel, and coal), and highway construction. He then prorated these quantities of moved earth for all countries, using their gross domestic product (GDP), energy consumption, and agriculture's effect on river sediment loads. From these he calculated a global estimate for human earth moving.

Hooke estimated that humans, as a geomorphic agent, annually move 40–45 billion tons (40–45 Gt/yr) of the planet's surface. Compare this quantity with natural river sediment transfer (14 Gt/yr), movement through stream meandering (39 Gt/yr), haulage by glaciers (4.3 Gt/yr), movement due to wave action and erosion (1.25 Gt/yr), wind transport (1 Gt/yr), sediment movement by continental and oceanic mountain building (34 Gt/yr), or deep-ocean sedimentation (7 Gt/yr). As Hooke stated,

> *Homo sapiens* has become an impressive geomorphic agent. Coupling our earth-moving prowess with our inadvertent adding of sediment load to rivers and the visual impact of our activities on the landscape, one is compelled to acknowledge that, for better or for worse, this biogeomorphic agent may be the premier geomorphic agent of our time.*

*R. L. Hooke, "On the efficacy of humans as geomorphic agents," *GSA Today*, The Geological Society of America, 4, no. 9 (September 1994): 217–226.

Summary and Review—Weathering, Karst Landscapes, and Mass Movement

▸ *Define* the science of geomorphology.

Geomorphology is the science that analyzes and describes the origin, evolution, form, and spatial distribution of landforms. The exogenic system, powered by solar energy and gravity, tears down the landscape through processes of landmass **denudation** involving weathering, mass movement, erosion, transportation, and deposition. Different rocks offer differing resistance to these weathering processes and produce a pattern on the landscape of **differential weathering**.

Agents of change include moving air, water, waves, and ice. W. M. Davis's *geomorphic cycle model* characterized landscapes as evolving through stages from youth to old age. Since the 1960s, research and understanding of the processes of denudation have moved toward the **dynamic equilibrium model**, which considers slope and landform stability to be consequences of the resistance of rock materials to the attack of denudation processes.

geomorphology (p. 402)
denudation (p. 402)
differential weathering (p. 402)
dynamic equilibrium model (p. 403)

1. Define geomorphology, and describe its relationship to physical geography.
2. Define landmass denudation. What processes are included in the concept?
3. What is the interplay between the resistance of rock structures and weathering variabilities?
4. Give a brief overview of the geomorphic cycle model. What was W. M. Davis's principal contribution to the models of landmass denudation?
5. What are the principal considerations in the dynamic equilibrium model?

▸ *Illustrate* the forces at work on materials residing on a slope.

Slopes are shaped by the relation between rate of weathering and breakup of slope materials and the rate of mass movement and erosion of those materials. A slope is considered stable if it is stronger than these denudation processes; it is unstable if it is weaker. In this struggle against gravity, a slope may reach a **geomorphic threshold**—the point at which there is enough energy to overcome resistance against movement. **Slopes** that form the boundaries of landforms have several general components: *waxing slope*, *free face*, *debris slope*, and *waning slope*. Slopes seek an angle of equilibrium among the operating forces.

> geomorphic threshold (p. 403)
> slopes (p. 403)

6. Describe conditions on a hillslope that is right at the geomorphic threshold. What factors might push the slope beyond this point?

7. Given all the interacting variables, do you think a landscape ever reaches a stable, old-age condition? Explain.

8. What are the general components of an ideal slope?

9. Relative to slopes, what is meant by an "angle of equilibrium"? Can you apply this concept to the photograph in Figure 13.2?

♦ *Define* weathering, and *explain* the importance of parent rock and joints and fractures in rock.

Weathering processes disintegrate both surface and subsurface rock into mineral particles or dissolve them in water. The upper layers of surface material undergo continual weathering and create broken-up rock called **regolith**. Weathered **bedrock** is the *parent rock* from which regolith forms. The unconsolidated, fragmented material that develops after weathering is **sediment**, which along with weathered rock forms the **parent material** from which soil evolves.

Important in weathering processes are **joints**, the fractures and separations in the rock. Jointing opens up rock surfaces on which weathering processes operate. Factors that influence weathering include character of the bedrock (hard or soft, soluble or insoluble, broken or unbroken), climatic elements (temperature, precipitation, freeze–thaw cycles), position of the water table, slope orientation, surface vegetation and its subsurface roots, and time.

> weathering (p. 405)
> regolith (p. 405)
> bedrock (p. 405)
> sediment (p. 405)
> parent material (p. 405)
> joints (p. 405)

10. Describe weathering processes operating on an open expanse of bedrock. How does regolith develop? How is sediment derived?

11. Describe the relationship between mesoscale climatic conditions and rates of weathering activities.

12. What is the relation among parent rock, parent material, regolith, and soil?

13. What role do joints play in the weathering process? Give an example from one of the illustrations in this chapter.

♦ *Describe* frost action, crystallization, pressure-release jointing, and the role of freezing water as physical weathering processes.

Physical weathering refers to the breakup of rock into smaller pieces with no alteration of mineral identity. The physical action of water when it freezes (expands) and thaws (contracts) is a powerful agent in shaping the landscape. This **frost action** may break apart any rock. Working in joints, expanded ice can produce *joint-block separation* through the process of

frost-wedging. Frost action loosens rock that falls from a steep cliff, producing a **talus slope** of poorly sorted debris at the base of the slope. Another process of physical weathering is *crystallization*; as crystals in rock grow and enlarge over time, they force apart mineral grains and break up rock.

As overburden is removed from a granitic batholith, the pressure of deep burial is relieved. The granite slowly responds with *pressure-release jointing*, with layer after layer of rock peeling off in curved slabs or plates. As these slabs weather, they slip off in a process called **sheeting**. This *exfoliation process* creates an arch-shaped or dome-shaped feature on the exposed landscape, forming an **exfoliation dome**.

> physical weathering (p. 406)
> frost action (p. 407)
> talus slope (p. 407)
> sheeting (p. 409)
> exfoliation dome (p. 409)

14. What is physical weathering? Give an example.

15. Why is freezing water such an effective physical weathering agent?

16. What weathering processes produce a granite dome? Describe the sequence of events.

♦ *Describe* the susceptibility of different minerals to the chemical weathering processes called hydration, hydrolysis, oxidation, carbonation, and solution.

Chemical weathering is the chemical decomposition of minerals in rock. It can cause **spheroidal weathering**, in which chemical weathering occurs in cracks in the rock. As cementing and binding materials are removed, the rock begins to disintegrate, and sharp edges and corners become rounded.

Hydration occurs when a mineral absorbs water and expands, thus creating a strong mechanical force that stresses rocks. **Hydrolysis** breaks down silicate minerals in rock, as in the chemical weathering of feldspar into clays and silica. Water actively participates in chemical reactions. **Oxidation** is the reaction of oxygen with certain metallic elements, the most familiar example being the rusting of iron, producing iron oxide. The dissolving of materials into *solution* is considered chemical weathering. For instance, a mild acid such as carbonic acid in rainwater will cause **carbonation**, wherein carbon combines with certain minerals, such as calcium, magnesium, potassium, and sodium.

> chemical weathering (p. 409)
> spheroidal weathering (p. 409)
> hydration (p. 409)
> hydrolysis (p. 410)
> oxidation (p. 411)
> carbonation (p. 411)

17. What is chemical weathering? Contrast this set of processes to physical weathering.

18. What is meant by the term *spheroidal weathering*? How does spheroidal weathering occur?

19. What is hydration? What is hydrolysis? Differentiate between these processes. How do they affect rocks?

20. Iron minerals in rock are susceptible to which form of chemical weathering? What characteristic color is associated with this type of weathering?

21. With what kind of minerals do carbon compounds react, and under what circumstances does this reaction occur? What is this weathering process called?

▶ *Review* the processes and features associated with karst topography.

Karst topography refers to distinctively pitted and weathered limestone landscapes. Surface circular **sinkholes** form and may extend to form a *karst valley*. A sinkhole may collapse through the roof of an underground cavern, forming a *collapse sinkhole*. The formation of caverns is part of karst processes and groundwater erosion. Limestone caves feature many unique erosional and depositional features, producing a dramatic subterranean world.

> karst topography (p. 412)
> sinkholes (p. 412)

22. Describe the development of limestone topography. What is the name applied to such landscapes? From what area was this name derived?

23. Differentiate among sinkholes, karst valleys, and cockpit karst. Within which form is the radio telescope at Arecibo, Puerto Rico?

24. In general, how would you characterize the region southwest of Orleans, Indiana?

25. What are some of the unique erosional and depositional features you find in a limestone cavern?

▶ *Portray* the various types of mass movements, and *identify* examples of each in relation to moisture content and speed of movement.

Any movement of a body of material, propelled and controlled by gravity, is **mass movement**, also called **mass wasting**. The **angle of repose** of loose sediment grains represents a balance of driving and resisting forces on a slope. Mass movement of Earth's surface produces some dramatic incidents, including: **rockfalls** (a volume of rock that falls); **debris avalanches** (a mass of tumbling, falling rock, debris, and soil moving at high speed); **landslides** (a large amount of material failing simultaneously); **mudflows** (material in motion with a high moisture content); and, **soil creep** (a persistent movement of individual soil particles that are lifted by the expansion of soil moisture as it freezes, by cycles of wetness and dryness, temperature variations, or the impact of grazing animals). In addition, human mining and construction activities have created massive **scarification** of landscapes.

> mass movement (p. 418)
> mass wasting (p. 418)
> angle of repose (p. 418)
> rockfall (p. 419)
> debris avalanche (p. 419)
> landslide (p. 419)
> mudflows (p. 421)
> soil creep (p. 421)
> scarification (p. 423)

26. Define the role of slopes in mass movements, using the terms *angle of repose*, *driving force*, *resisting force*, and *geomorphic threshold*.

27. What events occurred in the Madison River Canyon in 1959?

28. What are the classes of mass movement? Describe each briefly and differentiate among these classes.

29. Name and describe the type of mudflow associated with a volcanic eruption.

30. Describe the difference between a landslide and what happened on the slopes of Nevado Huascarán.

31. What is scarification, and why is it considered a type of mass movement? Give several examples of scarification. Why are humans a significant geomorphic agent?

NetWork

The *Geosystems* Home Page provides online resources for this chapter on the World Wide Web. To begin: Once on the Home Page, click on this textbook, scroll the Table of Contents menu, select this chapter, and click "Begin." You will find self-tests that are graded, review exercises, specific updates for items in the chapter, and many links to interesting related pathways on the Internet under "Destinations." *Geosystems* is at http://www.prenhall.com/christopherson.

Critical Thinking

A. Locate a slope, possibly near campus, near your home, or a local roadcut. Using Figure 13.3a and b, can you identify the forces and forms of a hillslope at your site? How would you go about assessing the stability of the slope? Is there any evidence of the mass wasting of materials, soil creep, or other processes discussed in this chapter?

B. The USGS has completed a landslide hazards potential map for the conterminous United States. You can check out this resource at http://landslides.usgs.gov/html_files/landslides/nationalmap/national.html. Note the zoom-in capability to look at specific areas in greater detail. The map legend rates landslide incidence and susceptibility. How does the map portray landslide incidences? In what way does the map rate landslide susceptibility to an area? Are you able to find a location that you have visited and determine its vulnerability?

C. Go to the *Geosystems* Home Page and to Critical Thinking in Chapter 13. Study the two remote-sensing images. The Himalayan Mountains are the primary source of the sediment transported by the rivers. Imagine yourself as a piece of rock at the very top of Mount Everest. Describe a scenario by which you join a collection of sediments in the Bay of Bengal. Discuss the changes in weathering processes acting upon you between the high, cold reaches of the Himalayas and the hot, humid mudflats of the Ganges Delta.

Gregory A. Pope, Associate Geography Professor

Gregory Pope lists physical geography, weather, geomorphology, weathering processes, GIS, and global environmental change as areas of special interest. He assumed his teaching post at Montclair State in 1996 and is the advisor for students in the geography program. Greg's Master's and Ph.D. degrees were completed at Arizona State University. His dissertation developed a boundary-layer model for interpreting spatial variation in weathering. He has published several articles derived from his dissertation (Figure 13.1.1).

Earlier in this chapter you can see Greg's influence in my mention of the importance of scale in analyzing weathering processes. He found both chemical and physical weathering at small-scale reaction sites on rock, even in the driest landscapes.

Greg did his undergraduate work at the University of Colorado in his hometown of Colorado Springs. As a child, he camped with his family in Colorado and other parts of the West. Especially vivid to him are camping trips to the Rockies: "I didn't know what geomorphology or geology were, but the rocks and mountain formations fascinated me." Beginning when he was only 4 or 5, during automobile trips to Illinois, he was the family navigator, using his own highway map collection.

Greg remembers, "We had a geographic encyclopedia set that I read and used for homework throughout my lower grades. I also read all the geography and geology books in my junior high before I finished eighth grade; they called it social studies, not geography, but it was really geography. I convinced my teacher to let me do an urban planning project about the unplanned growth in Colorado Springs. I asked the teacher to let me design a new city plan, as if Colorado Springs wasn't there. I did this one whole semester instead of what the rest of the class did."

FIGURE 13.1.1 Gregory A. Pope, Associate Geography Professor and Department Chair, Montclair State University, Upper Montclair, New Jersey.
Greg, along with geography professor Patricia Beyer, led a 2001 field trip through New York's Central Park analyzing the diverse types of stones used in the park's many bridges. [Photo by Bobbé Christopherson.]

"Following graduation from UCCS, I received a National Geographic Society internship and worked in Washington, DC, for half a year. I received a lot of valuable research experience working on *World*, the NGS children's magazine." He also worked as a contributing writer for the NGS National Geography Bee.

As a teaching assistant in graduate school at Arizona State, Greg, along with his advisor Ronald Dorn, began studying the White Mountains of eastern California, which fostered his interest in desert geomorphology. He knew he wanted to teach. "When I taught the labs, something all teaching assistants do, I felt comfortable. I knew teaching had to be part of my future."

Greg is studying the weathering of ancient to modern architectural structures in Portugal and collaborating with geoarcheologists who study weathering of petroglyphs. Also, he is looking at weathering in some stonework, specifically isolating pre–Clean Air Act (1970) contamination and weathering, examining tombstones near pollution sources. With his students he examines the rates of building-stone weathering around New York City—his take on urban geology—and then maps the spatial results. In addition, Greg is doing research on the tectonic geomorphology of the Rocky Mountains, a lifelong subject of interest. And with an NSF grant he is studying the effects of recent wildfires, and the intense temperatures they generate, on weathering processes in western soils in Colorado.

In 2001, Greg, along with geographer Dr. Patricia Beyer of Bloomsburg University, led us on a field trip through Central Park in New York City—called "Cultural Stones of Central Park: Outcrops, Arches, and Statues." They produced a wonderful field guide for our expedition. The popular park went through many stages of design and construction over the last 150 years, including a variety of arches and bridges to accommodate different forms of traffic. Thirty-six bridges remain of the original thirty-nine, built from a great variety of rock.

Greg said, "The stones used by the designers and builders in the art and architecture of the park are truly 'cultural stones.' Sources for all this building stone cover a wide geography, from many locales across the Northeast and into Canada." Greg and Patricia showed us how these materials presented a diverse array of surfaces on which physical and chemical weathering processes could work (Figure 13.1.2).

What lies ahead for geography: global change studies will be important; the integration of environmental studies is another possible avenue. Greg feels that "some discipline must emerge to pull all the disparate parts together—geography should be it. In the direction I am going—weathering studies, geoarcheology, and work in building my department's program—I am having fun! I am inventing stuff and combining subjects that interest me. And being a geographer, I get to synthesize all this into coherent patterns! I get to work with people from so many different fields. My advice is to have fun with it, have joy in what you choose to do."

FIGURE 13.1.2 Weathered sandstone in Central Park, New York City.
Here on the side of one of the many arches, constructed in 1859, we see the ravages of both physical and chemical weathering. The patterns etched into this Alberta sandstone, found in New Brunswick, are completely gone in several squares. Encrusted salts, setting up the crystallization process, are visible. [Photo by Bobbé Christopherson.]

The Mississippi River separates Minnesota from the bluffs of Wisconsin in the distance. This is a portion of the 8th largest river system on Earth ranked by discharge and 4th ranked by length. The Mississippi is a workhorse carrying much commerce. On the river in the photo you see a *tow,* which is one towboat pushing several barges, in this case covered dry-cargo barges perhaps filled with farm products, fertilizer, or dry cement. The *tow* must navigate through 29 locks and dams on the upper river between Granite City, Illinois, near St. Louis, Missouri, and St. Anthony Falls, Minnesota. Transit is possible because each dam holds back a navigable pool of water, maintained at a minimum depth of 2.7 m (9 ft) for the type of barges used on the Upper Mississippi. [Photo by Bobbé Christopherson.]

14

River Systems and Landforms

▶ Key Learning Concepts

After reading the chapter, you should be able to:

▶ **Define** the term *fluvial,* and **outline** the fluvial processes: erosion, transportation, and deposition.

▶ **Construct** a basic drainage basin model, and **identify** different types of drainage patterns and internal drainage, with examples.

▶ **Describe** the relation among velocity, depth, width, and discharge, and **explain** the various ways that a stream erodes and transports its load.

▶ **Develop** a model of a meandering stream, including point bar, undercut bank, and cutoff, and **explain** the role of stream gradient in these flow characteristics.

▶ **Define** a floodplain, and **analyze** the behavior of a stream channel during a flood.

▶ **Differentiate** the several types of river deltas, and **detail** each.

▶ **Explain** flood probability estimates, and **review** strategies for mitigating flood hazards.

arth's rivers and waterways form vast arterial networks that drain the continents. They also shape the landscape by removing the products of weathering, mass movement, and erosion and transporting them downstream. To call rivers "Earth's lifeblood" is no exaggeration, inasmuch as rivers redistribute mineral nutrients important for soil formation and plant growth and serve society in many ways.

Rivers not only provide essential water supplies, they also process waste (diluting and transporting it), provide critical cooling water for manufacturing and power generation, and form essential transportation networks. Rivers have been important in the geography of human history, influencing where settlements were built, where livelihoods were made, and where borders were drawn. This chapter discusses the dynamics of river systems and their landforms.

At any moment, approximately 1250 km³ (300 mi³) of water is flowing through Earth's waterways. Even though this volume is only 0.003% of all freshwater, the work performed by this energetic flow makes it a dominant agent of landmass denudation. Of the world's rivers, those with the greatest *discharge* (streamflow past a point in a given unit of time) are the Amazon of South America (Figure 14.1), the Congo (Zaire) of Africa, the Chàng Jiang (Yangtze) of Asia, and the Orinoco of South America (Table 14.1). In North America, the greatest discharges are from the Missouri–Ohio–Mississippi, Saint Lawrence, and Mackenzie River systems.

Hydrology is the science of water, its global circulation, distribution, and properties, specifically water at and below Earth's surface. For hydrology links on the web, see **http://www.worldwater.org/links.htm** or the Global Hydrology and Climate Center at **http://www.ghcc. msfc.nasa.gov/**; also see the Amazon River at **http:// boto.ocean.washington.edu/eos/index.html**.

In this chapter: We begin with a look at the largest rivers on Earth. Essential fluvial concepts of base level, drainage basin, and drainage density and patterns follow. Factors that affect streamflow characteristics and the work performed by flowing water, including erosion and transport, are discussed. Stream depositional features such as deltaic forms are illustrated with a detailed look at the Mississippi River Delta. Human response to floods and floodplain management are important aspects of river management and conclude the chapter.

Fluvial Processes and Landscapes

Stream-related processes are termed **fluvial** (from the Latin *fluvius*, meaning "river"). Geographers seek to describe stream patterns and the fluvial processes that created them. Fluvial systems, like all natural systems, have characteristic processes and produce predictable landforms. Yet a stream system can behave with randomness

FIGURE 14.1 Mouth of the Amazon River.
The mouth of the Amazon River discharges a fifth of all the freshwater that enters the world's oceans. The mouth of the Amazon is 160 km (100 mi) wide. Millions of tons of sediments are derived from the Amazon's drainage basin, which is as large as the Australian continent. Large islands of sediment are left where the river's discharge leaves the mouth and flows into the Atlantic Ocean. [*Terra* MISR sensor image courtesy of NASA/GSFC/JPL and the MISR Team.]

Table 14.1   Largest Rivers on Earth Ranked by Discharge

Rank by Volume	Average Discharge at Mouth in Thousands of m³/s (cfs)	River (with Tributaries)	Outflow/Location	Length km (mi)	Rank by Length
1	180 (6350)	Amazon (Ucayali, Tambo, Ene, Apurimac)	Atlantic Ocean/ Amapá-Pará, Brazil	6570 (4080)	2
2	41 (1460)	Congo, also known as the Zaire (Lualaba)	Atlantic Ocean/ Angola, Congo	4630 (2880)	10
3	34 (1201)	Yangtze (Chàng Jiang)	East China Sea/ Kiangsu, China	6300 (3915)	3
4	30 (1060)	Orinoco	Atlantic Ocean/ Venezuela	2737 (1700)	27
5	21.8 (779)	La Plata estuary (Paraná)	Atlantic Ocean/ Argentina	3945 (2450)	16
6	19.6 (699)	Ganges (Brahmaputra)	Bay of Bengal/India	2510 (1560)	23
7	19.4 (692)	Yenisey (Angara, Selenga or Selenge, Ider)	Gulf of Kara Sea/ Siberia	5870 (3650)	5
8	18.2 (650)	Mississippi (Missouri, Ohio, Tennessee, Jefferson, Beaverhead, Red Rock)	Gulf of Mexico/ Louisiana	6020 (3740)	4
9	16.0 (568)	Lena	Laptev Sea/Siberia	4400 (2730)	11
17	9.7 (348)	St. Lawrence	Gulf of St. Lawrence/ Canada and United States	3060 (1900)	21
36	2.83 (100)	Nile (Kagera, Ruvuvu, Luvironza)	Mediterranean Sea/ Egypt	6690 (4160)	1

and disorder. The term *river* is applied to a trunk stream or an entire river system. *Stream* is a more general term not necessarily related to size. There is some overlap in usage between the two terms.

Insolation and gravity power the hydrologic cycle and are the driving forces of fluvial systems. Individual streams vary greatly, depending on the climate in which they operate, the composition of the surface, topography over which they flow, the nature of vegetation and plant cover, and the length of time they have been operating in a specific setting.

Water dislodges, dissolves, or removes surface material in the **erosion** process. Streams produce *fluvial erosion*, in which weathered sediment is picked up for transport to new locations. Thus, a stream is a mixture of water and solids; the solids are carried in solution, suspension, and by mechanical **transport** (movement of material). Materials are laid down by another process, **deposition**. **Alluvium** is the general term for the clay, silt, sand, gravel, or other unconsolidated rock and mineral fragments deposited by running water as sorted or semisorted sediment on a floodplain, delta, or streambed.

Base Level of Streams

American geologist and ethnologist John Wesley Powell (1834–1902) was a director of the U.S. Geological Survey, first director of the U.S. Bureau of Ethnology, explorer of the Colorado River, and a pioneer in understanding the landscape (Figure 14.2). In 1875 he put forward the idea of **base level**, or a level below which a stream cannot

erode its valley. In general, the *ultimate base level* is sea level, the average level between high and low tides. As shown in Figure 14.3, you can imagine the base level as a surface extending inland from sea level, inclined gently upward under the continents. Ideally, this is the lowest practical level for all denudation processes.

Of course, Powell recognized that not every landscape has degraded all the way to sea level; clearly, other intermediate base levels are in operation. A *local base level*, or temporary one, may control the lower limit of local streams. The local base level may be a river, a lake, hard and resistant rock, or a human-made dam (Figure 14.3). The local base level set by Glen Canyon Dam in Arizona backed up Lake Powell into sinuous canyons. The multi-year western drought has brought down the reservoir to record low levels. In this arid landscape, Glen Canyon Dam is the local base level for the impounded Colorado River above the dam.

Over time, the work of streams modifies the landscape dramatically. Landforms are produced by two basic processes: (1) erosive action of flowing water, and (2) deposition of stream-transported materials. Let us begin our study by examining a basic fluvial unit—the drainage basin.

Drainage Basins

Figure 14.4 illustrates drainage basin concepts. Every stream has a **drainage basin**, ranging in size from tiny to vast. Every drainage basin is defined by ridges that form *drainage divides*; that is, the ridges are the dividing lines

FIGURE 14.2 Powell's journey to base level.
John Wesley Powell developed his base-level concept during extensive exploration of the West. He first ventured down the Colorado River in 1869 in heavy oak boats, drifting down the Green River, joining the Colorado River in what is now Canyon-lands National Park, Utah. Continuing on through river canyons that are now flooded by Lake Powell—a local base level not present at the time of Powell's expedition." [Photo by Bobbé Christopherson.]

that control into which basin precipitation drains. Drainage divides define a **watershed**, the catchment (water-receiving) area of the drainage basin. In any drainage basin, water initially moves downslope in a thin film called **sheetflow**, or *overland flow*. High ground that separates one valley from another and directs sheetflow is termed an *interfluve*. Surface runoff concentrates in *rills*, or small-scale downhill grooves, which may develop into deeper *gullies* and then into a stream course in the valley.

Drainage Divides and Basins Several high drainage divides, called **continental divides**, are situated in the United States and Canada. These are extensive mountain and highland regions separate drainage basins, sending flows to the Pacific, the Gulf of Mexico, the Atlantic, Hudson Bay, or the Arctic Ocean. The principal drainage divides and drainage basins in the United States and Canada are mapped in Figure 14.5. These divides form water-resource regions and provide a spatial framework for water-management planning.

A major drainage basin system is made up of many smaller drainage basins. Each drainage basin gathers and delivers its precipitation and sediment to a larger basin, concentrating the volume into the main stream. A good example from Figure 14.5 is the great Mississippi–Missouri–Ohio River system, draining some 3.1 million km^2 (1.2 million mi^2), or 41% of the continental United States.

Consider the travels of rainfall in north-central Pennsylvania. This water feeds hundreds of small streams that flow into the Allegheny River. At the same time, rainfall in southern Pennsylvania feeds hundreds of streams that flow into the Monongahela River. The two rivers then join at Pittsburgh to form the Ohio River. The Ohio flows southwestward and at Cairo, Illinois, connects with the Mississippi River, which eventually flows on past New Orleans into the Gulf of Mexico. Each contributing tributary, large or small, adds its discharge, pollution, and sediment load to the larger river.

(a)

(b)

FIGURE 14.3 Ultimate and local base levels.
(a) The concepts of ultimate base level (sea level) and local base level (natural, such as a lake, or artificial, such as a dam). Note how base level curves gently upward from the sea as it is traced inland; this is the theoretical limit for stream erosion. (b) A local base level was introduced to the landscape above Glen Canyon Dam when it closed off the canyon. Because of the western drought, Lake Powell is lowering in level, as you can tell by the light-colored rock ringing the lake. [Photo by Bobbé Christopherson.]

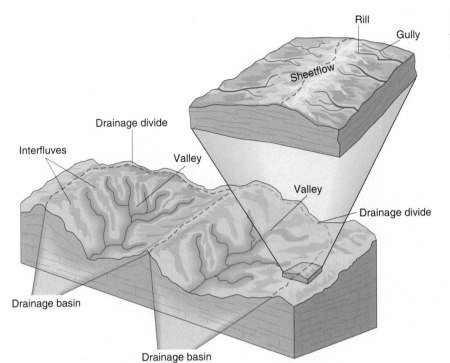

FIGURE 14.4 A drainage basin.
A drainage divide separates the drainage basin and its watershed from other basins.

In our example, sediment weathered and eroded in north-central and southern Pennsylvania is transported thousands of kilometers and accumulates on the floor of the Gulf of Mexico, where it forms the Mississippi River delta.

Imagine the complexity of an international drainage basin, such as the Danube River in Europe, that flows 2850 km (1770 mi) from western Germany's Black Forest to the Black Sea. The river crosses or forms the borders between nine countries (Figure 14.6). A total area of 817,000 km² (315,000 mi²) fall within the drainage basin, including some 300 tributaries.

The Danube serves many economic functions: commercial transport, municipal water source, agricultural irrigation, fishing, and hydroelectric production. An international struggle is underway to save the river from its burden of industrial and mining wastes, sewage, chemical discharge, agricultural runoff, and drainage from ships. The many shipping canals actually spread pollution and worsen overall conditions in the river. All of this pollution passes through Romania and the deltaic ecosystems in the Black Sea. The river is widely regarded as one of the most polluted on Earth.

Political changes in Europe in 1989 allowed the first scientific analysis of the entire river system. The United Nations Environment Programme (UNEP) and the European Union, along with other organizations, are dedicated to clearing the Danube, saving the delta, and restoring the environment of this valuable resource. The river delta is one of the international Biosphere Reserves; see http://www.unesco.org/mab/. More on the Black Sea environment is posted at http://www.blacksea-environment.org/.

Drainage Basins as Open Systems Drainage basins are open systems in which inputs include precipitation and the minerals and rocks of the regional geology.

Energy and materials are redistributed as the stream constantly adjusts to its landscape. System outputs of water and sediment exit through the mouth of the river, into a lake, another river, or the ocean.

Change that occurs in any portion of a drainage basin can affect the entire system, because the stream adjusts to carry the appropriate load of sediment relative to its discharge. If a river system is brought to a threshold where it can no longer maintain its present form, the relations within the drainage basin system are destabilized, initiating a transition period to a more stable condition. A stream drainage system constantly struggles toward equilibrium among the interacting variables of discharge, transported load, channel shape, and channel steepness.

Delaware River Basin Let us look at the Delaware River basin, within the Atlantic Ocean drainage region (Figure 14.7, p. 438). The Delaware River headwaters are in the Catskill Mountains of New York. This basin encompasses 33,060 km² (12,890 mi²) and includes parts of five states in the river's length, 595 km (370 mi) from headwaters to the mouth. The river system ends at Delaware Bay, which eventually enters the Atlantic Ocean. Topography varies from low-relief coastal plains to the Appalachian Mountains in the north. The entire basin lies within a humid, temperate climate and receives an average annual precipitation of 120 cm (47.2 in.).

The river provides water for an estimated 20 million people, not only within the basin but to cities outside the basin as well. Several major conduits export water from the Delaware River. Note on the map the Delaware Aqueduct to New York City (in the north) and the Delaware & Raritan Canal (near Trenton). Several reservoirs in the drainage basin allow some control over water flow and storage for dry periods. The sustainability of this water resource is critical to the entire region. Clearly, planning

FIGURE 14.5 Drainage basins and continental divides.
Continental divides (blue lines) separate the major drainage basins that empty into the Pacific,
Atlantic, Gulf of Mexico, and to the north through Canada into Hudson Bay and the Arctic
Ocean. Subdividing these large-scale basins are major river basins. [After U.S. Geological Survey;
The National Atlas of Canada, 1985, Energy, Mines, and Resources Canada; and Environment
Canada, *Currents of Change—Inquiry on Federal Water Policy—Final Report 1986.*]

for a drainage basin requires regional cooperation and
careful spatial analysis of all variables.

Internal Drainage Most streamwater finds its way to
progressively larger rivers and eventually into the ocean.
In some regions, however, stream drainage does not reach
the ocean. Instead, the water leaves the drainage basin by
means of evaporation or subsurface gravitational flow.
Such streams terminate in areas of **internal drainage**.

Portions of Asia, Africa, Australia, Mexico, and the
western United States have regions with such internal

drainage patterns. Internal drainage dominates the region
of the Great Basin of the western United States, as shown
on maps in Figures 14.5 and 15.23a. For example, the
Humboldt River flows across Nevada to the west, where
evaporation and seepage losses make it disappear into
the Humboldt "sink." Many streams and creeks flow into
the Great Salt Lake, but its only outlet is evaporation,
so it exemplifies a region of internal drainage. In the
Middle East, the Dead Sea region, and in Asia, the areas
around the Aral Sea and Caspian Sea, have no outlet to
the ocean.

Danube River delta

FIGURE 14.6 An international drainage basin—the Danube River.
The Danube crosses or forms the border of nine countries as it flows across Europe to the Black Sea. The river spews polluted discharge into the Black Sea through its arcuate-form delta. [*Terra* MODIS image, June 15, 2002, courtesy of MODIS Land Rapid Response Team, NASA/GSFC.]

Drainage Density and Patterns

A primary feature of any drainage basin is its drainage density. **Drainage density** is determined by dividing the total length of all stream channels in the basin by the area of the basin. The number and length of channels in a given area express the landscape's regional topography and surface appearance. For example, Figure 14.8 reveals a very high drainage density in a humid climate. In contrast, the typical desert has a very low drainage density.

The **drainage pattern** is the arrangement of channels in an area. Patterns are quite distinctive, for they are determined by the combination of regional steepness, variable rock resistance, variable climate, variable hydrology, relief of the land, and structural controls imposed by the underlying rocks. Consequently, the drainage pattern of any land area on Earth is a remarkable visual summary of every characteristic—geologic and climatic—of that region.

The *Landsat* image and topographic map in Figure 14.8 are of the Ohio River drainage near the junction of West Virginia, Ohio, and Kentucky. The high-density drainage pattern and intricate dissection of the land occur because the region has generally level, easily eroded sandstone, siltstone, and shale strata and a humid mesothermal climate. Fluvial action and other denudation processes are responsible for this dissected topography.

Common Drainage Patterns   The seven most common drainage patterns are shown in Figure 14.9. A most familiar pattern is *dendritic drainage* (Figure 14.9a). This treelike pattern (Greek *dendron*, or "tree") is similar to that of many natural systems, such as capillaries in the human circulatory system, the vein patterns in leaves, and tree roots. Energy expended by this drainage system is efficient because the overall length of the branches is minimized. Figure 14.8 shows dendritic drainage; on the satellite image and topographic map, you can trace the branching pattern of streams.

The *trellis drainage* pattern (Figure 14.9b) is characteristic of dipping or folded topography. Such drainage exists in the nearly parallel mountain folds of the Ridge and Valley Province in the eastern United States. Refer to Figure 12.18, where a satellite image of this region shows its distinctive drainage pattern. Here, drainage patterns are influenced by rock structures of variable resistance and folded strata. The principal streams are directed by the parallel folded structures, whereas smaller dendritic tributary streams are at work on nearby slopes, joining the main streams at right angles, like a plant trellis.

The inset sketch to Figure 14.9b suggests that a headward-eroding part of one stream (to the lower right of the inset) could break through a drainage divide and *capture* the headwaters of another stream in the next valley, and indeed this does happen. The dotted line is the abandoned former channel. The sharp bends in two of the streams in the illustration are called *elbows of capture* and are evidence that one stream has breached a drainage divide. This type of capture, or *stream piracy*, can also occur in other drainage patterns.

The remaining drainage patterns in Figure 14.9 are responses to other specific structural conditions:

- A *radial* drainage pattern (c) results when streams flow off a central peak or dome, such as occurs on a volcanic mountain.
- *Parallel* drainage (d) is associated with steep slopes.
- A *rectangular* pattern (e) is formed by a faulted and jointed landscape, which directs stream courses in patterns of right-angle turns.
- *Annular* patterns (f) are produced by structural domes, with concentric patterns of rock strata guiding stream courses. Figure 12.10c provides an example of annular drainage on a dome structure.

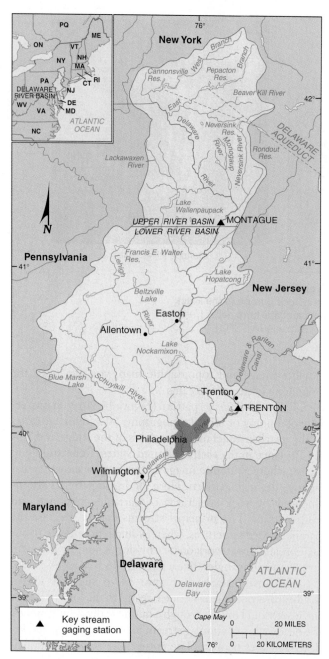

FIGURE 14.7 The Delaware River drainage basin.
This basin is under study by the USGS to assess the potential impact of global warming on a river system. [After U.S. Geological Survey, 1986, "Hydrologic events and surface water resources," *National Water Summary 1985,* Water Supply Paper 2300 (Washington, DC: Government Printing Office), p. 30.]

FIGURE 14.8 A stream-dissected landscape.
(a) A highly dissected topography (featuring a dendritic drainage pattern shown in Figure 14.9) around the junction of West Virginia, Ohio, and Kentucky. All of these borders are formed by rivers. (b) A portion of the USGS topographic map covering the area north of Huntington, West Virginia, reveals the intricate complexity of dissected landscapes. (The image and map are not at the same scale.) [(a) *Landsat* image from NASA; (b) Huntington Quadrangle, USGS.]

- In areas having disrupted surface patterns, such as the glaciated shield regions of Canada, northern Europe, and some parts of Michigan and other states, a *deranged* pattern (g) is in evidence, with no clear geometry in the drainage and no true stream valley pattern (see Figure 17.15a).

The structure and relief of the land dictate these seven drainage patterns. In Figure 14.9h, you see two distinct drainage patterns. Of the seven types illustrated, which two patterns are most like those in the aerial photo, made in central Montana?

Drainage patterns also occur that are discordant with the landscape through which they flow. For example, a drainage system may flow in apparent conflict with older, buried structures that have been uncovered by erosion, so that the streams appear to be *superimposed.* Where an

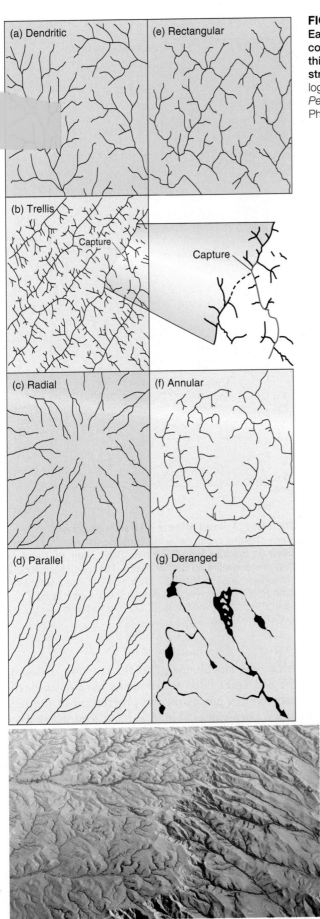

(a) Dendritic

(e) Rectangular

(b) Trellis

Capture

Capture

(c) Radial

(f) Annular

(d) Parallel

(g) Deranged

(h)

FIGURE 14.9 The seven most common drainage patterns.
Each pattern is a visual summary of all the geologic and climatic conditions of its region. (h) Two of these drainage patterns dominate this scene from central Montana, in response to local relief and rock structure. [(a) through (g) after A. D. Howard, "Drainage analysis in geological interpretation: A summation," *Bulletin of American Association of Petroleum Geologists* 51 (1967): 2248, adapted by permission. (h) Photo by Bobbé Christopherson.]

existing stream flows as rocks are uplifted, the stream keeps its original course, cutting into the rock in a pattern contrary to its structure. Such a stream is a *superposed stream* (the stream cuts across weak and resistant rocks alike). A few examples include Wills Creek, cutting a water gap through Haystack Mountain at Cumberland, Maryland; the Columbia River through the Cascade Mountains of Washington; and the River Arun that cuts across the Himalayas.

Streamflow Characteristics

A mass of water positioned above base level in a stream has potential energy. As the water flows downslope, or downstream, under the influence of gravity, this energy becomes kinetic energy. The rate of this conversion from potential to kinetic energy depends on the steepness of the stream channel.

Stream channels vary in *width* and *depth*. The streams that flow in them vary in *velocity* and in the *sediment load* they carry. All of these factors increase with increasing **discharge**, or the stream's volume of flow per unit of time. Discharge is calculated by multiplying the velocity of the stream by its width and depth for a specific cross section of the channel, as stated in the simple expression:

$$Q = wdv$$

where Q = discharge, w = channel width, d = channel depth, and v = stream velocity. As Q increases, some combination of channel width, depth, and stream velocity increases. Discharge is expressed either in cubic meters per second (m^3/s) or cubic feet per second (cfs).

Figure 14.10 illustrates the relation of discharge to width, depth, and velocity. The graphs show that mean velocity increases with greater discharge, despite the common misperception that downstream flow becomes more sluggish. (The increased velocity downstream often is masked by the apparent smooth, quiet flow of the water.) Not surprisingly, discharge also increases with width and depth.

Given the interplay of channel width and depth and stream velocity, the cross section of a stream varies over time, especially during heavy floods. Figure 14.11 shows changes in the San Juan River channel in Utah that occurred during a flood. Greater discharge increases the velocity and therefore the capacity of the river to transport sediment as the flood progresses. As a result, the river's ability to scour materials from its bed is enhanced. Such scouring represents a powerful clearing action, especially in the excavation of alluvium. Scouring might provide new recreational areas and wildlife habitats, as

FIGURE 14.11 A flood affects a stream channel.
(a) Stream-channel cross sections showing the progress of a 1941 flood on the San Juan River near Bluff, Utah. (b) Detail of stream-channel profile on September 9, 1941. [After L. Leopold and T. Maddock, Jr., *The Hydraulic Geometry of Stream Channels and Some Physiographic Implications,* U.S. Geological Survey Professional Paper 252 (Washington, DC: Government Printing Office, 1953), p. 32.]

FIGURE 14.10 Effects of stream discharge.
(a) Relation of stream width, depth, and velocity to stream discharge of the Powder River at Locate, Montana. Discharge is shown in cubic meters per second (m³/s) and cubic feet per second (cfs). (b) The Powder River area near the Wyoming border in southeastern Montana. [(a) After L. Leopold and T. Maddock, Jr., *The Hydraulic Geometry of Stream Channels and Some Physiographic Implications,* U.S. Geological Survey Professional Paper 252 (Washington, DC: Government Printing Office, 1953), p.7; (b) photo by Joyce Wilson.]

explored by the Grand Canyon experiment in News Report 14.1.

You can see in Figure 14.11 that the San Juan River's channel was deepest on October 14, when floodwaters were highest (blue line). Then, as the discharge returned

to normal, the kinetic energy of the river was reduced, and the bed again filled as sediments were redeposited. You can see this process in the progression of the red, green, and purple lines. The flood and scouring process graphed in Figure 14.11 moved a depth of about 3 m (10 ft) of sediment from this cross section of the stream channel. Such adjustments in a stream channel occur as the stream system continuously works toward equilibrium, in an effort to balance discharge, velocity, and sediment load.

Exotic Streams

Most streamflows increase discharge downstream because the area being drained increases. The Mississippi River is typical. It starts as many small brooks and grows to a mighty river pouring into the Gulf of Mexico. In contrast, a stream can originate in a humid region and subsequently flow through an arid region. In that case, the *discharge usually decreases with distance,* because of high

News Report 14.1

Scouring the Grand Canyon for New Beaches and Habitats

Glen Canyon Dam sealed the Colorado River gorge north of the Grand Canyon, near Lees Ferry and the Utah-Arizona border (see the map in Focus Study 15.1, Figure 15.1.1). Impoundment of water began in 1963, as did the inundation by the new Lake Powell of many canyons upstream from the dam. The lake began collecting the tremendous sediment load of the Colorado River, as dam operations reduced the volume of water in the river downstream, eliminating the natural seasonal fluctuations in river discharge. The production of hydroelectricity at Glen Canyon further affected the Grand Canyon with highly variable water releases keyed to electrical turbine operations. Water in the canyon rose and fell as much as 4.3 m (14 ft) in a 24-hour period as the distant lights and air-conditioning

of Las Vegas and Phoenix went on and off.

All of these changes affected the canyon. Over the years, the beaches were starved for sand, channels filled with sediment, fisheries were disrupted, and backwater channels became depleted of nutrients. In 1996, an unprecedented experiment on the Colorado River took place. The Grand Canyon was artificially flooded with a 1260-m³/s (45,000-cfs) release from Glen Canyon (Figure 14.1.1). The flow lasted for seven days and then was reduced to 224 m³/s (8000 cfs). Flows were decreased gradually to allow the newly formed beaches to drain, in contrast to the rapid opening of the pipes that initiated the flood. Lake Powell dropped 1 m (3.3 ft), and Lake Mead, downstream behind Hoover Dam, rose 0.8 m (2.6 ft).

The results were initially thought to be positive: 35% more beach area was created by the scouring of sand and sediments from the channel. Approximately 80% of the aggradation (building up of beaches) took place in the first 40 hours and was completed by the 100-hour mark. Numerous backwater channels were created, flush with fresh nutrients for the humpback chub and other endangered fish species. However, benefits turned out to be limited, and negatives, in the form of existing ecosystem disruption and immediate losses of some sediment deposits, marred the experiment. Present thinking is that the lack of sediment to supply the floodwaters for downstream deposition and beach building was the problem. Some of the initial beach deposits quickly disappeared. For more on this experiment, see http://water.usgs.gov/pubs/FS/FS-060-99/.

In November 2004, hydrologists conducted a second test. Scientists timed the artificial flood with natural peak runoff that supplies fresh sediment to the system. Also, instead of seven days, the artificial flood releases from Glen Canyon lasted 60 hours (2.5 days). The release coincided with local storms that fed tributaries, such as the Paria River, with sediment to accomplish the task of rebuilding beaches downstream. Tributaries deliver more than three million tonnes of sediment to the Colorado River and the experiment was another attempt at sediment distribution. (See Grand Canyon Monitoring and Research Center for progress reports and analysis at http://www.gcmrc.gov/.)

FIGURE 14.1.1 Glen Canyon Dam floods the Grand Canyon, on purpose. Bypass tubes dramatically discharge water from Lake Powell in a 1996 experiment to correct the damaging effects of the dam on the Grand Canyon. Note the workers on the walkway for scale, experiencing a deafening roar and vibration. Another modified experiment took place in 2004. [Photo by Tom Smart, Deseret News.]

---

potential evapotranspiration rates in the arid area. Such a stream is called an **exotic stream** (*exotic* means "of foreign origin").

The Nile River exemplifies exotic streams. This great river, Earth's longest, drains much of northeastern Africa. But as it courses through the deserts of Sudan and Egypt,

it loses water instead of gaining it, because of evaporation and withdrawal for agriculture. By the time it empties into the Mediterranean Sea, the Nile's flow has dwindled so much that it ranks only 36th in discharge.

The United States has exotic streams too, notably the Colorado River. Its flow decreases with distance from its

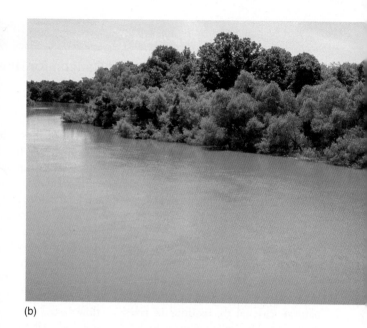

(b)

(a)

FIGURE 14.12 Stream velocity and discharge increase together.
(a) A low-discharge, low-velocity (but turbulent) mountain stream in the White Mountains, New Hampshire. (b) The high-discharge, high-velocity (but laminar, or smooth water flow) portion of the Ocmulgee River near Jacksonville, Georgia. [Photos by Bobbé Christopherson.]

source; in fact, the river no longer produces enough discharge to reach its mouth in the Gulf of California! The exotic Colorado River is depleted not only by passage across dry, desert lands but also by upstream removal of water for agriculture and municipal uses—see Focus Study 15.1 in Chapter 15 for a detailed discussion of this river and a satellite image of the river's former mouth.

Stream Erosion

A stream's erosional turbulence and abrasion carve and shape the landscape through which it flows. **Hydraulic action** is the work of flowing water alone. Running water causes hydraulic squeeze-and-release action that loosens and lifts rocks. As this debris moves along, it mechanically erodes the streambed further, through the process of **abrasion**, with rock particles grinding and carving the streambed like liquid sandpaper.

The upstream tributaries in a drainage basin usually have small and irregular discharges, and most of the stream's energy is expended in turbulent eddies. As a result, hydraulic action in these upstream sections is at a maximum, whereas the coarse-textured load of such a stream is small. The downstream portions of a river, however, move much larger volumes of water past a given point and carry larger suspended loads of sediment (Figure 14.12). Stream velocity determines rates of erosion and deposition. Sediment particles are deposited onto the streambed at slower velocities, whereas they are eroded at higher velocities.

Stream Transport

You may have watched a river or creek after a rainfall, the water colored brown by the heavy sediment load being transported. The amount of material available to a stream depends on topographic relief, the nature of rock and soil through which the stream flows, climate, vegetation, and human activity in a drainage basin. *Competence*, which is a stream's ability to move particles of a specific size, is a function of stream velocity. *Capacity* is the total possible load that a stream can transport. Four processes transport eroded materials: solution, suspension, saltation, and traction; each is shown in action in Figure 14.13.

Solution refers to the **dissolved load** of a stream, especially the chemical solution derived from minerals such as limestone or dolomite or from soluble salts. The main contributor of material in solution is chemical weathering. Sometimes the undesirable salt content that hinders human use of some rivers comes from dissolved rock formations and from springs in the stream channel; as an example, the San Juan and Little Colorado Rivers that flow into the Colorado River near the Utah–Arizona border add dissolved salts to the system.

The **suspended load** consists of fine-grained, clastic particles (bits and pieces of rock). They are held aloft in the stream, with the finest particles not deposited until the stream velocity slows nearly to zero. Turbulence in the water, with random upward motion, is an important mechanical factor in holding a load of sediment in suspension.

FIGURE 14.13 Fluvial transport.
Fluvial transportation of eroded materials through suspension, solution, saltation, and traction.

Bed load refers to coarser materials that are dragged, rolled, or pushed along the streambed by **traction** or by **saltation**, where particles bounce along in short hops and jumps (from the Latin *saltim*, which means "by leaps or jumps"). Particles transported by saltation are too large to remain in suspension but are not limited to the sliding and rolling motion of traction. These processes relate directly to a stream's velocity and its ability to retain particles in suspension. With increased kinetic energy, parts of the bed load are rafted upward and become suspended load. Saltation is also a process in wind transport of materials (see Chapter 15).

The early explorers who visited the Grand Canyon reported in their journals that they were kept awake at night by the thundering sound of the river and rapids. Imagine the tremendous quantity of material being moved along by the natural Colorado River before any dams were built. Dams and reservoirs now trap sediments that formerly contributed to the river's bed load.

If the load (bed and suspended) exceeds a stream's capacity, sediments accumulate as **aggradation** (the opposite of degradation) and the stream channel builds up through deposition. With excess sediment, a stream becomes a maze of interconnected channels that form a **braided stream** pattern (Figure 14.14). Braiding often occurs when reduced discharge lowers a stream's transporting ability, such as after flooding, or when a landslide occurs upstream, or from increased load where weak banks of sand or gravel exist. Locally, braiding also may result from a new sediment load from glacial meltwaters, as in three examples: Alaska's Chitina River, a river in Tibet, and a stream on Spitsbergen Island in the Arctic Ocean. Glacial action and erosion produced the materials that exceed stream capacity in all three locations.

Flow and Channel Characteristics

Streams have two general types of flow, laminar and turbulent. *Laminar flow* is a streamlined flow of water in which individual clay and other fine particles move along evenly in generally parallel paths. Natural streams have stretches of laminar flow only when the water is deep and the channel surfaces are smooth.

Flow becomes *turbulent* in shallow streams, or where the channel is rough, as in a section of rapids. Small eddies are caused by friction between streamflows and the channel sides and bed. Complex turbulent flows propel sand, pebbles, and even boulders, increasing the suspension, traction, and saltation.

Flow characteristics of a stream are best seen in a cross-sectional view. The greatest velocities in a stream are near the surface at center channel (Figure 14.15), corresponding to the deepest part of the stream channel. Velocities decrease closer to the sides and bottom of the channel because of the frictional drag on the water flow. The portion of the stream flowing at maximum velocity moves diagonally across the stream from bend to bend.

Laminar and turbulent flow form three types of stream channels: *braided, straight,* and *meandering.* Where slope is gradual, stream channels develop a sinuous (snakelike) form, weaving across the landscape. This action produces a **meandering stream**. The term *meander* comes from the ancient Greek Maiandros River in Asia Minor (the present-day Menderes River in Turkey), which had what came to be known as a meandering channel pattern. The tendency to meander is evidence of a river system's struggle to operate with least effort, between self-organizing order (equilibrium) and chaotic disorder in nature.

The outer portion of each meandering curve is subject to the fastest water velocity and therefore the greatest scouring erosive action; it can be the site of a steep bank called an **undercut bank**, or *cutbank* (Figures 14.15 and 14.16). In contrast, the inner portion of a meander experiences the slowest water velocity and thus receives sediment fill, forming a deposit called a **point bar**. As meanders develop, these scour-and-fill features gradually work their way downstream. As a result, the landscape near a meandering river bears meander scars of residual deposits from previous river channels (see Figure 14.22). The photograph of the Itkillik River in Alaska, Figure 14.16a, shows both meanders and meander scars.

Meandering streams create a remarkable looping pattern on the landscape. Meanders gradually form loops, as shown in the four-part sequence in Figure 14.16b. In (1), the stream erodes its outside bank as the curve migrates

(a)

(b)

(c)

FIGURE 14.14 Braided streams.
(a) Braided stream pattern in Chitina River, Wrangell–Saint Elias National Park, Alaska. (b) A 15-km (9.3-mi) stretch of the braided Brahmaputra River channel some 35 km (22 mi) south of Lhasa, Tibet, in a narrow valley south of the Tibetan Plateau. (c) A braided stream of meltwaters on Spitsbergen Island in the Arctic Ocean. Each of these streams reflects excessive sediment load associated with glacial meltwaters filled with fine sediments, or "glacial flour." [Photos by (a) Tom Bean; (b), International Space Station astronaut, October 13, 2001, courtesy of Earth Science and Image Analysis Lab, JSC/NASA; (c) Bobbé Christopherson.]

downstream, forming a neck. In (2), the narrowing neck of land created by the looping meander eventually erodes through and forms a *cutoff* in (3). A cutoff marks an abrupt change in the stream's lateral movements—the stream becomes straighter.

When the former meander becomes isolated from the rest of the river, the resulting **oxbow lake** (4) may gradually fill with silt or may again become part of the river when it floods. The Mississippi River is many miles shorter today than it was in the 1830s because of artificial cutoffs that were dredged across meander necks to improve navigation and safety. How many of these stream features—meanders, oxbow lakes, cutoffs—can you spot in the aerial photo in Figure 14.17, p. 447, looking east across northern Argentina? News Report 14.2 looks at the difficulty of using a meandering stream as a political boundary.

Stream Gradient

Every stream runs downhill under the pull of gravity. In the course of this process, each stream develops its own **gradient**, which is the rate of elevation decline from its headwaters to its mouth. This decline is far from linear; characteristically, the *longitudinal profile* (side view) of a stream features a steeper slope upstream and a gentler slope downstream, forming an uneven concave shape (Figure 14.18). This gradient is concave for complex reasons related to the stream's having just enough energy to transport the sediment load it receives. The longitudinal profile of streams can be expressed mathematically, enabling scientists to categorize and predict stream behavior.

An important fluvial concept is that of the graded stream, nicely defined by J. H. Mackin, a geomorphologist:

A graded stream is one in which, over a period of years, slope is delicately adjusted to provide, with available discharge and with prevailing channel characteristics, just the velocity required for transportation of the load supplied from the drainage basin.*

*J. H. Mackin, "Concept of the graded river," *Geological Society of America Bulletin* 59 (1948): 463.

Stream Processes, Floodplains, Oxbow Lake Formation

ANIMATION

Areas of maximum velocity

Maximum velocity

Point bar deposit

Undercut bank erosion

Pool (deep)

FIGURE 14.15 Meandering stream profile.
Aerial view and cross sections of a meandering stream, showing the location of maximum flow velocity, point-bar deposits, and areas of undercut bank erosion.

News Report

News Report 14.2

Rivers Make Poor Political Boundaries

Streams often serve the role of natural political boundaries. Commonly, where a stream forms a boundary, the boundary line is drawn down the middle of the stream. It is easy to see how boundary disputes might arise when they are based on river channels that shift their positions quite rapidly during times of flood. The boundaries based on the original stream channels are not altered, and existing land ownership remains set.

Carter Lake, Iowa, provides a fascinating example (Figure 14.2.1). The Nebraska-Iowa border originally was placed midchannel in the Missouri River. But in 1877, the meander loop that curved around the town of Carter Lake in Iowa was cut off by the river, leaving the town "captured" by Nebraska! The old boundary marked along the former meander bend still is used as the state line. The oxbow lake created by the cutoff was named Carter Lake.

This event illustrates why boundaries always should be fixed by surveys independent of river locations. Such surveys have been completed along the Rio Grande near El Paso, Texas, and along the Colorado River between Arizona and California, permanently establishing political boundaries separate from shifting river channels. In the latter example, a midpoint between the bluffs on either side of the floodplain is used as the state line. Meanwhile, Carter Lake is still in Iowa.

FIGURE 14.2.1 A stranded town.
Carter Lake, Iowa, sits within the curve of a former meander that was cut off by the Missouri River. The city and oxbow lake remain part of Iowa even though they are stranded within Nebraska.

445

(a)

(b)

FIGURE 14.16 Meandering stream development.
(a) Itkillik River in Alaska. (b) Development of a river meander and oxbow lake simplified
in four stages. [(a) U.S. Geological Survey photo.]

In other words, a **graded stream** is one that attains a graded condition; the term does not mean that the stream is at its lowest gradient. Rather, graded represents a present balance, a dynamic equilibrium among erosion, transportation, and deposition over time along a portion of the stream. Both high-gradient and low-gradient streams can achieve a graded condition. The longitudinal profile of a graded stream is called a *profile of equilibrium*—a parabolic curve, gently flattening toward the mouth. Stream dynamics can then be compared against this ideal balance.

One problem with applying the graded-stream concept is that an individual stream can have both graded and ungraded portions and may have graded sections without having an overall graded slope. Variations and interruptions are the rule rather than the exception. A profile of equilibrium may not be smooth throughout its course and cannot exist for long, for it represents a theoretical perfect balance. With streams, as in all of nature, change is the only constant.

Stream gradient may be affected by tectonic uplift of the landscape, which changes the base level. If tectonic

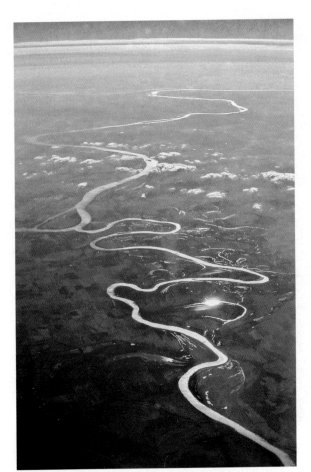

FIGURE 14.17 Meandering stream features.
Meandering streams catch the morning light over northern
Argentina. Identify where former meanders were cut off,
oxbow lakes, and former stream channels. Several inversion
layers are visible in the far distance, as the burning of tropical
rain forest to the north fills the air with smoke. [Photo by
Bobbé Christopherson.]

Nickpoints When the longitudinal profile of a stream
shows an abrupt change in gradient, such as at a waterfall
or an area of rapids, the point of interruption is termed a
nickpoint (also spelled *knickpoint*). At a nickpoint, the
conversion of potential energy to concentrated kinetic
energy works to eliminate the point of interruption and
smooth out the gradient. Figure 14.20a shows a stream
with two such interruptions. The photos show nickpoints
and falls in New Hampshire and South Dakota.

Nickpoints can result when a stream flows across a
zone of hard, resistant rock or from various tectonic uplift
episodes, such as might occur along a fault line. Temporary blockage in a channel, caused by a landslide or a logjam, also could be considered a nickpoint; when the
logjam breaks, the stream quickly readjusts its channel to
its former grade.

Waterfalls are interesting and beautiful gradient
breaks (Figure 14.20b and c). At the edge of a fall, a
stream is free-falling, moving at high velocity under the
acceleration of gravity, causing increased abrasion on the
channel below. The increased abrasion and hydraulic action generally undercut the waterfall. Eventually the excavation will cause the rock ledge at the lip of the fall to
collapse, and the waterfall will shift a bit farther upstream.
The height of the waterfall is gradually reduced as debris
accumulates at its base. Thus, a nickpoint migrates upstream, sometimes for kilometers, until it becomes a series
of rapids and is eventually eliminated.

At Niagara Falls on the Ontario–New York border,
glaciers advanced over the region and then receded. In
doing so, they exposed resistant rock strata that are underlain by less-resistant shales. This formation is a *cuesta*,
which is a ridge with a cliff on one side and beds gently
sloping away on the other side. This Niagara escarpment
actually stretches across more than 700 km (435 mi) from
east of the falls, northward through Ontario, Canada, the
Upper Peninsula of Michigan, curving south through
Wisconsin along the western shore of Lake Michigan and
the Door Peninsula. As this less-resistant material continues to weather away, the overlying rock strata collapse,
allowing Niagara Falls to erode farther upstream toward
Lake Erie (Figure 14.21a).

Niagara Falls is a place where natural processes labor
to eliminate a nickpoint, reducing this portion of the river
to a series of mere rapids. In fact, the falls have retreated

forces slowly lift the landscape, the stream gradient will
increase, stimulating renewed erosional activity. The meandering stream flowing through the uplifted landscape
becomes *rejuvenated;* that is, the river actively returns to
downcutting and eventually forms *entrenched meanders* in
the landscape. Figure 14.19 depicts actual rejuvenated
landscapes. Such rejuvenation also is occurring in
Figure 14.2.

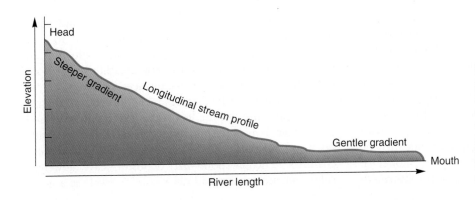

**FIGURE 14.18 An ideal longitudinal
profile.**
Idealized longitudinal profile of a
stream, showing its gradient. Upstream segments have a steeper gradient; downstream, the gradient is
gentler. The middle and lower portions
in the illustration appear graded, or in
dynamic equilibrium.

(a)

(b)

FIGURE 14.19 Entrenched meanders. (a) and (b) The San Juan River near Mexican Hat, Utah, cuts down into the uplifted Colorado Plateau landscape, producing entrenched meanders called the Goosenecks of the San Juan. (c) Entrenched meanders merge in central Montana. [Photos (a) and (c) by Bobbé Christopherson; (b) by Randall M. Christopherson.]

(c)

more than 11 km (6.8 mi) from the steep face of the Niagara escarpment during the last 12,000 years. A nickpoint is a relatively temporary and mobile feature on the landscape. In the past, engineers have used control facilities upstream to reduce flows over the American Falls at Niagara for inspection of the cliff (Figure 14.21b).

Stream Gradient and Landscape Forms In Chapter 13, we introduced William Morris Davis, a geomorphologist who founded the Association of American Geographers in 1904. His evolutionary concepts of erosion also included fluvial processes. Drawing from G. K. Gilbert, an important geomorphologist and his contemporary, Davis incorporated the graded-stream concept into his model, identifying erosion stages in a cyclic model he called *youth, maturity,* and *old age*—old age is approached as the floodplain broadens and a low stream gradient produces a wide, meandering flow pattern.

Today, a *functional model of dynamic equilibrium* is supported by geomorphologists. The dynamic equilibrium model emphasizes the effects of individual processes interacting on streams and hillslope systems. Stream form and behavior result from complex interactions of slope, discharge, and sediment load, all of which are variable within different climates and with different rock types. Landscapes simply do not provide enough clear evidence to support a cyclic model of evolution. Regardless, Davis's work was a breakthrough

in understanding landscapes, and many of his terms are still in use.

As suggested by S. A. Schumm and R. W. Lichty, two modern geomorphologists, the validity of cyclic or functional landscape models may depend on the *time frame*. Let us consider three time frames: geologic time, graded time, and steady time. Over the long span of *geologic time*, cyclic models of evolutionary development might explain the disappearance of entire mountain ranges through, for example, denudation. At the other extreme of time, *steady time* applies to short-term adjustments ongoing in a drainage basin. *Graded time* is between the two time frames, and within it lies the realm of dynamic equilibrium conditions.

Stream Deposition

After weathering, mass movement, erosion, and transportation, deposition is the next logical event in a sequence. In *deposition*, a stream deposits alluvium, or unconsolidated sediments, thereby creating depositional landforms, such as floodplains, terraces, or deltas.

As discussed earlier, stream meanders tend to migrate downstream through the landscape. Over time, the landscape near a meandering river comes to bear meander scars of residual deposits from former, abandoned channels. Former point-bar deposits leave low-lying ridges, creating a *bar-and-swale relief* (a swale is a gentle

(a)

(b)

(c)

FIGURE 14.20 Nickpoints interrupt a stream profile.
(a) Longitudinal stream profile showing nickpoints produced by resistant rock strata. Potential energy is converted into kinetic energy and concentrated at the nickpoint, accelerating erosion, which will eventually eliminate the feature. (b) A nickpoint and granite pothole interrupts stream gradient on the Pemigewassey River, Franconia Notch Park, New Hampshire. (c) Falls and rapids break the flow of the Big Sioux River in Sioux Falls, South Dakota. [Photos by Bobbé Christopherson.]

low area). The *Landsat* image in Figure 14.22 exhibits characteristic meandering scars: meander bends, oxbow lakes, natural levees, point bars, and undercut banks.

Floodplains The flat, low-lying area flanking a stream channel that is subjected to recurrent flooding is a **floodplain**. It is formed when the river overflows its channel during times of high flow. Thus, when floods occur, the floodplain is inundated. When the water recedes, it leaves behind alluvial deposits that generally mask the underlying rock with their accumulating thickness. The present river channel is embedded in these alluvial deposits. Figure 14.23 illustrates a characteristic floodplain, related features in photos, and a representative topographic map of an area near Philipp, Mississippi.

On either bank of some streams, **natural levees** develop as by-products of flooding. When floodwater arrives, the river overflows its banks, loses velocity as it

spreads out, and drops a portion of its sediment load to form the levees. Larger, sand-sized particles drop out first, forming the principal component of the levees, with finer silts and clays deposited farther from the river. Successive floods increase the height of the levees (*levée* is French for "raising"). The levees may grow in height until the river channel becomes elevated, or *perched*, above the surrounding floodplain.

On the topographic map (Figure 14.23d), you can see the natural levees represented by several contour lines that run immediately adjacent to the Tallahatchie River. These contour lines (5-ft interval) denote a height of 10–15 ft (3–4.5 m) above the river and the adjoining floodplain. Next time you have an opportunity to see a river and its floodplain, look for levees (they may be low and subtle).

Notice in Figure 14.23a an area labeled backswamp and a stream called a yazoo tributary. The natural levees and elevated channel of the river prevent this **yazoo**

(a)

(b)

FIGURE 14.21 Retreat of Niagara Falls.
(a) Headward retreat of Niagara Falls from the Niagara escarpment. It has taken the falls about 12,000 years to reach this position at a pace of about 1.3 m (4.3 ft) per year. (b) Niagara Falls, with the American Falls portion almost completely shut off by upstream controls for inspection. Horseshoe Falls in the background is still flowing. Such inspections allow engineers to assess the progress of natural processes that are working to eliminate the Niagara Falls nickpoint.
[(a) After W. K. Hamblin, *Earth's Dynamic Systems,* 6th ed. (Upper Saddle River, NJ: Macmillan Publishing, an imprint of Prentice Hall, Inc. © 1992), Figure 12.15, p. 246. Used by permission. (b) Photo courtesy of the New York Power Authority.]

NOTEBOOK Niagara Falls

(a)

FIGURE 14.22 Meander scars.
The Mississippi River forms a portion of the Mississippi–Arkansas border near Senatobia, Mississippi. Characteristic meander patterns and scars of former channels are visible in the *Landsat* image. The image of northwestern Mississippi shows the portion of the Mississippi River that forms the Mississippi–Arkansas border.
[Image by GEOPIC, Earth Satellite Corporation. Used by permission.]

(b)

(a) Typical floodplain landscape and related landscape features.

Meander scar

Meandering, graded stream

Oxbow lake

Meander scars

Cutoff

Point bar

Yazoo tributary

Bluffs

Undercut bank

(b)

Alluvial deposits

(a)

Natural levees

Backswamp

(c)

(d)

FIGURE 14.23 A floodplain.
(a) Typical floodplain landscape and related landscape features. (b) An oxbow lake. (c) Levees, oxbow lakes, farmland in a floodplain. (d) A portion of the Philipp, Mississippi, topographic map quadrangle. [(b) and (c) photos by Bobbé Christopherson; (d) topographic map from USGS.]

ANIMATION

Stream Processes, Floodplains, Oxbow Lake Formation

FIGURE 14.24 Flooding river and threatened levees.
Emergency levee repairs to prevent further erosion by raindrop impact and seepage from the river. The floodplain to
the left is below the level of the Sacramento River.
[Photo by California Department of Water Resources.]

tributary from joining the main channel, so it flows parallel to the river and through the **backswamp** area. (The
name comes from the Yazoo River in the southern part of
the Mississippi floodplain.)

People build cities on floodplains despite the threat of
flooding because floodplains are nearly level and they are
next to water. People often are encouraged by government assurances of artificial protection from floods and
disaster assistance if floods occur. Government assistance
may provide for the building of artificial levees on top of
natural levees. Artificial levees do increase the capacity in
the channel, but they also lead to even greater floods
when they are overtopped by floodwaters or when they
fail (Figure 14.24). Are there river or creek floodplains
where you live? If so, what is your impression of present
land-use patterns, local planning and zoning, and people's
hazard perception overall?

The catastrophic floods along the Mississippi River
and its tributaries in 1993 illustrate the risk of building
settlements on floodplains. Damage estimates from those
floods, detailed in News Report 14.3, exceeded $30 bil-
lion. In 2001, some two-thirds of overall disaster losses
worldwide were attributable to floods. Tropical storm
Allison left $6 billion in damage in its wandering visit to
Texas in June 2001; most damage occurred from flooding
along occupied floodplains.

Stream Terraces   As explained earlier, several factors
may rejuvenate stream energy and stream–landscape relations so that a stream can scour downward with renewed
vigor and increased erosion. The resulting entrenchment
of the river deeper into its own floodplain produces
alluvial terraces on either side of the valley, which look
like topographic steps above the river. Alluvial terraces
generally appear paired at similar elevations on each side
of the valley (Figure 14.25). If more than one set of paired

terraces is present, the valley probably has undergone
more than one episode of rejuvenation. The flat terrace
areas, above the lowest section of floodplain along the
river, have always been a location for settlement.

If the terraces on either side of the valley do not
match in elevation, then entrenchment actions must have
been continuous as the river meandered from side to side,
with each meander cutting a terrace slightly lower in
elevation—a condition of *unpaired terraces*. Thus, alluvial
terraces represent what originally was a depositional feature (a floodplain) that subsequently has been eroded by
the rejuvenated stream.

River Deltas   The mouth of a river is where it reaches a
base level. The river's forward velocity rapidly decelerates
as it enters a larger body of water. The reduced velocity
causes the sediment load to quickly exceed the river's
ability to transport it. Coarse sediments such as sand and
gravel drop out first and are deposited closest to the river's
mouth. Finer clays are carried farther and form the extreme end of the deposit. The depositional plain that
forms at the mouth of a river is called a **delta** for its characteristic triangular shape, named after the Greek letter

Alluvial terraces (paired)

(a)

(b)

FIGURE 14.25 Alluvial stream terraces.
(a) Alluvial terraces are formed as a stream cuts into a valley.
(b) Alluvial terraces along the Rakaia River in New Zealand.
[(a) After W. M. Davis, *Geographical Essays* (New York: Dover,
1964 [1909]), p. 515; (b) photo by Bill Bachman/Photo
Researchers, Inc.]

News Report 14.3

The 1993 Midwest Floods

The Mississippi and Missouri Rivers are no strangers to flooding. The damage a flood might cause increases as more people settle on the vulnerable floodplains. The widespread floods of 1993 in the upper Mississippi and lower Missouri River basins exceeded peak discharge records at nearly 100 gaging (stream-measuring) stations, making them some of the greatest floods in U.S. history.

In spring 1993, a series of low-pressure systems (with their counterclockwise winds) stalled in the West, and high pressure (with its clockwise winds) dominated the Eastern Seaboard. These systems combined to produce a region of sustained convergence, instability, and thundershowers over the Midwest. Precipitation was 150%–200% above normal for most cities in the flooded area (Figure 14.3.1a).

By late June, many reservoirs were filled and soils were saturated by record rainfall over the region. Then came July, when some stations received 75 cm (30 in.) of precipitation in the first 3 weeks! Three remarkable aspects of "The Great Flood of 1993" were that the flood stage along some rivers was sustained for more than a month, flood crests set new historical records, and many locations experienced multiple crests.

As 10,000 km (6200 mi) of levees were overtopped, more than 1000 levees were breached (broken through). Water reached almost 10 m (32 ft) above flood stage in some areas, flooding many cities and towns and 6 million hectares (14 million acres) of forest and farmland in the Missouri and Mississippi River drainage basins (Figure 14.3.1b). The result was a Presidential Disaster Declaration for 534 counties in Illinois, Iowa, Kansas, Minnesota, Missouri, the Dakotas, Nebraska, and Wisconsin. Overall, damage estimates topped $30 billion.

Some cities had prepared for the disaster through improved levee construction and hazard zoning of susceptible floodplains. Many others had postponed raising local taxes for such action and had done little to prevent flood damage. This unevenness between areas created political friction. This flood event was a painful reminder of nature's unpredictability and the need for improved hazard perception, preparation, and avoidance strategies. Yet, we find settlement on the floodplain and on former sandbars in the Mississippi River (Figure 14.3.2a). In Figure 14.3.2b we see new development sites being claimed in 2003 from the same river. For imagery of these 1993 floods, see **http://rsd.gsfc.nasa.gov/rsd/images/Flood_cp.html**, and for a brief paper, see **http://www.nwrfc.noaa.gov/floods/papers/oh_2/great.htm**.

(a)   (b)

FIGURE 14.3.1 Summer 1993 weather map summary.
(a) Weather pattern summarized for the summer of 1993. Atmospheric conditions produced a steady flow of moisture-rich, unstable, maritime tropical air from the Gulf of Mexico for a prolonged period of time. (b) A farm 32 km (20 mi) south of Des Moines, Iowa, is inundated during the Midwest floods of 1993. [Photo by Les Stone/Sygma.]

(continued)

News report 14.3 *(continued)*

(a)   (b)

FIGURE 14.3.2 Living and building in a floodplain.
Near Harpers Ferry, Iowa, (a) homes are located on a forested sandbar in the Mississippi River. (b) New fill is being dumped into the river to extend the land for further development. The river is at a high water level in the photo. [Photos by Bobbé Christopherson.]

delta Δ. The significance of the Nile River delta to food production was perceived by Herodotus in ancient times (see News Report 14.4).

Each flood stage deposits a new layer of alluvium over the surface of the delta so that it grows outward. At the same time, river channels divide into smaller courses known as *distributaries*, which appear as a reverse of the dendritic drainage pattern of tributary streams discussed earlier. Here are a few examples:

- The Ganges River delta features an intricate pattern of distributaries in a *braided delta*. Bountiful alluvium carried from deforested slopes upstream provides excess sediment that is deposited to form many deltaic islands (Figure 14.26). The combined Ganges–Brahmaputra River delta complex is the largest in the world.
- The Nile River delta is an *arcuate* (arc-shaped) delta (Figure 14.27). Also arcuate are the Danube River delta in Romania, where it enters the Black Sea, and the Indus River delta. (See News Report 14.4 for an update on the condition of the disappearing Nile delta.)
- The Tiber River in Italy has an *estuarine delta*, one that is in the process of filling an **estuary**, which is the seaward mouth of a river where the river's freshwater encounters seawater.

Mississippi River Delta   The Mississippi River delta has an interesting history. Over the past 120 million years, the Mississippi has collected sediments throughout its vast basin and deposited them into the Gulf of Mexico. During the past 5000 years, the river has formed a succession of seven distinct deltaic complexes along the Louisiana coast (Figure 14.28a).

Each generalized lobe in the illustration reflects distinct course changes in the Mississippi River, probably

where the river broke through its natural levees during episodes of severe flooding, thus changing the configuration of the delta. The seventh and current delta has been building for at least 500 years and is a classic example of a *bird-foot delta*—a long channel with many distributaries and sediments carried beyond the tip of the delta into the Gulf of Mexico.

The Mississippi River delta clearly is dynamic over time as sediments accumulate on the floor of the Gulf of Mexico and the distributaries shift (Figure 14.28b). The main channel persists because of much effort and expense directed at maintaining the artificial levee system. The 3.25-million-km² (1.25-million-mi²) Mississippi drainage basin produces enough sediment to extend the Louisiana coast 90 m (295 ft) a year—550 million metric tonnes a year—however, several factors are actually causing losses to the delta each year.

Compaction and the tremendous weight of the sediments in the Mississippi River create isostatic adjustments in Earth's crust. These adjustments are causing the entire region of the delta to subside, thereby placing ever-increasing stress on natural and artificial levees and other structures along the lower Mississippi. The many canals and waterways plowed through the delta by the oil and gas industry and shipping interests have worked to deprive the delta of new alluvial building materials. Also, the pumping of tremendous quantities of oil and gas from thousands of onshore and offshore wells is thought to be a cause of regional land deflation and lowering.

The city of New Orleans is now almost entirely below river level, with sections of the city below sea level. Severe flooding is a certainty for existing and planned settlements unless further intervention or urban relocation occurs. The building of multiple flood-control structures and extensive reclamation efforts by the U.S. Army Corps

News Report 14.4

The Nile Delta Is Disappearing

People along the Nile River have depended on its regular flow and annual floods for millennia. Herodotus noted in the 5th century B.C. how the people farmed the fields in the floodplain and delta regions. After harvesting their crops, they retreated from the area to their homes. They would await the annual floods that brought fresh silt and nutrients for next year's planting. This cycle of fertility continued until the completion of the Aswân High Dam in 1964. This structure caused a decrease in the supply of sediment to the delta, and partially as a result the delta coastline continues to actively recede. Herodotus stated in *The History*, Book Two, "In the part called the Delta, it seems to me that if the Nile no longer floods . . . for all time to come, the Egyptians will suffer."

J. Stanley, an oceanographer at the Smithsonian Institution, has proposed an intriguing explanation for the Nile Delta's recession, and one that goes beyond the impacts of the Aswân High Dam. Over the centuries, more than 9000 km (5500 mi) of canals were built in the delta to augment the natural distributary system. As the river discharge enters the network of canals, flow velocity is reduced, stream competence and capacity are lost, and sediment load is deposited far short of where the delta touches the Mediterranean Sea. River flows no longer effectively reach the sea.

The Nile delta is receding from the coast at an alarming 50 to 100 m (165 ft to 330 ft) per year. Seawater is intruding farther inland in both surface water and groundwater. Human action and reaction to this evolving situation will no doubt determine the delta's future. If Herodotus could only see the delta he described as it looks today!

FIGURE 14.26 The Ganges River enters the Bay of Bengal.
The complex distributary pattern in the "many mouths" of the Ganges River delta in Bangladesh and extreme eastern India from the *Terra* satellite. [*Terra* MODIS sensor image courtesy of MODIS Land Team, NASA.]

FIGURE 14.27 The Nile River delta.
The arcuate Nile River delta. Intensive agricultural activity and small settlements are visible on the delta and along the Nile River floodplain in this true-color image. Cairo is at the apex of the delta. You can see the two main distributaries: Damietta to the east and Rosetta to the west.
[January 30, 2001, *Terra* MISR sensor image courtesy of MISR Team, NASA/GSFC/JPL.]

of Engineers apparently have only delayed the peril, as demonstrated by recent flooding.

An additional problem for the lower Mississippi Valley is the possibility, in a worst-case flood, that the river could break from its existing channel and seek a new route to the Gulf of Mexico. If you examine the map in Figure 14.28c and look at the sediment plume to the west of the main delta in the satellite image, an obvious alternative to the Mississippi's present channel is the Atchafalaya River (its delta is shown in Figure 14.28d). The Atchafalaya would provide a much shorter route to the Gulf of Mexico, less than one-half the present distance, and it has a steeper gradient than the Mississippi. Presently, this alternative-route distributary carries about 30% of the Mississippi's total discharge. For the Mississippi to bypass New Orleans entirely would be a blessing, for it would remove the flood threat. However, this shift would be a financial disaster, as a major U.S. port would silt in and seawater would intrude into freshwater resources.

At present, artificial barriers block the Atchafalaya from reaching the Mississippi at the point shown; without the floodgates the two rivers do connect. The Old River Control Project (1963) maintains three structures and a lock about 320 km (200 mi) from the Mississippi's mouth

to keep these rivers in their channels (Figure 14.28e). A major flood is only a matter of time, one that might cause the river-channel change. A case study in News Report 14.5 about Bayou Lafourche, a former major distributary of the Mississippi until 1904, highlights many of these issues.

FIGURE 14.28 The Mississippi River delta. ➡
(a) Evolution of the present delta, from 5000 years ago (1) to present (7). (b) The bird-foot delta of the Mississippi River receives a continuous supply of sediments, focused by controlling levees. The delta extends ever farther into the Gulf of Mexico, although subsidence of the delta and rising sea level have diminished the overall surface area. (c) Location map of the Old Control Structures and potential capture point (arrow) where the Atchafalaya River may one day divert the present channel. (d) Where the Atchafalaya River enters the Gulf. (e) Old River Control Auxillary Structure, one of the dams to keep the Mississippi in its channel. (f) The end of the bird-foot delta stretches far out in the Gulf. (g) Mississippi Delta waterscape, note the house with the raised first floor on stilts.
[(a) Adapted from C. R. Kolb and J. R. Van Lopik, "Depositional environments of the Mississippi River deltaic plain," in *Deltas in Their Geologic Framework* (Houston: Houston Geological Society, 1966). (b) *Terra* MODIS sensor image, March 5, 2001, courtesy of Liam Gumley, Space Science and Engineering Center, University of Wisconsin, and the MODIS Science Team, NASA. Photos (d) through (g) by Bobbé Christopherson.]

(a)

[Shaded areas denote areas of previous deltas.]

(b)

(c)

(d)

(e)

(f)

(g)

FIGURE 14.28

News Report 14.5

What Once Was Bayou Lafourche—Analysis of a Photo

Bayou Lafourche was the main distributary of the Mississippi River between the Civil War and 1904. A dam stopped the flow of this distributary and led to water losses in the bayou and stagnation of the former marshlands. The bayou was closed to navigation by 1930. In 1955, attempts began to reverse the losses to this bayou by restoring some water flow with pumps. Meanwhile, the overall losses to the delta continue.

Bayou is a general term for several water features in the Lower Mississippi River system, including creeks and secondary waterways. These sometimes stagnant, winding watercourses pass through coastal marshlands and swamps and allow tidal waters access into deltaic lowlands.

Human impact on Bayou Lafourche's 175-km (110-mi) stretch of the delta was severe and is mirrored by similar impacts across southern

Louisiana. In fact, most of coastal Louisiana, which represents about 40% of the coastal marshes in the United States, is profoundly altered and disrupted by dam and levee construction, flow alterations, oil and gas exploration, wells, pumping, pipelines, and dredging for navigation, logging, and industry needs. Approximately 65 km^2 (25 mi^2) of these Louisiana coastal lowlands are reclaimed by the Gulf of Mexico in an average year.

(a)

(b)

FIGURE 14.5.1 A hurricane protection levee compound in the Mississippi delta.
(a) Development of settlements, farms, and industry within a 65-km (40-mi) hurricane-protection levee system. Bayou Lafourche is the waterway that runs along the highway, with locks at the north and south ends of the levee system to seal off the interior in times of storm surges. (b) A map of this general region. [Photo by Bobbé Christopherson.]

FIGURE 14.5.2 Agriculture in the bayou. New growth sprouts in a field of sugar cane, grown in long-lots along Bayou Lafourche. [Photo by Bobbé Christopherson.]

FIGURE 14.5.3 Port Fourchon. Port Fourchon is the service center for the offshore oil industry, with about 6000 workers using the port as a depot for commuting to offshore drilling rigs.

Figure 14.5.1 illustrates a portion of Bayou Lafourche in 2003. The towns of Galliano and Golden Meadow and a total population of more than 10,000 people are protected from rising water, tidal fluctuations, and hurricane storm surge within the compound surrounded by levees. The 65-km (40-mi) hurricane protection levee system, some 4 m (13 ft) in height, 28-m (91-ft) wide at the base, was begun in the late 1970s. You can see the long-lot land survey system dividing lands at right angles to Bayou Lafourche and Highway 1, some plots planted in sugar cane (Figure 14.5.2). Floodgates at either end of this enclave are closed when the flood threat is imminent. Water tables in the protected area have decreased since the levee enclosure was completed and surface water flows reduced.

With delta subsidence and rising sea levels, increasing levee height is an ongoing process—known as a levee "lift." Present thinking is to add another meter in an attempt to maintain 100-year storm-surge protection. The real irony is that the former coastal wetlands were a buffer between the Gulf storm surges and interior lands, yet today because of all the human disruptions there is much open water between Golden Meadow and the former shoreline along the margins of the delta. Disruptions include reduc-tion in sediment supplies to the delta, a rising sea level, and land deflation due to oil and gas removal through thousands of wells. Only a few marsh mitigation and restoration projects are underway. One is at Port Fourchon at the extreme southern end of the for-mer coast (south of the Figure 14.5.1 photo). This is a major oil and gas in-dustry transshipment port and heli-copter base where workers connect for their jobs on drilling rigs in the Gulf of Mexico (Figure 14.5.3).

Meanwhile, Galliano is presently 122 cm (48 in.) above sea level. The Mississippi delta lands are slowly sub-siding and sea level is continuing to rise. The story continues to unfold.

Rivers Without Deltas   The Amazon River, Earth's highest-discharge stream, exceeds 175,000 m³/s (6.2 mil-lion cfs) in discharge and carries sediments far into the deep Atlantic offshore. Yet the Amazon lacks a true delta. Its mouth, 160 km (100 mi) wide, has formed an underwa-ter deltaic plain deposited on a sloping continental shelf. As a result, the Amazon's mouth is braided into a broad maze of islands and channels (see Figure 14.1).

Other rivers also lack deltaic formations if they do not produce significant sediment or if they discharge into strong erosive currents. The Columbia River of the U.S. Northwest lacks a delta because offshore currents remove sediment before it can accumulate into a delta.

Floods and River Management

ANIMATION   Stream Processes, Floodplains

Throughout history, civilizations have settled floodplains and deltas, especially since the agricultural revolution of 10,000 years ago, when the fertility of floodplain soils was discovered. Early villages generally were built away from the area of flooding, or on stream terraces, because the floodplain was dedicated exclusively to farming. However, as commerce grew, competition for sites near rivers grew, because these locations were important for transporta-tion. Port and dock facilities were built, as were river bridges. Because water is a basic industrial raw material used for cooling and for diluting and removing wastes, waterside industrial sites became desirable. These com-peting human activities on vulnerable flood-prone lands place lives and property at risk during floods.

The abuse and misuse of river floodplains brought catastrophe to North Carolina in 1999. In short succes-sion during September and October, Hurricanes Dennis, Floyd, and Irene delivered several feet of precipitation to the state; each storm falling on already saturated ground. About 50,000 people were left homeless and at least 50 died, while more than 4000 homes were lost and an equal amount were badly damaged (Figure 14.29a). The dollar estimate for the ongoing disaster now exceeds $10 billion. However, the real tragedy will unfold for years to come.

Hogs, in factory farms, outnumber humans in North Carolina. More than 10 million hogs, each producing two

(a)

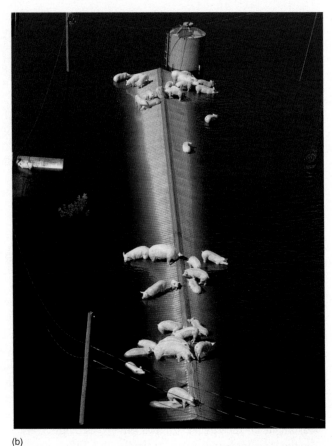

(b)

FIGURE 14.29 North Carolina 1999 floodplain disaster.
(a) Three hurricanes deluged North Carolina with several feet of rain during September and October 1999, Hurricane Floyd being the worst. (b) Hundreds of thousands of livestock were killed in the floods, which also washed out hundreds of animal sewage lagoons into wetlands, streams, and the ocean. Many of these factory farms and lagoons were sited on floodplains. [(b) Photo by Mel Nathanson, *Raleigh News & Observer*.]

tons of waste per year, were located in about 3000 agricultural factories. These generally unregulated operations collect almost 20 million tons of manure into nearly 500 open lagoons, many set on river floodplains. The hurricane downpour flushed out these waste lagoons, spewing hundreds of millions of gallons of untreated sewage into wetlands, streams, and eventually Pamlico Sound and the ocean—a spreading "dead zone." Add to this waste hundreds of thousands of hog, poultry, and other livestock carcasses, industrial toxins, floodplain junkyard oil, and municipal waste, and you have an environmental catastrophe (Figure 14.29b).

Final assessment of overall damage will take a decade, if not longer. There has been little progress toward resolving these waste/floodplain issues since these events in 1999. This raises great concern since the Atlantic is in an active hurricane cycle, exemplified by the category 5 Hurricane Isabel that hit North Carolina in 2003.

Catastrophic floods continue to be a threat, especially in poor nations. Bangladesh is perhaps the most persistent example. Bangladesh is one of the most densely populated countries on Earth, and more than *three-fourths* of its land area is a floodplain! The country's vast alluvial plain sprawls over an area the size of Alabama (130,000 km², or 50,000 mi²).

The flooding severity in Bangladesh is magnified as a consequence of human economic activities. Excessive forest harvesting in the upstream portions of the Ganges–Brahmaputra River watersheds increased runoff. Over time, the increased sediment load carried by the

river was deposited in the Bay of Bengal, creating new islands (see Figure 14.26). These islands, barely above sea level, became sites for new farming villages. As a result, about 150,000 people perished in the 1988 and 1991 floods. (For information on worldwide floods, see http://www.dartmouth.edu/~floods/Resources.html/ or a daily flood summary at http://iwin.nws.noaa.gov/iwin/us/nationalflood.html.)

Rating Floodplain Risk

A **flood** is a high water level that overflows the natural riverbank along any portion of a stream. Both floods and the floodplains they might occupy are rated statistically for the expected time intervals between floods. Thus, you hear about "10-year floods," "50-year floods," and so on. A *10-year flood* is the greatest level of flooding that is likely to occur once every 10 years. This also means that such flooding has only a 10% likelihood of occurring in any one year and is likely to occur about 10 times each century. For any given floodplain, such a frequency indicates a moderate threat.

A 50-year or 100-year flood is of greater and perhaps catastrophic consequence, but it is also less likely to occur in a given year. These probability ratings of flood levels are mapped for an area and the defined floodplains that result are then labeled as a "50-year floodplain" or a "100-year floodplain."

These statistical estimates are probabilities that events will occur randomly during any single year of the specified

FIGURE 14.30 Streamflow measurement.
(a) A typical streamflow measurement installation may use a variety of devices: staff gauge, stilling well with recording instrument, and suspended current meter. (b) An automated hydrographic station and (c) stilling well send telemetry to a satellite for collection by the USGS.
[Photos (b) courtesy of California Department of Water Resources; (c) by Bobbé Christopherson.]

period. Of course, two decades might pass without a 50-year flood, or a 50-year level of flooding could occur 3 years in a row. The record-breaking Mississippi River Valley floods in 1993 easily exceeded a 1000-year flood probability. See Focus Study 14.1 for more about flood-plain hazards and management strategies.

Streamflow Measurement

Flood patterns in a drainage basin are as complex as the weather, for floods and weather are equally variable and both include a level of unpredictability. Measuring and analyzing the behavior of each large watershed and stream enables engineers and concerned parties to develop the best possible flood-management strategy. Unfortunately, reliable data often are not available for small basins or for the changing landscapes of urban areas.

The key to flood avoidance or management is to possess extensive measurements of *streamflow*, a stream's discharge, and how it performs during a precipitation event.

(Figure 14.30a). Once the cross section of a stream is fully measured, only the stream level is needed to determine discharge (using the calculation: discharge = width × depth × velocity). A *staff gauge* (a pole marked with water levels) is placed in a stream, as shown in the figure, to measure stream level. Another method involves a *stilling well* on the stream bank with a gauge mounted in it to measure stream level. A movable current meter can be used to sample stream velocity at various locations.

Approximately 11,000 stream gaging stations (hydrologists use this spelling) are in use in the United States (an average of more than 200 per state). Of these, 7000 are operated by the U.S. Geological Survey and have continuous recorders for stage (level) and discharge (see **http://water.usgs.gov/pubs/circ/circ1123/**). Many of these stations automatically send telemetry data to satellites, from which information is retransmitted to regional centers (Figure 14.30b and c). Environment Canada's Water Survey of Canada maintains more than 3000 gaging stations (see **http://www.msc.ec.gc.ca/wsc/**).

Focus Study 14.1

Floodplain Strategies

Detailed measurements of stream-flows and flood records have been kept rigorously in the United States for only about 100 years, in particular since the 1940s. At any selected location along any given stream, the *probable maximum flood* (PMF) is a hypothetical flood of such a magnitude that there is virtually no possibility it will be exceeded. Because the collection and concentration of rainfall produce floods, hydrologists speak of a corollary, the *probable maximum precipitation* (PMP) for a given drainage basin, which is an amount of rainfall so great that it will never be exceeded.

These parameters are used by hydrologic engineers to establish a design flood against which to take protective measures. For urban areas near creeks, planning maps often include survey lines for a 50-year or a 100-year floodplain; such maps have been completed for most U.S. urban areas. The design flood usually is used to enforce planning restrictions and special insurance requirements. Restrictive zoning using these floodplain designations is an effective way of avoiding potential damage.

Such political action is not always enforced, and the scenario sometimes goes like this: (1) Minimal zoning precautions are not carefully supervised, (2) a flooding disaster occurs, (3) the public is outraged at being caught off guard, (4) businesses and homeowners are surprisingly resistant to stricter laws and enforcement, (5) eventually another flood refreshes the memory and promotes more planning meetings and questions. As strange as it seems, there is little indication that our risk perception improves as the risk increases. (The floodplain managers' organization at http://www.floods.org/ is a respected source of information.)

Bypass Channels as a Planning Strategy

A planning strategy used in some large river systems is to develop artificial floodplains; this is done by

(a)

(b)

(c)

FIGURE 14.1.1 Weir and bypass channels to divert floodwaters.
(a) The Bonnet Carré spillway (a weir), on lower Mississippi, New Orleans District, is opened in flood events to let water flow to Lake Pontchartrain and bypass New Orleans to the south. (b) The Sacramento weir awaits floodwaters from the Sacramento River, which it will direct through a channel into the Yolo Bypass visible in the distance in (c). When not flooded, the bypass is used for farmland crops. [Photos by Bobbé Christopherson.]

constructing *bypass channels* to accept seasonal or occasional floods. When not flooded, the bypass channel can serve as farmland, often benefiting from the occasional soil-replenishing inundation. When the river reaches flood stage, large gates called *weirs* are opened, allowing the water to enter the bypass channel. This alternate route relieves the main channel of the burden of carrying the entire discharge.

Along the Lower Mississippi north of New Orleans, the Bonnet Carré spillway is such a weir, in operation in the photo (Figure 14.1.1a). In high-water flood events, water is let through to a floodway into a bypass between a pair of guide levees to relieve the flood discharge flow. The bypass diverts the flood 9.7 km (6 mi) into Lake Pontchartrain, as it has done 7 times since 1937, to protect New Orleans. For more details see http://www.mvn.usace.army.mil/pao/bcarre/bcarre.htm.

In the Sacramento Valley, California, such weirs lead to bypass channels for taking floodwaters past the metropolitan areas. In Figure 14.1.1b we see the Sacramento weir, and in Figure 14.1.1c the bypass is filled with floodwaters. Farther downstream in the Sacramento–San Joaquin delta, artificial levees and pumps attempt to keep dry farmlands that are actually below river level (Figure 14.1.2). To

make matters worse, the area in the aerial photo is subsiding due to oxidation of peat in the soils, withdrawal of groundwater, and compaction. One of these delta islands flooded in a levee break in 2004.

Reservoir Considerations

Dams (an artificial structure placed in a river channel) and reservoirs (an impoundment of water behind a dam; a human-made lake) are common streamflow-control methods within a watershed. For conservation purposes, a dam holds back seasonal peak flows for distribution during low-water periods. In this way, streamflows are regulated to assure year-round water supplies. Dams also are constructed for flood control, to hold back excess flows for later release at more moderate discharge levels. Adding hydroelectric power production to these functions of conservation and flood control can define a modern multipurpose reclamation project. Of all types there are approximately 45,000 large dams in the world (at least 15 m high; or if between 5 and 15 m high, that have reservoir volume of at least 3 million m³, 106 million ft³) and innumerable smaller ones. The World Commission on Dams is at http://www.dams.org/.

The function of reservoir impoundment is to provide flexible storage capacity within a watershed to

regulate river flows, especially in a region with variable precipitation. Figure 14.1.3 shows one reservoir during drought conditions and during a time of wetter weather 6 years later.

Unfortunately, the multipurpose benefits of reservoir construction are countered by some negative consequences. The area upstream from a dam becomes permanently drowned. In mountainous regions, this may mean loss of white-water rapids and recreational sections of a river. In agricultural areas, the ironic end result may be that a hectare of farmland is inundated upstream to preserve a hectare of farmland downstream.

Furthermore, dams built in warm and arid climates lose substantial water to evaporation, compared with the free-flowing streams they replace. Reservoirs in the southwestern United States can lose 3–4 meters (10–13 ft) of water a year. Also, sedimentation can reduce the effective capacity of a reservoir and can shorten a dam's life span, as mentioned earlier regarding Glen Canyon Dam.

Vast environmental disruption for the sake of economic gain no longer appears popular with the public. The James Bay Project in central and northern Québec is a case in point. Launched in 1970 by Hydro-Québec and only one-third complete at this time, the project might eventually include 215 dams, 25 power stations, and 20 river diversions. Many unexpected environmental problems have arisen because no environmental impact studies were completed at the outset. The early stages remain a huge experiment with fragile ecosystems. Well into the planning phase, the public learned that corporate interests sought inexpensive, publicly generated electric power, and that much of it was for export to the United States—all at public expense.

The second major phase of the James Bay Project, the Great Whale project (named for a local river), may never be completed because of the success of conservation programs begun by utilities in the northeastern United States and because of court challenges to assess impacts before

FIGURE 14.1.2 A delta island.
Levees surround this delta island and its farms. The central portion of the island is below the river level. Note the former water channel and developing oxbow on the land. [Photo by Bobbé Christopherson.]

(continued)

Focus Stucy 14.1 *(continued)*

(a)

(b)

FIGURE 14.1.3 Reservoir extremes.
Comparative photographs of the New Hogan reservoir, central California, during (a) dry and (b) wet weather conditions. Reservoirs help regulate runoff variability. [Photos by author.]

further construction. In addition, the state of New York in 1992 withdrew its offer to buy 1000 megawatts of power from the Hydro-Québec project. For more on this project from a local perspective, see the First Nations 1999 resolution at http://www.afn.ca/resolutions/1990/sca/res14.htm; also see http://www.carc.org/pubs/v20no2/1.htm.

A Final Thought About Floods

The benefit of any levee, bypass, or other project intended to prevent flood destruction is measured in avoided damage and is used to justify the cost of the protection facility. Thus, ever-increasing damage leads to the justification of ever-increasing flood-control structures. All such strategies are subjected to cost–benefit analysis, but bias is a serious drawback because such an analysis usually is prepared by an agency or bureau with a vested interest in building more flood-control projects.

As suggested in an article titled "Settlement Control Beats Flood Control,"* published over 50 years ago, there are other ways to protect populations than with enormous, expensive, sometimes environmentally disruptive projects. Strictly zoning the floodplain is one approach. However, the flat, easily developed floodplains near pleasant rivers are desirable for housing and thus weaken political resolve. A reasoned zoning strategy would set aside the floodplain for farming or passive recreation, such as a riverine park, golf course, or plant and wildlife sanctuary, or for other uses that are not hurt by natural floods. This study concludes that "urban and industrial losses would be largely obviated by set-back levees and zoning and thus cancel the biggest share of the assessed benefits which justify big dams."

*Walter Kollmorgen, *Economic Geography* 29, no. 3 (July 1953): 215.

Scientists hope that adequate funding to maintain such hydrologic monitoring by stream gaging networks will continue, for the data are essential to hazard assessment and proper planning, yet we find river monitoring systems fighting for their budgets.

Hydrographs A graph of stream discharge over time for a specific place is a **hydrograph**. The hydrograph in Figure 14.31a shows the relation between precipitation input (the bar graph) and stream discharge (the curves). During dry periods, at low-water stages, the flow is described as base flow and is largely maintained by input from local groundwater (dark blue line).

When rainfall occurs in some portion of the watershed, the runoff collects and is concentrated in streams and tributaries. The amount, location, and duration of the rainfall episode determine the *peak flow*. Also important is the nature of the surface in a watershed; for example, a hydrograph for a specific portion of a stream changes after a forest fire or following urbanization of the watershed.

Human activities have enormous impact on water flow in a basin. The effects of urbanization are quite dramatic, both increasing and hastening peak flow, as you can see by comparing preurban streamflow (purple curve) and urbanized streamflow (light blue) in Figure 14.31a. In fact, urban areas produce runoff patterns quite similar

(b)

FIGURE 14.31 Urban flooding.
(a) Effect of urbanization on a typical stream hydrograph. Normal base flow is indicated with a dark blue line. The purple line indicates discharge after a storm, before urbanization. Following urbanization, stream discharge dramatically increases, as shown by the light blue line. (b) Severe flooding of an urban area in Linda, California, after a levee break on the Sacramento River in 1986; and it flooded again in 1997. [(b) Photo from California Department of Water Resources.]

to those of deserts. The sealed surfaces of the city drastically reduce infiltration and soil-moisture recharge; their effect is similar to that of the hard, nearly barren surfaces of the desert. A significant part of urban flooding occurs because of the alteration in surfaces and leads to shortened concentration times, for peak flows strike with little warning. These issues will intensify as urbanization of vulnerable areas continues. The flooded region in Figure 14.31b was hit in 1986 and again in 1997.

Summary and Review—River Systems and Landforms

▶ *Define* the term fluvial, and *outline* the fluvial processes: erosion, transportation, and deposition.

River systems, fluvial processes and landscapes, floodplains, and river control strategies are important to human populations as demands for limited water resources increase. **Hydrology** is the science of water, its global circulation, distribution, and properties, specifically water at and below Earth's surface. Stream-related processes are called **fluvial**. Water dislodges, dissolves, or removes surface material in the process called **erosion**. Streams produce *fluvial erosion*, in which weathered sediment is picked up for **transport**, movement to new locations. Sediments are laid down by another process, **deposition**. **Alluvium** is the general term for the clay, silt, sand, gravel, or other unconsolidated rock and mineral fragments deposited by running water as sorted or semisorted sediment on a floodplain, delta, or streambed.

Base level is the lowest elevation limit of stream erosion. A *local base level* occurs when something interrupts the stream's ability to achieve base level, such as is created by a dam or a landslide that blocks a stream channel.

hydrology (p. 432)
fluvial (p. 432)
erosion (p. 433)
transport (p. 433)
deposition (p. 433)
alluvium (p. 433)
base level (p. 433)

1. What role is played by rivers in the hydrologic cycle?
2. What are the five largest rivers on Earth in terms of discharge? Relate these to the weather patterns in each area and to regional potential evapotranspiration (POTET) and precipitation (PRECIP).
3. Define the term *fluvial*. What is a fluvial process?
4. What is the sequence of events that takes place as a stream dislodges material?
5. Explain the base-level concept. What happens to a local base level when a reservoir is constructed?

▶ *Construct* a basic drainage basin model, and *identify* different types of drainage patterns and internal drainage, with examples.

The basic fluvial system is a **drainage basin**, which is an open system. *Drainage divides* define the **watershed** catchment (water-receiving) area of the drainage basin. In any drainage basin, water initially moves downslope in a thin film called **sheetflow**, or *overland flow*. This surface runoff concentrates in *rills*, or small-scale downhill grooves, which may develop into deeper *gullies* and a stream course in a valley. High ground that separates one valley from another and directs sheetflow is termed an *interfluve*. Extensive mountain and highland regions act as **continental divides** that separate major drainage basins. Some regions, such as the Great Salt Lake Basin, have **internal drainage** that does not reach the ocean, the only outlets being evaporation and subsurface gravitational flow.

Drainage density is determined by the number and length of channels in a given area and is an expression of a landscape's topographic surface appearance. **Drainage pattern** refers to the arrangement of channels in an area as determined by the steepness, variable rock resistance, variable climate, hydrology, relief of the land, and structural controls imposed by the landscape. Seven basic drainage patterns are generally found in nature: dendritic, trellis, radial, parallel, rectangular, annular, and deranged.

> drainage basin (p. 433)
> watershed (p. 434)
> sheetflow (p. 434)
> continental divides (p. 434)
> internal drainage (p. 436)
> drainage density (p. 437)
> drainage pattern (p. 437)

6. What is the spatial geomorphic unit of an individual river system? How is it determined on the landscape? Define the several relevant key terms used.

7. In Figure 14.5, follow the Allegheny–Ohio–Mississippi River systems to the Gulf of Mexico. Analyze the pattern of tributaries, and describe the channel. What role do continental divides play in this drainage?

8. Describe drainage patterns. Define the various patterns that commonly appear in nature. What drainage patterns exist in your hometown? Where you attend school?

▶ *Describe* the relation among velocity, depth, width, and discharge, and *explain* the various ways that a stream erodes and transports its load.

Stream channels vary in *width* and *depth*. The streams that flow in them vary in velocity and in the *sediment load* they carry. All of these factors may increase with increasing discharge. **Discharge** is calculated by multiplying the velocity of the stream by its width and depth for a specific cross section of the channel. Most streams increase discharge downstream. But, some streams originate in a humid region and flow through an arid region, such that discharge decreases with distance. Such an **exotic stream** is exemplified by the Nile River or the Colorado River.

Hydraulic action is the work of *turbulence* in the water. Running water causes hydraulic squeeze-and-release action to loosen and lift rocks and sediment. As this debris moves along, it mechanically erodes the streambed further, through a process of **abrasion**.

Solution refers to the **dissolved load** of a stream, especially the chemical solution derived from minerals such as limestone or dolomite or from soluble salts. The **suspended load** consists of fine-grained, clastic particles held aloft in the stream, with the finest particles not deposited until the stream velocity slows nearly to zero. **Bed load** refers to coarser materials that are dragged and pushed and rolled along the streambed by **traction** or that bounce and hop along by **saltation**. If the load in a stream exceeds its capacity, sediments accumulate as **aggradation** as the stream channel builds through deposition. With excess sediment, a stream becomes a maze of interconnected channels that form a **braided stream** pattern.

> discharge (p. 439)
> exotic stream (p. 441)
> hydraulic action (p. 442)
> abrasion (p. 442)
> dissolved load (p. 442)
> suspended load (p. 442)
> bed load (p. 443)
> traction (p. 443)
> saltation (p. 443)
> aggradation (p. 443)
> braided stream (p. 443)

9. What was the impact of flood discharge on the channel of the San Juan River near Bluff, Utah? Why did these changes take place?

10. How does stream discharge do its erosive work? What are the processes at work in the channel?

11. Differentiate between stream competence and stream capacity.

12. How does a stream transport its sediment load? What processes are at work?

▶ *Develop* a model of a meandering stream, including point bar, undercut bank, and cutoff, and *explain* the role of stream gradient in these flow characteristics.

Where the slope is gradual, stream channels develop a sinuous form called a **meandering stream**. The outer portion of each meandering curve is subject to the fastest water velocity and can be the site of a steep **undercut bank**. On the other hand, the inner portion of a meander experiences the slowest water velocity and forms a **point-bar** deposit. When a meander neck is cut off as two undercut banks merge, the meander becomes isolated and forms an **oxbow lake**.

Every stream develops its own **gradient** and establishes a longitudinal profile. A portion of the stream is designated a **graded stream** when the stream is measured by available discharge, channel characteristics, its velocity, and the load supplied from the drainage basin. An interruption in a stream's longitudinal profile is called a **nickpoint**. A nickpoint can occur as the stream flows across hard, resistant rock or after tectonic uplift episodes.

> meandering stream (p. 443)
> undercut bank (p. 443)
> point bar (p. 443)
> oxbow lake (p. 444)
> gradient (p. 444)
> graded stream (p. 446)
> nickpoint (p. 447)

13. Describe the flow characteristics of a meandering stream. What is the pattern of flow in the channel? What are the erosional and depositional features and the typical landforms created?

14. Explain these statements: (a) All streams have a gradient, but not all streams are graded. (b) Graded streams may have ungraded segments.

15. Why is Niagara Falls an example of a nickpoint? Without human intervention, what do you think would eventually take place at Niagara Falls?

16. What is meant by "the validity of cyclic or equilibrium models depends on which of three time frames is being considered"? Explain and discuss.

FIGURE 14.1.1 Julie Dian-Reed, Senior Service Hydrologist and Weather Forecaster.
National Weather Service, Wilmington, Ohio. [Photo by National Weather Service.]

Careers in this field are bright and involve the GIS revolution. "My background in geography helps me bring together diverse subjects to complement our weather and river forecasting. We are beginning to incorporate GIS information as part of our suite of tools for both weather and river forecasting, and the improvement of warnings to the public.

For instance, recent advancements in river and flash flood hydrology are in the Advanced Hydrologic Prediction Services (AHPS) program. Julie says, "Recent technology allows forecasters to gather data from smaller stream basins in contrast to previous broad-scale, generalized information. What is evolving is a capability for specific flash flood forecasting and warnings to the public. GIS models are critical to this process."

A further benefit of the AHPS is a more probabilistic approach to predicting river flood crests by more accurately assessing precipitation amounts and locations within a basin. Julie explains, "We will forecast the likelihood that a river will reach a specific height and the timing of the crest. Better informed emergency and water resource managers will be able to make accurate risk assessments during an impending flood. GIS is part of this revolution in river forecasting, with online graphical displays of flood forecasts and probable inundation maps."

"I enjoy the variety of weather we receive in this region, and following the weather is a hobby of mine. Even while not working, I track the weather and try to take photos of a flodding event." Julie is enthusiastic about her integrated vocation and avocation!

The Loess Hills of western Iowa give evidence of the last ice age, as the finely ground sediments left by the glaciers were windblown to this setting. Such deposition by wind is an eolian process. The loess layers in Iowa date from 159,000 to 12,500 years ago. The loess deposits reach heights of about 61 m (200 ft) above the nearby prairie farmlands (background in photo) and Missouri River and run north and south more than 322 km (200 mi); see Figure 15.14 for the Midwest location. Only China has deposits that exceed these dimensions. Because of the nature of loess soils, they are susceptible to high erosion rates, producing a rough terrain of hills, gullies, and oversteepened slopes. [Photos by Bobbé Christopherson.]

CHAPTER 15

Eolian Processes and Arid Landscapes

▶ Key Learning Concepts

After reading the chapter, you should be able to:

- ▶ **Characterize** the unique work accomplished by wind and eolian processes.
- ▶ **Describe** eolian erosion, including deflation, abrasion, and the resultant landforms.
- ▶ **Describe** eolian transportation, and **explain** saltation and surface creep.
- ▶ **Identify** the major classes of sand dunes, and **present** examples within each class.
- ▶ **Define** loess deposits and their origins, locations, and landforms.
- ▶ **Portray** desert landscapes, and **locate** these regions on a world map.

Wind is an agent of geomorphic change. Like moving water, moving air (wind) causes erosion, transportation, and deposition of materials. Like moving water, moving air is a fluid and it behaves similarly, although it has a lower viscosity (it is "thinner") than water. Although lacking the lifting ability of water, wind processes can modify and move quantities of materials in deserts, along coastlines, and elsewhere.

Wind processes modify and move sediment in deserts and along coastlines in a variety of climates. Wind may contribute to soil formation in places far distant from the point of origin, bringing fine material from regions where glaciers deposited it. Elsewhere, fallow fields (those not planted) give up their soil resource to destructive wind erosion. Scientists are only now getting an accurate picture of the amount of windblown dust that fills the atmosphere and crosses the oceans between continents. Chemical fingerprints and satellites trace windblown dust from African soils to South America and Asian landscapes to Europe. Winds even spread living organisms. One study found related mosses, liverworts, and lichens distributed thousands of kilometers between islands in the Southern Ocean.

471

Earth's dry lands stand out in stark contrast on the water planet. Arid landscapes display unique landforms and life forms: "Instead of finding chaos and disorder the observer never fails to be amazed at a simplicity of form, an exactitude of repetition and a geometric order in the desert."* The polar regions are deserts as well, so there should be no surprise that Yuma, Arizona, and weather stations in Antarctica receive the same amount of annual precipitation.

In this chapter: We examine the work of wind, associated erosion, transport, and depositional processes, and resulting landforms. Windblown fine particles form vast loess deposits, the basis for rich agricultural soils. For convenience of organization, we include the discussion of arid lands in this chapter. Most deserts are rocky and covered with desert pavement, whereas other dry landscapes are covered in sand dunes. In desert environments, an overall lack of moisture and stabilizing vegetation sets the landscapes in sharp, sun-baked relief. Water is the major erosional agent in the desert, yet water is the limiting resource for human development. The troubled Colorado River and the Western drought is the subject of an important focus study.

The Work of Wind

The work of the wind—erosion, transportation, and deposition—is called **eolian** (also spelled *aeolian*). The word comes from Aeolus, ruler of the winds in Greek mythology. British Army major Ralph Bagnold, while stationed in Egypt in 1925, accomplished much eolian research. An engineering officer who spent time in the deserts west of the Nile, Bagnold measured, sketched, and developed hypotheses about the wind and desert forms. Imagine the scene in the 1920s: A Model-T Ford chugs across the desert west of Cairo taking Bagnold to his research area. He laid rolls of chicken wire over treacherous stretches of sand to prevent getting stuck! Bagnold's often-cited work, *The Physics of Blown Sand and Desert Dunes*, was published in 1941.

The actual ability of wind to move materials is small compared with that of other transporting agents such as water and ice, because air is so much less dense than those other media. Yet, over time, wind accomplishes enormous work. Bagnold studied the ability of wind to transport sand over the surface of a dune. Figure 15.1 shows that a steady wind of 50 kmph (30 mph) can move approximately one-half ton of sand per day over a square meter section of dune. The graph also demonstrates how rapidly the amount of transported sand increases with wind speed.

Grain size is important in wind erosion. Intermediate-sized grains move (bounce along) most easily. It is the largest and the smallest sand particles that require the strongest winds to move. The large particles are heavier and thus require stronger winds. Small particles are difficult to move because they exhibit a mutual cohesiveness and because they usually present a smooth surface

*R. A. Bagnold, *The Physics of Blown Sand and Desert Dunes* (London: Methuen, 1941).

FIGURE 15.1 Sand movement and wind velocity.
Sand movement relative to wind velocity, as measured over a square meter cross section of ground surface. [After R. A. Bagnold, *The Physics of Blown Sand and Desert Dunes* (London: Methuen, 1941). Adapted by permission.]

(aerodynamic) to the wind. Yet, the finest dust, once aloft, is carried from continent to continent (Figure 15.2). In addition, wind can prune and shape vegetation, especially where winds are strong in a consistent direction (see the chapter-opening photo for Chapter 6).

Eolian Erosion

Two principal wind-erosion processes are **deflation**, the removal and lifting of individual loose particles, and **abrasion**, the grinding of rock surfaces with a "sandblasting" action by particles captured in the air. Deflation and abrasion produce a variety of distinctive landforms and landscapes.

Deflation Deflation literally blows away loose or noncohesive sediment. Fine materials are eroded away by wind deflation and moving water, leaving behind a concentration of pebbles and gravel called **desert pavement**. Resembling a cobblestone street, desert pavement protects underlying sediment from further deflation and water erosion (Figure 15.3). Desert pavements are so common that many provincial names are used for them—for example, *gibber plain* in Australia; *gobi* in China; and in Africa, *lag gravels* or *serir*, or *reg* desert if some fine particles remain. Water is an important factor in the formation of desert pavement, washing away fine materials and concentrating and cementing the remaining rock pieces.

Heavy recreational activity damages fragile desert landscapes, especially in the arid lands of the United States, where more than 15 million off-road vehicles

(a)

(b)

FIGURE 15.2 The work of the wind.
(a) Streaks of windblown sands cover bedrock south of Atar, in central Mauritania. Note the circular Richat dome structure, just east of the Western Sahara political border corner (refer to Figure 12.10c). (b) A dust storm blowing off the Sudan coast at 21° N over the Red Sea (north is to the left, south to the right). [Images (a) *Terra* MODIS sensor image courtesy of NASA Earth Observatory, R. Stockli, R. Simmon, and B. Montgomery; and (b) astronaut photo from Shuttle STS-43 courtesy of JSC/NASA.]

(ORVs) are now in use. Such vehicles crush plants and animals; disrupt desert pavement, leading to greater deflation; and create ruts that easily concentrate sheetwash (water flowing across surfaces) to form gullies. Measures to restrict ORV use to specific areas, preserving the remaining desert, are controversial.

Military activities can also threaten desert landscapes. A serious environmental impact of the 1991 Persian Gulf War was the disruption of desert pavement. Thousands of square kilometers of stable desert pavement were shattered by bombardment with more than 68,000 tons of explosives and were disrupted by the movement of heavy vehicles. The resulting loosened sand and silt was made available for deflation, plaguing cities and farms with increased dust and sand accumulations that continue into this century.

Wherever wind encounters loose sediment, deflation may remove enough material to form basins called **blowout depressions**. These depressions range from small indentations less than a meter wide up to areas hundreds of meters wide and many meters deep. Chemical weathering, although slow in the desert owing to the lack of water, is important in the formation of a blowout, for it

removes the cementing materials that give particles their cohesiveness. (In arid climates chemical weathering is active on the surfaces of particles at the microscopic level, utilizing hygroscopic and capillary water in reactions.)

Large depressions in the Sahara Desert are at least partially formed by deflation. The enormous Munkhafad el Qaṭṭâra (Qaṭṭâra Depression) just inland from the Mediterranean Sea in the Western Desert of Egypt, which covers 18,000 km² is now about 130 m (427 ft) below sea level at its lowest point.

Abrasion   You may have seen work crews sandblasting surfaces on buildings and bridges to clean them or on streets to remove unwanted markings. Sandblasting uses a stream of compressed air filled with sand grains to quickly and smoothly abrade a surface. Abrasion by windblown particles is nature's slower version of sandblasting, and it is especially effective at polishing exposed rocks when the abrading particles are hard and angular.

Variables that affect the rate of abrasion include the hardness of surface rocks, wind velocity, and wind constancy. Abrasive action is restricted to the area

Time

(a)

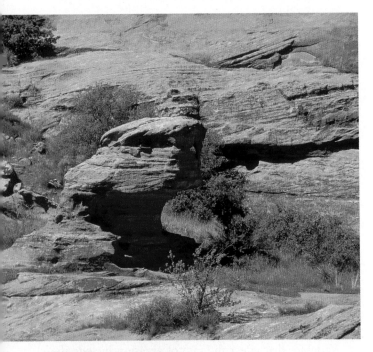

(b)

FIGURE 15.3 Desert pavement.
(a) Desert pavement is formed from larger rocks and fragments left after deflation and sheetwash (water flowing across the surface). (b) A typical desert pavement. [Photo by Bobbé Christopherson.]

immediately above the ground, usually no more than a meter or two in height, because sand grains are lifted only a short distance.

Rocks exposed to eolian abrasion appear pitted, grooved, or polished. They usually are aerodynamically shaped in a specific direction, according to the consistent flow of airborne particles carried by prevailing winds. Rocks that bear such evidence of eolian erosion are **ventifacts** (literally, "artifacts of the wind").

On a larger scale, deflation and abrasion are capable of streamlining rock structures that are aligned parallel to the most effective wind direction, leaving behind distinctive, elongated ridges or formations called **yardangs**. These wind-sculpted features can range from meters to kilometers in length and up to many meters in height. On Earth, some yardangs are large enough to be detected on satellite imagery. The Ica Valley of southern Peru contains yardangs reaching 100 m (330 ft) in height and several kilometers in length, and yardangs in the Lūt Desert of Iran attain 150 m (490 ft) in height. Abrasion is concentrated on the windward end of each yardang, with deflation operating on the leeward portions (Figure 15.4). The Sphinx in Egypt was perhaps partially formed as a yardang, suggesting a head and body. Some scientists think this shape led the ancients to complete the bulk of the sculpture artificially with masonry.

Eolian Transportation

As mentioned in Chapter 6, atmospheric circulation can transport fine material, such as volcanic debris, fire soot and smoke, and dust, worldwide within days. Wind exerts a drag, or frictional pull, on surface particles until they become airborne, just as water in a stream picks up sediment (again, think of air as a fluid). The distance that wind is capable of transporting particles in suspension varies greatly with particle size. Only the finest dust particles travel significant distances, so the finer material suspended in a dust storm is lifted much higher than the coarser particles of a sandstorm, which may be lifted only about 2 m (6.5 ft).

People living in areas of frequent dust storms are faced with infiltration of very fine particles into their homes and businesses through even the smallest cracks. (Figure 3.8 illustrates such dust storms in the Nevada desert, a dust plume from orbit, and alkali dust near Mono Lake, California.) People living in desert regions and along sandy beaches, where frequent sandstorms occur, contend with the sandblasting of painted surfaces and etched window glass.

Both human and natural sand erosion and transport from a beach are slowed by conservation measures such as the introduction of stabilizing native plants, the use of fences, and the restriction of pedestrian traffic to walkways (Figure 15.5). As human settlement further encroaches on coastal dunes, sand transport becomes problematic. As an example, at Nags Head, North Carolina, the Jocky's Ridge dune, some 43 m (140 ft) in height, is actively migrating over roads and yards.

The term *saltation* was used in Chapter 14 to describe movement of particles by water. The term also describes

FIGURE 15.4 A yardang.
A small wind-sculpted rock formation in Snow Canyon outside St. George, Utah. [Photo by Bobbé Christopherson.]

FIGURE 15.5 Preventing sand transport.
(a) Popham Beach State Park, Maine. Further erosion and transport of coastal dunes can be controlled by stabilizing strategies, planting native plants, and by confining pedestrian traffic to walkways. (b) Tufts of grass are planted to stabilize this Atlantic City, New Jersey, beach with a fence installed to prevent human foot traffic. [Photos by Bobbé Christopherson.]

(a)

(b)

the wind transport of grains along the ground, grains usually larger than 0.2 mm (0.008 in.). About 80% of wind transport of particles is accomplished by this skipping and bouncing action (Figure 15.6). Compared with fluvial transport, in which saltation is accomplished by hydraulic lift, eolian saltation is executed by aerodynamic lift, elastic bounce, and impact (compare Figure 15.6a with Figure 14.13). Grains lift much higher in air than in water because of air's lower viscosity.

On impact, grains hit other grains and knock them into the air. Saltating particles crash into other particles, knocking them both loose and forward. This type of movement is **surface creep**, which slides and rolls particles too large for saltation and affects about 20% of the material being transported. Once in motion, particles continue to be transported by lower wind velocities. In a desert or along a beach, you can hear a slight hissing sound produced by the myriad saltating grains of sand, almost like steam escaping, as they bounce along and collide with surface particles.

Through processes of weathering, erosion, and transportation, mineral grains are removed from parent rock and redistributed elsewhere. In Figure 13.5c, you saw the relation between the composition and color of the sandstone in the background and the derived sandy surface in the foreground. Wind action is not significant in the weathering process that frees individual grains of sand from the parent rock, but it is active in relocating the weathered grains.

(a)

(b)

FIGURE 15.6 How the wind moves sand.
(a) Eolian suspension, saltation, and surface creep are mechanisms of sediment transportation. Compare with the saltation and traction that occur in another fluid, water, in Figure 14.13. (b) Sand grains saltating along the surface in the Stovepipe Wells dune field, Death Valley. [Photo by author.]

FIGURE 15.7 Sand ripples.
Sand ripple patterns later may
become lithified into fixed patterns
in rock. The area in the photo looks
vast but is only about 1 m (3.3 ft)
wide. [Photo by author.]

Eolian Depositional Landforms

The smallest features shaped by individual saltating grains
are ripples (Figure 15.7). Ripples form in crests and
troughs, positioned transversely (at a right angle) to the
direction of the wind. Their formation is influenced by
the length of time particles are airborne, with each windblown bounce. Eolian ripples are similar to fluvial ripples,
although the impact of saltating grains is slight in water.

Many people have never been in a desert; they know
this arid landscape only from the movies, which leave the
impression that most deserts are covered by sand. Instead,
desert pavements predominate across most subtropical
arid landscapes; only about 10% of desert areas are
covered with sand. Sand grains generally are deposited as
transient ridges or hills called dunes.

A **dune** is a wind-sculpted accumulation of sand. An
extensive area of dunes, such as that found in North
Africa, is characteristic of an **erg desert**, or **sand sea**.
The Grand Erg Oriental in the central Sahara exceeds
1200 m (4000 ft) in depth and covers 192,000 km²
(75,000 mi²), comparable to the area of Nebraska. This
sand sea has been active for more than 1.3 million years
and has average dune heights of 120 m (400 ft). The
Sahara Marzūq sand sea shown in Figure 15.8a is wider
than 300 km (185 mi). Eolian processes are at work on the
Martian surface forming similar parallel ridges along the
floor of Melas Chasma in Figure 15.8b (note the dust
devil in the lower left corner of the image). Similar sand
seas, such as the Grand Ar Rub'al Khālī Erg, are active in
Saudi Arabia.

Dune fields, whether in arid regions or along coastlines,
tend to migrate in the direction of strong prevailing winds.
Strong seasonal winds or winds from a passing storm may
sometimes prove more effective than average prevailing

winds. When saltating sand grains encounter small patches
of sand, their kinetic energy (motion) is dissipated and they
start to accumulate; a dune is born. As height increases above
30 cm (12 in.), a steeply sloping slipface on the lee side of the
dune and characteristic dune features form.

By studying the dune model in Figure 15.9 you can see
that winds characteristically create a gently sloping *windward
side* (stoss side), with a more steeply sloped **slipface** on the
leeward side. A dune usually is asymmetrical in one or more
directions. The angle of a slipface is the steepest angle at
which loose material is stable—its *angle of repose*. Thus, the
constant flow of new material makes a slipface a type of
avalanche slope. Sand builds up as it moves over the crest of the
dune to the brink; then it avalanches, falling and cascading as
the slipface continually adjusts, seeking its angle of repose
(usually 30° to 34°). In this way, a dune migrates downwind,
as suggested by the successive dune profiles in Figure 15.9.

Dunes have many wind-shaped styles that make classification difficult. We can simplify dune forms into three
classes—*crescentic* (crescent, curved shape), *linear* (straight
forms), and *star dunes* (each is summarized in Figure 15.10). The ever-changing form of these eolian deposits is part of their beauty, eloquently described by one
author: "I see hills and hollows of sand like rising and
falling waves. Now at midmorning, they appear paper
white. At dawn they were fog gray. This evening they will
be eggshell brown."*

Star dunes are the mountainous giants of the sandy
desert. They form in response to complicated, changing
wind patterns and have multiple slipfaces. They are
pinwheel-shaped, with several radiating arms rising and

*J. E. Bowers, *Seasons of the Wind* (Flagstaff, AZ: Northland Press, 1985),
p. 1.

(a)

(b)

FIGURE 15.8 A sand sea.
(a) The Sahara Marzūq, an erg desert that dominates southwestern Libya. Effective northwesterly winds shape the pattern and direction of the transverse and barchanoid (series of connected barchans) dunes. This sand sea exceeds 300 km across.
(b) A similar pattern of dunes appear on the Martian surface in the southern area of Melas Chasma in Valles Marineris. Note the dust devil in the lower left. The area covered is about 2 km wide. [(a) *Terra* MODIS sensor image courtesy of MODIS Land Rapid Response Team, NASA/GSFC, November 9, 2001. (b) Mars Global Surveyor, Mars Orbiter Camera image courtesy of NASA/JPL/Malin Space Science Systems, July 11, 1999.]

FIGURE 15.9 Dune cross section.
Successive slipfaces exhibit a distinctive pattern as the dune migrates in the direction of the effective wind.

Effective wind direction

Previous slipfaces

Windward (stoss) slope

Slipface

Leeward slope

Direction of dune movement

Successive slipfaces created as dune migrates

Class	Type	Description
Crescentic	Barchan	Crescent-shaped dune with horns pointed downwind. Winds are constant with little directional variability. Limited sand available. Only one slipface. Can be scattered over bare rock or desert pavement or commonly in dune fields.
	Transverse	Asymmetrical ridge, transverse to wind direction (right angle). Only one slipface. Results from relatively ineffective wind and abundant sand supply.
	Parabolic	Role of anchoring vegetation important. Open end faces upwind with U-shaped "blowout" and arms anchored by vegetation. Multiple slipfaces, partially stabilized.
	Barchanoid ridge	A wavy, asymmetrical dune ridge aligned transverse to effective winds. Formed from coalesced barchans; look like connected crescents in rows with open areas between them.

Barchan

Transverse

Parabolic

Barchanoid ridge

FIGURE 15.10 Major dune forms.
Arrows show wind direction. [Adapted from E. D. McKee, *A Study of Global Sand Seas*, U.S. Geological Survey Professional Paper 1052 (Washington, DC: U.S. Government Printing Office, 1979).]

joining to form a common central peak. The best examples of star dunes are in the Sahara and the Namib Desert, where they approach 200 m (650 ft) in height (Figure 15.11).

The map in Figure 15.12 shows the correlation of active sand regions with deserts (tropical, continental interior, and coastal). Important is the limited extent of desert area covered by active sand dunes—only about 10%

of all continental land between 30° N and 30° S. Also noted on the map are dune fields in humid climates such as along coastal Oregon, the south shore of Lake Michigan (Figure 15.12b), along the Gulf and Atlantic coastlines, in Europe, and elsewhere. The remarkable coastal desert sands of Namibia are pictured in Figure 15.12c.

These same dune-forming principles and terms (for example, *dune* and *slipface*) apply to snow-covered

Class	Type	Description
Linear	Longitudinal	Long, slightly sinuous, ridge-shaped dune, aligned parallel with the wind direction; two slipfaces. Average 100 m high and 100 km long, and the "draa" at the extreme in size can reach to 400 m high. Results from strong effective winds varying in one direction.
	Seif	After Arabic word for "sword"; a more sinuous crest and shorter than longitudinal dunes. Rounded toward upwind direction and pointed downwind. (Not illustrated.)
Star dune		The giant of dunes. Pyramidal or star-shaped with three or more sinuous, radiating arms extending outward from a central peak. Slipfaces in multiple directions. Results from effective winds shifting in all directions. Tend to form isolated mounds in high effective winds and connected sinuous arms in low effective winds.
Other	Dome	Circular or elliptical mound with no slipface. Can be modified into barchanoid forms.
	Reversing	Asymmetrical ridge form intermediate between star dune and transverse dune. Wind variability can alter shape between forms.

Longitudinal

Star

Dome

Reversing

landscapes. *Snow dunes* are formed as wind deposits snow in drifts. In semiarid farming areas, capturing drifting snow with fences and tall stubble left in fields contributes significantly to soil moisture when the snow melts.

Loess Deposits

Approximately 15,000 years ago, in several episodes, Pleistocene glaciers retreated in many parts of the world, leaving behind large glacial outwash deposits of fine-grained clays and silts. These materials were blown great distances by the wind and redeposited in unstratified, homogeneous (evenly mixed) deposits. Peasants working along the Rhine River Valley in Germany gave these deposits the name **loess** (pronounced "luss"). No specific landforms were created; instead, loess covered existing landforms with a thick blanket of material that assumed the general topography of the existing landscape.

FIGURE 15.11 Mountains of the desert.
Star dune in the Namib Desert in southwestern Africa.
[Photo by Comstock.]

Because of its own binding strength, loess weathers and erodes into steep bluffs, or vertical faces. At Xi'an, Shaanxi Province, China, a loess wall has been excavated for dwelling space (Figure 15.13a). When a bank is cut into a loess deposit, it generally will stand vertically, although it can fail if saturated (Figure 15.13b). The hills and gullies in the Loess Hills of Iowa in the chapter opener demonstrate the erosion potential of these soils.

Figure 15.14 shows the worldwide distribution of loess deposits. Significant accumulations throughout the Mississippi and Missouri River valleys form continuous deposits 15–30 m (50–100 ft) thick. Loess deposits also occur in eastern Washington State and Idaho. This silt explains the fertility of the soils in these regions, for loess deposits are well drained and deep and have excellent moisture retention. Loess deposits also cover much of Ukraine, central Europe, China, the Pampas–Patagonia regions of Argentina, and lowland New Zealand. The

(a)

(b)

(c)

FIGURE 15.12 Sandy regions of the world.
(a) Worldwide distribution of active and stable sand regions. (b) Sand dunes along the shore of Lake Michigan in Indiana Dunes State Park, Indiana. (c) Sandy area in the Namib-Naukluft Park, Namib Desert, Namibia, Africa. [(a) After R. E. Snead, *Atlas of World Physical Features*, p. 134, © 1972 by John Wiley & Sons. Adapted by permission. (b) Photo by Bobbé Christopherson; (c) photo by Nigel J. Dennis/Photo Researchers, Inc.]

(a)

(b)

FIGURE 15.13 Example of loess deposits.
(a) Loess formation in Xi'an, Shaanxi Province, China, has sufficient structural strength to permit excavation for dwelling rooms. (b) A loess bluff in western Iowa. [(a) Photos by Betty Crowell; (b) photo by Bobbé Christopherson.]

soils derived from loess are some of Earth's "breadbasket" farming regions. Transport of loess soils occurred in the catastrophic Dust Bowl in the 1930s in the United States; see News Report 15.1.

In Europe and North America, loess is thought to be derived mainly from glacial and periglacial sources. The vast deposits of loess in China, covering more than 300,000 km² (116,000 mi²) are thought to be derived from windblown desert sediment rather than glacial sources. Accumulations in the Loess Plateau of China exceed 300 m (1000 ft) in thickness, forming complex weathered badlands and some good agricultural land. These wind-blown deposits are interwoven with much of Chinese history and society.

Overview of Desert Landscapes

Dry climates occupy about 26% of Earth's land surface. If all semiarid climates are considered, perhaps as much as 35% of all land area tends toward dryness, constituting the largest single climatic region on Earth (see Figures 10.4 and 10.5 for the location of these *arid deserts* and *semiarid steppe* climate regions and Figure 20.3 for the distribution of these desert environments).

News Report 15.1

The Dust Bowl

Deflation and wind transport of loess soils produced a catastrophe in the American Great Plains in the 1930s—the Dust Bowl. More than a century of overgrazing and intensive agriculture left soil susceptible to drought and eolian processes. Reduced precipitation amounts and above-normal temperatures triggered the multiyear disaster. The deflation of many centimeters of soil occurred in southern Nebraska, Kansas, Oklahoma, Texas, eastern Colorado, and even in southern Canada and northern Mexico.

Fine sediments were lifted by winds to form severe dust storms. The transported dust darkened the skies of Midwestern cities and drifted over farmland. Streetlights were left on throughout the day in Kansas City, St. Louis, and other Midwestern cities and towns. Such episodes can devastate economies, cause tremendous loss of topsoil, and even bury farmsteads. Southeastern Australia experienced severe dust storms in 1993 that included consequences similar to those of the American Dust Bowl. For more on the Dust Bowl, see **http://www. pbs.org/wgbh/amex/dustbowl/**.

Scientists are seeking answers as to why this took place when it did.

Tree-ring analysis shows that a drought of that intensity happened about twice a century over the past 400 years. One such drought in what is now called Nebraska in the 13th century appears to have been as intense as that in the 1930s; however, it lasted 38 years. The Dust Bowl weather conditions seem to correlate with cool Pacific Ocean sea-surface temperatures (SST) and warm Atlantic Ocean SST. Chapter 6 discussed the Pacific Decadal Oscillation (PDO) and its relation to drought occurrences in the United States.

FIGURE 15.14 Worldwide loess deposits.
Dots represent small, scattered loess formations. [After R. E. Snead, *Atlas of World Physical Features*, p. 138, © 1972 by John Wiley & Sons. Adapted by permission.]

Deserts occur worldwide as topographic plains, such as the Great Sandy and Simpson Deserts of Australia, the Arabian and Kalahari Deserts of Africa, and portions of the extensive Taklimakan Desert, which covers some 270,000 km² (105,000 mi²) in the central Tarim Basin of China. Deserts also are found in mountainous regions: interior Asia, from Iran to Pakistan, and in China and Mongolia. In South America, lying between the ocean and the Andes, is the rugged Atacama Desert. And Earth's deserts are expanding, as we will discuss shortly. Now, let us look at the link between climate and Earth's deserts.

Desert Climates

The spatial distribution of dry lands is related to three climatological settings: to subtropical high-pressure cells between 15° and 35°, both N and S latitudes (see Figures 6.11 and 6.13), to the rain shadow on the lee side of mountain ranges (see Figure 8.9), and to areas at great distance from moisture-bearing air masses, such as central Asia. Figure 15.15 portrays this distribution according to the climate classification used in this text and presents photographs of four major desert regions.

Desert areas possess unique landscapes created by the interaction of intermittent precipitation events, weathering processes, and wind. Rugged, hard-edged desert landscapes of cliffs and scarps contrast sharply with the vegetation-covered, rounded, and smoothed slopes characteristic of humid regions.

The daily surface energy balance for El Mirage, California, presented in Figure 4.20a and b, highlights the high sensible heat conditions and intense ground heating in the desert. Such areas receive a high input of insolation through generally clear skies, and they experience high radiative heat losses at night. A typical desert water balance experiences high potential evapotranspiration demand, low precipitation supply, and prolonged summer water deficits (see, for example, Figure 9.13e for Phoenix, Arizona). Fluvial processes in the desert generally are characterized by intermittent running water, with hard, poorly vegetated surfaces yielding high runoff during rainstorms (Figure 15.16).

Desert Fluvial Processes

Precipitation events in a desert may be rare indeed, a year or two apart, but when they do occur, a dry streambed can fill with a torrent called a **flash flood**. These channels may fill in a few minutes and surge briefly during and after a storm. Depending on the region, such a dry streambed is known as a **wash**, an *arroyo* (Spanish), or a *wadi* (Arabic). A desert highway that crosses a wash usually has signs posted to warn drivers not to proceed if rain is in the vicinity, for a flash flood can suddenly arise and sweep away anything in its path.

When washes fill with surging flash flood waters, a unique set of ecological relationships quickly develops. Crashing rocks and boulders break open seeds that respond to the timely moisture and germinate. Other plants and animals also spring into brief life cycles as the water irrigates their limited habitats.

At times of intense rainfall, remarkable scenes fill the desert. Figure 15.17 shows two photographs taken just one month apart in a sand dune field in Death Valley, California. A rainfall event produced 2.57 cm (1.01 in.) of precipitation in one day, in a place that receives only 4.6 cm (1.83 in.) in an average year. The stream in the photograph continued to run for hours and then collected

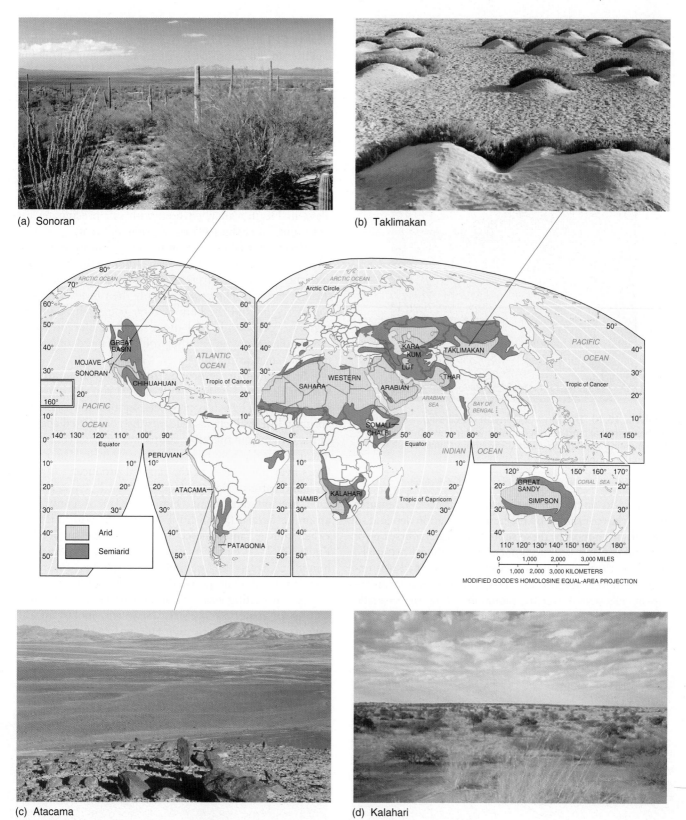

(a) Sonoran

(b) Taklimakan

(c) Atacama

(d) Kalahari

FIGURE 15.15 The world's dry regions.
Worldwide distribution of arid lands (*arid desert* climates) and semiarid lands (*semiarid steppe* climates). (a) Sonoran Desert in the American Southwest. (b) Taklimakan Desert in central Asia. (c) Atacama Desert in subtropical Chile near Baquedano. (d) Kalahari Desert in south-central Africa.
[(a) Photo by author; (b) photo by Wu Chunzhan/New China Pictures Co./East Photo; (c) photo by Jacques Jangoux/Photo Researchers, Inc.; (d) photo by Nigel J. Dennis/Photo Researchers, Inc.]

FIGURE 15.16 The land in Canyon de Chelly.
In this marginal land, undependable water flows between the towering sandstone walls surrounding Chinle Wash in Canyon de Chelly, Arizona. [Photo by author.]

in low spots on hard, underlying clay surfaces. The water was quickly consumed by the high evaporation demand so that, in just a month, these short-lived watercourses were dry and covered with accumulations of alluvial materials.

The 1985 precipitation event just described was exceeded by one August 15, 2004, when several thunderstorm cells stalled over the mountains along the east side of Death Valley. Rainfall totals reached 6.35 cm (2.5 in.) in little more than an hour. Cars were swept away from the

Furnace Creek Inn parking lot, about 4.8 km (3 mi) of Highways 190 and 178 washed out, and there were two fatalities in the Zabriskie Point area. One annex building had a 3-m- (10-ft-) high water mark on it. Again, water is sparse in the deserts, yet it is the major erosional force—sometimes dramatically so. For recent information and maps of Death Valley, see **http://www.nps.gov/deva/**.

As runoff water evaporates, salt crusts may be left behind on the desert floor. This intermittently wet and dry low area in a region of closed drainage is a **playa**, site of an *ephemeral lake* when water is present. Accompanying our earlier discussion of evaporites, Figure 11.10 shows such a playa in Death Valley, covered with salt precipitate just one month after the 1985 record rainfall event.

Permanent lakes and continuously flowing rivers are uncommon features in the desert, although the Nile River and the Colorado River are notable exceptions. Both these rivers are *exotic streams*, having their headwaters in a wetter region and the bulk of their course through arid regions. Focus Study 15.1 discusses the Colorado River, the ongoing drought, and the problem of increasing water demands in an arid land.

In arid climates, a prominent landform is an **alluvial fan**, which occurs at the mouth of a canyon where it exits into a valley. The fan is produced by flowing water that abruptly loses velocity as it leaves the constricted channel of the canyon and therefore drops layer upon layer of sediment along the base of the mountain block. Water then flows over the surface of the fan and produces a braided drainage pattern, shifting from channel to channel with each precipitation event (Figure 15.18). A continuous apron, or **bajada** (Spanish for "slope"), may form if individual alluvial fans coalesce into one sloping surface (see Figure 15.23c). Fan formation of any sort is reduced in humid climates because perennial streams constantly carry away sediment, preventing its deposition.

An interesting aspect of an alluvial fan is the natural sorting of materials by size. Near the mouth of the canyon at the apex of the fan, coarse materials are

(a)

(b)

FIGURE 15.17 An improbable river in Death Valley.
The Stovepipe Wells dune field of Death Valley, California, shown (a) the day after a 2.57-cm (1.01 in.) rainfall and (b) one month later (identical location). This 1985 rainfall event was exceeded by one in 2004. [Photos by author.]

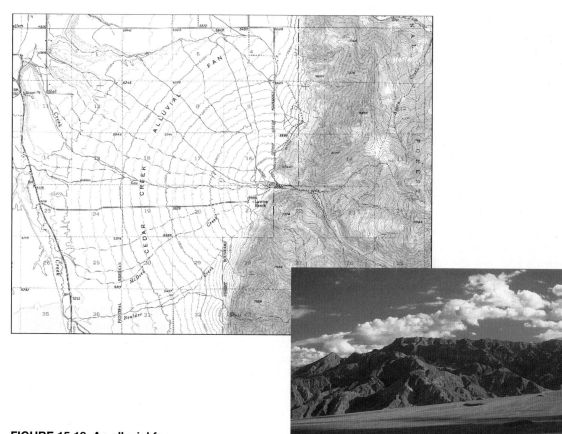

FIGURE 15.18 An alluvial fan.
The photo shows an alluvial fan in a desert land-
scape. The topographic map shows the Cedar
Creek alluvial fan. (Topographic map is the Ennis
Quadrangle, 15-minute series, scale 1:62,500, con-
tour interval = 40 ft; latitude/longitude coordinates
for mouth of canyon are 45°2′ N 111°35′ W.)
[Photo by Bobbé Christopherson; USGS map.]

deposited, grading slowly to pebbles and finer gravels
with distance out from the mouth. Then sands and silts
are deposited, with the finest clays and salts carried in
suspension and solution all the way to the valley floor.
Dissolved minerals accumulate as evaporite deposits on
the valley floor left after evaporation of the water from
the playa.

Well-developed alluvial fans also can be a major
source of groundwater. Some cities—San Bernardino,
California, for example—are built on alluvial fans and
extract their municipal water supplies from them. In
other parts of the world, such water-bearing alluvial fans
are known as *qanat* (Iran), *karex* (Pakistan), and *foggara*
(western Sahara).

Focus Study 15.1

The Colorado River: A System Out of Balance

The headwaters of an exotic stream
rise in a humid region of water surpluses but the stream flows mostly
through arid lands for the rest of its
journey to the sea. Exotic streams
have few incoming tributaries. Consequently, an exotic stream has a discharge pattern that is different than a
typical stream: Instead of discharge
increasing downstream, it decreases

(see Chapter 14). The Nile and
Colorado Rivers are prominent examples. These natural conditions are
further affected by human impacts of
water withdrawals and reservoir construction with high evaporation rates.

In the case of the Nile, the East
African mountains and plateaus provide a humid source area. The Nile
first rises as the remote headwaters of

the Kagera River in the eastern portion of the Lake Plateau country of
East Africa. On its way to Lake Victoria, it forms the partial boundary of
Tanzania, Rwanda, and Uganda. The
Nile itself then flows from the lake and
continues on its 6650 km (4132 mi)
course, with decreasing flows all the
way to its mouth on the Mediterranean Sea near Cairo.

(continued)

Focus Study 15.1 (continued)

The Colorado River Basin

The Colorado rises on the high slopes of Mount Richthofen (3962 m, or 13,000 ft) in Rocky Mountain National Park, Colorado (Figure 15.1.1a) and flows almost 2317 km (1440 mi) to where a trickle of water disappears in the sand, kilometers short of its former mouth in the Gulf of California.

Orographic precipitation totaling 102 cm (40 in.) per year falls mostly as snow in the Rockies, feeding the Colorado headwaters. But at Yuma, Arizona, near the river's end, annual precipitation is a scant 8.9 cm (3.5 in.), an extremely small amount when compared with the annual potential evapotranspiration demand in the Yuma region of 140 cm (55 in.).

From its source region, the Colorado River quickly leaves the humid Rockies and spills out into the arid desert of western Colorado and eastern Utah. At Grand Junction, Colorado, near the Utah border, annual precipitation is only 20 cm (8 in.) (Figure 15.1.1b). After carving its way through the intricate labyrinth of canyonlands in Utah, the river enters Lake Powell, behind Glen Canyon Dam (Figure 15.1.1c). The Colorado then flows through the Grand Canyon chasm, formed by the river's own erosive power.

West of the Grand Canyon, the river turns southward, tracing its final 644 km (400 mi) as the Arizona–California border. Along this stretch sits Hoover Dam, just east of Las Vegas (Figure 15.1.1d); Davis Dam, built to control the releases from Hoover (Figure 15.1.1e); Parker Dam, for the water needs of Los Angeles; three more dams for irrigation water (Palo Verde, Imperial, and Laguna, Figure 15.1.1f); and finally, Morelos Dam at the Mexican border (Figure 15.1.1g). Mexico owns the end of the river and whatever water is left, although the river no longer reaches its mouth in the Gulf of California (Figure 15.1.1i).

Figure 15.1.1j shows the annual water discharge and suspended sediment load for the Colorado River at Yuma, Arizona, from 1905 to 1964. The completion of Hoover Dam in the 1930s dramatically reduced suspended sediment. Addition of Glen Canyon Dam upstream from Hoover Dam in 1963 further reduced streamflows. In fact, Lake Powell, which formed upstream behind the artificial base level of Glen Canyon Dam, is forecast to fill with sediment over the next 100 years, if it remains in full operation.

Overall, the drainage basin encompasses 641,025 km^2 (247,500 mi^2) of mountain, basin-and-range, plateau, canyon, and desert landscapes in parts of seven states and two countries. A discussion of the Colorado River is included in this chapter because of its crucial part in the history of the Southwest and its role in the future of this arid and semiarid region.

Dividing Up the Colorado's Dammed Water

John Wesley Powell (1834–1902), the first person of record to successfully navigate the Colorado River through the Grand Canyon, was the first director of the U.S. Bureau of Eth-nology and later director of the U.S. Geological Survey (1881–1892). Powell perceived that the challenge of the West was too great for individual efforts and believed that solutions to problems such as water availability could be met only through private cooperative efforts. His 1878 study (reprinted 1962), *Report of the Lands of the Arid Region of the United States*, is a conservation landmark.

Today, Powell probably would be skeptical of the intervention by government agencies in building large-scale reclamation projects. Lake Powell is named after him. An anecdote in Wallace Stegner's book *Beyond the Hundredth Meridian* relates that, at an 1893 international irrigation conference held in Los Angeles, development-minded delegates bragged that the entire West could be conquered and reclaimed from nature and that "rain will certainly follow the plow." Powell spoke against that sentiment: "I tell you, gentlemen, you are piling up a heritage of conflict and litigation over water rights, for there is not sufficient water to supply the land."* He was booed from the hall, but history has shown Powell to be correct.

The Colorado River Compact was signed by six of the seven basin states in 1923. (The seventh, Arizona, signed in 1944, the same year as the Mexican Water Treaty.) With this compact, the Colorado River basin was divided into an upper basin and a lower basin, arbitrarily separated for

*W. Stegner, *Beyond the Hundredth Meridian* (Boston: Houghton Mifflin, 1954), 343.

FIGURE 15.1.1 The Colorado River drainage basin.
The Colorado River basin, showing division of the upper and lower basins near Lees Ferry in northern Arizona. (a) Headwaters of the Colorado River near Mount Richthofen in the Colorado Rockies. (b) The river near Moab, Utah. (c) Glen Canyon Dam, a regulatory, administrative facility near Lees Ferry, Arizona. (d) Hoover Dam spillways in rare operation during 1983 floods. (e) Davis Dam in full release during flood. (f) Irrigation canal and cropland irrigated with Colorado River water in the Palo Verde district. (g) Morelos Dam at the Mexican border is the final stop as the river dwindles to a mere canal. (h) Central Arizona Project aqueduct west of Phoenix. (i) The Colorado River (far upper left) stops short of its former delta. The bluish-purple water in the former channel is actually an inlet for Gulf of California water; the gray deposits are sediments in mud flats along this inlet. (j) Dam construction affects river discharge and sediment yields. [(a) Photo by Bobbé Christopherson; (b through g) photos by author; (h) photo by Tom Bean/DRK Photo; (i) *Terra* ASTER (thermal emission and reflection) image courtesy of NASA/GSFC/MITI/ERSDAC/JAROS and the U.S./Japan ASTER Science Team, September 8, 2000; and, (j) data from USGS, 1985, *National Water Summary 1984*, Water Supply Paper 2275 (Washington, DC: Government Printing Office), p. 55).]

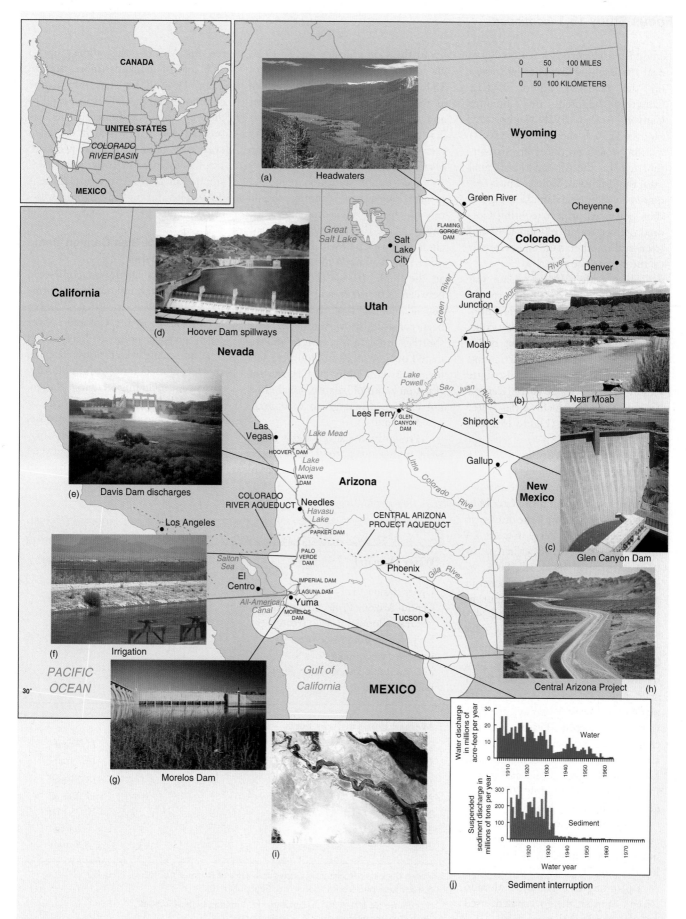

(a) Headwaters

(b) Near Moab

(c) Glen Canyon Dam

(d) Hoover Dam spillways

(e) Davis Dam discharges

(f) Irrigation

(g) Morelos Dam

(h) Central Arizona Project

(i)

(j) Sediment interruption

(continued)

Focus Study 15.1 *(continued)*

administrative purposes at Lees Ferry near the Utah–Arizona border (noted on the map in Figure 15.1.1). Congress adopted the Boulder Canyon Act in 1928, authorizing Hoover Dam as the first major reclamation project on the river. Also authorized was the All-American Canal into the Imperial Valley, which required an additional dam. Los Angeles then began its project to bring Colorado River water 390 km (240 mi) from still another dam and reservoir on the river to the city.

Shortly after Hoover Dam was finished and downstream enterprises were thus offered flood protection, the other projects were quickly completed. There are now eight major dams on the river and many irrigation works. The latest effort to redistribute Colorado River water is the Central Arizona Project, which carries water to the Phoenix area (Figure 15.1.1h).

Highly Variable River Flows
The flaw in all this planning and water distribution is that exotic streamflows are highly variable, and the Colorado is no exception. In 1917, the discharge measured at Lees Ferry totaled 24 million acre-feet (maf), whereas in 1934 it dropped by nearly 80%, to only 5.03 maf. In 1977, the discharge dropped again to 5.02 maf, but in 1984 it rose to an all-time high of 24.5 maf. Yet, in contrast, in 2000 to 2004 river discharge fell to exceptionally low levels, with 2002 dropping to a record low of 3.1 maf, and 2003 to 6.4 maf! The 2002 to 2004 water years marked the lowest 3-year average flow in the total record. Please revisit the chapter-opening pair of photos for Chapter 9 showing the comparison of Lake Mead water levels at Hoover Dam in July 1983 and July 2003 as evidence of this variability.

The average flows between 1906 and 1930 were almost 18 maf a year, but averages dropped to 14.1 maf during the past 70 years (1930 to 2003). As a planning basis for the Colorado River Compact, the government used average river discharges from 1914

up to the Compact signing in 1923, an exceptionally high average of 18.8 maf. That amount was perceived as more than enough for the upper and lower basins, each to receive 7.5 maf, and, later, for Mexico to receive 1.5 maf in the 1944 Mexican Water Treaty.

We might question whether proper long-range planning should rely on the providence of high variability. Tree-ring analyses of past climates have disclosed that the only other time Colorado discharges were at the high 1914–1923 level was between A.D. 1606 and 1625! The dependable flows of the river have been consistently overestimated. This shortfall problem is shown in an estimated budget of 20 million acre-feet for the river (Table 1): Clearly, the situation is out of balance, for there is not enough discharge to meet budgeted demands.

Presently, the seven states *ideally* want rights to a total as high as 25.0 maf. When added to the guaran-

tee for Mexico, this comes up to 26.5 maf of demands (more than in Table 1). And six states share one common opinion: California's right to the water must be limited to a court-ordered 4.4 maf, which it exceeded every year until California's access to surplus Colorado River water was stopped in 2002. At present, there is no surplus during the ongoing western drought.

Water Loss at Glen Canyon Dam and Lake Powell
Glen Canyon Dam was completed and began water impoundment, forming Lake Powell, in 1963, 27.4 km (17 mi) north of the Colorado River Basin division point at Lees Ferry. The advancing water slowly flooded the deep, fluted, inner gorges of many canyons, including Glen, Navajo, Labyrinth, and Cathedral. Glen Canyon Dam's primary purpose, according to the Bureau of Reclamation, was to regulate flows between the upper and lower basins. An

Table 1   Estimated Colorado River Budget through 2003

Water Demand	Quantity (maf)[a]
Upper Basin (7.5) Lower Basin (7.5)[b]	15.0
Central Arizona Project (rising to 2.8 maf)	1.0
Mexican allotment (1944 Treaty)	1.5
Evaporation from reservoirs	1.5
Bank storage at Lake Powell	0.5
Phreatophytic losses (water-demanding plants)	0.5
Budgeted total demand	20.0 maf

Average Flows at Lees Ferry	
1906–1930	17.7 maf
1930–2003 average flow of the river	14.1 maf
1990–2002	13.25 maf
2000	11.06 maf
2001	10.75 maf
2002	3.1 maf
2003	6.4 maf

Source: Bureau of Reclamation states and states of Arizona, California, Nevada.
A water year runs from October 1 to September 30.
[a] 1 million acre-feet = 325,872 gallons; 1.24 million liters; 43,560 ft³.
[b] In-basin consumptive uses 75% agricultural.

additional benefit is the production of hydroelectric power, which is sold wholesale at a low rate to utilities across the Southwest. Also, there is a growing recreation and tourism industry around Lake Powell, as previously inaccessible desert scenery now can be reached by boat.

Many thought that Lake Mead, behind Hoover Dam, could have served the primary administrative function of flow regulation, especially considering the serious water-loss problems with the Glen Canyon Dam and the absorbent rocks that contain its reservoir. Porous Navajo sandstone underlies most of the Lake Powell reservoir at Glen Canyon Dam. This sandstone absorbs an estimated 0.5 maf of the river's overall annual discharge as bank storage—the higher the lake level, the greater the loss into the sandstone. Second, Lake Powell is an open body of water in an arid desert, where hot, dry winds accelerate evaporative losses. Another 0.5 maf of the Colorado's overall discharge is lost annually from the reservoir in this manner, about a third of all Colorado River reservoir evaporation losses. Third, now-permanent sand bars and banks have stabilized along the regulated river, allowing water-demanding plants called phreatophytes to establish and extract an additional 0.5 maf of the river flow. To some extent, similar water losses result from the presence of other dams along the river as well.

Intense precipitation and heavy snowpack in the Rockies, attributable to the 1982–1983 El Niño (see Focus Study 10.1), led to record-high discharge rates on the Colorado, testing the controllability of one of the most regulated rivers in the world. Federal reservoir managers were not prepared for the high discharge, since they had set aside Lake Mead's primary purpose (flood control) in favor of competing water and power interests. What followed was the 1983 human-caused flood on the most engineered river in the world!

The only time the spillways at Hoover Dam ever operated was more than 40 years earlier, when the reservoir capacity was artificially raised for a test; now they were opened to release the floodwaters (Figure 15.1.1d). Davis Dam, which regulates releases from Hoover Dam, was within 30 cm (1 ft) of overflow, a real problem for a structure made partially of earth fill (Figure 15.1.1e). In addition, Glen Canyon Dam was over capacity and at risk and was damaged by the volume of discharge tearing through its spillways. The decision to increase releases from Hoover to protect the dam doomed towns and homeowners along the river to the south to floodwaters, especially inundating subdivisions near Needles, California.

The Western Drought and Beyond

The severity of the western drought from 1999 to 2005, in its sixth year, surpasses anything in the historical record and is approaching the driest in the record according to tree-ring analyses. Satellites can spot the symptoms of this western drought: reduced snow pack in the Rockies, reduced soil moisture, the drought-stressed conditions of vegetation, the prevalence of wildfires, and lowered reservoir levels.

Since A.D. 1226, nine droughts lasting 10 to 20 years and four lasting more than 20 years occurred. Whether the current drought will be of this duration or be more like the three droughts of the 20th century that ran 4 to 11 years is unknown. However, this is the first drought to occur in the Colorado River system in the presence of a human demand for water that is increasing at such record levels.

Water inflow to Lake Powell in the summer of 2004 was 31% of the 30-year average (1961–1990)—2003 was 53% and 2002 was just 25%. Lake Powell by 2005 had dropped below levels not seen since it began filling more than 30 years ago—only 38% of capacity. Lake Mead dropped to 54% of capacity by the end of 2004. The maximum reservoir elevation of Lake Mead is 375 m (1229 ft), and by July 2004 Lake Mead dropped to 343 m (1126.7 ft). Forecasts place the level at 341 m (1118 ft) by December 2005. The production of hydroelectric power has fallen to just 31% of capacity from these dams.

An official "shortage" declaration has not been issued and managers think they can continue to meet Lower Basin needs from Lake Mead's storage. The official status in 2005 placed the system in a "drought condition." There is speculation that Glen Canyon Dam might cease operations before the end of the decade because of these low-water conditions, leading some environmental groups to suggest decommissioning, possible removal, and restoration of the canyons to predevelopment status—an improbable outcome politically.

The worst appears yet to come. As of 2005, the total system stands at 50% of capacity and engineers estimate that it will take 13 normal winters of snowpack in the Rockies and basin-wide precipitation to "reset" the system. This recovery is much slower than previous drought recoveries because of the rapid increase in water demand across the Southwest. Arizona's population alone grew 36% between 1990 and 2000 (3.6 million to 5.55 million in 2004). Las Vegas went from 368,000 in 1985 to 2 million by the end of 2005 (540% increase!). Present probabilities for the period after 2016 place only a 20% chance of surplus occurring in any given year. Imagine the population after 2016 in the region if no steps are taken to control growth!

Government stopgap measures include, among others: buying water from farmers to meet contractual obligations for Colorado River water; reopening the desalination plant in Yuma, Arizona, to treat agricultural runoff water for its return to the river to meet international agreements with Mexico; and increasing storage, when possible, in upstream reservoirs in the system where evaporation rates are lower than at Lake Powell. Conservation (using less water) and efficiency (using water more effectively) are not yet being stressed as ways to reduce the tremendous demand for water. However, southern Nevada has

(continued)

Focus Study 15.1 *(continued)*

launched a campaign to replace lawns with drought-tolerant xeroscaping—desert landscaping. Limiting or halting further metropolitan construction and population growth in the Colorado River Basin are not part of present strategies to lower demand. Despite the conditions along the river, we still seem to be "supply" focused in our approach and not focusing on "demand." Remember, it is the tremendous increase in the demand for this water that stands in the way of resetting the Colorado River system in the foreseeable future.

In a speech in December 2003 in Las Vegas to the Colorado River Water Users Association, the director of the Bureau of Reclamation, in an obvious understatement about the water shortfall that could continue for a decade or more, stated:

> That's why the Colorado River must be carefully managed and used. That is why we'll keep encouraging all Colorado River water users to pay attention to their water use. And that is why Reclamation . . . will continue to manage the river more intensely than we ever had in the past.

We might wonder what John Wesley Powell would think if he were alive today to witness such errant attempts to control the mighty and variable Colorado. He foretold such a "heritage of conflict and litigation."

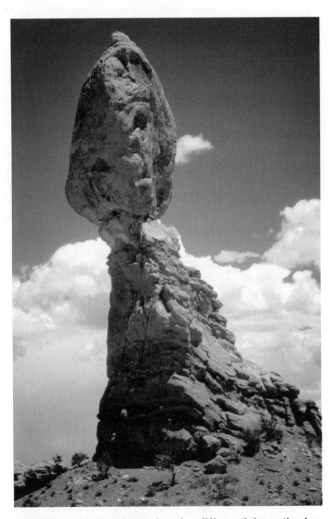

FIGURE 15.19 A balanced rock—differential weathering. Balanced Rock in Arches National Park, Utah, where writer-naturalist Edward Abbey (quoted in text) worked as a ranger years before it became a park. The overall feature is 39 m (128 ft) tall and composed of Entrada sandstone. The balanced-rock portion is 17 m (55 ft) tall and weighs 3255 metric (3577 short) tons. [Photo by author.]

Desert Landscapes

Contrary to popular belief, deserts are not wastelands, for they abound in specially adapted plants and animals. Moreover, the limited vegetation, intermittent rainfall, intense insolation, and distant vistas produce starkly beautiful landscapes (see Death Valley photographs in Focus Study 10, Figure 10.1.1). And all deserts are not the same: For example, North American deserts have more vegetation cover than do the generally barren Asian desert expanses, as we saw in Figure 15.15.

The shimmering heat waves and related mirage effects in the desert are products of light refraction through layers of air that have developed a temperature gradient near the hot ground. The desert's enchantment is captured in the book *Desert Solitaire*:

> Around noon the heat waves begin flowing upward from the expanses of sand and bare rock. They shimmer like transparent, filmy veils between my sanctuary in the shade and all the sun-dazzled world beyond. Objects and forms viewed through this tremulous flow appear somewhat displaced or distorted. . . . The great Balanced Rock floats a few inches above its pedestal, supported by a layer of superheated air. The buttes, pinnacles, and fins in the windows area bend and undulate beyond the middle ground like a painted backdrop stirred by a draft of air.*

The buttes, pinnacles, and mesas of arid landscapes are resistant horizontal rock strata that have eroded differentially. Removal of the less-resistant sandstone strata produces unusual desert sculptures—arches, windows, pedestals, and delicately balanced rocks (Figure 15.19). Specifically, the upper layers of sandstone along the top

*E. Abbey, *Desert Solitaire* (New York: McGraw-Hill, 1968), p. 154. Copyright © 1968 by Edward Abbey.

(a)

(b)

FIGURE 15.20 Monument Valley landscape.
(a) Mitten Buttes, Merrick Butte, and rainbow in Monument Valley, Navajo Tribal Park, along the Utah–Arizona border. (b) A schematic of the tremendous removal of material by weathering, erosion, and transport. [Photo by author.]

of an arch or butte are more resistant to weathering and protect the sandstone rock beneath.

The removal of all surrounding rock through differential weathering leaves enormous buttes as residuals on the landscape. If you imagine a line intersecting the tops of the Mitten Buttes shown in Figure 15.20, you can gain some idea of the quantity of material that has been removed. These buttes exceed 300 m (1000 ft) in height, similar to the Chrysler Building in New York City or First Canadian Place in Toronto.

Desert landscapes are places where stark erosional remnants can stand above the surrounding terrain as knobs or hills. Such a bare, exposed rock, called an *inselberg* (island mountain), is exemplified by Uluru (Ayers) Rock in Australia (Figure 15.21).

In a desert area, weak surface material may weather to a complex, rugged topography, usually of relatively low and varied relief. Such a landscape is called a *badland*, probably so named because in the American West it offered little economic value and was difficult to traverse in 19th-century wagons. The Badlands region of the Dakotas and north-central Arizona (the Painted Desert) are of this form.

Sand dunes that existed in some ancient deserts have lithified, forming sandstone structures that bear the imprint of *cross-stratification*. When such a dune was accumulating, sand cascaded down its slipface and distinct bedding planes (layers) were established that remained after the dune lithified (Figure 15.22). Ripple marks, animal tracks, and fossils also are found preserved in these

FIGURE 15.21 Australian landmark.
Uluru (Ayers) Rock in Northern Territory, Australia, is an isolated mass of weathered rock. The formation is 348 m (1145 ft) high and is 2.5 km long by 1.6 km wide. Uluru Rock is sacred to Aboriginal peoples and has been protected in Uluru National Park since 1950. [Photo by Porterfield/Chickering.]

FIGURE 15.22 Cross-bedding in sedimentary rocks.
The bedding pattern, called cross-stratification, in these sandstone rocks tells us about patterns that were established in the dunes before lithification (hardening into rock). [Photo by Bobbé Christopherson.]

sandstones, which originally were eolian-deposited sand dunes.

Basin and Range Province

A province is a large region that is characterized by several geologic or physiographic traits. Characterizing the **Basin and Range Province** of the western United States are alternating basins (valleys) and mountain ranges that lie in the rain shadow of mountains to the west (Figure 15.23). The physiography and geography combine to give the province a dry climate, few permanent streams, and *internal drainage patterns*—drainage basins that lack any outlet to the ocean (see Figure 14.5, "Great Basin" and internal drainage).

The vast Basin and Range Province—almost 800,000 km² (300,000 mi²)—was a major barrier to early settlers in their migration westward. The combination of desert climate and north–south trending mountain ranges

presented harsh challenges. Today, when you traverse U.S. Highway 50 across Nevada, you cross five passes (horsts) of more than 1950 m (6400 ft) and numerous basins (grabens). Throughout the drive you are reminded where you are by prideful signs that plainly state: "The Loneliest Road in America." It is difficult to imagine crossing this topography with wagons and oxen (Figure 15.23e).

The Basin and Range formed as a result of tectonic processes. As the North American plate lumbered westward, it overrode former oceanic crust and hot spots at such a rapid pace that slabs of subducted material literally were run over. The crust was stretched by tensional forces, creating a landscape fractured by many faults. The present landscape consists of nearly parallel sequences of *horsts* (upward-faulted blocks, which are the "ranges") and *grabens* (downward-faulted blocks, which are the "basins," or valleys). Figure 15.23c shows this pattern of normal faults.

John McPhee captured the feel of this desert province in his book *Basin and Range*:

> Supreme over all is silence. Discounting the cry of the occasional bird, the wailing of a pack of coyotes, silence—a great spatial silence—is pure in the Basin and Range. It is a soundless immensity with mountains in it. You stand . . . and look up at a high mountain front, and turn your head and look fifty miles down the valley, and there is utter silence.*

Basin-and-range relief is abrupt, and rock structures are angular and rugged. As the ranges erode, transported materials accumulate to great depths in the basins, gradually producing extensive desert plains. The basin's elevation averages roughly 1200–1500 m (4000–5000 ft) above sea level, with mountain crests rising higher by some 900–1500 m (3000–5000 ft). Death Valley, California, is the lowest of these basins, with an elevation of −86 m (−282 ft). However, to the west of the valley, the Panamint Range rises to 3368 m (11,050 ft) at Telescope Peak—almost 3.5 vertical kilometers (2.2 mi) of desert mountain relief!

In Figure 15.23c and d, note the **bolson**, a slope-and-basin area between the crests of two adjacent ridges in a dry region of internal drainage. Death Valley provides a dramatic example of these arid-land features. Figure 15.23c also identifies a *playa* (central salt pan), a *bajada* (coalesced alluvial fans), and a mountain front in retreat from weathering and erosional attack. A *pediment* is an area of bedrock that is layered with a thin veneer, or coating, of alluvium. It is an erosional surface, as opposed to the depositional surface of the bajada.

Vast arid and semiarid lands remain an enigma on the water planet. They challenge our technology to access water, our courage, and personal need for water. Yet, these lands hold a mysterious fascination, perhaps because they are so lacking in the moisture that infuses our lives.

---

*J. McPhee, *Basin and Range* (New York: Farrar, Straus, Giroux, 1981), p. 46.

FIGURE 15.23 Basin and Range Province in the western United States.

(a) Map and (b) *Landsat* image of the area. Recent scientific discoveries demonstrate that this province extends south through northern and central Mexico. (c) A bolson in the mountainous desert landscape of the Basin and Range Province. Parallel normal faults produce a series of horsts (ranges) and grabens (basins). (d) Death Valley features a central playa, parallel mountain ranges, alluvial fans, and *bajada* along the base of the ranges. (e) "The Loneliest Road in America," Highway U.S. 50, slashes westward through the province. [(b) Image from NASA; (d) photo by author; (e) photo by Bobbé Christopherson.]

flash flood (p. 482)
wash (p. 482)
playa (p. 484)
alluvial fan (p. 484)
bajada (p. 484)
Basin and Range Province (p. 492)
bolson (p. 492)
desertification (p. 494)

14. Characterize desert energy and water-balance regimes. What are the significant patterns of occurrence for arid landscapes in the world?

15. How would you describe the water budget of the Colorado River? What was the basis for agreements regarding distribution of the river's water? Why has thinking about the river's discharge been so optimistic? Overview current conditions of the Colorado River system in light of the ongoing drought and increasing water demands.

16. Describe a desert bolson from crest to crest. Draw a simple sketch with the components of the landscape labeled.

17. Where is the Basin and Range Province? Briefly describe its appearance and character.

18. What is meant by desertification? Using the maps in Figures 15.15 and 15.24, and the text description, locate several of the regions affected by desert expansion.

NetWork

The *Geosystems* Home Page provides online resources for this chapter on the World Wide Web. To begin: Once on the Home Page, click on this textbook, scroll the Table of Contents menu, select this chapter, and click "Begin." You will find self-tests that are graded, review exercises, specific updates for items in the chapter, and many links to interesting related pathways on the Internet under "Destinations." *Geosystems* is at http://www.prenhall.com/christopherson.

Critical Thinking

A. "Water is the major erosion and transport medium in the desert." Respond to this quotation. How is it possible for this to be true when deserts are so dry? What factors have you learned from this chapter that prove this statement true?

B. Where are the nearest eolian features (coastal, lakeshore, or desert dunes, or loess deposits) to your present location? Which causative factors discussed in this chapter explain the features you identified? Are you able to visit the site?

C. Relative to the Colorado River system, define, compare, and contrast what you think is meant by a "supply strategy" or a "demand strategy" in dealing with the present Colorado River water budget (Table 1 in Focus Study 15.1). Which strategy seems focused on centralized or decentralized solutions? What political and economic forces do you think are at play in planning for the river? Who are the stakeholders? With your analysis, briefly speculate what strategies you think would be most effective in dealing with this situation.

Basaltic rocks that hardened from centuries of lava flows confront the ocean along the south coast of the Reykjanes Peninsula, south of Iceland's capital Reykjavík. This island country sits astride the Mid-Atlantic Ridge and its sea-floor spreading process. Magma from a hot spot anchored in Earth's mantle feeds the island's volcanoes, as Iceland continues to grow. The scene is at approximately 63.8° N latitude. [Photo by Bobbé Christopherson.]

The Oceans, Coastal Processes, and Landforms

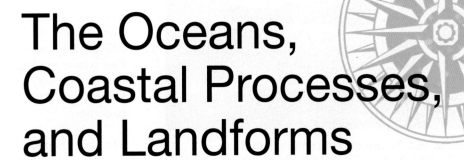

▶ Key Learning Concepts

After reading the chapter, you should be able to:

▶ *Describe* the chemical composition of seawater and the physical structure of the ocean.

▶ *Identify* the components of the coastal environment, and *list* the physical inputs to the coastal system, including tides and mean sea level.

▶ *Describe* wave motion at sea and near shore, and *explain* coastal straightening as a product of wave refraction.

▶ *Identify* characteristic coastal erosional and depositional landforms.

▶ *Describe* barrier islands and their hazards as they relate to human settlement.

▶ *Assess* living coastal environments: corals, wetlands, salt marshes, and mangroves.

▶ *Construct* an environmentally sensitive model for settlement and land use along the coast.

Walk along a shoreline and you witness the dramatic interaction of Earth's vast oceanic, atmospheric, and lithospheric systems. At times, the ocean attacks the coast in a stormy rage of erosive power; at other times, the moist sea breeze, salty mist, and repetitive motion of the water are gentle and calming (Figure 16.1). The coastlines are areas of dynamic change and beauty, where oceans and seas meet the land. Few have captured this confrontation between land and sea as well as biologist Rachel Carson:

The edge of the sea is a strange and beautiful place. All through the long history of Earth it has been an area of unrest where waves have broken heavily against the land, where the tides have pressed forward over the continents, receded, and then returned. For no two successive days is the shoreline precisely the same. Not only do the tides advance and retreat in their eternal rhythms, but the level of the sea

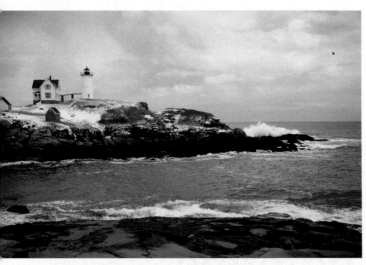

FIGURE 16.1 Where the land confronts the sea.
The Nubble Light on Cape Neddick, Maine, was built in 1879, near the town of York. Its 27-m (88-ft) tower has withstood storms and winds for almost 130 years. At such a place we feel the dramatic interaction of Earth's oceanic, atmospheric, and lithospheric systems. A photo of Nubble light was on the recorded disk of planetary scenes and sounds that is on the *Voyager* spacecraft, launched into deep space in 1977.
[Photo by Bobbé Christopherson.]

itself is never at rest. It rises or falls as the glaciers melt or grow, as the floors of the deep ocean basins shift under its increasing load of sediments, or as the earth's crust along the continental margins warps up or down in adjustment to strain and tension. Today a little more land may belong to the sea, tomorrow a little less. Always the edge of the sea remains an elusive and indefinable boundary.*

Despite such variability between sea and land, many people live and work near the ocean because of commerce, shipping, fishing, and tourism. A 1995 scientific assessment estimates that about 40% of Earth's population lives within 100 km (62 mi), and 49% within 200 km (145 mi), of coastlines. In the United States, about 50% of the people live in areas designated as *coastal* (this includes the Great Lakes). Therefore, an understanding of coastal processes and landforms is important to nearly half the world's population. And because these processes along coastlines often produce dramatic change, they are essential to consider in planning and development. A World Resources Institute study found as much as 50% of the world's coastlines at some risk of loss through things like erosion and rising sea level or disruption from pollution.

The ocean is a vast ecosystem, intricately linked to life on the planet and to life-sustaining systems in the atmosphere, the hydrosphere, and the lithosphere. In the

recent *Atlas of the Oceans—The Deep Frontier,* Jean-Michel Cousteau on "The Future of the Ocean" states:

> Today, we are coming to better appreciate the extent to which our actions affect an ecosystem—and the people who depend on it—thousands of miles away. The reef fisherman in Fiji is not undone by the local poacher, but by global warming intensified by the driving of a car in downtown Toronto, Canada. Yet these connections are not all bad news. The web of interdependence is built with strands of responsibility and hope. Our ever-expanding ability to communicate across borders and oceans is helping to drive a truly global dialogue about the planet's most pressing environmental challenges.†

In this chapter: We begin the chapter with a brief look at our global oceans and seas—1998 was celebrated as The International Year of the Ocean by all United Nations countries (see **http://www.yoto98.noaa.gov/**). The physical and chemical properties of the sea distinguish it from the waters of the continent. Our coverage includes discussions about tides, waves, coastal erosional and depositional landforms, beaches, barrier islands, and organic processes, including corals, wetlands, salt marshes, and mangroves. A systems framework of specific inputs (components and driving forces), actions (movements and processes), and outputs (results and consequences) organize our discussion of coastal processes. We conclude with a look at the considerable human impact on coastal environments.

Global Oceans and Seas

The ocean is one of Earth's last great scientific frontiers and is of great interest to geographers. Remote sensing from orbit, aircraft, surface vessels, submersibles, and the ocean floor is providing a wealth of data and a new capability to understand the oceanic system. The pattern of sea-surface temperatures is presented in Figure 5.11 and ocean currents in Figure 6.22. The world's oceans, their area, volume, and depth, are listed in a table in Figure 7.3. The locations of oceans and major seas are shown and listed alphabetically in Figure 16.2.

Chemical Composition of Seawater

Water is called the "universal solvent," dissolving at least 57 of the 92 elements found in nature. In fact, most natural elements and the compounds they form are found in the seas as dissolved solids, or *solutes.* Thus, seawater is a solution, and the concentration of dissolved solids is called **salinity**.

---

*"The Marginal World," in *The Edge of the Sea* by Rachel Carson. © 1955 by Rachel Carson, © renewed 1983 by R. Christie (Boston: Houghton Mifflin), p. 11.

†*Atlas of the Oceans—The Deep Frontier* by Sylvia Earle. © 2001 by National Geographic Society, text © 2001 by Sylvia Earle (Washington: National Geographic Society), p. 171.

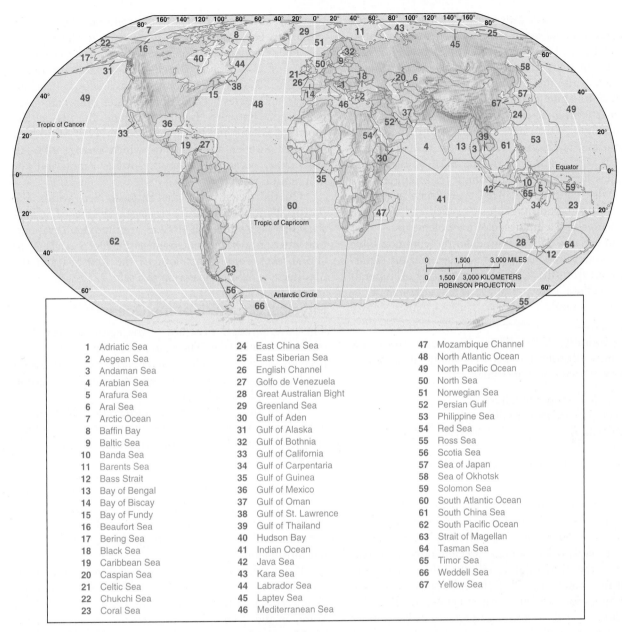

FIGURE 16.2 Principal oceans and seas of the world.
A sea is generally smaller than an ocean and is near a landmass; sometimes the term refers to a large, inland, salty body of water. Match the alphabetized name and number with its location on the map.

1	Adriatic Sea	24	East China Sea	47	Mozambique Channel
2	Aegean Sea	25	East Siberian Sea	48	North Atlantic Ocean
3	Andaman Sea	26	English Channel	49	North Pacific Ocean
4	Arabian Sea	27	Golfo de Venezuela	50	North Sea
5	Arafura Sea	28	Great Australian Bight	51	Norwegian Sea
6	Aral Sea	29	Greenland Sea	52	Persian Gulf
7	Arctic Ocean	30	Gulf of Aden	53	Philippine Sea
8	Baffin Bay	31	Gulf of Alaska	54	Red Sea
9	Baltic Sea	32	Gulf of Bothnia	55	Ross Sea
10	Banda Sea	33	Gulf of California	56	Scotia Sea
11	Barents Sea	34	Gulf of Carpentaria	57	Sea of Japan
12	Bass Strait	35	Gulf of Guinea	58	Sea of Okhotsk
13	Bay of Bengal	36	Gulf of Mexico	59	Solomon Sea
14	Bay of Biscay	37	Gulf of Oman	60	South Atlantic Ocean
15	Bay of Fundy	38	Gulf of St. Lawrence	61	South China Sea
16	Beaufort Sea	39	Gulf of Thailand	62	South Pacific Ocean
17	Bering Sea	40	Hudson Bay	63	Strait of Magellan
18	Black Sea	41	Indian Ocean	64	Tasman Sea
19	Caribbean Sea	42	Java Sea	65	Timor Sea
20	Caspian Sea	43	Kara Sea	66	Weddell Sea
21	Celtic Sea	44	Labrador Sea	67	Yellow Sea
22	Chukchi Sea	45	Laptev Sea		
23	Coral Sea	46	Mediterranean Sea		

The ocean remains a remarkably homogeneous mixture. The ratio of individual salts does not change, despite minor fluctuations in overall salinity. In 1874, the British HMS *Challenger* sailed around the world, taking surface and depth measurements and collecting samples of seawater. Analyses of those samples first demonstrated the uniform composition of seawater.

Although ocean chemistry was thought to be fairly constant during the Phanerozoic Eon (the past 542 million years), recent evidence suggests that seawater chemistry has varied over time within a narrow range. The variations are consistent with changes in sea-floor spreading rates, volcanism, and sea level. Evidence is gathered from fluid inclusions in marine formations such as in limestones and evaporites, which contain ancient seawaters. The ocean reflects conditions in Earth's environment. As an example, the high-latitude oceans have been freshening over the past decade in response to the large volume of freshwater coming from glacial ice melt and increased precipitation in response to warmer temperatures, increasing discharge from the Russian rivers that drain into the Arctic Ocean.

Ocean Chemistry Ocean chemistry is a result of complex exchanges among seawater, the atmosphere, minerals, bottom sediments, and living organisms. In addition, significant flows of mineral-rich water enter the ocean through hydrothermal (hot water) vents in the ocean floor. (These vents are called "black smokers" for the dense, black, mineral-laden water that spews from them, Figure 11.11b.) The uniformity of seawater results from complementary chemical reactions and continuous mixing—after all, the ocean basins interconnect, and water circulates among them.

Seven elements account for more than 99% of the dissolved solids in seawater. They are (with their ionic form) chlorine (as chloride, Cl^-), sodium (as Na^+), magnesium (as Mg^{2+}), sulfur (as sulfate, SO_4^{2-}), calcium (as Ca^{2+}), potassium (as K^+), and bromine (as bromide, Br^-). Seawater also contains dissolved gases (such as carbon dioxide, nitrogen, and oxygen), suspended and dissolved organic matter, and a multitude of trace elements.

Commercially, only sodium chloride (common table salt), magnesium, and bromine are extracted in any significant amount from the ocean. Future mining of minerals from the seafloor is technically feasible, although it remains uneconomical.

Average Salinity: 35‰   There are several ways to express salinity (dissolved solids by volume) in seawater, using the worldwide average value:

- 3.5% (% parts per hundred)
- 35,000 ppm (parts per million)
- 35,000 mg/l
- 35 g/kg
- 35‰ (‰ parts per thousand), the most common notation

Salinity worldwide normally varies between 34‰ and 37‰; variations are attributable to atmospheric conditions above the water and to the volume of freshwater inflows. In equatorial water, precipitation is great throughout the year, diluting salinity values to slightly lower than average (34.5‰). In subtropical oceans—where evaporation rates are greatest because of the influence of hot, dry, subtropical high-pressure cells—salinity is more concentrated, increasing to 36‰. Figure 16.3 plots the difference between evaporation and precipita-

tion and salinity by latitude to illustrate this slight spatial variability. Can you answer the question in the caption?

The term **brine** is applied to water that exceeds the average of 35‰ salinity. **Brackish** applies to water that is less than 35‰ salts. In general, oceans are lower in salinity near landmasses because of freshwater runoff and river discharges. Extreme examples include the Baltic Sea (north of Poland and Germany) and the Gulf of Bothnia (between Sweden and Finland), which average 10‰ or less salinity because of heavy freshwater runoff and low evaporation rates.

On the other hand, the Sargasso Sea, within the North Atlantic subtropical gyre, averages 38‰. The Persian Gulf has a salinity of 40‰ as a result of high evaporation rates in a nearly enclosed basin. Deep pockets, or "brine lakes," along the floor of the Red Sea and the Mediterranean Sea register up to a salty 225‰.

Physical Structure of the Ocean

The basic physical structure of the ocean is layered, as shown in Figure 16.4. The figure also graphs four key aspects of the ocean, each of which varies with increasing depth: average temperature, salinity, dissolved carbon dioxide, and dissolved oxygen.

The ocean's surface layer is warmed by the Sun and is wind-driven. Variations in water temperature and solutes are blended rapidly in a *mixing zone* that represents only 2% of the oceanic mass. Below the mixing zone is the *thermocline transition* zone, a more than 1-km-deep region of decreasing temperature gradient that lacks the motion of the surface. Friction at these depths dampens the effect of surface currents. In addition, colder water temperatures at the lower margin tend to inhibit any convective movements.

From a depth of 1–1.5 km (0.6–0.9 mi) to the ocean floor, temperature and salinity values are quite uniform. Temperatures in this *deep cold zone* are near 0°C (32°F). Water in the deep cold zone does not freeze, however, because of its salinity and intense pressures at those depths; seawater freezes at about −2°C (28.4°F) at the

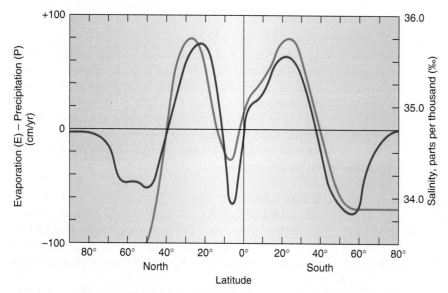

FIGURE 16.3 Variation in ocean salinity by latitude.
Salinity (green line) is principally a function of climatic conditions. Specifically important is the moisture relation expressed by the difference between evaporation and precipitation (E – P) (purple line). Why is salinity higher in the subtropics and lower along the equator? [G. Wüst, 1936.]

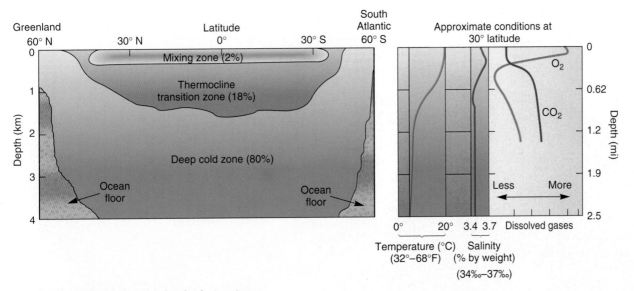

FIGURE 16.4 The ocean's physical structure.
Schematic of average physical structure observed throughout the ocean's vertical profile as sampled along a line from Greenland to the South Atlantic. Temperature, salinity, and dissolved gases are shown plotted by depth.

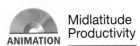

Midlatitude Productivity

surface. Generally, the coldest water is along the bottom. Now let us shift to the edge of the sea and examine Earth's coastlines. Many scientific activities related to the ocean are coordinated and conducted by the National Ocean Service. You can find information about these activities at http://www.nos.noaa.gov/.

Coastal System Components

We know that the continents were formed over many millions of years. However, most of Earth's coastlines are relatively new, existing in their present state as the setting for continuous change. A dynamic equilibrium exists among the energy of waves, tides, wind, and currents, the supply of materials, the slope of the coastal terrain, and the fluctuation of relative sea level. These interactions produce an infinite variety of erosional and depositional features and coastlines of diverse beauty.

Inputs to the Coastal System

Inputs to the coastal environment include many elements we have already discussed:

- *Solar energy* input drives the atmosphere and the hydrosphere. The conversion of insolation to kinetic energy produces prevailing winds, weather systems, and climate.
- *Atmospheric winds*, in turn, generate ocean currents and waves, key inputs to the coastal environment.
- *Climatic regimes*, which result from insolation and moisture, strongly influence coastal geomorphic processes.
- The nature of *coastal geomorphology* (rock type, landforms, tectonic activity) is important in determining rates of erosion and sediment production.
- *Human activities* are an increasingly significant input to coastal change.

All of these inputs occur within the ever-present influence of gravity's pull, not only from Earth but also from the Moon and Sun. Gravity provides the potential energy of position for materials in motion and generates the tides.

The Coastal Environment and Sea Level

The coastal environment is called the **littoral zone**, from the Latin word for "shore." Figure 16.5 illustrates the littoral zone and includes specific components discussed later in the chapter. The littoral zone spans some land as well as water. Landward, it extends to the highest waterline that occurs on shore during a storm. Seaward, it extends to where water is too deep for storm waves to move sediments on the seafloor—usually around 60 m, or 200 ft, in depth. The specific contact line between the sea and the land is the *shoreline*, although this line shifts with tides, storms, and sea-level adjustments. The *coast* continues inland from high tide to the first major landform change and may include areas considered to be part of the coast in local usage.

Because the level of the ocean varies, the littoral zone naturally shifts position from time to time. A rise in sea level causes submergence of land, whereas a drop in sea level exposes new coastal areas. In addition, uplift and subsidence of the land itself initiates changes in the littoral zone.

Sea level is an important concept. Every elevation you see in an atlas or on a map is referenced to mean sea level. Yet this average sea level changes daily with the tides and over the long term with changes in climate, tectonic plate movements, and glaciation. Thus, *sea level* is a relative term. At present, there exists no international system to determine exact sea level over time. The Global Sea Level Observing System (GLOSS) is an international group actively working on sea-level issues. (For copies of their newsletters and other scientific discussion, see http://www.pol.ac.uk/psmsl/gb.html. GLOSS is part of

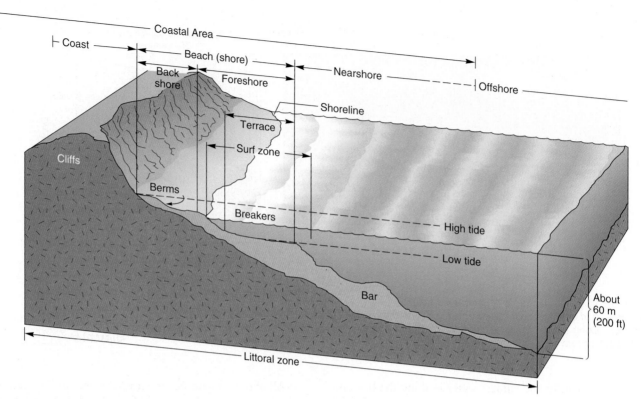

FIGURE 16.5 The littoral zone.
The littoral zone includes the coast, beach, and nearshore environments.

the larger Permanent Service for Mean Sea Level, which you can find at http://www.pol.ac.uk/psmsl/programmes/.)

Mean sea level (MSL) is a value based on average tidal levels recorded hourly at a given site over many years. MSL varies spatially because of ocean currents and waves, tidal variations, air temperature and pressure differences, ocean temperature variations, slight variations in Earth's gravity, and changes in oceanic volume. At present, sea level is rising at a historic rate related to global climate change. For more information, see News Report 16.1.

At present, the overall U.S. MSL is calculated at approximately 40 locations along the coastal margins of the continent. These sites are being upgraded with new equipment in the Next Generation Water Level Measurement System, using next-generation tide gauges, specifically along the U.S. and Canadian Atlantic coasts, Bermuda, and the Hawaiian Islands. The *NAVSTAR* satellites that make up the Global Positioning System (GPS) make possible the correlation of data within a network of ground- and ocean-based measurements.

Remote-sensing technology augments these measurements, including the *TOPEX/Poseidon* satellite launched in 1992 and still in service (see http://topex-www.jpl.nasa.gov/). This satellite has two radar altimeters that measure changes in mean sea level at any one location every 10 days between 66° N and 66° S latitudes. These measurements are made to an astonishing precision of 4.2 cm, 1.7 in. (Figure 16.6)! Spectacular *TOPEX/Poseidon* portraits of El Niño and La Niña in the Pacific are in Focus Study 10.1. Another ocean surface

topography satellite launched in December 2001, named *Jason-1*, is extending the science of determining mean sea level, ocean topography, atmosphere–ocean interactions, and ocean circulation.

Coastal System Actions

The coastal system is the scene of complex tidal fluctuation, winds, waves, ocean currents, and the occasional impact of storms. These forces shape landforms ranging from gentle beaches to steep cliffs, and they sustain delicate ecosystems.

Tides

Tides are complex daily oscillations in sea level, ranging worldwide from barely noticeable to several meters. They are experienced to varying degrees along every ocean shore around the world. Tidal action is a relentless energy agent for geomorphic change. As tides flood (rise) and ebb (fall), the daily migration of the shoreline landward and seaward causes significant changes that affect sediment erosion and transportation.

Tides also are important in human activities, including navigation, fishing, and recreation. Tides are especially important to ships because the entrance to many ports is limited by shallow water, and thus high tide is required for passage. Tall-masted ships may need a low tide to clear overhead bridges. Tides also exist in large lakes, but because the tidal range is small, tides are difficult to distinguish from changes caused by wind. Lake Superior, for instance, has a tidal variation of only about 5 cm (2 in.).

News Report 16.1

Sea-Level Variations and the Present MSL Increase

Sea level varies along the full extent of North American shorelines. The mean sea level (MSL) of the U.S. Gulf Coast is about 25 cm (10 in.) higher than that of Florida's east coast, which is the lowest in North America. MSL rises northward along the eastern coast, to 38 cm (15 in.) higher in Maine than in Florida. Along the U.S. western coast, MSL is higher than Florida's by about 58 cm (23 in.) in San Diego and by about 86 cm (34 in.) in Oregon.

Overall, North America's Pacific coast MSL averages about 66 cm (26 in.) higher than the Atlantic coast MSL. MSL is affected by differences in ocean currents, air pressure and wind patterns, water density, and water temperature.

Over the long term, sea-level fluctuations expose a range of coastal landforms to tidal and wave processes. As average global temperatures cycle through cold or warm climatic spells, the quantity of ice locked up in the ice sheets of Antarctica and Greenland and in hundreds of mountain glaciers can increase or decrease and result in

sea-level changes accordingly. At the peak of the most recent Pleistocene glaciation about 18,000 B.P. (years before the present), sea level was about 130 m (430 ft) lower than it is today. On the other hand, if Antarctica and Greenland ever became ice-free (ice sheets fully melted), sea level would rise at least 65 m (215 ft) worldwide.

Just 100 years ago sea level was 38 cm (15 in.) lower along the coast of southern Florida. Venice, Italy, has experienced a rise of 25 cm (10 in.) since 1890. During the last century, sea level rose 10–20 cm (4–8 in.), a rate 10 times higher than the average rate during the last 3000 years. The present sea-level rise is spatially uneven; for instance, the rate along the coast of Argentina is nearly 10 times the rate along the coast of France.

Given these trends and the predicted climatic change, sea level will continue to rise and be potentially devastating for many coastal locations. A rise of only 0.3 m (1 ft) would cause shorelines worldwide to move inland an average of 30 m (100 ft)!

This elevated sea level would inundate some valuable real estate along coastlines worldwide. Some 20,000 km² (7800 mi²) of land along North American shores alone would be drowned, at a staggering loss of $650 billion. A 95-cm (3.1-ft) sea-level rise could inundate 15% of Egypt's arable land, 17% of Bangladesh, and many island nations and communities. However, uncertainty exists in these forecasts, and the pace of the rise should be slow.

The 2001 IPCC forecast for global mean sea-level rise this century, given regional variations, is a range from 0.09–0.88 m (3.5–34.7 in.). Review the "Global Climate Change" section in Chapter 10 for specific forecasts. Despite any uncertainty, planning should start now along coastlines worldwide because preventive strategies are cheaper than recovery costs from possible destruction. Insurance underwriters have begun the process by refusing coverage for shoreline properties vulnerable to rising sea level.

Causes of Tides   Tides are produced by the gravitational pull of both the Sun and the Moon. (Earth's astronomical relation to the Sun and the Moon is discussed in Chapter 2.) The Sun's influence is only about half that of the Moon's because of the Sun's greater distance from Earth, although it is a significant force. Figure 16.7 illustrates the relation among the Moon, the Sun, and Earth and the generation of variable tidal bulges on opposite sides of the planet.

The gravitational pull of the Moon tugs on Earth's atmosphere, oceans, and lithosphere. The same is true for the Sun, to a lesser extent. Earth's solid and fluid surfaces all experience some stretching as a result of this gravitational pull. The stretching raises large *tidal bulges* in the atmosphere (which we can't see), smaller tidal bulges in the ocean, and very slight bulges in Earth's rigid crust. Our concern here is the tidal bulges in the ocean.

Gravity and inertia are essential elements in understanding tides. *Gravity* is the force of attraction between two bodies. *Inertia* is the tendency of objects to stay still if motionless or to keep moving in the same direction if in motion. The gravitational effect on the side of Earth facing the Moon or Sun is greater than that experienced by

the far side, where inertial forces are slightly greater. This difference exists because gravitational influences decrease with distance. It is this difference in the net force of gravitational attraction and inertia that generates the tides.

Figure 16.7a shows the Moon and the Sun in *conjunction* (lined up with Earth), a position in which the sum of their gravitational forces produces a large tidal bulge. The corresponding tidal bulge on Earth's opposite side is primarily the result of the farside water's remaining in position (being left behind) because its inertia exceeds the gravitational pull of the Moon and Sun. In effect, from this inertial point of view, as the nearside water and Earth are drawn toward the Moon and Sun, the farside water is left behind because of the slightly weaker gravitational pull. This arrangement produces the two opposing tidal bulges on opposite sides of Earth.

Tides appear to move in and out along the shoreline, but they do not actually do so. Instead, Earth's surface rotates into and out of the relatively "fixed" tidal bulges as Earth changes its position in relation to the Moon and Sun. Every 24 hours and 50 minutes, any given point on Earth rotates through two bulges as a direct result of this rotational positioning. Thus, every day, most coastal

FIGURE 16.6 Ocean topography as revealed by satellite.
Sea-level data recorded by the radar altimeter aboard the *TOPEX/Poseidon* satellite, October 3–12, 1992. The color scale is given in centimeters above or below Earth's geoid. The overall relief portrayed in the image is about 2 m (6.6 ft). The maximum sea level is in the western Pacific Ocean (white). The minimum is around Antarctica (blue and purple). [Image from JPL and NASA/GSFC; *TOPEX/Poseidon* is a joint U.S.–French Mission.]

locations experience two high (rising) tides, known as **flood tides**, and two low (falling) tides, known as **ebb tides**. The difference between consecutive high and low tides is considered the *tidal range*.

Spring and Neap Tides The combined gravitational effect of the Sun and Moon is strongest in the conjunction alignment and results in the greatest tidal range between high and low tides, known as **spring tides**. (*Spring* means to "spring forth"; it has no relation to the season of the year.) Figure 16.7b shows the other alignment that gives rise to spring tides, when the Moon and Sun are at *opposition*. In this arrangement, the Moon and Sun cause separate tidal bulges, affecting the water nearest to each of them. In addition, the left-behind water resulting from the pull of the body on the opposite side augments each bulge.

When the Moon and the Sun are neither in conjunction nor in opposition but are more or less in the positions shown in Figure 16.7c and d, their gravitational influences are offset and counteract each other, producing a lesser tidal range known as **neap tide**. (*Neap* means "without the power of advancing.")

Tides also are influenced by other factors, including ocean basin characteristics (size, depth, and topography), latitude, and shoreline shape. These factors cause a great variety of tidal ranges. For example, some locations may experience almost no difference between high and low tides. The highest tides occur when open water is forced into partially enclosed gulfs or bays. The Bay of Fundy in

Nova Scotia records the greatest tidal range on Earth, a difference of 16 m (52.5 ft) (Figure 16.8a, b). (For more on tides and tide prediction, contact the Scripps Institution of Oceanography library at **http://scilib.ucsd.edu/sio/tide/**.)

Tidal Power The fact that sea level changes daily with the tides suggests an opportunity: Could these predictable flows be harnessed to generate electricity? The answer is yes, given the right conditions. Bays and estuaries tend to focus tidal energy, concentrating it in a smaller area than in the open ocean. This circumstance provides an opportunity to construct a dam with water gates, locks to let ships through, and turbines to generate power.

Only 30 locations in the world are suited for tidal power generation. At present, only three are producing electricity. Two are outside North America—a 4-megawatt-capacity station in Russia in operation since 1968 (at Kislaya-Guba Bay on the White Sea) and a facility in France operating since 1967 (on the Rance River estuary on the Brittany coast). The tides in the Rance estuary fluctuate up to 13 m (43 ft), and power production has been almost continuous there, providing an electrical-generating capacity of a moderate 240 megawatts (about 20% of the capacity of Hoover Dam). The third area is the Bay of Fundy in Nova Scotia, Canada. At one of several favorable sites on the Bay, the Annapolis Tidal Generating Station was built in 1984. Nova Scotia Power Incorporated operates this 20-megawatt plant (Figure 16.8c). According

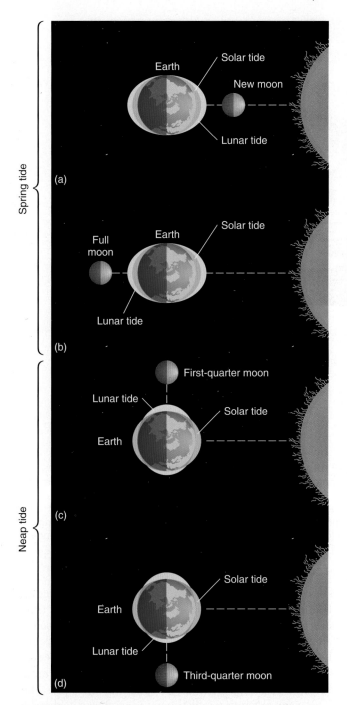

FIGURE 16.7 **The cause of tides.**
Gravitational relations of Sun, Moon, and Earth combine to produce spring tides (a, b) and neap tides (c, d). (Tides are greatly exaggerated for illustration.)

ANIMATION   Monthly Tidal Cycles

to the Canadian government, tidal power generation at ideal sites is economically competitive with fossil fuel plants.

Another kind of tidal power taking form in Norway involves sea-floor devices that harness the motion of coastal currents. Windmill-like turbines are set in motion by the tides and currents, as are paddle-driven turbines. Electrical production using this method began in 2003.

Waves

Friction between moving air (wind) and the ocean surface generates undulations of water called **waves**. Waves travel in groups, called *wave trains*. Waves vary widely in scale: On a small scale, a moving boat creates a wake of small waves; at a larger scale, storms generate large groups of wave trains. At the extreme is the wind wake produced by the presence of the Hawaiian Islands, traceable westward across the Pacific Ocean surface for 3000 km (1865 mi). The islands disrupt the steady trade winds and produce related surface temperature and wind changes.

A stormy area at sea is called a *generating region* for large wave trains, which radiate outward in all directions. The ocean is crisscrossed with intricate patterns of waves traveling in all directions. The waves seen along a coast may be the product of a storm center thousands of kilometers away.

Regular patterns of smooth, rounded waves, called **swells**, are the mature undulations of the open ocean. As waves leave the generating region, wave energy continues to run in these swells, which can range from small ripples to very large flat-crested waves. A wave leaving a deep-water generating region tends to extend its wavelength horizontally for many meters. Tremendous energy occasionally accumulates to form unusually large waves. One moonlit night in 1933, the U.S. Navy tanker *Ramapo* reported a wave in the Pacific higher than their mainmast, at about 34 m (112 ft)!

As you watch waves in open water, it appears that water is migrating in the direction of wave travel, but only a slight amount of water is actually advancing. It is the *wave energy* that is moving through the flexible medium of water. Water within a wave in the open ocean is simply transferring energy from molecule to molecule in simple cyclic undulations, called *waves of transition* (Figure 16.9). Individual water particles move forward only slightly, forming a vertically circular pattern.

The diameter of the paths formed by the orbiting water particles decreases with depth. As a deep-ocean wave approaches the shoreline and enters shallower water (10–20 m, or 30–65 ft), the orbiting water particles are vertically restricted. This restriction causes more-elliptical, flattened orbits to form near the bottom. This change from circular to elliptical orbits slows the entire wave, although more waves continue arriving. The resultant effects are closer-spaced waves, growing in height and steepness, with sharper wave crests. As the crest of each wave rises, a point is reached when its height exceeds its vertical stability and the wave falls into a characteristic **breaker**, crashing onto the beach (Figure 16.9b).

In a breaker, the orbital motion of transition gives way to elliptical *waves of translation* in which both energy and water move toward shore. The slope of the shore determines wave style. Plunging breakers indicate a steep bottom profile, whereas spilling breakers indicate a gentle, shallow bottom profile. In some areas, unexpected high waves can arise suddenly. It is a good idea to learn to recognize severe wave conditions before you venture along the shore in these areas.

(a)

(b)

(c)

FIGURE 16.8 Tidal range and tidal power.
Tidal range is great in some bays and estuaries, such as Halls Harbor near the Bay of Fundy at flood tide (a) and ebb tide (b). The flow of water between high and low tide is ideal for turning turbines and generating electricity, as is done at the Annapolis Tidal Generating Station (c), near the Bay of Fundy in Nova Scotia, in operation since 1984. [Photos by Jeff Newbery.]

When the backwash of water flows to the ocean from the beach in a concentrated column, usually at a right angle to the line of breakers, it is a *rip current*. A person caught in one of these can be swept offshore, but usually only a short distance. These brief, short torrents of water can be dangerous (Figure 16.9c).

As various wave trains move along in the open sea, they interact by *interference*. These interfering waves sometimes align so that the wave crests and troughs from one wave train are in phase with those of another. When this in-phase condition occurs, the height of the waves is increased, sometimes dramatically. The resulting "killer waves" or "sleeper waves" can sweep in unannounced and overtake unsuspecting victims. Signs along portions of the California, Oregon, Washington, and British Columbia coastline warn beachcombers to watch for "killer waves." On the other hand, out-of-phase wave trains will dampen wave energy at the shore. When you observe the breakers along a beach, the changing beat of the surf actually is produced by the patterns of *wave interference* that occurred in far-distant areas of the ocean.

Wave Refraction In general, wave action tends to straighten a coastline. Where waves approach an irregular coast, they bend around headlands, which are protruding landforms generally composed of resistant rocks (Figure 16.10). The submarine topography refracts, or bends, approaching waves. The refracted energy is focused around headlands and dissipates energy in coves, bays, and the submerged coastal valleys between headlands. Thus, headlands receive the brunt of wave attack along a coastline. This **wave refraction** redistributes wave energy so that different sections of the coastline vary in erosion potential, with the long-term effect of straightening the coast.

Figure 16.11 shows waves approaching the coast at an angle, for they usually arrive at some angle other than parallel. As the waves enter shallow water, they are refracted as the shoreline end of the wave slows. The wave portion in deeper water moves faster in comparison, thus producing a current parallel to the coast, zigzagging in the prevalent direction of the incoming waves. This **longshore current**, or *littoral current*, depends on wind direction and resultant wave direction. A longshore current is generated only in the surf zone and works in combination with wave action to transport large amounts of sand, gravel, sediment, and debris along the shore as *longshore drift*, or, to use the more comprehensive term, **littoral drift**.

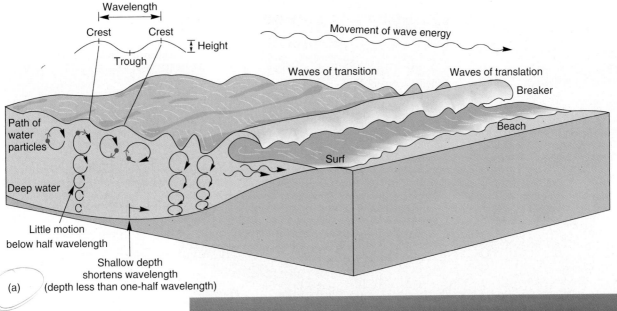

(a) Shallow depth shortens wavelength (depth less than one-half wavelength)

(b)

(c)

ANIMATION Wave Motion/Wave Refraction

FIGURE 16.9 Wave formation and breakers.
(a) The orbiting tracks of water particles change from circular motions and swells in deep water (waves of transition) to more elliptical orbits near the bottom in shallow water (waves of translation). (b) Cascades of waves attack the shore along Baja California, Mexico. (c) A dangerous rip current interrupts approaching breakers. Note the churned-up water where the rip current enters the surf. [Photos by Bobbé Christopherson.]

Particles on the beach also are moved along as **beach drift**, shifting back and forth between water and land with each *swash* and *backwash* of surf. Individual sediment grains trace arched paths along the beach. You have perhaps stood on a beach and heard the sound of myriad sand grains and seawater in the backwash of surf. These dislodged materials are available for transport and even-

tual deposition in coves and inlets and can represent a significant volume.

Tsunami, or Seismic Sea Wave An occasional wave that momentarily but powerfully influences coastlines is the tsunami. **Tsunami** is Japanese for "harbor wave," named for its devastating effect where its energy is

FIGURE 16.10 Coastal straightening.
(a) The process of coastal straightening is brought about by wave refraction. Wave energy is concentrated as it converges on headlands (b) and is diffused as it diverges in coves and bays (c). Headlands are frequent sites for lighthouses such as the light at Point Arena, California (d). [(b), (c) Photos by author; (d) photo by Bobbé Christopherson.]

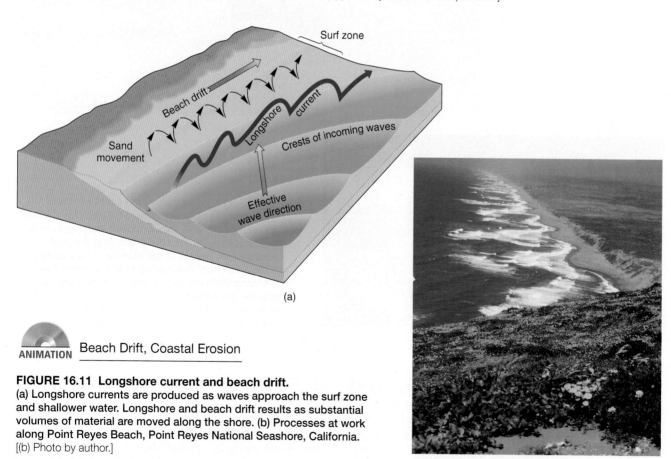

ANIMATION Beach Drift, Coastal Erosion

FIGURE 16.11 Longshore current and beach drift.
(a) Longshore currents are produced as waves approach the surf zone and shallower water. Longshore and beach drift results as substantial volumes of material are moved along the shore. (b) Processes at work along Point Reyes Beach, Point Reyes National Seashore, California. [(b) Photo by author.]

focused in harbors. Often tsunami are reported incorrectly as "tidal waves," but they have no relation to the tides. They are formed by sudden, sharp motions in the seafloor, caused by earthquakes, submarine landslides, or eruptions of undersea volcanoes. Thus, they properly are called *seismic sea waves*.

A tsunami typically begins when a large undersea disturbance generates a solitary wave of great wavelength, or sometimes a group of two or three long waves. These waves generally exceed 100 km (60 mi) in wavelength, but they are only a meter or so in height. Because of their great wavelength, tsunami are affected by the topography of the deep-ocean floor and are refracted by rises and ridges. They travel at great speeds in deep-ocean water—velocities of 600 to 800 kmph (375 to 500 mph) are not uncommon—but they often pass unnoticed on the open sea because their great length makes the slow rise and fall of water hard to observe.

As a tsunami approaches a coast, however, the shallow water forces the wavelength to shorten. As a result, the wave height may increase up to 15 m (50 ft) or more. Such a wave has the potential to devastate a coastal area, causing property damage and death. For example, in 1992 the citizens of Casares, Nicaragua, were surprised by a 12-m (39-ft) tsunami that took 270 lives. A 1998 Papua New Guinea tsunami, launched by a massive undersea landslide some 4 km^3 (1 mi^3), killed 2000. During the 20th century, there were 141 damaging tsunami and perhaps 900 smaller ones, with a total death toll of about 70,000. On December 26, 2004, a 9.0 magnitude earthquake struck off the west coast of northern Sumatra, triggering a massive tsunami across the Indian Ocean. Total deaths from the quake and tsunami exceeded 150,000. No warning system was in place as it is in the Pacific Ocean.

Hawai'i is vulnerable to tsunami because of its position in the open central Pacific, surrounded by the ring of fire that outlines the Pacific Basin. The U.S. Army Corps of Engineers has reported 41 damaging tsunami in Hawai'i during the past 142 years—statistically, 1 every 3.5 years. Figure 16.12 illustrates tsunami travel times across the Pacific Ocean to Honolulu; for instance, it would take a tsunami 10 hours to reach Hawai'i from the Philippines.

Because tsunami travel such great distances so quickly and are undetectable in the open ocean, accurate forecasts are difficult and occurrences are often unexpected. A warning system now is in operation for nations surrounding the Pacific, where the majority of tsunami occur. A warning is issued whenever seismic stations detect a significant quake or landslide under water, where it might generate a tsunami. A new tsunami hazard-mitigation network began operating in 2004 with the deployment of six pressure sensors on the ocean floor with accompanying surface buoys—this is the Deep-ocean Assessment and Reporting of Tsunamis (DART). Three instruments are along the Aleutian Islands, south of Alaska, two along the U.S. West Coast, and one near the equator off South America.

There is continuing threat from volcanic island collapses, such as the closely watched Cumbre Vieja volcano

FIGURE 16.12 Tsunami travel times to Honolulu, Hawai'i. [After NOAA.]

in the Canary Islands that could affect the Atlantic Basin. Forecasters watch the Kīlauea region of the southeast coast of Hawai'i, where a 40-km by 20-km by 2000-m (25-mi × 12-mi × 6560-ft) thick portion of relatively new and unstable crustal basalt could affect the Pacific Basin.

Warnings always should be heeded, despite the many false alarms, for the causes lie hidden beneath the ocean and are difficult to monitor in any consistent manner, although remote-sensing capabilities are improving accuracy. (For the tsunami research program, see http://www.pmel.noaa. gov/tsunami/. The tsunami home page is at http://www. geophys.washington.edu/tsunami/welcome.html.)

Coastal System Outputs

As you can see, coastlines are active places, with energy and sediment being continuously delivered to a narrow environment. The action of tides, currents, wind, waves, and changing sea level produces a variety of erosional and depositional landforms. We look first at erosional coastlines, then at depositional coastlines.

Erosional Coastal Processes and Landforms

The active margin of the Pacific Ocean along North and South America is a typical erosional coastline. *Erosional coastlines* tend to be rugged, of high relief, and tectonically active, as expected from their association with the leading edge of drifting lithospheric plates (review the plate tectonics discussion in Chapter 11). Figure 16.13 presents features commonly observed along an erosional coast.

Sea cliffs are formed by the undercutting action of the sea. As indentations slowly grow at water level, a sea cliff becomes notched and eventually will collapse and retreat. Other erosional forms evolve along cliff-dominated coastlines, including *sea caves*, *sea arches*, and *sea stacks*. As erosion continues, arches may collapse, leaving isolated stacks in the water (Figure 16.13b, c). The coasts of southern England and Oregon are prime examples of such erosional landscapes.

Stack, arch

(b)

Collapsing cliffs

(d)

Terrace

Former sea cliffs

Sea cave

Wave-cut platform (terrace)

Sea arch

Sea cliff

Landslides

Sea stack

Notched cliff

Wave-cut platform

(a)

Wave-built terrace

(c)

(e)

FIGURE 16.13 Erosional coastal features.
(a) Characteristic coastal erosional landforms: caves, arches (b), stacks (c), collapsing cliffs (d),
and wave-cut platforms near Bixby Bridge along the Cabrillo Highway, California (e). [(b), (c) Photos
by author; (d) photo by Lowell Georgia/Photo Researchers, Inc.; (e) photo by Bobbé Christopherson.]

Wave action can cut a horizontal bench in the tidal zone, extending from a sea cliff out into the sea. Such a structure is called a **wave-cut platform**, or *wave-cut terrace*. If the relation between the land and sea level has changed over time, multiple platforms or terraces may rise like stairsteps back from the coast. These marine terraces are remarkable indicators of a changing relation between the land and sea, with some terraces more than 370 m (1200 ft) above sea level. A tectonically active region, such as the California coast, has many examples of multiple wave-cut platforms, which at times can be unstable and vulnerable to failure (Figure 16.13d, e).

Depositional Coastal Processes and Landforms

Depositional coasts generally are along land of gentle relief, where sediments from many sources are available.

Such is the case with the Atlantic and Gulf coastal plains of the United States, which lie along the relatively passive, trailing edge of the North American lithospheric plate. Erosional processes and inundation, particularly during storm activity, influence depositional coasts.

Figure 16.14 illustrates characteristic landforms deposited by waves and currents. One notable deposition landform is the **barrier spit**, which consists of material deposited in a long ridge extending out from a coast. It partially crosses and blocks the mouth of a bay. Classic examples of a barrier spit include Sandy Hook, New Jersey (south of New York City), and Cape Cod, Massachusetts. The Little Sur River in California (16.14a) and Prion Bay in Southwestern National Park in Tasmania (16.14b) show barrier spits forming partway across the mouths of their respective rivers.

If a spit grows to completely cut off the bay from the ocean and form an inland lagoon, it is called a **bay barrier**, or *baymouth bar*. Spits and barriers are made up of

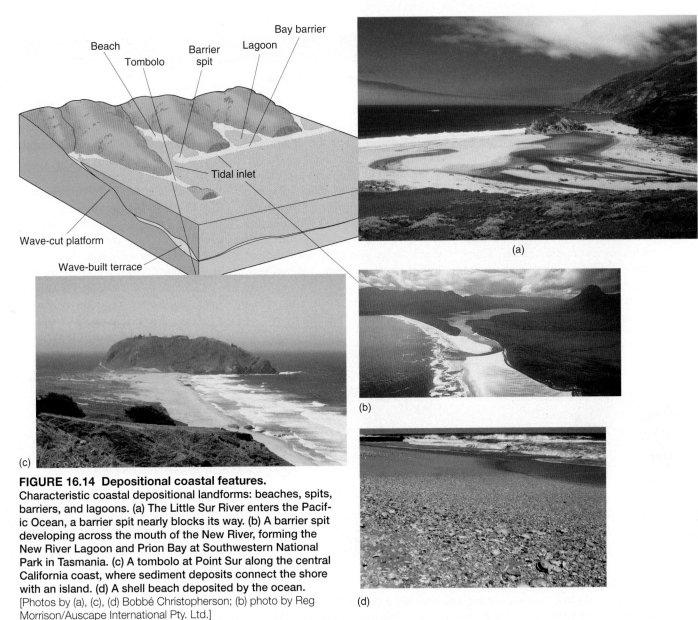

FIGURE 16.14 Depositional coastal features.
Characteristic coastal depositional landforms: beaches, spits, barriers, and lagoons. (a) The Little Sur River enters the Pacific Ocean, a barrier spit nearly blocks its way. (b) A barrier spit developing across the mouth of the New River, forming the New River Lagoon and Prion Bay at Southwestern National Park in Tasmania. (c) A tombolo at Point Sur along the central California coast, where sediment deposits connect the shore with an island. (d) A shell beach deposited by the ocean.
[Photos by (a), (c), (d) Bobbé Christopherson; (b) photo by Reg Morrison/Auscape International Pty. Ltd.]

materials that have been eroded and transported by *littoral drift* (considered beach and longshore drift combined). For much sediment to accumulate, offshore currents must be weak; strong currents carry material away before it can be deposited. Tidal flats and salt marshes are characteristic low-relief features wherever tidal influence is greater than wave action. If these deposits completely cut off the bay from the ocean, an inland **lagoon** is formed.

A **tombolo** occurs when sediment deposits connect the shoreline with an offshore island or sea stack (Figure 16.14c). The tombolo forms when sediments accumulate on an underwater wave-built terrace. Coastal erosion and depositional features operate through a complex history as the relation between the land and the ocean change. High Latitude Connection 16.1 examines isostatic rebound of the crust, causing such shoreline features to rise in relation to sea level.

Not all the beaches of the world are composed of sand, for they can be shingles (beach gravel), and shells, among other materials (Figure 16.14d). Let's examine these transient coastal deposits in more detail—let's go to the beach.

Beaches Of all the features associated with a depositional coastline, beaches probably are the most familiar. Beaches vary in type and permanence, especially along coastlines dominated by wave action. Technically, a **beach** is that place along a coast where sediment is in motion, deposited by waves and currents. Material from the land temporarily resides on the beach while it is in active transit along the shore. You may have experienced a beach at some time, along a sea coast, a lake shore, or even along a stream. Perhaps you have even built your own "landforms" in the sand, only to see them washed away by the waves: a lesson in erosion.

On average, the beach zone spans from about 5 m (16 ft) above high tide to 10 m (33 ft) below low tide (see Figure 16.5). However, the specific definition varies greatly along individual shorelines. Worldwide, quartz (SiO_2) dominates beach sands because it resists weathering and therefore remains after other minerals are removed. In volcanic areas, beaches are derived from wave-processed lava. Hawai'i and Iceland, for example, feature some black-sand beaches. Many beaches, such as those in southern France and western Italy, lack sand and are

High Latitude Connection 16.1

A Rebounding Shoreline and Coastal Features

We learned in Chapter 11, Figure 11.4, that the lighter crust rests on denser materials beneath—the principal of buoyancy applies. Where the load is greater, owing to glaciers, sediment, or mountains, the crust tends to sink, or ride lower. When a glacier melts, for instance, the crust rides higher with the decreased load in a recovery uplift known as *isostatic rebound*. Thus, the entire crust is in a constant state of compensating adjustment.

As Earth systems came out of the last ice age and the massive glaciers

and continental ice retreated, the unburdened crust began its rebound, or *postglacial uplift*. The last continental ice retreated about 6000 years ago. The floor of Hudson Bay, Canada, for example is rising at about 15 mm a year. Recently, as we see in Chapter 17, southeastern Alaska is experiencing an accelerated rise beyond the postglacial uplift associated with the last ice age—a rebound rate of 36 mm/year, discussed in Chapter 11. This new uplift is thought to be a result of the substantial losses of glacial ice related

to increasing temperatures across Alaska.

In the Svalbard Archipelago in the Arctic Ocean north of Norway, along Duve Fjord on Nordaustlandet Island, the land is rising at approximately 2.5 m (8.3 ft) per century. Figure 16.1.1 shows landscape features caused by this phenomenon. The former shoreline appears in successive terraces. Where a depositional tombolo had formed, the neck of land connecting to the offshore island is now stranded above sea level (Figure 16.1.1b).

(a)

(b)

FIGURE 16.1.1 A rebounding coast.
Along Duve Fjord, Nordaustlandet Island, Arctic Ocean: (a) Each of the curved terraces of cobbles along this coast is a former shoreline, formed as the island isostatically rebounded following the retreat of the glaciers. (b) A raised tombolo on the uplifting coast. [Photos by Bobbé Christopherson.]

composed of pebbles and cobbles—a type of *shingle beach*. Some shores have no beaches at all; scrambling across boulders and rocks may be the only way to move along the coast. The coasts of Maine and portions of Canada's Atlantic provinces are classic examples. These coasts, composed of resistant granite rock, are scenically rugged and have few beaches.

A beach acts to stabilize a shoreline by absorbing wave energy, as is evident by the amount of material that is in almost constant motion (see "sand movement" in Figure 16.11). Some beaches are stable. Others cycle seasonally: They accumulate during the summer, are moved offshore by winter storm waves, forming a submerged bar, and are redeposited onshore the following summer. Protected areas along a coastline tend to accumulate sediment, which can lead to large coastal sand dunes. Prevailing winds often drag such coastal dunes inland, sometimes burying trees and highways.

Maintaining Beaches Changes in coastal sediment transport can disrupt human activities—beaches are lost, harbors are closed, and coastal highways and beach houses can be inundated with sediment. Thus, people use various strategies to interrupt longshore drift and beach drift. The goal is either to halt sand accumulation or to force accumulation in a desired way through construction of engineered structures, called "hard" shoreline protection.

Figure 16.15 illustrates common approaches: a *jetty* to block material from harbor entrances, a *groin* to slow drift action along the coast, and a *breakwater* to create a zone of still water near the coastline. However, interrupting the littoral drift that is the natural replenishment for beaches may lead to unwanted changes in sediment distribution downcurrent. Careful planning and impact assessment should be part of any strategy for preserving or altering a beach.

Beach nourishment refers to the artificial replacement of sand along a beach (see News Report 16.2). Through

(a)

(b)

(c)

FIGURE 16.15 Interfering with the littoral drift of sand.
(a) Various constructions attempt to control littoral drift (beach drift and longshore drift) along a coast: breakwater, jetty, and groin. (b) Aerial photograph of coastal constructions. Note the disruption of sediment movement along the coast at Bald Head Island, North Carolina, looking west across the Cape Fear River mouth.
(c) A groin interrupts sand transport along a coast. [Photos by (b) Robert H. Goslee; (c) Bobbé Christopherson.]

ANIMATION   Coastal Stabilization Structures

such efforts, a beach that normally experiences a net loss of sediment will instead show a net gain. Years of human effort and expense to build beaches can be erased by a single storm, as occurred when Hurricane Camille completely eliminated offshore islands along the Gulf Coast in 1969, or in 2004 when Hurricanes Charley, Frances, Ivan, and Jeanne devastated coastal Florida beaches and barrier islands.

Barrier Formations   Barrier chains are long, narrow, depositional features, generally of sand, that form offshore roughly parallel to the coast. Common forms are **barrier beaches** or the broader, more extensive landform, **barrier islands**. Tidal variation in the area usually is moderate to low, with adequate sediment supplies coming from nearby coastal plains. Figure 16.16 illustrates the many features of barrier chains, using North Carolina's famed Outer Banks as an example, including Cape Hatteras, across Pamlico Sound from the mainland. The

area presently is designated as one of three national seashore reserves supervised by the National Park Service.

On the landward side of a barrier formation are tidal flats, marshes, swamps, lagoons, coastal dunes, and beaches, visible in Figure 16.16. Barrier beaches appear to adjust to sea level and may naturally shift position from time to time in response to wave action and longshore currents. A break in a barrier forms an inlet that connects a bay with the ocean. The name *barrier* is appropriate, for these formations take the brunt of storm energy and actually shield the mainland.

Because of the continuing loss to the barrier island, the famous Cape Hatteras lighthouse was moved inland in 1999 to safer ground (Figure 16.16c). Its new position is about 488 m (1600 ft) from the ocean, or approximately the distance it was in 1870 when it was built—that much sand was lost back to the sea. The change in latitude and longitude coordinates for this lighthouse is the opening photo for Chapter 1. (See http://www.ncsu.edu/coast/chl/ for

News Report 16.2

Engineers Nourish a Beach

The city of Miami, Florida, and surrounding Dade County have spent almost $70 million since the 1970s in a continuing effort to rebuild their beaches. Sand is transported to the replenishment area from a source area. To maintain a 200-m- (660-ft) wide beach, planners determine net sand loss per year and set a schedule for replenishment. In Miami Beach, an 8-year replenishment cycle is maintained. During Hurricane Andrew in 1992, the replenished Miami Beach is thought to have prevented millions of dollars in shoreline structural damage. The cost of beach replenishment following the 2004 hurricane onslaught and sand loss is still undetermined.

Unforeseen environmental impact may accompany the addition of sand to a beach, especially if the sand is from an unmatched source, in terms of its ecological traits. If the new sands do not physically and chemically match the existing varieties, disruption of coastal marine life is possible. The U.S. Army Corps of Engineers, which operates the Miami replenishment program, is running out of "borrowing areas" for sand that matches the natural sand of the beach. A proposal

to haul a different type of sand from the Bahamas is being studied as to possible environmental consequences.

In contrast to the term *hard structures*, this hauling of sand to replenish a beach is considered "soft" shoreline protection (Figure 16.2.1). Enormous energy and material must be committed to counteract the relentless energy

that nature expends along the coast. The four hurricanes that hit Florida in 2004 removed enormous quantities of sand from beaches and barrier islands. Each loss makes the coast that much more vulnerable. For more on beach nourishment through Duke University, see **http://www.nicholas.duke.edu/psds/nourishment.htm**.

FIGURE 16.2.1 Beach sand replenishment.
Efforts to create protection for the beach and former primary dune must be continuous. Barrier islands are a dynamic system of rapid change; increasing storm activity and rising sea level place these coastal environments at risk. Note the sign to help keep the sand in place. One bad storm can wipe out these efforts. [Photo by Bobbé Christopherson.]

details of the extraordinary effort to save this landmark.) Vulnerable settlements fill the Outer Banks and Cape Hatteras and are repeatedly slammed by tropical storms (Hurricanes Emily, 1993, Dennis, Floyd, and Irene, 1999, and Isabel in 2003) causing damage and beach erosion (Figure 16.16d). And previous damage is visible, such as the salt-spray-burned vegetation from Hurricane Dennis (Figure 16.16e)

Barrier beaches and islands are quite common worldwide, lying offshore of nearly 10% of Earth's coastlines. Examples are found off the shores of Africa, India's eastern coast, Sri Lanka, Australia, Alaska's northern slope, and the shores of the Baltic and Mediterranean Seas. Earth's most extensive chain of barrier islands is along the U.S. Atlantic and Gulf Coasts, extending some 5000 km (3100 mi) from Long Island to Texas and Mexico.

Barrier Island Origin and Hazards Various hypotheses have been proposed to explain the formation of barrier islands. They may begin as offshore bars or low ridges of

submerged sediment near shore and then gradually migrate toward shore as sea level rises.

Because many barrier islands seem to be migrating landward, they are an unwise choice for homesites or commercial building. Nonetheless, they are a common choice, even though they take the brunt of storm energy. The hazard represented by the settlement of barrier islands was made graphically clear when Hurricane Hugo assaulted South Carolina in 1989.

In the Charleston area, Hugo swept away beachfront houses, barrier-island developments, and millions of tons of sand; the hurricane destroyed up to 95% of the single-family homes in one community. The southern portion of one island was torn away. With increased development and continuing real estate appreciation, each future storm can be expected to cause ever-increasing capital losses.

Figure 16.17a shows an aerial photo of development along the Texas Gulf Coast, near Galveston. Portions of the barrier island chain along the Texas coast are experiencing beach erosion rates averaging 2 m (6.5 ft) per year as sea

FIGURE 16.16 Barrier-island chain.
(a) *Landsat* image of barrier-island chain along the North Carolina coast. You can see key depositional forms: spit, island, beach, lagoon, and inlet. A sound is a large inlet of the ocean; Pamlico Sound is an ideal example. (b) View of Cape Hatteras lighthouse at its old location illustrates the narrow strand of sand that stands between the ocean and the mainland. (c) The path along which the Cape Hatteras lighthouse was moved inland in 1999 to safer ground. Answer to Chapter 1 opening photo question: lighthouse moved South 12", West 48". (d) Construction continues along Hatteras and the Outer Banks ocean front despite the great risk. (e) Salt spray from Hurricane Dennis "burned" these trees south of the lighthouse. [(a) *Terra* MODIS image courtesy of NASA/GSFC; photos by (b) Eric Horan/Liaison Agency, Inc.; (c) (d) (e) by Bobbé Christopherson.]

level continues to rise. Just south of the developed area in Figure 16.17b near Corpus Christi, is the Padre Island National Seashore, which is the longest stretch of protected barrier island in the world (177 km, 110 mi), encompassing more than 130,000 acres (52,610 hectares).

The barrier islands off the Louisiana shore are disappearing at rates approaching 20 m (65 ft) per year, as hurricanes have taken their toll. Also, they are affected by subsidence through compaction of Mississippi delta sediments and a changing sea level that is rising at 1 cm (0.4 in.)

(a)

(b)

(c)

FIGURE 16.17 Coastal barrier islands along the Texas Gulf Coast.
(a) A barrier island stands between Texas and the Gulf of Mexico. Note the Intercoastal Waterway for shipping along Bolivar Island, near Galveston. (b) In the portion of Padre Island not set aside in the National Seashore, houses line up behind the primary dunes and people drive vehicles on the beach. (c) Typical construction places houses on stilts to deal with storm surges. [Photos by Bobbé Christopherson.]

(a)

(b)

Hurricane Georges

FIGURE 16.18 Hurricane Georges (1998) takes its toll.
This 1998 hurricane eroded huge amounts of sand from the Chandeleur Islands, just off the Louisiana and Mississippi Gulf Coast. Compare the *before* topographic map (a) with the *after* aerial photo composite (b). (c) *GOES-8* image of Hurricane Georges; note the islands just south of the hurricane's eye. [(b) Photo by Aerial Data Service, Earth Imaging, a USGS service company; topographic map provided by USGS; (c) Image courtesy of NOAA.]

(c)   Hurricane Georges

(a)   (b)   (c)

FIGURE 16.19 Florida Gulf Coast barrier islands.
(a) Marinas and development along the barrier islands near Clearwater, Florida. (b) A house on the
Cedar Key barrier island. (c) All island residents are concerned about how to get inland in the
event of a hurricane—evacuation signs mark certain highways. [Photos by Bobbé Christopherson.]

per year in the region. In 1998, Hurricane Georges destroyed large tracts of the Chandeleur Islands (30 to 40 km, 19 to 25 mi, from the Louisiana and Mississippi Gulf Coast), leaving the mainland with reduced protection (Figure 16.18). The regional office of the USGS predicts that in a few decades the barrier islands may be gone. Louisiana's increasingly exposed wetlands are disappearing at rates of 65 km² (25 mi²) per year, made worse by regional subsidence and a rising sea level.

The same questions about hazard perception and risk arise with the barrier islands along the Florida coast. Figure 16.19a is an aerial photo of the chain along the Gulf Coast near Clearwater. North along the Florida coast, near Cedar Key, a residence sits on the barrier island at sea level (Figure 16.19b). One of the problems of barrier-island living is being able to evacuate when a hurricane is on the way: Access is restricted, traffic snarls are common, and roadways and bridges may be washed out (Figure 16.19c).

Biological Processes: Coral Formations

Not all coastlines form by purely physical processes. Some form as the result of biological processes, such as coral growth. A **coral** is a simple marine animal with a small, cylindrical, saclike body called a polyp; it is related to other marine invertebrates, such as anemones and jellyfish. Corals

secrete calcium carbonate ($CaCO_3$) from the lower half of their bodies, forming a hard, calcified external skeleton.

Corals live in a *symbiotic* relationship with algae: They live together in a mutually helpful arrangement, each dependent on the other for survival. Corals cannot photosynthesize, but they do obtain some of their own nourishment. Algae perform photosynthesis and convert solar energy to chemical energy in the system, providing the coral with about 60% of its nutrition and assisting the coral with the calcification process. In return, corals provide the algae with nutrients. Coral reefs are the most diverse marine ecosystems. Preliminary estimates of coral species range upward to 9 million worldwide, yet, as in most ecosystems in water or on land, biodiversity is declining in these communities.

Figure 16.20 shows the distribution of currently living coral formations. Corals mostly thrive in warm tropical oceans, so the difference in ocean temperature between the western coasts and eastern coasts of continents is critical to their distribution. Western coastal waters tend to be cooler, thereby discouraging coral activity, whereas eastern coastal currents are warmer and thus enhance coral growth.

Living colonial corals range in distribution from about 30° N to 30° S. Corals occupy a very specific ecological zone: 10–55 m (30–180 ft) depth, 27‰–40‰ (parts per

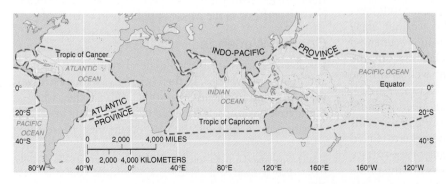

FIGURE 16.20 Worldwide distribution of living coral formations.
Yellow areas include prolific reef growth and atoll formation. The red dotted line marks the geographical limits of coral activity. Colonial corals range in distribution from about 30° N to 30° S. [After J. L. Davies, *Geographical Variation in Coastal Development,* Essex, England: Longman House, 1973. Adapted by permission.]

thousand) salinity, and 18° to 29°C (64°–85°F) water temperature—30°C (86°F) is the upper threshold; above that temperature the corals begin to bleach and die. However, an interesting exception to this are unique species of cool corals that exist in cold, dark water and deep ocean, at temperatures as low as 4°C (39°F) and in depths to 2000 m (6562 ft), beyond the expected range. These remarkable oddities are studied by scientists for they do not rely on algae but harvest nutrients from plankton and particulate matter.

Corals require clear, sediment-free water and consequently do not locate near the mouths of sediment-charged freshwater streams. For example, note the lack of these structures along the U.S. Gulf Coast.

Coral Reefs Both solitary and colonial corals exist. It is the colonial corals that produce enormous structures. Their skeletons accumulate, forming a coral rock. Through many generations, live corals near the ocean's surface build on the foundation of older coral skeletons, which in turn may rest upon a volcanic seamount or some other submarine feature built up from the ocean floor. *Coral reefs* form by this process. Thus, a coral reef is a biologically derived sedimentary rock. It can assume one of several distinctive shapes.

In 1842, Charles Darwin hypothesized an evolution of reef formation. He suggested that, as reefs developed around a volcanic island and the island itself gradually subsided, an equilibrium between the subsidence of the island and the upward growth of the corals is maintained. This idea, generally accepted today, is portrayed in Figure 16.21. Note the specific examples of each reef stage: *fringing reefs* (platforms of surrounding coral rock), *barrier reefs* (forming enclosed lagoons), and *atolls* (circular, ring-shaped).

Earth's most extensive fringing reef is the Bahamian platform in the western Atlantic (Figure 16.20d), covering some 96,000 km² (37,000 mi²). The largest barrier reef, the Great Barrier Reef along the shore of the state of Queensland, Australia, exceeds 2025 km (1260 mi) in length, is 16–145 km (10–90 mi) wide, and includes at least 700 coral-formed islands and keys (coral islets or barrier islands).

Coral Bleaching A troubling phenomenon is occurring among corals around the world as normally colorful corals turn stark white by expelling their own nutrient-supplying, colorful algae (red-brown to green), a phenomenon known as *bleaching*. Exactly why the corals eject their symbiotic partner is unknown, for without algae the corals die. Scientists are tracking their unprecedented bleaching and dying worldwide. Locations in the Caribbean Sea and the Indian Ocean, as well as off the shores of Australia, Indonesia, Japan, Kenya, Florida, Texas, and Hawai'i, are experiencing this phenomenon.

Possible causes include local pollution, disease, sedimentation, and changes in salinity. One acknowledged cause is the 1 to 2 C° (1.8 to 3.6 F°) warming of sea-surface temperatures, as stimulated by greenhouse warming of the atmosphere. In a report, the *Status of Coral Reefs of the*

World: 2000, from the Global Coral Reef Monitoring Network, warmer water was found to be a greater threat to corals than local pollution or other environmental problems (see http://www.coris.noaa.gov/).

Coral bleaching worldwide is continuing as average ocean temperatures climb higher, thus linking the issue of climate change to the health of all living coral formations. By the end of 2000, approximately 30% of reefs were lost, especially following the record El Niño event of 1998. As sea-surface temperatures continue to rise, forecasts state that coral losses will continue. For more information and Internet links, see http://www.usgs.gov/coralreef.html. The *World Atlas of Coral Reefs* in 2001 summarized:

> Humans are thus bringing new pressures to bear on the world's coral reefs and driving more profound changes, more rapidly, than any natural impact has ever done. Overfishing has become so widespread that there are few, if any, reefs in the world which are not threatened. . . . From onshore a much greater suite of damaging activities is taking place. Often remote from reefs, deforestation, urban development, and intensive agriculture are now producing vast quantities of sediments and pollutants which are pouring into the sea and rapidly degrading coral reefs. . . . A further specter overshadowing the world of coral reefs is that of global climate change.[*]

Wetlands, Salt Marshes, and Mangrove Swamps

Some coastal areas have great *biological productivity* (plant growth, spawning grounds for fish, shellfish, and other organisms) stemming from trapped organic matter and sediments. Such a rich coastal marsh environment can greatly outproduce a wheat field in raw vegetation per acre. Thus, coastal marshes can support rich wildlife habitats. Unfortunately, these wetland ecosystems are quite fragile and are threatened by human development.

Wetlands are saturated with water enough of the time to support *hydrophytic vegetation* (plants that grow in water or wet soil). Wetlands usually occur on poorly drained soils. Geographically, they occur not only along coastlines but also as northern bogs (peatlands with high water tables), as potholes in prairie lands, as cypress swamps (with standing or gently flowing water), as river bottomlands and floodplains, and as arctic and subarctic environments that experience permafrost during the year. We studied the loss to coastal wetlands in the Mississippi Delta in Chapter 14.

Coastal Wetlands

Coastal wetlands are of two general types—salt marshes and mangrove swamps. In the Northern Hemisphere, **salt marshes** tend to form north of the 30th parallel, whereas **mangrove swamps** form equatorward of that line. This distribution is dictated by the occurrence of freezing

---

[*]M. D. Spalding, C. Ravilious, and E. P. Green, and UNEP/WCMC, *World Atlas of Coral Reefs* (Berkeley: University of California Press, 2001), p. 11.

Fringing   Barrier   Atoll

Tahiti
(18° S 149° W)
Hawai'i
(20° N 156° W)
Grand Comoro
(12° S 44° E)

O'ahu
(22° N 158° W)
Rarotonga
(21° S 160° W)

Mayotte
(13° S 45° E)
Santa Cruz
(11° S 166° E)
Bora Bora
(16° S 151° W)

Truk
(7° N 152° E)
Clipperton
(10° N 109° W)
Aitutaki
(19° S 160° W)

Bikini
(12° N 165° E)
Eniwetok
(12° N 162° E)
Kwajalein
(9° N 167° E)

(a)

(b)

(c)

(d)

FIGURE 16.21 Coral forms.
(a) Common coral formations in a sequence of reef growth formed around a subsiding volcanic island: fringing reefs, barrier reefs, and an atoll. (b) Satellite image of a portion of the Maldive Islands, in the Indian Ocean (5° N 75° E). (c) Aerial photograph of the atolls in Bora Bora, Society Islands (16° S 152° W). (d) The Bahamas are the largest fringing reef in the world. The lighter areas feature ocean depths to 10 m (33 ft). The darker blues indicate depths dropping off to 4000 m (13,100 ft). The name Bahamas is from the Spanish *baja mar,* for "shallow sea." [(a) After D. R. Stoddart, *The Geographical Magazine* 63 (1971): 610; (b) *Landsat-7* image courtesy of NASA; (c) photo by Harvey Lloyd/Stock market; (d) *Terra* MODIS sensor image courtesy of NASA/GSFC and MODIS Land Response Team, April, 18, 2000.]

conditions, which control the survival of mangrove seedlings. Roughly the same latitudinal limits apply in the Southern Hemisphere.

Salt marshes usually form in estuaries and behind barrier beaches and spits. An accumulation of mud produces a site for the growth of *halophytic* (salt-tolerant) plants. This vegetation then traps additional alluvial sediments and adds to the salt marsh area. Because salt marshes are in the intertidal zone (between the farthest reaches of high and low tides), sinuous, branching channels are pro-

duced as tidal waters flood into and ebb from the marsh (Figure 16.22).

Sediment accumulation on tropical coastlines provides the site for mangrove trees, shrubs, and other small trees. The prop roots of the mangrove are constantly finding new anchorages. The roots are visible above the waterline but reach below the water surface, providing a habitat for a multitude of specialized life forms. Mangrove swamps often secure and fix enough material to form islands (Figure 16.23).

FIGURE 16.22 Coastal salt marsh.
Salt marshes are productive ecosystems commonly occurring poleward of 30° latitude in both hemispheres. This is Gearheart Marsh, part of the Arcata Marsh system on the Pacific Coast in northern California. [Photo by Bobbé Christopherson.]

(a)

(b)

FIGURE 16.23 Mangroves.
Mangroves tend to grow equatorward of 30° latitude. (a) Mangroves along the East Alligator River (12° S latitude) in Kakadu National Park, Northern Territory, Australia. (b) Mangroves retain sediments and can form anchors for island formations, such as Aldabra Island (9° S), Seychelles. (c) Mangroves in the Florida Keys. [(a) Photo by Belinda Wright/ DRK Photo; (b) photo by Wolfgang Kaehler Photography; (c) photo by author.]

(c)

Marco Island The development of Florida's Marco Island is an example of mangrove loss due to urbanization. Marco Island is on the southwestern Florida coast about 24 km (15 mi) south of Naples. Figure 16.24a shows Marco Island as it was in 1952: approximately 5300 acres of subtropical mangrove habitat and barrier-island ter-

rain, formed from river sediments and old shell and reef fragments.

Development began in 1962, despite tropical storms that occasionally assault the area, as Hurricane Donna did in 1960. Development included artificial landfill for housing sites and general urbanization of the entire

FIGURE 16.24 Mangrove loss at Marco Island.
Digitized inventory of Marco Island, Florida, showing dramatic development from (a) 1952 survey map and (b) 1984 color-infrared photograph of the island. This type of analysis is similar to GIS (Geographic Information System) studies and is invaluable in assessing human impacts. For more on GIS, see Figure 1.28. [From S. Patterson, *Mangrove Community Boundary Interpretation and Detection of Areal Changes on Marco Island, Florida,* Biological Report 86 (10), for National Wetlands Research Center, U.S. Fish and Wildlife Service, August 1986, pp. 23, 49.]

island. Scientists regarded the island and estuarine habitat as highly productive and unique in its mixture of mangrove species, bird species, and a productive aquatic environment. Consequently, numerous challenges and court actions were initiated in attempts to halt development. But construction continued, and by 1984 Marco Island was completely developed (Figure 16.23b). Compare the 1952 digitized inventory and 1984 color-infrared aerial photograph to see the impact of urban development on this island. Note that the only remaining natural community is restricted to a very limited portion along the extreme perimeter. Today, geographic information system models greatly facilitate such comparative analyses.

The World Resources Institute and the U.N. Environment Programme estimate that from preagricultural times to today, mangrove losses are running between 40% (examples, Cameroon and Indonesia) and nearly 80% (examples, Bangladesh and Philippines). Deliberate removal was a common practice by many governments in the early

days of settlement because of a falsely conceived fear of disease or pestilence in these swamplands.

Human Impact on Coastal Environments

Development of modern societies is closely linked to estuaries, wetlands, barrier beaches, and coastlines. Estuaries are important sites of human settlement because they provide natural harbors, a food source, and convenient sewage and waste disposal. Society depends on daily tidal flushing of estuaries to dilute the pollution created by waste disposal. Thus, the estuarine and coastal environment is vulnerable to abuse and destruction if development is not carefully planned. We are approaching a time when most barrier islands will be developed and occupied.

Barrier islands and coastal beaches do migrate over time. Houses that were more than a mile from the sea in

(a)   (b)

FIGURE 16.25 Beach erosion dooms houses in Massachusetts and New York.
(a) Two homes on stilts sit atop a primary berm made of cobbles. The ocean is just beyond the houses along this Massachusetts shore. (b) Placing houses on stilts along the shore offers little protection from an attacking storm surge, here on Long Island. [Photos by (a) Bobbé Christopherson; (b) Mark Wexler/Woodfin Camp & Associates.]

the Hamptons along the southeastern shore of Long Island, New York, are now within only 30 m (100 ft) of it. The time remaining for these homes can be quickly shortened by a single hurricane or by a sequence of storms (Figure 16.25).

Despite our understanding of beach and barrier-island migration, the effect of storms, and warnings from scientists and government agencies, coastal development proceeds. Society behaves as though beaches and barrier islands are stable, fixed features, or as though they can be engineered to be permanent. Experience has shown that

severe erosion generally cannot be prevented. Shoreline planning is the topic of Focus Study 16.1.

A reasonable conclusion seems to be that the key to protective environmental planning and zoning is to *allocate responsibility and cost in the event of a disaster.* An ideal system places a hazard tax on land, based on assessed risk, and restricts the government's responsibility to fund reconstruction or an individual's right to reconstruct on frequently damaged sites. Comprehensive mapping of erosion-hazard areas would help avoid the ever-increasing costs from recurring disasters.

Focus Study 16.1

An Environmental Approach to Shoreline Planning

Coastlines are places of wonderful opportunity. They also are zones of specific constraints. Poor understanding of this resource and a lack of environmental analysis often go hand in hand. The late ecologist and landscape architect Ian McHarg, in *Design with Nature*, discusses the New Jersey shore. He shows how proper understanding of a coastal environment could have avoided problems from major storms and coastal development. Much of what is presented here applies to coastal areas elsewhere. Figure 16.1.1 illustrates the New Jersey shore from ocean to back bay. Let us walk across this landscape and discover how it should be treated under ideal conditions.

Beaches and Dunes—Where to Build?

We begin our walk at the water's edge and proceed across the beach. Sand beaches are the primary natural defense against the ocean; they act as bulwarks against the pounding of a stormy sea. The shoreline tolerates recreation, but not construction, because of its shifting, changing nature during storms, daily tidal fluctuations, and the potential effects of rising sea level. Beaches are susceptible to pollution and require environmental protection to control nearshore dumping of dangerous materials. High water levels in the Great Lakes and along the southern California coast and the Atlantic hurricane

intensity demonstrate the vulnerability of shorelines to erosion and inundation by high water and storm surges. Wave erosion attacks cliff formations and undermines, bit by bit, the foundations of houses and structures built too close to the edge (Figure 16.1.2).

Walking inland, we encounter the primary dune. It is even more sensitive than the beach: It is fragile, easily disturbed, and vulnerable to erosion, and it cannot tolerate the passage of people trekking to the beach. Some vegetation is present, but it is easily disturbed and has only a small effect on sand stabilization. Primary dunes are like human-made dikes in the Netherlands: They are the primary defense

FIGURE 16.1.1 Planning along a coast.
(a) Coastal environment—a planning perspective from ocean to bay along the New Jersey shore. Note the placement of each letter on the illustration to identify photo location. (b) Light trucks and SUVs drive along the beach in the protected Island Beach State Park, whereas they are banned from the stretches of beach that are developed with houses and business. (c) Houses once distant from the surf are hit by waves. Sand is brought in to strengthen defenses. (d) The primary dune in a somewhat natural state. (e) An undeveloped trough behind the primary dune. (f) Development along the primary dunes challenges the natural setting—an amusement park. (g) The former trough, paved and developed, barely above high-tide levels. (h) Development and commerce in the bayshore and bay. [(a) After *Design with Nature* by Ian McHarg. Copyright © 1969 by Ian L. McHarg. Adapted by permission. All photos by Bobbé Christopherson.]

against the sea, so development or heavy traffic should not disturb them. Carefully controlled access points to the beach should be designated, and restricted access should be enforced, for even foot traffic can cause destruction. (See two examples of beach stabilization in Figure 15.5.)

The trough behind the primary dune is relatively tolerant of limited recreation and building. The plants that fix themselves to the surface send roots down to fresh groundwater reserves and anchor the sand in the process. Thus, if construction should inhibit the surface recharge of that water supply, the natural protective ground cover could die and destabilize the environment. Or subsequent saltwater intrusion might contaminate well water. Clearly, groundwater resources and the location of recharge aquifers must be considered in planning. The continuing use of septic sewage systems in the trough is not sustainable in these environments.

Behind the trough is the secondary dune, a second line of defense against the sea. It, too, is tolerant of some use yet is vulnerable to destruction. Next is the backdune, more suitable for

(continued)

Focus Study 16.1 (continued)

(a)

(b)

FIGURE 16.1.2 Examples of poor planning.
(a) Coastal erosion and a failed house foundation. (b) Seemingly futile efforts to halt collapse of coastal cliffs. [Photos by (a) author; (b) Bobbé Christopherson.]

development than any zone between it and the sea. Further inland are the bayshore and the bay, where no dredging or filling and only limited dumping of treated wastes and toxics should be permitted. This zone is tolerant to intensive recreation. In reality, of course, the opposite of such careful assessment and planning prevails.

A Scientific View and a Political Reality

Before an intensive government study of the New Jersey shore was completed in 1962, no analysis of coastal hazards had been done, outside of academic circles. What is common knowledge to geographers, botanists, biologists, and ecologists in the classroom and laboratory still has not filtered through to the general planning

and political processes. As a result, on the New Jersey shore and along much of the Atlantic and Gulf Coasts, improper development of the fragile coastal zone (primary dunes, trough, and secondary dunes) led to extensive destruction during storms in 1962, 1992, and numerous other times.

Scientists estimate that the coastlines will continue their retreat, in some cases 10s of meters in a few decades. Such estimates persist along the entire East Coast around to the Gulf Coast. Society must reconcile ecology and economics if these coastal environments are to be sustained.

South Carolina enacted the Beach Management Act of 1988 (modified 1990) to apply some of McHarg's principles. In its first 2 years, more than 70 lawsuits protested the act as

an invalid seizure of private property without compensation.

Destructive or threatening hurricane activity is helping to drive the importance of such cooperative planning and proactive laws. Similar measures in other states have faced the same difficult path. Implementation of any planning process is problematic, political pressure is intense, and results are mixed. Ian McHarg voiced an optimistic hope: "May it be that these simple ecological lessons will become known and incorporated into ordinance [law] so that people can continue to enjoy the special delights of life by the sea."*

*From *Design with Nature* by Ian McHarg, p. 17. © 1969 by Ian L. McHarg. Published by Bantam Doubleday Dell Publishing Group, Inc.

Summary and Review—The Oceans, Coastal Processes, and Landforms

▶ *Describe* the chemical composition of seawater and the physical structure of the ocean.

Water is called the "universal solvent," dissolving at least 57 of the 92 elements found in nature. Most natural elements and the compounds they form are found in the seas as dissolved solids. Seawater is a solution, and the concentration of dissolved solids is called **salinity**. **Brine** exceeds the average 35‰ (parts per

thousand) salinity; **brackish** applies to water that is less than 35‰. The ocean is divided by depth into a narrow mixing zone at the surface, a thermocline transition zone, and the deep cold zone.

salinity (p. 500)
brine (p. 502)
brackish (p. 502)

1. Describe the salinity of seawater: its composition, amount, and distribution.

2. Analyze the latitudinal distribution of salinity as shown in Figure 16.3. Why is salinity less along the equator and greater in the subtropics?

3. What are the three general zones relative to physical structure within the ocean? Characterize each by temperature, salinity, dissolved oxygen, and dissolved carbon dioxide.

▶ *Identify* the components of the coastal environment, and *list* the physical inputs to the coastal system, including tides and mean sea level.

The coastal environment is called the **littoral zone** and exists where the tide-driven, wave-driven sea confronts the land. Inputs to the coastal environment include solar energy, wind and weather, climatic variation, the nature of coastal geomorphology, and human activities.

Mean sea level (MSL) is based on average tidal levels recorded hourly at a given site over many years. MSL varies spatially because of ocean currents and waves, tidal variations, air temperature and pressure differences, ocean temperature variations, slight variations in Earth's gravity, and changes in oceanic volume.

Tides are complex daily oscillations in sea level, ranging worldwide from barely noticeable to many meters. Tides are produced by the gravitational pull of both the Moon and the Sun. Most coastal locations experience two high (rising) **flood tides** and two low (falling) **ebb tides** every day. The difference between consecutive high and low tides is the tidal range. **Spring tides** exhibit the greatest tidal range, when the Moon and Sun are either in conjunction or opposition. **Neap tides** produce a lesser tidal range.

littoral zone (p. 503)
mean sea level (MSL) (p. 504)
tide (p. 504)
flood tide (p. 506)
ebb tide (p. 506)
spring tide (p. 506)
neap tide (p. 506)

4. What are the key terms used to describe the coastal environment?

5. Define mean sea level. How is this value determined? Is it constant or variable around the world? Explain.

6. What interacting forces generate the pattern of tides?

7. What characteristic tides are expected during a new Moon or a full Moon? During the first-quarter and third-quarter phases of the Moon? What is meant by a flood tide? An ebb tide?

8. Is tidal power being used anywhere to generate electricity? Explain briefly how such a plant would utilize the tides to produce electricity. Are there any sites in North America? Where are they?

▶ *Describe* wave motion at sea and near shore, and *explain* coastal straightening as a product of wave refraction.

Friction between moving air (wind) and the ocean surface generates undulations of water that we call **waves**. Wave energy in the open sea travels through water, but the water itself stays in place. Regular patterns of smooth, rounded waves, called **swells**, are the mature undulations of the open ocean. Near shore, the restricted depth of water slows the wave, forming *waves of*

translation, in which both energy and water actually move forward toward shore. As the crest of each wave rises, the wave falls into a characteristic **breaker**.

Wave refraction redistributes wave energy so that different sections of the coastline vary in erosion potential. Headlands are eroded, whereas coves and bays receive materials, with the long-term effect of straightening the coast. As waves approach a shore at an angle, refraction produces a **longshore current** of water moving parallel to the shore. This current produces the *longshore drift* of sand, sediment, and gravel and assorted materials moving along the beach as **beach drift**. **Littoral drift** is transport of materials along the shore; a more comprehensive term, it considers *beach drift* and *longshore drift* combined.

A **tsunami** is a seismic sea wave triggered by an undersea landslide or earthquake. It travels at great speeds in the open sea and gains height as it comes ashore, posing a coastal hazard.

wave (p. 507)
swell (p. 507)
breaker (p. 507)
wave refraction (p. 508)
longshore current (p. 508)
littoral drift (p. 508)
beach drift (p. 509)
tsunami (p. 509)

9. What is a wave? How are waves generated, and how do they travel across the ocean? Does the water travel with the wave? Discuss the process of wave formation and transmission.

10. Describe the refraction process that occurs when waves reach an irregular coastline. Why is the coastline straightened?

11. Define the components of beach drift and the longshore current and longshore drift.

12. Explain how a seismic sea wave attains such tremendous velocities. Why is it given a Japanese name?

▶ *Identify* characteristic coastal erosional and depositional landforms.

An *erosional coast* features wave action that cuts a horizontal bench in the tidal zone, extending from a sea cliff out into the sea. Such a structure is called a **wave-cut platform**, or *wave-cut terrace*. In contrast, *depositional coasts* generally are located along land of gentle relief, where depositional sediments are available from many sources. Characteristic landforms deposited by waves and currents are a **barrier spit** (material deposited in a long ridge extending out from a coast); a **bay barrier**, or *baymouth bar* (a spit that cuts off the bay from the ocean and forms an inland **lagoon**); a **tombolo** (where sediment deposits connect the shoreline with an offshore island or sea stack); and a **beach** (land along the shore where sediment is in motion, deposited by waves and currents). A beach helps to stabilize the shoreline, although it may be unstable seasonally.

wave-cut platform (p. 512)
barrier spit (p. 512)
bay barrier (p. 512)
lagoon (p. 513)
tombolo (p. 513)
beach (p. 513)

13. What is meant by an erosional coast? What are the expected features of such a coast?

14. What is meant by a depositional coast? What are the expected features of such a coast?

15. How do people attempt to modify littoral drift? What strategies are used? What are the positive and negative impacts of these actions?

16. Describe a beach—its form, composition, function, and evolution.

17. What success has Miami had with beach replenishment? Is it a practical strategy?

▶ *Describe* barrier islands and their hazards as they relate to human settlement.

Barrier chains are long, narrow, depositional features, generally of sand, that form offshore roughly parallel to the coast. Common forms are **barrier beaches** and the broader, more extensive **barrier islands**. Barrier formations are transient coastal features, constantly on the move, and they are a poor, but common, choice for development.

barrier beach (p. 515)
barrier island (p. 515)

18. On the basis of the information in the text and any other sources at your disposal, do you think barrier islands and beaches should be used for development? If so, under what conditions? If not, why not?

19. After the Grand Strand off South Carolina was destroyed by Hurricane Hazel in 1954, settlements were rebuilt, only to be hit by Hurricane Hugo 35 years later, in 1989. Why does this type of recurring damage happen to human populations?

▶ *Assess* living coastal environments: corals, wetlands, salt marshes, and mangroves.

A **coral** is a simple marine invertebrate that forms a hard, calcified, external skeleton. Over generations, corals accumulate in large reef structures. Corals live in a *symbiotic* (mutually helpful) relationship with algae; each is dependent on the other for survival.

Wetlands are lands saturated with water that support specific plants adapted to wet conditions. They occur along coastlands and inland in bogs, swamps, and river bottomlands. Coastal wetlands form as **salt marshes** poleward of the 30th parallel in each hemisphere and as **mangrove swamps** equatorward of these parallels.

coral (p. 519)
wetlands (p. 520)
salt marsh (p. 520)
mangrove swamp (p. 520)

20. How are corals able to construct reefs and islands?

21. Describe a trend in corals that is troubling scientists, and discuss some possible causes.

22. Why are the coastal wetlands poleward of 30° N and S latitude different from those that are equatorward? Describe the differences.

▶ *Construct* an environmentally sensitive model for settlement and land use along the coast.

Coastlines are zones of specific constraints. Poor understanding of this resource and a lack of environmental analysis often go hand in hand, producing frequent disasters to coastal ecosystems and real property losses. Society must reconcile ecology and economics if these coastal environments are to be sustained.

23. Describe the condition of Marco Island, Florida. Was a rational model used to assess the environment prior to development? What economic and political forces were involved?

24. What type of environmental analysis is needed for rational development and growth in a region like the New Jersey shore? Evaluate South Carolina's approach to coastal hazards and protection.

NetWork

The *Geosystems* Home Page provides online resources for this chapter on the World Wide Web. To begin: Once on the Home Page, click on this textbook, scroll the Table of Contents menu, select this chapter, and click "Begin." You will find self-tests that are graded, review exercises, specific updates for items in the chapter, and many links to interesting related pathways on the Internet under "Destinations." *Geosystems* is at **http://www.prenhall.com/christopherson**.

Critical Thinking

A. This chapter includes the following statement: "The key to protective environmental planning and zoning is to *allocate responsibility and cost in the event of a disaster.* An ideal system places a hazard tax on land, based on assessed risk, and restricts the government's responsibility to fund reconstruction or an individual's right to reconstruct at frequently damaged sites." What do you think about this as a policy statement? How would you approach implementing such a strategy? In what way could you use geographic information systems (GIS), as described in Chapter 1, to survey, assess, list owners, and follow taxation status for a vulnerable stretch of coastline?

B. Under "Destinations" in Chapter 16 of the *Geosystems* Home Page there is a link called "Coral Reefs." Sample some of the links on this page. Do you find any information about the damage to and bleaching of coral reefs reported in 1998? Which places in the world? Are there some suspect causes presented in "Bleaching Hot Spots" or any of the "Coral Reef Alliance" references?

These peaks, known as nunataks, rise above the Greenland Ice Sheet. There are no clouds in the photo, for this is all ice at approximately 2500 m (8200 ft) elevation—an accumulation of ice perhaps 100,000 years in the making. Note the crevasses, or cracks, in the ice near the top of the frame. Greenland is 81% covered by ice, an amount second only to Antarctica. The Greenland Ice Sheet is thinning at about 1 m per year as ice mass is lost. [Photo by Bobbé Christopherson.]

Glacial and Periglacial Processes and Landforms

▶ **Key Learning Concepts** _____

After reading the chapter, you should be able to:

▶ *Differentiate* between alpine and continental glaciers, and *describe* their principal features.

▶ *Describe* the process of glacial ice formation, and *portray* the mechanics of glacial movement.

▶ *Describe* characteristic erosional and depositional landforms created by alpine glaciation and continental glaciation.

▶ *Analyze* the spatial distribution of periglacial processes, and *describe* several unique landforms and topographic features related to permafrost and frozen ground phenomena.

▶ *Explain* the Pleistocene ice-age epoch and related glacials and interglacials, and *describe* some of the methods used to study paleoclimatology.

A bout 77% of Earth's freshwater is frozen, with the bulk of that ice sitting restlessly in just two places—Greenland and Antarctica. The remaining ice covers various mountains and fills some alpine valleys. A volume of more than 32.7 million km^3 (7.8 million mi^3) of water is tied up as ice in Greenland (2.38 million km^3), Antarctica (30.1 million km^3), and ice caps and mountain glaciers worldwide (180,000 km^3). For comparison, the Vatnajökull ice cap in Figure 17.5 has a volume of 3100 km^3.

These deposits provide an extensive frozen record of Earth's climatic history over the past several million years and perhaps some clues to its climatic future. Worldwide, glacial ice is in retreat, melting at rates exceeding anything in the ice record. In the European Alps alone some 75% of the glaciers have receded in the past 50 years, losing more than 50% of their ice mass since 1850. At this rate, the European Alps will have only 20% of their preindustrial glacial ice left by 2050.

In this chapter: We focus on Earth's extensive ice deposits—their formation, movement, and the ways in which they produce various erosional and depositional landforms. Glaciers, transient landforms themselves, leave in their wake a variety of landscape features. The fate of glaciers is intricately tied to change in global temperature, which ultimately concerns us all. We discuss the methods used to decipher past climates—the science of *paleoclimatology*—and the clues for understanding future climate patterns. Exciting discoveries are arising from ice cores taken from Earth's two ice sheets, one dating back 740,000 years from the present.

We examine the cold world of permafrost and periglacial geomorphic processes. These periglacial environments may be separated from actual glacial ice across time (glaciers from past ages) and space (physical distance in the present). Approximately 25% of Earth's land area is subject to freezing conditions and frost action characteristic of periglacial regions, including areas that once were dominated by periglacial processes and have specific related features of relict, or past, permafrost. These areas near former and existing glaciers contain evidence of the last ice age.

Rivers of Ice

A **glacier** is a large mass of ice resting on land or floating as an ice shelf in the sea adjacent to land. Glaciers are not frozen lakes or groundwater ice. Instead, they form by the continual accumulation of snow that recrystallizes under its own weight into an ice mass. Glaciers are not stationary; they move slowly under the pressure of their own great weight and the pull of gravity. In fact, they move slowly in streamlike patterns, merging as tributaries into large rivers of ice, as you can see in the satellite image and aerial photo in Figure 17.1.

Today, about 11% of Earth's land area is dominated by these slowly flowing rivers of ice. But during colder episodes in the past, as much as 30% of continental land was covered by glacial ice. Through these "ice ages," below-freezing temperatures prevailed at lower latitudes more than they do today, allowing snow to accumulate year after year.

Glaciers form in areas of permanent snow, both at high latitudes and at high elevations at any latitude. A **snowline** is the lowest elevation where snow can survive year-round; specifically, it is the lowest line where winter snow accumulation persists throughout the summer. Glaciers form on some high mountains along the equator, such as in the Andes Mountains of South America and on Mount Kilimanjaro in Tanzania, Africa. On equatorial mountains, the snowline is around 5000 m (16,400 ft); on midlatitude mountains, such as the European Alps,

snowlines average 2700 m (8850 ft); and in southern Greenland, snowlines are as low as 600 m (1970 ft). (For Internet links to Global Land Ice Measurements from Space go to **http://www.glims.org/**, which includes an inventory of world glaciers, or the National Snow and Ice Data Center at **http://www-nsidc.colorado.edu/**. *Landsat*-7 images are listed at **http://www.emporia.edu/ earthsci/gage/glacier7.htm**.)

Glaciers are as varied as the landscape itself. They fall within two general groups, based on their form, size, and flow characteristics: alpine glaciers and continental glaciers.

Alpine Glaciers

With few exceptions, a glacier in a mountain range is called an **alpine glacier**, or *mountain glacier*. The name comes from the Alps of central Europe, where such glaciers abound. Alpine glaciers form in several subtypes. One prominent type is a *valley glacier*, literally a river of ice

(a)

(b)

FIGURE 17.1 Rivers of ice.
(a) Alpine glaciers merge from adjoining glacial valleys in the northeast region of Ellesmere Island in the Canadian Arctic.
(b) Numerous glaciers carve the coast of Greenland. [(a) *Terra* ASTER image courtesy of University of Alberta, NASA/GSFC/ ERSDAC/JAROS and the U.S./Japan ASTER Science Team, July 31, 2000. (b) photo by Bobbé Christopherson.]

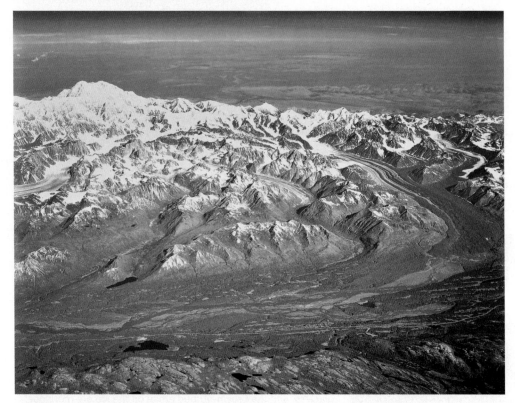

FIGURE 17.2 Glaciers in south-central Alaska.
Oblique infrared (false-color) image of Eldridge and Ruth Glaciers, with Mount McKinley at upper left, in the Alaska Range of Denali National Park. Photo made at 18,300 m (60,000 ft).
[Alaska High Altitude Aerial Photography from EROS Data Center, USGS.]

confined within a valley that originally was formed by stream action. Such glaciers range in length from only 100 m (325 ft) to more than 100 km (60 mi).

In Figure 17.2, at least a half dozen valley glaciers are identifiable in the high-altitude photograph of the Alaska Range. Several are named on the map, specifically the Eldridge and Ruth Glaciers, which fill valleys as they flow from source areas near Mount McKinley.

As a valley glacier flows slowly downhill, the mountains, canyons, and river valleys beneath its mass are profoundly altered by its erosive passage. Some of the debris created by the glacier's excavation is transported on the ice, visible as dark streaks and bands being transported for deposition elsewhere; other portions of its debris load are carried within or along its base (see Figures 17.1 and 17.7c).

Most alpine glaciers originate in a mountain *snowfield* that is confined in a bowl-shaped recess. This scooped-out erosional landform at the head of a valley is called a **cirque**. A glacier that forms in a cirque is called a *cirque glacier*. Several cirque glaciers may jointly feed a valley glacier, as shown on the topographic map and photograph in Figure 17.3. The Kuskulana Glacier, which flows off the map to the west, is supplied by numerous cirque glaciers feeding several tributary valley glaciers. In the aerial photos in Figure 17.1, how many valley glaciers join the main glacier in (a) and (b)?

Wherever several valley glaciers pour out of their confining valleys and coalesce at the base of a mountain range, a *piedmont glacier* is formed and spreads freely over the lowlands. Malaspina Glacier is an excellent example of a piedmont glacier (Figure 17.4a). The debris deposits on

(a)

(b)

FIGURE 17.3 Topographic map from southeastern Alaska.
(a) Numerous cirque glaciers depicted on a topographic map of southeastern Alaska (McCarthy C-7 Quadrangle, 1:63,360 scale, 100-ft contour interval). The blue areas represent active glaciers, and the brown-striped areas on the ice are moraines. (b) Aerial photo of the active Kuskulana glacier in the Wrangell Range. [(a) Courtesy of U.S. Geological Survey; (b) photo by Steve McCutcheon/Visuals Unlimited.]

the surface of the ice form beautiful streaked patterns as the glacier fans out over the coastal plain and into Yakutat Bay. A *tidewater glacier*, or *tidal glacier*, ends in the sea, calving (breaking off) to form floating ice called **icebergs** (Figure 17.4b and c). Icebergs usually form wherever glaciers meet the ocean. In Figure 17.36 at the end of the chapter you can enjoy the beauty of myriad iceberg shapes in a photo gallery.

Continental Glaciers

On a much larger scale than individual alpine glaciers, a continuous mass of ice is a **continental glacier**. In its most extensive form, it is an **ice sheet**. Most of Earth's glacial ice exists in the ice sheets that blanket 81% of Greenland,

1,756,000 km^2 (678,000 mi^2) of ice, and 90% of Antarctica, 14.2 million km^2 (5.48 million mi^2) of ice. Antarctica alone has 92% of all the glacial ice on the planet.

The Antarctic and Greenland ice sheets have such enormous mass that large portions of each landmass beneath the ice are isostatically depressed (pressed down by weight) below sea level. Each ice sheet reaches depths of more than 3000 m (10,000 ft), with average depths around 2000 m, burying all but the highest peaks.

Two additional types of continuous ice cover associated with mountain locations are *ice caps* and *ice fields*. An **ice cap** is roughly circular and, by definition, covers an area of less than 50,000 km^2 (19,300 mi^2). An ice cap completely buries the underlying landscape. The volcanic island of Iceland features several ice caps, such as the

(a)

(b)

(c)

FIGURE 17.4 Piedmont glacier and ice calving into the sea.
(a) Malaspina Glacier in southeastern Alaska. This piedmont glacier is nearly the size of Rhode Island and covers some 5000 km² 1950 mi². About 70% of Malaspina's ice comes from the Seward Glacier (upper center), which is fed by ice fields in the St. Elias Mountain Range. (b) When glaciers reach the sea, large pieces begin calving off, forming icebergs, such as these massive 30-m- (98-ft) high blocks ready to fall at Neko Harbor, Antarctica. (c) A tidewater glacier calving numerous icebergs into the ocean in southern Greenland. [(a) High-altitude photo courtesy of AeroMap U.S., Inc., Anchorage, Alaska; photos (b) and (c) by Bobbé Christopherson.]

Vatnajökull Ice Cap in the *Landsat* image in Figure 17.5a. Volcanoes lie beneath these icy surfaces. Iceland's Grímsvötn Volcano erupted in 1996, and again in 2004, producing large quantities of melted glacial water and floods, a flow called *jökulhlaup* by Icelanders.

An **ice field** is not extensive enough to form the characteristic dome of an ice cap; instead, it extends in a characteristic elongated pattern in a mountainous region. A fine example is the Patagonian ice field of Argentina and Chile, one of Earth's largest. It attains only 90 km (56 mi) width, but stretches 360 km (224 mi), from 46° to 51° S latitude (equivalent in latitude from South Dakota to

central Manitoba in the Northern Hemisphere). In an ice field, ridges and peaks are visible above the buried terrain; the term *nunatak* refers to these peaks, visible in Figure 17.5b.

Continuous ice sheets or ice caps are drained by rapidly moving, solid *ice streams* that form around their periphery, moving to the sea or to lowlands. Such frozen ice streams flow from the edges of Greenland and Antarctica, through stationary and slower-moving ice. An *outlet glacier* flows out from an ice sheet or ice cap but is constrained by a mountain valley or pass (Figure 17.6).

(a)

(b)

FIGURE 17.5 Ice cap and ice field.
(a) The Vatnajökull ice cap in southeastern Iceland (*jökull* means "ice cap" in Danish). Note the location of the Grímsvötn Volcano on the map and the path of the floods to the sea. (b) The southern Patagonian ice field of Argentina. [(a) *Landsat* image from NASA; (b) photo by Cosmonauts G. M. Greshko and Yu V. Romanenko, *Salyut 6.*]

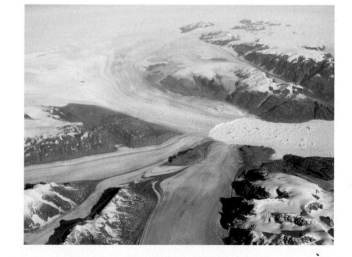

FIGURE 17.6 Greenland's east coast.
Ice streams from the ice sheet form outlet glaciers flowing to the sea along the southeast coast of Greenland. The center of the image is near the Arctic Circle (66.5° N). [Photo by Bobbé Christopherson.]

Glacial Processes

A glacier is a dynamic body, moving relentlessly downslope at rates that vary within its mass, excavating the landscape through which it flows. The mass is dense ice that is formed from snow and water through a process of compaction, recrystallization, and growth. A glacier's *mass budget* consists of net gains or losses of this glacial ice, which determine whether the glacier expands or retreats. Let us now look at glacial ice formation, mass balance, movement, and erosion before we discuss the fascinating landforms produced by these processes.

Formation of Glacial Ice

Consider for a moment the nature of ice. You may be surprised to learn that ice is both a mineral (an inorganic natural compound of specific chemical makeup and crystalline structure) and a rock (a mass of one or more minerals). Ice is a frozen fluid, a trait that it shares with igneous rocks. The accumulation of snow in layered deposits is similar to sedimentary rock formations. To give birth to a glacier, snow and ice are transformed under pressure, recrystallizing into a type of metamorphic rock. Glacial ice is a remarkable material!

The essential input to a glacier is snow that accumulates in a snowfield, a glacier's accumulation zone (Figure 17.7a and 17.7d). Snowfields typically are at the highest elevation of an ice sheet, ice cap, or head of a valley glacier, usually in a cirque. Avalanches from surrounding mountain slopes can add to the snowfield.

As the snow accumulation deepens in sedimentary-like layers, the increasing thickness results in increased weight and pressure on underlying ice. Rain and summer snowmelt then contribute water, which stimulates further melting, and that meltwater seeps down into the snowfield and refreezes.

Snow that survives the summer and into the following winter begins a slow transformation into glacial ice. Air spaces among ice crystals are pressed out as snow packs to a greater density. The ice recrystallizes and consolidates under pressure. In a transition step to glacial ice, snow becomes **firn**, which has a compact, granular texture.

As this process continues, many years pass before dense glacial ice is produced. Formation of **glacial ice** is analogous to metamorphic processes: Sediments (snow and firn) are pressured and recrystallized into a dense metamorphic rock (glacial ice). In Antarctica, glacial ice formation may take 1000 years because of the dryness of the climate (minimal snow input), whereas in wet climates the time is reduced to just a few years because of rapid, constant snow input to the system.

Glacial Mass Balance

A glacier is an open system, with *inputs* of snow and *outputs* of ice, meltwater, and water vapor. At its upper end, a glacier is fed by snowfall and other moisture in the *accumulation zone* (Figure 17.7b). This area ends at the **firn line**, indicating where the winter snow and ice accumulation survived the summer melting season. Toward a glacier's lower end,

it is wasted (reduced) through several processes: melting on the surface, internally, and at its base; ice removal by deflation (wind); the calving of ice blocks; and sublimation (recall from Chapter 7 that this is the direct evaporation of ice). Collectively, these losses are called **ablation**.

The zone where accumulation gain balances ablation loss is the *equilibrium line* (Figure 17.7a). This area of a glacier generally coincides with the firn line. A glacier achieves *positive net balance* of mass—grows larger—during cold periods with adequate precipitation. In warmer times, the equilibrium line migrates up-glacier, and the glacier retreats—grows smaller—because of its *negative net balance*. Internally, gravity continues to move a glacier forward even though its lower terminus might be in retreat owing to ablation. The mass-balance losses from the South Cascade glacier and Alaskan glaciers provide a case in point (News Report 17.1).

Highways of ice that flow from accumulation areas high in the mountains are marked by trails of transported debris called *moraines* (Figure 17.7c and e). A lateral moraine accumulates along the sides as rock is loosened by abrasion and plucking from the valley walls. A medial moraine forms down the middle when two glaciers merge and their lateral moraines combine, forming an elongated deposit on and within the combined glacier.

Tributary valley glaciers merge to form a *compound valley glacier*. The flowing movement of a compound valley glacier is different from that of a river with tributaries. Tributary glaciers flow into a compound glacier and merge alongside one another by extending and thinning rather than by blending, as do rivers. Each tributary maintains its own patterns of transported debris (dark streaks, visible in the photos in Figure 17.7c and f; also see Figure 17.13c).

Glacial Movement

Like all minerals, ice has specific properties of hardness, color, melting point (quite low in the case of ice), and brittleness. We know the properties of ice best from those brittle little cubes in the freezer. But glacial ice has different properties, depending on its location in a glacier. In a glacier's depths, glacial ice behaves in a plastic manner, distorting and flowing in response to weight and pressure from above and the degree of slope below. In contrast, the glacier's upper portion is more like the everyday ice we know, quite brittle.

A glacier's rate of flow ranges from almost nothing to a kilometer or two per year on a steep slope. The rate of snow accumulation in the formation area is critical to the pace of glacial movement.

Glaciers are not rigid blocks that simply slide downhill. The greatest movement within a valley glacier occurs *internally*, below the rigid surface layer, where the underlying zone moves plastically forward (Figure 17.8a). At the same time, the base creeps and slides along, varying its speed with temperature and the presence of any lubricating water or saturated sediment beneath the ice. This *basal slip* usually is much slower than the internal plastic flow of the glacier, so the upper portion of the glacier flows ahead of the lower portion. The difference in speed stretches the glacier's brittle surface ice.

FIGURE 17.9 Glacial sandpapering polishes rock.
Glacial polish and striations are examples of glacial abrasion and erosion. The polished, marked surface is seen beneath a glacial erratic—a rock left behind by a retreating glacier.
[Photo by Bobbé Christopherson.]

Glacial Erosion   The way in which a glacier erodes the land is similar to a large excavation project, with the glacier hauling debris from one site to another for deposition. The passing glacier mechanically plucks rock material and carries it away. Debris is carried on its surface and is also transported internally, or *englacially*, embedded within the glacier itself. There is evidence that rock pieces actually freeze to the basal layers of the glacier in a *glacial plucking*, or picking up, process and, once embedded, enable the glacier to scour and sandpaper the landscape as it moves—a process called **abrasion**. This abrasion and gouging produce a smooth surface on exposed rock, which shines with *glacial polish* when the glacier retreats. Larger rocks in the glacier act much like chisels, gouging the underlying surface and producing glacial striations parallel to the flow direction (Figure 17.9).

Glacial Landforms

Glacial erosion and deposition produce distinctive landforms that differ greatly from the way the land looked before the ice came and went. You might expect all glaciers to create the same landforms, but alpine and continental glaciers each generate their own characteristic landscapes. We look first at erosional landforms created by alpine glaciers, then at their depositional landforms. Finally, we examine the landscape that results from continental glaciation.

Erosional Landforms Created by Alpine Glaciation

Alpine glaciers create spectacular, dramatic landforms that bring to mind the Canadian Rockies, the Swiss Alps, or vaulted Himalayan peaks. Geomorphologist William Morris Davis depicted the stages of a valley glacier in drawings published in 1906 and redrawn here in Figure 17.10.

Study of these figures reveals the handiwork of ice as sculptor:

- In (a), you see typical stream-cut valleys as they exist before glaciation. Note the prominent V-shaped valley.
- In (b), you see the same landscape during subsequent glaciation. Glacial erosion and transport actively remove much of the regolith (weathered bedrock) and the soils that covered the stream-valley landscape. As the *cirque* walls erode away, sharp ridges form, dividing adjacent cirque basins. These **arêtes** ("knife-edge" in French) become the *sawtooth*, serrated ridges in glaciated mountains. Two eroding cirques may reduce an arête to a saddlelike depression or pass, called a **col**. A **horn** (pyramidal peak) results when several cirque glaciers gouge an individual mountain summit from all sides. Most famous is the Matterhorn in the Swiss Alps, but many others occur worldwide. A **bergschrund** forms when a crevasse or wide crack opens along the headwall of a glacier, most visible in summer when covering snow is gone.
- In (c), you see the same landscape at a time of warmer climate when the ice has retreated. The glaciated valleys now are U-shaped, greatly changed from their previous stream-cut V form. You can see the steep sides and the straightened course of the valleys. Physical weathering from the freeze-thaw cycle has loosened rock along the steep cliffs, where it has fallen to form *talus slopes* along the valley sides. Transported rocks are left behind by the retreating ice as *erratics*.

Note in Figure 17.10c that in the cirques where the valley glaciers originated, small mountain lakes called **tarns** have formed. One cirque contains small, circular, stair-stepped lakes called **paternoster** ("our father") **lakes** for their resemblance to rosary (religious) beads. Paternoster lakes may have formed from the differing resistance of rock to glacial processes or from damming by glacial deposits.

The valleys carved by tributary glaciers are left stranded high above the valley floor, because the primary glacier eroded the valley floor so deeply. These *hanging valleys* are the sites of spectacular waterfalls.

In Figure 17.11 see how many of the erosional forms from Figure 17.10 (arêtes, col, horn, cirque, cirque glacier, sawtooth ridge, U-shaped valleys, erratics, tarn, truncated spurs, among others) you can identify in these two photographs of glaciers at work. In terms of net mass

FIGURE 17.10 The geomorphic handiwork ➡ of alpine glaciers.
(a) A preglacial landscape with V-shaped stream-cut valleys. (b) The same landscape filled with valley glaciers. Note inset photos of a *horn*, a *cirque basin, sawtooth ridge,* and *bergschrund.* (c) When the glaciers retreat, the new landscape is unveiled. Note inset photos of *glacial erratics,* a *hanging valley* and waterfall, *tarn* (lake), and characteristic U-shaped glacial valley (trough). [After W. M. Davis, in Tarbuck, Lutgens, and Pinzke, *Applications and Investigations in Earth Science* (New York: Macmillan, an imprint of Prentice Hall, Inc., 1994), p. 85. Photos: waterfall by author; all other photos by Bobbé Christopherson.]

(a) Preglacial

V-shaped valley

Horn
Col
Sawtooth
Arête
Truncated spurs
Cirque
Medial moraine
Bergschrund
Main glacier

(b) Glacial

Glacial erratics

Arête
Col Horn
Paternoster lakes
Cirques
U-shaped glacial trough
Hanging valley

(c) Postglacial

Hanging waterfall
Tarn
U-shaped valley

NOTEBOOK   Teton Glacier

(a)

(b)

FIGURE 17.11 Erosional features of Alpine glaciation. (a) Many of the erosional glacial features appear in this scene from southern Greenland. How many of the erosional features illustrated in Figure 17.10 can you find in the photo? (b) These merging glaciers along Beverlysundet, Nordaust-landet Island, Arctic Ocean, form a medial moraine and haul erratic boulders on their surfaces that have cracked off the cliffs through frost action. [Photos by Bobbé Christopherson.]

balance, what is your assessment of the status of the merging glaciers in Figure 17.11b?

Where a glacial trough intersects the ocean, the glacier can continue to erode the landscape, even below sea level. As the glacier retreats, the trough floods and forms a deep **fjord** in which the sea extends inland, filling the lower reaches of the steep-sided valley (Figure 17.12). The fjord may be flooded further by rising sea level or by changes in the elevation of the coastal region. All along the glaciated coast of Alaska, glaciers now are in retreat, thus opening many new fjords that previously were blocked by ice. Coastlines with notable fjords include those of Norway, Greenland, Chile, the South Island of New Zealand, Alaska, and British Columbia.

Depositional Landforms Created by Alpine Glaciation

You have just seen how glaciers excavate tremendous amounts of material and create fascinating landforms in

FIGURE 17.12 Norwegian fjord. The beautiful Fjærfjorden, Norway, where an inlet of the sea fills a U-shaped, glacially carved valley. [Photo by Bobbé Christopherson.]

(a)

(b)

(c)

FIGURE 17.13 Depositional features of alpine glaciation.
Medial, lateral, and ground moraine deposits are evident in eastern and southern Greenland. (a) As
these tributary glaciers merge, their lateral moraines combine to form a medial moraine. (b) Retreating
ice leaves behind these lateral moraines. (c) Note how each lateral moraine remains identifiable as
they combine in this valley glacier. This is unlike a stream where currents quickly mix waters from
adjoining streams. How many moraines can you count? Note the medial moraine flowing from the
upper-left corner of the photo. [All photos by Bobbé Christopherson.]

the process. Glaciers produce a different set of distinctive
landforms when they melt and deposit their debris cargo
at the glacier's *terminus*. Figures 17.7a and e show that after
the glacier melts, debris accumulates in moraines to mark
the former margins of the glacier, both its end and sides.

Glacial Drift   The general term for all glacial deposits,
both unsorted and sorted, is **glacial drift**. Sediments deposited by glacial meltwater are sorted by size and are
termed **stratified drift**. Direct ice deposits leave unstratified and unsorted debris called **till**.

As a glacier flows to a lower elevation, a wide assortment of rock fragments becomes *entrained* (carried
along) on its surface or embedded within its mass or in
its base. As the glacier melts, this unsorted cargo is
deposited on the ground surface. Such till is poorly
sorted and is difficult to cultivate for farming, but the
clays and finer particles can provide a basis for soil
development.

Retreating glaciers leave behind large rocks (sometimes house-sized), boulders, and cobbles that are "foreign" in composition and origin from the ground on
which they were deposited. These *glacial erratics*, lying in
strange locations with no obvious means of transport,
were an early clue that blankets of ice once had covered
the land (pictured in Figure 17.10c and 17.11b).

Moraines   The deposition of glacial sediments produces a specific landform called a **moraine**. Several types
of moraines are exhibited in Figure 17.13. A **lateral
moraine** forms along each side of a glacier. If two glaciers
with lateral moraines join, a **medial moraine** may form
(see Figures 17.1 and 17.7f). Both medial and lateral
moraine forms are clearly visible in Figure 17.13. A deposition of till that is generally spread across a surface is
called a *ground moraine*, or *till plain*, and may hide the

former landscape. Such plains are found in portions of
the U.S. Midwest.

Eroded debris that is dropped at the glacier's farthest
extent is called a **terminal moraine**. However, there also
may be *end moraines*, formed at other points where a glacier paused after reaching a new equilibrium between
growth and ablation. Lakes may form behind terminal
and end moraines after a glacier's retreat, with the
moraine acting as a dam.

All of these till types are unsorted and unstratified. In
contrast, streams of glacial meltwater can carry and deposit sorted and stratified glacial drift beyond a terminal
moraine. Meltwater-deposited material downvalley from
a glacier is called a *valley train deposit*. Peyto Glacier in Alberta, Canada, produces such a valley train that continues
into Peyto Lake (Figure 17.14). Distributary stream channels appear braided across its surface. The picture also

FIGURE 17.14 A valley train deposit.
Peyto Glacier in Alberta, Canada. Note valley train, braided
stream, and milky-colored glacial meltwater. [Photo by author.]

shows the milky meltwater associated with glaciers, laden with finely ground "rock flour." Meltwater is produced by glaciers at all times, not just when they are retreating.

Erosional and Depositional Landforms Created by Continental Glaciation

The extent of the most recent continental glaciation in North America and Europe, 18,000 years ago, is portrayed several pages ahead in Figure 17.27. When these huge sheets of ice advanced and retreated, they produced some of the erosional and depositional features characteristic of alpine glaciation. However, because continental glaciers form under different circumstances—not in mountains, but across broad, open landscapes—the intricately carved alpine features, lateral moraines, and medial moraines all are lacking in continental glaciation. Table 17.1 compares the erosional and depositional features of alpine (valley) and continental glaciers.

Table 17.1	Features of Valley Glaciation and Continental Glaciation Compared	
Features and Landforms	**Alpine (Valley) Glacier**	**Continental Glaciation**
Erosional		
Striations, polish, etc.	Common	Common
Cirques	Common	Absent
Horns, arêtes, cols	Common	Absent
U-shaped valleys, truncated spurs, hanging valleys	Common	Rare
Fjords	Common	Absent
Roche moutonnée	Common	Common
Depositional		
Till	Common	Common
Terminal moraines	Common	Common
Recessional moraines	Common	Common
Ground moraines	Common	Common
Lateral moraines	Common	Absent
Medial moraines	Common, easily destroyed	Absent
Drumlins	Rare or absent	Locally common
Erratics	Common	Common
Stratified drift	Common	Common
Kettles	Common	Common
Eskers, crevasse fillings	Rare	Common
Kames	Common	Common
Kame terraces	Common	Present in hilly country

Source: Adapted from L. D. Leet, S. Judson, and M. Kauffman, *Physical Geology*, 5th ed., © 1978, p. 317. Reprinted by permission of Prentice Hall, Inc., Upper Saddle River, NJ.

Figure 17.15 illustrates some of the most common erosional and depositional features associated with the retreat of a continental glacier. A **till plain** forms behind an end moraine; it features *unstratified* coarse till, has low and rolling relief, and has a deranged drainage pattern (Figure 17.15a, and see Figure 14.9). Beyond the morainal deposits lies the **outwash plain** of *stratified drift* featuring stream channels that are meltwater-fed, braided, and overloaded with sorted and deposited materials.

Figure 17.15b shows a sinuously curving, narrow ridge of coarse sand and gravel called an **esker**. It forms along the channel of a meltwater stream that flows beneath a glacier, in an ice tunnel, or between ice walls. As a glacier retreats, the steep-sided esker is left behind in a pattern roughly parallel to the path of the glacier. The ridge may not be continuous and in places may even appear to be branched, following the path set by the subglacial watercourse. Commercially valuable deposits of sand and gravel are quarried from some eskers.

Sometimes an isolated block of ice, perhaps more than a kilometer across, remains in a ground moraine, an outwash plain, or valley floor after a glacier has retreated. As much as 20 to 30 years is required for it to melt. In the interim, material continues to accumulate around the melting ice block. When the block finally melts, it leaves behind a steep-sided hole. Such a feature then frequently fills with water. This feature is called a **kettle**. Thoreau's famous Walden Pond, mentioned in the quotation in Chapter 7, is such a glacial kettle and is pictured in Figure 17.15c.

Another feature of outwash plains is a **kame**, a small hill, knob, or mound of poorly sorted sand and gravel that is deposited directly by water, by ice in crevasses, or in ice-caused indentations in the surface (Figure 17.15d). Kames also can be found in deltaic forms and in terraces along valley walls.

Glacial action also forms two types of streamlined hills. One is erosional, called a *roche moutonnée*, and the other is depositional, called a *drumlin*. A **roche moutonnée** ("sheep rock" in French) is an asymmetrical hill of exposed bedrock. Its gently sloping upstream side (stoss side) has been polished smooth by glacial action, whereas its downstream side (lee side) is abrupt and steep where the glacier plucked rock pieces (Figure 17.16).

A **drumlin** is deposited till that has been streamlined in the direction of continental ice movement, blunt end upstream and tapered end downstream (the opposite of a roche moutonnée). "Swarms" of drumlins occur across the landscape in portions of New York and Wisconsin, among other areas. Sometimes their shape is that of an elongated teaspoon bowl, lying face down. They attain lengths of 100–5000 m (300 ft to more than 3 mi) and heights up to 200 m (650 ft).

Figure 17.17 shows a portion of a topographic map for the area south of Williamson, New York, which experienced continental glaciation during the last ice age. In studying the map, can you identify the numerous drumlins?

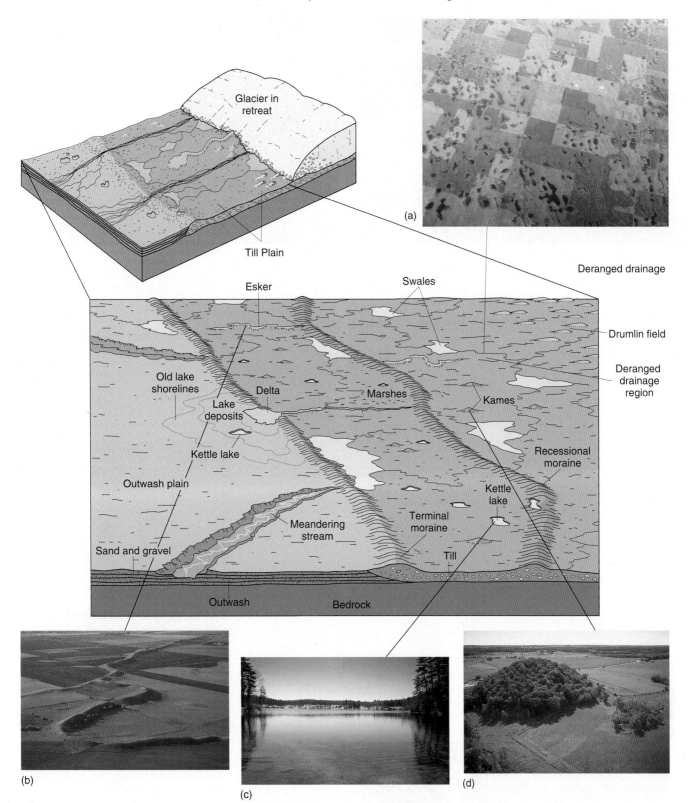

FIGURE 17.15 Continental glacier depositional features.
Common depositional landforms produced by glaciers. (a) Deranged drainage tells us continental glaciers advanced and retreated across the prairies of central Saskatchewan, Canada. (b) An esker through farmland near Dahlen, North Dakota. (c) Walden Pond is a kettle surrounded by mixed forest near Boston, Massachusetts. (d) A kame covered by a woodlot near Campbellsport, Wisconsin. [Illustration from R. M. Busch, ed., *Laboratory Manual in Physical Geology,* 3rd ed. (New York: Macmillan, an imprint of Prentice Hall, Inc., 1993), p. 188. Photos (a) and (c) by Bobbé Christopherson; (b) and (d) Tom Bean/DRK.]

(a)

FIGURE 17.16 Glacial erosion streamlined rock.
Roche moutonnée, as exemplified by Lembert Dome
in the Tuolumne Meadows area of Yosemite National
Park, California. [Photo by author.]

(b)

(a)

(b)

FIGURE 17.17 Glacially deposited streamlined features.
(a) Topographic map south of Williamson, New York, featuring
numerous drumlins. (7.5-minute series quadrangle map,
originally produced at a 1:24,000 scale, 10-ft contour interval.)
(b) A drumlin field, Snare Lake Canada. [(a) USGS map; (b) photo
by National Air Photo Library, NRC, Ottawa, Canada.]

In what direction do you think the continental glaciers moved across this region? Look for gentle, tapered slopes in the downstream direction of each drumlin, left behind as the glacier retreated.

Periglacial Landscapes

In 1909, Polish geologist W. Lozinski coined the term **periglacial** to describe frost weathering and freeze-thaw rock shattering in the Carpathian Mountains. He expanded the concept in 1910 to include several cold-climate processes, landforms, and topographic features that exist along the margins of glaciers, past and present. Periglacial regions occupy more than 20% of Earth's land surface and include features formed under nonglacial conditions. Under these conditions, a unique combination of frost action and periglacial processes operate, including permafrost, frost action, and ground ice.

Climatologically, these regions are in *subarctic* and *polar* climates, especially *tundra* climate. Such climates occur either at high latitude (tundra and boreal forest environments) or at high elevation in lower-latitude mountains (alpine environments). These periglacial regions are dominated by processes that are related to physical weathering, mass movement (Chapter 13), climate (Chapter 10), and soil (Chapter 18).

Geography of Permafrost

When soil or rock temperatures remain below 0°C (32°F) for at least 2 years, **permafrost** ("permanent frost") develops. An area of permafrost that is not covered by glaciers is considered periglacial, with the largest extent of such lands in Russia. Approximately 80% of Alaska has permafrost beneath its surface. Canada, China, Scandinavia, Greenland, and Antarctica, in addition to alpine mountain regions of the world, also are affected.

Note that this criterion is based solely on *temperature* and has nothing to do with how much or how little water is present. Two other factors also contribute to permafrost: the presence of fossil permafrost from previous ice-age conditions and the insulating effect of snow cover or vegetation that inhibits heat loss.

Continuous and Discontinuous Zones Permafrost regions are divided into two general categories: continuous and discontinuous, which merge along a general transition zone. *Continuous permafrost* is the region of severest cold and is perennial, roughly poleward of the −7°C (19°F) mean annual temperature isotherm (purple area in Figure 17.18). Continuous permafrost affects all surfaces except those beneath deep lakes or rivers. The depth of continuous permafrost may exceed 1000 m (3300 ft), averaging approximately 400 m (1300 ft).

Unconnected patches of *discontinuous permafrost* gradually coalesce poleward toward the continuous zone. Permafrost becomes scattered or sporadic until it gradually disappears equatorward of the −1°C (30.2°F) mean annual temperature isotherm (dark blue area on map). In the discontinuous zone, permafrost is absent on sun-exposed south-facing slopes, areas of warm soil, or areas insulated

FIGURE 17.18 Permafrost distribution.
Distribution of permafrost in the Northern Hemisphere. Alpine permafrost is noted except for small occurrences in Hawai'i, Mexico, Europe, and Japan. Sub-sea permafrost occurs in the ground beneath the Arctic Ocean along the margins of the continents, as shown. Note the towns of Resolute and Coppermine (Kugluktuk) in Nunavut (formerly part of NWT) and Hotchkiss in Alberta. A cross section of the permafrost beneath these towns is shown in **Figure 17.19.** [Adapted from T. L. Péwé, "Alpine permafrost in the contiguous United States: A review," *Arctic and Alpine Research* 15, no. 2 (May 1983): 146. © University of Colorado. Used by permission.]

by snow. In the Southern Hemisphere, north-facing slopes experience increased warmth.

Areas of discontinuous permafrost feature a mixture of *cryotic* (frozen) and *noncryotic* ground. In addition to these two types of ground, zones of high-altitude *alpine permafrost* extend to lower latitudes, as shown on the map. Microclimatic factors such as slope orientation and snow cover are important in the alpine environment.

In the Mackenzie Mountains of Canada (62° N), continuous permafrost extends down to an elevation of 1200 m (4000 ft), and discontinuous permafrost occurs throughout the range. The Colorado Rockies (40° N) experience continuous permafrost down to an elevation of 3400 m (11,150 ft) and discontinuous permafrost to 1700 m (5600 ft).

Behavior of Permafrost   We have looked at the spatial distribution of permafrost; now let us examine how permafrost behaves. Figure 17.19 is a stylized cross section from approximately 75° N to 55° N, using the three sites located on the map in Figure 17.18. The **active layer** is the zone of seasonally frozen ground that exists between the subsurface permafrost layer and the ground surface.

The active layer is subjected to consistent daily and seasonal freeze-thaw cycles. This cyclic melting of the active layer affects as little as 10 cm (4 in.) depth in the north (Ellesmere Island, 78° N), up to 2 m (6.6 ft) in the southern margins (55° N) of the periglacial region, and 15 m (50 ft) in the alpine permafrost of the Colorado Rockies (40° N).

The depth and thickness of the active layer and permafrost zone change slowly in response to climatic change. Higher temperatures degrade (reduce) permafrost and increase the thickness of the active layer; lower temperatures gradually aggrade (increase) permafrost depth and reduce active-layer thickness. Although somewhat sluggish in response, the active layer is a dynamic open system driven by energy gains and losses in the subsurface environment. As you might expect, most permafrost exists in disequilibrium with environmental conditions and therefore actively adjusts to inconstant climatic conditions.

The present trend in concert with global warming is for a deepening of the active layer at high latitudes. Across Canada and Siberia this warming is causing a large release of carbon from peat-rich thawed ground that in turn influences the global greenhouse—a real-time positive feedback mechanism. As described in Chapter 10, the thaw in the active layer in the Alaskan tundra has shortened the number of days that oil exploration equipment can venture out from 200 to only 100 days a year. The weight of the trucks requires a frozen, more solid surface to operate.

A *talik* is unfrozen ground that may occur above, below, or within a body of discontinuous permafrost or beneath a water body in the continuous region. Taliks occur beneath deep lakes and may extend to bedrock and noncryotic soil beneath large, deep lakes (see Figure 17.19). Taliks form connections between the active layer and groundwater, whereas in continuous permafrost groundwater is essentially cut off from surface water. In this way, permafrost disrupts aquifers and taliks, leading to water-supply problems.

Ground Ice and Frozen Ground Phenomena

In regions of permafrost, frozen subsurface water is termed *ground ice*. The moisture content of areas with ground ice varies from nearly none in drier regions to almost 100% in saturated soils. From the area of maximum energy loss, freezing progresses through the ground along a *freezing front*, or boundary between frozen and unfrozen soil. The presence of frozen water in the soil initiates geomorphic processes associated with *frost action* and the expansion of water as it freezes (discussed in Chapters 7 and 13).

Ground ice may occur as:

- Common *pore ice* (subsurface water frozen in the soil's pore spaces)
- *Lenses* (horizontal bodies) and *veins* (channels extending in any direction)
- *Segregated ice* (layers of buried ice that increase in mass by accreting water as the ground freezes, producing layers of relatively pure ice)
- *Intrusive ice* (the freezing of water injected under pressure, as in a pingo, discussed shortly)
- *Wedge ice* (surface water entering a crack and freezing)

Various aspects of frost action as a geomorphic agent are discussed in Chapter 13 ("Physical Weathering Processes" and the section "Classes of Mass Movements," which describes soil creep). Some forms of ground ice may occur at the surface during episodes of *icing*, in which a river or spring forms freezing layers of surface ice. Such icings can occur on slopes as well as level ground. Spring flooding along rivers can result from the presence of icings; it is a particular problem in rivers draining into the Arctic Ocean.

Frost Action Processes   The 9% expansion of water as it freezes produces strong mechanical forces. Such frost action shatters rock, producing angular pieces that form a *block field*, or *felsenmeer*. The felsenmeer accumulates as part of the arctic and alpine periglacial landscape, particularly on mountain summits and slopes.

If sufficient water freezes, the saturated soil and rocks are subjected to *frost-heaving* (vertical movement) and *frost-thrusting* (horizontal movement). Boulders and rock slabs may be thrust to the surface. Soil horizons (layers) may be disrupted by frost action and appear to be stirred or churned, a process termed *cryoturbation*. Frost action also can produce contractions in soil and rock, opening up cracks for ice wedges to form. Also, there is a tremendous increase in pressure in the soil as ice expands, particularly if there are multiple freezing fronts trapping unfrozen soil and water between them.

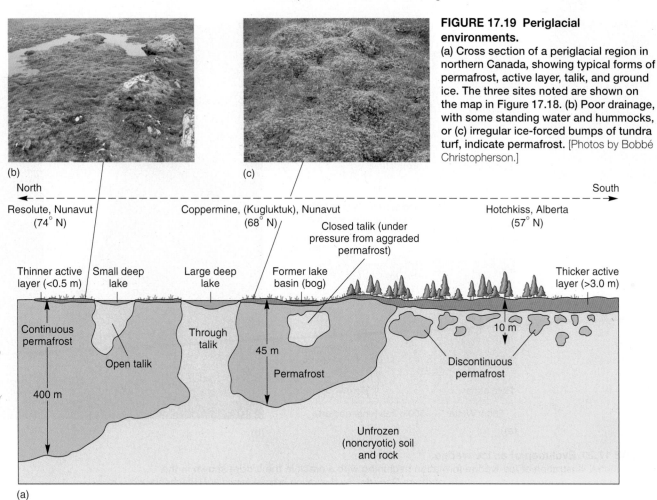

FIGURE 17.19 Periglacial environments.
(a) Cross section of a periglacial region in northern Canada, showing typical forms of permafrost, active layer, talik, and ground ice. The three sites noted are shown on the map in Figure 17.18. (b) Poor drainage, with some standing water and hummocks, or (c) irregular ice-forced bumps of tundra turf, indicate permafrost. [Photos by Bobbé Christopherson.]

An *ice wedge* develops when water enters a crack in the permafrost and freezes (Figure 17.20). Thermal contraction in ice-rich soil forms a tapered crack—wider at the top, narrowing toward the bottom. Repeated seasonal freezing and melting progressively enlarge the wedge, which may widen from a few millimeters to 5–6 m (16–20 ft) and up to 30 m (100 ft) in depth. Widening may be small each year, but after many years the wedge can become significant, like the sample shown in Figure 17.20b.

Thin layers of sediment that form foliations in the wedge mark the annual thickness of ice added. Thus, the age of a wedge can be determined by counting the foliations, like counting annual accumulations recorded in an ice core. In summer, when the active layer thaws, the wedge itself may not be visible. However, the presence of a wedge beneath the surface is noticeable where the ice-expanded sediments form raised, upturned ridges.

Frost-Action Landforms Large areas of frozen ground (soil-covered ice) can develop a heaved-up, circular, ice-cored mound called a *pingo*. It rises above the flat landscape, occasionally exceeding 60 m (200 ft) in height. Pingos rise when freezing water expands, sometimes as a result of pressure developed by artesian water injected into permafrost (Figure 17.21).

A *palsa*, which comes from the Swedish word for "elliptical," is a rounded or elliptical mound of peat that contains thin perennial ice lenses rather than an ice core, as in a pingo. Palsas can be 2–30 m (6–100 ft) wide by 1–10 m (3–30 ft) high and usually are covered by soil or vegetation over a cracked surface.

The expansion and contraction of frost action results in the transport of stones and boulders. As the water-ice volume changes and an ice wedge deepens, coarser particles are moved toward the surface. An area with a system of ground ice and frost action develops sorted and unsorted accumulations of rock at the surface that take the shape of polygons called **patterned ground**.

Patterned ground may include polygons of sorted rocks that coalesce into stone polygon nets (Figure 17.22a and b). Various terms are in use to describe such ice-wedge and stone polygon forms: nets, circles, vegetation-covered hummocks, and stripes (elongated polygons formed on a hillslope, Figure 17.22c). Ice wedges sometimes are found beneath the perimeter of each cell; however, questions still exist as to how such patterned ground actually develops and the degree to which such ground forms are the vestiges of past eras. Such features may take centuries to form.

Such polygon nets in patterned ground provide vivid evidence of frozen subsurface water on Mars (Figure 17.22d).

FIGURE 17.20 Evolution of an ice wedge.
(a) Sequential illustration of ice-wedge formation beginning with a crack in the tundra shown in the
photo. (b) An ice wedge and ground ice in northern Canada. [(a) Illustration adapted from A. H. Lachenbruch,
"Mechanics of thermal contraction and ice-wedge polygons in permafrost," *Geological Society of America Bulletin
Special Paper 70* (1962). Photos by (a) Bobbé Christopherson; (b) H. M. French.]

FIGURE 17.21 A pingo.
An ice-cored pingo resulting from hydraulic pressures that
pushed the mound upward above the landscape. Coastal erosion exposed the ice core. This pingo is near Tuktoyaktuk,
Mackenzie Delta, NWT, Canada. [Photo by H. M. French.]

In this 1999 image, the Mars Global Surveyor captured a
region of these features on the Martian northern plains.
To date, some 600 different sites on Mars demonstrate
such patterned ground.

Hillslope Processes: Gelifluction and Solifluction

Soil drainage is poor in areas of permafrost and ground
ice. The active layer of soil and regolith is saturated with
soil moisture during the thaw cycle (summer), and the
whole layer commences to flow from higher to lower elevation if the landscape is even slightly inclined. This flow
of soil is generally called *solifluction* and may occur in all
climate conditions. In the presence of ground ice or permafrost, the more specific term *gelifluction* is applied. In
this ice-bound type of soil flow, movement up to 5 cm
(2 in.) per year can occur on slopes as gentle as a degree
or two.

The cumulative effect of this landflow can be an overall flattening of a rolling landscape, with identifiable sag-
ging surfaces and scalloped and lobed patterns in the
downslope soil movements (Figure 17.23). Other types of
periglacial mass movement include failure in the active
layer, producing translational and rotational slides and
rapid flows associated with melting ground ice. Periglacial
mass movement processes are related to slope dynamics
and processes discussed in Chapter 13.

Thermokarst Landscapes

As ground ice melts, ir-
regular features develop across the landscape, creating
thermokarst topography. These forms result from thermal subsidence and erosion caused by ice-wedge melting
and poor drainage. (Note that the term refers only to
the topographic style and has nothing to do with the

(a)   (b)

(c)   (d)

FIGURE 17.22 Patterned ground phenomena.
(a) Patterned ground in the Dry Valleys of Antarctica. (b) Rock nets form polygons and circles (about a meter across) near Duvefjord, Nordaustlandet Island, Arctic Ocean. (c) Patterned ground in stripes, near Sundbukta, Barents Island, Arctic Ocean. (d) Polygons in the Martian northern plains; each cell averages a little more than 100 m across (300 ft). [Photos by (a) Galen Rowell/Mountain Light Photography, Inc.; (b) and (c) Bobbé Christopherson; and (d) image from the *Mars Global Surveyor,* Mars Orbiter Camera, courtesy of NASA/JPL/Malin Space Science Systems, May 1999.]

FIGURE 17.23 Soil flowage in periglacial environments.
Gelifluction lobes of soil underlain by frozen ground in permafrost conditions, on a hillside near Sundbukta, Barents Island, Arctic Ocean. [Photo by Bobbé Christopherson.]

solution processes and chemical weathering that cause limestone karst.)

Thermokarst topography is hummocky, marked by cave-ins, bogs, small depressions, pits, standing water, and small lakes. More thermokarst landforms are found in Siberia and Scandinavia than in North America. In Canada and Alaska, rounded thaw lakes, or cave-in lakes, are thermokarst features. With the incredibly warm temperatures recorded in the Canadian Arctic over the past 15 years, more thermokarst disruption of surfaces is occurring—leading to highway, railway, and building damage.

Humans and Periglacial Landscapes

In areas of permafrost and frozen-ground phenomena, people face several related problems. Because thawed ground above the permafrost zone frequently shifts, highways and rail lines become warped, twisted, and fail

FIGURE 17.24 Permafrost melting and structure collapse.
Building failure due to improper construction and the melting of permafrost, south of Fairbanks, Alaska. [Adapted from U.S. Geological Survey. Photo by Steve McCutcheon, illustration based on U.S. Geological Survey pamphlet "Permafrost" by L. L. Ray.]

and utility lines are disrupted. In addition, any building placed directly on frozen ground will "melt" into the defrosting soil, creating subsidence in structures (Figure 17.24).

Construction in periglacial regions dictates placing structures above the ground to allow air circulation beneath. This airflow allows the ground to cycle through its normal annual temperature pattern. Utilities such as water and sewer lines must be built aboveground in "utilidors" to protect them from freezing and thawing

ground (Figure 17.25). Likewise, the Trans-Alaska oil pipeline was constructed aboveground on racks for 675 of its 1285-km (420 of its 800 mi) length to avoid melting the frozen ground, causing shifting that could rupture the line. The pipeline that is underground uses a cooling system to keep the permafrost around the pipeline stable.

The effects of global warming are already showing up in Siberia, where the active layer is now more than twice the depth it was in the past. Buildings constructed with

(a)

Pipeline
1.2 m diameter

Average height
1.5 to 3.0 m

(b)

FIGURE 17.25 Special structures for permafrost.
(a) Proper construction in periglacial environments requires raising of buildings aboveground and running water and sewage lines in elevated "utilidors," here in Barentsburg, a Russian settlement on Spitsbergen Island. (b) Supporting the Trans-Alaska oil pipeline on racks protects the permafrost from heat. [Photos by (a) Bobbé Christopherson; (b) Galen Rowell/Mountain Light Photography, Inc.]

shallow pilings for support will certainly fail if this trend continues. In an era of global change, the regions of permafrost may provide further climatic indicators.

The Pleistocene Ice-Age Epoch

Imagine almost a third of Earth's land surface buried beneath ice sheets and glaciers—most of Canada, the northern Midwest, England, and northern Europe, and many mountain ranges, beneath thousands of meters of ice! This is how it was at the height of the Pleistocene Epoch of the late Cenozoic Era. In addition, periglacial regions along the margins of the ice during the last ice age covered about twice their present areal extent.

The Pleistocene is thought to have begun about 1.65 million years ago and is one of the more prolonged cold periods in Earth's history. It featured not just one glacial advance and retreat, but at least 18 expansions of ice over Europe and North America, each obliterating and confusing the evidence from the one before. Apparently, glaciation can take about 90,000 years, whereas deglaciation is rapid, requiring less than about 10,000 years to melt away the accumulation.

The term *ice age* is applied to any extended period of cold (not a single brief cold spell). An **ice age** is a period of generally cold climate that includes one or more *glacials*, interrupted by brief warm spells known as *interglacials*. Each glacial and interglacial is given a name that is usually based on the location where evidence of the episode is prominent—for example, "Wisconsinan glacial."

Modern research techniques to understand past climates include the examination of ancient ratios of oxygen isotopes, depths of coral growth in the tropics, analysis of ocean and lake sediments worldwide, and analysis of the latest ice cores from Greenland and Antarctica, including the Dome C ice core that plunges the record back 740,000 years (as of 2004). These techniques have opened the way for a new chronology and understanding of past climates and for putting perspective on the present warming trends. The record is clear and correlates across all proxy methods—present levels of (CO_2) and methane (CH_4) are the highest in 440,000 years. Earth systems have no experience in dealing with levels this high during this time span, and these concentrations are increasing.

Glaciologists currently recognize the Illinoian glacial and Wisconsinan glacial periods, with the Sangamon interglacial between them. These events span the 300,000-year period prior to our present Holocene Epoch (Figure 17.26). The chart shows that the Illinoian glacial actually consisted of two glacials occurring during Marine (oxygen) Isotope Stages (MIS) 6 and 8, as did the Wisconsinan (MIS 2 and 4), which are dated at 10,000 to 35,000 years ago. The oxygen isotope glacial/interglacial stages on the chart are numbered back to MIS 23 at approximately 900,000 years ago. To overcome local bias in core records, investigators correlate oxygen isotope data with other indicators worldwide. (For web links on glaciers and the Pleistocene, see http://research.umbc.edu/~miller/geog111/glacierlinks.htm.)

Changes in the Landscape

The continental ice sheets covered portions of Canada, the United States, Europe, and Asia about 18,000 years ago, as illustrated on the polar map projection in Figure 17.27. The ice sheets ranged in thickness to more than 2 km (1.2 mi). In North America, the Ohio and Missouri River systems mark the southern terminus of continuous ice at its greatest extent during the Pleistocene. The ice sheet disappeared by 7000 years ago.

As both alpine and continental glaciers retreated, they exposed a drastically altered landscape: the rocky soils of New England, the polished and scarred surfaces of Canada's Atlantic Provinces, the sharp crests of the Sawtooth Range and Tetons of Idaho and Wyoming, the scenery of the Canadian Rockies and the Sierra Nevada, the Great Lakes of the United States and Canada, the Matterhorn of Switzerland, and much more. In the Southern Hemisphere, there is evidence of this ice age in the form of fjords and sculpted mountains in New Zealand and Chile.

The continental glaciers came and went several times over the region we know as the Great Lakes (Figure 17.28). The ice enlarged and deepened stream valleys to form the basins of the future lakes. This complex history produced five lakes that today cover 244,000 km^2 (94,000 mi^2) and hold some 18% of all the lake water on Earth. Figure 17.28 shows the final formation of the Great Lakes, which involved two advancing and two retreating stages—between 13,200 and 10,000 years before the present. During the final retreat, tremendous quantities of glacial meltwater flowed into the gouged, isostatically (weight of the ice) depressed basins. Drainage at first was to the Mississippi River via the Illinois River, to the St. Lawrence River via the Ottawa River, and to the Hudson River in the east. In recent times, drainage has shifted through the St. Lawrence system.

Study of these glaciated landscapes is important, for we can better understand paleoclimatology (past climates) and discover the mechanisms that produce ice ages and climatic change. This knowledge can then act as a key to unlock an understanding of present conditions and perhaps empower an ability to forecast what lies ahead.

Lowered Sea Levels and Lower Temperatures

Sea levels 18,000 years ago were approximately 100 m (330 ft) lower than they are today because so much of Earth's water was frozen and tied up in the glaciers instead of being in the ocean. Imagine the coastline of New York being 100 km farther east; Alaska and Russia connected by land across the Bering Straits, and England and France joined by a land bridge. In fact, sea ice extended

FIGURE 17.26 Temperature record of the past 2 million years.
Pleistocene temperatures, determined by oxygen isotope fluctuations in fossil planktonic foraminifera (tiny marine organisms having a calcareous shell) from deep-sea cores. Twenty-three Marine (oxygen) Isotope Stages (MIS) cover 900,000 years, with names assigned for the past 300,000 years. See Focus Study 17.1 for application of this scale to ice-core data. [After N. J. Shackelton and N. D. Opdyke, *Oxygen-Isotope and Paleomagnetic Stratigraphy of Pacific Core V28-239, Late Pliocene to Latest Pleistocene.* Geological Society of America Memoir 145. ©1976 by the GSA. Adapted by permission.]

southward into the North Atlantic and Pacific and northward in the Southern Hemisphere about 50% farther than it does today.

Sea-surface temperatures 18,000 years ago averaged between 1.4 C° and 1.7 C° (2.5 F° and 3.1 F°) lower than today. During the coldest portion of the Pleistocene ice age, air temperatures were as much as 12 C° (22 F°) colder than today's average, although milder periods ranged to within 5 C° (9 F°) of present air temperatures.

Paleolakes

Figure 17.29 portrays the American West dotted with large lakes 12,000–30,000 years ago. Except for the Great Salt Lake in Utah (a remnant of the former Lake Bonneville noted on the map) and a few smaller lakes, only dry basins, ancient shorelines, and lake sediments remain today. These ancient lakes are called **paleolakes**. The three photos in Figure 17.29 show what these paleolake sites look like today: the Bonneville Salt Flats in Utah, Mono Lake in California, and Severe Dry Lake, Utah.

The term *pluvial* (Latin: "rain") describes any period of wet conditions, such as occurred during the Pleistocene Epoch. During pluvial periods, lake levels increased in arid regions. The drier periods between pluvials are called *interpluvials*. Interpluvials are marked with *lacustrine deposits*, which are lake sediments that form terraces along former shorelines.

Are paleolakes of glacial origin like the Great Lakes? Earlier researchers attempted to correlate pluvial and glacial ages, given their coincidence during the Pleistocene. However, few sites actually demonstrate such a simple relation. For example, in the western United States, the estimated volume of melted ice from glaciers is only a small portion of the actual water volume that was in the paleolakes. Also, these lakes tend to predate glacial times and are correlated instead with periods of wetter climate, or periods thought to have had lower evaporation rates. The term *paleolake* is increasingly used in the scientific literature to describe these lakes and to separate their occurrence from specific glacial stages.

Paleolakes existed in North and South America, Africa, Asia, and Australia. Today, the Caspian Sea in Kazakstan and southern Russia has a level of 30 m (100 ft) below mean world sea level, but ancient shorelines are visible about 80 m (265 ft) above the present lake level.

In North America, the two largest late Pleistocene paleolakes were in the Basin and Range Province of the West—Lake Bonneville and Lake Lahontan. These two lakes were, respectively, 8 times and 6 times the size of their present-day remnants. Figure 17.29 shows these and other paleolakes at their highest level and the few remaining modern lakes in light blue. Much research is still taking place since many of the ancient lakeshores are not yet field mapped.

The Great Salt Lake, near Salt Lake City, Utah, and the Bonneville Salt Flats (Figure 17.29b) in western Utah

FIGURE 17.27 Pleistocene glaciation.
Extent of Pleistocene glaciation in the Northern Hemisphere 18,000 years ago: North America, a polar perspective, and 9500 years ago (inset map). Earlier episodes produced continental glaciation of slightly greater extent. Note the depth of the continental ice sheets in meters. [From A. McIntyre, *CLIMAP* (Climate: Long-Range Investigation, Mapping, and Prediction) Project, Lamont-Doherty Earth Observatory. © 1981 by the GSA. Adapted by permission.]

18,000 years ago

9500 years ago

International Date Line

North Pole

Prime Meridian

Continental ice
Sea ice
Continents
Glacial maximum
Oceans (0°–24°)

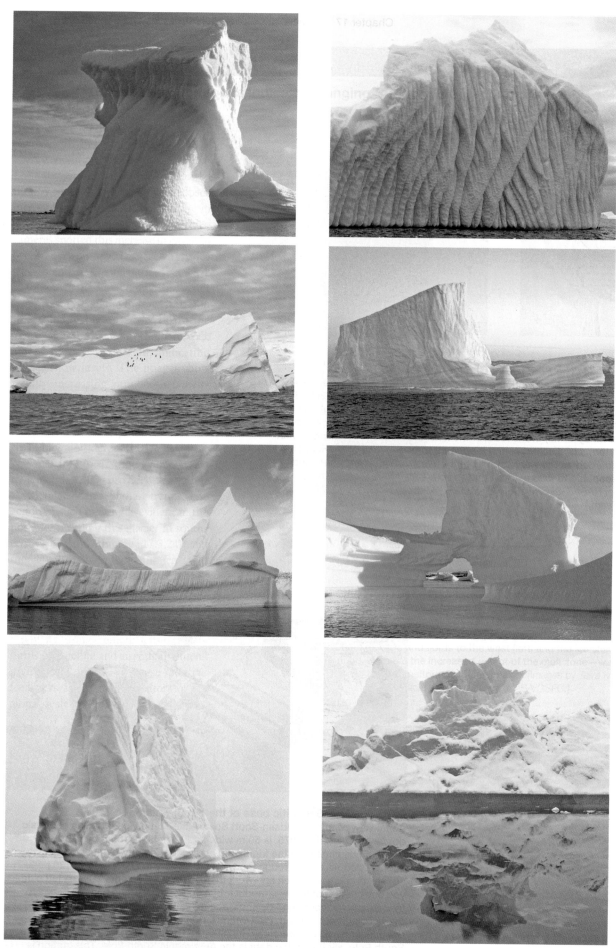

FIGURE 17.36 Arctic and Antarctic iceberg gallery.
[Photos by Bobbé Christopherson.]

High Latitude Connection 17.1

Climate Change Impacts an Arctic Ice Shelf

By late fall 2003, dramatic news spread through the climate science community that the Ward Hunt Ice Shelf—largest in the Arctic, on the north coast of Ellesmere Island in Nunavut (noted on the map in Figure 17.33a)—was breaking up. This was shocking news for the shelf had been stable for at least 4500 years. What scientists observed was that after three decades of mass losses, the ice shelf reached a systems threshold and rapidly broke up between 2000 and 2003. The breakup was captured by a sensor on the *Terra* satellite (Figure 17.1.1).

Backed up behind the Ward Hunt Ice Shelf was an *epishelf lake*, an extensive freshwater accumulation some 43 m deep, sitting on about 360 m of salty seawater in the Disraeli Fiord. As the ice-shelf dam failed, the freshwater epishelf lake drained. This was the largest such epishelf lake in the Northern Hemisphere. A unique biological community of freshwater and brackish water plankton was lost and delicate ecosystems disrupted.

Scientists noted that regional temperatures increased between 1967 and 2002 to an average July surface temperature of 1.3°C (34.3°F), which is above the threshold for ice-shelf failure established for Antarctica. Some other factors that pushed this threshold in Ward Hunt were freeze-thaw cycles, ocean tides and wind strength, sea-surface temperatures, and variations in salinity.

In 1907, Robert Perry noted that a continuous ice shelf guarded the entire north coast of Ellesmere Island. Most of this ice shelf was gone by 1982. Scientists attribute the loss of the Ellesmere Island Ice Shelf and the Ward Hunt Ice Shelf to "the cumulative effects of long-term warming since the 19th century." These conditions appeared to be accelerating during the latter third of the 20th century.

Crack in ice shelf

FIGURE 17.1.1 Ward Hunt Ice Shelf breakup.
Ellesmere Island, Nunavut, dominates this *Terra* satellite image made September 10, 2003. The northwest coast of Greenland is to the east and Nansen Sound and Greely Fiord are to the west. The center of the image is at 80° north latitude. The Ward Hunt Ice Shelf is near center on the north coast, with a large crack down its center and losses along its seaward front.[*Terra* MODIS sensor image courtesy of MODIS Land Rapid Response Team, NASA/GSFC, September 10, 2003.]

Summary and Review—Glacial and Periglacial Processes and Landforms

▶ *Differentiate* between alpine and continental glaciers, and *describe* their principal features.

More than 77% of Earth's freshwater is frozen. Ice covers about 11% of Earth's surface, and periglacial features occupy another 20% of ice-free but cold-dominated landscapes. A **glacier** is a mass of ice sitting on land or floating as an ice shelf in the ocean next to land. Glaciers form in areas of permanent snow. A **snowline** is the lowest elevation where snow occurs year-round and its altitude varies by latitude—higher near the equator, lower poleward.

A glacier in a mountain range is an **alpine glacier**. If confined within a valley, it is termed a *valley glacier*. The area of origin is a snowfield, usually in a bowl-shaped erosional landform called a **cirque**. Where alpine glaciers flow down to the sea, they calve and form **icebergs**. A **continental glacier** is a continuous mass of ice on land. Its most extensive form is an **ice sheet**; a smaller, roughly circular form is an **ice cap**; and the least extensive form, usually in mountains, is an **ice field**.

glacier (p. 532)　　　continental glacier (p. 534)
snowline (p. 532)　　　ice sheet (p. 534)
alpine glacier (p. 532)　ice cap (p. 534)
cirque (p. 533)　　　　ice field (p. 535)
iceberg (p. 534)

1. Describe the location of most freshwater on Earth today.
2. What is a glacier? What is implied about existing climate patterns in a glacial region?
3. Differentiate between an alpine glacier and the three types of continental glaciers. Which occurs in mountains? Which covers Antarctica?

4. How are icebergs generated? Describe their buoyancy characteristics from this discussion and the section on ice in Chapter 7. Why do you think an iceberg overturns from time to time as it melts?

▶ *Describe* the process of glacial ice formation, and *portray* the mechanics of glacial movement.

Snow becomes glacial ice through accumulation, increasing thickness, pressure on underlying layers, and recrystallization. Snow progresses through transitional steps from **firn** (compact, granular) to a denser **glacial ice** after many years.

A glacier is an open system with inputs and outputs that can be analyzed through observation of the growth and wasting of the glacier itself. A **firn line** is the lower extent of a fresh snow-covered area. A glacier is fed by snowfall and is wasted by **ablation** (losses from its upper and lower surfaces and along its margins). Accumulation and ablation achieve a mass balance in each glacier.

As a glacier moves downhill, vertical **crevasses** may develop. Sometimes a glacier will move rapidly in a **glacier surge**. The presence of water along the basal layer appears to be important in glacial movements. As a glacier moves, it plucks rock pieces and debris, incorporating them into the ice, and this debris scours and sandpapers underlying rock through **abrasion**.

firn (p. 537)
glacial ice (p. 537)
firn line (p. 537)
ablation (p. 537)
crevasse (p. 541)
glacier surge (p. 541)
abrasion (p. 542)

5. Trace the evolution of glacial ice from fresh fallen snow.
6. What is meant by glacial mass balance? What are the basic inputs and outputs underlying that balance?
7. What is meant by a glacier surge? What do scientists think produces surging episodes?

▶ *Describe* characteristic erosional and depositional landforms created by alpine glaciation and continental glaciation.

Extensive valley glaciers have profoundly reshaped mountains worldwide, carving V-shaped stream valleys into U-shaped glaciated valleys, producing many distinctive erosional and depositional landforms. As cirque walls erode away, sharp **arêtes** (sawtooth, serrated ridges) form, dividing adjacent cirque basins. Two eroding cirques may reduce an arête to a saddlelike **col**. A **horn** results when several cirque glaciers gouge an individual mountain summit from all sides, forming a pyramidal peak. A **bergschrund** forms when a crevasse or wide crack opens along the headwall of a glacier, most visible in summer when covering snow is gone. An ice-carved rock basin left as a glacier retreats may fill with water to form a **tarn**; tarns in a string separated by moraines are called **paternoster lakes**. Where a glacial valley trough joins the ocean, and the glacier retreats, the sea extends inland to form a **fjord**.

All glacial deposits, whether ice-borne or meltwater-borne, constitute **glacial drift**. Glacial meltwater deposits are sorted and are called **stratified drift**. Direct deposits from ice, called **till**, are unstratified and unsorted. Specific landforms produced by the deposition of drift are **moraines**. A **lateral moraine** forms along each side of a glacier; merging glaciers with lateral moraines form a **medial moraine**; and eroded debris dropped at the glacier's terminus is a **terminal moraine**.

Continental glaciation leaves different features than does alpine glaciation. A **till plain** forms behind end moraines, featuring unstratified coarse till, low and rolling relief, and deranged drainage. Beyond the morainal deposits, **outwash plains** of stratified drift feature stream channels that are meltwater-fed, braided, and overloaded with debris that is sorted and deposited across the landscape. An **esker** is a sinuously curving, narrow ridge of coarse sand and gravel that forms along the channel of a meltwater stream beneath a glacier. An isolated block of ice left by a retreating glacier becomes surrounded with debris; when the block finally melts, it leaves a steep-sided **kettle**. A **kame** is a small hill, knob, or mound of poorly sorted sand and gravel that is deposited directly by water or by ice in crevasses.

Glacial action forms two types of streamlined hills: the erosional **roche moutonnée** is an asymmetrical hill of exposed bedrock, gently sloping upstream and abruptly sloping downstream; the depositional **drumlin** is deposited till, streamlined in the direction of continental ice movement (blunt end upstream and tapered end downstream).

arête (p. 542)
col (p. 542)
horn (p. 542)
bergschrund (p. 542)
tarn (p. 542)
paternoster lakes (p. 542)
fjord (p. 544)
glacial drift (p. 545)
stratified drift (p. 545)
till (p. 545)
moraine (p. 545)
lateral moraine (p. 545)
medial moraine (p. 545)
terminal moraine (p. 545)
till plain (p. 546)
outwash plain (p. 546)
esker (p. 546)
kettle (p. 546)
kame (p. 546)
roche moutonnée (p. 546)
drumlin (p. 546)

8. How does a glacier accomplish erosion?
9. Describe the evolution of a V-shaped stream valley to a U-shaped glaciated valley. What features are visible after the glacier retreats?
10. How is an arête formed? A col? A horn? Briefly differentiate between them.
11. Differentiate between two forms of glacial drift—till and outwash.
12. What is a morainal deposit? What specific moraines are created by alpine and continental glaciers?
13. What are some common depositional features encountered in a till plain?
14. Contrast a roche moutonnée and a drumlin regarding appearance, orientation, and the way each forms.

▶ *Analyze* the spatial distribution of periglacial processes, and *describe* several unique landforms and topographic features related to permafrost and frozen ground phenomena.

The term **periglacial** describes cold-climate processes, landforms, and topographic features that exist along the margins of glaciers, past and present. Periglacial regions occupy more than 20% of Earth's land surface. These areas have either near-permanent ice or are at high elevation, and the ground is seasonally snow-free. When soil or rock temperatures remain below 0°C (32°F) for at least 2 years, **permafrost** ("permanent frost") develops. An area of permafrost that is not covered by glaciers is considered periglacial. Note that this criterion is based solely on temperature and has nothing to do with how much or how little water is present. The **active layer** is the zone of seasonally frozen ground that exists between the subsurface permafrost layer and the ground surface. **Patterned ground** forms in the periglacial environment where freezing and thawing of the ground create polygonal forms of arranged rocks at the surface; these forms can be circles, polygons, stripes, nets, and steps.

periglacial (p. 549)   active layer (p. 550)
permafrost (p. 549)   patterned ground (p. 551)

15. In terms of climatic types, describe the areas on Earth where periglacial landscapes occur. Include both higher latitude and higher altitude climate types.

16. Define two types of permafrost, and differentiate their occurrence on Earth. What are the characteristics of each?

17. Describe the active zone in permafrost regions, and relate the degree of development to specific latitudes.

18. What is a talik? Where might you expect to find taliks, and to what depth do they occur?

19. What is the difference between permafrost and ground ice?

20. Describe the role of frost action in the formation of various landform types in the periglacial region, such as a pingo and patterned ground.

21. Relate some of the specific problems humans encounter in developing periglacial landscapes.

▶ *Explain* the Pleistocene ice-age epoch and related glacials and interglacials, and *describe* some of the methods used to study paleoclimatology.

An **ice age** is any extended period of cold. The late Cenozoic Era featured pronounced ice-age conditions in an epoch called the Pleistocene. During this time, alpine and continental glaciers covered about 30% of Earth's land area in at least 18 glacials, punctuated by interglacials of milder weather. Beyond the ice, **paleolakes** formed because of wetter conditions. Evidence of ice-age conditions is gathered from ice cores drilled in Greenland and Antarctica, from ocean sediments, from coral growth in relation to past sea levels, and from rock. The study of past climates is *paleoclimatology*.

The apparent pattern followed by these low-temperature episodes indicates cyclic causes. A complicated mix of interacting variables appears to influence long-term climatic trends: celestial relations, solar variability, tectonic factors, atmospheric variables, and oceanic circulation.

ice age (p. 555)
paleolake (p. 556)

22. What is paleoclimatology? Describe Earth's past climatic patterns. Are we experiencing a normal climate pattern in this era, or have scientists noticed any significant trends?

23. Define an ice age. When was the most recent? Explain "glacial" and "interglacial" in your answer.

24. Summarize what science has learned about the causes of ice ages by listing and explaining at least four possible factors in climate change.

25. Describe the role of ice cores in deciphering past climates. What record do they preserve? Where were they drilled?

26. Explain the relationship between the criteria defining the Arctic and Antarctic regions. Is there any coincidence in the Arctic criteria and the distribution of Northern Hemisphere forests on the continents?

27. Based on information in this chapter and Chapter 10 in the "Global Climate Change" section, summarize a few of the changing conditions underway in each polar region. Are there any conditions that seem of greatest concern to you after studying them?

NetWork

The *Geosystems* Home Page provides online resources for this chapter on the World Wide Web. To begin: Once on the Home Page, click on this textbook, scroll the Table of Contents menu, select this chapter, and click "Begin." You will find self-tests that are graded, review exercises, specific updates for items in the chapter, and many links to interesting related pathways on the Internet under "Destinations." *Geosystems* is at http://www.prenhall.com/christopherson.

Critical Thinking

A. After checking back issues of the *New South Polar Times*, 1993–1998 at http://www.spotsylvania.k12.va.us/nspt/home.htm, imagine duty there for yourself. Of the 28 people who winter over at the station, some serve as scientists, technicians, and support staff. The station commander for the 1998 season was Katy McNitt-Jensen, in her third tour at the pole. Remember, the last airplane leaves mid-February and the first airplane lands mid-October—such is the isolation. What do you see as the positives and negatives to such service? How would you combat the elements? The isolation? The cold and dark conditions?

B. Analyze and compare the initial results from ice-core drilling efforts in Greenland and Antarctica as described in this chapter (Focus Study 17.1 and text). What do they tell us about past climates? How far back do they go in time? Have they found any evidence of humans on Earth in the cores?

C. What is the relationship between knowing about paleoclimates and being able to forecast future climates? Is there a link? Have scientists discovered any cyclic behavior to Earth's past climates that might repeat in the future? Explain.

Karl Birkeland, Avalanche Scientist, Forest Service National Avalanche Center

When snow scientist Karl Birkeland was in second grade, he wrote a little essay called "How Glaciers Form." It reads (spelling corrected): "Glaciers form by large snows, where snow does not melt. The reason glaciers are covered with dirt is because the glaciers shake the canyon and the dirt falls on them. Glaciers must be bigger than the Empire State Building. If they are near the water big icebergs might fall in." Karl adds, "So I was thinking about this way back then. I have this framed above my desk, misspellings and all."

Karl completed his undergraduate work at the University of Colorado in environment, population, and biology, and even thought about medical school. His father was a professor of geology at Colorado. A skier since childhood, while finishing his degree he was a member of the ski patrol. He began to realize that studying the environment, specifically snow science, was what he wanted to do: "Snow science became a focus. Drawing on my ski patrol work where we had to deal with avalanches, I felt I had a good start."

Karl finished his master's at Montana State University, where his thesis topic was "Spatial Variability of Snow Resistance on Potential Avalanche Slopes." He began work as an avalanche specialist at the Gallatin National Forest, Bozeman, Montana, in 1990, where he established the Avalanche Center. He completed his Ph.D. at Arizona State in 1997. In his dissertation, he looked at slopes throughout the Bridger Mountains, Montana, and developed models for assessing variations in snow stability. This was the first time that anyone had tried to map avalanche conditions over a mountain range in one day (Figure 17.1.1).

Karl states, "In the Rockies, we have better preserved weak layers within the snowpack. These weak layers can keep the snow packs unstable for long periods of time, even if there hasn't been fresh snowfall." This regional characteristic makes the location of his National Avalanche Center ideal.

As to his present responsibilities, Karl comments, "I'm in charge of transferring new and emerging technologies to the avalanche specialists at the regional avalanche centers around the country. I'm also involved in keeping track of any recent scientific developments in our field that should be conveyed to our avalanche workers." Karl assists the director of the Forest Service National Avalanche Center in coordinating all the avalanche centers.

Karl and his team continue to examine synoptic climatology and analyze air masses to understand resulting snow accumulations. Reconstructing past weather patterns is key. These efforts will produce a classification scheme that categorizes past data into coastal, continental, or intermountain characteristic types. "We can tie this in with weather patterns and predict the kind of avalanches we might expect on a given day." He says the big challenge is to "take scientific work and distill out the day-to-day practical significance for avalanche forecasters." In other words, what is included in this research that can help avalanche forecasters right now (Figure 17.1.2)?

I kept hearing "spatial" in his work, getting a sense of the geographer in him. "That's me!" he answers, "There is such a need to see things spatially, regionally, with maps as the medium of communication." Geography is in the family, as Karl's wife is a Ph.D. geographer studying rivers and riparian vegetation across the Colorado Plateau and southwestern deserts.

Karl participates in the biennial International Snow Science Workshop (see http://www.issw.net/). The key challenge is merging theory and practice. More than 700 avalanche forecasters, ski patrollers, helicopter ski guides, and snow scientists, attended the 2004 meeting in

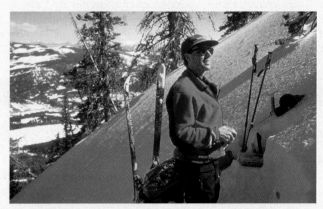

FIGURE 17.1.1 Karl Birkeland, Avalanche Scientist.
Karl works at the U.S. Forest Service National Avalanche Center.
[Photo by Ginger Birkeland.]

FIGURE 17.1.2 Investigating the origin of an avalanche.
Avalanche forecaster Chris Lundy investigates the crown face of a slab avalanche in Glacier National Park, Montana. This is the front along which an avalanche broke loose. [Photo by Karl Birkeland.]

FIGURE 17.1.3 The calving of nearby glacial ice shook the ground and air with a thunderous crack and triggered this ice fall avalanche along the coast of Antarctica.
[Photo by Bobbé Christopherson.]

Wyoming. The group is working to increase the global database, accident statistics, and forecasting efforts of many scientists in this growing field of interest. The future involves utilization of GIS modeling techniques to better tie together snowpack variables for spatial analysis.

"We want to bring together the operational avalanche workers and the avalanche researchers. The Internet has done a great job at linking this small community of individuals," Karl says. "This communication is needed because the avalanche risk is growing, owing to the increased development and construction in the mountains that are avalanche prone. All in all, the future is quite exciting in snow science!" And, Karl declares, "I continue to be very pro-geography and I like calling myself a geographer! Much of my current research focuses on spatial variations of snowpack properties, a clearly geographic topic that has critical real-world applications for avalanche forecasters."

PART IV

Soils, Ecosystems, and Biomes

Earth is the home of the only known biosphere in the Solar System—a unique, complex, and interactive system of abiotic (nonliving) and biotic (living) components working together to sustain a tremendous diversity of life. Energy enters the biosphere through conversion of solar energy by photosynthesis in the leaves of plants. Soil is the essential link among the lithosphere, plants, and the rest of Earth's physical systems. Thus, soil helps sustain life.

Life is organized into a feeding hierarchy from producers to consumers, ending with decomposers. Taken together, the soils, plants, animals, and all abiotic components produce aquatic and terrestrial ecosystems, generally grouped together in various biomes. Today we face crucial issues, principally the preservation of the diversity of life in the biosphere and the survival of the biosphere itself. Patterns of land and ocean temperatures, precipitation, weather phenomena, and stratospheric ozone, among many elements, are changing as global climate systems shift. The resilience of the biosphere, as we know it, is being tested in a real-time, one-time experiment. These important issues of biogeography are considered in Part 4.

Forests and farms sweep down to meet the long, narrow Vagavatnet Lake near the town of Vagamo in south-central Norway. The interaction of atmosphere, hydrosphere, lithosphere, and the passage of continental glaciation during the last ice age produce such a serene and productive biosphere at 62° N latitude. [Photo by Bobbé Christopherson.]

Light rains grace the fertile soils of southwestern Minnesota on a warm, humid Midwestern afternoon. Former prairies have sustained a productive agriculture since the early 19th century. Here feed grains are grown in rich Mollisols of the Soil Taxonomy classification. [Photo by Bobbé Christopherson.]

18

The Geography of Soils

▶ Key Learning Concepts _____

After reading the chapter, you should be able to:

▶ *Define* soil and soil science, and *describe* a pedon, polypedon, and typical soil profile.

▶ *Describe* soil properties of color, texture, structure, consistence, porosity, and soil moisture.

▶ *Explain* basic soil chemistry, including cation-exchange capacity, and *relate* these concepts to soil fertility.

▶ *Evaluate* the principal soil formation factors, including the human element.

▶ *Describe* the 12 soil orders of the Soil Taxonomy classification system, and *explain* their general occurrence.

E arth's landscape generally is covered with soil. **Soil** is a dynamic natural material composed of fine particles in which plants grow, and it contains both mineral fragments and organic matter. The soil system includes human interactions and supports all human, other animal, and plant life. If you have ever planted a garden, tended a houseplant, or been concerned about famine and soil loss, this chapter will interest you. A knowledge of soil is at the heart of agriculture and food production.

You kneel down and scoop a handful of prairie soil, compressing it and breaking it apart with your fingers. You are holding a historical object, one that bears the legacy of the last 15,000 years or more. This lump of soil contains information about the last ice age and intervening warm periods, about distinct and distant source materials, about several physical processes. We are using and abusing this legacy at rates much faster than it was formed. Soils do not reproduce, nor can they be re-created.

Soil science is interdisciplinary, involving physics, chemistry, biology, mineralogy, hydrology, taxonomy, climatology, and cartography. Physical geographers are interested in the spatial patterns formed by soil types and the environmental factors that interact to produce them. As an integrative science, physical geography is well suited to the study of soils.

Pedology concerns the origin, classification, distribution, and description of soil (*ped* from the Greek *pedon*, meaning "soil" or "earth"). Pedology is at the center of learning about soil as a natural body, but it does not dwell on its practical uses. *Edaphology* (from the Greek *edaphos*, meaning "soil" or "ground") focuses on soil as a medium for sustaining higher plants. Edaphology emphasizes plant growth, fertility, and the differences in productivity among soils. Pedology gives us a general understanding of soils and their classification, whereas edaphology reflects society's concern for food and fiber production and the management of soils to increase fertility and reduce soil losses.

In many locales, an *agricultural extension service* can provide specific information and perform a detailed analysis of local soils. Soil surveys and local soil maps are available for most counties in the United States and for the Canadian provinces. Your local phone book may list the U.S. Department of Agriculture, Natural Resources Conservation Service (http://www.nrcs.usda.gov/), or Agriculture Canada's Soil Information System (http://sis.agr.gc.ca/cansis/intro.html, with more information at http://www.metla.fi/info/vlib/soils/old.htm). Or, you may contact the appropriate department at a local college or university. See the National Soil Survey Center's site at http://soils.usda.gov/, and for links related to world soil resources, see http://www.nhq.nrcs.usda.gov/.

In this chapter: The geography of soils deals spatially with a complex substance, the characteristics of which vary from kilometer to kilometer, and even centimeter to centimeter. We begin with soil characteristics and the basic soil sampling and soil mapping units. The soil profile is a dynamic structure, mixing and exchanging materials and moisture across its horizons. Properties of soil include texture, structure, porosity, moisture, and chemistry—all integrating to form soil types. The chapter discusses both natural and human factors that affect soil formation. A global concern exists over the loss of soils to erosion, mistreatment, and conversion to other uses. The chapter concludes with a brief examination of the Soil Taxonomy, the 12 principal soil orders, and their spatial distribution.

Soil Characteristics

Classifying soils is similar to classifying climates because both involve interacting variables. Before we look at soil classification, let us examine the physical properties that distinguish soils as they develop through time in response to climate, relief, and topography.

Soil Profiles

Just as a book cannot be judged by its cover, so soils cannot be evaluated at the surface only. Instead, a soil profile should be studied from the surface to the deepest extent of plant roots, or to where regolith or bedrock is encountered. Such a profile, called a **pedon**, is a hexagonal column measuring 1 to 10 m^2 in top surface area (Figure 18.1). At the sides of the pedon, the various layers of the soil profile are visible in cross section and are labeled with letters. *A pedon is the basic sampling unit used in soil surveys.*

FIGURE 18.1 Soil sampling and mapping units. A soil pedon (sampling unit) is derived from a polypedon (mapping unit). Shown are typical O, A, E, B, C, and R soil horizons within a developed soil pedon. The true soil, or solum, includes the O, A, E, and B horizons.

Many pedons together in one area make up a **polypedon**, which has distinctive characteristics differentiating it from surrounding polypedons. A polypedon is an essential soil individual, comprising an identifiable series of soils in an area. It can have a minimum dimension of about m² and no specified maximum size. *The polypedon is the basic mapping unit used in preparing local soil maps.*

Soil Horizons

Each distinct layer exposed in a pedon is a **soil horizon**. A horizon is roughly parallel to the pedon's surface and has characteristics distinctly different from horizons directly above or below. The boundary between horizons usually is distinguishable when viewed in profile, such as along a road cut, on the basis of the properties of color, texture, structure, consistence (meaning soil consistency or cohesiveness), porosity, the presence or absence of certain minerals, moisture, and chemical processes (Figure 18.2). Soil horizons are the building blocks of soil classification.

FIGURE 18.2 A typical soil profile.
This is a Mollisol pedon in southeastern South Dakota. The parent material is glacial till, and the soil is well drained. The dark O and A horizons above the #1, transition into an E horizon. Distinct carbonate nodules are visible in the lower B and upper C horizons. [Photo from the Marbut Collection, Soil Science Society of America.]

At the top of the soil profile is the *O* (organic) *horizon*, named for its organic composition, derived from plant and animal litter that was deposited on the surface and transformed into humus. **Humus** is not just a single material; it is a mixture of decomposed and synthesized organic materials, usually dark in color. Microorganisms work busily on this organic debris, performing a portion of the *humification* (humus-making) process. The O horizon is 20%–30% or more organic matter, which is important because of its ability to retain water and nutrients and for the way it acts in a complementary manner to clay minerals.

At the bottom of the soil profile is the *R* (rock) *horizon*, consisting of either unconsolidated (loose) material or consolidated bedrock. When bedrock physically and chemically weathers into regolith, it may or may not contribute to overlying soil horizons. The A, E, B, and C horizons mark differing mineral strata between O and R. These middle layers are composed of sand, silt, clay, and other weathered by-products. (Table 11.3 presents a description of grain sizes for these weathered soil particles.)

In the *A horizon*, humus and clay particles are particularly important, for they provide essential chemical links between soil nutrients and plants. This horizon usually is richer in organic content, and hence darker, than lower horizons. Here is where human disruption through plowing, pasturing, and other uses takes place. The A horizon grades into the *E horizon*, made up of coarse sand, silt, and resistant minerals; generally, the E horizon is lighter in color.

From the lighter-colored E horizon, silicate clays and oxides of aluminum and iron are leached (removed by water) and carried to lower horizons with the water as it percolates through the soil. This process of removal of fine particles and minerals by water, leaving behind sand and silt, is termed **eluviation**—thus the E designation for this horizon. As precipitation increases, so does the rate of eluviation.

In contrast to the A and E horizons, *B horizons* accumulate clays, aluminum, and iron. B horizons are dominated by **illuviation**, a depositional process. (Eluviation is an erosional removal process; illuviation is depositional.) B horizons may exhibit reddish or yellowish hues because of the illuviated presence of minerals (silicate clays, iron and aluminum, carbonates, gypsum) and organic oxides.

Some materials occurring in the B horizon may have formed in place from weathering processes rather than arriving there by *translocation*, or migration. In the humid tropics, these layers often develop to some depth. Likewise, clay losses in an A horizon may be caused by destructive processes and not eluviation. Research to better understand erosion and deposition of clays between soil horizons is one of the challenges in modern soil science.

The combination of the A and E horizons and the B horizon is designated the **solum**, considered the true definable soil of the pedon. The A, E, and B horizons experience active soil processes (labeled on Figure 18.1).

Below the solum is the *C horizon* of weathered bedrock or weathered parent material. This zone is identified as *regolith* (although the term sometimes is used to include the solum as well). The C horizon is not much affected by soil operations in the solum and lies outside the biological influences experienced in the shallower horizons. Plant roots and soil microorganisms are rare in the C horizon. It lacks clay concentrations and generally is made up of carbonates, gypsum, or soluble salts, or of iron and silica, which form cemented soil structures. In dry climates, calcium carbonate commonly forms the cementing material of these hardened layers.

Soil scientists using the U.S. classification system employ letter suffixes to further designate special conditions within each soil horizon. A few examples include Ap (A horizon has been plowed), Bt (B horizon is of illuviated clay), Bf (permafrost or frozen soil), and Bh (illuviated humus as a dark coating on sand and silt particles).

Soil Properties

Soils are complex and varied, as this section reveals. Observing a real soil profile will help you identify color, texture, structure, and other soil properties. A good opportunity to observe soil profiles is at a construction site or excavation, perhaps on your campus, or at a road cut along a highway. The USDA Natural Resources Conservation Service *Soil Survey Manual* (U.S. Department of Agriculture Handbook No. 18, October 1993) presents information on all soil properties. The NRCS publication *Soil Survey Laboratory Methods Manual* (U.S. Department of Agriculture Soil Survey Investigations Report No. 42, v. 3.0, January 1996) details specific methods and practices in conducting soil analysis.

Soil Color

Color is important, for it sometimes suggests composition and chemical makeup. If you look at exposed soil, color may be the most obvious trait. Among the many possible hues are the reds and yellows found in soils of the southeastern United States (high in iron oxides), the blacks of prairie soils in portions of the U.S. grain-growing regions and Ukraine (richly organic), and white-to-pale hues found in soils containing silicates and aluminum oxides. However, color can be deceptive: Soils of high humus content are often dark, yet clays of warm–temperate and tropical regions with less than 3% organic content are some of the world's blackest soils.

To standardize color descriptions, soil scientists describe a soil's color by comparing it with a *Munsell Color Chart* (developed by artist and teacher Albert Munsell in 1913). These charts display 175 colors arranged by *hue* (the dominant spectral color, such as red), *value* (degree of darkness or lightness), and *chroma* (purity and saturation of the color, which increase with decreasing grayness). A Munsell notation identifies each color by a name, so soil scientists can make worldwide comparisons of soil color. Soil color is checked against the chart at various depths within a pedon (Figure 18.3).

Soil Texture

Soil texture, perhaps a soil's most permanent attribute, refers to the mixture of sizes of its particles and the

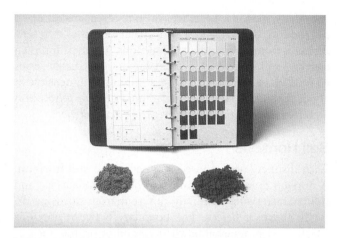

FIGURE 18.3   A Munsell Soil Color Chart page.
A soil sample is viewed through the hole to match it with a color on the chart. Hue, value, and chroma are the characteristics of color assessed by this system. [Photo courtesy of Gretag Macbeth, Munsell Color.]

proportion of different sizes. Individual mineral particles are called *soil separates.* All particles smaller in diameter than 2 mm (0.08 in.), such as very coarse sand, are considered part of the soil. Larger particles, such as pebbles, gravel, or cobbles, are not part of the soil. (Sands are graded from coarse, to medium, to fine, down to 0.05 mm; silt to 0.002 mm; and clay to less than 0.002 mm.)

Figure 18.4 is a *soil texture triangle* showing the relation of sand, silt, and clay concentrations in soil. Each corner of the triangle represents a soil consisting solely of the particle size noted (although rarely are true soils composed of a single separate). Every soil on Earth is defined somewhere in this triangle.

Figure 18.4 includes the common designation **loam**, which is a balanced mixture of sand, silt, and clay that is beneficial to plant growth. Farmers consider ideal a sandy loam with clay content below 30% (lower left) because of its water-holding characteristics and ease of cultivation. Soil texture is important in determining water-retention and water-transmission traits.

To see how the soil texture triangle works, consider a soil type in Indiana called the *Miami silt loam.* Samples from this soil type are plotted on the soil texture triangle as points 1, 2, and 3. A sample taken near the surface in the A horizon is recorded at point 1, in the B horizon at point 2, and in the C horizon at point 3. Textural analyses of these samples are summarized in Table 18.1 and in the three pie

Table 18.1	Textural Analysis of Miami Silt Loam		
Sample Points	**% Sand**	**% Silt**	**% Clay**
1 = A horizon	21.5	63.4	15.0
2 = B horizon	31.1	25.0	43.4
3 = C horizon	42.4	34.0	23.5

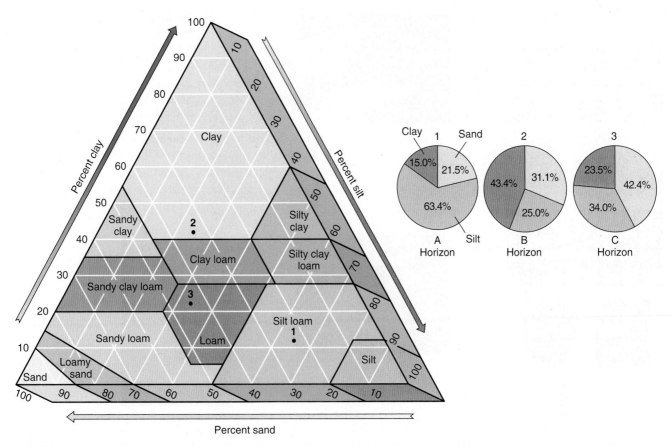

FIGURE 18.4 Soil texture triangle.
Measures of the ratio of clay, silt, and sand determine soil texture. As an example, points 1 (horizon A), 2 (horizon B), and 3 (horizon C) designate samples taken at three different horizons in the Miami silt loam in Indiana. Note the ratio of sand to silt to clay shown in the three pie diagrams. [After U.S. Department of Agriculture, Natural Resources Conservation Service, *Soil Survey Manual,* Agricultural Handbook No. 18, p. 138 (Washington, DC: U.S. Government Printing Office, 1993).]

diagrams in Figure 18.4. Note that silt dominates the surface, clay the B horizon, and sand the C horizon. The *Soil Survey Manual* presents guidelines for estimating soil texture by feel, a relatively accurate method when used by an experienced person. However, laboratory methods using graduated sieves and separation by mechanical analysis in water allow more precise measurements.

Soil Structure

Soil *texture* describes the size of soil particles, but soil *structure* refers to their *arrangement.* Structure can partially modify the effects of soil texture. The smallest natural lump or cluster of particles is a *ped.* The shape of soil peds determines which of the structural types the soil exhibits: crumb or granular, platy, blocky, prismatic, or columnar (Figure 18.5).

Peds separate from each other along zones of weakness, creating voids, or pores, that are important for moisture storage and drainage. Rounded peds have more pore space between them and greater permeability than other shapes. They are therefore better for plant growth than are blocky, prismatic, or platy peds, despite comparable fertility. Terms used to describe soil structure include fine, medium, or coarse. Adhesion among peds ranges from weak to strong.

Soil Consistence

In soil science, the term *consistence* is used to describe the consistency of a soil or cohesion of its particles. Consistence is a product of texture (particle size) and structure (ped shape). Consistence reflects a soil's resistance to breaking and manipulation under varying moisture conditions:

- A *wet soil* is sticky between the thumb and forefinger, ranging from a little adherence to either finger, to sticking to both fingers, to stretching when the fingers are moved apart. *Plasticity*, the quality of being moldable, is roughly measured by rolling a piece of soil between your fingers and thumb to see whether it rolls into a thin strand.

- A *moist soil* is filled to about half of field capacity (the usable water capacity of soil), and its consistence grades from loose (noncoherent), to *friable* (easily pulverized), to firm (not crushable between thumb and forefinger).

- A *dry soil* is typically brittle and rigid, with consistence ranging from loose, to soft, to hard, to extremely hard.

Soil particles are sometimes cemented together, to some degree. Soils are described as weakly cemented, strongly cemented, or *indurated* (hardened). Calcium carbonate, silica, and oxides or salts of iron and aluminum

Crumb or
granular

Platy

Blocky

Prismatic
or columnar

FIGURE 18.5 Types of soil structure.
Structure is important because it controls drainage, rooting of plants, and how well the soil delivers nutrients to plants. The shape of individual peds, shown here, controls a soil's structure. [Photos by National Soil Survey Center, Natural Resources Conservation Service, USDA, Soil Survey Staff.]

all can serve as cementing agents. The cementation of soil particles that occurs in various horizons is a function of consistence and may be continuous or discontinuous.

Soil Porosity

Soil porosity, permeability, and moisture storage are discussed in Chapter 9. Pores in the soil horizon control the movement of water—its intake, flow, and drainage—and air ventilation. Important porosity factors are pore *size*, pore *continuity* (whether they are interconnected), pore *shape* (whether they are spherical, irregular, or tubular), pore *orientation* (whether pore spaces are vertical, horizontal, or random), and pore *location* (whether they are within or between soil peds).

Porosity is improved by the biotic actions of plant roots, animal activity such as the tunneling of gophers or worms, and human intervention through soil manipulation (plowing, adding humus or sand, or planting soil-building crops). Much of a farmer's soil preparation work before planting, and for the home gardener as well, is done to improve soil porosity.

Soil Moisture

Reviewing Figures 9.9 and 9.10 (soil moisture types and availability) will help you understand this section. Plants operate most efficiently when the soil is at *field capacity*, which is the maximum water availability for plant use after large pore spaces have drained of gravitational water. Field capacity is determined by soil type. The depth to which a plant sends its roots determines the amount of

soil moisture to which the plant has access. If soil moisture is removed below field capacity, plants must exert increased energy to obtain available water. This moisture removal inefficiency worsens until the plant reaches its wilting point. Beyond this point, plants are unable to extract the water they need, and they die.

Soil moisture regimes and their associated climate types shape the biotic and abiotic properties of the soil more than any other factor (Table 18.2). The U.S. Natural Resources Conservation Service recognizes five soil moisture regimes based on Thornthwaite's water-balance principles (see "The Soil-Water-Budget Concept" in Chapter 9).

Soil Chemistry

Recall that soil pores may be filled with air, water, or a mixture of the two. Consequently, soil chemistry involves both air and water. The atmosphere within soil pores is mostly nitrogen, oxygen, and carbon dioxide. Nitrogen concentrations are about the same as in the atmosphere, but oxygen is less and carbon dioxide is greater because of ongoing respiration processes.

Water present in soil pores is called the *soil solution*. It is the medium for chemical reactions in soil. This solution is critical to plants as their source of nutrients, and it is the foundation of *soil fertility*. Carbon dioxide combines with the water to produce carbonic acid, and various organic materials combine with the water to produce organic acids. These acids are then active participants in soil processes, as are dissolved alkalies and salts.

Table 18.2   Principal Soil Moisture Regimes

Regime	Description
Aquic (L. *aqua*, "water")	*The groundwater table lies at or near the surface*, so the soil is almost constantly wet, as in bogs, marshes, and swamps, a reducing environment with virtually no dissolved oxygen present. Commonly, groundwater levels fluctuate seasonally; a small borehole will produce standing, stagnant water.
Aridic (torric) (L. *aridis*, "dry," and L. *torridus*, "hot and dry")	*Soils in this regime are dry more than half the time*, with soil temperatures at a depth of 50 cm (19.7 in.) above 5°C (41°F). In some or all areas, soils are never moist for as long as 90 consecutive days. This regime occurs mainly in arid climates, although where surface structure inhibits infiltration and recharge or where soils are thin and therefore dry, this regime occurs in semiarid regions as well.
Udic (L. *udus*, "humid")	*Soils have little or no moisture deficiency throughout the year*, specifically during the growing season. The water balance exhibits soil moisture surpluses that flush through the soils during one season of the year. If the water balance exhibits a moisture surplus in all months of the year, the regime is called *perudic*, with adequate soil moisture always available to plants.
Ustic (L. *ustus*, "burnt," implying dryness)	*This regime is intermediate between aridic and udic regimes;* it includes the semiarid and tropical wet–dry climates. Moisture is available but is limited, with a prolonged deficit period following the period of soil moisture utilization during the early portion of the growing season. Temperature is important in determining the moisture efficiency of available water in this borderline moisture regime.
Xeric (L. *xeros*, "dry")	*This regime applies to those few areas that experience a Mediterranean climate:* dry and warm summer, rainy and cool winter. There are at least 45 consecutive dry days during the 4 months following the summer solstice. Winter rains effectively leach the soils.

Source: U.S. Department of Agriculture, *Soil Taxonomy*, Agricultural Handbook No. 436 (Washington, DC: U.S. Government Printing Office, 1975).

To understand how the soil solution behaves, let us go through a quick chemistry review. An *ion* is an atom, or group of atoms, that carries an electrical charge (examples: Na^+, Cl^-, HCO_3^-). An ion has either a positive charge or a negative charge. For example, when NaCl (sodium chloride) dissolves in solution, it separates into two ions: Na^+, a *cation* (positively charged ion), and Cl^-, an *anion* (negatively charged ion). Some ions in soil carry single charges, whereas others carry double or even triple charges (e.g., sulfate, SO_4^{2-}; and aluminum, Al^{3+}).

Ions in soil are retained by **soil colloids**. These tiny particles of clay and organic material (humus) carry a negative electrical charge and consequently attract any positively charged ions in the soil (Figure 18.6). The positive ions, many metallic, are critical to plant growth. If it were not for the negatively charged soil colloids, the positive ions would be leached away by the soil solution and thus would be unavailable to plant roots.

Individual clay colloids are thin and platelike, with parallel surfaces that are negatively charged (see Figure 18.6). They are more chemically active than silt and sand particles but less active than organic colloids. Metallic cations attach to the surfaces of the colloids by *adsorption* (not *absorption*, which means "to enter"). Colloids can exchange cations between their surfaces and

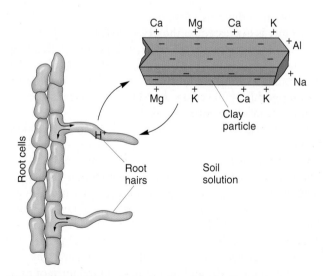

FIGURE 18.6 Soil colloids and cation-exchange capacity (CEC).
This typical soil colloid retains mineral ions by adsorption to its surface (opposite charges attract). This process holds the ions until they are absorbed by root hairs.

ANIMATION   Soil Ion Exchange: Soil Particles and Soil Water

FIGURE 18.7 pH scale.
The pH scale measures acidity (lower pH) and alkalinity (higher pH). (The complete pH scale ranges between 0 and 14.)

the soil solution, an ability called **cation-exchange capacity (CEC)**, which is the measure of soil fertility. A high CEC means that the soil colloids can store or exchange more cations from the soil solution, an indication of good soil fertility (unless there is a complicating factor, such as a soil that is too acid).

Therefore, **soil fertility** is the ability of soil to sustain plants. Soil is fertile when it contains organic substances and clay minerals that absorb water and adsorb certain elements needed by plants. Billions of dollars are expended to create fertile soil conditions, yet the future of Earth's most fertile soils is threatened because soil erosion is on the increase worldwide.

Soil Acidity and Alkalinity

A soil solution may contain significant hydrogen ions (H^+), the cations that stimulate acid formation. The result is a soil rich in hydrogen ions, or an *acid soil*. On the other hand, a soil high in base cations (calcium, magnesium, potassium, sodium) is a *basic* or *alkaline soil*. Such acidity or alkalinity is expressed on the pH scale (Figure 18.7).

Pure water is nearly neutral, with a pH of 7.0. Readings below 7.0 represent increasing acidity. Readings above 7.0 indicate increasing alkalinity. Acidity usually is regarded as strong at 5.0 or lower on the pH scale, whereas 10.0 or above is considered strongly alkaline.

The major contributor to soil acidity in this modern era is acid precipitation (rain, snow, fog, or dry deposition). Acid rain actually has been measured below pH 2.0—an incredibly low value for natural precipitation, as acid as lemon juice. Increased acidity in the soil solution accelerates the chemical weathering of mineral nutrients and increases their depletion rates. Because most crops are sensitive to specific pH levels, acid soils below pH 6.0 require treatment to raise the pH. This soil treatment is accomplished by the addition of bases in the form of minerals that are rich in base cations, usually lime (calcium carbonate, $CaCO_3$).

Soil Formation Factors and Management

Soil is an open system involving physical inputs and outputs. Soil-forming factors are both *passive* (parent material, topography and relief, and time) and *dynamic* (climate, biology, and human activities). These factors work together as a system to form soils. The roles of these factors are considered here and in the soil-order discussions that follow.

Natural Factors

Physical and chemical weathering of rocks in the upper lithosphere provides the raw mineral ingredients for soil formation. These rocks provide the parent materials, and their composition, texture, and chemical nature help determine the type of soil that forms. Clay minerals are the principal weathered by-products in soil.

Climate types correlate closely with soil types worldwide. The moisture, evaporation, and temperature regimes of climates determine the chemical reactions, organic activity, and eluviation rates of soils. Not only is the present climate important but many soils exhibit the imprint of past climates, sometimes over thousands of years. Most notable is the effect of glaciations. Among other contributions, glaciation produced the loess soil materials that have been windblown thousands of kilometers to their present locations (discussed in Chapter 15).

Vegetation, animal, and bacterial activity determine the organic content of soil, along with all that is living in soil—algae, fungi, worms, and insects. The chemical makeup of the vegetation contributes to the acidity or alkalinity of the soil solution. For example, broadleaf trees tend to increase alkalinity, whereas needleleaf trees tend to produce higher acidity. Thus, when civilization moves into new areas and alters the natural vegetation by logging or plowing, the affected soils are likewise altered, often permanently.

Topography also affects soil formation. Slopes that are too steep cannot have full soil development because gravity and erosional processes remove materials. Lands that are nearly level inhibit soil drainage and can become waterlogged. The compass orientation of slopes is important because it controls exposure to sunlight. In the Northern Hemisphere, a south-facing slope is warmer overall through the year because it receives direct sunlight. Water-balance relations are affected because north-facing slopes are colder, causing slower snowmelt and slower evaporation, providing more moisture for plants than is available on south-facing slopes, which tend to dry faster.

News Report 18.1

Soil Is Slipping Through Our Fingers

- The U.S. General Accounting Office estimates that from 3 to 5 million acres of prime farmland are lost each year in the United States through mismanagement or conversion to nonagricultural uses. About half of all cropland in the United States and Canada is experiencing excessive rates of soil erosion—these countries are two of the few that monitor loss of topsoil.

- The Canadian Environmental Advisory Council estimated that the organic content of cultivated prairie soils has declined by as much as 40% compared with non-cultivated native soils. In Ontario and Québec, losses of organic content increased to as much as 50%, and losses are even higher in the Atlantic Provinces, which were naturally low in organic content before cultivation.

- A 1995 study completed at Cornell University concluded that soil erosion is a major environmental threat to the sustainability and productive capacity of agriculture worldwide.

- Since 1950, as much as 38% of the world's farmable land has been lost to soil erosion, and the rate continues at 5 to 6 million hectares (about 12 to 15 million acres) per year—560 million hectares, 1380 million acres, to date (World Resources Institute and UNEP, 1997).

- The causes for degraded soils, in order of severity, include: overgrazing, vegetation removal, agricultural activities, overexploitation, and industrial and bioindustrial use (UNEP, 1997).

- The world's human population is growing at the rate of 6.9 million people a month (net increase), increasing the demand for food and agricultural productivity. Especially significant is that proportion of the global population that is adopting a meat-centered diet as American tastes and food outlets spread worldwide—as we see in the next chapter, meat production is an inefficient use of the grain supply.

All of the identified natural factors in soil development (parent material, climate, biological activity, landforms and topography) require *time* to operate. Over geologic time plate tectonics has redistributed landscapes and thus subjected soil-forming processes to diverse conditions.

The Human Factor

Human intervention has a major impact on soils. Millennia ago, farmers in most cultures learned to plant slopes "on the contour"—to make rows or mounds around a slope at the same elevation, not vertically up and down the slope. Planting on the contour prevents water from flowing straight down the slope and thus reduces soil erosion. It was common to plant and harvest a floodplain but to live on higher ground nearby. Floods were celebrated as blessings that brought water, nutrients, and more soil to the land. Society is drifting away from these common-sense strategies.

A few centimeters' thickness of prime farmland soil may require *500 years* to mature. Yet, this same thickness is being lost annually through soil erosion when the soil-holding vegetation is removed and the land is plowed regardless of topography. Flood control structures block sediments and nutrients from replenishing floodplain soils, leading to additional soil losses. Over the same period, exposed soils may be completely leached of needed cations, thereby losing their fertility. Unlike living species, soils do not reproduce nor can they be re-created.

Some 35% of farmlands are losing soil faster than it can form—a loss exceeding 23 billion metric tonnes (25 billion tons) per year. Soil depletion and loss are at record levels from Iowa to China, Peru to Ethiopia, the Middle East to the Americas. The impact on society is potentially disastrous as population and food demands increase (see News Report 18.1).

Soil erosion can be compensated for in the short run by using more fertilizer, increasing irrigation, and by planting higher-yielding strains. But the potential yield from prime agricultural land will drop by as much as 20% over the next 20 years if only moderate erosion continues. One study tabulated the market value of lost nutrients and other variables in the most comprehensive soil-erosion study to date. The sum of direct damage (to agricultural land) and indirect damage (to streams, society's infrastructure, and human health) was estimated at more than $25 billion a year in the United States and hundreds of billions of dollars worldwide. (Of course, this is a controversial assessment in the agricultural industry.) The cost to bring erosion under control in the United States is estimated at approximately $8.5 billion, or about 30 cents on every dollar of damage and loss. Figure 18.8 maps regions of soil loss.

Our spatial impact on precious soil resources across each agricultural region should be recognized and protective measures taken. There is a need for cooperative international action. Critical to sustaining soil fertility is a long-term assessment of the *cost* of continued behavior, that is, business as usual, compared to the *benefits* of mitigating damage and preserving the market value of prime soils through conservation actions—a cost–benefit analysis. In sustaining natural systems, the benefits of doing things right far outweigh the continued costs of doing things wrong, over the long term.

(a)

(b)

FIGURE 18.8 Soil degradation.
(a) Approximately 1.2 billion hectares (3.0 billion acres) of Earth's soils suffer degradation through erosion caused by human misuse and abuse. (b) Typical loss through soil sheet erosion on a northeastern Wisconsin farm. One millimeter of soil lost from an acre weighs about **5 tons.** [(a) *A Global Assessment of Soil Degradation,* adapted from United Nations Environment Programme, International Soil Reference and Information Centre, "Map of Status of Human-Induced Soil Degradation," Sheet 2, Nairobi, Kenya, 1990; (b) photo by D. P. Burnside/Photo Researchers, Inc.]

Soil Classification

Classification of soils is complicated by the continuing interaction of physical properties and processes just discussed. This interaction creates thousands of distinct soils, with well over 15,000 soil series identified in the United States and Canada alone. Not surprisingly, different classification systems are in use worldwide. The United States, Canada, the United Kingdom, Germany, Australia, Russia, and the United Nations Food and Agricultural Organization each have their own soil classification system. Each system reflects the environment of its country. For example, the National Soil Survey Committee of Canada developed a system suited to its great expanses of boreal forest, tundra, and cool climatic regimes.

Soil Taxonomy

The U.S. soil classification system, *Soil Taxonomy—A Basic System of Soil Classification for Making and Interpreting Soil Surveys,* was published in 1975 and revised in a second edition in 1999. Soil scientists refer to it as **Soil Taxonomy**. Over the years, various revisions and clarifications in the system were published in *Keys to the Soil Taxonomy,* now in its 9th edition (2003, http://soils.usda.gov/technical/classification/tax_keys/), which includes all the revisions to the 1975 Soil Taxonomy. Major revisions include the addition of two new soil orders: Andisols (volcanic soils) in 1990 and Gelisols (cold and frozen soils) in 1998. Much of the information in this chapter is derived from these two keystone publications.

Soil properties and morphology (appearance, form, and structure) actually seen in the field are key to the Soil Taxonomy system. Thus, it is open to addition, change, and modification as the sampling database grows. The system recognizes the importance of interactions between humans and soils and the changes that humans have introduced, both purposely and inadvertently.

The classification system divides soils into six categories, creating a hierarchical sorting system (Table 18.3). The smallest, most detailed category is the soil series, which ideally includes only one polypedon but may include adjoining polypedons. In sequence from smallest

Table 18.3 **U.S. Soil Taxonomy**	
Soil Category	**Number of Soils Included**
Orders	12
Suborders	47
Great groups	230
Subgroups	1,200
Families	6,000
Series	15,000

category to the largest, the Soil Taxonomy recognizes *soil series, soil families, soil subgroups, soil great groups, soil suborders, and soil orders.*

Pedogenic Regimes Prior to the Soil Taxonomy system, **pedogenic regimes** were used to describe soils. These regimes keyed specific soil-forming processes to climatic regions. Although each pedogenic process may be active in several soil orders and in different climates, we discuss them within the soil order where they commonly occur. Such climate-based regimes are convenient for relating climate and soil processes. However, *the Soil Taxonomy system recognizes the great uncertainty and inconsistency in basing soil classification on such climatic variables.* Aspects of several pedogenic processes are discussed with appropriate soil orders:

- **Laterization**: a leaching process in humid and warm climates, discussed with Oxisols
- **Salinization**: a process that concentrates salts in soils in climates with excessive potential evapotranspiration (POTET) rates, discussed with Aridisols
- **Calcification**: a process that produces an illuviated accumulation of calcium carbonates in continental climates, discussed with Mollisols and Aridisols
- **Podzolization**: a process of soil acidification associated with forest soils in cool climates, discussed with Spodosols
- **Gleization**: a process that includes an accumulation of humus and a thick, water-saturated gray layer of clay beneath, usually in cold, wet climates and poor drainage conditions

Diagnostic Soil Horizons

To identify a specific soil series within the Soil Taxonomy, the U.S. Natural Resources Conservation Service describes diagnostic horizons in a pedon. A *diagnostic horizon* reflects a distinctive physical property (color, texture, structure, consistence, porosity, moisture) or a dominant soil process (discussed with the soil types).

In the solum (A, E, and B horizons), two diagnostic horizons may be identified: the epipedon and the subsurface. The presence or absence of either of these diagnostic horizons usually distinguishes a soil for classification.

- The **epipedon** (literally, "over the soil") is the diagnostic horizon at the surface where most of the rock structure has been destroyed. It may extend downward through the A horizon, even including all or part of an illuviated B horizon. It is visibly darkened by organic matter and sometimes is leached of minerals. Excluded from the epipedon are alluvial deposits, eolian deposits, and cultivated areas, because soil-forming processes have lacked the time to erase these relatively short-lived characteristics.
- The **diagnostic subsurface horizon** originates below the surface at varying depths. It may include part of the A or B horizon or both. Many diagnostic subsurface horizons have been identified.

The 12 Soil Orders of the Soil Taxonomy

At the heart of the Soil Taxonomy are 12 general soil orders, listed in Table 18.4. Their worldwide distribution is shown in Figure 18.9 and in individual maps with each description. Please consult this table and these maps as you read the following descriptions. Because the Soil Taxonomy evaluates each soil order on its own characteristics, there is no priority to the classification. However, you will find a progression in this discussion, for the 12 orders are arranged loosely by latitude, beginning with Oxisols along the equator as in Chapters 10 (climates) and 20 (terrestrial biomes).

Oxisols The intense moisture, temperature, and uniform daylength of equatorial latitudes profoundly affect soils. These generally old landscapes, exposed to tropical conditions for millennia or hundreds of millennia, are deeply developed. Soil minerals are highly altered (except in certain newer volcanic soils in Indonesia—the Andisols). Oxisols are among the most mature soils on Earth. Distinct horizons usually are lacking where these soils are well drained (Figure 18.10a). Related vegetation is the luxuriant and diverse tropical and equatorial rain forest. Oxisols include five suborders.

Oxisols (tropical soils) are so called because they have a distinctive horizon of iron and aluminum oxides. The concentration of oxides results from heavy precipitation, which leaches soluble minerals and soil constituents from the A horizon. Typical Oxisols are reddish (from the iron oxide) or yellowish (from the aluminum oxides), with a weathered claylike texture, sometimes in a granular soil structure that is easily broken apart. The high degree of eluviation removes basic cations and colloidal material to lower illuviated horizons. Thus, Oxisols are low in CEC (cation-exchange capacity) and fertility, except in regions augmented by alluvial or volcanic materials.

To have the lush rain forests in the same regions as soils poor in inorganic nutrients seems an irony. However, this forest system relies on the recycling of nutrients from soil organic matter to sustain fertility, although this nutrient recycling ability is quickly lost when the ecosystem is disturbed.

FIGURE 18.9 Soil Taxonomy.
Worldwide distribution of the Soil
Taxonomy's 12 soil orders. [Adapted
from maps prepared by World Soil
Resources Staff, Natural Resources
Conservation Service, USDA, 1999.]

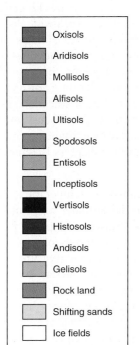

Oxisols
Aridisols
Mollisols
Alfisols
Ultisols
Spodosols
Entisols
Inceptisols
Vertisols
Histosols
Andisols
Gelisols
Rock land
Shifting sands
Ice fields

Consequently, Oxisols have a diagnostic subsurface horizon that is highly weathered, contains iron and aluminum oxides, is at least 30 cm (12 in.) thick, and lies within 2 m (6.5 ft) of the surface (see Figure 18.10). Figure 18.11 illustrates *laterization*, the leaching process that operates in well-drained soils in warm, humid, tropical and subtropical climates. If Oxisols are subjected to repeated wetting and drying, an *ironstone hardpan* develops (this is a hardened soil layer in the lower A or in the B horizon—iron-rich and humus-poor clay with quartz and other minerals). It is called a *plinthite* (from the Greek *plinthos*, meaning "brick"). This form of soil, also called a *laterite*, can be quarried in blocks and used as a building material (Figure 18.12).

Agricultural activities can be conducted with care in these soils. Early cultivation practices, called *slash-and-burn*

shifting cultivation, were adapted to these soil conditions and formed a unique style of crop rotation. The scenario went like this: People in the tropics cut down (slashed) and burned the rain forest in small tracts and then cultivated the land with stick and hoe. Mineral nutrients in the organic material and short-lived fertilizer input from the fire ash would quickly be exhausted. After several years, the soil lost fertility through leaching by intense rainfall, so the people shifted cultivation to another tract and repeated the process. After many years of moving from tract to tract, the group returned to the first patch to begin the cycle again. This practice protected the limited fertility of the soils somewhat, allowing periods of recovery.

The invasion of foreign plantation interests, development by local governments, vastly increased population pressures, and conversion of vast forest tracts to pasturage

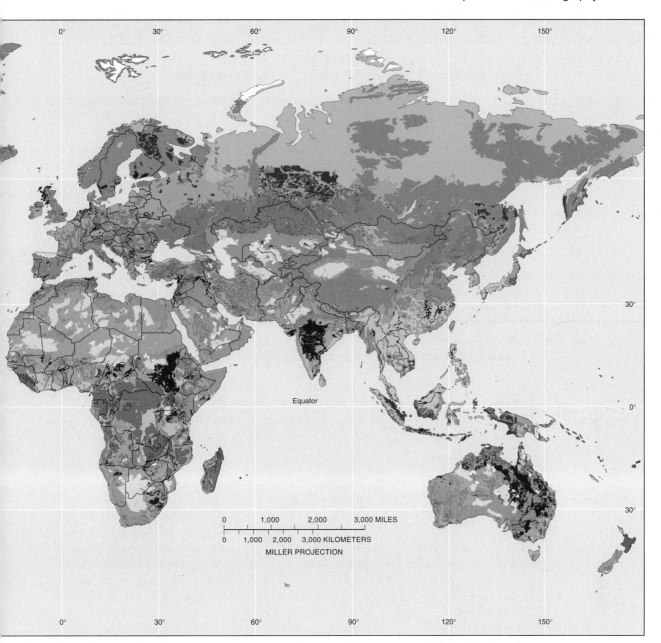

disrupted this orderly land rotation. Permanent tracts of cleared land, taken out of the former rotation mode, put tremendous pressure on the remaining forest and brought disastrous consequences. When Oxisols are disturbed, soil loss can exceed a thousand tons per square kilometer per year, not to mention the greatly increased extinction rate of plant and animal species that accompanies such soil depletion and rain forest destruction. The regions dominated by the Oxisols and rain forests are rightfully the focus of much worldwide environmental attention.

Aridisols   The largest single soil order occurs in the world's dry regions. **Aridisols** (desert soils) occupy approximately 19% of Earth's land surface and some 12% of U.S. land (see Figure 18.9). A pale, light soil color near the surface is diagnostic (Figure 18.13a).

Not surprisingly, the water balance in Aridisol regions has periods of soil-moisture deficit and generally inadequate soil moisture for plant growth. High potential evapotranspiration and low precipitation produce very shallow soil horizons. Usually there is no period greater than 3 months when the soils have adequate moisture. Lacking water and therefore lacking vegetation, Aridisols also lack organic matter of any consequence. Low precipitation means infrequent leaching, yet Aridisols are leached easily when exposed to excessive water, for they lack a significant colloidal structure.

Salinization is common in Aridisols, resulting from excessive potential evapotranspiration rates in deserts and semiarid regions. Salts dissolved in soil water migrate to surface horizons and are deposited there as the water evaporates. These deposits appear as subsurface salty

Table 18.4 Soil Taxonomy Soil Orders

Order	Derivation of Term	Marbut, 1938* (Canadian System**)	General Location and Climate	Description
Oxisols	Fr. *oxide*, "oxide" Gr. *oxide*, "acid or sharp"	Latosols lateritic soils	Tropical soils; hot, humid areas	Maximum weathering of Fe and Al and eluviation, continuous plinthite layer
Aridisols	L. *aridos*, "dry"	Reddish desert, gray desert, sierozems	Desert soils; hot, dry areas	Limited alteration of parent material, low climate activity, light color, low humus content, subsurface illuviation of carbonates
Mollisols	L. *mollis*, "soft"	Chestnut, chernozem (Chernozemic)	Grassland soils; subhumid, semiarid lands	Noticeably dark with organic material, humus rich, base saturation, high, friable surface with well-structured horizons
Alfisols	Invented syllable	Gray-brown podzolic, degraded chernozem (Luvisol)	Moderately weathered forest soils; humid temperate forests	B horizon high in clays, moderate to high degree of base saturation, illuviated clay accum., no pronounced color change with depth
Ultisols	L. *ultimus*, "last"	Red-yellow podzolic, reddish yellow lateritic	Highly weathered forest soils; subtropical forests	Similar to Alfisols, B horizon high in clays, generally low amount of base saturation, strong weathering in subsurface horizons, redder than Alfisols
Spodosols	Gr. *spodos* or L. *spodus*, "wood ash"	Podzols, brown podzolic (Podzol)	Northern conifer forest soils; cool, humid forests	Illuvial B horizon of Fe/Al clays, humus accum., without structure, partially cemented, highly leached, strongly acid, coarse texture of low bases
Entisols	Invented syllable from *recent*	Azonal soils, tundra	Recent soils, profile undeveloped, all climates	Limited development, inherited properties from parent material, pale color, low humus, few specific properties, hard and massive when dry
Inceptisols	L. *inceptum*, "beginning"	Ando, subarctic brown forest lithosols, some humic gleys (Brunisol, Cryosol with permafrost, Gleysol wet)	Weakly developed soils; humid regions	Intermediate development, embryonic soils but few diagnostic features, further weathering possible in altered or changed subsurface horizons
Vertisols	L. *verto*, "to turn"	Grumusols (1949) tropical black clays	Expandable clay soils; subtropics, tropics; sufficient dry period	Forms large cracks on drying, self-mixing action, contains >30% in swelling clays, light color, low humus content
Histosols	Gr. *histos*, "tissue"	Peat, muck, bog (Organic)	Organic soils, wet places	Peat or bog, >20% organic matter, much with clay >40 cm thick, surface, organic layers, no diagnostic horizons
Andisols	L. *ando*, "volcanic ash"	—	Areas affected by frequent volcanic activity (formerly within Inceptisols and Entisols)	Volcanic parent materials, particularly ash and volcanic glass; weathering and mineral transformation important; high CEC and organic content, generally fertile
Gelisols	L. *gelatio*, "freezing"	Formerly Inceptisols and Entisols (Cryosols, some Brunisols)	High latitudes in Northern Hemisphere, southern limits near tree line	Permafrost within 100 cm of the soil surface; evidence of cryoturbation (frost churning) and/or an active layer; patterned-ground

* C.F. Marbut, USDA soil scientist, developed the first American system of soil classification in the 1930s, first published in the USDA *Yearbook of Agriculture* in 1938.

** For comparison, soil orders from the Canadian System of Soil Classification (CSSC), last revised in 1998.

(a)

FIGURE 18.10 Oxisols.
(a) Deeply weathered Oxisol profile in central Puerto Rico.
(b) General map showing worldwide distribution of these
tropical soils. [Photo from the Marbut Collection, Soil Science
Society of America.]

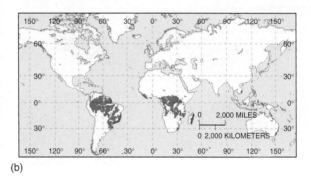

(b)

Warm and humid climates

Very little
organic
matter

Oxic
horizon

A Residual iron
and aluminum
–silica (SiO$_2$)
removed

B Accumulations
of iron and
aluminum
–plinthite
(ironstone)
formation

C Much soluble
material to
water table

Water table

To
streams

FIGURE 18.11 Laterization.
The laterization process is characteristic of moist tropical
and subtropical climate regimes.

FIGURE 18.12 Oxisols are used for bricks.
Here plinthite is being quarried in India for building material.
[Photo by Henry D. Foth.]

(a)

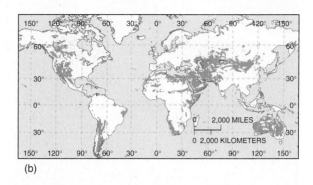

(b)

FIGURE 18.13 Aridisols.
(a) Soil profile from central Arizona. (b) General map showing worldwide distribution of these desert soils. [Photo from the Marbut Collection, Soil Science Society of America.]

horizons, which will damage or kill plants when the horizons occur near the root zone. The salt accumulation associated with a desert playa is an example of extreme salinization (see Chapter 15).

Obviously, salinization complicates farming in Aridisols. The introduction of irrigation water may either waterlog poorly drained soils or lead to salinization. Nonetheless, vegetation does grow where soils are well drained and low in salt content. If large capital investments are made in water, drainage, and fertilizers, Aridisols possess much agricultural potential (Figure 18.14). In the Nile and Indus River valleys, for example, Aridisols are intensively farmed with a careful balance of these environmental factors, although thousands of acres of once-productive land, not so carefully treated, now sit idle and salt-encrusted. In California, the Kesterson Wildlife Refuge was reduced to a toxic waste dump in the early 1980s by contaminated agricultural drainage. Focus Study 18.1 elaborates on the Kesterson tragedy. News Report 18.2 describes efforts to deal with salinization.

FIGURE 18.14 Agriculture in arid lands.
The Coachella Valley of southeastern California is a desert in agricultural bloom; asparagus is growing in the foreground. Providing drainage for excessive water application in such areas often is necessary to prevent salinization of the rooting zone. [Photo by author.]

Selenium Concentration in Western Soils

Irrigated agriculture has increased greatly since 1800, when only 8 million hectares (about 20 million acres) were irrigated worldwide. Today, approximately 255 million hectares (about 630 million acres) are irrigated, and this figure is on the increase (USDA, FAO, 1996). Representing about 16% of Earth's agricultural land, irrigated land accounts for nearly 36% of the harvest.

Two related problems common in irrigated lands are salinization and water logging, especially in arid lands that are poorly drained. In many areas, production has decreased and even ended because of salt buildup in the soils. Examples include areas along the Tigris and Euphrates Rivers, the Indus River valley, sections of South America and Africa, and the western United States.

Irrigation in the West

About 95% of the irrigated acreage in the United States lies west of the 98th meridian. This region is increasingly troubled with salinization and waterlogging problems. In addition, at least nine sites in the West, particularly California's western San Joaquin Valley, are experiencing related contamination of a more serious nature—increasing selenium concentrations. Toxic effects of selenium were reported during the 1980s in some domestic animals grazing on grasses grown in selenium-rich soils in the Great Plains. In California, as parent materials weathered, selenium-rich alluvium washed into the semiarid valley, forming the soils that needed only irrigation water to become productive.

Drainage of agricultural wastewater poses a particular problem in semiarid and arid lands, where river discharge is inadequate to dilute and remove field runoff. One solution to prevent salt accumulations and water-clogged soils is to place field drains beneath the soil to collect gravitational water from fields that have been purposely over watered to keep

salts away from the effective rooting depth of the crops. But agricultural drain water must go somewhere, and for the San Joaquin Valley of central California this problem triggered a 15-year controversy.

Death of Kesterson

Central California's potential drain outlets are to the ocean, San Francisco Bay, or the central valley. But all these suggested destinations failed to pass environmental-impact assessments under Environmental Policy Act requirements. Nonetheless, by the late 1970s, about 80 miles of a drain were finished, even though no formal plan or adequate funding had been completed—a drain was built with no outlet! In the absence of any plan, large-scale irrigation continued, supplying the field drains with salty, selenium-laden runoff that made its way to the Kesterson National Wildlife Refuge in the northern portion of the San Joaquin Valley east of San Francisco. The unfinished drain abruptly stopped at the boundary to the refuge.

The selenium-tainted drainage took only 3 years to destroy the wildlife refuge, which was officially declared a contaminated toxic waste site (Figure 18.1.1). Aquatic life forms (e.g., marsh plants, plankton, and insects) had taken in the selenium, which then made its way up the food chain and into the diets of higher life forms in the refuge. According to

FIGURE 18.1.1 Soil contamination in the wildlife refuge.
Salt-encrusted soil and plants at the contaminated Kesterson National Wildlife Refuge. [Photo by Gary R. Zahm/DRK Photo.]

U.S. Fish and Wildlife Service scientists, the toxicity moved through the food chain and genetically damaged and killed wildlife, including all varieties of birds that nested at Kesterson; approximately 90% of the exposed birds perished or were injured. Because this wildlife refuge was a major migration flyway and stopover point for birds from throughout the Western Hemisphere, the destruction of the refuge also violated several multinational wildlife protection treaties.

The field drains were sealed and removed in 1986, following a court order that forced the federal government to uphold existing laws. Irrigation water then immediately began backing up in the corporate farmlands, producing both waterlogging and selenium contamination.

Since 1985, more than 0.6 million hectares (1.5 million acres) of irrigated Aridisols and Alfisols have gone out of production in California, marking the end of several decades of irrigated farming in climatically marginal lands. Severe cutbacks in irrigated acreage no doubt will continue, underscoring the need to preserve prime farmlands in wetter regions and to understand essential soil processes.

Frustrated agricultural interests have asked the federal government to finish the drain, either to San Francisco Bay or to the ocean. However, neither option appears capable of passing an environmental-impact analysis. Another strategy is to allow irrigated lands to start pumping again into the former wildlife refuge because it is now a declared toxic dump site anyway! There are nine such threatened sites in the West; Kesterson was simply the first of these to fail from contamination. Meanwhile, the perimeter fence surrounding portions of Kesterson is posted with signs that read, "Area Beyond This Sign CLOSED, All Public Access Prohibited!" Such damage to a wildlife refuge presents a real warning to human populations—remember where we are at the top of the food chain.

News Report

News Report 18.2

Drainage Tiles, But Where to Go?

In the California valleys of Imperial, Coachella, and San Joaquin, and in the Wellton–Mohawk district of extreme southern Arizona, the fields have been excavated, drainage tiles (perforated ceramic pipe) laid in place, and the fields replaced. The field drains carry away excess irrigation water, which has been purposely applied to leach salts away from plant roots. The problem has been to find a place to dump the salt- and chemical-laden drainage water.

In the Imperial and Coachella Valleys, the drainage water was sent to the Salton Sea (Figure 18.2.1)—a large inland sea formed by an irrigation accident in 1906 (a broken canal levee allowed the Colorado River to flow into the desert of southern

California for 18 months). For the Wellton–Mohawk area, the contaminated drainage water was pumped back into the Colorado River just above the Mexican border. By the late

FIGURE 18.2.1 Fields drain into canals. Soil drainage canal collects contaminated water from field drains and directs it into the Salton Sea. [Photo by author.]

1950s, salty Colorado River water was killing every plant it touched in Mexico.

Today, the Wellton–Mohawk contaminated drainage water passes through a bypass drain near Yuma, Arizona. A desalination plant went into brief service in late 1993, pumping desalted water back to the river and saline sludge directly to the ocean, but it was shut down after less than a year of service. With the Western drought, there is serious discussion about reactivating the plant in 2005. The irony is that it would be cheaper for the government to buy the subsidized irrigated orchards and fields and allow the land to revert to desert than it would be to operate the $484 million desalinization works.

Mollisols Some of Earth's most significant agricultural soils are **Mollisols** (grassland soils). There are seven recognized suborders, which vary in fertility. The dominant diagnostic horizon is a dark, organic surface layer some 25 cm (10 in.) thick (Figure 18.15). As the Latin name implies (words using the same root, *mollis*, are "mollify" and "emollient"), Mollisols are soft, even when dry. They have granular or crumbly peds, loosely arranged when dry. These humus-rich soils are high in basic cations (calcium, magnesium, and potassium) and have a high cation-exchange capacity and therefore, high fertility. In soil moisture, these soils are intermediate between humid and arid.

Soils of the world's steppes and prairies belong to this soil group: the North American Great Plains, the Pampas of Argentina, and the region from Manchuria in China westward to Europe. Agriculture ranges from large-scale commercial grain farming to grazing along the drier portions of the soil order. With fertilization or soil-building practices, high crop yields are common. The "fertile triangle" of Ukraine, Russia, and western portions of the Russian Federation is of this soil type. The region stretches from north of the Caspian Sea and Kazakstan westward in a widening triangle toward Central Europe (Figure 18.16). The soils in the fertile triangle are known as the *chernozem* in other classification systems. The remainder of the Russian landscape presents varied soils of lower fertility and productivity.

In North America, the Great Plains straddle the 98th meridian, which is coincident with the 51-cm (20-in.)

isohyet of annual precipitation—wetter to the east and drier to the west. The Mollisols here mark the historic division between the short- and tall-grass prairies (Figure 18.17).

Calcification is a soil process characteristic of some Mollisols and adjoining marginal areas of Aridisols. Calcification is the accumulation of calcium carbonate or magnesium carbonate in the B and C horizons. Calcification by calcium carbonate ($CaCO_3$) forms a diagnostic subsurface horizon that is thickest along the boundary between dry and humid climates (Figure 18.18).

When cemented or hardened, these deposits are called *caliche*, or *kunkur*; they occur in widespread soil formations in central and western Australia, the Kalahari region of interior Southern Africa, and the High Plains of the west-central United States, among other places. (Soils dominated by the processes of calcification and salinization were formerly known as pedocals.)

Alfisols Spatially the most widespread of the soil orders are **Alfisols** (moderately weathered forest soils), extending in five suborders from near the equator to high latitudes. Representative Alfisol areas include Boromo and Burkina Faso (interior western Africa); Fort Nelson, British Columbia; the states near the Great Lakes; and the valleys of central California. Most Alfisols are grayish brown to reddish and are considered moist versions of the Mollisol soil group. Moderate eluviation is present, as well as a subsurface horizon of illuviated clays and clay formation because of a pattern of increased precipitation (Figure 18.19).

(a)

(b)

(c)

(d)

FIGURE 18.15 Mollisols.
(a) Soil profile from central Iowa. (b) General map showing worldwide distribution of these grassland soils. (c) Mollisols in northwestern Iowa. (d) In eastern Washington State, wheat flourishes in the Palouse. These fertile Mollisols will be increasingly valuable in the future. [Photos (a) from the Marbut Collection, Soil Science Society of America; (c) and (d) by Bobbé Christopherson.]

Alfisols have moderate to high reserves of basic cations and are fertile. However, productivity depends on moisture and temperature. Alfisols usually are supplemented by a moderate application of lime and fertilizer in areas of active agriculture. Some of the best farmland in the United States stretches from Illinois, Wisconsin, and eastern Minnesota through Indiana, Michigan, and Ohio to Pennsylvania and New York (see Figure 18.17c). This land produces grains, hay, and dairy products. The soil here is an Alfisol subgroup called Udalfs, which are characteristic of humid continental, hot summer climates.

The Xeralfs, another Alfisol subgroup, are associated with the moist winter, dry summer pattern of the Mediterranean climate. These naturally productive soils are farmed intensively for subtropical fruits, nuts, and special

crops that can grow in only a few locales worldwide (e.g., California grapes, olives, citrus, artichokes, almonds, figs).

Ultisols   Farther south in the United States are the **Ultisols** (highly weathered forest soils) and their five suborders. An Alfisol might degenerate into an Ultisol, given time and exposure to increased weathering under moist conditions. These soils tend to be reddish because of residual iron and aluminum oxides in the A horizon (Figure 18.20).

The relatively high precipitation in Ultisol regions causes greater mineral alteration and more eluvial leaching than in other soils. Therefore, the level of basic cations is lower and the soil fertility is lower. Fertility is further reduced by certain agricultural practices and the

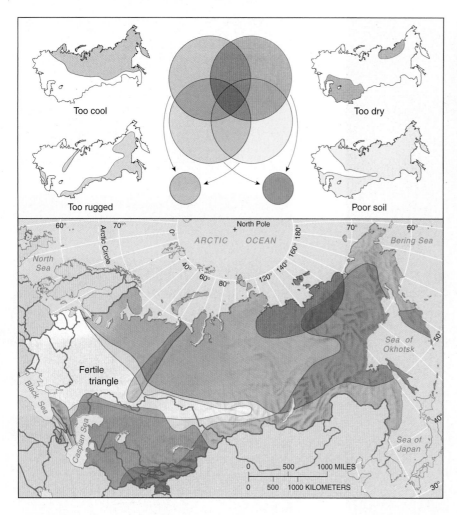

FIGURE 18.16 Soil fertility in Eurasia. Top: Four main factors influence soil fertility. Bottom: The fertile triangle of central Eurasia. The white area is highest in soil fertility (no negative factors). Areas shown with any color or color combination are limited for the reasons shown. The least-fertile area in Russia is northeastern Siberia. [After P. W. English and J. A. Miller, *World Regional Geography: A Question of Place*, p. 183, ©1989. Adapted by permission of John Wiley & Sons, Inc.]

effect of soil-damaging crops such as cotton and tobacco, which deplete nitrogen and expose soil to erosion. These soils respond well if subjected to good management—for example, crop rotation restores nitrogen, and certain cultivation practices prevent sheetwash and soil erosion. Peanut plantings assist in nitrogen restoration in these soils. Much needs to be done to achieve sustainable management of these soils.

Spodosols   The **Spodosols** (northern coniferous forest soils) and their four suborders occur generally to the north and east of the Alfisols. They are in cold and forested moist regimes (*humid continental mild summer* climates) in northern North America and Eurasia, Denmark, The Netherlands, and southern England. Because there are no comparable climates in the Southern Hemisphere, this soil type is not identified there. Spodosols form from sandy parent materials, shaded under evergreen forests of spruce, fir, and pine. Spodosols with more moderate properties form under mixed or deciduous forests (Figure 18.21).

Spodosols lack humus and clay in the A horizon. An eluviated horizon of sandy and leached clays and iron lies in the A horizon, above a B horizon of illuviated organic matter and iron and aluminum oxides (Figure 18.21c). The surface horizon receives organic litter from base-poor, acid-rich evergreen trees, which contribute to acid accumulations in the soil. The solution in acidic soils effectively leaches clays, iron, and aluminum, which are passed to the upper diagnostic horizon. An ashen-gray color is common in these subarctic forest soils and is characteristic of a formation process called *podzolization*. In the Canadian system, Spodosols fall within the Podzolic Great Group, as seen in the temperate rain forests of Vancouver Island, British Columbia, in Figure 18.21d and the rich forest soils of south-central Norway in Figure 18.21e.

When agriculture is attempted, the low basic cation content of Spodosols requires the addition of nitrogen, phosphate, and potash (potassium carbonate), and perhaps crop rotation as well. A soil *amendment* such as limestone can significantly increase crop production by raising the pH of these acidic soils. For example, the yields of several crops (corn, oats, wheat, and hay) grown in specific Spodosols in New York State were increased up to a third with the application of 1.8 metric tonnes (2 tons) of limestone per 0.4 hectare (1.0 acre) during each 6-year rotation.

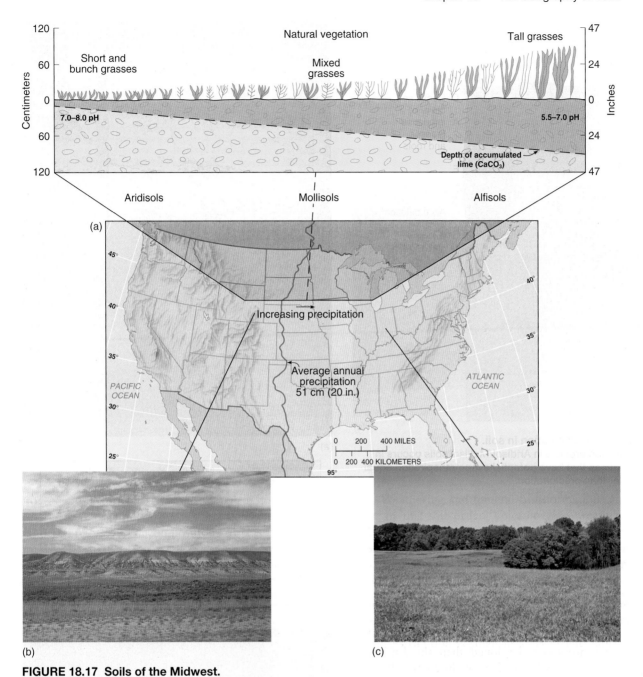

FIGURE 18.17 Soils of the Midwest.
(a) Aridisols (to the west), Mollisols (central), and Alfisols (to the east)—a soil continuum in the
north-central United States and southern Canadian prairies. Graduated changes that occur in
soil pH and the depth of accumulated lime are shown. (b) Bunch grasses and shallow soils of
Wyoming. (c) Farmlands and Alfisols south of Bedford, Indiana. [(a) Illustration adapted from N. C.
Brady, *The Nature and Properties of Soils,* 10th ed., © 1990 by Macmillan Publishing Company,
adapted by permission; photos by (b) author; and (c) Bobbé Christopherson.]

Entisols   The **Entisols** (recent, undeveloped soils) lack
vertical development of their horizons. The five suborders of this soil group are based on differences in parent
materials and climatic conditions, although the presence
of Entisols is not climate-dependent, for they occur in
many climates worldwide. Entisols are true soils, but
they have not had sufficient time to generate the usual
horizons.

Entisols generally are poor agricultural soils, although
those formed from river silt deposits are quite fertile. The
same conditions that inhibit complete development also
prevent adequate fertility—too much or too little water,
poor structure, and insufficient accumulation of weathered nutrients. Active slopes, alluvium-filled floodplains,
poorly drained tundra, tidal mudflats, dune sands and erg
(sandy) deserts, and plains of glacial outwash all are

(a)

(b)

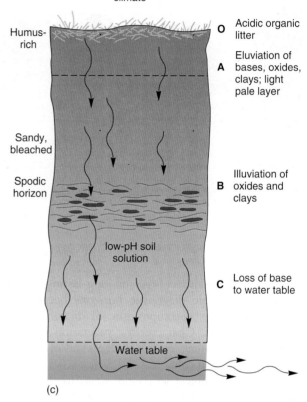

(c)

FIGURE 18.21 Spodosols.
(a) Soil profile from northern New York. (b) General map showing worldwide distribution of these northern coniferous forest soils. (c) The podzolization process is typical in cool and moist climatic regimes. (d) Characteristic temperate forest and Spodosols in the cool, moist climate of central Vancouver Island. (e) Spodosols and farms growing feed grains in central **Norway along a fjord.** [Photos (a) from the Marbut Collection, Soil Science Society of America; (d) and (e) by Bobbé Christopherson.]

(d)

(e)

FIGURE 18.22 Entisols.
A characteristic Entisol: young, undeveloped soil forming from a shale parent material in the desert near Zabriskie Point in Death Valley, California.
[Photo by author.]

Vertisol clays are black when wet, but not because of organics: Rather, the blackness is due to specific mineral content. They range from brown to dark gray. These deep clays swell when moistened and shrink when dried. In the drying process, vertical cracks may form, as wide as 2–3 cm

(0.8–1.2 in.) and up to 40 cm (16 in.) deep. Loose material falls into these cracks, only to disappear when the soil again expands and the cracks close. After many such cycles, soil contents tend to invert or mix vertically, bringing lower horizons to the surface (Figure 18.25).

(a)

(b)

(c)

FIGURE 18.23 Gelisols.
(a) General map showing worldwide distribution of these cold and frozen soils.
(b) Gelisols and tundra green in the brief summer season on Spitsbergen Island.
(c) Turning over the upper organic layer exposes the fibrous organic content and slow decomposition in this Arctic Ocean location. [Photos by Bobbé Christopherson.]

(a)

(b)

FIGURE 18.24 Andisols in agricultural production.
Fertile Andisols planted with sugar cane on Kaua'i, Hawai'i. The 50th state was the second-largest U.S.
producer of sugar cane after Florida in 1998. (b) Pasture and feed grains for raising sheep grow in
Andisols in southwestern Iceland. [Photos by (a) Wolfgang Kaehler Photography; (b) Bobbé Christopherson.]

(a)

(b)

(c)

FIGURE 18.25 Vertisol.
(a) Soil profile in the Lajas Valley of Puerto Rico. (b) General map
showing worldwide distribution of these expandable clay soils.
(c) Vertisols in the Texas coastal plain, northeast of Palacios near
the Tres Palacios River, planted with a commercial sorghum crop.
Note the dark soil color indicative of Vertisols, wet and shiny from
the rains of Tropical Storm Allison, June 2001. [Photos (a) from the
Marbut Collection, Soil Science Society of America; and (c) by Bobbé
Christopherson.]

(a)

FIGURE 18.26 Histosols.
(a) A bog in coastal Maine, near Popham Beach State Park. (b) General map showing worldwide distribution of these organic soils. [Photo by Bobbé Christopherson.]

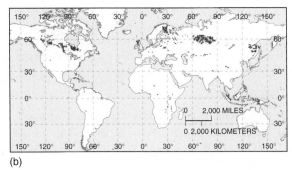

(b)

Despite the fact that clay soils are plastic and heavy when wet, with little available soil moisture for plants, Vertisols are high in bases and nutrients and thus are some of the better farming soils where they occur. For example, they occur in a narrow zone along the coastal plain of Texas (Figure 18.25c) and in a section along the Deccan region of India. Vertisols often are planted with grain sorghums, corn, and cotton.

Histosols Accumulations of thick organic matter can form **Histosols** (organic soils), including four suborders.

In the midlatitudes, when conditions are right, beds of former lakes may turn into Histosols, with water gradually replaced by organic material to form a bog and layers of peat (Figure 18.26). (Lake succession and bog or marsh formation are discussed in Chapter 19.) Histosols also form in small, poorly drained depressions, with conditions ideal for significant deposits of sphagnum peat to form. This material can be cut, baled, and sold as a soil amendment. In some locales, dried peat has served for centuries as a low-grade fuel.

Summary and Review—The Geography of Soils

▶ *Define* soil and soil science, and *describe* a pedon, polypedon, and typical soil profile.

Soil is the portion of the land surface in which plants can grow. It is a dynamic natural body composed of fine materials and contains both mineral and organic matter. **Soil science** is the interdisciplinary study of soils involving physics, chemistry, biology, mineralogy, hydrology, taxonomy, climatology, and cartography. *Pedology* deals with the origin, classification, distribution, and description of soil. *Edaphology* specifically focuses on the study of soil as a medium for sustaining the growth of plants.

The basic sampling unit used in soil surveys is the **pedon**. The **polypedon** is the soil unit used to prepare local soil maps and may contain many pedons. Each discernible layer in an exposed pedon is a **soil horizon**. The horizons are designated O (contains **humus**, a complex mixture of decomposed and synthesized organic materials), A (rich in humus and clay, darker), E (zone of **eluviation**, the removal of fine particles and minerals by water), B (zone of **illuviation**, the deposition of clays and minerals translocated from elsewhere), C (*regolith*, weathered bedrock), and R (bedrock). Soil horizons A, E, and B experience the most active soil processes and together are designated the **solum**.

soil (p. 577)
soil science (p. 577)
pedon (p. 578)
polypedon (p. 579)
soil horizon (p. 579)
humus (p. 579)
eluviation (p. 579)
illuviation (p. 579)
solum (p. 579)

1. Soils provide the foundation for animal and plant life and therefore are critical to Earth's ecosystems. Why is this true?

2. What are the differences among soil science, pedology, and edaphology?

3. Define polypedon and pedon, the basic units of soil.

4. Characterize the principal aspects of each soil horizon. Where does the main accumulation of organic material occur? Where does humus form? Explain the difference between the eluviated layer and the illuviated layer. Which horizons constitute the solum?

◆ *Describe* soil properties of color, texture, structure, consistence, porosity, and soil moisture.

We use several physical properties to classify soils. Color suggests composition and chemical makeup. Soil texture refers to the size of individual mineral particles and the proportion of different sizes. For example, **loam** is a balanced mixture of sand, silt, and clay. Soil structure refers to the arrangement of soil peds, which are the smallest natural cluster of particles in a soil. The cohesion of soil particles to each other is called soil consistence. Soil porosity refers to the size, alignment, shape, and location of spaces in the soil. Soil moisture refers to water in the soil pores and its availability to plants.

loam (p. 580)

5. Soil color is identified and compared using what technique?
6. Define a soil separate. What are the various sizes of particles in soil? What is loam? Why is loam regarded so highly by agriculturists?
7. What is a quick, hands-on method for determining soil consistence?
8. Summarize the five soil-moisture regimes common in mature soils.

◆ *Explain* basic soil chemistry, including cation-exchange capacity, and *relate* these concepts to soil fertility.

Particles of clay and organic material form negatively charged **soil colloids** that attract and retain positively charged mineral ions in the soil. The capacity to exchange ions between colloids and roots is an ability called the **cation-exchange capacity (CEC).** CEC is a measure of **soil fertility,** the ability of soil to sustain plants. Fertile soil contains organic substances and clay minerals that absorb water and retain certain elements needed by plants.

soil colloids (p. 583)
cation-exchange capacity (CEC) (p. 584)
soil fertility (p. 584)

9. What are soil colloids? How are they related to cations and anions in the soil? Explain cation-exchange capacity.
10. What is meant by the concept of soil fertility?

◆ *Evaluate* the principal soil formation factors, including the human element.

Environmental factors that affect soil formation include parent materials, climate, vegetation, topography, and time. Human influence is having great impact on Earth's prime soils. Essential soils for agriculture and their fertility are threatened by mismanagement, destruction, and conversion to other uses. Much soil loss is preventable through the application of known technologies, improved agricultural practices, and government policies.

11. Briefly describe the contribution of the following factors and their effect on soil formation: parent material, climate, vegetation, landforms, time, and humans.
12. Explain some of the details that support the concern over loss of our most fertile soils. What cost estimates have been placed on soil erosion?

◆ *Describe* the 12 soil orders of the Soil Taxonomy classification system, and *explain* their general occurrence.

The **Soil Taxonomy** classification system is used in the United States and is built around an analysis of various diagnostic hori-

zons and 12 soil orders, as actually seen in the field. The system divides soils into six hierarchical categories: series, families, subgroups, great groups, suborders, and orders.

Specific soil-forming processes keyed to climatic regions (not a basis for classification) are called **pedogenic regimes: laterization** (leaching in warm and humid climates), **salinization** (collection of salt residues in surface horizons in hot, dry climates), **calcification** (accumulation of carbonates in the B and C horizons in drier continental climates), **podzolization** (soil acidification in forest soils in cool climates), and **gleization** (humus and clay accumulation in cold, wet climates with poor drainage).

The Soil Taxonomy system uses two diagnostic horizons to identify soil: the **epipedon,** or the surface soil, and the **diagnostic subsurface horizon,** or the soil below the surface at various depths. (For an overview and definition of the 12 soil orders, please refer to Table 18.4.) The 12 soil orders are **Oxisols** (tropical soils), **Aridisols** (desert soils), **Mollisols** (grassland soils), **Alfisols** (moderately weathered, temperate forest soils), **Ultisols** (highly weathered, subtropical forest soils), **Spodosols** (northern conifer forest soils), **Entisols** (recent, undeveloped soils), **Inceptisols** (weakly developed, humid-region soils), **Gelisols** (cold soils underlain by permafrost), **Andisols** (soils formed from volcanic materials), **Vertisols** (expandable clay soils), and **Histosols** (organic soils).

Soil Taxonomy (p. 586)
pedogenic regimes (p. 587)
laterization (p. 587)
salinization (p. 587)
calcification (p. 587)
podzolization (p. 587)
gleization (p. 587)
epipedon (p. 587)
diagnostic subsurface horizon (p. 587)
Oxisols (p. 587)
Aridisols (p. 589)
Mollisols (p. 594)
Alfisols (p. 594)
Ultisols (p. 595)
Spodosols (p. 596)
Entisols (p. 597)
Inceptisols (p. 598)
Gelisols (p. 598)
Andisols (p. 599)
Vertisols (p. 599)
Histosols (p. 603)

13. Summarize the basis of soil classification described in this chapter. What have soil scientists revised in the latest Soil Taxonomy classification system?
14. What is the basis of the Soil Taxonomy system? How many orders, suborders, great groups, subgroups, families, and soil series are there?
15. Define an epipedon and a diagnostic subsurface horizon. Give a simple example of each.
16. Locate each soil order on the world map and on the U.S. map as you give a general description of it.
17. How was slash-and-burn shifting cultivation, as practiced in the past, a form of crop and soil rotation and conservation of soil properties?
18. Describe the salinization process in arid and semiarid soils. What associated soil horizons develop?

19. Which of the soil orders are associated with Earth's most productive agricultural areas?

20. What is the significance to plants of the 51-cm (20-in.) isohyet in the Midwest relative to soils, pH, and lime content?

21. Describe the podzolization process associated with northern coniferous forest soils. What characteristics are associated with the surface horizons? What strategies might enhance these soils?

22. What former Inceptisols now form a new soil order? Describe these soils as to location, nature, and formation processes. Why do you think they were separated into their own order?

23. Why has a selenium contamination problem arisen in western U.S. soils? Explain the impact of agricultural practices, and tell why you think this is or is not a serious problem.

 NetWork

The *Geosystems* Home Page provides online resources for this chapter on the World Wide Web. To begin: Once on the Home Page, click on this textbook, scroll the Table of Contents menu, select this chapter, and click "Begin." You will find self-tests that are graded, review exercises, specific updates for items in the chapter, and many links to interesting related pathways on the Internet under "Destinations." *Geosystems* is at **http://www. prenhall.com/christopherson.**

Critical Thinking

A. Select a small soil sample from your campus or near where you live. Using the sections in this chapter on soil characteristics, properties, and formation, describe as completely as possible this sample, within these limited constraints. Using the general soil map and any other sources available (e.g., local agriculture extension agent, Internet, related department on campus), are you able to roughly place this sample in one of the soil orders?

B. Using the local phone directory or Internet, see if you can locate an agency or extension agent that provides information about local soils and advice for soil management. Is there a place where you can get soil tested?

C. Refer to "Critical Thinking" for Chapter 18 on the *Geosystems* Home Page item 1. Please complete the analysis of the three photographs for environments of deciduous forests of eastern North America, through evergreen forests, and into the arctic tundra. What soil order would you expect in these regions? Why?

A Rockhopper penguin colony perches on the rocky coast of New Island, Falkland Islands, in the South Atlantic (52° S, 62° W). The Rockhopper (*Eudyptes chrysocome*) feed on krill, squid, octopus, and fish in the ocean and then climb these cliffs to nest and lay eggs and fledge their chicks. The photo was made on December 12th, at the beginning of this process. The pairs mate for life. The Antarctic (brown) Skua (*Catharacta antarctica,* upper-left inset) is a predator and scavenger that watches over the colony waiting for a chance to steal an egg or pluck a chick from a nest. [Photo by Bobbé Christopherson.]

CHAPTER 19

Ecosystem Essentials

▶ Key Learning Concepts

After reading the chapter, you should be able to:

▶ **Define** ecology, biogeography, and the ecosystem concept.

▶ **Describe** communities, habitats, and niches.

▶ **Explain** photosynthesis and respiration, and **derive** net photosynthesis and the world pattern of net primary productivity.

▶ **List** abiotic ecosystem components, and **relate** those components to ecosystem operations.

▶ **Explain** trophic relationships in ecosystems.

▶ **Relate** how biological evolution led to the biodiversity of life on Earth.

▶ **Define** succession, and **outline** the stages of general ecological succession in both terrestrial and aquatic ecosystems.

Diversity is an impressive feature of the living Earth. The diversity of organisms is a response to the interaction of the atmosphere, hydrosphere, and lithosphere, all powered by solar energy. The evolution of life on Earth, in all its forms, is intertwined with the evolution of Earth's physical and chemical systems. This interaction produces a variety of conditions within which the biosphere exists. The diversity of life also results from the intricate interplay and coevolution of living organisms themselves. We, as part of this vast natural complex, seek to find our place and understand the feelings nature stimulates within us.

The physical beauty of nature is certainly among its most powerful appeals to the human animal. The complexity of the aesthetic response is suggested by its wide-ranging expression from the contours of a mountain landscape to the ambient

607

colors of a setting Sun to the fleeting vitality of a breaching whale. Each exerts a powerful aesthetic impact on most people, often accompanied by feelings of awe at the extraordinary physical appeal and beauty of the natural world.[*]

The biosphere, the sphere of life and organic activity, extends from the ocean floor to about 8 km (5 mi) altitude in the atmosphere. The biosphere includes myriad ecosystems from simple to complex, each operating within general spatial boundaries. An **ecosystem** is a self-sustaining association of living plants and animals and their nonliving physical environment. Earth's biosphere itself is a collection of ecosystems within the natural boundary of the atmosphere and Earth's crust.

Natural ecosystems are open systems for both solar energy and matter, with almost all ecosystem boundaries functioning as transition zones rather than as sharp demarcations. Distinct ecosystems—for example, forests, seas, mountaintops, deserts, beaches, islands, lakes, and ponds—make up the larger whole.

Ecology is the study of the relationships between organisms and their environment and among the various ecosystems in the biosphere. The word *ecology*, developed by German naturalist Ernst Haeckel in 1869, is derived from the Greek *oikos* ("household," or "place to live") and *logos* ("study of"). **Biogeography** is the study of the distribution of plants and animals, the diverse spatial patterns they create, and the physical and biological processes, past and present, that produce Earth's species richness.

The degree to which modern society understands Earth's biogeography and conserves Earth's living legacy will determine the extent of our success as a species and the long-term survival of a habitable Earth:

The time is ripe to step up and expand current efforts to understand the great interlocking systems of air, water, and minerals nourishing the Earth.... Moreover, without vigorous action toward that goal, nations will be seriously handicapped in trying to cope with proven and suspected threats to ecosystems and to human health and welfare resulting from alterations in the cycles of carbon, nitrogen, phosphorus, sulfur, and related materials.... Society depends upon this life-support system of planet Earth.[†]

Earth's most influential biotic (living) agents are the humans. This is not arrogance; it is fact, because we powerfully influence every ecosystem on Earth. From the time humans first developed agriculture, the raising of animals, and the use of fire, we began a process that has led to our dominance over Earth's physical systems. The

penguin colony in the chapter-opening photo has declined 24% in numbers since 1970 because of environmental changes we discuss in this chapter.

Since life arose on the planet, there have been six major extinctions. The fifth one was 65 million years ago, whereas the sixth is happening across the present decades. Of all these extinction episodes, this is the only one of biotic origin, caused for the most part by human activity. We are the greatest evolutionary force as we essentially dominate freshwater quantity and quality, croplands, ocean fishery production, nitrogen and other chemical budgets, the concentration of greenhouse gases, and the present extinction rate.

In this chapter: We explore ecosystems and the community, habitat, and niche concepts. Plants are the essential living component in the biosphere, translating solar energy into usable forms to energize life. The role of nonliving systems, including biogeochemical cycles, is examined. We cover the organization of living ecosystems along complex food chains and webs. The biodiversity of living organisms is seen as a product of biological evolution over the past 3.6 billion years. Ecosystem stability and resilience, and how living landscapes change over space and time through the process of succession, are important. Included is coverage of the effects of global change on ecosystems and rates of succession.

Ecosystem Components and Cycles

An ecosystem is a complex of many variables, all functioning independently yet in concert, with complicated flows of energy and matter (Figure 19.1). The key is to understand the linkages and interconnections within and between systems. As the quote in the caption says, "... The effects of what you do in the world will always spread out like ripples in a pond." The decrease in the Black-browed Albatross (*Diomedea melanophris*) colony evident from the empty nests in Figure 19.1b was the result of changes in regulations in the fishing industry around the Falkland Islands. An increase in quotas for the squid harvest, a primary food source for the albatross, led to the reduction in bird population.

An ecosystem includes both biotic (living) and abiotic (nonliving) components. Nearly all ecosystems depend on a direct input of solar energy; the few limited ones that exist in dark caves, in wells, or on the ocean floor depend on chemical reactions (chemosynthesis).

Ecosystems are divided into subsystems, with the biotic portion composed of producers (plants), consumers (animals), and detritus feeders (worms, mites, bacteria, fungi). The abiotic flows in an ecosystem include gaseous, hydrologic, and mineral cycles. Figure 19.2a illustrates these essential elements of an ecosystem. Figure 19.2b and c illustrates how biotic and abiotic ingredients operate together to form a delicate forest floor in the Northwest and the hearty lichens in the hostile environment of Antarctica.

[*]S. R. Kellert, "The biological basis for human values of nature," in S. R. Kellert and E. O. Wilson, eds., *The Biophilia Hypothesis* (Washington, DC: Island Press, 1993), p. 49.

[†]G. F. White and M. K. Tolba, *Global Life Support Systems*, United Nations Environment Programme Information, No. 47 (Nairobi, Kenya: United Nations, 1979), p. 1.

(a)

(b)

FIGURE 19.1 The web of life.
(a) Study the spider web as you read the following quotation:

Life devours itself: everything that eats is itself eaten; everything that can be eaten is eaten; every chemical that is made by life can be broken down by life; all the sunlight that can be used is used. . . . The web of life has so many threads that a few can be broken without making it all unravel, and if this were not so, life could not have survived the normal accidents of weather and time, but still the snapping of each thread makes the whole web shudder, and weakens it. . . . You can never do just one thing: the effects of what you do in the world will always spread out like ripples in a pond.

(b) Abandoned Black-browed Albatross nests indicate a change in government fishing regulations in the surrounding ocean—". . . snapping each thread makes the whole web shudder. . . ." [(a) Photo by author; quotation from Friends of the Earth and Amory Lovins, The United Nations Stockholm Conference, *Only One Earth* (London: Earth Island Limited, 1972), p. 20. (b) Photos by Bobbé Christopherson.]

Communities

A convenient biotic subdivision within an ecosystem is a community. A **community** is formed by interactions among populations of living animals and plants at a particular time. An ecosystem is the interaction of many communities with the abiotic physical components of its environment.

Some examples help to clarify these concepts. In a forest ecosystem, a specific community may exist on the forest floor, whereas another community functions in the canopy of leaves high above (Figure 19.3). Not all forests appear in familiar structure, as the dwarf willow forest in Greenland demonstrates (Figure 19.3c). Similarly, within a lake ecosystem, the plants and animals that flourish in the bottom sediments form one community, whereas

those near the surface form another. A community is identified in several ways—by its physical appearance, the species present and the abundance of each, the complex patterns of their interdependence, and the trophic (feeding) structure of the community.

Within a community, two concepts are important: habitat and niche. **Habitat** is the type of environment where an organism resides or is biologically adapted to live. In terms of physical and natural factors, most species have specific habitat requirements with definite limits and a specific regimen of sustaining nutrients.

Niche (French *nicher*, "to nest") refers to the function, or occupation, of a life form within a given community. It is the way an organism obtains and sustains the physical, chemical, and biological factors it needs to survive. A niche

(a)

(b)

(c)

FIGURE 19.2 Biotic and abiotic components of ecosystems.
(a) Solar energy is the input that drives the biotic and abiotic components. Heat energy and biomass are the outputs from the biosphere. (b) Biotic and abiotic ingredients operate together to form this temperate forest-floor ecosystem in British Columbia, with mosses beginning the process of community development. (c) Two species of lichen in the photo, perhaps more, function under extremely hostile climate conditions in Antarctica. Each little indentation provides some advantage to the darker lichen. [Photos by Bobbé Christopherson.]

has several facets. Among these are a habitat niche, a trophic (food) niche, and a reproductive niche. For example, the Red-winged Blackbird (*Agelaius phoeniceus*) occurs throughout the United States and most of Canada in habitats of meadow, pastureland, and marsh. This species nests in blackberry tangles and thick vegetation in freshwater marshes, sloughs, and fields. Its trophic niche is weed seeds and cultivated seed crops throughout the year, and during the nesting season it adds insects to its diet—an aspect of its reproductive niche. These birds disperse seeds of many plants during their travels.

Other habitats produce comparable niches. In a stable community, no niche is left unfilled. The *competitive exclusion principle* states that no two species can occupy the same niche (food or space) successfully in a stable community. Thus, closely related species are spatially separated. In other words, each species operates to reduce competition and to maximize its own reproduction rate—literally, species survival depends on successful reproduction. This strategy in turn leads to greater diversity as species shift and adapt to fill each niche. Figure 19.4 shows eight different community environments and representative life forms.

Some species are *symbiotic*, an arrangement where two or more species exist together in an overlapping relationship. One type of symbiosis, *mutualism*, occurs when each organism benefits and is sustained over an extended period by the relation. For example, lichen (pronounced "liken") is made up of algae and fungi living together (Figure 19.2c). The alga is the producer and food source for the fungus, and the fungus provides structure and physical support. Their mutualism allows the two to occupy a niche in which neither could survive alone. Lichen developed from an earlier parasitic relationship in which the fungi broke into the alga cells. Today, the two organisms have evolved into a supportive harmony and symbiotic relationship. The partnership of corals and algae discussed in Chapter 16 is another example of mutualism in a symbiotic relationship.

By contrast, another form of symbiosis is a *parasitic* relationship, which may eventually kill the host, thus destroying the parasite's own niche and habitat. An example is parasitic mistletoe (*Phoradendron*), which lives on and may kill various kinds of trees. Some scientists are questioning whether our human society and the physical systems of Earth constitute a global-scale symbiotic

(a)

(b)

(c)

FIGURE 19.3 Different forest communities.
(a) Pioneer Mothers' Memorial Forest near Paoli, Indiana, is 36 hectares (88 acres) of old-growth forest, virtually undisturbed since around 1816, featuring black walnut, white oak, yellow poplar, white ash, and beech trees, among others. (b) Turtles bask on a log in a subtropical swamp (low, waterlogged ground) at Juniper Springs Recreation Area, Florida. (c) Fall colors highlight this dwarf willow forest in Greenland. Musk Oxen are foraging for food and drinking meltwater. The inset photo shows a 13-year-old willow, dated by the small annual growth segments below the leaf. Some dwarf willows can range to over 300 years in age. [Photos by Bobbé Christopherson.]

relationship of mutualism, which is sustainable, or a parasitic one, which is unsustainable.

Plants: The Essential Biotic Component

Plants are the critical biotic link between solar energy and the biosphere. Ultimately, the fate of all members of the biosphere, including humans, rests on the success of plants and their ability to turn sunlight into food.

Land plants (and animals) became common about 430 million years ago, according to fossilized remains. **Vascular plants** developed conductive tissues and true roots for internal transport of fluid and nutrients. (*Vascular* is from a Latin word for "vessel-bearing," referring to the conducting cells.)

In the present day, about 270,000 species of plants are known to exist, and most are vascular. Many more species have yet to be identified. They represent a great untapped resource base. Only about 20 species of plants provide 90% of the world's food—just three, wheat, maize (corn), and rice, comprise half of the food supply. Plants are a major source of new medicines and chemical compounds that benefit humanity. Plants also are the core of healthy, functioning ecosystems that sustain all life.

Leaves are solar-powered chemical factories, wherein photochemical reactions take place. Veins in the leaf bring in water and nutrient supplies and carry off the sugars (food) produced by photosynthesis. The veins in each leaf connect to the stems and branches of the plant and to the main circulation system.

Flows of carbon dioxide, water, light, and oxygen enter and exit the surface of each leaf (see Figure 1.4). Gases flow into and out of a leaf through small pores called **stomata** (singular: *stoma*), which usually are most numerous on the lower side of the leaf. Each stoma is surrounded by guard cells that open and close the pore, depending on the plant's changing needs.

Water that moves through a plant exits the leaves through the stomata and evaporates from leaf surfaces, thereby assisting the plant's temperature regulation. As water evaporates from the leaves, a pressure deficit is created that allows atmospheric pressure to push water up through the plant all the way from the roots, in the same manner that a soda straw works. We can only imagine the complex operation of a 100-m (330-ft) tree!

Photosynthesis and Respiration

Powered by energy from certain wavelengths of visible light, **photosynthesis** unites carbon dioxide and hydrogen (hydrogen is derived from water in the plant). The term is descriptive: *photo-* refers to sunlight, and *-synthesis*

FIGURE 19.4 Plants and animals fit specific niches.
(a) Elephant heads (*Pedicularis groenlandica*), a wildflower that grows above 1800 m (6000 ft) in wet mountain meadows. (b) A western yellow-bellied racer (*Coluber constrictor*) lives in prairies and meadows and feeds on insects, lizards, and mice. (c) A female polar bear (*Ursus maritimus*) with two 8-month-old cubs on pack ice in the Arctic Ocean. (d) A Svalbard reindeer (*Rangifer tarandus platyrhynchus*) on Spitsbergen Island, Arctic Ocean, losing velvet from his antlers. (e) Killdeer chicks (*Charadrius vociferus*) begin life roughing it on an exposed rocky nest, with protective parents creating a noisy diversion nearby. (f) Coral mushroom (*Ramaria* sp.), a fungal decomposer at work on the organic matter accumulated on the forest floor. (g) A large barrel cactus flourishes in the rocky soils and harsh conditions of the desert. (h) A leopard seal (*Hydrurga leptonyx*) resting on a bergy bit (small iceberg) off the Antarctic Peninsula. Fur is stained with blood from a recent meal. [Photos (a, c, d, f, h) by Bobbé Christopherson; (b, e, g) by author.]

FIGURE 19.5 How plants live and grow.
The balance between photosynthesis and respiration determines net photosynthesis and plant growth. [Temperate rain forest photo by Bobbé Christopherson.]

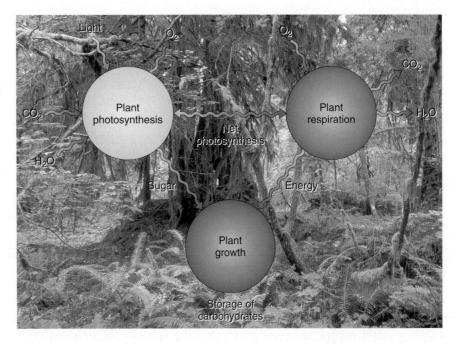

describes the "manufacturing" of starches and sugars through reactions within plant leaves. The process releases oxygen and produces energy-rich food for the plant.

The largest concentration of light-responsive, photosynthetic structures (known as *organelles*) in a leaf rests below the leaf's upper layers. These organelle units within cells are called *chloroplasts*, and within each resides a green, light-sensitive pigment called **chlorophyll**. Within this pigment, light stimulates photochemistry. Consequently, competition for light is a dominant factor in the formation of plant communities. This competition is expressed in the height, orientation, distribution, and structure of plants.

Only about one-quarter of the light energy arriving at the surface of a leaf is useful to the light-sensitive chlorophyll. Chlorophyll absorbs only the orange-red and violet-blue wavelengths for photochemical operations, and it reflects predominantly green hues (and some yellow). That is why trees and other vegetation look green.

Photosynthesis essentially follows this equation:

$$6CO_2 + 6H_2O + Light \rightarrow C_6H_{12}O_6 + 6O_2$$
(carbon dioxide) (water) (solar energy) (glucose, carbohydrate) (oxygen)

From the equation, you can see that photosynthesis removes carbon (in the form of CO_2) from Earth's atmosphere. The quantity is enormous: approximately 91 billion metric tonnes (100 billion tons) of carbon dioxide per year. Carbohydrates, the organic result of the photosynthetic process, are combinations of carbon, hydrogen, and oxygen. They can form simple sugars, such as *glucose* ($C_6H_{12}O_6$). Plants use glucose to build starches, which are more complex carbohydrates and the principal food stored in plants.

Plants store energy for later use. They consume this energy as needed through respiration by converting the carbohydrates to energy for their other operations. Thus, **respiration** is essentially a reverse of the photosynthetic process:

$$C_6H_{12}O_6 + 6O_2 \rightarrow 6CO_2 + 6H_2O + energy$$
(glucose carbohydrate) (oxygen) (carbon dioxide) (water) (heat energy)

In respiration, plants oxidize carbohydrates (stored energy), releasing carbon dioxide, water, and energy as heat. The overall growth of a plant depends on a surplus of carbohydrates beyond what is lost through plant respiration. Figure 19.5 presents a simple schematic of this process, which produces plant growth.

The *compensation point* is the break-even point between the production and consumption of organic material. Each leaf must operate on the production side of the compensation point, or else the plant eliminates it—something each of us has no doubt experienced with a houseplant that received inadequate water or light. The difference between photosynthetic production and respiration loss is called *net photosynthesis*. The amount varies, depending on controlling environmental factors such as light, water, temperature, soil fertility, and the plant's site, elevation, and competition from other plants and animals.

The rapidly increasing carbon dioxide concentration in the atmosphere stimulates the photosynthetic processes. However, with higher temperatures of global warming occurring more at night than during the day in many regions, this results in higher respiration rates.

One of the key scientific findings is that this unnatural warming is emphasized during the night and during wintertime, whereas a natural warming cycle would be balanced day to night or summer to winter. Plants respire all the time but at night respiration processes are the

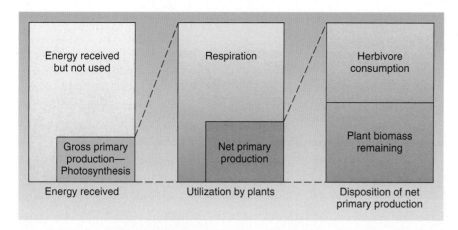

FIGURE 19.6 Energy budget of the biosphere.
Energy receipt, energy utilization, and energy disposition of net primary production by green plants.

plant's main operation. Studies of tropical rain forests in Costa Rica determined that increased respiration rates had made the forest a net carbon gain to the atmosphere.

Plant productivity increases as light availability increases—up to a point. When the light level is too high, *light saturation* occurs and most plants actually reduce their output in response. Some plants are adapted to shade, whereas others flourish in full sunlight. Crops such as rice, wheat, and sugarcane do well with high light intensity. Figure 19.6 portrays the general energy budget for green plants, showing energy receipt, energy utilization, and disposition of net primary production—how the energy is spent.

Net Primary Productivity   The net photosynthesis for an entire plant community is its **net primary productivity**. This is the amount of stored chemical energy that the community generates for the ecosystem. **Biomass** is the net dry weight of all this organic material and its stored chemical energy.

Net primary productivity is measured as fixed carbon per square meter per year. ("Fixed" means chemically bound into plant tissues.) Study the map and satellite images in Figure 19.7 and you can see that on land, net primary production tends to be highest between the Tropics of Cancer and Capricorn at sea level and decreases toward higher latitudes and altitudes. Precipitation also affects productivity, as evidenced by the correlations of abundant precipitation with high productivity adjacent to the equator and reduced precipitation with low productivity in the subtropical deserts. Even though deserts receive high amounts of solar radiation, other controlling factors limit productivity, namely, water availability and soil conditions.

In the oceans, differing nutrient levels control and limit productivity. Regions with nutrient-rich upwelling currents off western coastlines generally are the most productive. The map in Figure 19.7 shows that the tropical ocean and areas of subtropical high pressure are quite low in productivity.

In temperate and high latitudes, the rate at which carbon is fixed by vegetation varies seasonally. It increases in spring and summer as plants flourish with increasing solar input and, in some areas, with more available (nonfrozen) water, and it decreases in late fall and winter. Productivity rates in the tropics are high throughout the year, and turnover in the photosynthesis–respiration cycle is faster, exceeding by many times the rates experienced in a desert environment or in the far northern limits of the tundra. A lush hectare (2.5 acres) of sugarcane in the tropics might fix 45 metric tonnes (50 tons) of carbon in a year, whereas desert plants in an equivalent area might achieve only 1% of that amount.

Table 19.1 lists various ecosystems, their net primary productivity, and an estimate of net total biomass worldwide—170 billion metric tons of dry organic matter per year. Compare the various ecosystems, especially cultivated land (in *italics*) with most of the natural communities. Net productivity is generally regarded as the most important aspect of any type of community, and the distribution of productivity over Earth's surface is an important subject of biogeography.

Abiotic Ecosystem Components

Critical in each ecosystem are the flow of energy and the cycling of nutrients and water in life-supporting systems. These abiotic (nonliving) components set the stage for ecosystem operations.

Light, Temperature, Water, and Climate   Solar energy powers ecosystems, so the pattern of solar energy receipt is crucial. Solar energy enters an ecosystem by way of photosynthesis, and heat energy is dissipated from the system at many points. Of the total energy intercepted at Earth's surface and available for work, only about 1.0% is actually fixed by photosynthesis as chemical energy (energy stored as carbohydrates in plants).

The duration of Sun exposure is the *photoperiod.* Along the equator, days are essentially 12 hours long year-round; however, with increasing distance from the equator, seasonal effects become pronounced, as discussed in Chapter 2. Plants have adapted their flowering and seed germination to seasonal changes in insolation. Some seeds germinate only when daylength reaches a certain number of hours. A plant that responds in the opposite manner is the poinsettia (*Euphorbia pulcherrima*), which requires at least 2 months of 14-hour nights to start flowering.

(a)

(b)

(c)

FIGURE 19.7 Net primary productivity.
(a) Worldwide net primary productivity in grams of carbon per square meter per year (approximate values).
(b) Normalized difference vegetation index during 1984. False coloration indicates bare ground in browns, dense vegetation in blues. (c) *SeaWIFS* image of surface chlorophyll concentration for land and sea; intensity of green color represents higher levels of chlorophyll. This is the first continuous record over a 3-year period for the oceans. [(a) After D. E. Reichle, *Analysis of Temperate Forest Ecosystems* (Heidelberg: Springer, 1970). Adapted by permission. (b) NASA/GSFC. (c) *SeaWIFS* image by NASA/GSFC and ORBIMAGE, Dulles, VA. Used by permission.]

Other components are important to ecosystem processes. Air and soil temperatures determine the rates at which chemical reactions proceed (see Chapter 18). Significant temperature factors are seasonal variation and duration and the pattern of minimum and maximum temperatures (Chapter 5).

Operations of the hydrologic cycle and water availability depend on precipitation and evaporation rates and their seasonal distribution (see Chapters 7, 8, and 9). Water quality—its mineral content, salinity, and levels of pollution and toxicity—is important. Also, regional climates affect the pattern of vegetation and ultimately influence soil development. All of these factors work together to form the limits for ecosystems in a given location.

Figure 19.8 illustrates the general relationship among temperature, precipitation, and vegetation. Can you identify the characteristic vegetation type and related temperature and moisture regime that fits areas you know well, such as a place you have lived or an area around the school you now attend?

Life Zones   Alexander von Humboldt (1769–1859), an explorer, geographer, and scientist, observed that plants and animals recur in related groupings wherever similar

Table 19.1 Net Primary Productivity and Plant Biomass on Earth

Ecosystem	Area (10^6 km²)[a]	Net Primary Productivity per Unit Area (g/m²/yr)[b]		World Net Biomass (10^9/tons/yr)[c]
		Normal Range	Mean	
Tropical rain forest	17.0	1000–3500	2200	37.4
Tropical seasonal forest	7.5	1000–2500	1600	12.0
Temperate evergreen forest	5.0	600–2500	1300	6.5
Temperate deciduous forest	7.0	600–2500	1200	8.4
Boreal forest	12.0	400–2000	800	9.6
Woodland and shrubland	8.5	250–1200	700	6.0
Savanna	15.0	200–2000	900	13.5
Temperate grassland	9.0	200–1500	600	5.4
Tundra and alpine region	8.0	10–400	140	1.1
Desert and semidesert scrub	18.0	10–250	90	1.6
Extreme desert, rock, sand, ice	24.0	0–10	3	0.07
Cultivated land	*14.0*	*100–3500*	*650*	*9.1*
Swamp and marsh	2.0	800–3500	2000	4.0
Lake and stream	2.0	100–1500	250	0.5
Total continental	**149**	—	**773**	**115.17**
Open ocean	332.0	2–400	125	41.5
Upwelling zones	0.4	400–1000	500	0.2
Continental shelf	26.6	200–600	360	9.6
Algal beds and reefs	0.6	500–4000	2500	1.6
Estuaries	1.4	200–3500	1500	2.1
Total marine	**361.0**	—	**152**	**55.0**
Grand total	**510.0**	—	**333**	**170.17**

Source: R. H. Whittaker, *Communities and Ecosystems* (Heidelberg: Springer, 1975), p. 224. Reprinted by permission.
[a] 1 km² = 0.39 mi².
[b] 1 g per m² = 8.92 lb per acre.
[c] 1 metric ton (10^6 g) = 1.1023 tons.

conditions occur in the abiotic environment. He described the similarities and dissimilarities of vegetation and the associated uneven distribution of various organisms. After several years of study in the Andes Mountains of Peru, he described a distinct relation between altitude and plant communities as his *life zone concept.* As he climbed the mountains, he noticed that the experience was similar to that of traveling away from the equator toward higher latitudes (Figure 19.9a).

This zonation of plants with altitude is noticeable on any trip from lower valleys to higher elevations. Each **life zone** possesses its own temperature, precipitation, and insolation relations and therefore its own biotic communities.

The Grand Canyon in Arizona provides a good example. Plants and animals of the inner gorge at the bottom of the canyon (600 m, or 2000 ft, in elevation) are characteristic of the lower Sonoran Desert of northern Mexico. However, communities similar to those of southern Canadian forests dominate the north rim of the canyon (2100 m, or 7000 ft, in elevation). On the summits of the nearby San Francisco Mountains (3600 m, or 12,000 ft, in elevation), the vegetation is similar to that of the arctic tundra of northern Canada. Needleleaf forest

and alpine tundra in the Colorado Rockies at high elevation are pictured in Figure 19.9b and c.

Other Factors Beyond these general conditions, each ecosystem further produces its own *microclimate,* which is specific to individual sites. For example, in a forest the insolation reaching the ground is reduced and shade from its own trees blankets the forest floor. A pine forest cuts light by 20%–40%, whereas a birch–beech forest reduces it by as much as 50%–75%. Forests also are about 5% more humid than nonforested landscapes, have moderated temperatures (warmer winters and cooler summers), and experience reduced winds.

Slope orientation and Sun exposure are important, too, for they translate into differences in temperature and moisture efficiency, especially in middle and higher latitudes. With all other factors equal, slopes facing away from the Sun's rays tend to be moister and more vegetated than slopes facing toward the Sun. (In the Northern Hemisphere, these moister slopes face north, as evidenced by the apparent tendency of moss, a plant that needs high moisture and shade, to grow on the north side of tree trunks.) Such highly localized *microecosystems* are evident

FIGURE 19.8 How temperature and precipitation affect ecosystems.
(a) Climate controls (wetness, dryness, warmth, cold) and ecosystem types; (b) subtropical desert near Death Valley, California;
(c) Sonoran desert of Arizona; (d) cold desert of north-central Nevada; (e) dry tundra of East Greenland; (f) El Yunque tropical rain
forest of Puerto Rico; (g) deciduous trees in fall colors in Ohio; (h) needleleaf forest near Independence Pass, Colorado; and
(i) moist tundra of Spitsbergen, Arctic Ocean. [Photos (c, d) by author; (f) photo by Tom Bean; (b, e, g, h, i) by Bobbé Christopherson.]

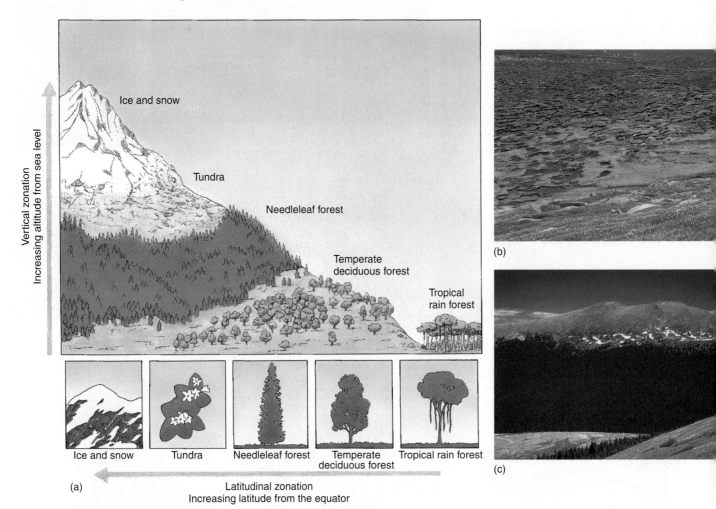

FIGURE 19.9 Vertical and latitudinal zonation of plant communities.
(a) Progression of plant community life zones with changing altitude or latitude. (b) Alpine tundra ecosystem in the Colorado Rockies. (c) The timberline for a needleleaf forest in the Rockies. The line of trees marks the limit of highest continuous forest, whereas the area above it marks the zone above which no trees grow. [Photos by Bobbé Christopherson.]

where changes in exposure and moisture occur related to geographic orientation.

Even Earth's magnetic field, another abiotic component, plays an interesting role in ecosystems. Read about turtle and animal navigation in News Report 19.1.

Elemental Cycles

The most abundant natural elements in living matter are hydrogen (H), oxygen (O), and carbon (C). Together, these elements make up more than 99% of Earth's biomass; in fact, all life (organic molecules) contains hydrogen and carbon. In addition, nitrogen (N), calcium (Ca), potassium (K), magnesium (Mg), sulfur (S), and phosphorus (P) are significant nutrients, elements necessary for the growth of a living organism.

Several key chemical cycles function in nature. Oxygen, carbon, and nitrogen each have *gaseous cycles*, part of which are in the atmosphere. Other elements have *sedimentary cycles*, which principally involve mineral and solid phases; major elements that have these cycles

include phosphorus, calcium, potassium, and sulfur. Some elements combine gaseous and sedimentary cycles. The recycling of gases and sedimentary (nutrient) materials forms Earth's **biogeochemical cycles**, so called because they involve chemical reactions necessary for growth and development of living systems. The chemical elements themselves recycle over and over again in life processes.

Oxygen and Carbon Cycles We consider these two cycles together because they are so closely intertwined through photosynthesis and respiration (Figure 19.10). The atmosphere is the principal reserve of available oxygen. Larger reserves of oxygen exist in Earth's crust, but they are unavailable, being chemically bound with other elements, especially the silicate (SiO_2) and carbonate (CO_3) mineral families. Unoxidized reserves of fossil fuels and sediments also contain oxygen.

The oceans are enormous pools of carbon—about 42,900 billion metric tonnes. (Metric tonnes times 1.10

Earth's Magnetic Field—An Abiotic Factor

The fact that birds and bees can detect Earth's magnetic field and use it for finding direction is well established. Small amounts of magnetically sensitive particles in the skull of the bird and the abdomen of the bee provide compass directions for bird migration and for alerting the hive to the location of the latest nectar find. For migrating birds, the timing and accuracy of an exact flight path is important, for the birds know the nutritional places they need to visit to build their fat storage for long journeys and where each "pit stop" is located along the way.

A team of biologists from the University of North Carolina, Catherine and Kenneth Lohmann, have found that sea turtles can detect magnetic fields of different strengths and the inclination (angle) of these magnetic fields. This means that the turtles have a built-in navigation system that helps them know where they are on Earth (like our global positioning system, which requires multimillion-dollar satellites). Using magnetic field strengths and inclination, the turtles evidently are aware of their global position—similar to knowing their latitude and longitude.

This magnetic map is evidently imprinted in their brain.

Loggerhead sea turtles hatch in Florida, crawl into the water, and spend the next 70 years traveling thousands of miles between North America and Africa around the subtropical high-pressure gyre in the Atlantic Ocean (see Figure 6.14). The females return to near where they were hatched to lay their eggs! The researchers think that the hatchlings are imprinted with magnetic data unique to the beach where they hatched and then develop a more global sense as they live a life swimming across the ocean.

---

equals short tons, or, here, 47,190 billion tons.) However, all of this carbon is bound chemically in carbon dioxide, calcium carbonate, and other compounds. The ocean initially absorbs carbon dioxide by means of the photosynthesis carried on by phytoplankton; it becomes part of the living organisms and is fixed in certain carbonate minerals, such as limestone ($CaCO_3$).

The atmosphere, which is the integrating link in the cycle, contains only about 700 billion tonnes of carbon (as carbon dioxide) at any moment. This is far less carbon than is stored in fossil fuels and oil shales (13,200 billion metric tonnes, as hydrocarbon molecules) or in living and dead organic matter (2500 billion metric tonnes, as carbohydrate molecules). Carbon dioxide is released into the atmosphere by the respiration of plants and animals, volcanic activity, and fossil fuel combustion by industry and transportation.

The carbon dumped into the atmosphere by human activity constitutes a vast geochemical experiment, using the real-time atmosphere as a laboratory. Since 1970, we have added to the atmospheric pool an amount of carbon equivalent to more than 25% of the total amount added since 1880; fourfold since 1950.

Global emissions of carbon from fossil fuels declined slightly from a 1997 high of 6.395 down to 6.299 billion metric tonnes (combined with oxygen in 18.9 billion metric tonnes of carbon dioxide) in 2000—400% more than the 1.6 billion metric tonnes of carbon in 1950. Emissions rose to more than 6.500 billion metric tonnes in 2003. About 50% of the carbon dioxide emitted since the beginning of the Industrial Revolution and not absorbed by oceans and organisms remains in the atmosphere, enhancing Earth's natural greenhouse effect. The greenhouse effect was covered more fully in Chapters 3, 5, and 10.

Nitrogen Cycle   Nitrogen, which accounts for 78.084% of each breath we take, is the major constituent of the atmosphere. Nitrogen also is important in the makeup of organic molecules, especially proteins, and therefore is essential to living processes. A simplified view of the nitrogen cycle is portrayed in Figure 19.11.

This vast atmospheric reservoir of nitrogen is not directly accessible to most organisms. *Nitrogen-fixing bacteria* provide the key link to life, live principally in the soil, and are associated with the roots of certain plants—for example, the *legumes* such as clover, alfalfa, soybeans, peas, beans, and peanuts. Colonies of these bacteria reside in nodules on the legume roots and chemically combine the nitrogen from the air in the form of nitrates (NO_3) and ammonia (NH_3).

Plants use these chemically bound forms of nitrogen to produce their own organic matter. Anyone or anything feeding on the plants thus ingests the nitrogen. Finally, the nitrogen in the organic wastes (sewage, manure) of these consuming organisms is freed by denitrifying bacteria, which recycle nitrogen back to the atmosphere.

To improve agricultural yields, many farmers use synthetic inorganic fertilizers, as opposed to soil-building organic fertilizers (manure and compost). Inorganic fertilizers are chemically produced through artificial nitrogen fixation at factories using the Haber–Bosch process developed in 1913, a process that takes inert nitrogen gas and hydrogen to form ammonia (NH_3, or as the ammonium ion NH_4^+), a reactive form of nitrogen that plants can use. The annual production of inorganic fertilizers far exceeds the ability of natural denitrification systems—and the present production of synthetic fertilizers now is doubling every 8 years; some 1.82 million tonnes (2 million tons) are produced per week, worldwide. Humans presently

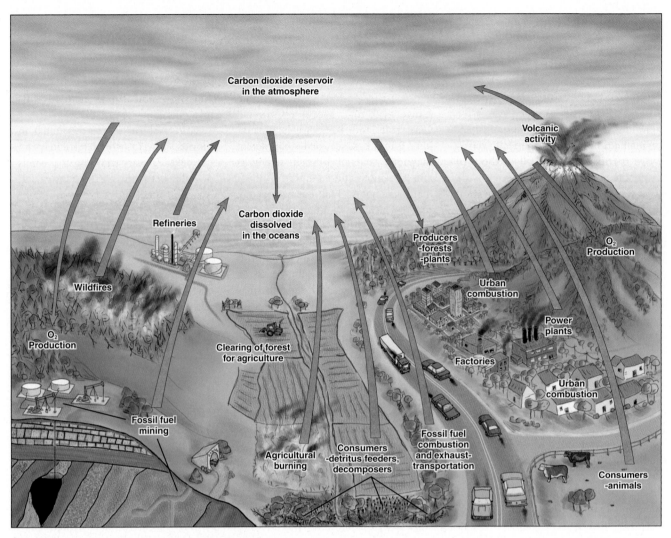

FIGURE 19.10 The carbon and oxygen cycles.
Carbon is fixed through photosynthesis and begins its passage through ecosystem operations. Respiration by living organisms, burning of forests and grasslands, and the combustion of fossil fuels releases carbon to the atmosphere. These cycles are greatly influenced by human activities.

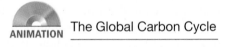 **ANIMATION** The Global Carbon Cycle

fix more nitrogen per year than all terrestrial sources combined. The cross-over point when anthropogenic sources of fixed nitrogen exceeded the normal range of natural fixed nitrogen was 1970.

A recent scientific study compared nitrogen content of woodland streams in more industrialized North America and Europe with more pristine streams in South America. The type of dissolved nitrogen exported by the streams from each ecosystem demonstrates the human dominance of the nitrogen cycle. Some 98% of the North American river export was in an inorganic nitrogen form. Yet, from the relatively pristine forests of Argentina and Chile, sampled streams contained 70% dissolved organic nitrogen. The conclusion points to the presence of inorganic nitrogen from industrial pollution and fossil fuel burning, rather than the previous assumptions that forests naturally produced inorganic nitrogen in runoff, since the pristine streams were naturally dominated by organic nitrogen.

This surplus of usable nitrogen accumulates in Earth's ecosystems. Some is present as excess nutrients, washed

from soil into waterways and eventually to the ocean. This excess nitrogen load begins a water pollution process that feeds an excessive growth of algae and phytoplankton, increases biochemical oxygen demand, diminishes dissolved oxygen reserves, and eventually disrupts the aquatic ecosystem. In addition, excess nitrogen compounds in air pollution are a component in acid deposition, further altering the nitrogen cycle in soils and waterways (see Chapter 3). For more on this problem, specific to North America, read News Report 19.2.

Limiting Factors

The term **limiting factor** identifies the one physical or chemical abiotic component that most inhibits biotic operations, through either its lack or its excess. Here are a few examples:

- Low temperatures limit plant growth at high elevations.
- Lack of water limits growth in a desert.

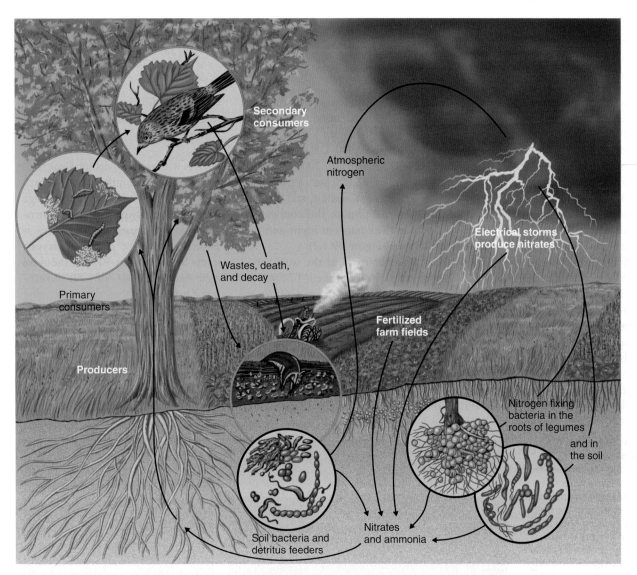

FIGURE 19.11 The nitrogen cycle.
The atmosphere is the essential reservoir of gaseous nitrogen. Atmospheric nitrogen gas is chemically fixed by bacteria to produce ammonia. Lightning and forest fires produce nitrates, and fossil fuel combustion forms nitrogen compounds that are washed from the atmosphere by precipitation. Plants absorb nitrogen compounds and produce organic material. [From T. Audesirk and G. Audesirk, *Biology: Life on Earth,* 4th ed., Figure 45-10, p. 903. © 1996 Prentice Hall Inc. Used by permission.]

- Excess water limits growth in a bog.
- Changes in salinity levels affect aquatic ecosystems.
- Lack of iron in ocean surface environments limits photosynthetic production.
- Low phosphorus content of soils limits plant growth.
- General lack of active chlorophyll above 6100 m (20,000 ft) limits primary productivity.

In most ecosystems, precipitation is the limiting factor. However, temperature, light levels, and soil nutrients certainly affect vegetation patterns.

Each organism possesses a range of tolerance for each limiting factor in its environment. This fact is illustrated vividly in Figure 19.12, which shows the geographic range for two tree species and two bird species. The coast redwood (*Sequoia sempervirens*) is limited to a narrow section of the Coast Ranges in California, covering barely 9500 km² (less than 4000 mi²), concentrated in areas that receive necessary summer advection fog. On the other hand, the red maple (*Acer rubrum*) thrives over a large area under varying conditions of moisture and temperature, thus demonstrating a broader tolerance to environmental variations.

The Mallard Duck (*Anas platyrhynchos*) and the Snail Kite (*Rostrhamus sociabilis*) also demonstrate a variation in tolerance and range (Figure 19.12b). The Mallard, a generalist, feeds from widely diverse sources, is easily domesticated, and is found throughout most of North America in at least one season of the year. By contrast, the Snail Kite is a specialist that feeds only on one specific type of snail. This single food source, then, is its limiting factor. Note its small habitat area near Lake Okeechobee in Florida.

(a)

(b)

(c)

(d)

FIGURE 19.24 Disturbances in ecosystems alter succession patterns.
(a) River flooding along the American River, California. (b) Community development proceeds in a burned forest, progressing perhaps by chance, as species with an adaptive edge, such as fireweed (*Epilobium* sp.), succeed. (c) Unwise practices cause extensive soil erosion on a hog farm in Iowa, 1999; a disrupted prairie landscape. (d) Drought-stressed Coulter pines are attacked by bark beetles, altering relationships in the ecosystem, now ripe for wildfire. [Photos (a) (b) (d) by Bobbé Christopherson; (c) by USDA/NRCS.]

the beginning and development of an ecosystem. Examples include new surfaces created by mass movement of land, areas exposed by a retreating glacier, cooled lava flows and volcanic eruption landscapes, or surface mining and clear-cut logging scars, or an area of sand dunes. Plants taking hold on new lava flows from the Kīlauea volcano, Hawai'i, illustrate primary succession in Figure 19.25. In

terrestrial ecosystems, succession begins with early species that form a **pioneer community**. You can see these plants taking hold in the photo. Primary succession often begins with lichens, mosses, and ferns growing on bare rock. These early inhabitants prepare the way for further succession to grasses, shrubs, and trees.

More common in nature is **secondary succession**, which begins if some aspects of a previously functioning community are present. An area where the natural community has been destroyed or disturbed, but where the underlying soil remains intact, may experience secondary succession. Examples of secondary succession are most of the areas affected by the Mount St. Helens eruption and blast in 1980 (Figure 19.26). Some soils, young trees, and plants were protected under ash and snow, so community development began almost immediately after the event. About 38,450 hectares (95,000 acres) of trees were blown down and burned, however, and as the photos show, the effects of the eruption were devastating and lingering. Of course, the areas completely destroyed near the Mount St. Helens volcano or those buried beneath the massive landslide north of the mountain became candidates for primary succession. An eruption sequence began unfolding in October 2004, although it is not expected to be anywhere near the 1980 catastrophic blast. Depending on the severity of the eventual dome eruption, the area will again be stimulated by such disturbance.

(b)

FIGURE 19.25 Primary succession.
Ferns take hold as primary succession on new lava flows that came from the Kīlauea volcano, Hawai'i. [Photo by Bobbé Christopherson.]

Landsat-7 8/22/1999

FIGURE 19.26 The pace of change in the region of Mount St. Helens.
(a) The area north of the volcano before the eruption (1979) and (b) after the eruption (1980) experienced profound change in all ecosystems. A series of four photo comparisons made in 1983 and 1999 of matching scenes illustrate the slow recovery of secondary succession: (c) and (d) at Meta Lake; (e) Spirit Lake panorama view (note the mat of floating logs after 19 years); (f) detail of the destroyed forest community. [(a) and all 1983 photos by author; (b) and all 1999 photos by Bobbé Christopherson; *Landsat-7* 8/22/99 satellite image courtesy of NASA/GSFC.]

A musk ox (*Ovibos moschatus*) bull and five cows graze near Sydkap Point, Scoresby Sund, Greenland. The bull weighs about 400 kg (880 lb) and stands about 150 cm (60 in.) high at the shoulder. Approximately 3500 to 5000 musk oxen are presently in Greenland. In the background, an armada of iceberg sculptures enters Scoresby Sund after a long sea voyage from north Greenland, passing by like city-block-sized pieces of Carrera marble, carved in infinite shapes. [Photo by Bobbé Christopherson.]

CHAPTER 20

Terrestrial Biomes

▶ Key Learning Concepts

After reading the chapter, you should be able to:

▶ **Define** the concept of biogeographic realms of plants and animals, and **define** ecotone, terrestrial ecosystem, and biome.

▶ **Define** six formation classes and the life-form designations, and **explain** their relationship to plant communities.

▶ **Describe** 10 major terrestrial biomes, and **locate** them on a world map.

▶ **Relate** human impacts, real and potential, to several of the biomes.

I n 1874, Earth was viewed as offering few limits to human enterprise. But visionary American diplomat and conservationist George Perkins Marsh perceived the need to manage and to conserve the environment:

> We have now felled forest enough everywhere, in many districts far too much. Let us now restore this one element of material life to its normal proportions, and devise means of maintaining the permanence of its relations to the fields, the meadows, and the pastures, to the rain and the dews of heaven, to the springs and rivulets with which it waters the Earth.*

Unfortunately, his warning was ignored. The clearing of the forests is perhaps the most fundamental change wrought by humans in our transformation of Earth. The development and expansion of civilization was fueled and built on the consumption of forests and other natural resources. This natural "work" supplied by Earth's physical, chemical, and biological systems has an estimated annual worth of some $35 trillion to the global economy.

*G. P. Marsh, *Physical Geography as Modified by Human Action* (New York: Scribner's, 1874), pp. 385–386.

649

Today, we see that Earth's natural systems do pose limits to human activity. Scientists are measuring ecosystems carefully, and they are building elaborate computer models to simulate our evolving real-time human-environment experiment—in particular, shifting patterns of environmental factors (temperatures and changing frost periods; precipitation timing and amounts; air, water, and soil chemistry; and a redistribution of nutrients). We know from history that the effect of climate change on ecosystems will be significant. More than a decade ago two Harvard researchers assessed the impact of these alterations on ecosystems:

Based on more than a decade of research, it is obvious that the CO_2-rich atmosphere of our future will have direct and dramatic effects on the composition and operation of ecosystems. According to the best scientific evidence, we see no reason to be sanguine [optimistic] about the response of these habitats to our changing environment.*

What will the quotations about Earth's environment be like 130 years from now, about the same time interval since G. P. Marsh made his assessment? Humans now have become the most powerful biotic agent on Earth, influencing all ecosystems on a planetary scale.

In this chapter: We explore Earth's major terrestrial ecosystems, conveniently grouped into 10 biomes. Each biome is an idealized assemblage because much alteration of the natural environment has occurred. The discussion covers biome appearance, structure, and location, and the present conditions of related plants, animals, and environment, and the state of biodiversity. Table 20.1 brings together most all the information about Earth's physical systems that is presented through the pages of this text. Through biomes, we synthesize the physical geography of place and region.

Biogeographic Realms

Figure 20.1 shows Earth's biogeographic realms. The upper map illustrates the botanical (plant) realms worldwide. The lower map shows the zoological (animal) realms. Each realm contains many distinct ecosystems that distinguish it from other realms.

A **biogeographic realm** is a geographic region where a group of plant and animal species evolved. As you can see, these realms correspond generally to continents. The main topographic barrier that separates these realms is the ocean. Every species attempts to migrate worldwide according to its niche requirements, reproductive success, and competition with other species but is constrained by climatic and topographic barriers. The collective effort of all species to seek optimal habitats and to maximize each personal niche requirement results in the generalized realms shown on the maps. Recognition that such distinct

regions of flora (plants) and fauna (animals) exist was an early beginning for *biogeography* as a discipline.

The Australian realm is unique, giving rise to 450 species of *Eucalyptus* among its plants and 125 species of marsupials, including four of the six orders (animals such as kangaroos, that carry their young in pouches, where gestation is completed; Figure 20.2). Also, monotremes, egg-laying mammals such as the platypus, add further distinctiveness to this realm. Australia's unique flora and fauna is the result of its early isolation from the other continents. During critical evolutionary times, Australia drifted away from Pangaea (see Chapter 11) and never again was reconnected by a land bridge, even when sea level lowered during repeated glacial ages. New Zealand's isolation from Australia also undoubtedly explains why no native marsupials exist in New Zealand.

Transition Zones

Alfred Wallace (1823–1913), the first scholar of *zoogeography*, used the stark contrast in animal species among islands in present-day Indonesia to delimit a boundary between the Oriental and Australian realms (Figure 20.1b). He thought that deep water in the straits had completely prevented species crossover. However, he later recognized that a boundary line actually is a wide transition zone, where one region grades into the other. Boundaries between natural systems are "zones of shared traits." Thus, despite distinctions among individual realms and our discussion of biomes, it is best to think of their borders as transition zones of mixed identity and composition, rather than as rigidly defined boundaries.

A boundary transition zone between adjoining ecosystem regions is an **ecotone**. Because ecotones are defined by different physical factors, they vary in width—think of a transition zone instead of a line. Climatic ecotones usually are more gradual than physical ecotones, where differences in soil or topography sometimes form abrupt boundaries. An ecotone between prairies and northern forests may occupy many kilometers of land. The ecotone is an area of tension, as similar species of plants and animals compete for the resource base.

Scientists have identified these boundary and edge areas as generating biodiversity (a species richness) as adaptation and natural selection favor the population with reproductive success. Organisms may compete, yet they may act to facilitate other organisms in a positive interaction in whatever strategies affect rates of ecosystem productivity, such as pollinating insects. A process of *coevolution* is possible where interacting populations in the tension zone can lead to genetic changes in each population (microevolution). Edges and ecotones are dynamic areas in the evolution of ecosystems and communities.

Terrestrial Ecosystems

A **terrestrial ecosystem** is a self-sustaining association of land-based plants and animals and their abiotic environment, characterized by specific plant formation classes.

*F. A. Bazzaz and E. D. Fajer, "Plant life in a CO_2-rich world," *Scientific American* 256, no. 1 (January 1992): 74.

(a)   BOTANICAL GEOGRAPHIC REALMS

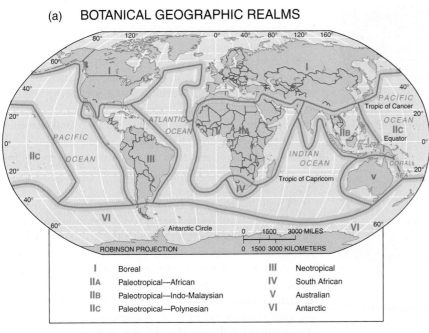

I	Boreal	III	Neotropical
IIA	Paleotropical—African	IV	South African
IIB	Paleotropical—Indo-Malaysian	V	Australian
IIC	Paleotropical—Polynesian	VI	Antarctic

(b)   ZOOLOGICAL GEOGRAPHIC REALMS

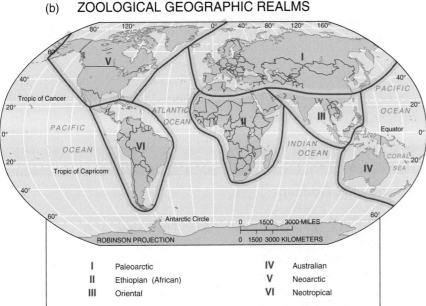

I	Paleoarctic	IV	Australian
II	Ethiopian (African)	V	Neoarctic
III	Oriental	VI	Neotropical

FIGURE 20.1 Biogeographic realms.
(a) Botanical geographic realms (after R. D. Good, 1947). (b) Zoological geographic realms (after L. F. deBeaufort, 1951). [From E. P. Odum, *Fundamentals of Ecology*, 3rd ed., reprinted with permission of Brooks/Cole an imprint of the Wadsworth Group, a div. of Thomson Learning.]

Plants are the most visible part of the biotic landscape, and they are key members of Earth's terrestrial ecosystems. In their growth, form, and distribution, plants reflect Earth's physical systems: its energy patterns; atmospheric composition; temperature and winds; air masses; water quantity, quality, and seasonal timing; soils; regional climates; geomorphic processes; and ecosystem dynamics. As discussed in the previous chapter, aquatic ecosystems are likewise important and result from the interaction of physical systems (see News Report 20.1).

A brief review of the ways that living organisms relate to the environment and to each other is helpful at this point. Interacting populations of plants and animals in an area form a *community*. An *ecosystem* involves the interplay between a community of plants and animals and their abiotic physical environment. Each plant and animal occupies an area in which it is biologically suited to live—

its *habitat*—and within that habitat it performs a basic operational function—its *niche* (see Chapter 19).

A **biome** is a large, stable terrestrial ecosystem characterized by specific plant and animal communities. Each biome usually is named for its *dominant vegetation*, because that is the single most easily identified feature. We can generalize Earth's wide-ranging plant species into six broad biomes: *forest, savanna, grassland, shrubland, desert,* and *tundra.* Because plant distributions respond to environmental conditions and reflect variations in climate and soil, the world climate map in Figure 10.5 is a helpful reference for this chapter. (The biome concept can also be applied to aquatic ecosystems: polar seas, temperate seas, tropical seas, seafloor, shoreline, and coral reef.)

More specific characteristics are used for the structural classification of plants themselves. *Life-form designations*

FIGURE 20.2 The unique Australian realm.
The flora and fauna of Australia form a special assemblage of communities. [Photo by Mark Newman/Photo Researchers, Inc.]

are based on the outward physical properties of individual plants or the general form and structure of a vegetation cover. These physical life forms include:

- *Trees* (large woody main trunk, perennial; usually exceeds 3 m, or 10 ft, in height)

- *Lianas* (woody climbers and vines)
- *Shrubs* (smaller woody plants; stems branching at the ground)
- *Herbs* (small plants without woody stems; includes grasses and other nonwoody vascular plants)
- *Bryophytes* (nonflowering, spore-producing plants; includes mosses and liverworts)
- *Epiphytes* (plants growing above the ground on other plants, using them for support)
- *Thallophytes* (lack true leaves, stems, or roots; includes bacteria, fungi, algae, lichens)

These plant assemblages in their general biomes are divided into more-specific vegetation units called **formation classes**, which refer to the structure and appearance of dominant plants in a terrestrial ecosystem. Examples are equatorial rain forest, northern needleleaf forest, Mediterranean shrubland, and arctic tundra. Each formation class includes numerous plant communities, and each community includes innumerable plant habitats. Within those habitats, Earth's diversity is expressed in 270,000 plant species. Despite this intricate complexity, we can generalize Earth's numerous formation classes into 10 global terrestrial biome regions, as portrayed in Figure 20.3 and detailed in Table 20.1.

Now let us go on a tour of Earth's biomes, keeping in mind that they synthesize all we have learned about the

News Report 20.1

Aquatic Ecosystems and the LME Concept

Oceans, estuaries, and freshwater bodies (streams and lakes) are home to aquatic ecosystems. Poor understanding of aquatic ecosystems has led to the demise of some species, such as herring in the Georges Bank prime fishing area of the Atlantic, and an overall decline in fisheries worldwide. One analytical tool is the designation of certain areas as large marine ecosystems (LMEs, http://www.edc.uri.edu/lme/). The LME was conceived by Kenneth Sherman and Lewis Alexander, scientists with the U.S. Marine Fisheries Service Laboratory and the University of Rhode Island, respectively, to encourage resource planners, managers, and researchers to consider complete ecosystems, not just a targeted species.

The Large Marine Ecosystems Strategy is a worldwide scientific and political effort to understand these aquatic ecosystems and move closer to sustainability of these resources. Involved is the World Conservation Union (IUCN), the Intergovernmental Oceanographic Commission of UNESCO (IOC) and other United Nations agencies, and the National Oceanic and Atmospheric Administration (NOAA).

An LME is a distinctive oceanic region having unique organisms, floor topography, currents, areas of nutrient-rich upwelling circulation, or areas of significant predation, including human. Examples of identified LMEs include the Gulf of Alaska, California Current, Gulf of Mexico, Northeast Continental Shelf, and the Baltic and Mediterranean Seas. Some 62 LMEs, each encompassing more than 200,000 km² (77,200 mi²) are presently defined worldwide.

The largest of 11 government-protected areas in North America is the Monterey Bay National Marine Sanctuary, established in 1992 within the California Current LME. (The Florida Keys Marine Sanctuary is second largest.) Stretching along 645 km (400 mi) of coastline and covering 13,500 km² (5300 mi²), the Monterey Bay sanctuary is home to 27 species of marine mammals, 94 species of shorebirds, 345 species of fish, and the largest sampling of invertebrates in any one place in the Pacific. This represents one of the most species-rich and diverse marine communities on Earth. In the United States, these protected areas are administered by NOAA and The National Marine Sanctuaries. Its web site is at http://www.sanctuaries.nos.noaa.gov/.

Table 20.1 Major Terrestrial Biomes and Their Characteristics

Biome and Ecosystems (map symbol)	Vegetation Characteristics	Soil Orders (Soil Taxonomy and CSSC)	Climate Type	Annual Precipitation Range	Temperature Patterns	Water Balance
Equatorial and Tropical Rain Forest (ETR) Evergreen broadleaf forest Selva	Leaf canopy thick and continuous; broadleaf evergreen trees, vines (lianas), epiphytes, tree ferns, palms	Oxisols Ultisols (on well-drained uplands)	Tropical	180–400 cm (>6 cm/mo)	Always warm (21–30°C; avg. 25°C)	Surpluses all year
Tropical Seasonal Forest and Scrub (TrSF) Tropical monsoon forest Tropical deciduous forest Scrub woodland and thorn forest	Transitional between rain forest and grasslands; broadleaf, some deciduous trees; open parkland to dense undergrowth; acacias and other thorn trees in open growth	Oxisols Ultisols Vertisols (in India) Some Alfisols	Tropical monsoon savanna	130–200 cm (<40 rainy days during 4 driest months)	Variable, always warm (>18°C)	Seasonal surpluses and deficits
Tropical Savanna (TrS) Tropical grassland Thorn tree scrub Thorn woodland	Transitional between seasonal forests, rain forests, and semiarid tropical steppes and desert; trees with flattened crowns, clumped grasses, and bush thickets; fire association	Alfisols (dry: Ultalfs) Ultisols Oxisols	Tropical savanna	9–150 cm, seasonal	No cold-weather limitations	Tends toward deficits, therefore fire- and drought-susceptible
Midlatitude Broadleaf and Mixed Forest (MBME) Temperate broadleaf Midlatitude deciduous Temperate needleleaf	Mixed broadleaf and needleleaf trees; deciduous broadleaf, losing leaves in winter; southern and eastern evergreen pines demonstrate fire association	Ultisols Some Alfisols (Podzols, red and yellow)	Humid subtropical warm summer Humid continental warm summer	75–150 cm	Temperate, with cold season	Seasonal pattern with summer maximum PRECIP and POTET (PET); no irrigation needed
Needleleaf Forest and Montane Forest (NF/MF) Taiga Boreal forest Other montane forests and highlands	Needleleaf conifers, mostly evergreen pine, spruce, fir; Russian larch, a deciduous needleleaf	Spodosols Histosols Inceptisols Alfisols (Boralfs: cold) (Gleysols) (Podzols)	Subarctic Humid continental cool summer	30–100 cm	Short summer, cold winter	Low POTET (PET), moderate PRECIP, moist soils, some waterlogged and frozen in winter; no deficits
Temperate Rain Forest (TeR) West coast forest Coast redwoods (U.S.)	Narrow margin of lush evergreen and deciduous trees on windward slopes; redwoods, tallest trees on Earth	Spodosols Inceptisols (mountainous environs) (Podzols)	Marine west coast	150–500 cm	Mild summer and mild winter for latitude	Large surpluses and runoff
Mediterranean Shrubland (MSh) Sclerophyllous shrubs Australian eucalyptus forest	Short shrubs, drought adapted, tending to grassy woodlands and chaparral	Alfisols (Xeralfs) Mollisols (Xerolls) (Luvisols)	Mediterranean dry summer	25–65 cm	Hot, dry summers, cool winters	Summer deficits, winter surpluses
Midlatitude Grasslands (MGr) Temperate grassland Sclerophyllous shrub	Tallgrass prairies and shortgrass steppes, highly modified by human activity; major areas of commercial grain farming; plains, pampas, and veld	Mollisols Aridisols (Chernozemic)	Humid subtropical Humid continental hot summer	25–75 cm	Temperate continental regimes	Soil moisture utilization and recharge balanced; irrigation and dry farming in drier areas
Warm Desert and Semidesert (DBW) Subtropical desert and scrubland	Bare ground graduating into xerophytic plants including succulents, cacti, and dry shrubs	Aridisols Entisols (sand dunes)	Desert Arid	<2 cm	Average annual temperature, around 18°C, highest temperatures on Earth	Chronic deficits, irregular precipitation events, PRECIP $< \frac{1}{2}$ POTET (PET)
Cold Desert and Semidesert (DBC) Midlatitude desert, scrubland, and steppe	Cold desert vegetation includes short grass and dry shrubs	Aridisols Entisols	Steppe Semiarid	2–25 cm	Average annual temperature around 18°C	PRECIP $> \frac{1}{2}$ POTET (PET)
Arctic and Alpine Tundra (AAT)	Treeless; dwarf shrubs, stunted sedges, mosses, lichens, and short grasses; alpine, grass meadows	Gelisols Histosols Entisols (permafrost) (Organic) (Cryosols)	Tundra Subarctic very cold	15–180 cm	Warmest months <10°C, only 2 or 3 months above freezing	Not applicable most of the year, poor drainage in summer
Ice			Ice sheet ice cap			

FIGURE 20.3 The 10 major global terrestrial biomes.

atmosphere, hydrosphere, lithosphere, and biosphere in the pages of this text. Here we bring it all together.

Earth's Major Terrestrial Biomes

Few natural communities of plants and animals remain; most biomes have been greatly altered by human intervention. Thus, the "natural vegetation" identified on many biome maps reflects *ideal potential* mature vegetation, given the environmental characteristics in a region. Even though human practices have greatly altered these ideal forms, it is valuable to study the natural (undisturbed) biomes to better understand the environment and to assess the extent of human-caused alteration.

Knowing the ideal in a region in addition guides us to a closer approximation of natural vegetation in the plants we introduce. In the United States and Canada, we

humans are perpetuating a type of *transition community*, somewhere between grasslands and a forest—an artificial successional stage, produced by large-scale interruption and disturbance. We plant trees and lawns and then must invest energy, water, and capital to sustain such artificial modifications. Or we graze animals and plant crops that perpetuate this disturbed transitional state.

When these activities cease, natural succession recovers, and we see the land slowly return to its various vegetation-cover potentials. Some irreversible damage might occur when plants, animals, and organisms are knowingly or accidentally brought into an ecosystem or biome in which they are not native. These exotic species pose an increasing concern to scientists and society—a subject of News Report 20.2.

The global distribution of Earth's major terrestrial biomes, based on vegetation formation classes, is portrayed in Figure 20.3. Table 20.1 describes each biome on

MODIFIED GOODE'S HOMOLOSINE EQUAL-AREA PROJECTION

the map and summarizes other pertinent information gathered from throughout *Geosystems*.

Now is a good time to refer to "The Living Earth" composite image that appears inside the front cover of this text. A computer artist using hundreds of thousands of satellite images produced this cloudless view of Earth in natural color typical of a local summer day. Compare the map in Figure 20.3 with this remarkable composite image and see what correlation you can make. Then compare the biomes with population distribution as indicated by the nighttime lighting display also on the inside front cover.

Equatorial and Tropical Rain Forest

Earth is girdled with a lush biome—the **equatorial and tropical rain forest**. In a climate of consistent year-round daylength (12 hours), high insolation, average annual temperatures around 25°C (77°F), and plentiful moisture,

plant and animal populations have responded with the most diverse expressions of life on the planet.

As Figure 20.3 shows, the Amazon region, also called the *selva*, is the largest tract of equatorial and tropical rain forest. In addition, rain forests cover equatorial regions of Africa, Indonesia, the margins of Madagascar and Southeast Asia, the Pacific coast of Ecuador and Colombia, and the east coast of Central America, with small discontinuous patches elsewhere. The cloud forests of western Venezuela are such tracts of rain forest at high elevation, perpetuated by high humidity and cloud cover. Undisturbed tracts of rain forest are rare.

Rain forests represent approximately one-half of Earth's remaining forests, occupying about 7% of the total land area worldwide. This biome is stable in its natural state, resulting from the long-term residence of these continental plates near equatorial latitudes and their escape from glacial activity.

News Report 20.2

Alien Invaders of Exotic Species

The title above sounds like that of a cult science fiction movie; instead, it refers to a real problem in the integrity of many ecosystems, both aquatic and terrestrial. Animal and plant species brought or somehow getting into biomes outside their native home is an important subject in biogeography and invasion biology. Niche takeovers by nonnative species can prove damaging to otherwise healthy ecosystems. These intruders are known as *exotic species*, or *alien* or *nonnative species*.

The Institute for Biological Invasions at the University of Tennessee at **http://invasions.bio.utk.edu/bio_invasions/index.html** is dedicated to these invasive plants, animals, and other organisms. Some states post warning signs at their borders (Figure 20.2.1a) and others offer web sites, such as Minnesota's "Prohibited and Noxious Plants by Scientific Name" (**http://www.dnr.state.mn.us/ecological_services/invasives.html**).

The African "killer bees" are frequently in the headlines, as are brown tree snakes in Guam, zebra mussels in the Great Lakes, and Eurasian cheatgrass in the Utah desert, to name but a few examples. Brought from Africa to the central coast of Brazil in 1957 to increase honey production, killer bees now have interbred with native bees and range to southern California, Arizona, New Mexico, Texas, and Puerto Rico. More than 1000 people have died from their attacks. The dangers are in their mass response to disturbances and that some people are allergic to even one bee sting. This is a classic case of geographical diffusion of an exotic species.

Probably 90% of exotic species fail when they try to move into established niches in a community. The 10% that succeed, some 4500 species documented in a 1993 Office of Technology Assessment report, prove damaging to as much as one-fifth of ecosystems they invade. Figures 20.2.1b and c show two such nonnative species: purple loosestrife (*Lythrum salicaria*) and kudzu (*Pueraria montana*).

Purple loosestrife was introduced from Europe in the 1800s as a desired ornamental and had some medicinal applications. The plant's seeds also arrived in ships that used soil for ballast. A hardy perennial, it got loose and invaded wetlands across the eastern portions of the United States and Canada, through the upper Midwest, and as far west as Vancouver Island, British Columbia, replacing plants on which native wildlife depend (Figure 20.2.1b). The plant is known to infect drier landscapes as well and poses a potential threat to agriculture.

The infamous kudzu was brought from Japan in 1876 for cattle feed, erosion control, and as an ornamental. It has spread to east Texas, southern Pennsylvania, and across the South into Florida. The plant can grow 0.3 m (1 ft) on a hot, humid

(a)

(b)

(c)

(d)

FIGURE 20.2.1 Exotic species.
(a) A sign in Wallowa County, Washington State, makes a plea for awareness in this agricultural country, listing noxious (nonnative, often problematic) weeds at the county line. (b) Invasive purple loosestrife near Lake Michigan and the Indiana Dunes National Lakeshore, Indiana (foreground and left center). (c) Kudzu overruns pasture and forest in western Georgia. (d) Gorse (yellow-flowering shrub), intended for barriers between pastures, is loose across the landscape in West Point, Falkland Islands. [All photos by Bobbé Christopherson.]

summer day and overtake forests and structures. Serious effort is going into finding some practical use for the prodigious plant, from pulp (treeless paper) to recipes for cooking to the telling of new kudzu jokes.

Gorse was introduced into the timber-poor Falkland Islands as a barrier plant, like a substitute fence. Its thorny stickers and dense growth were meant to keep grazing sheep and a few cattle under control. The plant got loose and spread beyond this intended use (Figure 20.2.1d).

In learning about biomes, we find out about native plant, animal, and organism locations and the unique place each species occupies. We must avoid knowingly or perhaps unconsciously adding to an escalating problem by bringing exotic species into nonnative sites where they do not naturally belong.

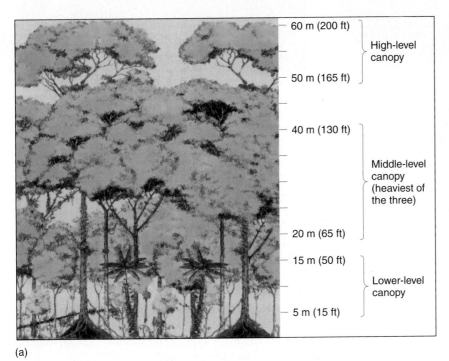

(a)

FIGURE 20.4 The three levels of a rainforest canopy.
(a) Lower, middle, and high levels of the rain forest. (b) Undisturbed rain forest on the Peninsula de Osa in Costa Rica. Note a few high-level canopy trees that extend above the dense, continuous cover of the middle canopy. [Photo by Barbara Cushman Rowell/Mountain Light Photography, Inc.]

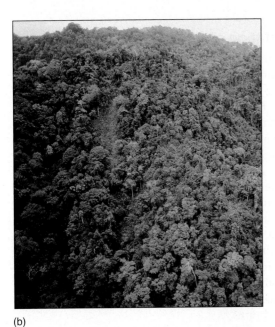

(b)

Rain forests feature ecological niches that are distributed vertically rather than horizontally because of the competition for light. Biomass in a rain forest is concentrated high up in the canopy, that dense mass of overhead leaves. The canopy is filled with a rich variety of plants and animals. Lianas (vines) stretch from tree to tree, entwining them with cords that can reach 20 cm (8 in.) in diameter. Epiphytes flourish there too—plants such as orchids, bromeliads, and ferns live entirely aboveground, supported physically but not nutritionally by the structures of other plants. Calm conditions on the forest floor make wind pollination difficult, so pollination is by insects, other animals, and self-pollination.

The rain forest canopy forms three levels—see Figure 20.4. The upper level is not continuous, but features emergent tall trees whose high crowns rise above the middle canopy. This upper level appears as a broken *overstory* of tall trees breaking through a middle canopy that is nearly continuous (Figure 20.4b). The broad leaves block much of the light and create a darkened *understory* area and forest floor. The lower level is composed of seedlings, ferns, bamboo, and the like, leaving the litter-strewn ground level in deep shade and fairly open.

Aerial photographs of a rain forest, or views along riverbanks covered by dense vegetation, or the false Hollywood-movie imagery of the jungle make it difficult to imagine the shadowy environment of the actual rainforest floor, which receives only about 1% of the sunlight arriving at the canopy (Figure 20.5a). The constant moisture, rotting fruit and moldy odors, strings of thin roots and vines dropping down from above, windless air, and

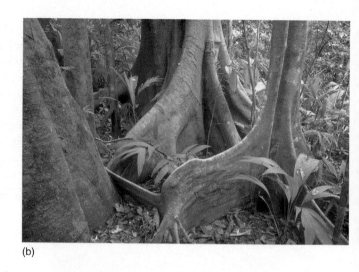

(a)

(b)

FIGURE 20.5 The rain forest.
(a) Equatorial rain forest is thick along the Amazon River where light breaks through to the surface, producing a rich gallery of vegetation. (b) The rainforest floor in Corcovado, Costa Rica, with typical buttressed trees and lianas. [(a) Photo by Wolfgang Kaehler Photography; (b) photo by Frank S. Balthis.]

echoing sounds of life in the trees together create a unique environment.

The smooth, slender trunks of rainforest trees are covered with thin bark and buttressed by large wall-like flanks that grow out from the trees to brace the trunks (Figure 20.5b). These buttresses form angular, open enclosures, a ready habitat for various animals. There are usually no branches for at least the lower two-thirds of the tree trunks.

The wood of many rainforest trees is extremely hard, heavy, and dense—in fact, some species will not even float in water. (Exceptions are balsa and a few others, which are light.) Varieties of trees include mahogany, ebony, and rosewood. Logging is difficult because individual species are widely scattered; a species may occur only once or twice per square kilometer. Selective cutting is required for species-specific logging, whereas pulpwood production takes everything. Conversion of the forest to pasture usually is done by setting destructive fires.

Rainforest soils, principally Oxisols, are essentially infertile, yet they support rich vegetation. The trees have adapted to these soil conditions with root systems able to capture nutrients from litter decay at the soil surface. High precipitation and temperatures work to produce deeply weathered and leached soils, characteristic of the laterization process, with a claylike texture sometimes breaking up into a granular structure. Oxisols lack nutrients and colloidal material. With much investment in fertilizers, pesticides, and machinery, these soils can be productive.

The animal and insect life of the rain forest is diverse, ranging from small decomposers (bacteria) working the surface to many animals living exclusively in the upper stories of the trees. These tree dwellers are referred to as *arboreal*, from the Latin for "tree," and include sloths, monkeys, lemurs, parrots, and snakes. Beautiful birds of many colors, tree frogs, lizards, bats, and a rich insect community that includes more than 500 species of butterflies are found in rain forests. Surface animals include pigs

(bushpigs and the giant forest hog in Africa, wild boar and bearded pig in Asia, and peccary in South America), species of small antelopes (bovids), and mammalian predators (the tiger in Asia, jaguar in South America, and leopard in Africa and Asia).

The present human assault on Earth's rain forests has put this diverse fauna and the varied flora at risk. It also jeopardizes an important recycling system for atmospheric carbon dioxide and a potential source of valuable pharmaceuticals and many types of new foods—and so much is still unknown and undiscovered.

Deforestation of the Tropics

More than half of Earth's original rain forest is gone, cleared for pasture, timber, fuel wood, and farming. Worldwide, an area nearly the size of Wisconsin is lost each year (169,000 km^2, 65,000 mi^2) and about a third more is disrupted by selective cutting of canopy trees, a damage that occurs along the *edges* of deforested areas.

When orbiting astronauts look down on the rain forests at night, they see thousands of human-set fires. During the day, the lower atmosphere in these regions is choked with the smoke. These fires are used to clear land for agriculture, which is intended to feed the domestic population as well as to produce accelerating cash exports of beef, rubber, coffee, soybeans, and other commodities. Edible fruits are not abundant in an undisturbed rain forest, but cultivated clearings produce bananas (plantains), mangos, jack fruit, guava, and starch-rich roots such as manioc and yams.

Because of the poor soil fertility, the cleared lands are quickly exhausted under intensive farming and are then generally abandoned in favor of newly burned and cleared lands, unless fertility is maintained artificially. Unfortunately, the dominant trees require from 100 to 250 years to reestablish themselves after major disturbances. Once cleared, the

(a)   1975

(b)   1986

(c)   1992

(e)

(f)   Brazil's deforestation
areas compared to North
America

(d)   2001

FIGURE 20.6 Amazon rain forest losses over three decades.
Satellite images of a portion of western Brazil called Rondônia, recorded in false color in (a) June 1975 (*Landsat 2*), (b) August 1986 (*Landsat 5*), (c) June 1992 (*Landsat 4*), and in true color (d) June 1, 2001 (*Terra* MODIS sensor; the area of the other three images is in the upper left). Highway BR364, the main artery in the region, passes through the towns of Jaru and Ji-Paraná near the center of image (d) at about 62° W 11° S. The branching pattern of feeder roads encroaches on the rain forest. (e) Surface view of burning rain forest. (f) Deforested areas in Brazil, cleared between 1972 and 2000, and for 2001–2003, and the record-setting 2004, are compared to North America. [*Landsat* images courtesy of NASA/USGS EROS Data Center; *Terra* MODIS image courtesy of NASA/MODIS Land Rapid Response Team; (e) photo by George Holton/Photo Researchers, Inc.]

former forest becomes a mass of low bushes intertwined with vines and ferns, slowing the return of the forests.

Figure 20.6a–c shows satellite false-color images of a portion of western Brazil called Rondônia, recorded in June 1975 (*Landsat 2*), August 1986 (*Landsat 5*), June 1992 (*Landsat 4*), and in (d) true color in June 2001 (*Terra*, MODIS). These images give a sense of the level of rain-forest destruction in progress. You can clearly see encroachment along new roads branching from highway BR364. The *edges* of every road and cleared area represent

a significant portion of the species habitat disturbance, population dynamic changes, and carbon losses to the atmosphere—perhaps as much as 1.5 times more impact occurs along these edges than in the tracts of clear-cut logging. The hotter, drier, and windier edge conditions penetrate the forest up to 100 m. Figure 20.6e shows a tract of former rain forest that has just been burned to begin the clearing and road-building process.

In 1998, despite all the efforts to introduce sustainable forestry practices, deforestation increased 27% over

1997—some 16,800 km² (6500 mi²)—and rose again to a 19,836 km² (7660 mi²) loss in 2000 (an area roughly equivalent to Connecticut, Rhode Island, and Delaware combined and noted on the map). Since 1972, this has brought total deforestation to more than 15% of the entire Amazon Basin—more than the area of the combined north-central states of Minnesota, Iowa, Missouri, North and South Dakota, Nebraska, and Kansas (Figure 20.6f), or equivalent to the area of France in 30 years! Remember, in 1970 only about 1% of the Brazilian Amazon had been deforested.

Estimates based on satellite imagery and ground surveys place the losses for 2001–2003 at a staggering 75,545 km² (29,170 mi²)—or an area equal to slightly less than the states of Vermont, New Hampshire, and Massachusetts combined. The losses in 2004 slightly exceeded the equivalent area of New Jersey (as noted on the map in Figure 20.6f). This is a highly charged issue, as can be seen by the growth of Brazil's cattle industry using these new pasture lands. Cattle herds increased from 25 million head to 57 million head during the period 1990–2002. Brazil is a major beef exporter and consumer of its own rainforest resource base.

The United Nations Food and Agricultural Organization (FAO) estimates that if this destruction to rain forests continues unabated, they will be completely removed by 2050! By continent, rainforest losses are estimated at more than 50% in Africa, more than 40% in Asia, and 40% in Central and South America.

To slow this continuing catastrophe of deforestation, in 1985 the FAO, the UN Development Programme, the World Bank, the World Resources Institute, and several nongovernmental organizations initiated the Tropical Forestry Action Plan. This plan, armed with $8 billion in development aid money, instituted an accurate worldwide survey of the rate of deforestation, using satellites to obtain remote measurements of land cover for building more-accurate GIS models. Critics charge that this effort has yet to show progress. (Among many references, see Tropical Rainforest Coalition at **http://www. rainforest.org/**, World Resources Institute publications on forests at **http://www.wri.org/wri/cat-frst.html**, the Rainforest Information Centre at **http://www. rainforestinfo.org.au/welcome.htm**, or the Rainforest Action Network at **http://www.ran.org/**.)

Focus Study 20.1 looks more closely at efforts to curb increasing rates of species extinction and loss of Earth's biodiversity, much of which is directly attributable to the loss of rain forests.

Tropical Seasonal Forest and Scrub

A varied biome on the margins of the rain forest is the **tropical seasonal forest and scrub**, which occupies regions of low and erratic rainfall. The shifting intertropical convergence zone (ITCZ) brings precipitation with the seasonally shifting high Sun of summer and then a season of dryness with the low Sun of winter, producing a seasonal pattern of moisture deficits, some leaf loss, and dry-season flowering. The term *semideciduous* applies to some of the broadleaf trees that lose their leaves during the dry season.

(a)

(b)

FIGURE 20.7 Kenyan landscapes.
Two views of the open thorn forest and savanna near and in the Samburu Reserve, Kenya, representing a broad ecotone between the two biomes. [(a) Photo by Gael Summer-Hebden; (b) photo by Stephen J. Krasemann/DRK Photo.]

Areas of this biome have fewer than 40 rainy days during their four consecutive driest months, yet heavy monsoon downpours characterize their summers (see Chapter 6, especially Figure 6.21). The climates *tropical monsoon* and *tropical savanna* apply to these transitional communities between rain forests and tropical grasslands.

Portraying such a varied biome is difficult. In many areas humans disturb the natural biome, so that the savanna grassland adjoins the rain forest directly. The biome does include a gradation from wetter to drier areas: monsoonal forests, to open woodlands and scrub woodland, to thorn forests, to drought-resistant scrub species (Figure 20.7). In South America, an area of transitional tropical deciduous forest surrounds an area of savanna in southeastern Brazil and portions of Paraguay.

The monsoonal forests average 15 m (50 ft) high with no continuous canopy of leaves, graduating into orchardlike parkland with grassy openings or into areas choked by dense undergrowth. In more open tracts, a

Focus Study 20.1

Biodiversity and Biosphere Reserves

When the first European settlers landed on the Hawaiian Islands in the late 1700s, they counted 43 species of birds. Today, 15 of those species are extinct and 19 more are threatened or endangered, with only one-fifth of the original species relatively healthy. Humans have great impact on animals and plants and global biodiversity.

Relatively pristine habitats around the world are being lost at unprecedented rates as an expanding human population converts them to agriculture, forestry, and urban centers. As these habitats are altered, untold numbers of species are disappearing before they have been recognized, much less studied, and the functioning of entire ecosystems is threatened. This loss of biodiversity, at the very time when the value of biotic resources is becoming widely recognized, has made it strikingly clear that current strategies for conservation are failing dismally.*

As more is learned about Earth's ecosystems and their related communities, more is known of their value to civilization and our interdependence with them. Natural ecosystems are a major source of new foods, new chemicals, new medicines, and specialty woods, and of course they are indicators of a healthy, functioning biosphere.

International efforts are under way to study and preserve specific segments of the biosphere, among others: the UN Environment Programme, World Resources Institute, The World Conservation Union, Natural World Heritage Sites, Wetlands of International Importance, and the Nature Conservancy. An important part of the effort is the setting aside of biosphere reserves and the focus

provided by the World Conservation Monitoring Centre and its IUCN Red List of global endangered species. (See IUCN Red List at http://www.redlist.org/, the IUCN at http://www.iucn.org/, the endangered species home page of the Fish and Wildlife Service at http://endangered.fws.gov/, and the World Wildlife Fund at http://www.panda.org/.)

The motivation to set aside natural sanctuaries is directly related to concern over the increase in the rate of species extinctions. We are facing a loss of genetic diversity that may be unparalleled in Earth's history, even compared with the major extinctions over the geologic record.

Species Threatened—An Example

Black rhinos (*Diceros bicornis*) and white rhinos (*Ceratotherium simum*) in Africa exemplify species in jeopardy. Rhinos once grazed over much of the savanna grasslands and woodlands.

Today, they survive only in protected districts in heavily guarded sanctuaries and are threatened even there (Figure 20.1.1). In 1998, 2599 black rhinos remained, 96% less than the 70,000 in 1960. Populations remained statistically stable through the last half of the 1990s, and this count is supported by South Africa's guarding about 50% of the remaining population. A slow recovery is underway as numbers of black rhinos rose to 3,600 in 2003, an increase of 500 over the 2001 count (source: IUCN/SSC/African Rhino Specialist Group). This is not the case with the western black rhino, a subspecies, which is down to its last few individuals.

There were 11 white rhinos (northern subspecies) in existence in 1984, increasing to an estimated 29 by 1998 (Congo 25, Côte d'Ivoire 4), although the IUNC suggests this number might actually be only 20 rhinos in 2004. Political unrest in the Congo and the area of the Garamba National Park has led to the deaths of an

FIGURE 20.1.1 The rhinoceros in Africa.
Black rhino and young in Tanzania, escorted by Oxpecker birds. These rhinos are quite nearsighted (they can see clearly only up to 10 m); the birds act as the rhino's early-warning system for disturbances in the distance. [Photo by Stephen J. Krasemann/DRK Photo.]

*T. D. Sisk, A. E. Launer, K. R. Switky, and P. R. Ehrlich, "Identifying extinction threats," *Bioscience* 44 (October 1994): 592.

(continued)

Focus Study 20.1 *(continued)*

Table 1 Known and Estimated Species on Earth

Categories of living organisms	Number of Known Species	Estimated Number of Species		Working Estimate (000)	Accuracy
		High (000)	Low (000)		
Viruses	4,000	1,000	50	400	Very poor
Bacteria	4,000	3,000	50	1000	Very poor
Fungi	72,000	27,000	200	1500	Moderate
Protozoa	40,000	200	60	200	Very poor
Algae	40,000	1,000	150	400	Very poor
Plants	270,000	500	300	320	Good
Nematodes	25,000	1,000	100	400	Poor
Arthropods					
Crustaceans	40,000	200	75	150	Moderate
Arachnids	75,000	1,000	300	750	Moderate
Insects	950,000	100,000	2,000	8,000	Moderate
Mollusks	70,000	200	100	200	Moderate
Chordates	45,000	55	50	50	Good
Others	115,000	800	200	250	Moderate
Total	**1,750,000**	**111,655**	**3,635**	**13,620**	**Very poor**

Source: United Nations Environment Programme, *Global Biodiversity Assessment* (Cambridge: Cambridge University Press, 1995), Table 3.1–2, p. 118.

unknown number of these few. The southern white rhino population reached 11,000 by 2002.

Rhinoceros horn sells for $29,000 per kilogram for use as an imagined aphrodisiac. These large land mammals are nearing extinction and will survive only as a dwindling zoo population. The limited genetic pool that remains complicates further reproduction.

Table 1 summarizes the numbers of known and estimated species on Earth. Scientists have classified only 1.75 million species of plants and animals of an estimated 13.6 million overall; this figure represents an increase in what scientists once thought to be the diversity of life on Earth. And those yet-to-be-discovered species represent a potential future resource for society—the wide range of estimates places the expected species count between a low of 3.6 million and a high of 111.7 million.

Estimates of annual species loss overall range between 1000 and 30,000, although this range might be

conservative. The possibility exists that more than half of Earth's present species could be extinct within the next 100 years. The effects of pollution, loss of wild habitat, excessive grazing, poaching, and collecting are at the root of such devastation. Approximately 60% of the extinctions are attributable to the clearing and loss of rain forests alone.

The Question of Medicine and Food

Wheat, maize (corn), and rice—just three grains—fulfill about 50% of human planetary food demands. About 7000 plant species have been gathered for food throughout human history, but more than 30,000 plant species have edible parts. Undiscovered potential food resources are in nature waiting to be found and developed. Biodiversity, if preserved, provides us a potential cushion for all future food needs, but only if species are identified, inventoried, and protected. The same is true for pharmaceuticals.

Nature's biodiversity is like a full medicine cabinet. Since 1959, 25% of all prescription drugs were originally derived from higher plants. In preliminary surveys, 3000 plants have been identified as having anticancer properties. The rosy periwinkle (*Catharanthus roseus*) of Madagascar contains two alkaloids that combat two forms of cancer. (Alkaloids are compounds found in certain plants that help the plant defend against insects and are potentially significant as human medicines; examples are atropine, quinine, and morphine.) Yet, less than 3% of flowering plants have been examined for alkaloid content. It defies common sense to throw away the medicine cabinet before we even open the door to see what is inside.

Genetic material from the developing world is already making an immense contribution to the food tables and medicine cabinets of the industrialized Northern Hemisphere. Agriculture in developed countries reaps an estimated $5 billion a year from Third World genetic material.... One Peruvian variety of tomato has been worth $8 million annually to U.S. tomato processors because it added soluble solid content to tomatoes.... And one in four prescription drugs is derived at least in part from plants.... Their value to the $60-billion-a-year U.S. pharmaceutical industry is considerable.*

Biosphere Reserves

Formal natural reserves are a possible strategy for slowing the loss of biodiversity and protecting this resource base. Setting up such a biosphere reserve involves principles of *island biogeography*. Island communities are special places for study because of their spatial isolation and the relatively small number of species present. Islands resemble natural experiments,

*J. W. King, "Breeding uniformity: Will global biotechnology threaten global biodiversity?" *The Amicus Journal* 15 (Spring 1993): 26.

because the impact of individual factors, such as civilization, can be more easily assessed on islands than over larger continental areas.

Studies of islands also can assist in the study of mainland ecosystems, for in many ways a park or biosphere reserve, surrounded by modified areas and artificial boundaries, is like an island. Indeed, a biosphere reserve is conceived as an ecological island in the midst of change. The intent is to establish a core in which genetic material is protected from outside disturbances, surrounded by a buffer zone that is, in turn, surrounded by a transition zone and experimental research areas. An important variable to consider in setting aside a biosphere reserve is any change that might occur in temperature and precipitation patterns as a result of global change. A carefully considered reserve could end up outside its natural range as ecotones shift.

Scientists predict that new, undisturbed reserves will not be possible in a decade, because pristine areas will be gone. The biosphere reserve program is coordinated by the Man and the Biosphere (MAB) Programme of UNESCO (http://www.unesco.org/mab/). Nearly 300 such biosphere reserves covering some 12 million hectares (30 million acres) are now operated voluntarily in 76 countries.

Not all protected areas are ideal bioregional entities. Some are simply imposed on existing park space and some remain in the planning stage, although they are officially designated. In Canada, there are six biosphere reserve sites; see the Canada/MAB Programme site at http://www.eman-rese.ca/partners/mab/intro.html. The United States has 90 biosphere reserves under various jurisdictions. Also, for other MAB links check http://www.mabnet.org/misc/links.html.

Some of the best biosphere reserve examples have been in operation since the late 1970s and range from the struggling Everglades National Park in Florida, to the Changbai Reserve in China, to the Tai Forest on the Côte d'Ivoire (Ivory Coast), to the Charlevoix Biosphere Reserve in Québec, Canada. Added to these efforts is the work of the Nature Conservancy, which acquires land for preservation as a valuable part of the reserve (http://nature.org/).

The ultimate goal, about half achieved, is to establish at least one reserve in each of the 194 distinctive biogeographic communities presently identified. UNESCO and the World Conservation Monitoring Centre compile the United Nations' list of national parks and protected areas.

Presently in these biogeographic communities, 6930 areas covering 657 million hectares (1.62 billion acres) are designated in some protective form, representing about 4.8% of national land area on the planet. Even though these are not all set aside to the degree of biosphere reserves, they do demonstrate progress toward preservation of our planet's plant and animal heritage.

The preservation of species diversity is a problem that must today be confronted by one species, *Homo sapiens*. . . . The diversity of species is worth preserving because it represents a wealth of knowledge that cannot be replaced. Moreover, today's extinctions are unlike those in previous eras, in which long periods of recovery could follow extinctions. The present situation is an inexorably irreversible one in which human overpopulation will destroy most species unless we plan for protection immediately. Accepting that the goal is worthwhile requires that more energy be devoted to planning and priorities and less to emotionalism and indignation [on all sides].*

*D. E. Koshland, Jr., "Preserving biodiversity," *Science* 253 (August 16, 1991): 717.

---

common tree is the acacia, with its flat top and usually thorny stems. These trees have branches that look like an upside-down umbrella (spreading and open skyward), as do trees in the tropical savanna.

Local names are given to these communities: the *caatinga* of the Bahia State of northeastern Brazil, the *chaco* area of Paraguay and northern Argentina, the *brigalow* scrub of Australia, and the *dornveld* of southern Africa. Figure 20.3 shows areas of this biome in Africa, extending from eastern Angola through Zambia to Tanzania and Kenya; in southeast Asia and portions of India, from interior Myanmar through northeastern Thailand; and in parts of Indonesia.

The trees throughout most of this biome make poor lumber, but some, especially teak, may be valuable for fine cabinetry. In addition, some of the plants with dry-season adaptations produce usable waxes and gums, such as carnauba and palm-hard waxes. Animal life includes the koalas and cockatoos of Australia and the elephants, large cats, rodents, and ground-dwelling birds in other occurrences of this biome.

Tropical Savanna

The **tropical savanna** is large expanses of grassland, interrupted by trees and shrubs. This is a transitional biome between the tropical forests and semiarid tropical steppes and deserts. The savanna biome also includes treeless tracts of grasslands, and in dry savannas, grasses grow discontinuously in clumps, with bare ground between them. The trees of the savanna woodlands are characteristically flat-topped, in response to light and moisture regimes.

Savannas covered more than 40% of Earth's land surface before human intervention but were especially modified by human-caused fire. Fires occur annually throughout the biome. The timing of these fires is important. Early in the dry season they are beneficial and increase tree cover; late in the season they are very hot and kill trees and seeds. Savanna trees are adapted to resist the "cooler" fires.

Forests and elephant grasses averaging 5 m (16 ft) high once penetrated much farther into the dry regions, for they are known to survive there when protected. Savanna grasslands are much richer in humus than the wetter tropics and are better drained, thereby providing a better base for agriculture and grazing. Sorghums, wheat, and groundnuts (peanuts) are common commodities.

Tropical savannas receive their precipitation during less than 6 months of the year, when they are influenced by the shifting ITCZ. The rest of the year they are under the drier influence of shifting subtropical high-pressure cells. Savanna shrubs and trees are frequently *xerophytic*, or drought-resistant, with various adaptations to protect them from the dryness: small, thick leaves; rough bark; or leaf surfaces that are waxy or hairy.

Africa has the largest area of this biome, including the famous Serengeti Plains of Tanzania and Kenya and the Sahel region, south of the Sahara. Sections of Australia, India, and South America also are part of the savanna biome. Some of the local names for these lands include the *Llanos* in Venezuela, stretching along the coast and inland east of Lake Maricaibo and the Andes; the *Campo Cerrado* of Brazil and Guiana; and the *Pantanal* of southwestern Brazil.

Particularly in Africa, savannas are the home of large land mammals (zebra, giraffe, buffalo, gazelle, wildebeest, antelope, rhinoceros, elephant). They graze on savanna grasses (Figure 20.8) or feed upon the grazers themselves (lion, cheetah). Birds include the Ostrich, Martial Eagle (largest of all eagles), and Secretary Bird. Many species of venomous snakes as well as the crocodile are present in this biome.

Unfortunately, in our lifetime we may see the reduction of these animal herds to zoo stock only, because of poaching and habitat losses, unless preservation efforts succeed. The loss of the rhino is discussed in Focus Study 20.1. Establishment of large tracts of savanna as biosphere reserves is critical for the preservation of this biome and its associated fauna.

Midlatitude Broadleaf and Mixed Forest

Moist continental climates support a mixed forest in areas of warm to hot summers and cool to cold winters. This **midlatitude broadleaf and mixed forest** biome includes several distinct communities in North America, Europe, and Asia. Along the Gulf of Mexico, relatively lush evergreen broadleaf forests occur. Northward are mixed deciduous and evergreen needleleaf stands associated with sandy soils and burning. When areas are given fire protection, broadleaf trees quickly take over. Pines (longleaf, shortleaf, pitch, loblolly) predominate in the southeastern and Atlantic coastal plains. Into New England and westward in a narrow belt to the Great Lakes, white and red pines and eastern hemlock are the principal evergreens, mixed with deciduous oak, beech, hickory, maple, elm, chestnut, and many others (Figure 20.9a).

These mixed stands contain valuable timber, but their distribution has been greatly altered by human activity. Native stands of white pine in Michigan and Minnesota were removed before 1910; only later reforestation sustains their presence today. In northern China these forests have almost disappeared as a result of centuries of harvest. The forest species that once flourished in China are similar to species in eastern North America: oak, ash, walnut, elm, maple, and birch.

This biome is quite consistent in appearance from continent to continent and at one time represented the principal vegetation of the *humid subtropical hot summer* and *marine west coast* (cool summer, winter drought) climatic regions of North America, Europe, and Asia.

A wide assortment of mammals, birds, reptiles, and amphibians is distributed throughout this biome. Representative animals (some migratory) include red fox, white-tail deer, southern flying squirrel, opossum, bear, and a great variety of birds, including Tanager and Cardinal (Figure 20.9b). To the north of this biome, poorer soils and

(a)

(b)

FIGURE 20.8 Animals and plants of the Serengeti Plains.
(a) Savanna landscape of the Serengeti Plains, with wildebeest, zebras, and thorn forest.
(b) Near Samburu, Kenya, Grevy's zebra and reticulated giraffe forage. [Photos by (a) Stephen F. Cunha; (b) Galen Rowell/Mountain Light Photography, Inc.]

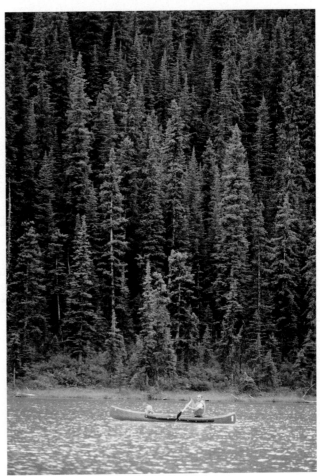

FIGURE 20.10 Boreal forest of Canada (boreal means "northern"). [Photo by author.]

FIGURE 20.9 The mixed broadleaf forest.
(a) A mixed broadleaf forest in the Midwest. (b) A young white-tailed deer grazes in the undergrowth of a mixed forest. [Photos by (a) Bobbé Christopherson; (b) by John Shaw/Tom Stack & Associates.]

colder climates favor stands of coniferous trees and a gradual transition to the northern needleleaf forests.

Needleleaf Forest and Montane Forest

Stretching from the east coast of Canada and the Atlantic provinces westward to Alaska and continuing from Siberia across the entire extent of Russia to the European Plain is the northern **needleleaf forest** biome, also called the **boreal forest** (Figure 20.10). A more open form of boreal forest, transitional to arctic and subarctic regions, is termed the **taiga**. The Southern Hemisphere, lacking *humid microthermal* climates except in mountainous locales, has no such biome. But **montane forests** of needleleaf trees exist worldwide at high elevation.

Boreal forests of pine, spruce, fir, and larch occupy most of the subarctic climates on Earth that are dominated by trees. Although these forests are similar in formation, individual species vary between North America and Eurasia.

The larch (*Larix*) is interesting because it is the rare needleleaf tree that loses its needles in the winter months, perhaps as a defense against the extreme cold of its native Siberia (see the Verkhoyansk climographs and photograph in Figure 10.20). Larches also occur in North America.

The Sierra Nevada, Rocky Mountains, Alps, and Himalayas have similar forest communities occurring at lower latitudes. Douglas fir and white fir grow in the western mountains in the United States and Canada. Economically, these forests are important for lumbering, with trees for lumber occurring in the southern margins of the biome and for pulpwood throughout the middle and northern portions. Present logging practices and the sustainability of these yields are issues of increasing controversy.

In the Sierra Nevada montane forests of California, the giant sequoias naturally occur in seven isolated groves. These trees are Earth's largest (in terms of biomass) living things, although they began as a small seed (Figure 20.11a). Some of these *Sequoia gigantea* exceed 8 m in diameter (28 ft) and 83 m (270 ft) in height. The largest of these is the General Sherman tree in Sequoia National Park (Figure 20.11c); it is estimated to be 3500 years old. The bark is fibrous, a half-meter thick, and lacks resins, so it effectively resists fire. Imagine the lightning strikes and fires that must have moved past the Sherman tree in 35 centuries!

(a)

(b)

(c)

FIGURE 20.11 Sequoia: the seed, the seedling, the giant.
(a) A *Sequoia gigantea* seed. About 300 seeds are in each sequoia cone. (b) Seedling at approximately 50 years of age. (c) The General Sherman tree in Sequoia National Park, California. This tree is probably wider than your classroom. The first branch is 45 m (150 ft) off the ground and is 2 m (6.5 ft) in diameter! [All photos by author.]

Standing among these giant trees is an overwhelming experience and creates a sense of the majesty of the biosphere.

Certain regions of the northern needleleaf biome experience permafrost conditions discussed in Chapter 17. When coupled with rocky and poorly developed soils, these conditions generally limit the existence of trees to those with shallow rooting systems. The summer thaw of surface active layers results in muskeg (moss-covered) bogs and Histosols (organic) soils of poor drainage and stability. Soils of the taiga are typically Spodosols (Podzolic, subject to podzolization), characteristically acidic and leached of humus and clays. Global warming is altering conditions in the high latitudes, causing increased melting and depth in the active layer. By the late 1990s, some affected forests were beginning to die in response to waterlogging in soils.

Representative fauna in this biome include wolf, moose (the largest in the deer family), bear, lynx, beaver, wolverine, marten, small rodents, and migratory birds during the brief summer season (Figure 20.12). Birds include hawks and eagles, several species of grouse, Pine Grosbeak, Clark's Nutcracker, and several owls. About 50 species of insects particularly adapted to the presence of coniferous trees inhabit the biome.

Temperate Rain Forest

Lush forests at middle and high latitudes are in the **temperate rain forest** biome. In North America, it occurs only along narrow margins of the Pacific Northwest (Figure 20.13a). Some similar types of forest exist in southern China, small portions of southern Japan, New Zealand, and a few areas of southern Chile.

This biome contrasts with the diversity of the equatorial and tropical rain forest in that only a few species make up the bulk of the trees. The rain forest of the Olympic Peninsula in Washington State is a mixture of broadleaf and needleleaf trees, huge ferns, and thick undergrowth. Precipitation approaching 400 cm (160 in.) per year on

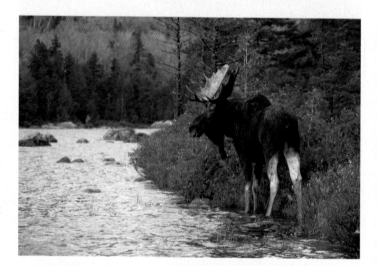

FIGURE 20.12 Animals of the northern needleleaf forest.
A bull moose with a full set of antlers grazes along the water's edge. [Photo by Jeremy Woodhouse/PhotoDisc, Inc.]

the western slopes, moderate air temperatures, summer fog, and an overall maritime influence produce this moist, lush vegetation community. Animals include bear, badger, deer, wild pig, wolf, bobcat, and fox. The trees are home to numerous bird species, including the famous Spotted Owl (Figure 20.13b).

The tallest trees in the world occur in this biome—the coastal redwoods (*Sequoia sempervirens*). Their distribution is shown on the map in Figure 19.12a. These trees can exceed 1500 years of age and typically range in height from 60 to 90 m (200 to 300 ft), with some exceeding 100 m (330 ft). Virgin stands of other representative trees, such as Douglas fir, spruce, cedar, and hemlock, have been reduced by timber harvests to a few remaining valleys in Oregon and Washington, less than 10% of the original forest. Replanting and secondary-growth forests are predominant. In similar forests in Chile, large-scale timber

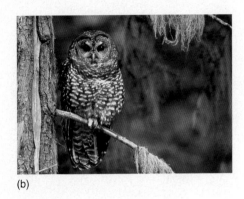

(a)   (b)

FIGURE 20.13 Temperate rain forest.
(a) An ancient temperate rain forest in the Pacific Northwest in the Gifford Pinchot National Forest features huge old-growth Douglas fir, redwoods, cedars, and a mix of deciduous trees, ferns, and mosses. Only a small percentage of these old-growth forests remain. (b) The Northern Spotted Owl is characteristic of the temperate rain forest. [Photos by (a) Bobbé Christopherson; (b) Greg Vaughn/Tony Stone Images.]

harvests and new mills began operations in 2000. U.S. corporations are shifting logging operations to these forests in the Chilean Lake District and northern Patagonia.

Many studies by the U.S. Forest Service and others noted the failing ecology of these forest ecosystems and suggested that timber management plans should include ecosystem preservation as a priority. The ultimate solution must be one of economic and ecological synthesis rather than continuing conflict and forest losses—the forests can't be all cut down, nor can they be all preserved. The opposing sides in this conflict between logging and society interests should combine their efforts in a sustainable forestry model.

Mediterranean Shrubland

The **Mediterranean shrubland** biome, also referred to as a temperate shrubland, occupies those regions poleward of the shifting subtropical high-pressure cells. As those cells shift poleward with the high summer Sun, they cut off available storm systems and moisture. Their stable high-pressure presence produces the characteristic *Mediterranean dry summer* and establishes conditions conducive to fire (Figure 19.27d).

Plant ecologists think that this biome is well adapted to frequent fires, for many of its characteristically deep-rooted plants have the ability to resprout from their roots after a fire. Earlier stands of evergreen woodlands (holm or evergreen oak) no longer dominate.

The dominant shrub formations that occupy these regions are stunted and able to withstand hot-summer drought. The vegetation is called *sclerophyllous* (from *sclero*, for "hard," and *phyllos*, for "leaf"). It averages a meter or two in height and has deep, well-developed roots, leathery leaves, and uneven low branches.

Typically, the vegetation varies between woody shrubs covering more than 50% of the ground and grassy woodlands covering 25%–60% of the ground. In California, the

Spanish word *chaparro* for "scrubby evergreen" gives us the word **chaparral** for this vegetation type (Figure 20.14). This scrubland includes species such as manzanita, toyon, red bud, ceanothus, mountain mahogany, blue and live oaks, and the dreaded poison oak.

A counterpart to the California chaparral in the Mediterranean region, called *maquis*, includes live and cork oak trees (source of cork), as well as pine and olive trees. The overall similar appearance of the California and Spanish oak savannas is demonstrated in Figure 10.16c and d. In Chile, such a region is called *mattoral*; in southwestern Australia, *mallee scrub*. In Australia, the bulk of the eucalyptus species is sclerophyllous in form and structure in whichever climate it occurs.

As described in Chapter 10, Mediterranean climates are important in commercial agriculture for subtropical fruits, vegetables, and nuts, with many food types produced only in this biome (e.g., artichokes, olives, almonds). Larger animals, such as several types of deer, are grazers and browsers, with coyote, wolf, and bobcat as predators. Many rodents, other small animals, and a variety of birds also proliferate.

Midlatitude Grasslands

Of all the natural biomes, the **midlatitude grasslands** are the most modified by human activity. Here are the world's "breadbaskets"—regions that produce bountiful grain (wheat and corn), soybeans, and livestock (hogs and cattle). Figure 20.15 shows a typical midlatitude grassland under cultivation. In these regions, the only naturally occurring trees were deciduous broadleafs along streams and other limited sites. These regions are called grasslands because of the predominance of grasslike plants before human intervention:

In this study of vegetation, attention has been devoted to grass because grass is the dominant feature of the Plains and is at the same time an index to their history. Grass is

FIGURE 20.14 Mediterranean chaparral.
Chaparral vegetation associated with the Mediterranean dry-summer climate in southern California. [Photo by Bobbé Christopherson.]

FIGURE 20.15 Grasslands of North America.
Midlatitude grasslands of Alberta, Canada, under cultivation. [Photo by Fotopic/Omni–Photo Communications, Inc.]

the visible feature which distinguishes the Plains from the desert. Grass grows, has its natural habitat, in the transition area between timber and desert. . . . The history of the Plains is the history of the grasslands.*

In North America, tallgrass prairies once rose to heights of 2 m (6.5 ft) and extended westward to about the 98th meridian, with shortgrass prairies in the drier lands farther west. The 98th meridian is roughly the location of the 51-cm (20-in.) isohyet, with wetter conditions to the east and drier conditions to the west (see Figure 18.17).

The deep, tough sod of these grasslands posed problems for the first European settlers, as did the climate. The self-scouring steel plow, introduced in 1837 by John Deere, allowed the interlaced grass sod to be broken apart, freeing the soils for agriculture. Other inventions were critical to opening this region and solving its unique

spatial problems: barbed wire (the fencing material for a treeless prairie); well-drilling techniques developed by Pennsylvania oil drillers but used for water wells; windmills for pumping; and railroads to conquer the distances.

Few patches of the original prairies (tall grasslands) or steppes (short grasslands) remain within this biome. For prairies alone, the reduction of natural vegetation went from 100 million hectares (250 million acres) down to a few areas of several hundred hectares each. The Tallgrass Prairie National Preserve is a 4410-hectare (10,894-acre) section of prairie located in the Flint Hills of Kansas. This is the only national park unit preserving this vestige of the former grasslands that covered much of the North American Great Plains. A state-designated 10-hectare (25-acre) remnant of the "prairie" is located 8 km north of Ames, Iowa, and the Hayden Prairie Preserve, also in Iowa, is pictured in News Report 19.4. Some of these patches of original grassland have never felt the plow. The map in Figure 20.3 shows the natural location of these former prairie and steppe grasslands.

Outside North America, the Pampas of Argentina and Uruguay and the grasslands of Ukraine are characteristic midlatitude grassland biomes. In each region of the world where these grasslands have occurred, human development of them was critical to territorial expansion.

This biome is the home of large grazing animals, including deer, pronghorn, and bison (the almost complete annihilation of the latter is part of American history, Figure 20.16). Grasshoppers and other insects feed on the grasses and crops as well, and gophers, prairie dogs, ground squirrels, Turkey Vultures, grouse, and Prairie Chickens are on the land. Predators include coyote, the nearly extinct black-footed ferret, badger, and birds of prey—hawks, eagles, and owls.

Deserts

Earth's **desert biomes** cover more than one-third of its land area, as is obvious in Figure 20.3. In Chapter 15, we examined desert landscapes and, in Chapter 10, desert climates. On a planet with such a rich biosphere, the deserts stand out as unique regions of fascinating adaptations for survival. Much as a group of humans in the desert might behave with short supplies, plant communities also compete for water and site advantage. Some desert plants, called *ephemerals*, wait years for a rainfall event, at which time their seeds quickly germinate and the plants develop, flower, and produce new seeds, which then rest again until the next rainfall event. The seeds of some xerophytic species open only when fractured by the tumbling, churning action of flash floods cascading down a desert arroyo, and, of course, such an event produces the moisture that a germinating seed needs.

Perennial desert plants employ other adaptive features to cope with the desert. Unique xerophytic features include

- Long, deep taproots (example, the mesquite); succulence (thick, fleshy, water-holding tissue, such as that of cacti)

*From *The Great Plains*, p. 32, ©1931 by Walter Prescott Webb, published by Ginn and Company, Needham Heights, MA.

FIGURE 20.16 Bison graze on the grasslands of Montana's National Bison range. [Photo by Lowell Georgia/CORBIS.]

- Spreading root systems to maximize water availability, waxy coatings and fine hairs on leaves to retard water loss, leafless conditions during dry periods (example, palo verde and ocotillo)
- Reflective surfaces to reduce leaf temperatures
- Tissue that tastes bad to discourage herbivores

The creosote bush (*Lorrea divaricata*) sends out a wide pattern of roots and contaminates the surrounding soil with toxins that prevent the germination of other creosote seeds, possible competitors for water. When a creosote bush dies, surrounding plants or germinating seeds will occupy the abandoned site, but they must wait for infrequent rains to remove the toxins, according to one hypothesis.

The faunas of both warm and cold deserts are limited by the extreme conditions and include few resident large animals. Exceptions are the camel, which can lose up to 30% of its body weight in water without suffering (for humans, a 10%–12% loss is dangerous), and the desert bighorn sheep, in scattered populations in inaccessible mountains and places such as the Grand Canyon along rocky outcrops and cliffs but not at the rim. In an effort to reestablish bighorn populations, several states are transplanting the animals to their former ranges. Some representative desert animals are the ring-tail cat, kangaroo rat, lizards, scorpions, and snakes. Most of these animals are quite secretive and become active only at night, when temperatures are lower. In addition, various birds have adapted to desert conditions and available food sources—for example, Roadrunners, Thrashers, Ravens, Wrens, Hawks, Grouse, and Nighthawks.

We are witnessing an unwanted expansion of the desert biome, discussed in Chapter 15 and Figure 15.24. This process, known as *desertification*, is now a worldwide phenomenon along the margins of semiarid and arid lands. Desertification is due principally to poor agricultural practices (overgrazing and agricultural activities that abuse soil structure and fertility), improper soil-moisture

management, erosion and salinization, deforestation, and the ongoing climatic change.

Earth's deserts are subdivided into desert and semi-desert associations, to distinguish those with expanses of bare ground from those covered by xerophytic plants of various types. The two broad associations are further separated into warm deserts, principally tropical and subtropical, and cold deserts, principally midlatitude.

Warm Desert and Semidesert Earth's **warm desert and semidesert** biomes are caused by the presence of dry air and low precipitation from subtropical high-pressure cells. These areas are very dry, as evidenced by the Atacama Desert of northern Chile, where only a minute amount of rain has ever been recorded—a 30-year annual average of only 0.05 cm (0.02 in.)! Like the Atacama, the true deserts are under the influence of the descending, drying, and stable air of high-pressure systems from 8 to 12 months of the year.

Remember that these dry regions are defined by low amounts of precipitation that fail to satisfy high amounts of potential evapotranspiration. Deserts receive precipitation that is less than one-half of potential evapotranspiration. *Semiarid steppe* climates receive precipitation that is more than one-half of annual potential evapotranspiration.

A few of the subtropical deserts—such as those in Chile, Western Sahara, and Namibia—are right on the seacoast and are influenced by cool offshore ocean currents. As a result, these true deserts experience summer fog that mists the plant and animal populations with needed moisture.

Vegetation ranges from almost none in the arid deserts to numerous xerophytic shrubs, succulents, and thorn tree forms. The lower Sonoran Desert of southern Arizona is a warm desert (Figure 20.17). This desert landscape features the unique saguaro cactus (*Carnegiea gigantea*), which, if undisturbed, grows to many meters in height and up to 200 years in age. First blooms do not appear until it is 50 to 75 years old!

The equatorward margin of the subtropical high-pressure cell is a region of transition to savanna, thorn tree, scrub woodland, and tropical seasonal forest. Poleward of the warm deserts, the subtropical cells shift to produce the Mediterranean dry-summer regime along west coasts and may grade into cool deserts elsewhere.

Cold Desert and Semidesert The **cold desert and semidesert** biomes tend to occur at higher latitudes. Here, seasonal shifting of subtropical high pressure is of some influence less than 6 months of the year. Interior locations are dry because of their distance from moisture sources or their location in rain-shadow areas on the lee side of mountain ranges, such as the Sierra Nevada of the western United States, the Himalayas, and the Andes. The combination of interior location and rain-shadow positioning produces the cold deserts of the Great Basin of western North America.

Winter snows occur in the cold deserts, but generally they are light. Summers are hot, with highs from 30°C to 40°C (86°F to 104°F). Nighttime lows, even in the

FIGURE 20.17 Sonoran Desert scene.
Characteristic vegetation in the Lower Sonoran Desert west of Tucson, Arizona (32° N; elevation 900 m, or 2950 ft). [Photo by author.]

summer, can cool 10–20 C° (18–36 F°) from the daytime high. The dryness, generally clear skies, and sparse vegetation lead to high radiative heat loss and cool evenings. Many areas of these cold deserts that are covered by sagebrush and scrub vegetation were actually dry shortgrass regions in the past, before extensive grazing altered their ecology. The deserts in the upper Great Basin today are the result of more than a century of such cultural practices.

Arctic and Alpine Tundra

NOTEBOOK

High-Latitude Animals
Geographic Scenes: East Greenland

The **arctic tundra** is found in the extreme northern area of North America, around the coast of Greenland, and in Russia, bordering on the Arctic Ocean and generally north of the 10°C (50°F) isotherm for the warmest month

(Figure 20.18). The opening photo for this chapter shows such tundra lands. Daylength varies greatly throughout the year, seasonally changing from almost continuous day to continuous night. Winters in this biome, *tundra* climate classification, are long and cold; cool summers are brief. The region, except for a few portions of Alaska and Siberia, was covered by ice during all of the Pleistocene glacial episodes.

Intensely cold continental polar air masses and stable high-pressure anticyclones govern tundra winters. A growing season of sorts lasts only 60–80 days, and even then frosts can occur at any time. Vegetation is fragile in this flat, treeless world; soils are poorly developed periglacial surfaces, which are underlain by permafrost. In the summer months, only the surface horizons thaw, thus producing a mucky surface of poor drainage. Roots can penetrate only to the depth of thawed ground, usually about a meter. The surface is shaped by freeze-thaw cycles that create the frozen-ground phenomena discussed in Chapter 17.

(a)

(b)

FIGURE 20.18 Arctic tundra.
(a) Tundra on the Kamchatka Peninsula, Russia. The Uzon Caldera is in the center-background mountain range. (b) Lush by tundra standards, flowers, grasses, mosses, and dwarf willow flourish in the cold climate of the Arctic. [Photos (a) by Wolfgang Kaehler Photography; (b) by Bobbé Christopherson.]

Tundra vegetation consists of low, ground-level herbaceous plant species, such as sedges, mosses, arctic meadow grass, snow lichen, and some woody dwarf willow, as shown in Figure 19.3c and here in Figure 20.18. Owing to the short growing season, some perennials form flower buds one summer and open them for pollination the next. Animals of the tundra include musk ox (see the chapter-opening photo), caribou, reindeer, rabbit, ptarmigans, lemmings, and other small rodents (important food for the larger carnivores), wolf, fox, weasel, Snowy Owls, polar bears (in nearby marine and pack ice environments), and, of course, mosquitoes. The tundra is an important breeding ground for geese, swans, and other waterfowl.

Alpine tundra is similar to arctic tundra, but it can occur at lower latitudes because it is associated with high elevation. This biome usually is described as above the timberline, that elevation above which trees cannot grow. Timberlines increase in elevation equatorward in both hemispheres. Alpine tundra communities occur in the Andes near the equator, the White Mountains of California, the Colorado and Canadian Rockies, the Alps, and Mount Kilimanjaro of equatorial Africa, as well as in mountains from the Middle East to Asia.

Alpine meadows feature grasses, herbaceous annuals (small plants), and stunted shrubs, such as willows and heaths. Because alpine locations are frequently windy sites, many plants appear to have been sculpted by the wind. Alpine tundra can experience permafrost conditions. Characteristic fauna include mountain goats, bighorn sheep, elk, and voles (Figure 20.19).

FIGURE 20.19 Alpine tundra conditions.
An alpine tundra and grazing mountain goats near Mount Evans, Colorado, at 3660 m (12,000 ft) elevation. [Photo by Bobbé Christopherson.]

Because the tundra biome is of such low productivity, it is fragile. Disturbances such as tire tracks, hydroelectric projects, and mineral exploitation leave marks that persist for hundreds of years. As development continues, the region will face even greater challenges from oil spills, contamination, and disruption (News Report 20.3).

News Report 20.3

ANWR Faces Threats

Planning continues for oil exploration and development of the Arctic National Wildlife Refuge (ANWR) on Alaska's North Slope. The May 2001 U.S. *National Energy Policy* recommends oil exploration and development in ANWR, using the latest environmentally sensitive techniques. However, the U.S. Senate defeated the plan in 2002 as a tentative step toward ANWR protection. Political and corporate pressure is fairly constant to exploit these expected reserves of fossil fuels.

This pristine wilderness is above the Arctic Circle, bordering on the Beaufort Sea and adjoining the Yukon Territory of Canada (Figure 20.3.1). The refuge area sustains almost 200,000 caribou, polar and grizzly

bears, musk oxen, and wolves. Some have referred to it as "America's Serengeti," given the annual migration of hundreds of thousands of large animals.

The ANWR remains the only portion of Alaska's Arctic coast that is not open to oil and gas exploration at this time—the other 90% is open. But controversy over this refuge continues. In 1995, Congress considered potential ANWR lease revenues in its Budget Resolution, implying a green light for development. A presidential veto stopped this Budget Reconciliation Act. A 1998 USGS assessment disclosed that the oil under the Arctic Refuge coastal plain is likely held in many small reservoirs. Recovering this oil would require extensive alteration

of the Arctic National Wildlife Refuge landscape, a point disputed in the *National Energy Policy*.

An ironic situation was described in High Latitude Connection 10.1, where the thawing permafrost and deepening active layer triggered by rising temperatures has reduced the number of days oil exploration heavy equipment can operate on lands just outside ANWR. Previously 200 days of work were possible, as the ground remained frozen and supported the heavy trucks and rigs. Now with regional-scale warming, such travel is restricted to just 100 days.

Oil extraction could be devastating to the fragile tundra, according to a

(continued)

News Report 20.3 *(continued)*

FIGURE 20.3.1 Arctic National Wildlife Refuge.
In Alaska's Arctic National Wildlife Refuge, Mount Chamberlin, the second-highest peak of the Brooks Range, overlooks tundra in the foreground. Is this region destined for petroleum exploration and development or for continued preservation as wilderness? [Photo by Scott T. Smith.]

USGS study released in 2002. The cost of development would price the oil at well over $30 per barrel. The estimated oil reserve in the ANWR, approximately 3.2 billion barrels (less than a 5-month supply at present U.S. demand levels), could be offset simply by a small increase in U.S. automobile efficiency. However, 2002 and 2003 marked years when mileage efficiency decreased. In 2002, federal funding toward improving automobile mileage efficiency was formally dropped and research monies for hybrid car technology stopped in favor of fuel cell research for the future. Clearly, economic ventures should be weighed against the ecosystem itself, its limitations and its uniqueness, and a total assessment of all costs to the consumer.

Summary and Review—Terrestrial Biomes

▶ *Define* the concept of biogeographic realms of plants and animals, and *define* ecotone, terrestrial ecosystem, and biome.

Earth is the only planet in the Solar System with a biosphere. An impressive feature of the living Earth is its diversity, which biogeographers categorize into discrete spatial biomes for analysis and study. The interplay among supporting physical factors within Earth's ecosystem determines the distribution of plant and animal communities. A **biogeographic realm** of plants and animals is a geographic region in which a group of species evolved. This recognition was a start at understanding distinct regions of flora and fauna and the broader pattern of terrestrial ecosystems. A boundary transition zone adjoining ecosystems is an **ecotone**.

A **terrestrial ecosystem** is a self-sustaining association of plants and animals and their abiotic environment that is characterized by specific plant formation classes. A **biome** is a large, stable ecosystem characterized by specific plant and animal communities. Biomes carry the name of the dominant vegetation because it is the most easily identified feature: forest, savanna, grassland, shrubland, desert, tundra.

    biogeographic realm (p. 650)
    ecotone (p. 650)
    terrestrial ecosystem (p. 650)
    biome (p. 651)

1. Reread the opening quotation and analysis in this chapter. What clue does it give you to the path ahead for Earth's forests? Is our future direction controllable? Explain.

2. What is a biogeographic realm? How is the world subdivided according to plant and animal types?

3. Describe a transition zone between two ecosystems. How wide is an ecotone? Explain.

4. Define *biome*. What is the basis of the designation?

▶ *Define* six formation classes and the life-form designations, and *explain* their relationship to plant communities.

Biomes are divided into more specific vegetation units called **formation classes**. The structure and appearance of the vegetation is described: rain forest, needleleaf forest, Mediterranean shrubland, arctic tundra, and so forth. Specific life-form designations include trees, lianas, shrubs, herbs, bryophytes, epiphytes (plants growing aboveground on other plants), and thallophytes (lacking true leaves, stems, or roots, including bacteria, fungi, algae, and lichens).

    formation classes (p. 652)

5. Distinguish between formation classes and life-form designations as a basis for spatial classification.

▶ *Describe* 10 major terrestrial biomes, and *locate* them on a world map.

Biomes are Earth's major terrestrial ecosystems, each named for its dominant plant community. The 10 major biomes are generalized from numerous formation classes that describe vegetation. Ideally, a biome represents a mature community of natural vegetation. In reality, few undisturbed biomes exist in the world, for most have been modified by human activity. Many of Earth's plant and animal communities are experiencing an accelerated rate of change that could produce dramatic alterations within our lifetime.

For an overview of Earth's 10 major terrestrial biomes and their vegetation characteristics, soil orders, climate-type designations, annual precipitation range, temperature patterns, and water balance characteristics, please review Table 20.1.

    equatorial and tropical rain forest (p. 655)
    tropical seasonal forest and scrub (p. 660)
    tropical savanna (p. 663)

6. Using the integrative chart in Table 20.1 and the world map in Figure 20.3, select any two biomes and study the correlation of vegetation characteristics, soil, moisture, and climate with their spatial distribution. Then, contrast the two using each characteristic.

7. Describe the equatorial and tropical rain forests. Why is the rainforest floor somewhat clear of plant growth? Why are logging activities for specific species so difficult there?

8. What issues surround the deforestation of the rain forest? What is the impact of these losses on the rest of the biosphere? What new threat to the rain forest has emerged?

9. What do *caatinga, chaco, brigalow,* and *dornveld* refer to? Explain.

10. Describe the role of fire or fire ecology in the tropical savanna biome and the midlatitude broadleaf and mixed forest biome.

11. Why does the northern needleleaf forest biome not exist in the Southern Hemisphere, except in mountainous regions? Where is this biome located in the Northern Hemisphere, and what is its relationship to climate type?

12. In which biome do we find Earth's tallest trees? Which biome is dominated by small, stunted plants, lichens, and mosses?

13. What type of vegetation predominates in the Mediterranean dry-summer climates? Describe the adaptations necessary for these plants to survive.

14. What is the significance of the 98th meridian in terms of North American grasslands? What types of inventions enabled humans to cope with the grasslands?

15. Describe some of the unique adaptations found in a desert biome.

16. What is desertification (review from Chapter 15 and this chapter)? Explain its impact.

17. What physical weathering processes are specifically related to the tundra biome? What types of plants and animals are found there?

▶ *Relate* human impacts, real and potential, to several of the biomes.

The equatorial and tropical rainforest biome is undergoing rapid deforestation. Because the rain forest is Earth's most diverse biome and is important to the climate system, such losses are creating great concern among citizens, scientists, and nations. Efforts are underway worldwide to set aside and protect remaining representative sites within most of Earth's principal biomes. These biosphere reserves are coordinated by the Man and the Biosphere (MAB) Programme of UNESCO. Nearly 300 such biosphere reserves, covering some 12 million hectares (30 million acres), are now operated voluntarily in 76 countries.

18. What is the relationship between island biogeography and biosphere reserves? Describe a biosphere reserve. What are the goals?

19. Compare the map in Figure 20.3 with the composite satellite image inside the front cover of this text. What correlations can you make between the local summertime portrait of Earth's biosphere and the biomes identified on the map?

20. From the background in News Report 20.3, what is your assessment of ANWR? Using credible Internet sources, what is the present status relative to ongoing, but halted, efforts to begin oil and gas exploration?

 Network

The *Geosystems* Home Page provides online resources for this chapter on the World Wide Web. To begin: Once on the Home Page, click on this textbook, scroll the Table of Contents menu, select this chapter, and click "Begin." You will find self-tests that are graded, review exercises, specific updates for items in the chapter, and many links to interesting related pathways on the Internet under "Destinations." *Geosystems* is at **http://www. prenhall.com/christopherson**.

Critical Thinking

A. Given the information presented in this chapter about deforestation in the tropics, assess the present situation for yourself. What are the main issues? What natural assets are at stake? Natural resources versus sovereign state rights? How are global biodiversity and greenhouse warming related to these issues? What is the perspective of the less-developed countries that possess most of the rain forest? What is the perspective of the more-developed countries and their transnational corporations and millions of environmentally active citizens? What kind of action plan would you develop to accommodate all parties? How would you proceed?

B. Using Figure 20.3 (biomes), Figure 10.5 (climates), Figures 10.2 and 9.6 (precipitation), and Figure 8.2 (air masses), and the printed graphic scales on these four maps, consider the following hypothetical. Assume a northward climatic shift in the United States and Canada of 500 km (310 mi) (in other words imagine moving North America 500 km south, to simulate climatic categories shifting north). Describe your analysis of conditions through the Midwest from Texas to the prairies of Canada. Describe your analysis of conditions from New York through New England and into the Maritime provinces. What economic dislocations and relocations do you envision?

Dr. Katrina A. Moser, Assistant Professor, Biogeography and Paleolimnology

A list of Dr. Moser's areas of interest brings the term "multidisciplinary" to mind: climate change, environmental change, paleolimnology, biogeography, aquatic ecology, and Quaternary environments. She teaches Earth systems science and subjects related to this list of interests.

Katrina explains, "Some of this topic diversity goes back to high school where I had an interest in geology, but so many other things as well." She describes enjoying hikes with her dad on family outings to a cottage on a lake in Québec, Canada, her birth country. "When I was five years old I got my first rock hammer as a present because I had started a little rock collection. I remember paddling out on the lake and wondering what was living in the lake and in the mud on the bottom of the lake. My parents had reference books in which I looked things up. The course of my present journey was set early in life!"

"In my undergraduate work at Queen's University, Kingston, Ontario, Canada, I began working on a project on Ellesmere Island in the far north. I did a paleoecological study of Pleistocene fossils in the Alexandra Fjord, in what is now Nunavut." She earned a B.Sc. degree and moved on to McMaster University, Hamilton, Ontario.

About geography, Katrina says, "The strength of the geographic approach is that it synthesizes in a way that you can see patterns and make connections. It simply doesn't have the same limits that restrict other disciplines." She went on to complete her M.Sc. geography degree and in the process developed several new techniques for paleoenvironmental research, including a new semi-automated method to count and analyze charcoal (the fire history) through an image analysis system. In addition, Katrina completed a B.Ed. degree (bachelor of education) back at Queen's before returning to

McMaster for her Ph.D. in Geography.

She directed her studies toward paleolimnology, a sub-discipline of paleoenvironmental studies—reconstructing life and environments from the past with emphasis on processes and conditions in ancient and modern lakes. "I studied the sediments in lakes and the geologic setting of the lake basin and its relation to conditions in the lake, among many other topics."

In a paper in *Nature*, by Katrina and coauthors, it stated that, "There was a strong limnological response to warming climate conditions during the mid Holocene." This led her to an interest in, ". . . diatoms as proxies of climatic variability." Katrina studied diatoms for a year in Dr. John P. Smol's lab at Queen's University, learning diatom taxonomy, research and statistical methods, and developing mathematical models to estimate past climate conditions using diatom fossils.

Katrina explains that, "Looking in a microscope there is a universe of microorganisms that tell me a story about the past. These sediments are like an ancient library and the key books in this library are the diatoms. Diatoms are microscopic single-celled plants that exhibit a wide-variety of forms; some 12,000 species in hundreds of genera have been identified. What we see are their opaline silica skeletons. We can relate their occurrence to particular past environmental conditions."

Finishing her Ph.D. in 1997, she accepted a teaching/research position at the University of Utah, where she is presently working. Katrina also is Director of the Environmental Change Observatory (ECO) at the university, which is developing new techniques for environmental analysis, establishing a record of climate change, providing insights into present climate change, as well as working on the causes of water quality

changes and resource issues. With physical geography as a base, she states, "I see Earth systems science as a way to analyze system changes in the past and present, thereby developing methods to understand and predict future trends. What my paleoenvironmental focus does is extend the observational record from recent history, far back in Earth's history. In Earth systems science we build models of systems based on all these observations."

Katrina's interest in water quality and resources is reflected by her work on an NSF research project. Of great concern is whether there is any precedent for a long-term drought lasting decades in the western United States. The present western drought is in its sixth year and scientists want to know how long is will last. Katrina uses her techniques of reconstructing climate change over long-time frames with lake sediments to provide clues. "Dramatic diatom changes during the past 10,000 years suggest that there were periods of much longer and more intense drought in the past."

The tools she employs in her paleoenvironmental research include: diatom analysis (mentioned earlier), pollen analysis, charcoal from fires, and sediment characteristics (isotopic make up, grain size, magnetic properties, geochemistry). "In setting up research projects, I find that because the problems are so interdisciplinary in nature, I try to put together a group of qualified scientists from various fields—this multidisciplinary approach is the key to Earth systems science."

Dr. Moser has published several peer-reviewed journal articles, a coauthored book titled *Ecology and Distribution of Diatoms from Boreal Forest Lakes in Wood Buffalo National Park, Northern Alberta and Northwest Territories, Canada*, and she has many articles in preparation for publication of her latest research. She has presented

56 papers at professional meetings (through 2004), has given numerous talks, and serves as a scientific reviewer for many journals. And, with all this activity she gave birth to two children in 1998 and 2003.

At the University of Utah, Katrina co-taught classes with the late Don Currey, a beloved and respected scientist and geographer who passed away in 2004. He inspired many with his drive to constantly research new areas and recheck old sites to take advantage of all the new technology available. Katrina remembers, "Don was so popular, he was never alone in the hospital in his final days, as a flow of undergraduate and graduate students visited. He was a joy to listen to and such a story teller—bigger than life. I can't imagine teaching without him." Katrina's teaching is influenced by his approach, as she explains, "I like to get the students out in the field to see things first hand, as in 'experiential education.' There is nothing better than a challenging discussion while making field observations, with all points of view heard and respected."

About Geography, Katrina says, "With many Earth systems displaying such dynamic trends, with ecosystems having components that are out-of-

FIGURE 20.1.1 Dr. Katrina A. Moser at work in the Unita Mountains, Utah. Katrina removes a core of lake sediment that contains the environmental history in this area for the past several millennia. Now, her paleolimnology detective work begins. [Photo provided by Katrina Moser.]

phase with climatic conditions, with many special variables and connections to be made across millennial-scale time frames, I see Geography in a key position to contribute to our understanding of Earth systems. When we look at the environment, we need to see the long-term perspective and seek to understand the spatial

expressions of temporal variation. This is why I like a 'Geosystems' way of thinking in an integrated, multidisciplinary, applied approach."

In the photo of Katrina at work (Figure 20.1.1), she is extracting a core from a lake in the Uinta Mountains of Utah. She explains, "We retrieved the uppermost sediments using a modified Livingston core. This preserves the sediment–water interface. We can see algae and worm tubes preserved in the top layers. Sediments are laid down year after year; here we have retrieved a sequence nearly 3000 years old. Now the detective work, as each layer contains geochemical and biological evidence and represents clues about environmental conditions in each of those past years. We see in the layers a record of climate, forest fire activity, changes in vegetation and soils, changes in atmospheric conditions, and water quality. A paleolimnologist is a real-time detective unlocking a mystery. Such a long-term perspective is critical to understanding natural variability of the system, which is essential to understanding present trends, and key to the formulation of sound policies to protect our environment." Katrina has such perspective!

Dr. Robert Corell, Chair of the ACIA Integration Team, welcomes nearly 400 scientists, managers, government and non-governmental representatives, student researchers, and indigenous people of the Arctic to a first-ever Arctic climate symposium, held in Reykjavik, Iceland, November 2004. The ACIA released their climate impact assessment to the Arctic Council and to a global audience. [Photo by Bobbé Christopherson.]

CHAPTER 21

Earth and the Human Denominator

▶ Key Learning Concepts

After reading the chapter, you should be able to:

▶ **Determine** an answer for Carl Sagan's question, "Who speaks for Earth?"

▶ **Describe** the growth in human population, and **speculate** on possible future trends.

▶ **Analyze** "An Oily Bird," and **relate** your analysis to energy consumption patterns in the United States and Canada.

▶ **List** the subjects of recent environmental agreements, conventions, and protocols, and **relate** them to physical geography and Earth systems science.

▶ **List** 12 paradigms for the 21st century.

▶ **Appraise** your place in the biosphere, and **realize** your relation to Earth systems.

> During my space flight, I came to appreciate my profound connection to the home planet and the process of life evolving in our special corner of the Universe, and I grasped that I was part of a vast and mysterious dance whose outcome will be determined largely by human values and actions.*

Earth can be observed from profound vantage points, as this astronaut experienced on the 1969 *Apollo IX* mission. Since November 2000, the International Space Station (ISS) has operated at an orbital altitude of about 400 km (250 mi), in the upper reaches of Earth's thermosphere (see Figure 3.2). Just

---

*Rusty Sweickart, "Our backs against the bomb, our eyes on the stars," *Discovery*, July 1987, p. 62.

FIGURE 21.1 Our orbital perspective—Earth's Space Station.
The International Space Station (ISS) began scientific operations in November 2000. More than 300 hours of human space walks went into the construction of the ISS through January 2003. All life-support supplies must be brought from Earth. Water recycling and purification techniques provide the bulk of the water needed. Oxygen is electrolytically derived from water and supplemented by compressed oxygen. Enormous solar-panel arrays generate electricity. Without plants, carbon dioxide must be scrubbed from the air the astronauts breathe. [Space Shuttle *Atlantis* photo made October 16, 2002, courtesy of NASA.]

imagine, a 16-nation coalition has built a scientific workstation, in orbit, that is 44.5 m long, 73 m wide, and 27.5 m tall (240 ft by 146 ft by 90 ft), and it is not yet finished (Figure 21.1). The 182-tonne (400,000 lb) complex has had more than 120 astronauts and scientists aboard.

More than 175 scientific investigations and experiments are completed or underway, to help us better understand Earth and life systems through research in the unique space environment. Problems with the Space Shuttle launch craft to service the station have delayed some progress. If completed as planned, the ISS will weigh 455 tonnes (1 million lb) and have the cabin volume of a 747 jumbo jet. This seems like an important fulfillment of our human nature: to question, to explore, to discover, to test the limits of the known, and to push beyond to the unknown. (See **http://spaceflight.nasa. gov/station/** for more information and updates.)

Our vantage point in this book is that of physical geography. We examine Earth's many systems: its energy, atmosphere, winds, ocean currents, water, weather, climate, endogenic and exogenic systems, soils, ecosystems, and biomes. This exploration has led us to an examination of the planet's most abundant large animal, *Homo sapiens*.

We stand in the first decade of the 21st century. The 21st century will be an adventure for the global society, historically unparalleled in experimenting with Earth's life-supporting systems. You will spend the majority of your life in this century. What preparations and "future thinking" can you do to understand all that is to occur?

In his 1980 book and public television series *Cosmos*, the late astronomer Carl Sagan asked:

> What account would we give of our stewardship of the planet Earth? We have heard the rationales offered by the nuclear superpowers. We know who speaks for the nations. But who speaks for the human species? Who speaks for Earth?*

Indeed, who does speak for Earth? We might answer: Perhaps we physical geographers, and other scientists who have studied Earth and know the operations of the global ecosystem, should speak for Earth. However, some might say that questions of technology, environmental politics, and future thinking belong outside of science, and that our job is merely to learn how Earth's processes work and to leave the spokesperson's role to others. Biologist-ecologist Marston Bates addressed this line of thought in 1960:

> Then we came to humans and their place in this system of life. We could have left humans out, playing the ecological game of "let's pretend humans don't exist." But this seems as unfair as the corresponding game of the economists, "let's pretend nature doesn't exist." The economy of nature and the ecology of humans are inseparable and attempts to separate them are more than misleading, they are dangerous. Human destiny is tied to nature's destiny and the arrogance of the engineering mind does not change this. Humans may be a very peculiar animal, but they are still a part of the system of nature.†

A fact of life is that Earth's more-developed countries (MDCs), through their economic dominance, speak for the billions who live in less-developed countries (LDCs). The gross state product (GSP) of California's 35.5 million people was $1.3 trillion in 2002, which was greater by more than $100 billion the gross domestic product (GDP) of China and its 1.3 billion people. The scale of disparity on our home planet is difficult to comprehend.

The fate of traditional modes of life may rest in some distant financial capital. But, economics aside, the reality is that the remote lands of Siberia are linked by Earth systems to the Pampas of Argentina to the Great Plains in North America and to those harvesting grain by hand in the remote Pamirs of Tajikistan (Figure 21.2).

To understand these linkages among Earth's myriad systems is our quest in *Geosystems* (Figure 21.3). We explored the atmosphere, hydrosphere, lithosphere, and biosphere, synthesizing operating systems into the web of life. We now see global linkages among physical and living systems and how actions in one place can affect change elsewhere. In this sense, Earth can be compared to a spaceship. In the same way the members of a crew

*C. Sagan, *Cosmos* (New York: Random House, 1980), p. 329.
†M. Bates, *The Forest and the Sea* (New York: Random House, 1960), p. 247.

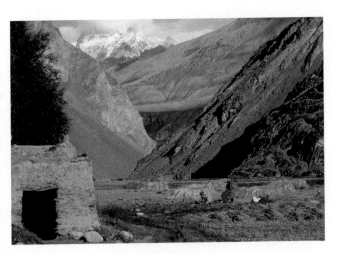

FIGURE 21.2 Lands distant from the global centers of power.
These lands were depopulated under Stalin when they were part of
the former USSR. Only in the last decade or so have people returned
to this traditional rural landscape in the Pamirs of Tajikistan, beginning life again, so distant from the more-developed countries.
[Photo by Stephen F. Cunha.]

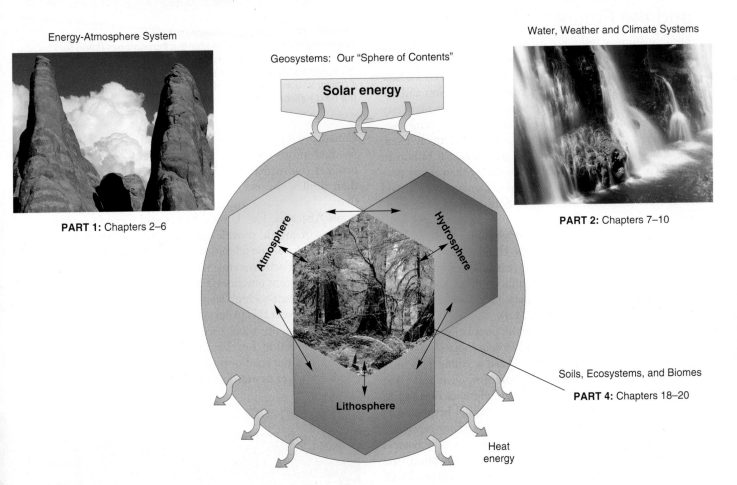

Energy-Atmosphere System

PART 1: Chapters 2–6

Geosystems: Our "Sphere of Contents"

Solar energy

Atmosphere

Hydrosphere

Lithosphere

Water, Weather and Climate Systems

PART 2: Chapters 7–10

Soils, Ecosystems, and Biomes

PART 4: Chapters 18–20

Heat
energy

Earth-Atmosphere Interface

PART 3: Chapters 11–17

FIGURE 21.3 The spheres within
Geosystems.
Through this text we covered the atmosphere, hydrosphere, lithosphere, and the
synthesizing biosphere—the culmination
of life-sustaining interactions. The systems approach shows us the flow of
energy and matter and the sequence of
events through time and across space.
Ultimately, *Geosystems* describes the
support systems of life. [Atmosphere and
hydrosphere photos by author; lithosphere
and biosphere photos by Bobbé Christopherson.]

aboard the Space Station are inextricably linked to each other's lives and survival, we too are each connected through the operation of planetary systems.

Growing international environmental awareness in the public sector is gradually prodding governments into action. People, when informed, generally favor environmental protection and public health progress over economic interests. A survey by Louis Harris and Associates confirmed these opinion trends. Majorities in 22 nations are willing to "endorse environmental protection at the risk of slowing down economic growth." The public seems to know they are saving money as consumers if things are done right with the environment.

The Human Count and the Future

Because human influence is pervasive, we consider the totality of our impact the *human denominator*. Just as the denominator in a fraction tells how many parts a whole is divided into, so the growing human population and the increasing demand for resources and rising planetary impact suggest how much the whole Earth system must adjust. Yet, Earth's resource base remains relatively fixed.

The human population of Earth passed 6 billion in August 1999; more people are alive today than at any previous point in the planet's long history, unevenly arranged in 192 countries and numerous colonies. In the year leading up to this milestone, 83,000,000 more people were added. Population growth has slowed since then to about 70 million more people per year—this is 192,000 a day, or a new U.S. population every 4.2 years. At the end of 2004, world population was at 6.4 billion. Over the span of human history these billion-mark milestones are occurring at closer intervals (Figure 21.4).

Thirty-seven percent of Earth's population live in just two countries (21% in China and 16% in India—2.4 billion people combined). Moreover, we are a young planetary population, with some 30% of those now alive under the age of 15 (2004 data from Population Reference Bureau and U.S. Bureau of Census *POPClock Projection:* **http://www. census.gov/cgi-bin/popclock**). *Virtually all new population growth is in the less-developed countries*, which now possess 81%, or about 5.1 billion, of the total population; the MDCs have the other 19%, or 1.3 billion (Figure 21.5). Table 21.1 illustrates this natural increase principle and the differences found in the MDCs and LDCs.

If you consider only the population count, the MDCs do not have a population growth problem. In fact, some European countries are actually declining in growth or are near replacement levels. However, people in the developed world produce a greater impact on the planet per person and therefore constitute a population impact crisis. An equation expresses this impact concept:

$$\text{Planetary impact } (I) = P \cdot A \cdot T$$

where P is *population*; A is *affluence*, or consumption per person; and T is *technology*, or the level of environmental impact per unit of production. The United States and

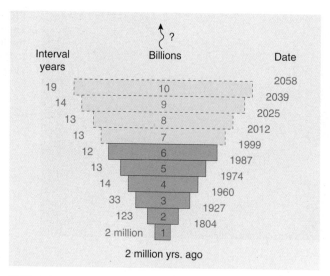

FIGURE 21.4 Human population growth.
The number of years required for the human population to add 1 billion to the count grows shorter and shorter. The 21st century should mark a slowing of this history of growth if policy actions are taken. Note the population forecasts for the next half-century.

Canada, with about 4.5% of the world's population, produce more than 50% of the world's gross domestic product (more than $10.5 trillion a year) and use more than double the energy per capita of Europeans, more than 7 times Latin Americans, 10 times Asians, and 20 times Africans. The United States and Canada produce 22 tons of carbon dioxide emissions per person per year.

Therefore, consideration by people in the MDCs of the state of Earth systems, natural resources, and sustainability of current practices is critical. Earth systems science is here to provide a high level of planetary monitoring and analysis to assist this process. The MDCs not only should effect change in their own planetary impact but also provide direction for the LDCs, which have a right to move along the road of progress and improve their living conditions. Let us examine this idea of global impact.

FIGURE 21.5 Dhaka, Bangladesh.
The Bangladesh capital typifies the plight of the less-developed world as millions converge on cities to find work and more modern lifestyles. [Photo by Bruce Bander/Photo Researchers, Inc.]

Table 21.1 Planetary Population Increase Rates				
	Birth Rate	Death Rate	Natural Increase (%)	Population Growth 2004–2050 (%)
World—2004	**21**	**9**	**1.3**	45%
More developed	**11**	**10**	**0.1**	4%
United States	14	8	0.6	43%
Japan	9	8	0.1	−21%
United Kingdom	12	10	0.1	10%
Less developed	**24**	**8**	**1.5**	55%
LDC exclu. China	**27**	**9**	**1.8**	69%
Mexico	25	5	2.1	41%
Nigeria	42	13	2.9	124%
India	25	8	1.7	50%

Source: World Population Data Sheet 2004. Population Reference Bureau, Washington, DC, http://www.prb.org.

An Oily Bird

At first glance, the chain of events that exposes wildlife to oil contamination seems to stem from a technological problem. An oil tanker splits open at sea and releases its petroleum cargo, which is moved by ocean currents toward shore, where it coats coastal waters, beaches, and animals. In response, concerned citizens mobilize and try to save as much of the spoiled environment as possible (Figure 21.6). But the real problem goes far beyond the physical facts of the spill.

In Prince William Sound off the southern coast of Alaska, in clear weather and calm seas, the *Exxon Valdez*, a single-hulled supertanker operated by Exxon Corporation, struck a reef. The tanker spilled 42 million liters (11 million gallons) of oil. It took only 12 hours for the *Exxon Valdez* to empty its contents, yet a complete cleanup is impossible, and costs and private claims exceeded

$15 billion. Scientists are still finding damage and residual oil spill. Eventually, more than 2400 km (about 1500 mi) of sensitive coastline was ruined for years to come, affecting three national parks and eight other protected areas. For perspective, had this spill occurred farther south, every beach and bay along the Pacific Coast from southern Oregon to the Mexican border would have been blackened (Figure 21.7).

The death toll of animals was massive: At least 5000 sea otters died, or about 30% of resident otters; about 300,000 birds and uncounted fish, shellfish, plants, and aquatic microorganisms also perished. Sublethal effects, namely mutations, now are appearing in fish. The Pacific herring is still in significant decline, as are the harbor seals; other species are in recovery, such as the Bald Eagle and Common Murre. More than a decade-and-a-half later, oil remains in mudflats and marsh soils still can be scooped from beneath rocks along shorelines.

The immediate problem of cleaning oil off a bird symbolizes a national and international concern with far-reaching spatial significance. And while we search for answers, oil slicks continue their contamination. In just 1 year following the *Exxon Valdez* disaster, nearly 76 million liters (20 million gallons) of oil were spilled in 10,000 accidents worldwide, as evidenced on the startling map in Figure 21.7a. This is an average of 27 accidents a day, ranging from a few disastrous spills to numerous small ones. There were 50 spills equal to the *Exxon Valdez* or larger since 1970. In addition to oceanic oil spills, people improperly dispose of crankcase oil from their automobiles in a volume that annually exceeds all of these tanker spills.

The immediate effect of global oil spills on wildlife is contamination and death. But the issues involved are much bigger than dead birds. Let's ask some fundamental questions about the *Exxon Valdez* accident:

- Why was the oil tanker there in the first place?
- Why are only a few of the 28 tankers that traverse the Prince William Sound of the safer double-hulled design?

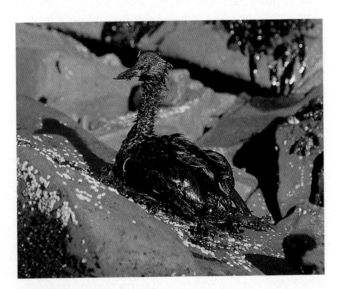

FIGURE 21.6 An oily bird.
A Western Grebe contaminated with oil from the *Exxon Valdez* tanker accident in Prince William Sound, Alaska. This oily bird is the result of a long chain of events and mistakes. [Photo by Geoffrey Orth/SIPA Press.]

FIGURE 21.7 Worldwide oil spills and the 1989 *Exxon Valdez* accident.
(a) Location of visible oil slicks worldwide in the 1990s. (b) Track of spreading oil for the first 56 days of the Alaskan spill. [(a) Data from Organization for Economic Cooperation and Development.]

- Why is petroleum imported into the continental United States from Alaska in such enormous quantities? Is the demand for petroleum products based on real need and the operation of efficient systems?
- The U.S. demand for oil is higher per capita than the demand of any other country. Efficiencies and mileage ratings in the U.S. transportation sector again fell in the 2004 fleet of cars and trucks, with SUVs (light trucks) leading in sales. Why?

A combination of economic growth, waste, low prices, and a lack of alternatives has spurred the demand for petroleum. In addition, our land-use policies continue to foster a diffuse sprawl of our population, thereby adding stress to transportation systems. All of these factors increased U.S. demand for oil to more than 8 billion barrels a year.

Yet, hypocrisy is apparent in our outrage over energy politics and oil-spill accidents. We continue to consume gasoline at record levels in inefficient vehicles, thus creating the demand for oil imports. The task of physical geography is to analyze all the spatial aspects of these events in the environment and the ironies that are symbolized by an oily bird.

The Need for International Cooperation

The idea for a global meeting on the environment was put forward at the 1972 U.N. Conference on the Human Environment held in Stockholm. The U.N. General Assembly in 1987 achieved a landmark in global planning by agreeing to hold the Earth Summit. *Our Common Future*, a book written at the time, set the tone for the 1992 Earth Summit:

> The Earth is one but the world is not. We all depend on one biosphere for sustaining our lives. Yet each community, each country, strives for survival and prosperity with little regard for its impact on others. Some consume the Earth's resources at a rate that would leave little for future generations. Others, many more in number, consume far too little and live with the prospect of hunger, squalor, disease, and early death.*

The setting for the 1992 Earth Summit in Rio de Janeiro was ironic, for many of the very problems discussed at the conference were evident on the city's streets, with their abundant air pollution, water pollution, toxics, noise, wealth and poverty, and daily struggle for health and education.

Five agreements were written at the Earth Summit: the Climate Change Framework; Biological Diversity Treaty; the Management, Conservation, and Sustainable Development of All Types of Forests; the Earth Charter (a nonbinding statement of 27 environmental and economic principles); and Agenda 21 for Sustainable Development. A product of the Earth Summit was the United Nations Framework Convention on Climate Change (FCCC), which led to a series of Conference of the Parties (COP) meetings. The Kyoto Protocol, agreed to in 1997 to reduce global carbon emissions, was finalized in Marrakech, Morocco, in 2001 at *COP*-7 (Figure 21.8). Subsequent

*World Commission on Environment and Development, *Our Common Future* (Oxford, UK: Oxford University Press, 1987), p. 27.

FIGURE 21.8 The Marrakech, Morocco, Climate Summit, November 2001.
The Marrakech Climate Summit (*COP-7*) voted to accept the terms of the Kyoto Protocol. A consensus was reached on the reduction of global carbon dioxide emissions and the Kyoto Rulebook. [Photo courtesy of IISD/ENB-Leila Mead.]

COP meetings were held in New Delhi, The Hague, Milan, and *COP-10* in Buenos Aires in December 2004. Following Russian ratification, the Kyoto Protocol and Rulebook became international law in March 2005, without United States or Australian participation.

Asking whether the Earth Summit "succeeded" or "failed" is the wrong question. The occurrence of this largest-ever official gathering is a remarkable accomplishment. From the Earth Summit emerged a new organization—the U.N. Commission on Sustainable Development—to oversee the promises made in the five agreements. This momentum lead to a second Earth Summit in 2002 (**http://www.earthsummit2002.org/**) in Johannesburg, South Africa, that completed an agenda including climate change, freshwater, gender issues, global public goods, HIV/AIDS, sustainable finance, and the five Rio Conventions.

A clear conclusion is that members of society must work to move the solutions for environmental and developmental problems off the bench and into play. We need international cooperation to consider our symbiotic relations with each other and with Earth's resilient, yet fragile, life-support systems (News Report 21.1).

News Report 21.1

Gaia Hypothesis Triggers Debate

Some scientists and philosophers view Earth as one vast, self-regulating organism. The concept is one of global symbiosis, or mutualism. This controversial concept is called the *Gaia hypothesis* (Gaia was the Earth-Mother goddess in ancient mythology). It was proposed in 1979 by James Lovelock, a British astronomer and inventor, and elaborated by American biologist Lynn Margulis.

Gaia is the ultimate synergistic relationship, in which the whole greatly exceeds the sum of the individual interacting components. The hypothesis contends that life processes control and shape Earth's inorganic physical and chemical processes, with the ecosphere so interactive that a very small mass can affect a very large mass. Thus, Lovelock and Margulis think that the material environment and the evolution of species are tightly joined; as species evolve through natural selection, they (including us) in turn affect their environment. The

present oxygen-rich composition of the atmosphere is given as proof of this coevolution of living and nonliving systems.

From the perspective of physical geography, the Gaia hypothesis permits a view of all Earth and the spatial interrelations among systems. In fact, such a perspective is necessary for analyzing specific environmental issues. Many variables interact synergistically, producing both wanted and unwanted results.

One disturbing aspect of this unity is that any biotic threat to the operation of an ecosystem tends to move toward extinction itself. This trend preserves the system overall. Earth-systems operation and feedback naturally tend to eliminate offensive members. The degree to which humans represent a planetary threat, then, becomes a topic of great concern, for Earth (Gaia) will prevail, regardless of the outcome of the human experiment.

The maladies of Gaia do not last long in terms of her life span. Anything that makes the world uncomfortable to live in tends to induce the evolution of those species that can achieve a new and more comfortable environment. It follows that, if the world is made unfit by what we do, there is the probability of a change in regime to one that will be better for life but not necessarily better for us.*

The debate is vigorous regarding the true applicability of this hypothesis to nature, or whether it is true science at all. Regardless, it remains philosophically intriguing in its portrayal of the relationship between humans and Earth.

*J. Lovelock, *The Ages of Gaia—A Biography of Our Living Earth* (New York: Norton, 1988), p. 178.

A critical corollary to these international efforts is the linkage of academic disciplines. A positive step in that direction is the Earth systems science approach embodied in physical geography that synthesizes content from across the disciplines to create a holistic perspective. Exciting progress toward an integrated understanding of Earth's physical and biological systems is in progress. Such a multidisciplinary effort began in 2000 at a Ministerial Meeting of the Arctic Council, when the Arctic Climate Impact Assessment (ACIA) was launched. The ACIA final report, *Climate Change in the Arctic*, was released in November 2004. The effort involved 300 scientists, eight countries, and six circumpolar indigenous peoples' organizations. For a look at this international triumph of cooperative science, see High Latitude Connection 21.1 (see http://www.acia.uaf.edu).

More than a decade ago, *Time* magazine gave Earth an interesting honor and is today a time capsule of activist suggestions (Figure 21.9). As if a victim of short-lived fame, we wonder where Earth awareness fits into the global scheme of things today. Is there a follow-up in pop culture to making the cover of *Time*?

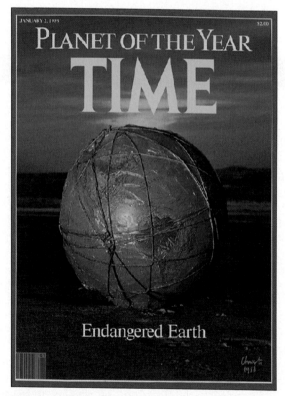

FIGURE 21.9 Earth made the cover of *Time*.
More than a decade ago, global concerns about environmental impacts prompted *Time* magazine to deviate from its 60-year tradition of naming a prominent citizen as its person of the year, instead naming Earth the "Planet of the Year." The magazine devoted 33 pages to Earth's physical and human geography. Importantly, *Time* also offered positive policy strategies for consideration—an interesting time capsule for comparison with real events over the years. [January 2, 1989 issue, Copyright © 1989 The *Time* Inc. Magazine Company. Reprinted by permission.]

Twelve Paradigms for the 21st Century

As we conclude, a brief list of the dominant themes and patterns of concern for the 21st century seems appropriate. We hope these *paradigms* will provide a useful framework for brainstorming and discussion of the central issues that will affect us in the new century. The paradigms:

Twelve Paradigms for the 21st Century

1. Population increases in the less-developed countries
2. Planetary impact per person (on the biosphere and resources; $I = P \cdot A \cdot T$)
3. Feeding the world's population
4. Global and national disparities of wealth and resource allocation
5. Status of women and children (health, welfare, rights)
6. Global climate change (temperatures, sea level, weather and climate patterns, disease, and diversity)
7. Energy supplies and energy demands; renewables and demand management
8. Loss of biodiversity (habitats, genetic wealth, and species richness)
9. Pollution of air, surface water (quality and quantity), groundwater, oceans, and land
10. The persistence of wilderness (biosphere reserves and biodiversity hot spots)
11. Globalization versus cultural diversity
12. Conflict resolution

Who Speaks for Earth?

Geographic awareness and education is an increasingly positive force on Earth. The National Geographic Society conducts an annual National Geography Bee to promote geography education and global awareness to millions of 6th to 8th graders, their parents, and teachers. The National Geography World Championship (formerly the International Geography Olympiad) is now an annual event (Figure 21.10). There are presently 60 geographic alliances in 47 states, which coordinate geographic education among teachers and students at all levels: K–12, community college, college, and university. People are learning more about Earth–human relations.

Yet, ideological and ethical differences still remain within society. This dichotomy was addressed by biologist Edward O. Wilson:

The evidence of swift environmental change calls for an ethic uncoupled from other systems of belief. Those committed by religion to believe that life was put on Earth in one divine stroke will recognize that we are destroying the Creation; and those who perceive biodiversity to be the product of blind evolution will agree. . . . Defenders of both premises seem destined to gravitate toward the same position on conservation. . . .

FIGURE 21.10 National Geographic World Championship.
Students from across the globe met in Vancouver, British Columbia, August 2001, for the fifth National Geographic World Championship (name changed from Geography Olympiad), sponsored by the National Geographic Society, with Alex Trebek of *Jeopardy!* fame as moderator. Here Alex congratulates the Canadian team on their second-place finish—the United States placed first, the Hungarian team third in the final standings. The 7th National Geographic World Championship will be held in 2005. The championship is an indicator of growing international geographic awareness. [Photo by O. Louis Mazzatenta, 2001 National Geographic Society.]

High Latitude Connection 21.1

Report from Reykjavik—Arctic Climate Impact Assessment

Temperature, precipitation, ocean salinity, spatial distribution of pack ice, and flora and fauna, are changing in the northern high latitudes. Some of the most rapid and severe climate change on Earth is occurring now in the Arctic Region. These dynamic conditions will accelerate over this century, resulting in major physical, ecological, social, and economic impacts, as scientists working on the Intergovernmental Panel on Climate Change (IPCC) *Third Assessment Report* (2001) initially described.

The Arctic Council decided to act in a first-ever regional scale scien-

tific investigation. The Arctic Council, a ministerial intergovernmental forum (Canada, Denmark/Greenland, Faroe Islands, Finland, Iceland, Norway, Russia, Sweden, and the United States; at **http://www.arctic-council.org/**), is joined by six Indigenous Peoples' organizations. Official observers include Germany, France, The Netherlands, Poland, United Kingdom, several non-governmental organizations, and several scientific international bodies.

The Arctic Council initiated the Arctic Climate Impact Assessment (ACIA,   **http://www.acia.uaf.edu**,

Figure 21.1.1a) in 2000 and charged its working groups to complete the research—Arctic Monitoring and Assessment Program (AMAP, **http://www.amap.no/**), the Conservation of Flora and Fauna (CAFF, **http://www.caff.is/**), along with the International Arctic Science Committee (IASC, **http://www.iasc.no/**), comprising national scientific organizations, including 18 national academies of science. The Arctic Council Ministers declared that the goal of ACIA is to ". . . evaluate and synthesize knowledge on climate variability and change and increased ultraviolet radiation,

(a)   (b)

FIGURE 21.1.1 ACIA process and scientific symposium.
(a) The logo of the Arctic Climate Impact Assessment. (b) Scientific papers, panels, and poster sessions delivered the findings of the ACIA process to the attendees at the Reykjavik symposium. [Photos by Bobbé Christopherson.]

(continued)

High Latitude Connection 21.1 *(continued)*

and support policy-making processes and the work of the IPCC, and address environmental, human health, social, cultural, and economic impacts and consequences, including policy recommendations."

Following fours years of research and field investigations by more than 300 scientists divided into 18 working groups, and an intensive review process, the ACIA report was forwarded to the Arctic Council and to the international science community and governments at a first-ever ACIA Symposium held in Reykjavik, Iceland, in November 2004 (see photo that opens this chapter and Figure 21.1.1b). The ACIA Synthesis report is *Impacts of A Warming Arctic–Arctic Climate Impact Assessment* (Cambridge: Cambridge University Press, 2004; ISBN: 0-521-61778-2). The full 1200-page science report was distributed in December.

The meeting began with the announcement by the Deputy Head of the Russian Roshydromet (Hydrometeorological Bureau) that Russia has ratified the Kyoto Protocol to limit emissions of greenhouse gases. An action, along with 130 countries, that makes the Protocol international law, without the participation of the United States and Australia.

ACIA science reinforces earlier discussion in this text on global climate change. Using the conservative "B2" computer model scenario of carbon dioxide and other greenhouse gases as a base line, and drawing information from five general circulation models (GCMs) from the IPCC in all, this state-of-the-art science work portrays worsening conditions in the circumpolar Arctic environment.

The ACIA report concludes that the region is warming more rapidly than previously thought, at nearly twice the rate of the rest of the globe. Dr. Robert Corell, Chair of the ACIA Integration Team stated, "The impacts of global warming are affecting

people now in the Arctic. The impacts of climate change on the region and the globe are projected to increase substantially in the years to come." Evelyn Hurwich, President of the Circumpolar Conservation Union, said, "Climate change is a weapon of mass destruction staring us in the face, threatening all of creation. The science is in and the evidence before us. There is now a moral duty to act on it."

A particular strength of the ACIA effort is the integration of traditional knowledge and experience from the region's indigenous people. Chief Gary Harrison of the Arctic Athabaskan Council summarized, "Everything is under threat, our homes are threatened by storms and melting permafrost, our livelihoods are threatened by changes to plants and animals we harvest. Even our lives are threatened, as traditional travel routes become dangerous."

ACIA findings affirm the 2001 IPCC Third Assessment Report, that greenhouse gases produced by the human combustion of fossil fuels are the root cause of the warming. These climate changes are exceeding that attributable to natural cycles and oscillations of climate. By 2100 temperatures in the region are forecasted to increase by 4–7 C° (7–13 F°), a large amount of temperature change (compare with IPCC forecasts in Chapter 10). Consider the past 50 years in Alaska, Western Canada, and Eastern Russia where average winter temperature increased 3–4 C° (5–7 F°).

Increased temperatures will accelerate the on-going loss of Arctic Ocean sea ice, melt a significant portion of the Greenland Ice Sheet, and continue the freshening of the northern oceans. In turn, these events will contribute to global sea-level rise and further affect the global climate engine, intensifying warming effects. Imagine a near complete ice free Arctic Ocean. If this opens the Arctic

to freighter and tanker traffic and resource exploitation such as oil and gas mining, then a whole new suite of environmental impacts looms ahead.

A summary of the *Key Findings* presented at the ACIA Symposium and detailed in the Synthesis report and final science report:

1. Arctic climate is now warming rapidly and much larger changes are projected.
2. Arctic warming and its consequences have worldwide implications.
3. Arctic vegetation zones are very likely to shift, causing wide-ranging impacts.
4. Animal species' diversity, ranges, and distribution will change.
5. Many coastal communities and facilities face increasing exposure to storms.
6. Reduced sea ice is very likely to increase marine transport and access to resources.
7. Thawing ground will disrupt transportation, buildings, and other infrastructure.
8. Indigenous communities are facing major economic and cultural impacts.
9. Elevated ultraviolet radiation levels will affect people, plants, and animals.
10. Multiple influences interact to cause impacts to people and ecosystems.

As Sheila Watt-Cloutier of the Inuit Circumpolar Conference concluded, "The Arctic is the early warning system for the rest of the world. We must all take what action we can to slow the pace of climate change." The ACIA process that is in motion is an important component in the preparation for the International Polar Year of science and research in the high latitudes of both polar regions, to be conducted from March 1, 2007 to March 1, 2009 (see http://www.ipy.org./).

For what, in the final analysis, is morality but the command of conscience seasoned by a rational examination of consequences? . . . An enduring environmental ethic will aim to preserve not only the health and freedom of our species, but access to the world in which the human spirit was born.*

United Nations Secretary-General Kofi Annan, recipient of the 2001 Nobel Peace Prize, spoke to the Association of American Geographers annual meeting (Figure 21.11) on March 1, 2001, and offered us this thought:

The idea of interdependence is old hat to geographers, but for most people it is a new garment they are only now trying on for size. Getting it to fit—and getting it imprinted on the mental maps that guide our voices and our choices—is one of the crucial projects of human geography for the 21st century. I look forward to working with you in that all-important journey.

We might ask the largest mammals on Earth as to who speaks for the planet and its functioning systems, but society has yet to decipher the complex languages of the whales, the *Cetaceans* (Figure 21.12). Society subjected whales to profligate slaughter from the 17th to 20th centuries. In the 50 years from 1925 to 1975, 1.5 million whales were killed. Of the 11 major species, this left 8 species at commercial extinction (not profitable), some threatened with physical extinction.

*E. O. Wilson, *The Diversity of Life* (Cambridge, MA: Harvard University Press, 1992), p. 351.

FIGURE 21.11 Secretary-General Kofi Annan addresses the AAG.
U.N. Secretary-General and Nobel Prize winner speaks to more than 3000 geographers at the Association of American Geographers Annual Meeting in New York City, March 1, 2001. [Photo courtesy of the Association of American Geographers, by Kevin J. McCormick.]

The largest are the blue whales (30-m, 100-ft, length), which numbered about 200,000 individuals before commercial whaling. They haven't been hunted now for 40 years and their numbers are still less than 10,000—perhaps as few as 3000 remain. We might say, who speaks for the whales?

FIGURE 21.12 Humpback whales in Antarctic waters.
(a) Three humpback whales swim about feeding on krill. They migrate between the Southern Ocean and the tropics but do not mix with the northern humpback population. Humpbacks are about half the size of the blue whale. (b) The tail fluke so gracefully enters the water with its black-and-white pattern and distinctive marks that act like a fingerprint. A central database is maintained using such marks to track their migrations. The orange coloration is algae that grows on the tail during the long trip. (c) Deception Island, Antarctica, a volcanic caldera, was an important whale processing center. There were many days when the ocean waters in the caldera were blood red from the slaughter. The industry collapsed in 1932 and the whalers left. These pieces of wood are staves for oil barrels that were abandoned more than 70 years ago, collapsed and forgotten in the frigid air, much like the business that went on here. [All photos by Bobbé Christopherson.]

In an ideal sense, we hold all of Earth's systems in common. In that spirit the world did come together and forced international controls to stop the loss of the whales. Through such realization, we gave the whales a voice to answer our question. Now, what about Sagan's question to us? The late Carl Sagan asked, "Who speaks for Earth?" He answered with this perspective:

We have begun to contemplate our origins: starstuff pondering the stars; organized assemblages of ten billion billion billion atoms considering the evolution of atoms; tracing the long journey by which, here at least, consciousness arose. Our loyalties are to the species and the planet. We speak for Earth. Our obligation to survive is owed not just to ourselves but also to that Cosmos, ancient and vast, from which we spring.*

May we all perceive our spatial importance within Earth's ecosystems and do our part to maintain a life-supporting and sustaining Earth for ourselves and countless generations in the future.

*C. Sagan, *Cosmos* (New York: Random House, 1980), p. 345.

Critical Thinking and Learning

A. What part do you think technology, politics, and thinking about the future should play in science courses?

B. Assess population growth issues: the count, the impact per person, and future projection. What strategies do you see as important?

C. According to the discussion in the chapter, what worldwide factors led to the *Exxon Valdez* accident? Describe the complexity of that event from a global perspective. In your analysis, examine both supply-side (corporations and utilities) and demand-side (consumers) issues, as well as environmental and strategic factors. And, what about the oily bird?

D. What is meant by the Gaia hypothesis? Describe several concepts from this text that might pertain to this hypothesis.

E. Relate the content of the various chapters in this text to the integrative Earth systems science concept. Which chapters help you to better understand Earth–human relations and human impacts?

F. After examining the list of 12 paradigm issues for the 21st century, suggest items that need to be added to the list, omitted from the list, or expanded in coverage. Rearrange and organize the list as needed to match your concerns.

G. This chapter states that we already know many of the solutions to the problems we face. Why do you think these solutions are not being implemented at a faster pace?

H. Who speaks for Earth?

NetWork

The *Geosystems* Home Page provides online resources for this chapter on the World Wide Web. To begin: Once on the Home Page, click on this textbook, scroll the Table of Contents menu, select this chapter, and click "Begin." You will find self-tests that are graded, review exercises, specific updates for items in the chapter, and many links to interesting related pathways on the Internet under "Destinations." *Geosystems* is at http://www.prenhall.com/christopherson.

Appendix A

Maps in This Text and Topographic Maps

Maps Used in This Text

Geosystems uses several map projections: Goode's homolosine, Robinson, and Miller cylindrical, among others. Each was chosen to best present specific types of data. **Goode's homolosine projection** is an interrupted world map designed in 1923 by Dr. J. Paul Goode of the University of Chicago. Rand McNally *Goode's Atlas* first used it in 1925. Goode's homolosine equal-area projection (Figure A.1) is a combination of two oval projections (*homolo*graphic and *sin*usoidal projections).

Two equal-area projections are cut and pasted together to improve the rendering of landmass shapes. A *sinusoidal projection* is used between 40° N and 40° S latitudes. Its central meridian is a straight line; all other meridians are drawn as sinusoidal curves (based on sine-wave curves) and parallels are evenly spaced. A *Mollweide projection*, also called a *homolographic projection*, is used from 40° N to the North Pole and from 40° S to the South Pole. Its central meridian is a straight line; all other meridians are drawn as elliptical arcs and parallels are unequally spaced—farther apart at the equator, closer together poleward. This technique of combining two projections preserves areal size relationships, making the projection excellent for mapping spatial distributions when interruptions of oceans or continents do not pose a problem.

We use Goode's homolosine projection throughout this book. Examples include the world climate map and smaller climate type maps in Chapter 10, topographic regions and continental shields maps (Figures 12.3 and 12.4), world karst map (Figure 13.14), world sand regions and loess deposits (Figures 15.12 and 15.14) and the terrestrial biomes map in Chapter 20.

Another projection we use is the **Robinson projection**, designed by Arthur Robinson in 1963 (Figure A.2). This projection is neither equal area nor true shape, but is a compromise between the two. The North and South Poles appear as lines slightly more than half the length of the equator; thus higher latitudes are exaggerated less than on other oval and cylindrical projections. Some of the Robinson maps employed include the latitudinal geographic zones map in Chapter 1 (Figure 1.13), daily net radiation map (Figure 2.11), the world temperature range map in Chapter 5 (Figure 5.17), the maps of lithospheric plates of crust and volcanoes and earthquakes in Chapter 11 (Figures 11.18 and 11.21), and the global oil spills map in Chapter 21 (Figure 21.7).

Another compromise map, the **Miller cylindrical projection**, is used in this text (Figure A.3). Examples of this projection include the world time zone map (Figure 1.17), global temperature maps in Chapter 5 (Figures 5.14 and 5.16), two global pressure maps in Figure 6.11, and the global soil maps and soil order maps in Chapter 18 (Figure 18.9). This projection is neither true shape nor true area, but is a compromise that avoids the severe scale distortion of the Mercator. The Miller projection frequently appears in world atlases. The American Geographical Society presented Osborn Miller's map projection in 1942.

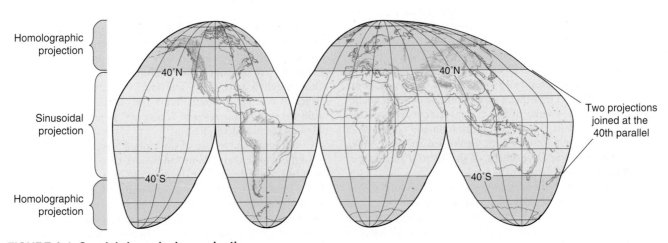

FIGURE A.1 Goode's homolosine projection.
An equal-area map. [Copyright by the University of Chicago. Used by permission of the University of Chicago Press.]

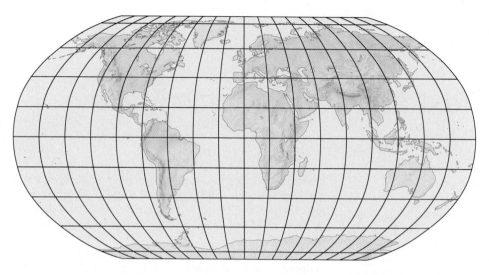

FIGURE A.2 Robinson projection.
A compromise between equal area and true shape. [Developed by Arthur H. Robinson, 1963.]

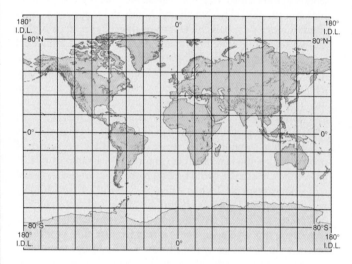

FIGURE A.3 Miller cylindrical projection.
A compromise map projection between equal area and true shape. [Developed by Osborn M. Miller, American Geographical Society, 1942.]

Mapping, Quadrangles, and Topographic Maps

The westward expansion across the vast North American continent demanded a land survey for the creation of accurate maps. Maps were needed to subdivide the land and to guide travel, exploration, settlement, and transportation. In 1785, the Public Lands Survey System began surveying and mapping government land in the United States. In 1836, the Clerk of Surveys in the Land Office of the Department of the Interior directed public-land surveys. The Bureau of Land Management replaced this Land Office in 1946. The actual preparation and recording of survey information fell to the U.S. Geological Survey (USGS), also a branch of the Department of the Interior (see **http://mapping.usgs.gov/**).

In Canada, the National Resources Canada conducts the national mapping program. Canadian mapping includes base maps, thematic maps, aeronautical charts, federal topographic maps, and the National Atlas of Canada, now in its fifth edition (see **http://atlas.gc.ca/**).

Quadrangle Maps

The USGS depicts survey information on quadrangle maps, so called because they are rectangular maps with four corner angles. The angles are junctures of parallels of latitude and meridians of longitude rather than political boundaries. These quadrangle maps utilize the Albers equal-area projection, from the conic class of map projections.

The accuracy of conformality (shape) and scale of this base map is improved by the use of not one but two standard parallels. (Remember from Chapter 1 that standard lines are where the projection cone touches the globe's surface, producing greatest accuracy.) For the conterminous United States (the "lower 48"), these parallels are 29.5° N and 45.5° N latitude (noted on the Alber's projection shown in Figure 1.21c). The standard parallels shift for conic projections of Alaska (55° N and 65° N) and for Hawai'i (8° N and 18° N).

Because a single map of the United States at 1:24,000 scale would be more than 200 m wide (more than 600 ft), some system had to be devised for dividing the map into a manageable size. Thus, a quadrangle system using latitude and longitude coordinates developed. Note that these maps are not perfect rectangles, because meridians converge toward the poles. The width of quadrangles narrows noticeably as you move north (poleward).

Quadrangle maps are published in different series, covering different amounts of Earth's surface at different scales. You see in Figure A.4 that each series is referred to by its angular dimensions, which range from 1° to 2° (1:250,000 scale) to 7.5′ × 7.5′ (1:24,000 scale). A map that is one-half a degree (30′) on each side

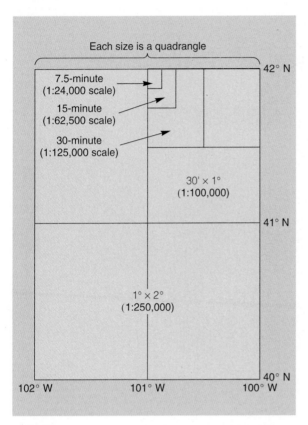

FIGURE A.4 Quadrangle system of maps used by the USGS.

is called a "30-minute quadrangle," and a map one-fourth of a degree (15′) on each side is a "15-minute quadrangle" (this was the USGS standard size from 1910 to 1950). A map that is one-eighth of a degree (7.5′) on each side is a 7.5-minute quadrangle, the most widely produced of all USGS topographic maps and the standard since 1950. The progression toward more-detailed maps and a larger-scale map standard through the years reflects the continuing refinement of geographic data and new mapping technologies.

The USGS National Mapping Program recently completed coverage of the entire country (except Alaska) on 7.5-minute maps (1 in. to 2000 ft, a large scale). It takes 53,838 separate 7.5-minute quadrangles to cover the lower 48 states, Hawai'i, and the U.S. territories. A series of smaller-scale, more-general 15-minute topographic maps offer Alaskan coverage.

In the United States, most quadrangle maps remain in English units of feet and miles. The eventual changeover to the metric system requires revision of the units used on all maps, with the 1:24,000 scale eventually changing to a scale of 1:25,000. However, after completing only a few metric quads, the USGS halted the program in 1991. In Canada, the entire country is mapped at a scale of 1:250,000, using metric units (1.0 cm to 2.5 km). About half the country also is mapped at 1:50,000 (1.0 cm to 0.50 km).

Topographic Maps

The most popular and widely used quadrangle maps are **topographic maps** prepared by the USGS. An example of such a map is a portion of the Cumberland, Maryland, quad shown in Figure A.5. You will find topographic maps throughout *Geosystems* because they portray landscapes so effectively. As examples, see Figures 13.15 and 13.16, karst landscapes and sinkholes near Orleans, Indiana, and Winter Park, Florida;

FIGURE A.5 An example of a topographic map from the Appalachians.
Cumberland, MD, PA, WV 7.5-minute quadrangle topographic map prepared by the USGS. Note the water gap through Haystack Mountain.

(a)

(b)

FIGURE A.6 Topographic map of a hypothetical landscape.
(a) Perspective view of a hypothetical landscape. (b) Depiction of that landscape on a topographic map. The contour interval on the map is 20 feet (6.1 m). [After the U.S. Geological Survey.]

Figure 14.8, river drainage patterns; Figure 14.23, river meander scars; Figure 15.18, an alluvial fan in Montana; Figure 17.3, glaciers in Alaska; and Figure 17.17, drumlins in New York.

A **planimetric map** shows the horizontal position (latitude/longitude) of boundaries, land-use aspects, bodies of water, and economic and cultural features. A highway map is a common example of a planimetric map.

A topographic map adds a vertical component to show topography (configuration of the land surface), including slope and relief (the vertical difference in local landscape elevation). These fine details are shown through the use of elevation contour lines (Figure A.6). A *contour line* connects all points at the same elevation. Elevations are shown above or below a vertical datum, or reference level, which usually is mean sea level. The contour interval is the vertical distance in elevation between two adjacent contour lines (20 ft, or 6.1 m in Figure 6b).

The topographic map in Figure 6b shows a hypothetical landscape, demonstrating how contour lines and intervals depict slope and relief, which are the three-dimensional aspects of terrain. The pattern of lines and the spacing between them indicate slope. The steeper a slope or cliff, the closer together the contour lines appear—in the figure, note the narrowly spaced contours that represent the cliffs to the left of the highway. A

wider spacing of these contour lines portrays a more gradual slope, as you can see from the widely spaced lines on the beach and to the right of the river valley.

In Figure A.7 are the standard symbols commonly used on these topographic maps. These symbols and the colors used are standard on all USGS topographic maps: black for human constructions, blue for water features, brown for relief features and contours, pink for urbanized areas, and green for woodlands, orchards, brush, and the like.

The margins of a topographic map contain a wealth of information about its concept and content. In the margins of topographic maps, you find the quadrangle name, names of adjoining quads, quad series and type, position in the latitude-longitude and other coordinate systems, title, legend, magnetic declination (alignment of magnetic north) and compass information, datum plane, symbols used for roads and trails, the dates and history of the survey of that particular quad, and more.

Topographic maps may be purchased directly from the USGS or Centre for Topographic Information, NRC (http://maps.nrcan.gc.ca/main.html). Many state geological survey offices, national and state park headquarters, outfitters, sports shops, and bookstores also sell topographic maps to assist people in planning their outdoor activities.

Control data and monuments
Vertical control

Third order or better, with tablet	$^{BM}\times_{16.3}$
Third order or better, recoverable mark	$\times_{120.0}$
Bench mark at found section corner	$^{BM}|_{18.6}$
Spot elevation	$\times_{5.3}$

Contours
Topographic

Intermediate	
Index	
Supplementary	
Depression	
Cut; fill	

Bathymetric

Intermediate	
Index	
Primary	
Index primary	
Supplementary	

Boundaries

National	
State or territorial	
County or equivalent	
Civil township or equivalent	
Incorporated city or equivalent	
Park, reservation, or monument	

Surface features

Levee	(Levee)
Sand or mud area, dunes, or shifting sand	(Sand)
Intricate surface area	(Strip mine)
Gravel beach or glacial moraine	(Gravel)
Tailings pond	(Tailings pond)

Mines and caves

Quarry or open pit mine	
Gravel, sand, clay, or borrow pit	
Mine dump	(Mine dump)
Tailings	(Tailings)

Vegetation

Woods	
Scrub	
Orchard	
Vineyard	
Mangrove	(Mangrove)

Glaciers and permanent snowfields

Contours and limits	
Form lines	

Marine shoreline
Topographic maps

Approximate mean high water	
Indefinite or unsurveyed	

Topographic-bathymetric maps

Mean high water	
Apparent (edge of vegetation)	

Coastal features

Foreshore flat	
Rock or coral reef	
Rock bare or awash	
Group of rocks bare or awash	
Exposed wreck	
Depth curve; sounding	
Breakwater, pier, jetty, or wharf	
Seawall	

Rivers, lakes, and canals

Intermittent stream	
Intermittent river	
Disappearing stream	
Perennial stream	
Perennial river	
Small falls; small rapids	
Large falls; large rapids	
Masonry dam	
Dam with lock	
Dam carrying road	
Perennial lake; Intermittent lake or pond	
Dry lake	(Dry lake)
Narrow wash	
Wide wash	(Wide wash)
Canal, flume, or aquaduct with lock	
Well or spring; spring or seep	

Submerged areas and bogs

Marsh or swamp	
Submerged marsh or swamp	
Wooded marsh or swamp	
Submerged wooded marsh or swamp	
Rice field	(Rice)
Land subject to inundation	Max pool 431

Buildings and related features

Building	
School; church	
Built-up area	
Racetrack	
Airport	
Landing strip	
Well (other than water); windmill	
Tanks	
Covered reservoir	
Gaging station	
Landmark object (feature as labeled)	
Campground; picnic area	
Cemetery: small; large	(Cem)

Roads and related features
Roads on Provisional edition maps are not classified as primary, secondary, or light duty. They are all symbolized as light duty roads.

Primary highway	
Secondary highway	
Light duty road	
Unimproved road	
Trail	
Dual highway	
Dual highway with median strip	

Railroads and related features

Standard gauge single track; station	
Standard gauge multiple track	
Abandoned	

Transmission lines and pipelines

Power transmission line; pole; tower	
Telephone line	Telephone
Aboveground oil or gas pipeline	
Underground oil or gas pipeline	Pipeline

FIGURE A.7 Standardized topographic map symbols used on USGS maps.
English units still prevail, although a few USGS maps are in metric. [From USGS, Topographic Maps, 1969.]

A.5

Appendix B

The Köppen Climate Classification System

The Köppen climate classification system was designed by Wladimir Köppen (1846–1940), a German climatologist and botanist, and is widely used for its ease of comprehension. The basis of any empirical classification system is the choice of criteria used to draw lines on a map to designate different climates. Köppen–Geiger climate classification uses *average monthly temperatures, average monthly precipitation,* and *total annual precipitation* to devise its spatial categories and boundaries. But we must remember that boundaries really are transition zones of gradual change. The trends and overall patterns of boundary lines are more important than their precise placement, especially with the small scales generally used on world maps.

Take a few minutes and examine the Köppen system, the criteria, and the considerations for each of the principal climate categories. The modified Köppen–Geiger system has its drawbacks, however. It does not consider winds, temperature extremes, precipitation intensity, quantity of sunshine, cloud cover, or net radiation. Yet the system is important because its correlations with the actual world are reasonable and the input data are standardized and readily available.

Köppen's Climatic Designations

Figure 10.4 in Chapter 10 shows the distribution of each of Köppen's six climate classifications on the land. This generalized map shows the spatial pattern of climate. The Köppen system uses capital letters (A, B, C, D, E, H) to designate climatic categories from the equator to the poles. The guidelines for each of these categories are in the margin of Figure B.1.

Five of the climate classifications are based on thermal criteria:

A Tropical (equatorial regions)
C Mesothermal (Mediterranean, humid subtropical, marine west coast regions)
D Microthermal (humid continental, subarctic regions)
E Polar (polar regions)
H Highland (compared to lowlands at the same latitude, highlands have lower temperatures—recall the normal lapse rate—and more efficient precipitation due to lower moisture demand)

Only one climate classification is based on moisture as well:

B Dry (deserts and steppes)

Within each climate classification additional lowercase letters are used to signify temperature and moisture conditions.

For example, in a tropical rain forest Af climate, the A tells us that the average coolest month is above 18°C (64.4°F, average for the month), and the f indicates that the weather is constantly wet, with the driest month receiving at least 6 cm (2.4 in.) of precipitation. (The designation f is from the German *feucht,* for "moist".) As you can see on the climate map, the tropical rain forest Af climate dominates along the equator and equatorial rain forest.

In a Dfa climate, the D means that the average warmest month is above 10°C (50°F), with at least one month falling below 0°C (32°F); the f says that at least 3 cm (1.2 in.) of precipitation falls during every month; and the a indicates a warmest summer month averaging above 22°C (71.6°F). Thus, a Dfa climate is a humid-continental, hot-summer climate in the microthermal category.

What's in a Boundary?

Originally, Köppen proposed that the isotherm boundary between mesothermal C and microthermal D climates be a coldest month of −3° C (26.6°F) or lower. That might be an accurate criterion for Europe, but for conditions in North America, the 0°C isotherm is considered more appropriate. The difference between the 0°C and −3°C isotherms for January covers an area about the width of the state of Ohio. Remember, these isotherm lines are really transition zones and do not mean abrupt change from one temperature to another.

A line denoting at least one month below freezing runs from New York City roughly along the Ohio River, trending westward until it meets the dry climates in the southeastern corner of Colorado. The line marking as the coldest month runs farther north along Lake Erie and the southern tip of Lake Michigan. From year to year, the position of the 0°C isotherm for January can shift back and forth several hundred kilometers as weather conditions vary. The map in Figure B.1 uses the 0°C isotherm for the C-D climate boundary.

Köppen Guidelines

The Köppen guidelines and map portrayal are in Figure B.1. First, check the guidelines for a climate type, then the color legend for the subdivisions of the type, then check out the distribution of that climate on the map. You may want to compare this with the climate map in Figure 10.5 that presents causal elements that produce these climates.

FIGURE B.1 World climates and their guidelines according to the Köppen classification system.

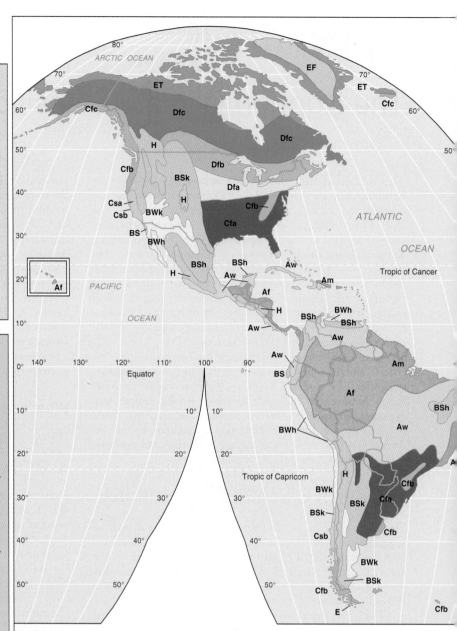

Köppen Guidelines
Tropical Climates — A

Consistently warm with all months averaging above 18°C (64.4°F); annual water supply exceeds water demand.

Af — Tropical rain forest:
 f = All months receive precipitation in excess of 6 cm (2.4 in.).

Am — Tropical monsoon:
 m = A marked short dry season with 1 or more months receiving less than 6 cm (2.4 in.) precipitation; an otherwise excessively wet rainy season. ITCZ 6–12 months dominant.

Aw — Tropical savanna:
 w = Summer wet season, winter dry season; ITCZ dominant 6 months or less, winter water-balance deficits.

Mesothermal Climates — C

Warmest month above 10°C (50°F); coldest month above 0°C (32°F) but below 18°C (64.4°F); seasonal climates.

Cfa, Cwa — Humid subtropical:
 a = Hot summer; warmest month above 22°C (71.6°F).
 f = Year-round precipitation.
 w = Winter drought, summer wettest month 10 times more precipitation than driest winter month.

Cfb, Cfc — Marine west coast, mild-to-cool summer:
 f = Receives year-round precipitation.
 b = Warmest month below 22°C (71.6°F) with 4 months above 10°C.
 c = 1–3 months above 10°C.

Csa, Csb — Mediterranean summer dry:
 s = Pronounced summer drought with 70% of precipitation in winter.
 a = Hot summer with warmest month above 22°C (71.6°F).
 b = Mild summer; warmest month below 22°C.

Microthermal Climates — D

Warmest month above 10°C (50°F); coldest month below 0°C (32°F); cool temperate-to-cold conditions; snow climates. In Southern Hemisphere, occurs only in highland climates.

Dfa, Dwa — Humid continental:
 a = Hot summer; warmest month above 22°C (71.6°F).
 f = Year-round precipitation.
 w = Winter drought.

Dfb, Dwb — Humid continental:
 b = Mild summer; warmest month below 22°C (71.6°F).
 f = Year-round precipitation.
 w = Winter drought.

Dfc, Dwc, Dwb — Subarctic:
Cool summers, cold winters.
 f = Year-round precipitation.
 w = Winter drought.
 c = 1–4 months above 10°C.
 d = Coldest month below −38°C (−36.4°F), in Siberia only.

Dry Arid and Semiarid Climates — B

Potential evapotranspiration* (natural moisture demand) exceeds precipitation (natural moisture supply) in all B climates. Subdivisions based on precipitation timing and amount and mean annual temperature.

Earth's arid climates.

BWh — Hot low-latitude desert

BWk — Cold midlatitude desert
 BW = Precipitation less than 1/2 natural moisture demand.
 h = Mean annual temperature >18°C (64.4°F).
 k = Mean annual temperature <18°C.

Earth's semiarid climates.

BSh — Hot low-latitude steppe

BSk — Cold midlatitude steppe
 BS = Precipitation more than 1/2 natural moisture demand but not equal to it.
 h = Mean annual temperature >18°C.
 k = Mean annual temperature <18°C.

Polar Climates — E

Warmest month below 10°C (50°F); always cold; ice climates.

ET — Tundra:
 Warmest month 0–10°C (32–50°F); precipitation exceeds small potential evapotranspiration demand*; snow cover 8–10 months.

EF — Ice cap:
 Warmest month below 0°C (32°F); precipitation exceeds a very small potential evapotranspiration demand; the polar regions.

EM — Polar marine:
 All months above −7°C (20°F), warmest month above 0°C; annual temperature range <17 C° (30 F°).

*Potential evapotranspiration = the amount of water that would evaporate or transpire if it were available — the natural moisture demand in an environment; see Chapter 9.

MODIFIED GOODE'S HOMOLOSINE EQUAL-AREA PROJECTION

A TROPICAL CLIMATES

Af Tropical rain forest climate
Am Tropical monsoon climate

Aw Tropical savanna climate

C MESOTHERMAL CLIMATES

Cfa Humid subtropical, without dry season, hot summers

Cwa
Cwb Humid subtropical, winter-dry

Cfb
Cfc Marine west coast, without dry season, warm to cool summers

Csa
Csb Mediterranean summer-dry

D MICROTHERMAL CLIMATES

Dfa
Dwa Humid continental, hot summers

Dfb
Dwb Humid continental, warm summers

Dfc
Dwc Subarctic, cool summers

Dfd
Dwd Subarctic, very cold winter

w Winter dry
f Without a dry season

B DRY ARID AND SEMIARID CLIMATES

BW Desert climate

BS Steppe climate

h Low-latitude hot **B** climates
k Midlatitude cold **B** climates

E POLAR CLIMATES
H HIGHLAND

ET Tundra climate

EF Ice cap and sheets
H Denotes cold climate due to elevation

Firn line (17)   The snow line that is visible on the surface of a glacier, where winter snows survive the summer ablation season; analogous to a snow line on land (see ablation).

Fjord (17)   A drowned glaciated valley, or glacial trough, along a seacoast.

Flash flood (15)   A sudden and short-lived torrent of water that exceeds the capacity of a stream channel; associated with desert and semi-arid washes.

Flood (14)   A high water level that overflows the natural riverbank along any portion of a stream.

Floodplain (14)   A flat, low-lying area along a stream channel, created by and subject to recurrent flooding; alluvial deposits generally mask underlying rock.

Flood tide (16)   Rising tide during the daily tidal cycle (compare ebb tide).

Fluvial (14)   Stream-related processes; from the Latin *fluvius* for "river" or "running water."

Fog (7)   A cloud, generally stratiform, in contact with the ground, with visibility usually reduced to less than 1 km (3300 ft).

Folding (12)   The bending and deformation of beds of rock strata subjected to compressional forces.

Food chain (19)   The circuit along which energy flows from producers (plants), which manufacture their own food, to consumers (animals); a one-directional flow of chemical energy, ending with decomposers.

Food web (19)   A complex network of interconnected food chains (see food chain).

Formation class (20)   That portion of a biome that concerns the plant communities only, categorized by size, shape, and structure of the dominant vegetation.

Friction force (6)   The effect of drag by the wind as it moves across a surface; may be operative through 500 m (1600 ft) of altitude. Surface friction slows the wind and therefore reduces the effectiveness of the Coriolis force.

Frost action (13)   A powerful mechanical force produced as water expands up to 9% of its volume as it freezes. Water freezing in a cavity in a rock can break the rock if it exceeds the rock's tensional strength.

Funnel cloud (8)   The visible swirl extending from the bottom side of a cloud, which may or may not develop into a tornado. A tornado is a funnel cloud that has extended all the way to the ground (see tornado).

Fusion (2)   The process of forcibly joining positively charged hydrogen and helium nuclei under extreme temperature and pressure; occurs naturally in thermonuclear reactions within stars, such as our Sun.

Gelifluction (17)   Refers to soil flow in periglacial environments, a progressive, lateral movement; a type of solifluction formed under periglacial conditions of permafrost and frozen ground (see solifluction).

Gelisols (18)   A new soil order in the Soil Taxonomy, added in 1998, describing cold and frozen soils at high latitudes or high elevations; characteristic tundra vegetation (see Canadian System of Soil Classification for similar types).

General circulation model (GCM) (10)   Complex, computer-based climate model that produces generalizations of reality and forecasts of future weather and climate conditions. Complex GCMs (three-dimensional models) are in use in the United States and in other countries.

Genetic classification (10)   A climate classification that uses causative factors to determine climatic regions; for example, an analysis of the effect of interacting air masses (compare empirical classification).

Geodesy (1)   The science that determines Earth's shape and size through surveys, mathematical means, and remote sensing (see geoid).

Geographic information system (GIS) (1)   A computer-based data processing tool or methodology used for gathering, manipulating, and analyzing geographic information to produce a holistic, interactive analysis.

Geography (1)   The science that studies the interdependence and interaction among geographic areas, natural systems, processes, society, and cultural activities over space—a spatial science. The five themes of geographic education are location, place, movement, regions, and human-Earth relationships.

Geoid (1)   A word that describes Earth's shape; literally, "the shape of Earth is Earth-shaped." A theoretical surface at sea level that extends through the continents; deviates from a perfect sphere.

Geologic cycle (11)   A general term characterizing the vast cycling that proceeds in the lithosphere. It encompasses the hydrologic cycle, tectonic cycle, and rock cycle.

Geologic time scale (11)   A depiction of eras, periods, and epochs that span Earth's history; shows both the sequence of rock strata and their absolute dates, as determined by methods such as radioactive isotopic dating.

Geomagnetic reversal (11)   A polarity change in Earth's magnetic field. With uneven regularity, the magnetic field fades to zero, then returns to full strength but with the magnetic poles reversed. Reversals have been recorded 9 times during the past 4 million years.

Geomorphic threshold (13)   The threshold up to which landforms change before lurching to a new set of relationships, with rapid realignments of landscape materials and slopes.

Geomorphology (13)   The science that analyzes and describes the origin, evolution, form, classification, and spatial distribution of landforms.

Geostrophic wind (6)   A wind moving between areas of different pressure along a path that is parallel to the isobars. It is a product of the pressure gradient force and the Coriolis force (see isobar, pressure gradient force, Coriolis force).

Geothermal energy (11, 12)   The energy in steam and hot water heated by subsurface magma near groundwater. Geothermal energy literally refers to heat from Earth's interior, whereas *geothermal power* relates to specific applied strategies of geothermal electric or geothermal direct applications. This energy is used in Iceland, New Zealand, Italy, and northern California, among other locations.

Glacial drift (17)   The general term for all glacial deposits, both unsorted (till) and sorted (stratified drift).

Glacial ice (17)   A hardened form of ice, very dense in comparison to normal snow or firn.

Glacier (17)   A large mass of perennial ice resting on land or floating shelflike in the sea adjacent to the land; formed from the accumulation and recrystallization of snow, which then flows slowly under the pressure of its own weight and the pull of gravity.

Glacier surge (17)   The rapid, lurching, unexpected forward movement of a glacier.

Glacio-eustatic (7)   Changes in sea level in response to changes in the amount of water stored on Earth as ice; the more water that is bound up in glaciers and ice sheets, the lower the sea level (compare eustasy).

Gleization (18)   A process of humus and clay accumulation in cold, wet climates with poor drainage.

Global positioning system (GPS) (1)   Latitude, longitude, and elevation are accurately calibrated using this handheld instrument that calibrates radio signals from satellites.

Goode's homolosine projection (Appendix A)   An equal-area projection formed by splicing together a sinusoidal and a homolographic projection.

Graben (12)   Pairs or groups of faults that produce downward-faulted blocks; characteristic of the basins of the interior western United States (compare horst; see Basin and Range Province).

Graded stream (14)   An idealized condition in which a stream's load and the landscape mutually adjust. This forms a dynamic equilibrium among erosion, transported load, deposition, and the stream's capacity.

Gradient (14)   The drop in elevation from a stream's headwaters to its mouth, ideally forming a concave slope.

Granite (11)   A coarse-grained (slow-cooling) intrusive igneous rock of 25% quartz and more than 50% potassium and sodium feldspars; characteristic of the continental crust.

Gravitational water (9)   That portion of surplus water that percolates downward from the capillary zone, pulled by gravity to the groundwater zone.

Gravity (2)   The mutual force exerted by the masses of objects that are attracted one to another and produced in an amount proportional to each object's mass.

Great circle (1)   Any circle drawn on a globe with its center coinciding with the center of the globe. An infinite number of great circles can be drawn, but only one parallel is a great circle—the equator (compare small circle).

Greenhouse effect (4)   The process whereby radiatively active gases (carbon dioxide, water vapor, methane, and CFCs) absorb insolation and emit the energy at longer wavelengths, which are retained longer, delaying the loss of infrared to space. Thus, the lower troposphere is warmed through the radiation and re-radiation of infrared wavelengths. The approximate similarity between this process and that of a greenhouse explains the name.

Greenwich Mean Time (GMT) (1)   Former world standard time, now reported as Coordinated Universal Time (UTC) (see Coordinated Universal Time).

Ground ice (17)   Subsurface water that is frozen in regions of permafrost. The moisture content of areas with ground ice may vary from nearly absent in regions of drier permafrost to almost 100% in saturated soils (compare permafrost).

Groundwater (9)   Water beneath the surface that is beyond the soil-root zone; a major source of potable water.

Groundwater mining (9)   Pumping an aquifer beyond its capacity to flow and recharge; an overuse of the groundwater resource.

Gulf Stream (5)   A strong, northward-moving, warm current off the east coast of North America, which carries its water far into the North Atlantic.

Habitat (19, 20)   A physical location to which an organism is biologically suited. Most species have specific habitat parameters and limits (compare to niche).

Hail (8)   A type of precipitation formed when a raindrop is repeatedly circulated above and below the freezing level in a cloud, with each cycle freezing more moisture onto the hailstone until it becomes too heavy to stay aloft.

Hair hygrometer (7)   An instrument for measuring relative humidity; based on the principle that human hair will change as much as 4% in length between 0% and 100% relative humidity.

Heat (3)   The flow of kinetic energy from one body to another because of a temperature difference between them.

Herbivore (19)   The primary consumer in a food web, which eats plant material formed by a producer (plant) that has photosynthesized organic molecules (compare carnivore).

Heterosphere (3)   A zone of the atmosphere above the mesopause, 80 km (50 mi) in altitude; composed of rarefied layers of oxygen atoms and nitrogen molecules; includes the ionosphere.

Histosols (18)   A soil order in the Soil Taxonomy. Formed from thick accumulations of organic matter, such as beds of former lakes, bogs, and layers of peat.

Homosphere (3)   A zone of the atmosphere from Earth's surface up to 80 km (50 mi), composed of an even mixture of gases, including nitrogen, oxygen, argon, carbon dioxide, and trace gases.

Horn (17)   A pyramidal, sharp-pointed peak that results when several cirque glaciers gouge an individual mountain summit from all sides.

Horst (12)   Upward-faulted blocks produced by pairs or groups of faults; characteristic of the mountain ranges of the interior of the western United States (see graben and Basin and Range Province).

Hot spot (11)   An individual point of upwelling material originating in the asthenosphere, or deeper in the mantle; tends to remain fixed relative to migrating plates; some 100 are identified worldwide, exemplified by Yellowstone National Park, Hawai'i, and Iceland.

Human–Earth relationships (1)   One of the oldest themes of geography (the human-land tradition); includes the spatial analysis of settlement patterns, resource utilization and exploitation, hazard perception and planning, and the impact of environmental modification and artificial landscape creation.

Humidity (7)   Water vapor content of the air. The capacity of the air to hold water vapor is mostly a function of the temperature of the air and the water vapor.

Humus (18)   A mixture of organic debris in the soil worked by consumers and decomposers in the humification process; characteristically formed from plant and animal litter deposited at the surface.

Hurricane (8)   A tropical cyclone that is fully organized and intensified in inward-spiraling rainbands; ranges from 160 to 960 km (100 to 600 mi) in diameter, with wind speeds in excess of 119 kmph (65 knots, or 74 mph); a name used specifically in the Atlantic and eastern Pacific (compare typhoon).

Hydration (13)   A chemical weathering process involving water that is added to a mineral, which initiates swelling and stress within the rock,

mechanically forcing grains apart as the constituents expand (contrast to hydrolysis).

Hydrology (14)   The science of water, its global circulation, distribution, and properties, specifically water at and below Earth's surface.

Hydraulic action (14)   The erosive work accomplished by the turbulence of water; causes a squeezing and releasing action in joints in bedrock; capable of prying and lifting rocks.

Hydrograph (14)   A graph of stream discharge (in cms or cfs) over a period of time (minutes, hours, days, years) at a specific place on a stream. The relationship between stream discharge and precipitation input is illustrated on the graph.

Hydrologic cycle (9)   A simplified model of the flow of water, ice, and water vapor from place to place. Water flows through the atmosphere, across the land, where it is also stored as ice, and within groundwater. Solar energy empowers the cycle.

Hydrolysis (13)   A chemical weathering process in which minerals chemically combine with water; a decomposition process that causes silicate minerals in rocks to break down and become altered (contrast with hydration).

Hydrosphere (1)   An abiotic open system that includes all of Earth's water.

Hygroscopic water (9)   That portion of soil moisture that is so tightly bound to each soil particle that it is unavailable to plant roots; the water, along with some bound capillary water, that is left in the soil after the wilting point is reached (see wilting point).

Ice age (17)   A cold episode, with accompanying alpine and continental ice accumulations, that has repeated roughly every 200 to 300 million years since the late Precambrian era (1.25 billion years ago); includes the most recent episode during the Pleistocene Ice Age, which began 1.65 million years ago.

Iceberg (17)   Floating ice created by ice calving (a large piece breaking off) and floating adrift; a hazard to shipping because about nine-tenths of the ice is submerged and can be irregular in form.

Ice cap (17)   A large, dome-shaped glacier, less extensive than an ice sheet ($<50,000$ km^2, $<19,300$ mi^2), although it buries mountain peaks and the local landscape.

Ice field (17)   The least extensive form of a glacier, with mountain ridges and peaks visible above the ice; less than an ice cap or ice sheet.

Icelandic low (6)   See subpolar low-pressure cell.

Ice sheet (17)   An enormous continuous continental glacier. The bulk of glacial ice on Earth covers Antarctica and Greenland in two ice sheets.

Ice wedge (17)   Formed when water enters a thermal contraction crack in permafrost and freezes. Repeated seasonal freezing and melting of the water progressively expands the wedge.

Igneous rock (11)   One of the basic rock types; it has solidified and crystallized from a hot molten state (either magma or lava). (Compare metamorphic rock, sedimentary rock.)

Illuviation (18)   The downward movement and deposition of finer particles and minerals from the upper horizon of the soil; a depositional process. Deposition usually is in the B horizon, where accumulations of clays, aluminum, carbonates, iron, and some humus occur (compare eluviation; see calcification).

Inceptisols (18)   A soil order in the Soil Taxonomy. Weakly developed soils that are inherently infertile; usually young soils that are weakly developed, although they are more developed than Entisols.

Industrial smog (3)   Air pollution associated with coal-burning industries; it may contain sulfur oxides, particulates, carbon dioxide, and exotics.

Infiltration (9)   Water access to subsurface regions of soil moisture storage through penetration of the soil surface.

Insolation (2)   Solar radiation that is incoming to Earth systems.

Interception (9)   Delays the fall of precipitation toward Earth's surface; caused by vegetation or other ground cover.

Internal drainage (14)   In regions where rivers do not flow into the ocean, the outflow is through evaporation or subsurface gravitational flow. Portions of Africa, Asia, Australia, and the western United States have such drainage.

International Date Line (1)   The 180° meridian, an important corollary to the prime meridian on the opposite side of the planet; established by the treaty of 1884 to mark the place where each day officially begins.

Intertropical convergence zone (ITCZ) (6)   See equatorial low-pressure trough.

Ionosphere (3)   A layer in the atmosphere above 80 km (50 mi) where gamma, X-ray, and some ultraviolet radiation is absorbed and converted into infrared and where the solar wind stimulates the auroras.

Isobar (6)   An isoline connecting all points of equal atmospheric pressure.

Isostasy (11)   A state of equilibrium in Earth's crust formed by the interplay between portions of the lithosphere and the asthenosphere and the principle of buoyancy. The crust depresses under weight and recovers with its removal, for example, the melting of glacial ice. The uplift is known as isostatic rebound.

Isotherm (5)   An isoline connecting all points of equal temperature.

Jet stream (6)   The most prominent movement in upper-level westerly wind flows; irregular, concentrated, sinuous bands of geostrophic wind, traveling at 300 kmph (190 mph). (See polar jet stream, subtropical jet stream.)

Joint (13)   A fracture or separation in rock that occurs without displacement of the sides; increases the surface area of rock exposed to weathering processes.

Kame (17)   A depositional feature of glaciation; a small hill of poorly sorted sand and gravel that accumulates in crevasses or in ice-caused indentations in the surface.

Karst topography (13)   Distinctive topography formed in a region of chemically weathered limestone with poorly developed surface drainage and solution features that appear pitted and bumpy; originally named after the Krš Plateau in Slovenia.

Katabatic winds (6)   Air drainage from elevated regions, flowing as gravity winds. Layers of air at the surface cool, become denser, and flow downslope; known worldwide by many local names.

Kettle (17)   Forms when an isolated block of ice persists in a ground moraine, an outwash plain, or valley floor after a glacier retreats; as the block finally melts, it leaves behind a steep-sided hole that frequently fills with water.

Kinetic energy (3)   The energy of motion in a body; derived from the vibration of the body's own movement and stated as temperature.

Köppen–Geiger climate classification (10, Appendix B)   An empirical classification system that uses average monthly temperatures, average monthly precipitation, and total annual precipitation to establish regional climate designations.

Lagoon (16)   An area of coastal seawater that is virtually cut off from the ocean by a bay barrier or barrier beach; also, the water surrounded and enclosed by an atoll.

Landfall (8)   The location along a coast where a storm moves onshore.

Land–sea breeze (6)   Wind along coastlines and adjoining interior areas created by different heating characteristics of land and water surfaces—onshore (landward) breeze in the afternoon and offshore (seaward) breeze at night.

Landslide (13)   A sudden rapid downslope movement of a cohesive mass of regolith and/or bedrock in a variety of mass-movement forms under the influence of gravity; a form of mass movement.

Land–water heating difference (5)   Differences in the degree and way that land and water heat, as a result of contrasts in transmission, evaporation, mixing, and specific heat capacities. Land surfaces heat and cool faster than water and have continentality, whereas water provides a marine influence.

Latent heat (7)   Heat energy is stored in one of three states—ice, water, or water vapor. The energy is absorbed or released in each phase change from one state to another. Heat energy is absorbed as the latent heat of melting, vaporization, or evaporation. Heat energy is released as the latent heat of condensation and freezing (or fusion).

Latent heat of condensation (7)   The heat energy released to the environment in a phase change from water vapor to liquid; under normal sea-level pressure, 540 calories are released from each gram of water vapor that changes phase to water at boiling; and 585 calories are released from each gram of water vapor that condenses at 20°C (68°F).

Latent heat of sublimation (7)   The heat energy absorbed or released in the phase change from ice to water vapor or water vapor to ice—no liquid phase. The change from water vapor to ice is also called deposition.

Latent heat of vaporization (7)   The heat energy absorbed from the environment in a phase change from liquid to water vapor at the boiling point; under normal sea-level pressure, 540 calories must be added to each gram of boiling water to achieve a phase change to water vapor.

Lateral moraine (17)   Debris transported by a glacier that accumulates along the sides of the glacier and is deposited along these margins.

Laterization (18)   A pedogenic process operating in well-drained soils that occur in warm and humid regions; typical of Oxisols. Plentiful precipitation leaches soluble minerals and soil constituents. Resulting soils usually are reddish or yellowish.

Latitude (1)   The angular distance measured north or south of the equator from a point at the center of Earth. A line connecting all points of the same latitudinal angle is called a parallel (compare longitude).

Lava (11, 12)   Magma that issues from volcanic activity onto the surface; the extrusive rock that results when magma solidifies (see magma).

Life zone (19)   A zonation by altitude of plants and animals that form distinctive communities. Each life zone possesses its own temperature and precipitation relations.

Lightning (8)   Flashes of light caused by tens of millions of volts of electrical charge heating the air to temperatures of 15,000° to 30,000°C.

Limestone (11)   The most common chemical sedimentary rock (nonclastic); it is lithified calcium carbonate; very susceptible to chemical weathering by acids in the environment, including carbonic acid in rainfall.

Limiting factor (19)   The physical or chemical factor that most inhibits biotic processes, either through lack or excess.

Lithification (11)   The compaction, cementation, and hardening of sediments into sedimentary rock.

Lithosphere (1)   Earth's crust and that portion of the upper-most mantle directly below the crust, extending down to about 70 km (45 mi). Some sources use this term to refer to the entire Earth.

Littoral drift (16)   Transport of sand, gravel, sediment, and debris along the shore; a more comprehensive term, considers *beach drift* and *longshore drift* combined.

Littoral zone (16)   A specific coastal environment; that region between the high-water line during a storm and a depth at which storm waves are unable to move sea-floor sediments.

Loam (18)   A soil that is a mixture of sand, silt, and clay in almost equal proportions, with no one texture dominant; an ideal agricultural soil.

Location (1)   A basic theme of geography dealing with the absolute and relative position of people, places, and things on Earth's surface.

Loess (15)   Large quantities of fine-grained clays and silts left as glacial outwash deposits; subsequently blown by the wind great distances and redeposited as a generally unstratified, homogeneous blanket of material covering existing landscapes; in China, loess originated from desert lands.

Longitude (1)   The angular distance measured east or west of a prime meridian from a point at the center of Earth. A line connecting all points of the same longitude is called a meridian.

Longshore current (16)   A current that forms parallel to a beach as waves arrive at an angle to the shore; generated in the surf zone by wave action, transporting large amounts of sand and sediment (see littoral current, beach drift).

Lysimeter (9)   A weather instrument for measuring potential and actual evapotranspiration; isolates a portion of a field so that the moisture moving through the plot is measured.

Magma (11)   Molten rock from beneath Earth's surface; fluid, gaseous, under tremendous pressure, and either intruded into existing country rock or extruded onto the surface as lava (see lava).

Magnetosphere (2)   Earth's magnetic force field, which is generated by dynamo-like motions within the planet's outer core; deflects the solar wind flow toward the upper atmosphere above each pole.

Mangrove swamp (16)   A wetland ecosystem between 30° N or S and the equator; tends to form a distinctive community of mangrove plants (compare salt marsh).

Mantle (11)   An area within the planet representing about 80% of Earth's total volume, with densities increasing with depth and averaging 4.5 g/cm^3; occurs between the core and the crust; is rich in iron and magnesium oxides and silicates.

Map (1, Appendix A)   A generalized view of an area, usually some portion of Earth's surface, as seen from above at a greatly reduced size (see scale and map projection).

Map projection (1, Appendix A) The reduction of a spherical globe onto a flat surface in some orderly and systematic realignment of the latitude and longitude grid.

Marine effect (5) A quality of regions that are dominated by the moderating effect of the ocean and that exhibit a smaller range of minimum and maximum temperature than continental stations (see continentality, land–water heating differences).

Mass movement (13) All unit movements of materials propelled by gravity; can range from dry to wet, slow to fast, small to large, and free-falling to gradual or intermittent.

Mass wasting (13) Gravitational movement of nonunified material downslope; a specific form of mass movement.

Meandering stream (14) The sinuous, curving pattern common to graded streams, with the energetic outer portion of each curve subjected to the greatest erosive action and the lower-energy inner portion receiving sediment deposits (see graded stream).

Mean sea level (MSL) (16) The average of tidal levels recorded hourly at a given site over a long period, which must be at least a full lunar tidal cycle.

Medial moraine (17) Debris transported by a glacier that accumulates down the middle of the glacier, resulting from two glaciers merging their lateral moraines; forms a depositional feature following glacial retreat.

Mediterranean shrubland (20) A major biome dominated by Mediterranean dry-summer climates and characterized by sclerophyllous scrub and short, stunted, tough forests (see chapparal).

Mercator projection (1) A true-shape projection, with meridians appearing as equally spaced straight lines and parallels appearing as straight lines that are spaced closer together near the equator. The poles are infinitely stretched, with the 84th north parallel and 84th south parallel fixed at the same length as that of the equator. It presents false notions of the size (area) of midlatitude and poleward landmasses but presents true compass direction (see rhumb line).

Mercury barometer (6) A device that measures air pressure using a column of mercury in a tube, one end of which is sealed, and the other end inserted in an open vessel of mercury (see air pressure).

Meridian (1) See longitude; a line designating an angle of longitude.

Mesocyclone (8) A large, rotating atmospheric circulation, initiated within a parent cumulonimbus cloud at midtroposphere elevation; generally produces heavy rain, large hail, blustery winds, and lightning; may lead to tornado activity.

Mesosphere (3) The upper region of the homosphere from 50 to 80 km (30 to 50 mi) above the ground; designated by temperature criteria; atmosphere extremely rarified.

Metamorphic rock (11) One of three basic rock types, it is existing igneous and sedimentary rock that has undergone profound physical and chemical changes under increased pressure and temperature. Constituent mineral structures may exhibit foliated or nonfoliated textures (compare igneous rock and sedimentary rock).

Meteorology (8) The scientific study of the atmosphere, including its physical characteristics and motions; related chemical, physical, and geological processes; the complex linkages of atmospheric systems; and weather forecasting.

Microclimatology (4) The study of local climates at or near Earth's surface, or that height above the Earth's surface where the effects of the surface are no longer of effect.

Midlatitude broadleaf and mixed forest (20) A biome in moist continental climates in areas of warm-to-hot summers and cool-to-cold winters; relatively lush stands of broadleaf forests trend northward into needleleaf evergreen stands.

Midlatitude cyclone (8) An organized area of low pressure, with converging and ascending airflow producing an interaction of air masses; migrates along storm tracks. Such lows or depressions form the dominant weather pattern in the middle and higher latitudes of both hemispheres.

Midlatitude grassland (20) The major biome most modified by human activity; so named because of the predominance of grasslike plants, although deciduous broadleafs appear along streams and other limited sites; location of the world's breadbaskets of grain and livestock production.

Mid-ocean ridge (11) A submarine mountain range that extends more than 65,000 km (40,000 mi) worldwide and averages more than 1000 km (600 mi) in width; centered along sea-floor spreading centers (see sea-floor spreading).

Milky Way Galaxy (2) A flattened, disk-shaped mass in space estimated to contain up to 400 billion stars; includes our Solar System.

Miller cylindrical projection (Appendix A) A compromise map projection that avoids the severe distortion of the Mercator projection (see map projection).

Mineral (11) An element or combination of elements that forms an inorganic natural compound; described by a specific formula and crystal structure.

Model (1) A simplified version of a system, representing an idealized part of the real world.

Mohorovičić discontinuity, or Moho (11) The boundary between the crust and the rest of the lithospheric upper mantle; named for the Yugoslavian seismologist Mohorovičić; a zone of sharp material and density contrasts; also known as the Moho.

Moist adiabatic rate (MAR) (7) The rate at which a saturated parcel of air cools in ascent; a rate of 6 C° per 1000 m (3.3 F° per 1000 ft). This rate may vary, with moisture content and temperature, from 4 C° to 10 C° per 1000 m (2 F° to 6 F° per 1000 ft) (see adiabatic; compare dry adiabatic rate).

Moisture droplet (7) A tiny water particle that constitutes the initial composition of clouds. Each droplet measures approximately 0.002 cm (0.0008 in.) in diameter and is invisible to the unaided eye.

Mollisols (18) A soil order in the Soil Taxonomy. These have a mollic epipedon and a humus-rich organic content high in alkalinity. Some of the world's most significant agricultural soils are Mollisols.

Moment magnitude scale (12) An earthquake magnitude scale. Considers the amount of fault slippage, the size of the area that ruptured, and the nature of the materials that faulted in estimating the magnitude of an earthquake—an assessment of the seismic moment. Replaces the Richter scale (amplitude magnitude); especially valuable in assessing larger-magnitude events.

Monsoon (6) An annual cycle of dryness and wetness, with seasonally shifting winds produced by changing atmospheric pressure systems; affects India, Southeast Asia, Indonesia, northern Australia, and portions of Africa. From the Arabic word *mausim*, meaning "season."

Montane forest (20) Needleleaf forest associated with mountain elevations (see needleleaf forest).

Moraine (17) Marginal glacial deposits (lateral, medial, terminal, ground) of unsorted and unstratified material.

Mountain–valley breeze (6) A light wind produced as cooler mountain air flows downslope at night and as warmer valley air flows upslope during the day.

Movement (1) A major theme in geography involving migration, communication, and the interaction of people and processes across space.

Mudflow (13) Fluid downslope flows of material containing more water than earthflows.

Natural levee (14) A long, low ridge that forms on both sides of a stream in a developed floodplain; they are depositional products (coarse gravels and sand) of river flooding.

Neap tide (16) Unusually low tidal range produced during the first and third quarters of the Moon, with an offsetting pull from the Sun (compare spring tide).

Needleleaf forest (20) Forests of pine, spruce, fir, and larch, stretching from the east coast of Canada westward to Alaska and continuing from Siberia westward across the entire extent of Russia to the European Plain; called the *taiga* (a Russian word) or the *boreal forest*; principally in the microthermal climates. Includes montane forests that may be at lower latitudes at higher elevation.

Negative feedback (1) Feedback that tends to slow or dampen response in a system; promotes self-regulation in a system; far more common than positive feedback in living systems (see feedback loop, compare to positive feedback).

Net primary productivity (19) The net photosynthesis (photosynthesis minus respiration) for a given community; considers all growth and all reduction factors that affect the amount of useful chemical energy (biomass) fixed in an ecosystem.

Net radiation (NET R) (4) The net all-wave radiation available at Earth's surface; the final outcome of the radiation balance process between incoming shortwave insolation and outgoing longwave energy.

Niche (19, 20) The basic function, or occupation, of a life form within a given community; the way an organism obtains its food, air, and water.

Nickpoint (knickpoint) (14) The point at which the longitudinal profile of a stream is abruptly broken by a change in gradient; for example, a waterfall, rapids, or cascade.

Nimbostratus (7) Rain-producing, dark, grayish, stratiform clouds characterized by gentle drizzle.

Nitrogen dioxide (3) A noxious (harmful) reddish-brown gas produced in combustion engines; can be damaging to human respiratory tracts and to plants; participates in photochemical reactions and acid deposition.

Noctilucent cloud (3) A rare, shining band of ice crystals that may glow at high latitudes long after sunset; formed within the mesosphere, where cosmic and meteoric dust act as nuclei for the formation of ice crystals.

Normal fault (12) A type of geologic fault in rocks. Tension produces strain that breaks a rock, with one side moving vertically relative to the other side along an inclined fault plane (compare reverse fault).

Normal lapse rate (3) The average rate of temperature decrease with increasing altitude in the lower atmosphere; an average value of 6.4 C° per km, or 1000 m (3.5 F° per 1000 ft). (Compare environmental lapse rate.)

Occluded front (8) In a cyclonic circulation, the overrunning of a surface warm front by a cold front and the subsequent lifting of the warm air wedge off the ground; initial precipitation is moderate to heavy.

Ocean basin (12) The physical container (a depression in the lithosphere) holding an ocean.

Omnivore (19) A consumer that feeds on both producers (plants) and consumers (meat)—a role occupied by humans, among other animals (compare consumer, producer).

Open system (1) A system with inputs and outputs crossing back and forth between the system and the surrounding environment. Earth is an open system in terms of energy (compare closed system).

Orogenesis (12) The process of mountain building that occurs when large-scale compression leads to deformation and uplift of the crust; literally, the birth of mountains.

Orographic lifting (8) The uplift of a migrating air mass as it is forced to move upward over a mountain range—a topographic barrier. The lifted air cools adiabatically as it moves upslope; clouds may form and produce increased precipitation.

Outgassing (7) The release of trapped gases from rocks, forced out through cracks, fissures, and volcanoes from within Earth; the terrestrial source of Earth's water.

Outwash plain (17) Glacial stream deposits of stratified drift of meltwater-fed, braided, and overloaded streams; occurs beyond a glacier's morainal deposits.

Overland flow (9) Surplus water that flows across the land surface toward stream channels. Together with precipitation and subsurface flows, it constitutes the total runoff from an area.

Oxbow lake (14) A lake that was formerly part of the channel of a meandering stream; isolated when a stream eroded its outer bank forming a cutoff through the neck of the looping meander (see meandering stream). In Australia, known as a billabong (Aboriginal for "dead river").

Oxidation (13) A chemical weathering process in which oxygen dissolved in water oxidizes (combines with) certain metallic elements to form oxides; most familiar is the "rusting" of iron in a rock or soil (Ultisols, Oxisols), which produces a reddish-brown stain of iron oxide .

Oxisols (18) A soil order in the Soil Taxonomy. Tropical soils that are old, deeply developed, and lacking in horizons wherever well drained; heavily weathered, low in cation-exchange capacity, and low in fertility.

Ozone layer (3) See ozonosphere.

Ozonosphere (3) A layer of ozone occupying the full extent of the stratosphere (20 to 50 km, or 12 to 30 mi) above the surface; the region of the atmosphere where ultraviolet wavelengths of insolation are extensively absorbed and converted into heat.

Pacific high (6) A high-pressure cell that dominates the Pacific in July, retreating southward in the Northern Hemisphere in January; also known as the Hawaiian high.

Pahoehoe (12) Basaltic lava that is more fluid than aa. Pahoehoe forms a thin crust that forms folds and appears "ropy," like coiled, twisted rope.

Paleoclimatology (10, 17) The science that studies the climates, and the causes of variations in climate, of past ages, throughout historic and geologic time.

Paleolake (17) An ancient lake, such as Lake Bonneville or Lake Lahonton, associated with former wet periods when the lake basins were filled to higher levels than today.

Palsa (17) A rounded or elliptical mound of peat that contains thin perennial ice lenses rather than an ice core, as in a pingo. Palsas can be 2 to 30 m wide by 1 to 10 m high and usually are covered by soil or vegetation over a cracked surface.

PAN (3) See peroxyacetyl nitrate.

Pangaea (11) The supercontinent formed by the collision of all continental masses approximately 225 million years ago; named in the continental drift theory by Wegener in 1912 (see plate tectonics).

Parallel (1) See latitude; a line, parallel to the equator, that designates an angle of latitude.

Parent material (13) The unconsolidated material, from both organic and mineral sources, that is the basis of soil development.

Particulate matter (PM) (3) Dust, dirt, soot, salt, sulfate aerosols, fugitive natural particles, or other material particles suspended in air.

Paternoster lake (17) One of a series of small, circular, stair-stepped lakes formed in individual rock basins aligned down the course of a glaciated valley; named because they look like a string of rosary (religious) beads.

Patterned ground (17) Areas in the periglacial environment where freezing and thawing of the ground create polygonal forms of arranged rocks at the surface; can be circles, polygons, stripes, nets, and steps.

Pedogenic regime (18) A specific soil-forming process keyed to a specific climatic regime: laterization, calcification, salinization, and podzolization, among others; not the basis for soil classification in the Soil Taxonomy.

Pedon (18) A soil profile extending from the surface to the lowest extent of plant roots or to the depth where regolith or bedrock is encountered; imagined as a hexagonal column; the basic soil sampling unit.

Percolation (9) The process by which water permeates the soil or porous rock into the subsurface environment.

Periglacial (17) Cold-climate processes, landforms, and topographic features along the margins of glaciers, past and present; periglacial characteristics exist on more than 20% of Earth's land surface; includes permafrost, frost action, and ground ice.

Perihelion (2) That point of Earth's closest approach to the Sun in its elliptical orbit; occurs on January 3 at 147,255,000 km (91,500,000 mi); variable over a 100,000-year cycle (compare aphelion).

Permafrost (17) Forms when soil or rock temperatures remain below 0°C (32°F) for at least 2 years in areas considered periglacial; criterion is based on temperature and not on whether water is present (compare frozen ground; see periglacial).

Permeability (9) The ability of water to flow through soil or rock; a function of the texture and structure of the medium.

Peroxyacetyl nitrate (PAN) (3) A pollutant formed from photochemical reactions involving nitric oxide (NO) and volatile organic compounds (VOCs). PAN produces no known human health effect, but it is particularly damaging to plants.

Phase change (7) The change in phase, or state, among ice, water, and water vapor; involves the absorption or release of latent heat (see latent heat).

Photochemical smog (3) Air pollution produced by the interaction of ultraviolet light, nitrogen dioxide, and hydrocarbons; produces ozone and PAN through a series of complex photochemical reactions. Automobiles are the major source of the contributive gases.

Photogrammetry (1) The science of obtaining accurate measurements from aerial photos and remote sensing; used to create and to improve surface maps.

Photosynthesis (19) The process by which plants produce their own food from carbon dioxide and water, powered by solar energy. The joining of carbon dioxide and hydrogen in plants, under the influence of certain wavelengths of visible light; releases oxygen and produces energy-rich organic material, sugars, and starches (compare respiration).

Physical geography (1) The science concerned with the spatial aspects and interactions of the physical elements and processes that make up the environment: energy, air, water, weather, climate, landforms, soils, animals, plants, microorganisms, and Earth.

Physical weathering (13) The breaking and disintegrating of rock without any chemical alteration; sometimes referred to as mechanical or fragmentation weathering.

Pingo (17) Large areas of frozen ground (soil-covered ice) that develop a heaved-up, circular, ice-cored mound, rising above a periglacial landscape as water freezes into ice and expands; sometimes results from pressure developed by freezing artesian water that is injected into permafrost; occasionally exceeds 60 m in height (200 ft).

Pioneer community (19) The initial plant community in an area; usually found on new surfaces or those that have been stripped of life, as in beginning primary succession; including lichens, mosses, and ferns growing on bare rock.

Place (1) A major theme in geography, focused on the tangible and intangible characteristics that make each location unique; no two places on Earth are alike.

Plane of the ecliptic (2) A plane (flat surface) intersecting all the points of Earth's orbit.

Planetesimal hypothesis (2) Proposes a process by which early protoplanets formed from the condensing masses of a nebular cloud of dust, gas, and icy comets; a formation process now being observed in other parts of the galaxy.

Planimetric map (Appendix A) A basic map showing the horizontal position of boundaries, land-use activities, and political, economic, and social outlines.

Plateau basalt (12) An accumulation of horizontal flows formed when lava spreads out from elongated fissures onto the surface in extensive sheets; associated with effusive eruptions; also known as flood basalts (see basalt).

Plate tectonics (11) The conceptual model and theory that encompasses continental drift, sea-floor spreading, and related aspects of crustal movement; accepted as the foundation of crustal tectonic processes (see continental drift).

Playa (15) An area of salt crust left behind by evaporation on a desert floor usually in the middle of a desert or semiarid bolson or valley; intermittently wet and dry.

Pluton (11) A mass of intrusive igneous rock that has cooled slowly in the crust; forms in any size or shape. The largest partially exposed pluton is a batholith (see batholith).

Podzolization (18) A pedogenic process in cool, moist climates; forms a highly leached soil with strong surface acidity because of humus from acid-rich trees.

Point bar (14) In a stream the inner portion of a meander, where sediment fill is redeposited.

Polar easterlies (6) Variable weak, cold, and dry winds moving away from the polar region; an anticyclonic circulation.

Polar front (6) A significant zone of contrast between cold and warm air masses; roughly situated between 50° and 60° N and S latitude.

Polar high-pressure cells (6) A weak, anticyclonic, thermally produced pressure system positioned roughly over each pole; the region of the lowest temperatures on Earth (see Antarctic high).

Polypedon (18) The identifiable soil in an area, with distinctive characteristics differentiating it from surrounding polypedons that form the basic mapping unit; composed of many pedons (see pedon).

Porosity (9) The total volume of available pore space in soil; a result of the texture and structure of the soil.

Positive feedback (1) Feedback that amplifies or encourages responses in a system (compare to negative feedback, feedback loop).

Potential evapotranspiration (9) POTET, or PE; the amount of moisture that would evaporate and transpire if adequate moisture were available; it is the amount lost under optimum moisture conditions, the moisture demand.

Potentiometric surface (9) A pressure level in a confined aquifer, defined by the level to which water rises in wells; caused by the fact that the water in a confined aquifer is under the pressure of its own weight; also known as a piezometric surface. This surface can extend above the surface of the land, causing water to rise above the water table in wells in confined aquifers (see artesian water).

Precipitation (9) Rain, snow, sleet, and hail—the moisture supply; called PRECIP, or P, in the water balance.

Pressure gradient force (6) Causes air to move from an area of higher barometric pressure to an area of lower barometric pressure due to the pressure difference.

Primary succession (19) Succession that occurs among plant species in an area of new surfaces created by mass movement of land, areas exposed by a retreating glacier, cooled lava flows and volcanic eruption landscapes, or surface mining and clear-cut logging scars, or an area of sand dunes, with no trace of a former community.

Prime meridian (1) An arbitrary meridian designated as 0° longitude, the point from which longitudes are measured east or west, at Greenwich, England, selected by international agreement in an 1884 treaty.

Process (1) A set of actions and changes that occur in some special order; analysis of processes is central to modern geographic synthesis.

Producer (19) Organism (plant) in an ecosystem that uses carbon dioxide as its sole source of carbon, which it chemically fixes through photosynthesis to provide its own nourishment; also called an autotroph (compare consumer).

Pyroclastic (12) An explosively ejected rock fragment launched by a volcanic eruption; sometimes described by the more general term tephra.

Radiation fog (7) Formed by radiative cooling of a land surface, especially on clear nights in areas of moist ground; occurs when the air layer directly above the surface is chilled to the dew-point temperature, thereby producing saturated conditions.

Rain gauge (9) A weather instrument; a standardized device that captures and measures rainfall.

Rain shadow (8) The area on the leeward slope of a mountain range, where precipitation receipt is greatly reduced compared to the windward slope on the other side (see orographic lifting).

Reflection (4) The portion of arriving insolation that is returned directly to space without being absorbed and converted into heat and without performing any work (see albedo).

Refraction (4) The bending effect on electromagnetic waves that occurs when insolation enters the atmosphere or another medium; the same process by which a crystal, or prism, disperses the component colors of the light passing through it.

Region (1) A geographic theme that focuses on areas that display unity and internal homogeneity of traits; includes the study of how a region forms, evolves, and interrelates with other regions.

Regolith (13) Partially weathered rock overlying bedrock, whether residual or transported.

Relative humidity (7) The ratio of water vapor actually in the air (content) compared to the maximum water vapor the air could hold (capacity) at that temperature; expressed as a percentage (compare vapor pressure, specific humidity).

Relief (12) Elevation differences in a local landscape; an expression of local height difference of landforms.

Remote sensing (1) Information acquired from a distance, without physical contact with the subject; for example, photography, orbital imagery, or radar.

Respiration (19) The process by which plants use food to derive energy for their operations; essentially, the reverse of the photosynthetic process; releases carbon dioxide, water, and heat energy into the environment (compare photosynthesis).

Reverse fault (12) Compressional forces produce strain that breaks a rock so that one side moves upward relative to the other side; also called a thrust fault (compare normal fault).

Revolution (2) The annual orbital movement of Earth about the Sun; determines the length of the year and the seasons.

Rhumb line (1) A line of constant compass direction, or constant bearing, which crosses all meridians at the same angle; a portion of a great circle.

Richter scale (12) An open-ended, logarithmic scale that estimates earthquake magnitude; designed by Charles Richter in 1935; now replaced by the moment magnitude scale (see moment magnitude scale).

Ring of fire (12) See circum-Pacific belt.

Robinson projection (Appendix A) A compromise (neither equal area nor true shape) oval projection developed in 1963 by Arthur Robinson.

Roche moutonnée (17) A glacial erosion feature; an asymmetrical hill of exposed bedrock; displays a gently sloping upstream side that has been smoothed and polished by a glacier and an abrupt, steep downstream side.

Rock (11) An assemblage of minerals bound together, or sometimes a mass of a single mineral.

Rock cycle (11) A model representing the interrelationships among the three rock-forming processes: igneous, sedimentary, and metamorphic; shows how each can be transformed into another rock type.

Rockfall (13) Free-falling movement of debris from a cliff or steep slope, generally falling straight or bounding downslope.

Rossby wave (6) An undulating horizontal motion in the upper-air westerly circulation at middle and high latitudes.

Rotation (2)   The turning of Earth on its axis averaging about 24 hours in duration; determines day–night relation; counterclockwise when viewed from above the North Pole and from above the equator, west to east or eastward.

Salinity (16)   The concentration of natural elements and compounds dissolved in solution, as solutes; measured by weight in parts per thousand (‰) in seawater.

Salinization (18)   A pedogenic process that results from high potential evapotranspiration rates in deserts and semiarid regions. Soil water is drawn to surface horizons, and dissolved salts are deposited as the water evaporates.

Saltation (14, 15)   The transport of sand grains (usually larger than 0.2 mm, or 0.008 in.) by stream or wind, bouncing the grains along the ground in asymmetrical paths.

Salt marsh (16)   A wetland ecosystem characteristic of latitudes poleward of the 30th parallel (compare to mangrove swamp).

Sand sea (15)   An extensive area of sand and dunes; characteristic of Earth's erg deserts (contrast to reg deserts).

Saturated (7)   Air that is holding all the water vapor that it can hold at a given temperature, known as the dew-point temperature.

Scale (1)   The ratio of the distance on a map to that in the real world; expressed as a representative fraction, graphic scale, or written scale.

Scarification (13)   Human-induced mass movements of Earth materials, such as large-scale open-pit mining and strip mining.

Scattering (4)   Deflection and redirection of insolation by atmospheric gases, dust, ice, and water vapor; the shorter the wavelength, the greater the scattering, thus skies in the lower atmosphere are blue.

Scientific method (1)   An approach that uses applied common sense in an organized and objective manner; based on observation, generalization, formulation, and testing of a hypothesis, and ultimately the development of a theory.

Sea-floor spreading (11)   As proposed by Hess and Dietz, the mechanism driving the movement of the continents; associated with upwelling flows of magma along the worldwide system of mid-ocean ridges (see mid-ocean ridge).

Secondary succession (19)   Succession that occurs among plant species in an area where vestiges of a previously functioning community are present; an area where the natural community has been destroyed or disturbed, but where the underlying soil remains intact.

Sediment (13)   Fine-grained mineral matter that is transported and deposited by air, water, or ice.

Sedimentary rock (11)   One of three basic rock types; formed from the compaction, cementation, and hardening of sediments derived from other rocks (compare igneous rock, metamorphic rock).

Seismic wave (11)   The shock wave sent through the planet by an earthquake or underground nuclear test. Transmission varies according to temperature and the density of various layers within the planet; provides indirect diagnostic evidence of Earth's internal structure.

Seismograph (12)   A device that measures seismic waves of energy transmitted throughout Earth's interior or along the coast.

Sensible heat (3)   Heat that can be measured with a thermometer; a measure of the concentration of kinetic energy from molecular motion.

Sheet flow (14)   Surface water that moves downslope in a thin film as overland flow, not concentrated in channels larger than rills.

Sheeting (13)   A form of weathering associated with fracturing or fragmentation of rock by pressure release; often related to exfoliation processes (see exfoliation dome).

Shield volcano (12)   A symmetrical mountain landform built from effusive eruptions (low-viscosity magma); gently sloped, gradually rising from the surrounding landscape to a summit crater; typical of the Hawaiian Islands (compare to effusive eruption; composite volcano).

Sinkhole (13)   Nearly circular depression created by the weathering of karst landscapes with subterranean drainage; also known as a doline in traditional studies; may collapse through the roof of an underground space (see karst topography).

Sleet (8)   Freezing rain, ice glaze, or ice pellets.

Sling psychrometer (7)   A weather instrument that measures relative humidity using two thermometers—a dry bulb and a wet bulb—mounted side-by-side.

Slipface (15)   On a sand dune, formed as dune height increases above 30 cm (12 in.) on the leeward side at an angle at which loose material is stable—its angle of repose (30 to 34°).

Slope (13)   A curved, inclined surface that bounds a landform.

Small circle (1)   A circle on a globe's surface that does not share Earth's center; for example, all parallels of latitude other than the equator (compare to great circle).

Snow line (17)   A temporary line marking the elevation where winter snowfall persists throughout the summer; seasonally, the lowest elevation covered by snow during the summer.

Soil (18)   A dynamic natural body made up of fine materials covering Earth's surface in which plants grow, composed of both mineral and organic matter.

Soil colloid (18)   A tiny clay and organic particle in soil; provides chemically active sites for mineral ion adsorption (see cation-exchange capacity).

Soil creep (13)   A persistent mass movement of surface soil where individual soil particles are lifted and disturbed by the expansion of soil moisture as it freezes or by grazing livestock or digging animals.

Soil fertility (18)   The ability of soil to support plant productivity when it contains organic substances and clay minerals that absorb water and certain elemental ions needed by plants through adsorption (see cation-exchange capacity, CEC).

Soil horizon (18)   The various layers exposed in a pedon; roughly parallel to the surface and identified as O, A, E, B, C, and R (bedrock).

Soil-moisture recharge (9)   Water entering available soil storage spaces.

Soil-moisture storage (9)   STRGE, the retention of moisture within soil; it is a savings account that can accept deposits (soil-moisture recharge) or experiences withdrawals (soil-moisture utilization) as conditions change.

Soil-moisture utilization (9)   The extraction of soil moisture by plants for their needs; efficiency of withdrawal decreases as the soil storage is reduced.

Soil science (18)   Interdisciplinary science of soils. Pedology concerns the origin, classification, distribution, and description of soil. Edaphology focuses on soil as a medium for sustaining higher plants.

Soil Taxonomy (18)   A soil classification system based on observable soil properties actually seen in the field; published in 1975 by the U.S. Soil Conservation Service and revised in 1990 and 1998 by the Natural Resources Conservation Service to include 12 soil orders.

Soil-water budget (9)   An accounting system for soil moisture using inputs of precipitation and outputs of evapotranspiration and gravitational water.

Solar constant (2)   The amount of insolation intercepted by Earth on a surface perpendicular to the Sun's rays when Earth is at its average distance from the Sun; a value of 1370 W/m^2, or 1.968 calories/cm^2 per minute; averaged over the entire globe at the thermopause.

Solar wind (2)   Clouds of ionized (charged) gases emitted by the Sun and traveling in all directions from the Sun's surface. Effects on Earth include auroras, disturbance of radio signals, and possible influences on weather.

Solifluction (17)   Gentle downslope movement of a saturated surface material (soil and regolith) in various climatic regimes where temperatures are above freezing (compare gelifluction).

Solum (18)   A true soil profile in the pedon; ideally, a combination of O, A, E, and B horizons (see pedon).

Spatial (1)   The nature or character of physical space, as in an area; occupying or operating within a space. Geography is a spatial science; spatial analysis its essential approach.

Spatial analysis (1)   The examination of spatial interactions, patterns, and variations over area and/or space; a key integrative approach of geography.

Specific heat (5)   The increase of temperature in a material when energy is absorbed; water has a higher specific heat (can store more heat) than a comparable volume of soil or rock.

Specific humidity (7)   The mass of water vapor (in grams) per unit mass of air (in kilograms) at any specified temperature. The maximum mass of water vapor that a kilogram of air can hold at any specified temperature is termed its maximum specific humidity (compare vapor pressure, relative humidity).

Speed of light (2)   Specifically, 299,792 kilometers per second (186,282 miles per second), or more than 9.4 trillion kilometers per year

(5.9 trillion miles per year)—a distance known as a light-year; at light speed Earth is 8 minutes and 20 seconds from the Sun.

Spheroidal weathering (13)   A chemical weathering process in which the sharp edges and corners of boulders and rocks are weathered in thin plates that create a rounded, spheroidal form.

Spodosols (18)   A soil order in the Soil Taxonomy classification that occurs in northern coniferous forests; best developed in cold, moist, forested climates; lacks humus and clay in the A horizons, with high acidity associated with podzolization processes.

Spring tide (16)   The highest tidal range, which occurs when the Moon and the Sun are in conjunction (at new Moon) or in opposition (at full Moon) stages (compare neap tide).

Squall line (8)   A zone slightly ahead of a fast-advancing cold front, where wind patterns are rapidly changing and blustery and precipitation is strong.

Stability (7)   The condition of a parcel, whether it remains where it is or changes its initial position. The parcel is stable if it resists displacement upward, unstable if it continues to rise.

Stationary front (8)   A frontal area of contact between contrasting air masses that shows little horizontal movement; winds in opposite direction on either side of the front flow parallel along the front.

Steady-state equilibrium (1)   The condition that occurs in a system when rates of input and output are equal and the amounts of energy and stored matter are nearly constant around a stable average.

Stomata (19)   A small opening on the undersides of leaves through which water and gasses pass.

Storm surge (8)   A large quantity of seawater pushed inland by the strong winds associated with a tropical cyclone.

Storm tracks (8)   Seasonally shifting paths followed by migrating low-pressure systems.

Stratified drift (17)   Sediments deposited by glacial meltwater that appear sorted; a specific form of glacial drift (compare till).

Stratigraphy (11)   A science that analyzes the sequence, spacing, geophysical and geochemical properties, and spatial distribution of rock strata.

Stratocumulus (7)   A lumpy, grayish, low-level cloud, patchy with sky visible, sometimes present at the end of the day.

Stratosphere (3)   That portion of the homosphere that ranges from 20 to 50 km (12.5 to 30 mi) above Earth's surface, with temperatures ranging from 257°C (270°F) at the tropopause to 0°C (32°F) at the stratopause. The functional ozonosphere is within the stratosphere.

Stratus (7)   A stratiform (flat, horizontal) cloud generally below 2000 m (6500 ft).

Strike-slip fault (12)   Horizontal movement along a faultline, that is, movement in the same direction as the fault; also known as a transcurrent fault. Such movement is described as right lateral or left lateral, depending on the relative motion observed across the fault (see transform fault).

Subduction zone (11)   An area where two plates of crust collide and the denser oceanic crust dives beneath the less dense continental plate, forming deep oceanic trenches and seismically active regions.

Sublimation (7)   A process in which ice evaporates directly to water vapor or water vapor freezes to ice (deposition).

Subpolar low-pressure cell (6)   A region of low pressure centered approximately at 60° latitude in the North Atlantic near Iceland and in the North Pacific near the Aleutians, as well as in the Southern Hemisphere. Airflow is cyclonic; it weakens in summer and strengthens in winter (refer to cyclone).

Subsolar point (2)   The only point receiving perpendicular insolation at a given moment—the Sun directly overhead (see declination).

Subtropical high-pressure cell (6)   One of several dynamic high-pressure areas covering roughly the region from 20° to 35° N and S latitudes; responsible for the hot, dry areas of Earth's arid and semiarid deserts (refer to anticyclone).

Sulfate aerosols (3)   Sulfur compounds in the atmosphere, principally sulfuric acid; principal sources relate to fossil fuel combustion; scatters and reflects insolation.

Sulfur dioxide (3)   A colorless gas detected by its pungent odor; produced by the combustion of fossil fuels, especially coal, that contain sulfur as an impurity; can react in the atmosphere to form sulfuric acid, a component of acid deposition.

Summer (June) solstice (2)   The time when the Sun's declination is at the Tropic of Cancer, at 23.5° N latitude, June 20–21 each year (compare to winter solstice).

Sunrise (2)   That moment when the disk of the Sun first appears above the horizon.

Sunset (2)   That moment when the disk of the Sun totally disappears.

Sunspot (2)   Magnetic disturbances on the surface of the Sun, occurring in an average 11-year cycle; related flares, prominences, and outbreaks produce surges in solar wind.

Surface creep (15)   A form of eolian transport that involves particles too large for saltation; a process whereby individual grains are impacted by moving grains and slide and roll.

Surplus (9)   (SURPL) The amount of moisture that exceeds potential evapotranspiration; moisture oversupply when soil moisture storage is at field capacity; extra or surplus water.

Suspended load (14)   Fine particles held in suspension in a stream. The finest particles are not deposited until the stream velocity nears zero.

Swell (16)   Regular patterns of smooth, rounded waves in open water; can range from small ripples to very large waves.

Syncline (12)   A trough in folded strata, with beds that slope toward the axis of the downfold (compare anticline).

System (1)   Any ordered, interrelated set of materials or items existing, separate from the environment, or within a boundary; energy transformations and energy and matter storage and retrieval occur within a system.

Taiga (20)   See needleleaf forest.

Talik (17)   An unfrozen portion of the ground that may occur above, below, or within a body of discontinuous permafrost or beneath a body of water in the continuous region, such as a deep lake; may extend to bedrock and noncryotic soil under large deep lakes.

Talus slope (13)   Formed by angular rock fragments that cascade down a slope along the base of a mountain; poorly sorted, cone-shaped deposits.

Tarn (17)   A small mountain lake, especially one that collects in a cirque basin behind risers of rock material or in an ice-gouged depression.

Temperate rain forest (20)   A major biome of lush forests at middle and high latitudes; occurs along narrow margins of the Pacific Northwest in North America, among other locations; includes the tallest trees in the world.

Temperature (5)   A measure of sensible heat energy present in the atmosphere and other media; indicates the average kinetic energy of individual molecules within a substance.

Temperature inversion (3)   A reversal of the normal decrease of temperature with increasing altitude; can occur anywhere from ground level up to several thousand meters; functions to block atmospheric convection and thereby trap pollutants.

Terminal moraine (17)   Eroded debris that is dropped at a glacier's farthest extent.

Terrane (12)   A migrating piece of Earth's crust, dragged about by processes of mantle convection and plate tectonics. Displaced terranes are distinct in their history, composition, and structure from the continents that accept them.

Terrestrial ecosystem (20)   A self-regulating association characterized by specific plant formations; usually named for the predominant vegetation and known as a biome when large and stable (see biome).

Thermal equator (5)   The isoline on an isothermal map that connects all points of highest mean temperature.

Thermohaline circulation (6)   Deep ocean currents produced by differences in temperature and salinity with depth; Earth's deep currents.

Thermokarst (17)   Topography of hummocky, irregular relief marked by cave-ins, bogs, small depressions, and pits formed as ground ice melts; an erosion process caused by ground ice melting; not related to solution processes and chemical weathering associated with limestone (karst).

Thermopause (2, 3)   A zone approximately 480 km (300 mi) in altitude that serves conceptually as the top of the atmosphere; an altitude used for the determination of the solar constant.

Thermosphere (3)   A region of the heterosphere extending from 80 to 480 km (50 to 300 mi) in altitude; contains the functional ionosphere layer.

Threshold (1)   A moment in which a system can no longer maintain its character, so it lurches to a new operational level, which may not be compatible with previous conditions.

Index

YOU SHOULD CAREFULLY READ THE TERMS AND CONDITIONS BEFORE USING THE CD-ROM PACKAGE. USING THIS CD-ROM PACKAGE INDICATES YOUR ACCEPTANCE OF THESE TERMS AND CONDITIONS.

Pearson Education, Inc. provides this program and licenses its use. You assume responsibility for the selection of the program to achieve your intended results, and for the installation, use, and results obtained from the program. This license extends only to use of the program in the United States or countries in which the program is marketed by authorized distributors.

LICENSE GRANT

You hereby accept a nonexclusive, nontransferable, permanent license to install and use the program ON A SINGLE COMPUTER at any given time. You may copy the program solely for backup or archival purposes in support of your use of the program on the single computer. You may not modify, translate, disassemble, decompile, or reverse engineer the program, in whole or in part.

TERM

The License is effective until terminated. Pearson Education, Inc. reserves the right to terminate this License automatically if any provision of the License is violated. You may terminate the License at any time. To terminate this License, you must return the program, including documentation, along with a written warranty stating that all copies in your possession have been returned or destroyed.

LIMITED WARRANTY

THE PROGRAM IS PROVIDED "AS IS" WITHOUT WARRANTY OF ANY KIND, EITHER EXPRESSED OR IMPLIED, INCLUDING, BUT NOT LIMITED TO, THE IMPLIED WARRANTIES OR MERCHANTABILITY AND FITNESS FOR A PARTICULAR PURPOSE. THE ENTIRE RISK AS TO THE QUALITY AND PERFORMANCE OF THE PROGRAM IS WITH YOU. SHOULD THE PROGRAM PROVE DEFECTIVE, YOU (AND NOT PEARSON EDUCATION, INC. OR ANY AUTHORIZED DEALER) ASSUME THE ENTIRE COST OF ALL NECESSARY SERVICING, REPAIR, OR CORRECTION. NO ORAL OR WRITTEN INFORMATION OR ADVICE GIVEN BY PEARSON EDUCATION, INC., ITS DEALERS, DISTRIBUTORS, OR AGENTS SHALL CREATE A WARRANTY OR INCREASE THE SCOPE OF THIS WARRANTY.

SOME STATES DO NOT ALLOW THE EXCLUSION OF IMPLIED WARRANTIES, SO THE ABOVE EXCLUSION MAY NOT APPLY TO YOU. THIS WARRANTY GIVES YOU SPECIFIC LEGAL RIGHTS AND YOU MAY ALSO HAVE OTHER LEGAL RIGHTS THAT VARY FROM STATE TO STATE.

Pearson Education, Inc. does not warrant that the functions contained in the program will meet your requirements or that the operation of the program will be uninterrupted or error-free. However, Pearson Education, Inc. warrants the CD-ROM(s) on which the program is furnished to be free from defects in material and workmanship under normal use for a period of ninety (90) days from the date of delivery to you as evidenced by a copy of your receipt. The program should not be relied on as the sole basis to solve a problem whose incorrect solution could result in injury to person or property. If the program is employed in such a manner, it is at the user's own risk and Pearson Education, Inc. explicitly disclaims all liability for such misuse.

LIMITATION OF REMEDIES

Pearson Education, Inc.'s entire liability and your exclusive remedy shall be: 1. the replacement of any CD-ROM not meeting Pearson Education, Inc.'s "LIMITED WARRANTY" and that is returned to Pearson Education, or 2. if Pearson Education is unable to deliver a replacement CD-ROM that is free of defects in materials or workmanship, you may terminate this agreement by returning the program.

IN NO EVENT WILL PEARSON EDUCATION, INC. BE LIABLE TO YOU FOR ANY DAMAGES, INCLUDING ANY LOST PROFITS, LOST SAVINGS, OR OTHER INCIDENTAL OR CONSEQUENTIAL DAMAGES ARISING OUT OF THE USE OR INABILITY TO USE SUCH PROGRAM EVEN IF PEARSON EDUCATION, INC. OR AN AUTHORIZED DISTRIBUTOR HAS BEEN ADVISED OF THE POSSIBILITY OF SUCH DAMAGES, OR FOR ANY CLAIM BY ANY OTHER PARTY. SOME STATES DO NOT ALLOW FOR THE LIMITATION OR EXCLUSION OF LIABILITY FOR INCIDENTAL OR CONSEQUENTIAL DAMAGES, SO THE ABOVE LIMITATION OR EXCLUSION MAY NOT APPLY TO YOU.

GENERAL

You may not sublicense, assign, or transfer the license of the program. Any attempt to sublicense, assign or transfer any of the rights, duties, or obligations hereunder

is void. This Agreement will be governed by the laws of the State of New York.

Should you have any questions concerning this Agreement, you may contact Pearson Education, Inc. by writing to:

ESM Media Development
Higher Education Division
Pearson Education, Inc.
1 Lake Street
Upper Saddle River, NJ 07458

Should you have any questions concerning technical support, you may write to:
New Media Production
Higher Education Division
Pearson Education, Inc.
1 Lake Street
Upper Saddle River, NJ 07458

YOU ACKNOWLEDGE THAT YOU HAVE READ THIS AGREEMENT, UNDERSTAND IT, AND AGREE TO BE BOUND BY ITS TERMS AND CONDITIONS. YOU FURTHER AGREE THAT IT IS THE COMPLETE AND EXCLUSIVE STATEMENT OF THE AGREEMENT BETWEEN US THAT SUPERSEDES ANY PROPOSAL OR PRIOR AGREEMENT, ORAL OR WRITTEN, AND ANY OTHER COMMUNICATIONS BETWEEN US RELATING TO THE SUBJECT MATTER OF THIS AGREEMENT.

PROGRAM INSTRUCTIONS

Windows:

Inserting this CD-ROM will automatically start the application, if you need to manually start the application follow the following steps:

● Insert the "Geosystems - Student Animations CD" CD into your CD-ROM drive.
● Browse to the "geosystems" CD-ROM using your Windows explorer window.
● Double-click the "GeoSys.exe" program file.

Macintosh:

Inserting this CD-ROM will automatically start the application, if you need to manually start the application follow the following steps:

● Insert the "Geosystems - Student Animations CD" CD into your CD-ROM drive.
● Double-click the "geosystems" CD icon on your desktop.
● Double-click the "GeoSys" application file.

MINIMUM SYSTEM REQUIREMENTS

Windows:

233 MHz Intel Pentium processor	*Optional components:*
Windows 98/ME/2000/NT4-sp3/XP	● Active internet connection
16 bit Sound Card	● Internet Browser:
32 MB or more of available RAM	● Netscape 6.2 or 7.0
800x600 monitor resolution set to 16 bit color	● Microsoft Internet Explorer 5.0, 5.5 or 6.0
Mouse or other pointing device	
4x CD-ROM Drive	

Macintosh:

PowerPC 200Mhz processor	For some early release versions of OS9
Operating System OS9.2.1/OSX	you may need
50 MB or more of available RAM	to install the "Carbonlib" system file –
(Turn on virtual memory if necessary, minimum setting 128MB)	see Read Me file for details
800x600 monitor resolution set to 16 bit color	*Optional components:*
	● Active internet connection
Mouse or other pointing device	● Internet Browser:
4x CD-ROM Drive	● Netscape 6.2 or 7.0
	● Microsoft Internet Explorer 5.1 or 5.2

TECHNICAL SUPPORT

If you are having problems with this software, call (800) 677-6337 between 8:00 a.m. and 8:00 p.m. EST, Monday through Friday, and 5:00 p.m. through Midnight EST on Sundays. You can also get support by filling out the web form located at : **http://247.prenhall.com/mediaform**

Our technical staff will need to know certain things about your system in order to help us solve your problems more quickly and efficiently. If possible, please be at your computer when you call for support. You should have the following information ready:

● Textbook ISBN	● hard disk space available
● CD-Rom/Diskette ISBN	● Sound card? Yes or No
● corresponding product and title	● printer make and model
● computer make and model	● network connection
● Operating System (Windows or Macintosh) and Version	● detailed description of the problem, including the exact wording of any error messages.
● RAM available	

NOTE: Pearson does not support and/or assist with the following:
● third-party software (i.e. Microsoft including Microsoft Office Suite, Apple, Borland, etc.)
● homework assistance

World – Physical

ARCTIC OCEAN

180° 160°W 140°W 120°W 100°W 80°W 60°W

80°N

Point Barrow
QUEEN ELIZABETH ISLANDS
Ellesmere Island
GREENLAND
Beaufort Sea
Victoria Island
Baffin Island
Baffin Bay

Bering Strait
Yukon R.
Mt. McKinley 20,320 ft. (6,194 m.)
MACKENZIE MTS.
Great Bear Lake
Great Slave L.
Hudson Bay
Davis Strait
Cape Farewell
Labrador
Labrador Sea
Reykjanes R.

60°N
Bering Sea
Aleutian Islands
Aleutian Trench
Gulf of Alaska
NORTH AMERICA
ROCKY MOUNTAINS
Saskatchewan R.
Lake Winnipeg
Canadian Shield
Great Lakes
Newfoundland

Vancouver I.
CASCADE RA.
Cape Mendocino
Great Basin
GREAT PLAINS
Missouri R.
Cape Cod
Sohm Plain
Azores Is.

40°N
Mendocino Fracture Zone
SIERRA NEVADA
Colorado R.
Ohio R.
APPALACHIAN MTS.
Cape Hatteras
Hatteras Plain
ATLANTIC OCEAN

Northeast Pacific Basin
Mississippi R.
Bermuda Rise
Mid-Atlantic Ridge
Cape Verde Plai

Murray Fracture Zone
SIERRA MADRE
Rio Grande
Bahama Is.

Tropic of Cancer
Hawaiian Ridge
Baja California
Gulf of Mexico
Cape Verde Is.

20°N
Hawaiian Is.
Molokai Fracture Zone
Cuba
West Indies
Puerto Rico Trench
Caribbean Sea

Johnston Atoll
Clarion Fracture Zone
Islas Revillagigedo
CENTRAL AMERICA
Cape Verde Is.

Central Pacific Basin
PACIFIC OCEAN
Middle America Trench
Demerara Plain

Line Islands
Clipperton Fracture Zone
Guiana Highlands
Orinoco R.

0° Equator
Kiritimati I.
Northeast Pacific Basin
Galápagos Is.
AMAZON BASIN
SOUTH AMERICA
Amazon R.
Cape São Roque

POLYNESIA
Phoenix Is.
Millennium I.
Parinas Point
Pernambu Plain

Marquesas Is.
ANDES
Brazilian Highlands

Samoa Is.
Tuamotu Archipelago
Mato Grosso Plateau

20°S
Tonga Is.
Cook Is.
Society Is. Tahiti
Sala-y-Gómez
Atacama Desert
Cape Frio

Tropic of Capricorn
Austral Islands
Pitcairn I.
Sala-y-Gómez Ridge
East Pacific Rise
Nazca Ridge
Gran Chaco
Paraná R.

Kermadec Tr.
Louisville Ridge
Easter I.
Mt. Aconcagua 22,834 ft. (6,960 m.)
Rio Grande Rise

Southwest Pacific Basin
Challenger Fracture Zone
Juan Fernández Is.
Pampas
Patagonia
Argentine Plain

40°S
Eltanin Fracture Zone
Southeast Pacific Basin
Humboldt Plain
Falkland Is.
South Georgia

Udintsev Fracture Zone
Strait of Magellan
Cape Horn
South Georgia Ridge
South Sandw

60°S
Pacific-Antarctic Ridge
Antarctic Circle
Drake Passage

80°S